# IN CASE OF ACCIDENT[1]

*In case of accident notify the laboratory instructor* **immediately.**

## FIRE

**Burning Clothing.** Prevent the person from running and fanning the flames. Rolling the person on the floor will help extinguish the flames and prevent inhalation of the flames. If a safety shower is nearby hold the person under the shower until flames are extinguished and chemicals washed away. Do not use a fire blanket if a shower is nearby. The blanket does not cool and smoldering continues. Remove contaminated clothing. Wrap the person in a blanket to avoid shock. Get prompt medical attention.

Do not, under any circumstances, use a carbon tetrachloride (toxic) fire extinguisher and be very careful using a $CO_2$ extinguisher (the person may smother).

**Burning Reagents.** Extinguish all nearby burners and remove combustible material and solvents. Small fires in flasks and beakers can be extinguished by covering the container with a fiberglass-wire gauze square, a big beaker, or a watch glass. Use a dry chemical or carbon dioxide fire extinguisher directed at the base of the flames. **Do not use water.**

**Burns, Either Thermal or Chemical.** Flush the burned area with cold water for at least 15 min. Resume if pain returns. Wash off chemicals with a mild detergent and water. Current practice recommends that no neutralizing chemicals, unguents, creams, lotions, or salves be applied. If chemicals are spilled on a person over a large area quickly remove the contaminated clothing while under the safety shower. Seconds count, and time should not be wasted because of modesty. Get prompt medical attention.

**CHEMICALS IN THE EYE:** Flush the eye with copious amounts of water for 15 min using an eyewash fountain or bottle or by placing the injured person face up on the floor and pouring water in the open eye. Hold the eye open to wash behind the eyelids. After 15 min of washing obtain prompt medical attention, regardless of the severity of the injury.

**CUTS: Minor Cuts.** This type of cut is most common in the organic laboratory and usually arises from broken glass. Wash the cut, remove any pieces of glass, and apply pressure to stop the bleeding. Get medical attention.

**Major Cuts.** If blood is spurting place a pad directly on the wound, apply firm pressure, wrap the injured to avoid shock, and get **immediate** medical attention. Never use a tourniquet.

**POISONS:** Call 800 information (1-800-555-1212) for the telephone number of the nearest Poison Control Center, which is usually also an 800 number.

---

[1]Adapted from *Safety in Academic Chemistry Laboratories,* prepared by the American Chemical Society Committee on Chemical Safety, March 1996.

# Macroscale and Microscale Organic Experiments

# Macroscale and Microscale Organic Experiments

**FIFTH EDITION**

**Kenneth L. Williamson**
*Mount Holyoke College, Emeritus*

**Robert Minard**
*Pennsylvania State University, Emeritus*

**Katherine M. Masters**
*Pennsylvania State University*

**Houghton Mifflin Company**
*Boston   New York*

**Publisher:** *Charles Hartford*
**Executive Sponsoring Editor:** *Richard Stratton*
**Development Editor:** *Kate Heinle*
**Senior Project Editor:** *Margaret Park Bridges*
**Manufacturing Manager:** *Karen B. Fawcett*
**Senior Marketing Manager:** *Katherine Greig*
**Marketing Assistant:** *Naveen Hariprasad*

**Cover image:** *© Jim Scherer Photography*

*Printed in the U.S.A.*

Library of Congress Control Number: 2005930127

Instructor's examination copy
ISBN-10: 0-618-73067-2
ISBN-13: 978-0-618-73067-4

For orders, use student text ISBNs
ISBN-10: 0-618-59067-6
ISBN-13: 978-0-618-59067-4

3456789-DOC-10 09 08 07

# Contents

## Oxidation and Reduction

## Aromatic Substitution and Elimination

## Reactions of Aldehydes and Ketones

## Reactions of Carboxylic Acids, Esters, and Amines

## The Diels–Alder and Related Reactions

## Derivatives of 1,2-Diphenylethane: A Multistep Synthesis

## Photochemistry

## Natural Product Chemistry and Biochemistry

**I**nnovation and exploration have always been the hallmarks of *Macroscale and Microscale Organic Experiments*. We are proud to have been part of a movement toward the increased use of microscale experiments in the undergraduate organic laboratory course.

As in previous editions, ease of use continues to be a chief attribute of the pedagogical features in this text. From the first edition, icons in the margin clearly indicate whether an experiment is to be conducted on a macroscale level or a microscale level. Wherever possible, we have expanded the popular introductory "In This Experiment" sections, which were introduced in the fourth edition, giving an overall view of the experimental work to be carried out without the detail that may obscure the end result. "Cleaning Up" sections at the end of almost every experiment focus students' attention on all the substances produced in a typical organic reaction and continue to highlight current laboratory safety rules and regulations.

In preparing the fifth edition, we have attempted to build on the strengths of previous editions while continuing to add innovative and new techniques, features, and experiments.

## New to This Edition

We welcome two new coauthors, Bob Minard and Katherine Masters, from Penn State University. In addition to original new experiments and revisions of previous ones, they have contributed a large-university perspective to maintaining an efficient and safe laboratory course.

We have enhanced the first 15 chapters of the text, focusing especially on instrumental methods. In keeping with this approach, we have eliminated from the text the chapter on qualitative organic analysis (which has been transferred to the website). Wet chemical methods of organic analysis, while an extremely valuable academic exercise in experimentation, observation, and deductive reasoning, are no longer used in the real world of organic chemistry. Spectrographic methods (infrared [IR], ultraviolet/visible [UV/Vis], and nuclear magnetic resonance [NMR] spectroscopy and mass spectrometry [MS]) have supplanted these methods and have received enhanced attention in this edition.

The place of organic chemistry laboratory work has been placed in a broader context in Chapter 1, along with a detailed description, keyed to the website, of the equipment involved. New information regarding working in a closed system as well as a discussion of laboratory courtesy has been included in the chapter on laboratory safety. A thorough discussion of intramolecular attractive forces unites the chapter on melting points and boiling points, which naturally leads to the fourth chapter on recrystallization.

Distillation methods (simple, fractional, steam, and vacuum) are now in one chapter, along with sublimation. We have introduced a bioassay of eugenol, isolated by steam distillation from cloves, for its antibacterial activity, an assay that can also be applied to sulfanilamide, which is synthesized in Chapter 45. To the chapter on extraction (Chapter 7) has been added an experiment on the isolation and purification of the components of analgesic tablets, which was taken from Chapter 44 of the fourth edition.

An extensive discussion of the principles of chromatography has been included in the chapter on thin-layer chromatography (TLC), which leads logically to the chapters on column and gas chromatography (GC), which now also includes new material on capillary GC as well as temperature programming.

The discussions of NMR theory and the interpretation of $^1$H NMR spectra have been revised, and diffuse reflectance IR analysis is introduced. The chapter on mass spectrometry (Chapter 13) is entirely new. Because magnetic sector instruments are rarely used, only time-of-flight and quadrupole mass analyzers are described. The interpretation of electron ionization spectra is given thorough treatment, and exact mass measurement is discussed. There are sections on GC-MS and computer-aided spectral identification. ESI and MALDI ionization are described, with high-mass examples provided.

Chapter 14 in this edition combines the material on UV spectroscopy with a section on refractive indices. The new chapter includes a compact section on organic qualitative analysis, including solubility tests. This section highlights a typical approach to unknown identification, using the spectroscopic methods described in Chapters 11–14.

Because of the increased availability of fast desktop computers and inexpensive software, the chapter on computational chemistry has been extensively revised to provide expanded coverage of ab initio and semiempirical models and calculations. The discussion emphasizes the reaction energetics that are not available by molecular mechanics methods.

## Green Chemistry

In an attempt to reduce the environmental impact of the chemical enterprise, close attention is paid to techniques and experiments that are nontoxic to living things and the environment. Microscale experiments, conducted on a scale of 1/100th to 1/1000th of that formerly used, vastly reduce but do not completely eliminate toxic reagents from the environment. To move to a "greener" laboratory, we are constantly on the watch for experiments and reagents that accomplish the same result with minimum damage to the environment.

Experiments of this type are indicated with a green chemistry icon, for instance the air oxidation of fluorene in Chapter 9 instead of a dichromate oxidation, or the benzoin condensation catalyzed by thiamine (an edible vitamin) as an alternative to potassium cyanide (Chapter 53). The use of household bleach to oxidize cyclohexanol is another example where a toxic oxidizing agent (chromium ion) is replaced with a much more benign substance (Chapter 22). And Chapter 67 deals with the chemistry of polymer recycling.

# Supplements

## *Online Study Center*

The new Online Study Center provides valuable resources intended to enhance students' organic chemistry laboratory experience and help prepare them to work more effectively and safely in the laboratory. Students can access the Online Study Center by visiting http://college.hmco.com/PIC/williamsonMMOE5e.

The Online Study Center icon appears throughout the text in the margin to direct students to the section on the website where appropriate resources appear. These resources include photos and videos that are intended to augment the descriptions and drawings in the text. The photographs amplify the figures in the text that illustrate apparatus and techniques, and the short videos demonstrate techniques that are more easily shown in motion than in static drawings or photos.

Several complete experiments are presented as videos on the website. The "Synthesis of Ferrocene," for example, illustrates in 3 min the operations needed for this complex procedure—a procedure that requires a minimum of 90 min in the laboratory.

In an effort to keep up with changes in the content and existence of relevant online resources, the "Surfing the Web" sections from previous editions have been moved to the website (General Resources). References to online resources can now be updated in real time as necessary.

Helpful supplemental materials, such as instructions on preparing a laboratory notebook and report and two samples of completed records, tables listing the physical properties of a number of functional group classes, and additional experiments are also be available on the website (General Resources).

*Online Study Center*
Video: How to Assemble
Apparatus

*Online Study Center*
**General Resources**
Web Links

## *Online Teaching Center*

The new Online Teaching Center centralizes important preparatory materials and resources for instructors in one convenient location. Instructors can access the Online Teaching Center for this text by visiting http://college.hmco.com/PIC/williamsonMMOE5e.

### PowerPoint Slides

New to this edition, PowerPoint slides supply a flexible, customizable instructor resource that is designed to enhance prelab discussions. For the technique chapters and other selected experiments, PowerPoint slides briefly outline the main techniques, underlying concepts, common problems, challenging manipulations, special safety precautions, and typical results. Relevant illustrations of apparatus, structures, reactions, and links to photos and videos of techniques are also provided.

### Instructor's Guide

The *Instructor's Guide* is an important adjunct to this text. It contains discussions about the time needed to carry out each experiment, an assessment of the relative difficulty of each experiment, problems that might be encountered, answers to end-of-chapter questions, a list of chemicals and apparatus required for each experiment (both per student and per 24-student laboratory), sources of supply for unusual items, and a discussion of hardware and software needed for running computational chemistry experiments. Note especially the comments regarding the transfer hydrogenation experiment in Chapter 25.

## Acknowledgments

We wish to express our thanks to the many people at Houghton Mifflin who have made this book possible: Richard Stratton, Executive Sponsoring Editor; Katherine Greig, Senior Marketing Manager; Kate Heinle, Development Editor; and Margaret Park Bridges, Senior Production Editor; to Rebecca Dodson, Editorial Project Manager, and Caryl Wenzel, the skilled copyeditor, at Schawk, Inc.; and to the reviewers of this text: Debra L. Bautista (Eastern Kentucky University), Edwin M. Bryant (Lansing Community College), Tina Choe (Loyola Marymount University), John M. Farrar (Indiana University, Purdue University, Fort Wayne), Nancy Faulk (Blinn College/Bryan Campus), Sam Leung (Washburn University), and Carl H. Snyder (University of Miami).

*Kenneth L. Williamson*
*Bob Minard*
*Katherine Masters*

# Organic Experiments and Waste Disposal

**A**n unusual feature of this book is the advice at the end of each experiment on how to dispose of its chemical waste. Waste disposal thus becomes part of the experiment, which is not considered finished until the waste products are appropriately taken care of. This is a valuable addition to the book for several reasons.

Although chemical waste from laboratories is less than 0.1% of that generated in the United States, its disposal is nevertheless subject to many of the same federal, state, and local regulations as is chemical waste from industry. Accordingly, there are both strong ethical and legal reasons for proper disposal of laboratory wastes. These reasons are backed up by a financial concern, because the cost of waste disposal can become a significant part of the cost of operating a laboratory.

There is yet another reason to include instructions for waste disposal in a teaching laboratory. Students will someday be among those conducting and regulating waste disposal operations and voting on appropriations for them. Learning the principles and methods of sound waste disposal early in their careers will benefit them and society later.

The basics of waste disposal are easy to grasp. Some innocuous water-soluble wastes are flushed down the drain with a large proportion of water. Common inorganic acids and bases are neutralized and flushed down the drain. Containers are provided for several classes of solvents, for example, combustible solvents and halogenated solvents. (The containers are subsequently removed for suitable disposal by licensed waste handlers.) Some toxic substances can be oxidized or reduced to innocuous substances that can then be flushed down the drain; for example, hydrazines, mercaptans, and inorganic cyanides can be thus oxidized by sodium hypochlorite solution, widely available as household bleach. Dilute solutions of highly toxic cations are expensive to dispose of because of their bulk; precipitation of the cation by a suitable reagent followed by its separation greatly reduces its bulk and cost. These and many other procedures can be found throughout this book.

One other principle of waste control lies at the heart of this book. Microscale experimentation, by minimizing the scale of chemical operations, also minimizes the volume of waste. Chromatography procedures to separate and purify products, spectroscopy methods to identify and characterize products, and well-designed small-scale equipment enable one to conduct experiments today on a tenth to a thousandth the scale commonly in use a generation ago.

Chemists often provide great detail in their directions for preparing chemicals so that the synthesis can be repeated, but they seldom say much about how to dispose of the hazardous byproducts. Yet the proper disposal of a chemical's byproducts is as important as its proper preparation. Dr. Williamson sets a good example by providing explicit directions for such disposal.

*Blaine C. McKusick*

# CHAPTER

# 1

# Introduction

**Online Study Center**

This icon will direct you to techniques, equipment setups, and online resources at http://college.hmco.com/PIC/williamsonMMOE5e.

**PRELAB EXERCISE:** Study the glassware diagrams presented in this chapter and be prepared to identify the reaction tube, the fractionating column, the distilling head, the filter adapter, and the Hirsch funnel.

**W**elcome to the organic chemistry laboratory! Here, the reactions that you learned in your organic lectures and studied in your textbook will come to life. The main goal of the laboratory course is for you to learn and carry out techniques for the synthesis, isolation, purification, and analysis of organic compounds, thus experiencing the experimental nature of organic chemistry. We want you to enjoy your laboratory experience and ask you to remember that safety always comes first.

## Synthesis–The Big Picture

Synthesis is a pervasive process in the universe. Science has shown us that our world consists of energy and matter and that both of these are being transformed from one form to another throughout the cosmos (see Fig. 1.1).

- Cosmosynthesis started roughly 14 billion years ago when the Big Bang transformed subatomic particles into hydrogen and helium. These elements condense into stars and are fused by nucleosynthetic reactions to give us the periodic table of elements. Most stars ultimately expand or explode and disperse their product elements into space where they can recondense to form new stars, some with planets around them. For the early Earth, these elements combined into minerals and other substances such as water, carbon dioxide, and nitrogen.
- Biosynthesis on planet Earth has been going on for over 2 billion years. In the process of growth and reproduction, even the simplest single cell organisms can synthesize incredibly complex structural, catalytic, informational, and

■ **FIG. 1.1**
The three types of synthesis and the steps by which synthetic chemists study the resulting products. Theory allows correlation of structure to properties and guides the cycle of synthetic design, synthesis, separation/purification, and analysis.

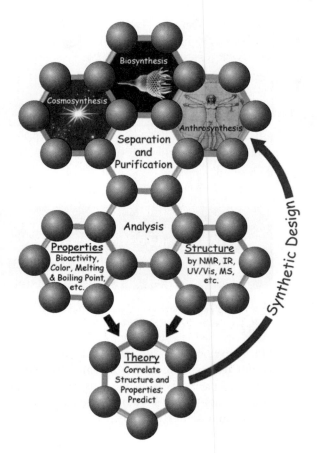

regulatory molecules from simple starting materials such as $CO_2$, $H_2O$, and $N_2$. Organic chemistry's beginnings are based on isolating and studying these natural products such as urea, ethanol, caffeine, or oil of wintergreen (methyl salicylate). The whole field of biochemistry is based on the isolation, structure determination, and reactivity of organic molecules found in living systems.

● Anthrosynthesis is a human activity that produces either known substances economically (e.g., iron, ammonia, and vitamin C) or new substances not found in the cosmos or biology (e.g., aspirin, Splenda, and Teflon). To date, chemists have synthesized or isolated from nature over 20 million different substances, and 99% of these are organic compounds. As you know, this great variety of organic compounds stems from carbon's unique ability to form chains that can be linear, branched, or cyclic and that can have other elements bonded to them to yield myriad structures.

Whether chemists study rocks from the earth, molecules in cells, or products of anthrosynthesis, they almost always separate and study the structure and properties of pure substances (e.g., silica, hemoglobin, or carbon nanotubes) rather than the mixtures (e.g., a shovel of dirt, blood, or soot) typically found in nature or synthetic reactions. Only pure substances will show reproducible data for physical, chemical, and biological properties. Chemists then try to understand the relationship between a compound's structure and its properties and devise a reasonable theory that relates these. If a theory is found to be true, it can be used to design the synthesis of a new substance with enhanced properties such as elasticity, antibiotic activity, or biodegradability. This cycle of Synthesis → Separation/Purification → Analysis → Theory → Synthetic Design is depicted in Figure 1.1. The cycle can be repeated to test and refine the theory and further improve the desired properties. This cycle is at the heart of progress in the pharmaceutical, polymer/plastic, and metallurgy industries.

Once you start your laboratory work, you will soon realize that synthesis involves much more than reacting chemicals. The techniques of separation, purification, and analysis often require considerable time and attention. Therefore, the first part of this lab text is designed to help you understand, apply, and master these techniques. They will then be used repeatedly in the dozens of synthesis and isolation experiments presented in later chapters.

## Experimental Organic Chemistry— What's in It for Me?

You are probably not a chemistry major. The vast majority of students in this laboratory course are majoring in the life sciences or engineering. Although you may never use the exact same techniques taught in this course, you will undoubtedly apply the skills taught here to whatever problem or question your ultimate career may present. Application of the scientific method involves the following steps:

1. Designing an experiment, therapy, or approach to solve a problem
2. Executing the plan or experiment
3. Observing the outcome to verify that you obtained the desired results
4. Recording the findings to communicate them both orally and in writing

You may not always get the desired result the first time when doing an experiment. Do not take this as a personal failure; it is evidence that carrying out an organic reaction can often be trickier than baking a cake (and we all probably had trouble baking our first cake!). Your laboratory instructor is there to provide guidance and advice but not to solve your problems. Learning to think for yourself is the most valuable skill you will take from this course. However, you should not be afraid to ask questions or bounce ideas off your instructor or fellow students. Just formulating the question or talking about your problem often helps you find the answer. To begin troubleshooting a problem, you will need to scrutinize carefully every detail of the approach, therapy, or experimental design; the execution of the experiment; and your observations to see where things might have gone wrong. This process will lead you to an idea or hypothesis about how to change things to make the experiment work. The sequence is then repeated, sometimes many times, until you solve the problem, answer the question, or prove/disprove the idea.

The teaching lab is more controlled than the real world. In this laboratory environment, you will be guided more than you would be on the job. Nevertheless, the experiments in this text are designed to be sufficiently challenging to give you a taste of experimental problem-solving methods practiced by professional scientists. We earnestly hope that you will find the techniques, the apparatus, and the experiments of just the right complexity, not too easy but not too hard, so that you can learn at a satisfying pace.

## Macroscale and Microscale Experiments

This laboratory text presents a unique approach for carrying out organic experiments; they can be conducted on either a *macroscale* or a *microscale*. Macroscale was the traditional way of teaching the principles of experimental organic chemistry and is the basis for all the experiments in this book, a book that traces its history to 1934 when the late Louis Fieser, an outstanding organic chemist and professor at Harvard University, was its author. Macroscale experiments typically involve the use of a few grams of *starting material*, the chief reagent used in the reaction. Most teaching institutions are equipped to carry out traditional macroscale experiments. Instructors are familiar with these techniques and experiments, and much research in industry and academe is carried out on this scale. For these reasons, this book has macroscale versions of most experiments.

For reasons primarily related to safety and cost, there is a growing trend toward carrying out microscale laboratory work, on a scale one-tenth to one-thousandth of that previously used. Using smaller quantities of chemicals exposes the laboratory worker to smaller amounts of toxic, flammable, explosive, carcinogenic, and teratogenic material. Microscale experiments can be carried out more rapidly than macroscale experiments because of rapid heat transfer, filtration, and drying. Because the apparatus advocated by the authors is inexpensive, more than one reaction may be set up at once. The cost of chemicals is, of course, greatly reduced. A principal advantage of microscale experimentation is that the quantity of waste is one-tenth to one-thousandth of that formerly produced. To allow maximum flexibility in the conduct of organic experiments, this book presents both macroscale and microscale procedures for the vast majority of the experiments. As will be seen, some of the equipment and techniques differ. A careful reading of both the microscale and macroscale procedures will reveal which changes and precautions must be employed in going from one scale to the other.

## Synthesis and Analysis

Synthesis and analysis are two major concerns of the organic chemist, and both are dealt with in this book. The typical sequence of activity in synthetic organic chemistry involves the following steps:

1. Designing the experiment based on knowledge of chemical reactivity, the equipment and techniques available, and full awareness of all safety issues
2. Setting up and running the reaction
3. Isolating the reaction product

4. Purifying the crude product, if necessary
5. Analyzing the product using chromatography or spectroscopy to verify purity and structure
6. Disposing of unwanted chemicals in a safe manner

Each of these steps uses certain laboratory techniques. The experiments at the beginning of this text are designed to teach you these techniques and to help you understand the molecular level principles that underlie them.

### 1. Designing the Experiment

Because the first step of experimental design often requires considerable experience, this part has already been done for you for most of the experiments in this introductory level book. Synthetic experimental design becomes increasingly important in an advanced course and in graduate research programs. Remember that safety is paramount, and therefore it is important to be aware of all possible personal and environmental hazards before running any reaction. These are more fully discussed in Chapter 2 and are highlighted in each experiment.

### 2. Running the Reaction

The rational synthesis of an organic compound, whether it involves the transformation of one functional group into another or a carbon-carbon bond-forming reaction, starts with a *reaction*. Organic reactions usually take place in the liquid phase and are *homogeneous*—the reactants are entirely in one phase. The reactants can be solids and/or liquids dissolved in an appropriate solvent to mediate the reaction. Some reactions are *heterogeneous*—that is, one of the reactants is a solid and requires stirring or shaking to bring the reactants in contact with one another. A few heterogeneous reactions involve the reaction of a gas, such as oxygen, carbon dioxide, or hydrogen, with material in solution. Examples of all these reactions are found in this book.

An *exothermic* reaction evolves heat. If it is highly exothermic with a low activation energy, one reactant is added slowly to the other, and heat is removed by external cooling. Most organic reactions are, however, mildly *endothermic*, which means the reaction mixture must be heated to overcome the activation barrier and to

Effect of temperature

increase the rate of the reaction. A very useful rule of thumb is that *the rate of an organic reaction doubles with a 10°C rise in temperature*. Louis Fieser introduced the idea of changing the traditional solvents of many reactions to high-boiling solvents to reduce reaction times. Throughout this book we will use solvents such as triethylene glycol, with a boiling point (bp) of 290°C, to replace ethanol (bp 78°C) and triethylene glycol dimethyl ether (bp 222°C) to replace dimethoxyethane (bp 85°C). Using these high-boiling solvents can greatly increase the rates of many reactions.

The progress of a reaction can be followed by observing a change in color or pH, the evolution of a gas, or the separation of a solid product or a liquid layer.

**Chapters 8–10:** Chromatography

Quite often, the extent of reaction can be determined by withdrawing tiny samples at certain time intervals and analyzing them by *thin-layer chromatography* or *gas chromatography* to measure the amount of starting material remaining and/or the amount of product formed. The next step, product isolation, should not be carried out until one is confident that the desired amount of product has been formed.

### 3. Product Isolation: Workup of the Reaction

Running an organic reaction is usually the easiest part of a synthesis. The real challenge lies in isolating and purifying the product from the reaction because organic reactions seldom give quantitative yields of a single pure substance. Any unwanted byproducts need to be removed.

In some cases the solvent and concentrations of reactants are chosen so that, after the reaction mixture has been cooled, the product will *crystallize* or *precipitate* if it is a solid. The product is then collected by *filtration*, and the crystals are washed with an appropriate solvent. If sufficiently pure at that point, the product is dried and collected; otherwise, it is purified by the process of *recrystallization* or, less commonly, by *sublimation*.

More typically, the product of a reaction does not crystallize from the reaction mixture and is often isolated by the process of *liquid/liquid extraction*. This process involves two liquids, a water-insoluble organic liquid such as dichloromethane and a neutral, acidic, or basic aqueous solution. The two liquids do not mix, but when shaken together, the organic materials and inorganic byproducts go into the liquid layer, organic or aqueous, that they are the most soluble in. After shaking, two layers again form and can be separated. Most organic products remain in the organic liquid and can be isolated by evaporation of the organic solvent.

If the product is a liquid, it is isolated by *distillation*, usually after extraction. Occasionally, the product can be isolated by the process of *steam distillation* from the reaction mixture.

### 4. Purification

When an organic product is first isolated, it will often contain significant impurities. This impure or crude product will need to be further purified or cleaned up before it can be analyzed or used in other reactions. Solids may be purified by recrystallization or sublimation and liquids by distillation or steam distillation. Small amounts of solids and liquids can also be purified by *chromatography*.

### 5. Analysis to Verify Purity and Structure

The purity of the product can be determined by melting point analysis for solids, boiling point analysis or refractive index for liquids, and chromatographic analysis for either solids or liquids. Once the purity of the product has been verified, structure determination can be accomplished by using one of the various spectroscopic methods, such as $^1H$ and $^{13}C$ nuclear magnetic resonance (NMR), infrared (IR), and ultraviolet/visible (UV/Vis) spectroscopies. Mass spectrometry (MS) is another tool that can aid in the identification of a structure.

### 6. Chemical Waste Disposal

All waste chemicals must be disposed of in their proper waste containers. Instructions on chemical disposal will appear at the end of each experiment. It is recommended that nothing be disposed of until you are sure of your product identity and purity; you do not want to accidentally throw out your product before the analysis is complete. Proper disposal of chemicals is essential for protecting the environment in accordance with local, state, and federal regulations.

**Chapter 4**: Recrystallization

**Chapter 7**: Liquid/Liquid Extraction

**Chapter 5**: Distillation

**Chapter 6**: Steam Distillation and Vacuum Distillation

**Chapters 11–14**: Structure Analysis

**CAUTION**: Never smell chemicals in an attempt to identify them.

# Equipment for Experimental Organic Chemistry

## A. Equipment for Running Reactions

Organic reactions are usually carried out by dissolving the reactants in a solvent and then heating the mixture to its boiling point, thus maintaining the reaction at that elevated temperature for as long as is necessary to complete the reaction. To keep the solvent from boiling away, the vapor is condensed to a liquid, which is allowed to run back into the boiling solvent.

Microscale reactions with volumes up to 4 mL can be carried out in a *reaction tube* (Fig. 1.2a). The mass of the reaction tube is so small and heat transfer is so rapid that 1 mL of nitrobenzene (bp 210°C) will boil in 10 s, and 1 mL of benzene (mp 5°C) will crystallize in the same period of time. Cooling is effected by simply agitating the tube in a small beaker of ice water, and heating is effected by immersing the reaction tube to an appropriate depth in an electrically heated sand bath. This sand bath usually consists of an electric 100-mL flask heater or heating mantle half filled with sand. The temperature is controlled by the setting on a variable voltage controller often called a Variac (Fig. 1.2b). The air above the heater is not hot. It is possible to hold a reaction tube containing refluxing solvents between the thumb and forefinger without the need for forceps or other protective device. Because sand is a fairly poor conductor of heat, there is a very large variation in temperature in the sand bath depending on its depth. The temperature of a 5-mL flask can be regulated by using a spatula to pile up or remove sand from near the flask's base. The heater is easily capable of producing temperatures in excess of 300°C; therefore, never leave the controller at its maximum setting. Ordinarily, it is set at 20%–40% of maximum.

Because the area of the tube exposed to heat is fairly small, it is difficult to transfer enough heat to the contents of the tube to cause the solvents to boil away. The reaction tube is 100 mm long, so the upper part of the tube can function as an efficient *air condenser* (Fig. 1.2a) because the area of glass is large and the volume of vapor is comparatively small. The air condenser can be made even longer by attaching the empty *distilling column* (Fig. 1.2c and 1.15o) to the reaction tube using the *connector with support rod* (Fig. 1.2c and Fig. 1.15m). The black connector is made of Viton, which is resistant to high-boiling aromatic solvents. The cream-colored connector is made of Santoprene, which is resistant to all but high-boiling aromatic solvents. As solvents such as water and ethanol boil, the hot vapor ascends to the upper part of the tube. These condense and run back down the tube. This process is called *refluxing* and is the most common method for conducting a reaction at a constant temperature, the boiling point of the solvent. For very low-boiling solvents such as diethyl ether (bp 35°C), a pipe cleaner dampened with water makes an efficient cooling device. A water-cooled condenser is also available (Fig. 1.3) but is seldom needed for microscale experiments.

A Petri dish containing sand and heated on a hot plate is not recommended for microscale experiments. It is too easy to burn oneself on the hot plate; too much heat wells up from the sand, so air condensers do not function well; the ⸱⸱

Turn on the sand bath about 20 min before you intend to use it. The sand heats slowly.

**CAUTION**: Never put a mercury thermometer in a sand bath! It will break, releasing highly toxic mercury vapor.

*Online Study Center*

Photos: Williamson Microscale Kit, Refluxing a Liquid in a Reaction Tube on a Sand Bath; Video: The Reaction Tube in Use

*Online Study Center*

Video: How to Assemble Apparatus

■ **FIG. 1.2**
(a) A reaction tube being heated on a hot sand bath in a flask heater. The area of the tube exposed to the heat is small. The liquid boils and condenses on the cool upper portion of the tube, which functions as an air condenser. (b) A variable voltage controller used to control the temperature of the sand bath. (c) The condensing area can be increased by adding a distilling column as an air condenser.

Cool part of tube

Refluxing liquid (Air condenser)

Sand

Heated area

Boiling liquid

Boiling chip

Electric flask heater

(a)

Distilling column

Wet pipe cleaner

Connector stir rod

(c)

(b)

**CAUTION**: Organic reactions should be conducted in a fume hood with the sash lowered.

dishes will break from thermal shock; and the ceramic coating on some hot plates will chip and come off.

Larger scale (macroscale) reactions involving volumes of tens to thousands of milliliters are usually carried out in large, round-bottom flasks that fit snugly (without sand!) into the appropriately sized flask heater or heating mantle (Fig. 1.4). The round shape can be heated more evenly than a flat-bottom flask or beaker. Heat transfer is slower than in microscale because of the smaller ratio of surface area to volume in a round-bottomed flask. Cooling is again conducted using an ice bath, but heating is sometimes done on a steam bath for low-boiling liquids. The narrow neck is necessary for connection via a *standard-taper ground glass joint* to a water-cooled *reflux condenser*, where the water flows in a jacket around the central tube.

## FIG. 1.3

Refluxing solvent in a 5-mL round-bottomed flask fitted with a water-cooled condenser.

Water out

Water in

## FIG. 1.4

A reflux apparatus for larger reactions. Liquid boils in the flask and condenses on the cold inner surface of the water-cooled condenser.

Central tube

Water out

Water in

Magnetic stirrer

The high heat capacity of water makes it possible to remove a large amount of heat in the larger volume of refluxing vapor (Fig. 1.4).

### Heating and Stirring

In modern organic laboratories, electric flask heaters (heating mantles), used alone or as sand baths, are used exclusively for heating. Bunsen burners are almost never used because of the danger of igniting flammable organic vapors. For solvents that boil below 90°C, the most common method for heating macroscale flasks is the *steam bath*.

Reactions are often stirred using a *magnetic stirrer* (Fig. 1.4) to help mix reagents and to promote smooth boiling. A Teflon-coated bar magnet (*stirring bar*) is placed in the reaction flask, and a magnetic stirrer is placed under the flask and flask heater. The stirrer contains a large, horizontally rotating bar magnet that attracts the Teflon-coated stirring bar magnet and causes it to turn. The speed of stirring can be adjusted on the front of the magnetic stirrer.

## B. Equipment for the Isolation of Products

### Filtration

If the product of a reaction crystallizes from the reaction mixture on cooling, the solid crystals are isolated by *filtration*. This can be done in several ways when using microscale techniques. If the crystals are large enough and in a reaction tube, insert a *Pasteur pipette* to the bottom of the tube while expelling the air an⸍

withdrawing the solvent (Fig. 1.5). Highly effective filtration occurs between the square, flat tip of the pipette and the bottom of the tube. This method of filtration has several advantages over the alternatives. The mixture of crystals and solvent can be kept on ice during the entire process. This minimizes the solubility of the crystals in the solvent. There are no transfer losses of material because an external filtration device is not used. This technique allows several recrystallizations to be carried out in the same tube with final drying of the product under vacuum. If you know the *tare* (the weight of the empty tube), the weight of the product can be determined without removing it from the tube. In this manner a compound can be synthesized, purified by crystallization, and dried all in the same reaction tube. After removal of material for analysis, the compound in the tube can then be used for the next reaction. This technique is used in many of this book's microscale experiments. When the crystals are dry, they are easily removed from the reaction tube. When they are wet, it is difficult to scrape them out. If the crystals are in more than about 2 mL of solvent, they can be isolated by filtration with a *Hirsch funnel*. The one that is in the microscale kit of apparatus is particularly easy to use because the funnel fits into the *filter flask* with no adapter and is equipped with a *polyethylene frit* for the capture of the filtered crystals (Fig. 1.6). The Wilfilter is especially good for collecting small quantities of crystals (Fig. 1.7).

■ **FIG. 1.5**
Filtration using a Pasteur pipette and a reaction tube.

Pipette bulb

Pasteur pipette

Reaction tube

Note square, flat tip

■ **FIG. 1.7**
A Wilfilter is placed upside down in a centrifuge tube and spun in a centrifuge.

■ **FIG. 1.6**
A Hirsch funnel with an integral adapter, a polyethylene frit, and a 25-mL filter flask.

Filter paper, 12 mm dia.

Polyethylene filter disk (frit), 10 mm dia.

Hirsch funnel

To aspirator

25-mL Filter flask

Crystals

Wilfilter

Filtrate

*Online Study Center*

Video: Macroscale Crystallization

Macroscale quantities of material can be recrystallized in conical *Erlenmeyer flasks* of the appropriate size. The crystals are collected in porcelain or plastic *Büchner funnels* fit with pieces of filter paper covering the holes in the bottom of the funnel (Fig. 1.8). A rubber *filter adapter* (*Filtervac*) is used to form a vacuum tight seal between the flask and the funnel.

### Extraction

**Chapter 7**: Extraction

*Online Study Center*

Photos: Extraction with Ether, Extraction with Dichloromethane; Videos: Extraction with Ether, Extraction with Dichloromethane

The product of a reaction will often not crystallize. It may be a liquid or a viscous oil, it may be a mixture of compounds, or it may be too soluble in the reaction solvent being used. In this case, an immiscible solvent is added, the two layers are shaken to effect *extraction*, and after the layers separate, one layer is removed. On a microscale, this can be done with a Pasteur pipette. The extraction process is repeated if necessary. A tall, thin column of liquid, such as that produced in a reaction tube, makes it easy to selectively remove one layer by pipette. This is more difficult to do in the usual test tube because the height/diameter ratio is small.

On a larger scale, a *separatory funnel* is used for extraction (Fig. 1.9a). The mixture can be shaken in the funnel and then the lower layer removed through the

**■ FIG. 1.9**
(a) A separatory funnel with a Teflon stopcock. (b) A microscale separatory funnel. Remove the polyethylene frit from the micro Büchner funnel before using.

**■ FIG. 1.8**
A suction filter assembly.

(a)    (b)

■ **FIG. 1.10**
A small-scale sublimation
apparatus.

15-mL Centrifuge
tube, to be filled
with ice

Rubber
adapter

To
vacuum

25-mL Filter flask

Material to
be sublimed

Sublimate

Heat source

stopcock after the stopper is removed. These funnels are available in sizes from 10 mL to 5000 mL. The chromatography column in the apparatus kit is also a *micro separatory funnel* (Fig. 1.9b). Remember to remove the frit at the column base of the micro Büchner funnel and to close the valve before adding liquid.

## C. Equipment for Purification

Many solids can be purified by the process of *sublimation*. The solid is heated, and the vapor of the solid condenses on a cold surface to form crystals in an apparatus constructed from a *centrifuge tube* fitted with a rubber adapter and pushed into a *filter flask* (Fig. 1.10). Caffeine can be purified in this manner. This is primarily a microscale technique, although sublimers holding several grams of solid are available.

Mixtures of solids and, occasionally, of liquids can be separated and purified by *column chromatography*. The *chromatography column* for both microscale and macroscale work is very similar (Fig. 1.11).

Some of the compounds to be synthesized in these experiments are liquids. On a very small scale, the best way to separate and purify a mixture of liquids is by *gas chromatography*, but this technique is limited to less than 100 mg of material on the usual gas chromatograph. For larger quantities of material, *distillation* is used. For this purpose, small distilling flasks are used. These flasks have a large surface area to allow sufficient heat input to cause the liquid to vaporize rapidly so that it can be distilled and then condensed for collection in a *receiver*. The apparatus (Fig. 1.12) consists of a *distilling flask*, a *distilling adapter* (which also functions as an air condenser on a microscale), a *thermometer adapter*, and a *thermometer*; for macroscale, a water-cooled *condenser* and *distilling adapter* are added to the apparatus (Fig. 1.13). *Fractional distillation* is carried out using a small, packed *fractionating column* (Fig. 1.14). The apparatus is very similar for both microscale and macroscale. On a microscale, 2 mL to 4 mL of a liquid can be

*Online Study Center*

Photo: Sublimation Apparatus

*Online Study Center*

Photo: Column Chromatography;
Videos: Extraction with Ether,
Extraction with Dichloromethane

**Chapter 9**: Column
Chromatography

⚠
**CAUTION**: Use mercury-free
thermometers whenever
possible.

*Online Study Center*

Photo: Simple Distillation
Apparatus; Video: How to
Assemble Apparatus; Photo:
Fractional Distillation Apparatus

■ **FIG. 1.11**
A chromatography column consisting of a funnel, a tube, a base fitted with a polyethylene frit, and a Leur valve.

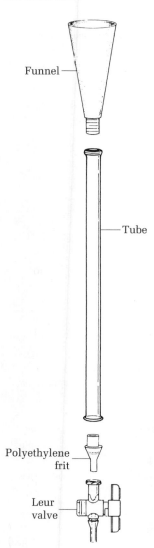

Funnel

Tube

Polyethylene frit

Leur valve

■ **FIG. 1.12**
A small-scale simple distillation apparatus. Note that the entire thermometer bulb is below the side arm of the distilling adapter.

Thermometer

Thermometer adapter

Distilling adapter and air condenser

Vial (receiver)

30-mL Beaker

Ice (if needed)

Distilling flask

Hot sand bath

Boiling chip

fractionally distilled, and 1 mL or more can be simply distilled. The usual scale in these experiments for macroscale distillation is about 25 mL.

Some liquids with a relatively high vapor pressure can be isolated and purified by *steam distillation*, a process in which the organic compound codistills with water at the boiling point of water. The microscale and macroscale apparatus for this process are shown in Chapter 6.

The collection of typical equipment used for microscale experimentation is shown in Figure 1.15 and for macroscale experimentation in Figure 1.16. Other equipment commonly used in the organic laboratory is shown in Figure 1.17.

## Check-in of Lab Equipment

Your first duty will be to check in to your assigned lab desk. The identity of much of the apparatus should already be apparent from the preceding outline of the experimental processes used in the organic laboratory.

Check to see that your thermometer reads about 22°C–25°C (71.6°F–77°F), which is normal room temperature. Examine the fluid column to see that it is

■ **FIG. 1.13**
An apparatus for simple distillation.

Thermometer

Thermometer adapter

Distillation head

Keck clamp

Clamp

Condenser

Distilling flask

Keck clamp

Boiling chip

Distillation adapter

Thermowell

Water out

Water in

Receiver (graduated cylinder)

■ **FIG. 1.14**
A microscale fractional distillation apparatus. The thermometer adapter is to be fitted with a thermometer.

Thermometer adapter (Santoprene)

Distilling head and air condenser

Receiver vial (1-dram)

Fractionating column with copper sponge

Rod

Elastomeric connector [Santoprene (white) or Viton (black)]

5-mL Round-bottomed flask

# ■ FIG. 1.15

Microscale apparatus kit.

(a) Pipette (1 mL), graduated in 1/1000ths.
(b) Chromatography column (glass) with polypropylene funnel and 20-μm polyethylene frit in base, which doubles as a micro Büchner funnel. The column, base, and stopcock are also used as a separatory funnel.
(c) Thermometer adapter.
(d) Connector only (Viton).
(e) Magnetic stirring bars (4 × 12 mm) in distillation receiver vial.

(f) Hirsch funnel (polypropylene) with 20-μm fritted polyethylene disk.
(g) Stopcock for chromatography column and separatory funnel.
(h) Claisen adapter/distillation head with air condenser.
(i) Filter flask, 25 mL.
(j) Distillation head, 105° connecting adapter.
(k) Rubber septa/sleeve stoppers, 8 mm.
(l) Syringe (polypropylene).

(m) Connector with support rod.
(n) Centrifuge tube (15 mL)/sublimation receiver, with cap.
(o) Distillation column/air condenser.
(p) Reaction tube, calibrated, 10 × 100 mm.
(q) Erlenmeyer flasks, 10 mL.
(r) Long-necked flask, 5 mL.
(s) Short-necked flask, 5 mL.
(t) Rubber adapter for sublimation apparatus.
(u) Tubing (polyethylene), 1/16-in. diameter.
(v) Spatula (stainless steel) with scoop end.

## ■ FIG. 1.16
Macroscale apparatus kit with 14/20 standard-taper ground-glass joints.

(a) Polyacetal Keck clamps, size 14.
(b) Hex-head glass stopper, 14/20 standard taper.
(c) Hirsch funnel (polypropylene) with 20-μm fritted polyethylene disk.
(d) Filter adapter for sublimation apparatus.
(e) Distilling head with O-ring thermometer adapter.
(f) Vacuum adapter.

(g) Centrifuge tube (15 mL)/sublimation receiver.
(h) Filter flask, 25 mL.
(i) Claisen adapter.
(j) Water-jacketed condenser.
(k) Chromatography column (glass) with polypropylene funnel and 20-μm polyethylene frit in base, which doubles as a micro Büchner funnel.

(l) Stopcock for chromatography column.
(m) Separatory funnel, 125 mL.
(n) Pear-shaped flask, 100 mL.
(o) Pear-shaped flask, 50 mL.
(p) Conical flask (15 mL) with side arm for inlet tube.
(q) Distilling column/air condenser.
(r) Conical reaction vial (5 mL)/distillation receiver.

unbroken and continuous from the bulb up. Replace any flasks that have star-shaped cracks. Remember that apparatus with graduations, stopcocks, or ground glass joints and anything porcelain are expensive. Erlenmeyer flasks, beakers, and test tubes are, by comparison, fairly cheap.

**FIG. 1.17**

Miscellaneous apparatus.
(a) 1.0 ± 0.01 mL graduated
pipette. (b) Septum. (c) 1.0-mL
syringe with blunt needle.
(d) Calibrated Pasteur pipette.
(e) Pipette pump. (f) Glass scorer.
(g) Filtervac. (h) Set of neoprene
filter adapters.

## Transfer of Liquids and Solids

Pasteur pipettes (Fig. 1.18) are very useful for transferring small quantities of liquid, adding reagents dropwise, and carrying out recrystallizations. Discard used Pasteur pipettes in the special container for waste glass. Surprisingly, the acetone used to wash out a dirty Pasteur pipette usually costs more than the pipette itself.

A plastic funnel that fits on the top of the reaction tube is very convenient for the transfer of solids to reaction tubes or small Erlenmeyer flasks for microscale experiments (Fig. 1.19). It is also the top of the chromatography column (Fig. 1.11). A special spatula with a scoop end (Fig. 1.17p) is used to remove solid material

■ **FIG. 1.17** (continued)
(i) Hirsch funnel with perforated plate in place. (j) Rubber thermometer adapter. (k) Powder funnel. (l) Polyethylene wash bottle. (m) Single-pan electronic balance with automatic zeroing and 0.001 g digital readout; 100 g capacity. (n) Electric flask heater. (o) Solid-state control for electric flask heater. (p) Stainless steel spatula. (q) Stirring bar. (r) Keck clamp. (s) Wilfilter.

from the reaction tube. On a large scale, a powder funnel is useful for adding solids to a flask (Fig. 1.17k). A funnel can also be fashioned from a sheet of weighing paper.

# Weighing and Measuring

The single-pan electronic balance (Fig. 1.17m), which is capable of weighing to ±0.001 g and having a capacity of at least 100 g, is the single most important instrument making microscale organic experiments possible. Most of the weighing measurements made in microscale experiments will use this type of balance. Weighing is fast and accurate with these balances. There should be one electronic balance for every 12 students. For macroscale experiments, a balance of such

2.0 mL

1.5 mL

1.0 mL

0.75 mL

0.50 mL

0.25 mL

Never pipette
by mouth!

0.10 mL

0.05 mL

**■ FIG. 1.18**
The approximate calibration
of Kimble 9" Pasteur pipette.

high accuracy is not necessary. Here, a balance with $\pm 0.01$ g accuracy would be satisfactory.

A container such as a reaction tube standing in a beaker or flask is placed on the pan. Set the digital readout to register zero and then add the desired quantity of the reagent to the reaction tube as the weight is measured periodically to the nearest milligram. Even liquids are weighed when accuracy is needed. It is much easier to weigh a liquid to 0.001 g than it is to measure it volumetrically to 0.001 mL.

It is often convenient to weigh reagents on glossy weighing paper and then transfer the chemical to the reaction container. The success of an experiment often depends on using just the right amount of starting materials and reagents. Inexperienced workers might think that if 1 mL of a reagent will do the job, then 2 mL will do the job twice as well. Such assumptions are usually erroneous.

Liquids can be measured by either volume or weight according to the following relationship:

$$\text{Volume (mL)} = \frac{\text{weight (g)}}{\text{density (g/mL)}}$$

Modern Erlenmeyer flasks and beakers have approximate volume calibrations fused into the glass, but these are *very* approximate. Better graduations are found on the microscale *reaction tube*. Somewhat more accurate volumetric measurements are made in 10-mL graduated cylinders. For volumes less than 4 mL, use a graduated pipette. **Never** apply suction to a pipette by mouth. The pipette can be fitted with a small plastic syringe using appropriately sized rubber tubing. A Pasteur pipette can be converted into a calibrated pipette with the addition of a plastic syringe (Fig. 1.17d). Figure 1.18 also shows the calibration marks for a 9-in. Pasteur pipette. You will find among your equipment a 1-mL pipette, calibrated in hundredths of a milliliter (Fig. 1.17a). Determine whether it is designed to *deliver* 1 mL or *contain* 1 mL between the top and bottom calibration marks. For our purposes, the latter is the better pipette.

Because the viscosity, surface tension, vapor pressure, and wetting characteristics of organic liquids are different from those of water, the so-called automatic pipette (designed for aqueous solutions) gives poor accuracy in measuring organic liquids. Syringes (Fig. 1.17c and Fig. 1.17d) and pipette pumps (Fig. 1.20), on the other hand, are quite useful, and these will be used frequently. Do not use a syringe that is equipped with a metal needle to measure corrosive reagents because these reagents will dissolve the metal in the needle. Because many reactions are "killed" by traces of moisture, many students' experiments are ruined by damp or wet apparatus. Several reactions that require especially dry or oxygen-free atmospheres will be run in systems sealed with a rubber septum (Fig. 1.17b). Reagents can be added to the system via syringe through this septum to minimize exposure to oxygen or atmospheric moisture.

Careful measurements of weights and volumes take more time than less accurate measurements. Think carefully about which measurements need to be made with accuracy and which do not.

■ **FIG. 1.19**
A funnel for adding solids and liquids to a reaction tube.

■ **FIG. 1.20**
Using a pipette pump to measure liquids to ±0.01 mL.

2-mL Pipette pump

1-mL Pipette graduated to ± 0.01 mL

Reaction tube

## Tares

Tare = weight of empty container

The tare of a container is its weight when empty. Throughout this laboratory course, it will be necessary to know the tares of containers so that the weights of the compounds within can be calculated. If identifying marks can be placed on the containers (e.g., with a diamond stylus), you may want to record tares for frequently used containers in your laboratory notebook.

To be strictly correct, we should use the word *mass* instead of *weight* because gravitational acceleration is not constant at all places on earth. But electronic balances record weights, unlike two-pan or triple-beam balances, which record masses.

## Washing and Drying Laboratory Equipment

### *Washing*

Clean apparatus immediately.

Considerable time may be saved by cleaning each piece of equipment soon after use, for you will know at that point which contaminant is present and be able to select the proper method for its removal. A residue is easier to remove

Both ethanol and acetone are very flammable.

Wash acetone goes in the organic solvents waste container; halogenated solvents go in the halogenated solvents waste container.

■ **FIG. 1.21**
A recrystallization solvent bottle and dispenser.

before it has dried and hardened. A small amount of organic residue can usually be dissolved with a few milliliters of an appropriate organic solvent. Acetone (bp 56.1°C) has great solvent power and is often effective, but it is extremely flammable and somewhat expensive. Because it is miscible with water and vaporizes readily, it is easy to remove from the vessel. Detergent and water can also be used to clean dirty glassware if an appropriate solvent is not found. Cleaning after an operation may often be carried out while another experiment is in process.

A *polyethylene bottle* (Fig. 1.17l) is a convenient wash bottle for acetone. Be careful not to store solvent bottles in the vicinity of the reaction where they can provide additional fuel for an accidental fire. The name, symbol, and formula of a solvent is written on a bottle with a marker or a wax pencil. For macroscale crystallizations, extractions, and quick cleaning of apparatus, it is convenient to have a bottle for each frequently used solvent—95% ethanol, ligroin or hexanes, dichloromethane, ether, and ethyl acetate. A pinhole opposite the spout, which is covered with the finger when in use, will prevent the spout from dribbling the solvent. For microscale work, these solvents are best dispensed from 25-mL or 50-mL bottles with an attached test tube containing a graduated (1-mL) polypropylene pipette (Fig. 1.21). Be aware of any potential hazards stemming from the reactivity of these wash solvents with chemical residues in flasks. Also, be sure to dispose of wash solvents in the proper container. Acetone and most other organic solvents do not contain halogens and can therefore go in the regular organic solvents waste container. However, if dichloromethane or another halogen-containing solvent is used, it must be disposed of in the halogenated solvents waste container.

Sometimes a flask will not be clean after a washing with detergent and acetone. At that point try an abrasive household cleaner. If still no success, try adding dilute acid or base to the dirty glassware, let it soak for a few minutes, and rinse with plenty of water and acetone.

### *Drying*

To dry a piece of apparatus rapidly, rinse with a few milliliters of acetone and invert over a beaker to drain. **Do not use compressed air**, which contains droplets of oil, water, and particles of rust. Instead, draw a slow stream of air through the apparatus using the suction of your water aspirator or house vacuum line.

## Miscellaneous Cleanup

If a glass tube or thermometer becomes stuck to a rubber connector, it can be removed by painting on glycerol and forcing the pointed tip of a small spatula between the rubber and the glass. Another method is to select a cork borer that fits snugly over the glass tube, moisten it with glycerol, and slowly work it through the connector. If the stuck object is valuable, such as a thermometer, the best policy is to cut the rubber with a sharp knife. Care should be taken to avoid force that could potentially cause a thermometer to break, causing injury and the release of mercury.

# The Laboratory Notebook

The laboratory notebook: What you did. How you did it. What you observed. Your conclusions.

A complete, accurate record is an essential part of all laboratory work. The failure to keep such a record means that laboratory labor is lost. An adequate record includes the procedure (what was done), observations (what happened), and conclusions (what the results mean).

Typically, laboratory personnel use a lined, $8.5 \times 11$ paperbound notebook and record all data in ink. Never record **anything** on scraps of paper to be recorded later in the notebook. Do not erase, remove, or obliterate notes. The use of whiteout is not good practice either. Simply draw a single line through incorrect entries.

When working in the laboratory, record everything you do and everything you observe **as it happens**. The recorded observations constitute the most important part of the laboratory record; they form the basis for the conclusions you will draw at the end of each experiment. One way to record observations is in a narrative form. Alternatively, the procedure can be written in outline form on the left-hand side of the page, with the observations recorded on the right-hand side.

In some colleges and universities, you will be expected to have all the relevant information about the running of an experiment entered in your notebook *before coming to the laboratory* so that your textbook will not be needed when you are conducting experiments. In industrial laboratories, a notebook may be designed so that carbon copies of all entries are kept. Notebooks are signed and dated by a supervisor, and the carbon copies are removed from the notebook each day and stored in a secure repository. Notebook records become critical legal documents, especially if a discovery turns out to be worth millions of dollars!

Record the physical properties of the product from your experiment, the yield in grams, and the percent yield. Analyze your results. When things did not turn out as expected, explain why. When your record of an experiment is complete, another chemist should be able to understand your account and determine what you did, how you did it, and what conclusions you reached. From the information in your notebook, this person should be able to repeat your work.

The detailed organization and format of your laboratory notebook will depend on your lab instructor's course design and varies from one instructor to another. On the publisher website, you will find an outline for preparing a typical laboratory record and two examples of completed laboratory records.

*Online Study Center*

**General Resources**
Laboratory Records

# CHAPTER

# 2

# Laboratory Safety, Courtesy, and Waste Disposal

**PRELAB EXERCISE:** Read this chapter carefully. Locate the emergency eyewash station, the safety shower, the fire extinguisher, and the emergency exits in your laboratory. Check your safety glasses or goggles for size and transparency. Learn which reactions must be carried out in the hood. Learn to use your laboratory fire extinguisher; learn how to summon help and how to put out a clothing fire. Learn first aid procedures for acid and alkali spills on the skin. Learn how to tell if your laboratory hood is working properly. Learn which operations under reduced pressure require special precautions. Check to see that compressed gas cylinders in your lab are firmly fastened to benches or walls. Learn the procedures for properly disposing of solid and liquid waste in your laboratory.

Small-scale (microscale) organic experiments are much safer to conduct than their macroscale counterparts that are run on a scale up to 100 times larger. However, for either microscale or macroscale experiments, the organic chemistry laboratory is an excellent place to learn and practice safety. The commonsense procedures practiced here also apply to other laboratories as well as to the shop, kitchen, and studio.

General laboratory safety information—particularly applicable to this organic chemistry laboratory course—is presented in this chapter. But it is not comprehensive. Throughout this text you will find specific cautions and safety information presented as margin notes printed in red. For a brief and thorough discussion of the topics in this chapter, you should read *Safety in Academic Chemistry Laboratories*.[1] There are also some specific admonitions regarding contact lenses (*see* "Eye Safety").

---

1. American Chemical Society Joint Board-Council Committee on Chemical Safety. *Safety in Academic Chemistry Laboratories, Vol. 1: Accident Prevention for College and University Students,* 7th ed.; American Chemical Society: Washington, DC, 2003.

# Important General Rules

- Know the safety rules of your particular laboratory.
- Know the locations of emergency eyewashes, fire extinguishers, safety showers, and emergency exits.
- Never eat, drink, smoke, or apply cosmetics while in the laboratory.
- Wear gloves and aprons when handling corrosive materials.
- Never work alone.

Dress sensibly.

- Perform no unauthorized experiments and do not distract your fellow workers; horseplay has no place in the laboratory.
- Dress properly for lab work. Do not wear open-toed shoes; your feet must be completely covered; wear shoes that have rubber soles and no heels or sneakers. Confine long hair and loose clothes. Do not wear shorts.
- Immediately report any accident to your instructor.
- If the fire extinguisher is used, report this to your instructor.
- Never use mouth suction to fill a pipette.
- Always wash your hands before leaving the laboratory.
- Do not use a solvent to remove a chemical from your skin. This will only hasten the absorption of the chemical through the skin.
- Do not use cell phones or tape, CD, MP3, or similar music players while working in the laboratory.
- Refer to the chemical supplier's hazard warning information or Material Safety Data Sheet (MSDS) when handling a new chemical for the first time.

Eye protection

## *Eye Safety*

**Eye protection is extremely important. Safety glasses of some type must be worn at all times.** It has been determined "that contact lenses can be worn in most work environments provided the same approved eye protection is worn as required of other workers in the area. Approved eye protection refers to safety glasses or goggles."[2] Chemical splash goggles are the preferred eye protection. One of the most important features of safety glasses/goggles is the brow bar. It is critical to have proper eye protection above the eyes; a brow bar satisfies this requirement for adequate splash protection.

Ordinary prescription eyeglasses do not offer adequate protection. Laboratory safety glasses should be constructed of plastic or tempered glass. If you do not have such glasses, wear goggles that afford protection from splashes and objects coming from the side as well as from the front. If plastic safety glasses are permitted in your laboratory, they should have side shields (Fig. 2.1). Eye safety cannot be overemphasized in the chemistry laboratory.

---

2. Ramsey, H.; and Breazeale, W. H. J. Jr. *Chem. Eng. News* **1998**, *76* (22), 6.

**■ FIG. 2.1**
(a) Chemical splash goggles.
(b) Safety glasses.

Brow bar        Brow bar

(a)                (b)

## Laboratory Courtesy

Please show up on time and be prepared for the day's work. Clean up and leave promptly at the end of the lab period. Clean your desktop and sink before you leave the lab. Be certain that no items such as litmus paper, used filter papers, used cotton, or stir bars collect in the sink. Dispose of all trash properly. Please keep the balances clean. Always replace the caps on reagent bottles after use.

## Working with Flammable Substances

Relative flammability of organic solvents.

Flammable substances are the most common hazard in the organic laboratory. Two factors can make today's organic laboratory much safer than its predecessor: (1) making the scale of the experiments as small as possible and (2) not using flames. Diethyl ether (bp 35°C), the most flammable substance you will usually work with in this course, has an ignition temperature of 160°C, which means that a hot plate at that temperature will cause it to burn. For comparison, *n*-hexane (bp 69°C), a constituent of gasoline, has an ignition temperature of 225°C. The flash points of these organic liquids—that is, the temperatures at which they will catch fire if exposed to a flame or spark—are below $-20°C$. These are very flammable liquids; however, if you are careful to eliminate all possible sources of ignition, they are not difficult to work with. Except for water, almost all of the liquids you will use in the laboratory are flammable.

**■ FIG. 2.2**
Solvent safety can.

Flammable vapors travel along bench tops.

**CAUTION**: Keep ignition sources away from flammable liquids.

Bulk solvents should be stored in and dispensed from safety cans (Fig. 2.2). These and other liquids will burn in the presence of the proper amount of their flammable vapors, oxygen, and a source of ignition (most commonly a flame or spark). It is usually difficult to remove oxygen from a fire, although it is possible to put out a fire in a beaker or a flask by simply covering the vessel with a flat object, thus cutting off the supply of air. Your lab notebook might do in an emergency. The best prevention is to pay close attention to sources of ignition—open flames, sparks, and hot surfaces. Remember, the vapors of flammable liquids are **always** heavier than air and thus will travel along bench tops, down drain troughs, and remain in sinks. For this reason all flames within the vicinity of a flammable liquid must be extinguished. Adequate ventilation is one of the best ways to prevent flammable vapors from accumulating. Work in an exhaust hood when manipulating large quantities of flammable liquids.

If a person's clothing catches fire and there is a safety shower close at hand, then shove the person under it and turn the shower on. Otherwise, push the person down and roll him or her over to extinguish the flames (stop, drop, and roll!). It is extremely important to prevent the victim from running or standing because the greatest harm comes from breathing the hot vapors that rise past the mouth. The safety shower

■ **FIG. 2.3**
Carbon dioxide fire
extinguisher.

Flammable vapors plus air
in a confined space are
explosive.

might then be used to extinguish glowing cloth that is no longer aflame. A so-called fire blanket should not be used because it tends to funnel flames past the victim's mouth, and clothing continues to char beneath it. It is, however, useful for retaining warmth to ward off shock after the flames are extinguished.

An organic chemistry laboratory should be equipped with a carbon dioxide or dry chemical (monoammonium phosphate) *fire extinguisher* (Fig. 2.3). To use this type of extinguisher, lift it from its support, pull the ring to break the seal, raise the horn, aim it at the base of the fire, and squeeze the handle. Do not hold on to the horn because it will become extremely cold. Do not replace the extinguisher; report the incident so the extinguisher can be refilled.

When disposing of certain chemicals, be alert for the possibility of *spontaneous combustion*. This may occur in oily rags; organic materials exposed to strong oxidizing agents such as nitric acid, permanganate ion, and peroxides; alkali metals such as sodium; or very finely divided metals such as zinc dust and platinum catalysts. Fires sometimes start when these chemicals are left in contact with filter paper.

## Working with Hazardous Chemicals

If you do not know the properties of a chemical you will be working with, it is wise to regard the chemical as hazardous. The *flammability* of organic substances poses the most serious hazard in the organic laboratory. There is a possibility that storage containers in the laboratory may contribute to a fire. Large quantities of organic solvents should not be stored in glass bottles; they should be stored in solvent safety cans. Do not store chemicals on the floor.

A flammable liquid can often be vaporized to form, with air, a mixture that is explosive in a confined space. The beginning chemist is sometimes surprised to learn that diethyl ether is more likely to cause a laboratory fire or explosion than a worker's accidental anesthesia. The chances of being confined in a laboratory with a concentration of ether high enough to cause a loss of consciousness are extremely small, but a spark in such a room would probably destroy the building.

The probability of forming an explosive mixture of volatile organic liquids with air is far greater than that of producing an explosive solid or liquid. The chief functional groups that render compounds explosive are the *peroxide, acetylide, azide, diazonium, nitroso, nitro,* and *ozonide* groups (Fig. 2.4). Not all members of

■ **FIG. 2.4**
Functional groups that can be
explosive in some compounds.

$$R - O - O - R \qquad R - C \equiv C - \text{Metal}$$

**Peroxide**                    **Acetylide**

$$R - N = N = N \qquad R - NO_2$$

**Azide**                    **Nitro**

$$R - N = O \qquad R - \overset{+}{N} \equiv N$$

**Nitroso**           **Diazonium salts**           **Ozonide**

Safety glasses or goggles must
be worn at all times.

these groups are equally sensitive to shock or heat. You would find it difficult to det-
onate trinitrotoluene (TNT) in the laboratory, but nitroglycerine is treacherously explosive. Peroxides present special problems that are discussed in the next section.

You will need to contend with the corrosiveness of many of the reagents you will
handle. The principal danger here is to the eyes. Proper eye protection is mandatory,
and even small-scale experiments can be hazardous to the eyes. It takes only a single
drop of a corrosive reagent to do permanent damage. Handling concentrated acids
and alkalis, dehydrating agents, and oxidizing agents calls for commonsense care to
avoid spills and splashes and to avoid breathing the often corrosive vapors.

Certain organic chemicals present acute toxicity problems from short-duration
exposure and chronic toxicity problems from long-term or repeated exposure.
Exposure can result from ingestion, contact with the skin, or, most commonly,
inhalation. Currently, great attention is being focused on chemicals that are
teratogens (chemicals that often have no effect on a pregnant woman but cause
abnormalities in a fetus), mutagens (chemicals causing changes in the structure of
the DNA, which can lead to mutations in offspring), and carcinogens (cancer-
causing chemicals). Small-scale experiments significantly reduce these hazards
but do not completely eliminate them.

## Working with Explosive Hazards

### 1. Peroxides

Certain functional groups can make an organic molecule become sensitive to
heat and shock, such that it will explode. Chemists work with these functional
groups only when there are no good alternatives. One of these functional groups,
the peroxide group (R—O—O—R), is particularly insidious because it can form
spontaneously when oxygen and light are present (Fig. 2.5). Ethers, especially
cyclic ethers and those made from primary or secondary alcohols (such as
tetrahydrofuran, diethyl ether, and diisopropyl ether), form peroxides. Other
compounds that form peroxides are aldehydes, alkenes that have allylic hydrogen atoms (such as cyclohexene), compounds having benzylic hydrogens on a

Ethers form explosive peroxides.

**■ FIG. 2.5**
Some compounds that form
peroxides.

**Tetrahydrofuran**    **Diisopropyl ether**    **Dioxane**    **Benzylic compounds**    **Ketones**

**Cyclohexene**    **Vinyl acetate**    **Allylic compounds**    **Aldehydes**

Never distill to dryness.

tertiary carbon atom (such as isopropyl benzene), and vinyl compounds (such as vinyl acetate). Peroxides are low-power explosives but are extremely sensitive to shock, sparks, light, heat, friction, and impact. The greatest danger from peroxide impurities comes when the peroxide-forming compound is distilled. The peroxide has a higher boiling point than the parent compound and remains in the distilling flask as a residue that can become overheated and explode. For this reason, one should never distill a liquid until the distilling flask is completely dry, and the distillation should be run in a hood with the sash down to help contain a possible explosion.

| The Detection of Peroxides | The Removal of Peroxides |
| --- | --- |
| To a solution of 0.01 g of sodium iodide in 0.1 mL of glacial acetic acid, add 0.1 mL of the liquid suspected of containing a peroxide. If the mixture turns brown, a high concentration of peroxide is present; if it turns yellow, a low concentration of peroxide is present. | Pouring the solvent through a column of activated alumina will remove peroxides and simultaneously dry the solvent. Do not allow the column to dry out while in use. When the alumina column is no longer effective, wash the column with 5% aqueous ferrous sulfate and discard the column as nonhazardous waste. |

Problems with peroxide formation are especially critical for ethers. Ethers (R-O-R′) form peroxides readily. Because ethers are frequently used as solvents, they are often used in quantity and then removed to leave reaction products. Cans of diethyl ether should be dated when opened. If opened cans are not used within one month, they should be treated for peroxides and disposed of.

$t$-Butyl methyl ether, $(CH_3)_3C-O-CH_3$, with a primary carbon on one side of the oxygen and a tertiary carbon on the other does not form peroxides easily. It is highly desirable to use this in place of diethyl ether for extraction. Refer to the discussion in Chapter 7.

You may have occasion to use *30% hydrogen peroxide*. This material causes severe burns if it contacts the skin and decomposes violently if contaminated with metals or their salts. Be particularly careful not to contaminate the reagent bottle.

## 2. Closed Systems

A closed system is defined as not being open to the atmosphere. Any sealed system is a closed system. If a closed system is not properly prepared, an explosion may result from the system being under pressure, caused from gas or heat evolution from the reaction or from applied heat to the system. One way to prevent an explosion of a closed system is to use glassware that can withstand the pressure and to evacuate the system under vacuum before it is closed to the atmosphere.

Most reactions are run in open systems; that is, they are run in apparatus that are open to atmosphere, either directly or through a nitrogen line hooked up to a bubbler, which is open to atmosphere.

# Working with Corrosive Substances

Handle strong acids, alkalis, dehydrating agents, and oxidizing agents carefully so as to avoid contact with the skin and eyes and to avoid breathing the corrosive vapors that attack the respiratory tract. All strong, concentrated acids attack the skin and eyes. *Concentrated sulfuric acid* is both a dehydrating agent and a strong acid and will cause very severe burns. *Nitric acid* and *chromic acid* (used in cleaning solutions) also cause bad burns. *Hydrofluoric acid* is especially harmful and causes deep, painful, and slow-healing wounds. It should be used only after thorough instruction. You should wear approved safety glasses or goggles, protective gloves, and an apron when handling these materials.

⚠️ Add $H_2SO4$, $P_2O_5$, CaO, and NaOH to water, not the reverse.

Sodium, potassium, and ammonium hydroxides are common bases that you will encounter. Sodium and potassium hydroxides are extremely damaging to the eyes, and ammonium hydroxide is a severe bronchial irritant. Like sulfuric acid, sodium hydroxide, phosphorous pentoxide, and calcium oxide are powerful dehydrating agents. Their great affinity for water will cause burns to the skin. Because they release a great deal of heat when they react with water, to avoid spattering they should always be added to water rather than water being added to them. That is, the heavier substance should always be added to the lighter one so that rapid mixing results as a consequence of the law of gravity.

You will receive special instructions when it comes time to handle metallic sodium, lithium aluminum hydride, and sodium hydride, three substances that can react explosively with water.

Among the strong oxidizing agents, *perchloric acid* ($HClO_4$) is probably the most hazardous. It can form heavy metal and organic *perchlorates* that are *explosive*, and it can react explosively if it comes in contact with organic compounds.

If one of these substances gets on the skin or in the eyes, wash the affected area with very large quantities of water, using the safety shower and/or eyewash station (Fig. 2.6) until medical assistance arrives. Do not attempt to neutralize the reagent chemically. Remove contaminated clothing so that thorough washing can take place. Take care to wash the reagent from under the fingernails.

Wipe up spilled hydroxide pellets *rapidly*.

Take care not to let the reagents, such as sulfuric acid, run down the outside of a bottle or flask and come in contact with your fingers. Wipe up spills immediately with a very damp sponge, especially in the area around the balances. Pellets of sodium and potassium hydroxide are very hygroscopic and will dissolve in the water they pick up from the air; they should therefore be wiped up very quickly. When handling large quantities of corrosive chemicals, wear protective gloves, a face mask, and a neoprene apron. The corrosive vapors can be avoided by carrying out work in a good exhaust hood.

Do not use a plastic syringe with a metal needle to dispense corrosive inorganic reagents, such as concentrated acids or bases.

# Working with Toxic Substances

Many chemicals have very specific toxic effects. They interfere with the body's metabolism in a known way. For example, the cyanide ion combines irreversibly with hemoglobin to form cyanometmyoglobin, which can no longer carry oxygen.

Aniline acts in the same way. Carbon tetrachloride and other halogenated compounds can cause liver and kidney failure. Carcinogenic and mutagenic substances deserve special attention because of their long-term insidious effects. The ability of certain carcinogens to cause cancer is very great; for example, special precautions are needed when handling aflatoxin $B_1$. In other cases, such as with dioxane, the hazard is so low that no special precautions are needed beyond reasonable, normal care in the laboratory.

Women of childbearing age should be careful when handling any substance of unknown properties. Certain substances are highly suspected as teratogens and will cause abnormalities in an embryo or fetus. Among these are benzene, toluene, xylene, aniline, nitrobenzene, phenol, formaldehyde, dimethylformamide (DMF), dimethyl sulfoxide (DMSO), polychlorinated biphenyls (PCBs), estradiol, hydrogen sulfide, carbon disulfide, carbon monoxide, nitrites, nitrous oxide, organolead and mercury compounds, and the notorious sedative thalidomide. Some of these substances will be used in subsequent experiments. Use care when working with these (and all) substances. Of course, the leading known cause of embryotoxic effects is ethyl alcohol in the form of maternal alcoholism. The amount of ethanol vapor inhaled in the laboratory or absorbed through the skin is so minute that it is unlikely to have morbid effects.

It is impossible to avoid handling every known or suspected toxic substance, so it is wise to know what measures should be taken. Because the eating of food or the consumption of beverages in the laboratory is strictly forbidden and because one should never taste material in the laboratory, the possibility of poisoning by mouth is remote. Be more careful than your predecessors—the hallucinogenic properties of LSD and **all** artificial sweeteners were discovered by accident. The two most important measures to be taken, then, are (1) avoiding skin contact by wearing the *proper* type of protective gloves (*see* "Gloves") and (2) avoiding inhalation by working in a good exhaust hood.

Many of the chemicals used in this course will be unfamiliar to you. Their properties can be looked up in reference books, a very useful one being the *Aldrich Handbook of Fine Chemicals*.[3] Note that 1,4-dichlorobenzene is listed as a "toxic irritant" and naphthalene is listed as an "irritant." Both are used as mothballs. Camphor, used in vaporizers, is classified as a "flammable solid irritant." Salicylic acid, which we will use to synthesize aspirin (Chapter 41), is listed as a "moisture-sensitive toxic." Aspirin (acetylsalicylic acid) is classified as an "irritant." Caffeine, which we will isolate from tea or cola syrup (Chapter 7), is classified as "toxic." Substances not so familiar to you, for example, 1-naphthol and benzoic acid, are classified, respectively, as "toxic irritant" and "irritant." To put things in perspective, nicotine is classified as "highly toxic." Pay attention to these health warnings. In laboratory quantities, common chemicals can be hazardous. Wash your hands carefully after coming in contact with laboratory chemicals. Consult the *Hazardous Laboratory Chemicals Disposal Guide*[4] for information on truly hazardous chemicals.

---

3. Free copies of this catalog can be obtained from http://www.sigmaaldrich.com/Brands/Aldrich.html.

4. Armour, M-A. *Hazardous Laboratory Chemicals Disposal Guide*, 3rd ed.; CRC Press LLC: Boca Raton, FL, 2003.

Because you have not had previous experience working with organic chemicals, most of the experiments you will carry out in this course will not involve the use of known carcinogens, although you will work routinely with flammable, corrosive, and toxic substances. A few experiments involve the use of substances that are suspected of being carcinogenic, such as hydrazine. If you pay proper attention to the rules of safety, you should find working with these substances no more hazardous than working with ammonia or nitric acid. The single, short-duration exposure you might receive from a suspected carcinogen, should an accident occur, would probably have no long-term consequences. The reason for taking the precautions noted in each experiment is to learn, from the beginning, good safety habits.

## Gloves

Be aware that protective gloves in the organic laboratory may not offer much protection. Polyethylene and latex rubber gloves are very permeable to many organic liquids. An undetected pinhole may bring with it long-term contact with reagents. Disposable polyvinyl chloride (PVC) gloves offer reasonable protection from contact with aqueous solutions of acids, bases, and dyes, but no one type of glove is useful as a protection against all reagents. It is for this reason that no less than 25 different types of chemically resistant gloves are available from laboratory supply houses. Some gloves are quite expensive and will last for years.

If disposable gloves are available, fresh nitrile gloves can be worn whenever handling a corrosive substance and disposed of once the transfer is complete. When not wearing gloves, it is advised that you wash your hands every 15 min to remove any traces of chemicals that might be on them.

## Using the Laboratory Hood

Modern practice dictates that in laboratories where workers spend most of their time working with chemicals, there should be one exhaust hood for every two people. This precaution is often not possible in the beginning organic chemistry laboratory, however. In this course you will find that for some experiments the hood must be used and for others it is advisable; in these instances, it may be necessary to schedule experimental work around access to the hoods. Many experiments formerly carried out in the hood can now be carried out at the lab desk because the concentration of vapors is significantly minimized when working at a microscale.

Keep the hood sash closed.

The hood offers a number of advantages when working with toxic and flammable substances. Not only does it draw off the toxic and flammable fumes, but it also affords an excellent physical barrier on all four sides of a reacting system when the sash is pulled down. If a chemical spill occurs, it may be contained within the hood.

It is your responsibility each time you use a hood to see that it is working properly. You should find some type of indicating device that will give you this information on the hood itself. A simple propeller on a cork works well. The hood is a backup device. Never use it alone to dispose of chemicals by evaporation; use an aspirator tube or carry out a distillation. Toxic and flammable fumes should be trapped or condensed in some way and disposed of in the prescribed manner. The sash should be pulled down unless you are actually carrying out manipulations on the experimental apparatus. The water, gas, and electrical controls should be on the outside of the hood so it is not necessary to open the hood to make adjustments. The ability of the hood to remove vapors is greatly enhanced if the apparatus is kept as close to the back of the hood as possible. Everything should be at least 15 cm back from the hood sash. Chemicals should not be permanently stored in the hood but should be removed to ventilated storage areas. If the hood is cluttered with chemicals, you will not achieve a good, smooth airflow or have adequate room for experiments.

## Working at Reduced Pressure

Implosion

Whenever a vessel or system is evacuated, an implosion could result from atmospheric pressure on the empty vessel. It makes little difference whether the vacuum is perfect or just 10 mm Hg; the pressure difference is almost the same (760 versus 750 mm Hg). An implosion may occur if there is a star crack in the flask or if the flask is scratched or etched. Only with heavy-walled flasks specifically designed for vacuum filtration is the use of a safety shield (Fig. 2.7) ordinarily unnecessary. The chances of implosion of the apparatus used for microscale experiments are remote.

Dewar flasks (thermos bottles) are often found in the laboratory without shielding. These should be wrapped with friction tape or covered with a plastic net to prevent the glass from flying about in case of an implosion (Fig. 2.8). Similarly, vacuum desiccators should be wrapped with tape before being evacuated.

■ **FIG. 2.7**
Safety shield.

■ **FIG. 2.8**
Dewar flask with safety net in place.

# Working with Compressed Gas Cylinders

Many reactions are carried out under an inert atmosphere so that the reactants and/or products will not react with oxygen or moisture in the air. Nitrogen and argon are the inert gases most frequently used. Oxygen is widely used both as a reactant and to provide a hot flame for glassblowing and welding. It is used in the oxidative coupling of alkynes (Chapter 24). Helium is the carrier gas used in gas chromatography. Other gases commonly used in the laboratory are ammonia, often used as a solvent; chlorine, used for chlorination reactions; acetylene, used in combination with oxygen for welding; and hydrogen, used for high- and low-pressure hydrogenation reactions.

The following rule applies to all compressed gases: Compressed gas cylinders should be firmly secured at all times. For temporary use, a clamp that attaches to the laboratory bench top and has a belt for the cylinder will suffice (Fig. 2.9). Eyebolts and chains should be used to secure cylinders in permanent installations. Flammable gases should be stored 20 ft from oxidizing gases.

A variety of outlet threads are used on gas cylinders to prevent incompatible gases from being mixed because of an interchange of connections. Both right- and left-handed external and internal threads are used. Left-handed nuts are notched to differentiate them from right-handed nuts. Right-handed threads are used on non-fuel and oxidizing gases, and left-handed threads are used on fuel gases, such as hydrogen. Never grease the threads on tank or regulator valves.

Cylinders come equipped with caps that should be left in place during storage and transportation. These caps can be removed by hand. Under these caps is a cylinder valve. It can be opened by turning the valve counterclockwise; however, because most compressed gases in full cylinders are under very high pressure (commonly up to 3000 lb/in.$^2$), a pressure regulator must be attached to the cylinder. This pressure regulator is almost always of the diaphragm type and has two gauges, one indicating the pressure in the cylinder, the other the outlet pressure (Fig. 2.10). On the outlet, low-pressure side of the regulator is a small needle valve and then the outlet connector. After connecting the regulator to the cylinder,

Always clamp gas cylinders.

■ **FIG. 2.9**
Gas cylinder clamp.

Hand-wheel valve

■ **FIG. 2.10**
Gas pressure regulator. Turn two-flanged diaphragm valve clockwise to increase outlet pressure.

*Clockwise* movement of diaphragm valve handle *increases* pressure.

Outlet pressure gauge

Cylinder pressure gauge

Needle valve

Outlet

Diaphragm valve

Notches indicating left-hand thread

■ **FIG. 2.11**
Gas cylinder cart.

Never attempt to identify an unknown organic compound by smelling it.

Clean up spills rapidly.

Mercury requires special measures—see your instructor.

unscrew the diaphragm valve (turn it counterclockwise) before opening the cylinder valve on the top of the cylinder. This valve should be opened only as far as necessary. For most gas flow rates in the laboratory, this will be a very small amount. The gas flow or pressure is increased by turning the two-flanged diaphragm valve **clockwise**. When the apparatus is not being used, turn off the cylinder valve (clockwise) on the top of the cylinder (Fig. 2.9). Before removing the regulator from the cylinder, reduce the flow or pressure to zero. Cylinders should never be emptied to zero pressure and left with the valve open because the residual contents will become contaminated with air. Empty cylinders should be labeled "empty." Their valves should be closed and capped, and the cylinders should be returned to the storage area and separated from full cylinders. Gas cylinders should never be dragged or rolled from place to place but should be fastened onto and moved in a cart designed for that purpose (Fig. 2.11). The cap should be in place whenever the cylinder is moved. If you detect even a small leak from any valve or connection, immediately seek the help of an instructor to remedy the problem. If there is a major leak of a corrosive or flammable gas, notify those around you to leave the area and seek help immediately.

## Odoriferous Chemicals

Some organic chemicals just smell bad. Among these are the thiols (organic derivatives of hydrogen sulfide), isonitriles, many amines (e.g., cadaverine), and butyric acid. Washing apparatus and, if necessary, hands in a solution of a quaternary ammonium salt may solve the problem. These compounds apparently complex with many odoriferous substances, allowing them to be rinsed away. Commercial products (e.g., Zephiran, Roccal, San-O-Fec, and others) are available at pet and farm supply stores.

## Waste Disposal–Cleaning Up

*Spilled solids* should simply be swept up and placed in the appropriate solid waste container. This should be done promptly because many solids are hygroscopic and become difficult if not impossible to sweep up in a short time. This is particularly true of sodium hydroxide and potassium hydroxide; these strong bases should be dissolved in water and neutralized with sodium bisulfate before disposal.

The method used to clean up spills depends on the type and amount of chemical spilled. If more than 1 or 2 g or mL of any chemical, particularly a corrosive or volatile one, is spilled, you should consult your instructor for the best way to clean up the spill. If a large amount of volatile or noxious liquid is spilled as might happen if a reagent bottle is dropped and broken, advise those in the area to leave the laboratory and contact your instructor immediately. If a spill involves a large amount of flammable liquid, be aware of any potential ignition sources and try to eliminate them. Large amounts of spilled acid can be neutralized with granular limestone or cement; large amounts of bases with solid sodium bisulfate, $NaHSO_4$. Large amounts of volatile liquids can be absorbed into materials such as vermiculite, diatomaceous earth, dry sand, kitty litter, or paper towels and these materials swept up and placed in a separate disposal container.

For spills of amounts less than 2 g of chemical, proceed as follows. Acid spills should be neutralized by dropping solid sodium carbonate onto them, testing the pH, wiping up the neutralized material with a sponge, and rinsing the neutral salt solution down the drain. Bases should be neutralized by sprinkling solid sodium bisulfate onto them, checking the pH, and wiping up with a sponge or towel. Do not use paper towels to wipe up spills of strong oxidizers such as dichromates or nitrates; the towels can ignite. Bits of sodium metal will also cause paper towels to ignite. Sodium metal is best destroyed with *n*-butyl alcohol. Always wear goves when cleaning up a spill.

### Cleaning Up

In the not-too-distant past it was common practice to wash all unwanted liquids from the organic laboratory down the drain and to place all solid waste in the trash basket. For environmental reasons, this is never a wise practice and is no longer allowed by law.

Organic reactions usually employ a solvent and often involve the use of a strong acid, a strong base, an oxidant, a reductant, or a catalyst. None of these should be washed down the drain or placed in the wastebasket. Place the material, classified as waste, in containers labeled for nonhazardous solid waste, organic solvents, halogenated organic solvents, or hazardous wastes of various types.

Waste containers:
Nonhazardous solid waste
Organic solvents
Halogenated organic solvents
Hazardous waste (various types)

Nonhazardous waste encompasses such solids as paper, corks, TLC plates, solid chromatographic absorbents such as alumina or silica that are dry and free of residual organic solvents, and solid drying agents such as calcium chloride or sodium sulfate that are also dry and free of residual organic solvents. These will ultimately end up in a sanitary landfill (the local dump). Any chemicals that are leached by rainwater from this landfill must not be harmful to the environment. In the *organic solvents* container are placed the solvents that are used for recrystallization and for running reactions, cleaning apparatus, and so forth. These solvents can contain dissolved, solid, nonhazardous organic solids. This solution will go to an incinerator where it will be burned. If the solvent is halogenated (e.g., dichloromethane) or contains halogenated material, it must go in the *halogenated organic solvents* container. Ultimately, this will go to a special incinerator equipped with a scrubber to remove HCl from the combustion gases. The organic laboratory should also have several other waste disposal containers for special hazardous, reactive, and noncombustible wastes that would be incompatible with waste organic solvents and other materials. For example, it would be dangerous to place oxidants in lysts with many organics. In particular, separate waste containers should be provided for toxic heavy metal wastes containing mercury, chromium, or lead salts, and so forth. The cleaning up sections throughout this text will call your attention to these special hazards.

Hazardous wastes such as sodium hydrosulfite (a reducing agent), platinum catalysts, and $Cr^{6+}$ (an oxidizing agent) cannot be burned and must be shipped to a secure landfill. To dispose of small quantities of a hazardous waste (e.g., solid mercury hydroxide), the material must be carefully packed in bottles and placed in a 55-gal ($\approx$208-L) drum called a lab pack, to which an inert material has been added. The lab pack is carefully documented and hauled off to a site where such waste is disposed of by a bonded, licensed, and heavily regulated waste disposal company. Formerly, many hazardous wastes were disposed of by burial in a secure landfill.

The kinds of hazardous waste that can be thus disposed of have become extremely limited in recent years, and much of the waste undergoes various kinds of treatment at the disposal site (e.g., neutralization, incineration, or reduction) to put it in a form that can be safely buried in a secure landfill or flushed to a sewer. There are relatively few places for approved disposal of hazardous waste. For example, there are none in New England, so most hazardous waste from this area is trucked to South Carolina. The charge to small generators of waste is usually based on the volume of waste. So, 1000 mL of a 2% cyanide solution would cost far more to dispose of than 20 g of solid cyanide, even though the total amount of this poisonous substance is the same. It now costs far more to dispose of most hazardous chemicals than it does to purchase them new.

American law states that a material is not a waste until the laboratory worker declares it a waste. So—for pedagogical and practical reasons—we want you to regard the chemical treatment of the byproducts of each reaction in this text as a part of the experiment.

In the section titled "Cleaning Up" at the end of each experiment, the goal is to reduce the volume of hazardous waste, to convert hazardous waste to less hazardous waste, or to convert it to nonhazardous waste. The simplest example is concentrated sulfuric acid. As a byproduct from a reaction, it is obviously hazardous. But after careful dilution with water and neutralization with sodium carbonate, the sulfuric acid becomes a dilute solution of sodium sulfate, which in almost every locale can be flushed down the drain with a large quantity of water. Anything flushed down the drain must be accompanied by a large excess of water. Similarly, concentrated bases can be neutralized, oxidants such as $Cr^{6+}$ can be reduced, and reductants such as hydrosulfite can be oxidized (by hypochlorite or household bleach). Dilute solutions of heavy metal ions can be precipitated as their insoluble sulfides or hydroxides. The precipitate may still be a hazardous waste, but it will have a much smaller volume.

One type of hazardous waste is unique: a harmless solid that is damp with an organic solvent. Alumina from a chromatography column and calcium chloride used to dry an ether solution are examples. Being solids, they obviously cannot go in the organic solvents container, and being flammable they cannot go in the nonhazardous waste container. A solution to this problem is to spread the solid out in the hood to let the solvent evaporate. You can then place the solid in the nonhazardous waste container. The savings in waste disposal costs by this operation are enormous. However, be aware of the regulations in your area as they may not allow evaporation of small amounts of organic solvents in a hood. If this is the case, special containers should be available for disposal of these wet solids.

Our goal in "Cleaning Up" is to make you more aware of *all* aspects of an experiment. Waste disposal is now an extremely important aspect. Check to be sure the procedure you use is permitted by law in your location. Three sources of information have been used as the basis of the procedures at the end of each experiment: the *Aldrich Catalog Handbook of Fine Chemicals*,[5] which gives brief

---

5. See footnote 3 on page 30.

disposal procedures for every chemical in their catalog; *Prudent Practices in the Laboratory: Handling and Disposal of Chemicals*[6]; and the *Hazardous Laboratory Chemicals Disposal Guide*.[7] The last title listed here should be on the bookshelf of every laboratory. This 464-page book gives detailed information about hundreds of hazardous substances, including their physical properties, hazardous reactions, physiological properties, health hazards, spillage disposal, and waste disposal. Many of the treatment procedures in "Cleaning Up" are adaptations of these procedures. *Destruction of Hazardous Chemicals in the Laboratory*[8] complements this book.

The area of waste disposal is changing rapidly. Many levels of laws apply—local, state, and federal. What may be permissible to wash down the drain or evaporate in the hood in one jurisdiction may be illegal in another, so before carrying out any waste disposal, check with your college or university waste disposal officer.

## Biohazards

The use of microbial growth bioassays is becoming common in chemistry laboratories. The use of infectious materials presents new hazards that must be recognized and addressed. The first step in reducing hazards when using these materials is to select infectious materials that are known not to cause illness in humans and are of minimal hazard to the environment. A number of procedures should be followed to make use of these materials safe: Individuals need to wash hands after they handle these materials and before they leave the laboratory; work surfaces need to be decontaminated at the end of each use; and all infectious materials need to be decontaminated before disposal.

## QUESTIONS

1. Write a balanced equation for the reaction between the iodide ion, a peroxide, and the hydrogen ion. What causes the orange or brown color?
2. Why does the horn of the carbon dioxide fire extinguisher become cold when the extinguisher is used?
3. Why is water *not* used to extinguish most fires in an organic laboratory?

6. National Research Council. *Prudent Practices in the Laboratory: Handling and Disposal of Chemicals* National Academy Press: Washington, DC, 1995.

7. See footnote 4 on page 30.

8. Lunn, G.; Sansone, E. B. *Destruction of Hazardous Chemicals in the Laboratory*; Wiley: New York, 1994.

# CHAPTER 3

# Melting Points and Boiling Points

*Online Study Center*

This icon will direct you to
techniques, equipment setups,
and online resources at
http://college. hmco.com/PIC/
williamsonMMOE5e.

**PRELAB EXERCISE:** Draw the structure of each of the following organic compounds, identify the intermolecular attractive forces for each, and list them in order of increasing boiling point as predicted by the relative strength of those intermolecular forces: (a) acetaldehyde, (b) sodium formate, (c) ethanol, and (d) propane.

## PART 1: Five Concepts for Predicting Physical Properties

In organic chemistry, structure is everything. A molecule's structure determines both its physical properties and its reactivity. Since the dawn of modern chemistry 200 years ago, over 20 million substances, most of them organic compounds, have been isolated, and their properties and reactions have been studied. It became apparent from these studies that certain structural features in organic molecules would affect the observed properties in a predictable way and that these millions of organic compounds could be organized into classes based on molecular size, composition, and the pattern of bonds between atoms. Chemists also saw trends in certain properties based on systematic changes in these structural features. This organized knowledge allows us to look at a compound's structure and to predict the physical properties of that compound.

Physical properties, such as melting point, boiling point, and solubility, are largely determined by *intermolecular attractive forces*. You learned about these properties in previous chemistry courses. Because a solid understanding of these concepts is critical to understanding organic chemistry, we will review the different types of forces in the context of structural organic chemistry. Using five simple concepts, you should be able to look at the structures of a group of different organic molecules and predict which might be liquids, gases, or solids and which might be soluble in water. You can often predict the boiling point, melting point, or solubility of one structure relative to other structures. In fact, as your knowledge grows, you may be able to predict a compound's approximate melting or boiling temperature based on its structure. Your understanding of intermolecular attractive forces will be very useful in this chapter's experiments on melting and boiling points and those in Chapters 5 and 6 that involve distillation and boiling points.

# 1. London Attractive Forces (Often Called Van der Waals Forces)

Organic molecules that contain only carbon and hydrogen (hydrocarbons) are weakly attracted to each other by London forces. Though weak, these attractive forces increase as molecular size increases. Thus, the larger the molecule, the greater the attractive force for neighboring molecules and the greater the energy required to get two molecules to move apart. This trend can be seen if we compare

$CH_4$

**Methane**
mp −182°C
bp −162°C
Gas at room temp.

**Hexane, $C_6H_{14}$**
mp −95°C
bp +69°C
Liquid at room temp.

**Tetracosane, $C_{24}H_{50}$**
mp +51°C
bp +391°C
Solid at room temp.

the melting points and boiling points of three hydrocarbons of different size: methane, hexane, and tetracosane.

We know that methane is called natural gas because methane's physical state at room temperature (20°C = 68°F) is a gas. Its London forces are so weak that methane must be cooled to −162°C at 1 atm of pressure before the molecules will stick together enough to form a liquid. Hexane is a very common liquid solvent found on most organic laboratory shelves. The intermolecular forces between its molecules are strong enough to keep them from flying apart, but the molecules are still able to flex and slide by each other to form a fluid. Hexane must be heated above 69°C, which is 231°C hotter than methane, to convert all its molecules to a gas. Tetracosane, a $C_{24}$ solid hydrocarbon, is four times larger than hexane, and its London forces are strong enough to hold the molecules rigidly in place at room temperature. Tetracosane is one of the many long-chain hydrocarbons found in candle wax, which must be heated in order to disrupt the intermolecular forces and melt the wax into a liquid. A lot of energy is required to convert liquid tetracosane to a gas, as evidenced by its extremely high boiling point (391°C).

## 2. Dipole-Dipole Attractive Forces

The attractive forces between molecules increases when functional groups containing electronegative atoms such as chlorine, oxygen, and nitrogen are present because these atoms are more electronegative than carbon. These atoms pull electrons toward themselves, making their end of the bond slightly negatively charged ($\delta^-$) and leaving the carbon slightly positively charged ($\delta^+$), as shown for isopropyl chloride and acetone.

**Isobutane**
mp −137°C
bp 0°C
Gas at room temp.

**Isopropyl chloride**
mp −117°C
bp +35°C
Liquid at room temp.

**Acetone**
mp −94°C
bp +56°C
Liquid at room temp.

A bond with a slight charge separation is termed a *polar* bond, and polar bonds often give a molecule a *dipole*: slightly positive and negative ends symbolized by an arrow in the direction of the negative charge (+→). Attraction of the positive end of one molecule's dipole to the negative end of another's dipole occurs between polar molecules, which increases the intermolecular attractive force. Dipole-dipole attractive forces are stronger than London forces, as demonstrated in the previous examples that show an increase in melting point and boiling point when a methyl group of isobutane is replaced by chlorine or oxygen.

## 3. Hydrogen Bonding

Hydrogen bonding is an even stronger intermolecular attractive force, as evidenced by the large increase in the melting point and boiling points of the alcohol methanol (MW = 32) relative to those of the hydrocarbon ethane (MW = 30), both of comparable molecular weight. Hydrogen bonding occurs with organic molecules containing O—H groups (for example, alcohols and carboxylic acids) or N—H groups (for example, amines or amides). The hydrogen in these groups is attracted to the unshared pair of electrons on the O or N of another molecule, forming a hydrogen bond, often symbolized by a dashed line, which is shown for methanol.

$CH_3$—$CH_3$

**Ethane**
mp −172°C
bp −88°C
Gas at room temp.

**Methanol**
mp −97°C
bp +65°C
Liquid at room temp.

As this example indicates, the hydrogen bonds extend throughout the liquid.

One can think of these hydrogen bonds as molecular Velcro that can be pulled apart if there is sufficient energy. Hydrogen bonding plays a major role in the special physical behavior of water and is a major determinant of the chemistry of proteins and nucleic acids in living systems.

## 4. Ionic Attractive Forces

Recall that ionic substances, such as table salt (NaCl), are usually solids with high melting points (>300°C) due to the strong attractive forces between positive and negative ions. Most organic molecules contain only covalent bonds and have no ionic attractive forces between them. However, there are three important exceptions involving acidic or basic functional groups that can form ionic structures as the pH is raised or lowered.

1.  The hydrogen on the —OH of the carboxyl group in carboxylic acids, such as acetic acid, is acidic ($H^+$ donating) and reacts with bases such as potassium hydroxide (KOH) and sodium bicarbonate ($NaHCO_3$) to form salts. The process is reversed by lowering the pH by adding an acid.

Potassium acetate (mp 306°C) ⇌ [KOH / H⁺] Acetic acid (mp 17°C) ⇌ [NaHCO₃ / H⁺] Sodium acetate (mp >300°C)

The dry salts are ionic and have very high melting points, which is expected for ionic substances. Note that this acidity is *not* observed for alcohols where the —OH group is attached to a singly bonded ($sp^3$ or saturated) carbon.

2. The hydrogen on an —OH group that is attached to an aromatic ring is weakly acidic and reacts with strong bases such as sodium hydroxide (NaOH) to form high melting ionic salts, as evidenced by the reaction of phenol to sodium phenolate. Again, the reaction is reversed by the addition of an acid.

**Phenol**
mp 41°C

**Sodium phenolate**
mp 382°C

Note again that this acidity is *not* observed for alcohols where the —OH group is attached to a singly bonded ($sp^3$ or saturated) carbon.

3. Amines (but not amides) are basic (H⁺ accepting) and will react with acid to form ionic amine salts with elevated melting points, as shown, for example, for isopropyl amine.

**Isopropyl amine**
mp −95°C

**Isopropyl amine hydrochloride**
mp 162°C

Amine salts can be converted back to amines by raising the pH by adding a base.

# 5. Competing Intermolecular Forces and Solubility

For pure compounds containing identical molecules, the total attractive force between molecules is the sum of all the attractive forces listed previously, both weak and strong. These forces tend to work together to raise melting and boiling points as the size of the molecule's hydrocarbon skeleton increases and as polar, hydrogen bonded, or ionic functional groups are incorporated into the molecule.

However, solubility involves the interaction of two different molecules, which may have different types of attractive forces. When we try to dissolve one substance in another, we have to disrupt the attractive forces in both substances to get the molecules of the two substances to intermingle. For example, to get water (polar with hydrogen bonding) and the hydrocarbons (nonpolar and no hydrogen bonding) in motor oil to mix and dissolve in one another, we would have to disrupt the London attractive forces between the oil molecules and the hydrogen bonds between the water molecules. Because London forces are weak, separating the oil molecules does not require much energy. However, breaking apart the much stronger hydrogen bonds by inserting oil molecules between the water molecules requires considerable energy and is unfavorable. Therefore, oil or even the simplest hydrocarbon, methane, is insoluble in water. This is the molecular basis of the old adage "Oil and water don't mix."

In addition to carbon and hydrogen, the majority of organic molecules contain other elements, such as nitrogen and oxygen, in functional groups that can be polar, exhibit hydrogen bonding, show ionic tendencies, or have any combination thereof. Can we predict the water solubility of these organic substances? Let's look at some examples and see.

Figure 3.1 shows a collection of small organic molecules of about the same size and molecular weight listed in order of increasing boiling point or melting point, which is consistent with the types of intermolecular forces discussed. With the exception of the hydrocarbon butane, all of these substances are very soluble in water—at least 100 g will dissolve in 1 L of water. It appears that the intermolecular attractive forces between these polar, hydrogen bonded, or ionic molecules and water compensates for the disruption of hydrogen bonding between water molecules so the organic molecules can move into and intermingle with the water molecules—in other words, dissolve.

Figure 3.2 shows a collection of larger organic molecules than those in Figure 3.1, again listed in order of increasing melting and boiling point, which is consistent with our knowledge of the strength of intermolecular attractive forces. The important difference for this group is that the hydrocarbon portion of each molecule is four carbons larger than for those in Figure 3.1. We might predict that the larger hydrocarbon portion makes them behave more like the water-insoluble hydrocarbon octane. Indeed, the larger hydrocarbon portion of these molecules greatly reduces their solubility in water to less than 5 g/L of water except for the two ionic compounds. These two, the amine hydrochloride and the sodium carboxylate compounds, have higher solubility—tens of grams per liter of water— proof that ionic charges can interact strongly with water molecules.

This trend in water solubility based on the size of the hydrocarbon portion continues for the set of even larger organic molecules shown in Figure 3.3 containing $C_{16}$ to $C_{18}$ hydrocarbon chains. Most are virtually insoluble in water; even the ionic forms have solubilities of less than 1 g/L of water.

In addition to water, many other liquid solvents are used in the organic laboratory to dissolve substances, including acetone, dichloromethane ($CH_2Cl_2$), ethanol ($CH_3CH_2OH$), diethyl ether ($CH_3CH_2OCH_2CH_3$), hexane, methanol, and toluene ($C_6H_5CH_3$). Predicting the solubility of different organic compounds,

■ **FIG. 3.1**

Some small organic molecules containing 4 atoms (carbon, oxygen, and nitrogen) in order of increasing melting or boiling points.

**C₄ Hydrocarbon, butane**
mp −138°C; bp 0°C
Insoluble in $H_2O$

**C₃ Amine**
mp −43°C; bp 48°C
Soluble in $H_2O$; Soluble in organic solvents

**C₃ Ketone, acetone**
mp −94°C; bp 56°C
Soluble in $H_2O$; Soluble in organic solvents

**C₃ Alcohol**
mp −127°C; bp 97°C
Soluble in $H_2O$; Soluble in organic solvents

**C₂ Carboxylic acid. acetic acid**
mp 17°C; bp 117°C
Soluble in $H_2O$; Soluble in organic solvents

**C₃ Amine hydrochloride**
mp 161°C
Soluble in $H_2O$; Insoluble in organic solvents

**Sodium salt of C₂ carboxylic acid**
mp 324°C
Soluble in $H_2O$; Insoluble in organic solvents

■ **FIG. 3.2**

Some organic molecules containing 8 atoms (carbon, oxygen, and nitrogen) in order of increasing melting or boiling points.

**C₈ Hydrocarbon, octane**
mp −57°C; bp 126°C
Insoluble in $H_2O$

**C₇ Amine**
mp −18°C; bp 154°C
Insoluble in $H_2O$; Soluble in organic solvents

**C₇ Ketone**
mp −35°C; bp 150°C
Insoluble in $H_2O$; Soluble in organic solvents

**C₇ Alcohol**
mp −34°C; bp 175°C
Insoluble in $H_2O$; Soluble in organic solvents

**C₆ Carboxylic acid**
mp −2°C; bp 200°C
Insoluble in $H_2O$; Soluble in organic solvents

**C₇ Amine hydrochloride**
mp 242°C
Soluble in $H_2O$; Insoluble in organic solvents

**Sodium salt of C₆ carboxylic acid**
mp 245°C
Soluble in $H_2O$; Insoluble in organic solvents

such as those shown in Figures 3.1, 3.2, and 3.3, in these solvents can be done using the intermolecular attractive force concepts discussed here. For example, because the molecules in Figure 3.3 have long hydrocarbon skeletons, you would predict that these would probably be soluble in the hydrocarbon hexane. This predictive rule can be summed up as "Like dissolves like." You might also predict

### ■ FIG. 3.3

Some organic molecules containing 18 atoms (carbon, oxygen, and nitrogen) in order of increasing melting or boiling points.

$C_{18}$ **Hydrocarbon**
mp 6°C
Insoluble in $H_2O$

$C_{17}$ **Ketone**
mp 48°C
Insoluble in $H_2O$; Soluble in organic solvents

$C_{17}$ **Amine**
mp 49°C
Insoluble in $H_2O$; Soluble in organic solvents

$C_{17}$ **Alcohol**
mp 53°C
Insoluble in $H_2O$; Soluble in organic solvents

$C_{16}$ **Carboxylic acid**
mp 63°C
Insoluble in $H_2O$; Soluble in organic solvents

$C_{17}$ **Amine hydrochloride**
mp 165°C
Soluble in $H_2O$; Slightly soluble in organic solvents

**Sodium salt of $C_{16}$ carboxylic acid**
mp 270°C
Soluble in $H_2O$; Slightly soluble in organic solvents

that the two ionic materials would be the least soluble in hexane because of the strong intermolecular forces in these solids, as evidenced by their high melting points, which are so unlike the weak London forces in hexane. The solubility of organic compounds in organic solvents and water at low, neutral, and high pH will be considered in more detail when you learn about recrystallization in Chapter 4 and extraction in Chapter 7.

# PART 2: Melting Points
## A. Thermometers

**T**here are a few types of thermometers that can be used to read melting point (and boiling point) temperatures: mercury-in-glass thermometers, non-mercury thermometers, and digital thermometers. Mercury-in-glass thermometers provide highly accurate readings and are ideal for use at high temperatures (260°C–400°C). Care should be taken not to break the thermometer, which will release the toxic mercury. Teflon-coated mercury thermometers are usable up to 260°C and are less likely to spill mercury if broken. If breakage does happen, inform your instructor immediately because special equipment is required to clean up mercury spills. A digital thermometer (Fig. 3.4) has a low heat capacity and a fast response time. It is more robust than a glass thermometer and does not, of course, contain mercury. Nonmercury thermometers may be filled with isoamyl benzoate (a biodegradable liquid) or with a custom organic red-spirit liquid instead of mercury. These thermometers give reasonably accurate readings at temperatures up to 150°C, but above this temperature they need to be carefully calibrated. These thermometers should be stored vertically to prevent thread separation.

⚠️
**CAUTION**: Mercury is toxic. Immediately report any broken thermometers to your instructor.

## B. Melting Points

The melting point of a pure solid organic compound is one of its characteristic physical properties, along with molecular weight, boiling point, refractive index, and density. A pure solid will melt reproducibly over a narrow range of temperatures, typically less than 1°C. The process of determining this melting point is done on a truly micro scale using less than 1 mg of material. The apparatus is simple, consisting of a thermometer, a capillary tube to hold the sample, and a heating bath.

Melting points—a micro technique

Melting points are determined for three reasons: (1) If the compound is a known one, the melting point will help to characterize the sample. (2) If the compound is new, then the melting point is recorded to allow future characterization by others. (3) The range of the melting point is indicative of the purity of the compound; an impure compound will melt over a wide range of temperatures. Recrystallization of the compound will purify it and decrease the melting point range. In addition, the entire range will be displaced upward. For example, an impure sample might melt from 120°C to 124°C and after recrystallization melt at 125°C–125.5°C. A solid is considered pure if the melting point does not rise after recrystallization.

Characterization

An indication of purity

A crystal is an orderly arrangement of molecules in a solid. As heat is added to the solid, the molecules will vibrate and perhaps rotate but still remain a solid. At a characteristic temperature the crystal will suddenly acquire the necessary energy to overcome the forces that attract one molecule to another and will undergo translational motion—in other words, it will become a liquid.

The forces by which one molecule is attracted to another include ionic attractions, London forces, hydrogen bonds, and dipole-dipole attractions. Most, but by no means all, organic molecules are covalent in nature and melt at temperatures below 300°C. Typical inorganic compounds are ionic and have much higher

■ **FIG. 3.4**
A digital thermometer.

Digital
thermometer

Surface
probe

■ **FIG. 3.5**

A melting point–composition diagram for mixtures of the solids X and Y.

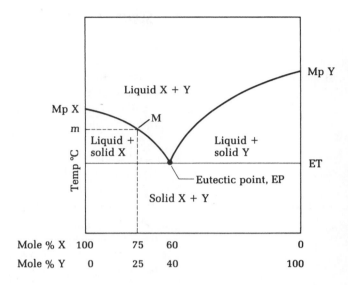

| Mole % X | 100 | | 75 | 60 | | 0 |
| Mole % Y | 0 | | 25 | 40 | | 100 |

melting points (e.g., sodium chloride melts at 800°C). Ionic organic molecules often decompose before melting, as do compounds having many strong hydrogen bonds, such as sucrose.

**Melting point generalizations**

Other factors being equal, larger molecules melt at higher temperatures than smaller molecules. Among structural isomers, the more symmetrical isomers will have the higher melting point. Among optical isomers, the *R* and *S* enantiomers will have the same melting points; but the racemate (a mixture of equal parts of *R* and *S*) will usually possess a different melting point. Diastereomers, another type of stereoisomer, will have different melting points. Molecules that can form hydrogen bonds will usually possess higher melting points than their counterparts of similar molecular weight.

**A phase diagram**

The melting point behavior of impure compounds is best understood by considering a simple binary mixture of compounds X and Y (Fig. 3.5). This melting point–composition diagram shows melting point behavior as a function of composition. The melting point of a pure compound is the temperature at which the vapor pressures of the solid and liquid are equal. But in dealing with a mixture, the situation is different. Consider the case of a mixture of 75% X and 25% Y. At a temperature below the eutectic temperature (ET), the mixture is solid Y and solid X. At ET, the solid begins to melt. The melt is a solution of Y dissolved in liquid X. The vapor pressure of the solution of X and Y together is less than that of pure X at the melting point; therefore, the temperature at which X will melt is lower when mixed with Y. This is an application of Raoult's law (*see* Chapter 5). As the temperature is increased, more and more of solid X melts until it is all gone at point **M** (temperature *m*). The melting point range is thus from ET to *m*. In practice it is very difficult to detect the ET when a melting point is determined in a capillary because it represents the point at which an infinitesimal amount of the solid mixture has begun to melt.

In this hypothetical example, the liquid solution becomes saturated with Y at the eutectic point (EP). This is the point at which X and Y and their liquid solutions are in equilibrium. A mixture of X and Y containing 60% X will appear to have a sharp melting point at the ET.

The melting point range of a mixture of compounds is generally broad, and the breadth of the range is an indication of purity. The chances of accidentally coming on the eutectic composition are small. Recrystallization will enrich the predominant compound while excluding the impurity and will, therefore, decrease the melting point range.

It should be apparent that the impurity must be soluble in the compound, so an insoluble impurity such as sand or charcoal will not depress the melting point. The impurity does not need to be a solid. It can be a liquid such as water (if it is soluble) or an organic solvent, such as the one used to recrystallize the compound; this advocates the necessity for drying the compound before determining its melting point.

Advantage is taken of the depression of melting points of mixtures to prove whether or not two compounds having the same melting points are identical. If X and Y are identical, then a mixture of the two will have the same melting point; if X and Y are not identical, then a small amount of X in Y or of Y in X will reduce the melting point.

*Melting point depression*

*The eutectic point*

*Mixed melting points*

## *Apparatus*

### Melting Point Capillaries

Before using a melting point apparatus, the sample needs to be prepared for analysis. Most melting point determinations require that the sample be placed in a capillary tube. The experiments in this book require capillary tubes for sample preparation. Capillaries may be obtained commercially or may be produced by drawing out 12-mm soft-glass tubing. The tubing is rotated in the hottest part of a Bunsen burner flame until it is very soft and begins to sag. It should not be drawn out during heating; it is removed from the flame and after a slight hesitation drawn steadily and not too rapidly to arm's length. With some practice it is possible to

■ **FIG. 3.6**
Sealing a melting point
capillary tube.

Samples that sublime

A small rubber band can be
made by cutting off a very
short piece of ¼" gum rubber
tubing.

Thomas–Hoover Uni-Melt

produce 10–15 good tubes in a single drawing. The long capillary tube can be cut into 100-mm lengths with a glass scorer. Each tube is sealed by rotating the end in the edge of a small flame, as seen in Figure 3.6.

**Filling Melting Point Capillaries.** The dry sample is ground to a fine powder on a watch glass or a piece of glassine paper on a hard surface using the flat portion of a spatula. It is formed into a small pile, and the open end of the melting point capillary is forced down into the pile. The sample is shaken into the closed end of the capillary by rapping sharply on a hard surface or by dropping it down a 2-ft length of glass tubing onto a hard surface. The height of the sample should be no more than 2–3 mm.

**Sealed Capillaries.** Some samples sublime (go from a solid state directly to the vapor phase without appearing to melt) or undergo rapid air oxidation and decompose at the melting point. These samples should be sealed under vacuum. This can be accomplished by forcing a capillary through a hole previously made in a rubber septum and evacuating the capillary using a water aspirator or a mechanical vacuum pump (Fig. 3.7). Using the flame from a small micro burner, the tube is gently heated about 15 mm above the tightly packed sample. This will cause any material in this region to sublime away. It is then heated more strongly in the same place to collapse the tube, taking care that the tube is straight when it cools. It is also possible to seal the end of a Pasteur pipette, add the sample, pack it down, and seal off a sample under vacuum in the same way.

**Melting Point Devices.** The apparatus required for determining an accurate melting point need not be elaborate; the same results are obtained on the simplest as on the most complex devices. The simplest setup involves attaching the sample-filled, melting point capillary to a thermometer using a rubber band and immersing them into a silicone oil bath (Fig. 3.8). This rubber band must be above the level of the oil bath; otherwise, it will break in the hot oil. The sample should be close to and on a level with the center of the thermometer bulb. This method can analyze compounds whose melting points go up to ~350°C. If determinations are to be done on two or three samples that differ in melting point by as much as 10°C, two or three capillaries can be secured to the thermometer together; the melting points can be observed in succession without removing the thermometer from the bath. As a precaution against the interchange of tubes while they are being attached, use some system of identification, such as one, two, and three dots made with a marking pencil.

More sophisticated melting point devices, some of which can attain temperatures of 500°C, will now be described.

The Thomas–Hoover Uni-Melt apparatus (Fig. 3.9) will accommodate up to seven capillaries in a small, magnified, lighted beaker of high-boiling silicone oil that is stirred and heated electrically. The heating rate is controlled with a variable transformer that is part of the apparatus. The rising mercury column of the thermometer can be observed with an optional traveling periscope device so the eye need not move away from the capillary. For industrial, analytical, and control

■ **FIG 3.7**
Evacuation of a melting point capillary prior to sealing.

■ **FIG 3.8**
A simple melting point apparatus.

■ **FIG. 3.9**
The Thomas–Hoover Uni-Melt melting point apparatus.

work the Mettler apparatus determines the melting point automatically and displays the result in digital form.

The Mel-Temp apparatus (Fig. 3.10) consists of an electrically heated aluminum block that accommodates three capillaries. The sample is illuminated through the lower port and observed with a six-power lens through the upper port. The heating rate can be controlled and, with a special thermometer, the apparatus can be used up to 500°C—far above the useful limit of silicone oil (which is about 350°C). For this melting point apparatus it is advisable to use a digital thermometer rather than the mercury-in-glass thermometer.

The Fisher-Johns melting point apparatus (Fig. 3.11) is used to determine the melting point of a single sample. Instead of a capillary tube, the sample is placed between two thin glass disks that are placed on an aluminum heating stage. Heating is controlled by a variable transformer, and melting is observed through a magnifier; the melting temperature is read from a mercury-in-glass thermometer. This apparatus can be used for compounds that melt between 20°C and 300°C.

## *Determining the Melting Point*

The accuracy of the melting point depends on the accuracy of the thermometer, so the first exercise in the following experiments will be to calibrate the thermometer. Melting points of pure, known compounds will be determined and deviations recorded so a correction can be applied to future melting points. Be forewarned, however, that thermometers are usually fairly accurate.

■ **FIG. 3.10**
The Mel-Temp melting point apparatus.

■ **FIG 3.11**
The Fisher-Johns melting point apparatus.

The rate of heating is the most important factor in obtaining accurate melting points. Heat no faster than 1°C per minute.

The most critical factor in determining an accurate melting point is the rate of heating. At the melting point the temperature increase should not be greater than 1°C per minute. This may seem extraordinarily slow, but it is necessary in order for heat from the oil bath or the heating block to be transferred equally to the sample and to the glass and mercury of the thermometer.

From your own experience you know the rate at which ice melts. Consider performing a melting point experiment on an ice cube. Because water melts at 0°C, you would need to have a melting point bath a few degrees below zero. To observe the true melting point of the ice cube, you would need to raise the temperature extraordinarily slowly. The ice cube would appear to begin to melt at 0°C and, if you waited for temperature equilibrium to be established, it would all be melted at 0.5°C. If you were impatient and raised the temperature too rapidly, the ice might appear to melt over a range of 0°C to 20°C. Similarly, melting points determined in capillaries will not be accurate if the rate of heating is too fast.

## EXPERIMENTS

## 1. Calibration of the Thermometer

Determine the melting point of standard substances (Table 3.1) over the temperature range of interest. The difference between the values found and those expected constitutes the correction that must be applied to future temperature readings. If the thermometer has been calibrated previously, then determine one or more melting points of known substances to familiarize yourself with the technique. If the

| TABLE 3.1 • Melting Point Standards | | |
|---|---|---|
| **Compound** | **Structure** | **Melting Point (°C)** |
| Naphthalene | (a) | 80–82 |
| Urea | (b) | 132.5–133 |
| Sulfanilamide | (c) | 164–165 |
| 4-Toluic acid | (d) | 180–182 |
| Anthracene | (e) | 214–217 |
| Caffeine (evacuated capillary) | (f) | 234–236.5 |

determinations do not agree within 1°C, then repeat the process. Both mercury-in-glass and digital thermometers will need to be calibrated; nonmercury thermometers are not typically used for melting point determination.

## 2. Melting Points of Pure Urea and Cinnamic Acid

$$\bigcirc\!\!-CH=CH-COOH$$

**Cinnamic acid**

Using a metal spatula, crush the sample to a fine powder on a hard surface such as a watch glass. Push the open end of a melting point capillary into the powder and force the powder down in the capillary by tapping the capillary or by dropping it through a long glass tube held vertically and resting on a hard surface. The column of solid should be no more than 2–3 mm in height and should be tightly packed.

Heat rapidly to within 20°C of the melting point.

Heat slowly (<1°C/min) near the melting temperature.

If the approximate melting temperature is known, the bath can be heated rapidly until the temperature is about 20°C below this point; the heating during the last 15°C–20°C should slow down considerably so *the rate of heating at the melting point is no more than 1°C per minute* while the sample is melting. As the melting point is approached, the sample may shrink because of crystalline structure changes. However, the melting process begins when the first drops of liquid are seen in the capillary and ends when the last trace of solid disappears. For a pure compound this whole process may occur over a range of only 0.5°C; hence, it is necessary to slowly increase the temperature during the determination.

Determine the melting point (mp) of either urea (mp 132.5°C–133°C) or cinnamic acid (mp 132.5°C–133°C). Repeat the determination; if the two determinations do not check within 1°C, repeat a third time.

## 3. Melting Points of Urea-Cinnamic Acid Mixtures

> **IN THIS EXPERIMENT** you will see dramatic evidence of the phenomenon of melting point depression, which will allow you to prepare a phase diagram like that shown in Figure 3.5 (on page 47).

Make mixtures of urea and cinnamic acid in the approximate proportions 1:4, 1:1, and 4:1 by putting side by side the correct number of equal-sized small piles of the two substances and then mixing them. Grind the mixture thoroughly for at least a minute on a watch glass using a metal spatula. Note the ranges of melting of the three mixtures and use the temperatures of complete liquefaction to construct a rough diagram of melting point versus composition.

## 4. Unknowns

Determine the melting point of one or more of the unknowns selected by your instructor and identify the substance based on its melting point (Table 3.2). Prepare two capillaries of each unknown. Run a very fast determination on the first sample to ascertain the approximate melting point. Cool the melting point bath to just below the melting point and make a slow, careful determination using the other capillary.

## 5. An Investigation: Determination of Molecular Weight Using Melting Point Depression

Before the mass spectrometer came into common usage, the molal freezing point depression of camphor was used to determine molecular weights. Whereas a 1% solid solution of urea in cinnamic acid will cause a relatively small melting point depression, a 1% by weight solid solution in camphor of any organic compound with a molecular weight of 100 will cause a 4.0°C depression in the melting point of the camphor. Quantitative use of this relationship can be used to determine the

| TABLE 3.2 • Melting Point Unknowns | |
|---|---|
| **Compound** | **Melting Point (°C)** |
| Benzophenone | 49–51 |
| Maleic anhydride | 52–54 |
| 4-Nitrotoluene | 54–56 |
| Naphthalene | 80–82 |
| Acetanilide | 113.5–114 |
| Benzoic acid | 121.5–122 |
| Urea | 132.5–133 |
| Salicylic acid | 158.5–159 |
| Sulfanilamide | 165–166 |
| Succinic acid | 184.5–185 |
| 3,5-Dinitrobenzoic acid | 205–207 |
| p-Terphenyl | 210–211 |

molecular weight of an unknown. You can learn more about the details of this technique in very old editions of *Organic Experiments*[1] or by searching the Web for "colligative properties, molal freezing point depression." Visit this book's website for more information.

# PART 3: Boiling Points

The boiling point of a pure organic liquid is one of its characteristic physical properties, just like its density, molecular weight, and refractive index, and the melting point of its solid state. The boiling point is used to characterize a new organic liquid, and knowledge of the boiling point helps to compare one organic liquid to another, as in the process of identifying an unknown organic substance.

A comparison of boiling points with melting points is instructive. The process of determining the boiling point is more complex than that for the melting point: It requires more material, and because it is less affected by impurities, it is not as good an indication of purity. Boiling points can be determined on a few microliters of a liquid, but on a small scale it is difficult to determine the boiling point *range*. This requires enough material to distill—about 1 to 2 mL. Like the melting point, the boiling point of a liquid is affected by the forces that attract one molecule to another—ionic attraction, London forces, dipole-dipole interactions, and hydrogen bonding, as discussed in Part 1 of this chapter.

## Structure and Boiling Point

In a homologous series of molecules the boiling point increases in a perfectly regular manner. The normal saturated hydrocarbons have boiling points ranging from $-162°C$ for methane to $330°C$ for $n\text{-}C_{19}H_{40}$, an increase of about 27°C for each

---

1. Fieser L. F. *Organic Experiments*, 2nd ed.; D.C. Heath: Lexington, MA, 1968; 38–42.

$CH_2$ group. It is convenient to remember that *n*-heptane with a molecular weight of 100 has a boiling point near 100°C (98.4°C). A spherical molecule such as 2,2-dimethylpropane has a lower boiling point than *n*-pentane because it does not have as many points of attraction to adjacent molecules. For molecules of the same molecular weight, those with dipoles, such as carbonyl groups, will have higher boiling points than those without, and molecules that can form hydrogen bonds will boil even higher. The boiling point of such molecules depends on the number of hydrogen bonds that can be formed. An alcohol with one hydroxyl group will boil at a lower temperature than an alcohol with two hydroxyl groups if they both have the same molecular weight. A number of other generalizations can be made about boiling point behavior as a function of structure; you will learn about these throughout your study of organic chemistry.

## Boiling Point as a Function of Pressure

Boiling points decrease about 0.5°C for each 10-mm decrease in atmospheric pressure.

Because the boiling point of a pure liquid is defined as the temperature at which the vapor pressure of the liquid exactly equals the pressure exerted on it, the boiling point will be a function of atmospheric pressure. At an altitude of 14,000 ft the boiling point of water is 81°C. At pressures near that of the atmosphere at sea level (760 mm), the boiling point of most liquids decreases about 0.5°C for each 10-mm decrease in atmospheric pressure. This generalization does not hold for greatly reduced pressures because the boiling point decreases as a nonlinear function of pressure (*see* Fig. 5.2 on page 91). Under these conditions a nomograph relating the observed boiling point, the boiling point at 760 mm, and the pressure in millimeters should be consulted (*see* Fig. 6.19 on page 128). This nomograph is not highly accurate; the change in boiling point as a function of pressure also depends on the type of compound (polar, nonpolar, hydrogen bonding, etc.). Consult the *CRC Handbook of Chemistry and Physics*[2] for the correction of boiling points to standard pressure.

CAUTION: Mercury is toxic. Immediately report any broken thermometers to your instructor.

Most mercury-in-glass laboratory thermometers have a mark around the stem that is 3 in. (76 mm) from the bottom of the bulb. This is the immersion line; the thermometer will record accurate temperatures if immersed to this line. If you break a mercury thermometer, immediately inform your instructor, who will use special apparatus to clean up the mercury. Mercury vapor is very toxic.

## Calibrating the Thermometer

If you have not previously carried out a calibration, test the 0°C point of your thermometer with a well-stirred mixture of crushed ice and distilled water. To check the 100°C point, put 2 mL of water in a test tube with a boiling chip to prevent bumping and boil the water gently over a hot sand bath with the thermometer in the vapor from the boiling water. Make sure that the thermometer does not touch the side of the test tube. Then immerse the bulb of the thermometer into the liquid and see if you can observe superheating. Check the atmospheric pressure to determine the true boiling point of the water.

2. Lide, D. R., ed. *CRC Handbook of Chemistry and Physics,* 86th ed.; CRC Press: Boca Raton, FL, 2005.

# Distillation Considerations

## Prevention of Superheating—Boiling Sticks and Boiling Stones

Superheating occurs when a very clean liquid in a very clean vessel is heated to a temperature above its boiling point without ever actually boiling. That is, if a thermometer is placed in the superheated liquid, the thermometer will register a temperature higher than the boiling point of the liquid. If boiling does occur under these conditions, it occurs with explosive violence. To avoid this problem, boiling stones or boiling sticks are always added to liquids before heating them to boiling— whether to determine a boiling point or to carry out a reaction or distillation. These boiling stones or sticks provide the nuclei on which the bubble of vapor indicative of a boiling liquid can form; be careful not to confuse the bubbling for boiling. Some boiling stones, also called boiling chips, are composed of porous unglazed porcelain. This material is filled with air in numerous fine capillaries. With heating, this air expands to form the fine bubbles on which even boiling can take place. Once the liquid cools, it will fill these capillaries and the boiling chip will become ineffectual, so another chip must be added each time the liquid is reheated to boiling. Wooden boiling sticks about 1.5 mm in diameter—often called applicator sticks—also promote even boiling and, unlike stones, are removed easily from the solution. None of these work well for vacuum distillation (*see* Chapter 6).

## Closed Systems

Distillations that are run at atmospheric pressure need to be open to the atmosphere to avoid pressure buildup, which could lead to an explosion. Therefore, always make sure that a distillation setup is not a closed system—unless, of course, you are running a vacuum distillation.

# Boiling Point Determination: Apparatus and Technique

## Boiling Point Determination by Distillation

All procedures involving volatile and/or flammable solvents should be conducted in a fume hood.

When enough material is available (at least 3 mL), the best method for determining the boiling point of a liquid is to distill it (*see* Chapter 5). Distillation allows the boiling range to be determined and thus gives an indication of purity. Bear in mind, however, that a constant boiling point is not a guarantee of homogeneity and thus purity. Constant-boiling azeotropes such as 95% ethanol are common.

### Boiling Point Determination Using a Digital Thermometer and a Reaction Tube

*Online Study Center*

Photo: Boiling Point Determination with a Digital Thermometer

Boiling points can be measured rapidly and accurately using an electronic digital thermometer, as depicted in Figure 3.12. Although digital thermometers are currently too expensive for each student to have, several of these in the laboratory can greatly speed up the determination of boiling points. Digital thermometers are

■ **FIG. 3.12**
Using a digital thermometer
for determining boiling points.

Digital
thermometer

°C

79

Surface
probe

Refluxing
vapor

Boiling chip

Boiling liquid

much safer to use because there is no danger from toxic mercury vapor if the thermometer is accidentally dropped.

The surface probe of the digital thermometer is the active element. Unlike the bulb of mercury at the end of a thermometer, this element has a very low heat capacity and a very fast response time, so boiling points can be determined very quickly in this apparatus. About 0.2 mL to 0.3 mL of the liquid and a boiling chip are heated on a sand bath until the liquid refluxes about 3 cm up the tube. The probe should not touch the side of the reaction tube but should be placed approximately 5 mm above the liquid. The boiling point is the highest temperature recorded by the thermometer and is maintained for about 1 min. The application

**■ FIG. 3.13**
A small-scale boiling-point
apparatus. Be sure the
thermometer does not touch
the tube.

Thermometer
adapter

Refluxing
vapor

Boiling chip          Boiling liquid

Smaller-scale boiling point
apparatus

of heat will drive tiny bubbles of air from the boiling chip; do not mistake these tiny bubbles for true boiling; this mistake can readily happen if the unknown has a very high boiling point.

### Boiling Point Determination in a Reaction Tube

If a digital thermometer is not available, use the apparatus shown in Figure 3.13. Using a distilling adapter at the top of a reaction tube allows access to the atmosphere. Place 0.3 mL of the liquid along with a boiling stone in a 10 × 100-mm reaction tube, clamp a thermometer so that the bulb is just above the level of the liquid, and then heat the liquid with a sand bath. It is *very important* that no part of the thermometer touch the reaction tube. Heating is regulated so that the boiling liquid refluxes about 3 cm up the thermometer but does not boil out of the apparatus. If you cannot see the refluxing liquid, carefully run your finger down the side of the reaction tube until you feel heat. This indicates where the liquid is refluxing. Droplets of liquid must drip from the thermometer bulb to thoroughly heat the mercury. The boiling point is the highest temperature recorded by the thermometer and is maintained over about a 1-min time interval.

The application of heat will drive out tiny air bubbles from the boiling chip. Do not mistake these tiny bubbles for true boiling. This can occur if the unknown has a very high boiling point. It may take several minutes to heat up the mercury in the thermometer bulb. True boiling is indicated by drops dripping from the thermometer, with a constant temperature recorded on the thermometer. If the temperature is not constant, then you are probably not observing true boiling.

### Boiling Point Determination Using a 3-mm to 5-mm Tube

For smaller quantities, a 3-mm to 5-mm diameter tube is attached to the side of the thermometer by a rubber band (Fig. 3.14) and heated with a liquid bath. The tube, which can be made from tubing 3 mm to 5 mm in diameter, contains a small inverted capillary. This is made by cutting a 6-mm piece from the sealed end of a melting point capillary, inverting it, and sealing the closed end to the capillary. A centimeter ruler is printed on the inside back cover of this book.

When the sample is heated in this device, the air in the inverted capillary will expand, and an occasional bubble will escape. At the true boiling point a continuous and rapid stream of bubbles will emerge from the inverted capillary. When this occurs, the heating is stopped, and the bath is allowed to cool. A time will come when bubbling ceases and the liquid just begins to rise in the inverted capillary. The temperature at which this happens is recorded. The liquid is allowed to partially fill the small capillary, and the heat is applied carefully until the first bubble emerges from the capillary. The temperature is recorded at that point. The two temperatures approximate the boiling point range for the liquid.

As the liquid was being heated, the air expanded in the inverted capillary and was replaced by vapor of the liquid. The liquid was actually slightly superheated when rapid bubbles emerged from the capillary, but on cooling a point was reached at which the pressure on the inside of the capillary matched the outside (atmospheric) pressure. This is, by definition, the boiling point.

**■ FIG. 3.14**
A smaller-scale boiling point apparatus.

Melting point capillary

Tube, 3 mm dia. × 80 mm long

Seal

Sample

Heating bath

**Cleaning Up.** Place the boiling point sample in either the halogenated or nonhalogenated waste container. Do not pour it down the sink.

## QUESTIONS

1. What effect would poor circulation of the melting point bath liquid have on the observed melting point?

2. What is the effect of an insoluble impurity, such as sodium sulfate, on the observed melting point of a compound?

3. Three test tubes, labeled A, B, and C, contain substances with approximately the same melting points. How could you prove the test tubes contain three different chemical compounds?

4. One of the most common causes of inaccurate melting points is too rapid heating of the melting point bath. Under these circumstances, how will the observed melting point compare to the true melting point?

5. Strictly speaking, why is it incorrect to speak of a melting *point*?

6. What effect would the incomplete drying of a sample (for example, the incomplete removal of a recrystallization solvent) have on the melting point?

7. Why should the melting point sample be finely powdered?

8. You suspect that an unknown is acetanilide (mp 113.5°C–114°C). Give a qualitative estimation of the melting point when the acetanilide is mixed with 10% by weight of naphthalene.

9. You have an unknown with an observed melting point of 90°C–93°C. Is your unknown compound A with a reported melting point of 95.5°C–96°C or compound B with a reported melting point of 90.5°C–91°C? Explain.

10. Why is it important to heat the melting point bath or block slowly and steadily when the temperature gets close to the melting point?

11. Why is it important to pack the sample tightly in the melting point capillary?

12. An unknown compound is suspected to be acetanilide (mp 113.5°C–114°C). What would happen to the melting point if this unknown were mixed with (a) an equal quantity of pure acetanilide? (b) an equal quantity of benzoic acid?

13. Which would be expected to have the higher boiling point, *t*-butyl alcohol (2-methyl-2-propanol) or *n*-butyl alcohol (1-butanol)? Explain.

14. What is the purpose of the side arm of the thermometer adapter in Figure 3.13?

*Online Study Center*

**General Resources**
Web Links

# REFERENCE

Weissberger, Arnold, and Bryant W. Rossiter (eds.). *Physical Methods of Chemistry*, Vol. 1, Part V. New York: Wiley-Interscience, 1971.

# CHAPTER
# 4

# Recrystallization

**Online Study Center**

This icon will direct you to techniques, equipment setups, and online resources at http://college.hmco.com/PIC/williamsonMMOE5e.

Recrystallization: the most important purification method for solids, especially for small-scale experiments.

**PRELAB EXERCISE:** Write an expanded outline for the seven-step process of recrystallization.

Recrystallization is the most important method for purifying solid organic compounds. It is also a very powerful, convenient, and efficient method of purification, and it is an important industrial technique that is still relevant in the chemical world today. For instance, the commercial purification of sugar is done by recrystallization on an enormous scale.

A pure, crystalline organic substance is composed of a three-dimensional array of molecules held together primarily by London forces. These attractive forces are fairly weak; most organic solids melt in the range between 22°C and 250°C. An impure organic solid will not have a well-defined crystal lattice because impurities do not allow the crystalline structure to form. The goal of recrystallization is to remove impurities from a solid to allow a perfect crystal lattice to grow.

There are four important concepts to consider when discussing the process of recrystallization:

1. Solubility
2. Saturation level
3. Exclusion
4. Nucleation

Recrystallization involves dissolving the material to be purified (the solute) in an appropriate hot solvent to yield a solution (*solubility*). As the solvent cools, the solution becomes saturated with respect to the solute (*saturation level*), which then recrystallizes. As the perfectly regular array of a crystal is formed, impurities are excluded (*exclusion*), and the crystal is thus a single pure substance. Soluble impurities remain in solution because they are not concentrated enough to saturate the solution. Recrystallization is initiated at a point of *nucleation*—a seed crystal, a speck of dust, or a scratch on the wall of the test tube if the solution is super-saturated with the solute.

In this chapter you will carry out the recrystallization process, one of the most important laboratory operations of organic chemistry, by following its seven steps. Then you will perform several actual recrystallization experiments.

## The Seven Steps of Recrystallization

The process of recrystallization can be broken into seven discrete steps: (1) choosing the solvent and solvent pairs; (2) dissolving the solute; (3) decolorizing the solution with pelletized Norit; (4) filtering suspended solids; (5) recrystallizing the solute; (6) collecting and washing the crystals; and (7) drying the crystals. A detailed description of each of these steps is given in the following sections.

## Step 1. Choosing the Solvent and Solvent Pairs

Similia similibus solvuntur.

In choosing the solvent, the chemist is guided by the dictum "like dissolves like." Even the nonchemist knows that oil and water do not mix and that sugar and salt dissolve in water but not in oil. Hydrocarbon solvents such as hexane will dissolve hydrocarbons and other nonpolar compounds, and hydroxylic solvents such as water and ethanol will dissolve polar compounds. It is often difficult to decide, simply by looking at the structure of a molecule, just how polar or nonpolar it is and which solvent would be best. Therefore, the solvent is often chosen by experimentation. If an appropriate single solvent cannot be found for a given substance, a solvent pair system may be used. The requirement for this solvent pair is miscibility; both solvents should dissolve in each other for use as a recrystallization solvent system.

**Online Study Center**

Video: The Reaction Tube in Use

The ideal solvent

The best recrystallization solvent (and none is ideal) will dissolve the solute when the solution is hot but not when the solution is cold; it will either not dissolve the impurities at all or it will dissolve them very well (so they do not recrystallize out along with the solute); it will not react with the solute; and it will be nonflammable, nontoxic, inexpensive, and very volatile (so it can be removed from the crystals).

Miscible: capable of being mixed

Some common solvents and their properties are presented in Table 4.1 in order of decreasing polarity of the solvent. Solvents adjacent to each other in the list will dissolve in each other; that is, they are miscible with each other, and each solvent will, in general, dissolve substances that are similar to it in chemical structure. These solvents are used both for recrystallization and as solvents in which reactions are carried out.

### Procedure

#### Choosing a Solvent

**Online Study Center**

Video: Picking a Solvent

To choose a solvent for recrystallization, place a few crystals of the impure solute in a small test tube or centrifuge tube and add a very small drop of the solvent. Allow the drop to flow down the side of the tube and onto the crystals. If the crystals

## TABLE 4.1 • Recrystallization Solvents

| Solvent | Boiling Point (°C) | Remarks |
|---|---|---|
| Water ($H_2O$) | 100 | It is the solvent of choice because it is cheap, nonflammable, and nontoxic and will dissolve a large variety of polar organic molecules. Its high boiling point and high heat of vaporization make it difficult to remove from crystals. |
| Acetic acid ($CH_3COOH$) | 118 | It will react with alcohols and amines, and it is difficult to remove from crystals. It is not a common solvent for recrystallizations, although it is used as a solvent when carrying out oxidation reactions. |
| Dimethyl sulfoxide (DMSO; $CH_3SOCH_3$) | 189 | It is not a commonly used solvent for recrystallization, but it is used for reactions. |
| Methanol ($CH_3OH$) | 64 | It is a very good solvent that is often used for recrystallization. It will dissolve molecules of higher polarity than other alcohols. |
| 95% Ethanol ($CH_3CH_2OH$) | 78 | It is one of the most commonly used recrystallization solvents. Its high boiling point makes it a better solvent for less polar molecules than methanol. It evaporates readily from crystals. Esters may undergo an interchange of alcohol groups on recrystallization. |
| Acetone ($CH_3COCH_3$) | 56 | It is an excellent solvent, but its low boiling point means there is not much difference in the solubility of a compound at its boiling point compared to about 22°C. |
| 2-Butanone; also known as methyl ethyl ketone (MEK; $CH_3COCH_2CH_3$) | 80 | It is an excellent solvent that has many of the most desirable properties of a good recrystallization solvent. |
| Ethyl acetate ($CH_3COOC_2H_5$) | 78 | It is an excellent solvent that has about the right combination of moderately high boiling point and the volatility needed to remove it from crystals. |
| Dichloromethane; also known as methylene chloride ($CH_2Cl_2$) | 40 | Although a common extraction solvent, dichloromethane boils too to make it a good recrystallization solvent. It is useful in a solvent pair with ligroin. |
| Diethyl ether; also known as ether ($CH_3CH_2OCH_2CH_3$) | 35 | Its boiling point is too low for recrystallization, although it is an extremely good solvent and fairly inert. It is used in a solvent pair with ligroin. |
| Methyl $t$-butyl ether ($CH_3OC(CH_3)_3$) | 52 | It is a relatively new and inexpensive solvent because of its large-scale use as an antiknock agent and oxygenate in gasoline. It does not easily form peroxides; it is less volatile than diethyl ether, but it has the same solvent characteristics. (*See also* Chapter 7.) |
| Dioxane ($C_4H_8O_2$) | 101 | It is a very good solvent that is not too difficult to remove from crystals; it is a mild carcinogen, and it forms peroxides. |
| Toluene ($C_6H_5CH_3$) | 111 | It is an excellent solvent that has replaced the formerly widely used benzene (a weak carcinogen) for the recrystallization of aryl compounds. Because of its boiling point, it is not easily removed from crystals. |

*(continued)*

| TABLE 4.1 • *(continued)* | | |
|---|---|---|
| **Solvent** | **Boiling Point (°C)** | **Remarks** |
| Pentane ($C_5H_{12}$) | 36 | It is a widely used solvent for nonpolar substances. It is not often used alone for recrystallization, but it is good in combination with several other solvents as part of a solvent pair. |
| Hexane ($C_6H_{14}$) | 69 | It is frequently used to recrystallize nonpolar substances. It is inert and has the correct balance between boiling point and volatility. It is often used as part of a solvent pair. (*See also* ligroin.) |
| Cyclohexane ($C_6H_{12}$) | 81 | It is similar in all respects to hexane. (*See also* ligroin.) |
| Petroleum ether | 30–60 | It is a mixture of hydrocarbons, of which pentane is the chief component. It is used interchangeably with pentane because it is cheap. Unlike diethyl ether or *t*-butyl methyl ether, it is not an ether in the modern chemical sense. |
| Ligroin | 60–90 | It is a mixture of hydrocarbons with the properties of hexane and cyclohexane. It is a very commonly used recrystallization solvent. It is also sold as "hexanes." |

**Note:** The solvents in this table are listed in decreasing order of polarity. Adjacent solvents in the list are, in general, miscible with each other.

dissolve instantly at about 22°C, that solvent cannot be used for recrystallization because too much of the solute will remain in solution at low temperatures. If the crystals do not dissolve at about 22°C, warm the tube on a hot sand bath and observe the crystals. If they do not go into solution, add 1 more drop of solvent. If the crystals go into solution at the boiling point of the solvent and then recrystallize when the tube is cooled, you have found a good recrystallization solvent. If not, remove the solvent by evaporation and try a different solvent. In this trial-and-error process it is easiest to try low-boiling solvents first because they can be easily removed. Occasionally, no single satisfactory solvent can be found, so mixed solvents, or *solvent pairs*, are used.

### Solvent Pairs

To use a mixed solvent pair, dissolve the crystals in the better solvent (more solubilizing) and add the poorer solvent (less solubilizing) to the hot solution until it becomes cloudy, and the solution is saturated with the solute. The two solvents must, of course, be miscible with each other. Some useful solvent pairs are given in Table 4.2.

| TABLE 4.2 • **Solvent Pairs** | |
|---|---|
| Acetic acid–water | Ethyl acetate–cyclohexane |
| Ethanol-water | Acetone-ligroin |
| Acetone-water | Ethyl acetate–ligroin |
| Dioxane-water | *t*-Butyl methyl ether–ligroin |
| Acetone-ethanol | Dichloromethane-ligroin |
| Ethanol–*t*-butyl methyl ether | Toluene-ligroin |

# Step 2. Dissolving the Solute

## *Microscale Procedure*

Once a recrystallization solvent has been found, the impure crystals are placed in a reaction tube, the solvent is added dropwise, the crystals are stirred with a microspatula or a small glass rod, and the tube is warmed on a steam bath or a sand bath until the crystals dissolve. Care must be exercised to use the minimum amount of solvent at or near boiling. Observe the mixture carefully as solvent is being added. Allow sufficient time for the boiling solvent to dissolve the solute and note the rate at which most of the material dissolves. When you believe most of the material has been dissolved, stop adding solvent. There is a possibility that your sample is contaminated with a small quantity of an insoluble impurity that never will dissolve. To hasten the solution process, crush large crystals with a stirring rod, taking care not to break the reaction tube.

If the solution contains no undissolved impurities and is not colored from impurities, you can simply let it cool, allowing the solute to recrystallize (step 5), and then collect the crystals (step 6). On the other hand, if the solution is colored, it must be treated with activated (decolorizing) charcoal and then filtered before recrystallization (step 3). If it contains solid impurities, it must be filtered before recrystallization takes place (step 4).

On a microscale, there is a tendency to use too much solvent so that on cooling the hot solution little or no material recrystallizes. This is not a hopeless situation. The remedy is to evaporate some of the solvent (by careful boiling) and repeat the cooling process. Inspect the hot solution.

A solution (solute dissolved in solvent) can become *superheated*; that is, heated above its boiling point without actually boiling. When boiling does suddenly occur, it can happen with almost explosive violence, a process called *bumping*. To prevent this from happening, a *wood applicator* stick can be added to the solution (Fig. 4.1). Air trapped in the wood comes out of the stick and forms the nuclei on which even boiling can occur. Porous porcelain *boiling chips* work in the same way. Never add a boiling chip or a boiling stick to a hot solution because the hot solutuion may be superheated and boil over or bump.

## *Macroscale Procedure*

Place the substance to be recrystallized in an Erlenmeyer flask (never use a beaker), add enough solvent to cover the crystals, and then heat the flask on a steam bath (if the solvent boils below 90°C) or a hot plate until the solvent boils. (Note: Adding a boiling stick or a boiling chip to the solution will promote even boiling. It is easy to superheat the solution; that is, heat it above the boiling point with no boiling taking place. Once the solution does boil, it does so with explosive violence; it bumps.) Never add a boiling chip or boiling stick to a hot solution.

Stir the mixture or, better, swirl it (Fig. 4.2) to promote dissolution. Add solvent gradually, keeping it at the boiling point, until all of the solute dissolves. A glass rod with a flattened end can sometimes be useful in crushing large particles of solute to speed up the dissolving process. Be sure no flames are nearby when working with flammable solvents.

Do not use too much solvent.

**Online Study Center**

Video: Recrystallization

Prevention of bumping

Do not use wood applicator sticks (boiling sticks) in place of boiling chips in a reaction. Use them only for recrystallization.

**Online Study Center**

Video: Macroscale Crystallization

All procedures involving volatile and/or flammable solvents should be conducted in a fume hood.

■ **FIG. 4.1**

A reaction tube being used for recrystallization. The wood applicator stick ("boiling stick") promotes even boiling and is easier to remove than a boiling chip. The Thermowell sand is cool on top and hotter deeper down, so it provides a range of temperatures. The reaction tube is long and narrow; it can be held in the hand while the solvent refluxes. Do not use a boiling stick in place of a boiling chip in a reaction.

Wood applicator stick

Cool at this point

Air condenser

Boiling solvent

Temperature controlled by depth in sand

■ **FIG. 4.2**

Swirling of a solution to mix contents and help dissolve material to be recrystallized.

Be careful not to add too much solvent. Note how rapidly most of the material dissolves; stop adding solvent when you suspect that almost all of the desired material has dissolved. It is best to err on the side of too little solvent rather than too much. Undissolved material noted at this point could be an insoluble impurity that never will dissolve. Allow the solvent to boil, and if no further material dissolves, proceed to step 4 to remove suspended solids from the solution by filtration or if the solution is colored, go to step 3 to carry out the decolorization process. If the solution is clear, proceed to step 5.

# Step 3. Decolorizing the Solution with Pelletized Norit

Activated charcoal = decolorizing carbon = Norit

The vast majority of pure organic chemicals are colorless or a light shade of yellow; consequently, this step is not usually required. Occasionally, a chemical reaction will produce high molecular weight byproducts that are highly colored. The impurities can be adsorbed onto the surface of activated charcoal by simply boiling the solution with charcoal. Activated charcoal has an extremely large surface area per gram (several hundred square meters) and can bind a large number of molecules to this surface. On a commercial scale, the impurities in brown sugar are adsorbed onto charcoal in the process of refining sugar.

Add a small amount (0.1% of the solute weight is sufficient) of pelletized Norit to the colored solution and then boil the solution for a few minutes. Be careful not to add the charcoal pieces to a superheated solution; the charcoal functions like hundreds of boiling chips and will cause the solution to boil over. Remove the Norit by filtration as described in step 4.

# Step 4. Filtering Suspended Solids

The filtration of a hot, saturated solution to remove solid impurities or charcoal can be performed in a number of ways. These processes include gravity filtration, pressure filtration, decantation, or removing the solvent with a Pasteur pipette. Vacuum filtration is not used because the hot solvent will cool during the process, and the product will recrystallize in the filter. Filtration can be one of the most vexing operations in the laboratory if the desired compound recrystallizes during filtration. Test the solution or a small portion of it before filtration to ensure that no crystals form at about 22°C. Like decolorization with charcoal, the removal of solid impurities by filtration is usually not necessary.

## *Microscale Procedure*

### (A) Removal of Solution with a Pasteur Pipette

If the solid impurities are large in size, they can be removed by filtration of the liquid through the small space between the square end of a Pasteur pipette and the bottom of a reaction tube (Fig. 4.3). Expel air from the pipette by squeezing the pipette bulb as the pipette is being pushed to the bottom of the tube. Use a small additional quantity of solvent to rinse the tube and pipette. Anhydrous calcium chloride, a drying agent, is easily removed in this way. The removal of very fine material, such as traces of charcoal, is facilitated by filtration of the solution through a small piece of filter paper (3 mm$^2$) placed in a reaction tube. This process is even easier if the filter paper is the thick variety, such as that from which Soxhlet extraction thimbles are made.[1]

### (B) Filtration in a Pasteur Pipette

To filter 0.1 mL to 2 mL of a solution, dilute the solution with enough solvent so that the solute will not recrystallize at about 22°C. Prepare a filter pipette by pushing a tiny bit of cotton into a Pasteur pipette, put the solution to be filtered into this filter pipette using another Pasteur pipette, and then force the liquid through the filter using air pressure from a pipette bulb (Fig. 4.4). Fresh solvent should be added to rinse the pipette and cotton. The filtered solution is then concentrated by evaporation. One problem encountered with this method is using too much cotton packed too tightly in the pipette so that the solution cannot be forced through it. To remove very fine impurities, such as traces of decolorizing charcoal, a 3-mm to 4-mm layer of Celite filter aid can be added to the top of the cotton.

### (C) Removal of Fine Impurities by Centrifugation

To remove fine solid impurities from up to 4 mL of solution, dilute the solution with enough solvent so that the solute will not recrystallize at about 22°C. Counterbalance the reaction tube and centrifuge for about 2 min at high speed in a laboratory centrifuge. The clear supernatant can be decanted (poured off) from the solid on the bottom of the tube. Alternatively, with care, the solution can be removed with a Pasteur pipette, leaving the solid behind.

*Online Study Center*

Video: Filtration of Crystals Using the Pasteur Pipette

*Online Study Center*

Photo: Preparation of a Filter Pipette

---

1. Belletire, J. L.; Mahmoodi, N. O. *J. Chem. Educ.* **1989**, *66*, 964.

■ **FIG. 4.3**
Filtration using a Pasteur
pipette and a reaction tube.

Pipette bulb

Pasteur pipette

Reaction tube

Note square,
flat tip

■ **FIG. 4.4**
Filtration of a solution in a
Pasteur pipette.

Solution
to be filtered

Cotton

Celite
(optional)

### (D) Pressure Filtration with a Micro Büchner Funnel

The technique applicable to volumes from 0.1 mL to 5 mL is to use a micro Büchner funnel. It is made of polyethylene and is fitted with a porous polyethylene frit that is 6 mm in diameter. This funnel fits in the bottom of an inexpensive disposable polyethylene pipette in which a hole is cut (Fig. 4.5). The solution to be filtered is placed in the pipette using a Pasteur pipette. The thumb covers the hole in the plastic pipette and pressure is applied to filter the solution. It is good practice to place a 6-mm-diameter piece of filter paper over the frit, which would otherwise become clogged with insoluble material.

    The glass chromatography column can be used in the same way. A piece of filter paper is placed over the frit. The solution to be filtered is placed in the chromatography column, and pressure is applied to the solution using a pipette bulb. In both procedures, dilute the solution to be filtered so that it does not recrystallize in the apparatus and use a small amount of clean solvent to rinse the apparatus. The filtered solution is then concentrated by evaporation.

### ■ FIG. 4.5

A pressure filtration apparatus. The solution to be filtered is added through the aperture, which is closed by a finger as pressure is applied.

— Filling aperture

— Polyethylene transfer pipette

— Polyethylene frit
— Büchner funnel

### ■ FIG. 4.6

Gravity filtration of hot solution through fluted filter paper.

 *Macroscale Procedure*

#### (A) Decantation

Decant: to pour off. A fast, easy separation procedure

On a large scale, it is often possible to pour off (decant) the hot solution, leaving the insoluble material behind. This is especially easy if the solid is granular like sodium sulfate. The solid remaining in the flask and the inside of the flask should be rinsed with a few milliliters of the solvent in order to recover as much of the product as possible.

#### (B) Gravity Filtration

The most common method for the removal of insoluble solid material is gravity filtration through fluted filter paper (Fig. 4.6). This is the method of choice for removing finely divided charcoal, dust, lint, and so on. The following equipment is needed for this process: three labeled Erlenmeyer flasks on a steam bath or a hot plate (flask A contains the solution to be filtered, flask B contains a few milliliters of solvent and a stemless funnel, and flask C contains several milliliters of the crystallizing solvent to be used for rinsing purposes), a fluted piece of filter paper, a towel for holding the hot flask and drying out the stemless funnel, and boiling chips for all solutions.

A piece of filter paper is fluted as shown in Figure 4.7 and is then placed in a stemless funnel. Appropriate sizes of Erlenmeyer flasks, stemless funnels, and

■ **FIG. 4.7**

Fluting a piece of filter paper.

⚠️

Be aware that the vapors of
low-boiling solvents can ignite
on an electric hot plate.

■ **FIG. 4.8**

Assemblies for gravity
filtration. Stemless funnels
have diameters of 2.5, 4.2,
5.0, and 6.0 cm.

filter paper are shown in Figure 4.8. The funnel is stemless so that the saturated
solution being filtered will not have a chance to cool and clog the stem with crystals. The filter paper should fit entirely inside the rim of the funnel; it is fluted to
allow rapid filtration. Test to see that the funnel is stable in the neck of flask B. If
not, support it with a ring attached to a ring stand. A few milliliters of solvent and
a boiling chip should be placed in flask B into which the solution is to be filtered.
This solvent is brought to a boil on the steam bath or hot plate along with the solution to be filtered.

The solution to be filtered (in flask A) should be saturated with the solute at
the boiling point. Note the volume and then add 10% more solvent (from flask C).
The resulting slightly dilute solution is not as likely to recrystallize in the funnel
during filtration. Bring the solution to be filtered to a boil, grasp flask A in a towel,
and pour the solution into the filter paper in the stemless funnel equipped in flask
B (Fig. 4.6). The funnel should be warm to prevent recrystallization from occurring in the funnel. This can be accomplished in two ways: (1) Invert the funnel
over a steam bath for a few seconds, pick up the funnel with a towel, wipe it
perfectly dry, place it on top of flask B, and then add the fluted filter paper; or
(2) place the stemless funnel in the neck of flask B and allow the solvent to reflux
into the funnel, thereby warming it.

Pour the solution to be filtered (in flask A) at a steady rate into the fluted filter paper (equipped in flask B). Check to see whether recrystallization is occurring
in the filter. If it does, add boiling solvent (from flask C heated on a steam bath or
a hot plate) until the crystals dissolve, dilute the solution being filtered, and carry
on. Rinse flask A with a few milliliters of boiling solvent (from flask C) and rinse
the fluted filter paper with this same solvent.

Because the filtrate has been diluted to prevent it from recrystallizing during
the filtration process, the excess solvent must now be removed by boiling the solution. This process can be speeded up somewhat by blowing a slow current of air
into the flask in the hood or using an aspirator tube to pull vapors into the aspirator (Fig. 4.9 and Fig. 4.10). However, the fastest method is to heat the solvent in
the filter flask on a sand bath while the flask is connected to the water aspirator.

**■ FIG. 4.9**
An aspirator tube in use. A boiling stick may be necessary to promote boiling.

Aspirator tube
To water aspirator
Clamp here
Tared flask
Boiling stick
Steam in
Outlet to sink
Steam bath

**■ FIG. 4.10**
A tube being used to remove solvent vapors.

To water aspirator or to nitrogen or air supply

Pasteur pipette

Boiling stick

Hot sand bath

**■ FIG. 4.11**
Evaporation of a solvent under a vacuum.

Hirsch funnel
25-mL Filter flask
To water aspirator

Be sure that you are wearing gloves when doing this step!

The vacuum is controlled with the thumb (Fig. 4.11).[2] Be sure that you are wearing gloves when doing this step! If your thumb is not large enough, put a one-holed rubber stopper into the Hirsch funnel or the filter flask and again control the vacuum with your thumb. If the vacuum is not controlled, the solution may boil over and go out the vacuum hose.

## Step 5. Recrystallizing the Solute

On both a macroscale and a microscale, the recrystallization process should normally start from a solution that is saturated with the solute at the boiling point. If it has been necessary to remove impurities or charcoal by filtration, the solution

2. See also Mayo, D. W.; Pike, R. M.; Butcher, S. M. *Microscale Organic Laboratory*; Wiley: New York, 1986; 97.

has been diluted. To concentrate the solution, simply boil off the solvent under an aspirator tube as shown in Figure 4.9 (macroscale) or blow off solvent using a gentle stream of air or, better, nitrogen in the hood as shown in Figure 4.10 (microscale). Be sure to have a boiling chip (macroscale) or a boiling stick (microscale) in the solution during this process but make sure you remove it before initiating recrystallization.

Once it has been ascertained that the hot solution is saturated with the compound just below the boiling point of the solvent, allow it to cool slowly to about 22°C. Slow cooling is a critical step in recrystallization. If the solution is not allowed to cool slowly, precipitation will occur, resulting in impurities "crashing out" of solution along with the desired solute; thus, no exclusion will occur. On a microscale, it is best to allow the reaction tube to cool in a beaker filled with cotton or paper towels, which acts as insulation, so cooling takes place slowly. Even insulated in this manner, the small reaction tube will cool to about 22°C within a few minutes. Slow cooling will guarantee the formation of large crystals, which are easily separated by filtration and easily washed free of adhering impure solvent. On a small scale, it is difficult to obtain crystals that are too large and occlude impurities. Once the tube has cooled to about 22°C without disturbance, it can be cooled in ice to maximize the amount of product that comes out of solution. On a macroscale, the Erlenmeyer flask is set atop a cork ring or other insulator and allowed to cool gradually to about 22°C. If the flask is moved during recrystallization, many nuclei will form, and the crystals will be small and have a large surface area. They will not be easy to filter and wash clean of the mother liquor. Once recrystallization ceases at about 22°C, the flask should be placed in ice to cool further. Make sure to clamp the flask in the ice bath so that it does not tip over.

With slow cooling, recrystallization should begin immediately. If not, add a seed crystal or scratch the inside of the tube with a glass rod at the liquid-air interface. Recrystallization must start on some nucleation center. A minute crystal of the desired compound saved from the crude material will suffice. If a seed crystal is not available, recrystallization can be started on the rough surface of a fresh scratch on the inside of the container.

# Step 6. Collecting and Washing the Crystals

Once recrystallization is complete, the crystals must be separated from the ice-cold mother liquor, washed with ice-cold solvent, and dried.

## *Microscale Procedure*

### (A) Filtration Using a Pasteur Pipette

The most important filtration technique used in microscale organic experiments employs a Pasteur pipette (Fig. 4.12). About 70% of the crystalline products from the experiments in this text can be isolated in this way. The others will be isolated by filtration on a Hirsch funnel.

The ice-cold crystalline mixture is stirred with a Pasteur pipette and, while air is being expelled from the pipette, forced to the bottom of the reaction tube. The

---

*Margin notes:*

A saturated solution

Slow cooling is important.

**Online Study Center**

Videos: Recrystallization, Microscale Crystallization

Add a seed crystal or scratch the tube.

**Online Study Center**

Video: Filtration of Crystals Using the Pasteur Pipette

■ **FIG. 4.12**
Filtration using a Pasteur pipette and a reaction tube.

Pipette bulb

Pasteur pipette

Reaction tube

Note square, flat tip

■ **FIG. 4.13**
Drying crystals under reduced pressure in a reaction tube.

To aspirator

Glass tube

Thermometer adapter

Reaction tube

Beaker of warm water

Crystals to be dried

bulb is released, and the solvent is drawn into the pipette through the very small space between the square tip of the pipette and the curved bottom of the reaction tube. When all the solvent has been withdrawn, it is expelled into another reaction tube. It is sometimes useful to rap the tube containing the wet crystals against a hard surface to pack them so that more solvent can be removed. The tube is returned to the ice bath, and a few drops of cold solvent are added to the crystals. The mixture is stirred to wash the crystals, and the solvent is again removed. This process can be repeated as many times as necessary. Volatile solvents can be removed from the damp crystals under vacuum (Fig. 4.13). Alternatively, the last traces of solvent can be removed by centrifugation using a Wilfilter (Fig. 4.14).

### (B) Filtration Using a Hirsch Funnel

When the volume of material to be filtered is greater than 1.5 mL, collect the material on a Hirsch funnel. The Hirsch funnel in the Williamson/Kontes kit[3] is unique. It is composed of polypropylene and has an integral molded stopper that fits the 25-mL filter flask. It comes fitted with a 20-μm polyethylene fritted disk, which is not meant to be disposable, although it costs only about twice as much as an 11-cm

*Online Study Center*

Video: Microscale Filtration on the Hirsch Funnel

---

3. The microscale kit is available through Kontes (www.kontes.com).

■ **FIG. 4.14**

The Wilfilter filtration apparatus. Filtration occurs between the flat face of the polypropylene Wilfilter and the top of the reaction tube.

(a)          (b)

■ **FIG. 4.15**

The Hirsch funnel being used for vacuum filtration. This unique design has a removable and replaceable 20-μm polyethylene frit. No adapter is needed because there is a vacuum-tight fit to the filter flask. Always use a piece of filter paper.

piece of filter paper (Fig. 4.15). Although products can be collected directly on this disk, it is good practice to place an 11- or 12-mm-diameter piece of no. 1 filter paper on the disk. In this way the frit will not become clogged with insoluble impurities. The disk of filter paper can be cut with a cork borer or leather punch. A piece of filter paper *must* be used on the old-style porcelain Hirsch funnels.

Clamp the clean, dry 25-mL filter flask in an ice bath to prevent it from falling over and place the Hirsch funnel with filter paper in the flask. (The reason for cooling the filter flask is to keep the mother liquor cold so it will not dissolve the crystals on the Hirsch funnel when fresh cold solvent is used to wash crystals from the container onto the funnel.) In a separate flask, cool ~10 mL of solvent in an ice bath; this solvent is used for washing the recrystallization flask and for washing the crystals. Wet the filter paper with the solvent used in the recrystallization, turn on the water aspirator (see "The Water Aspirator and the Trap"), and ascertain that the filter paper is pulled down onto the frit. Pour and scrape the crystals and

Break the vacuum, add a very small quantity of ice-cold wash solvent, and reapply vacuum.

mother liquor onto the Hirsch funnel and, as soon as the liquid is gone from the crystals, break the vacuum at the filter flask by removing the rubber hose.

Rinse the recrystallization flask with ice-cold fresh solvent. Pour this rinse through the Hirsch funnel and reapply vacuum to the filter flask. As soon as all the liquid has disappeared from the crystals, wash the crystals with a few drops of ice-cold solvent. Repeat this washing process as many times as necessary to remove colored material or other impurities from the crystals. In some cases, only one very small wash will be needed. After the crystals have been washed with ice-cold solvent, the vacuum can be left on to dry the crystals. Sometimes it is useful to press solvent from the crystals by using a cork.

### (C) Filtration with a Wilfilter (Replacing a Craig Tube)

The isolation of less than 100 mg of recrystallized material from a reaction tube (or any other container) is not easy. If the amount of solvent is large enough (1 mL or more), the material can be recovered by filtration on a Hirsch funnel. But when the volume of liquid is less than 1 mL, much product is left in the tube during transfer to a Hirsch funnel. The solvent can be removed with a Pasteur pipette pressed against the bottom of the tube, a very effective filtration technique, but scraping the damp crystals from the reaction tube results in major losses. If the solvent is relatively low boiling, it can be evaporated by connecting the tube to a water aspirator (*see* Fig. 4.13 on page 73). Once the crystals are dry, they are easily scraped from the tube with little or no loss. Some solvents—and water is the principal culprit—are not easily removed by evaporation. And even though removal of the solvent under vacuum is not terribly difficult, it takes time.

We have invented a filtration device that circumvents these problems: the Wilfilter. After recrystallization has ceased, most of the solvent is removed from the crystals using a Pasteur pipette in the usual way (*see* Fig. 4.12 on page 73). Then the polypropylene Wilfilter is placed on the top of the reaction tube followed by a 15-mL polypropylene centrifuge tube (*see* Fig. 4.14 on page 74). The assembly is inverted and placed in a centrifuge such as the International Clinical Centrifuge that holds twelve 15-mL tubes. The assembly, properly counterbalanced, is centrifuged for about 1 min at top speed. The centrifuge tube is removed from the centrifuge, and the reaction tube is then removed from the centrifuge tube. The three fingers on the Wilfilter keep it attached to the reaction tube. The filtrate is left in the centrifuge tube.

Filtration with the Wilfilter occurs between the top surface of the reaction tube and the flat surface of the Wilfilter. Liquid will pass through that space during centrifugation; crystals will not. The crystals will be found on top of the Wilfilter and inside the reaction tube. The very large centrifugal forces remove all the liquid, so the crystals will be virtually dry and thus easily removed from the reaction tube by shaking or scraping with the metal spatula.

The Wilfilter replaces an older device known as a Craig tube (Fig. 4.16), which consists of an outer tube of 1-, 2-, or 3-mL capacity with an inner plunger made of Teflon (expensive) or glass (fragile). The material to be recrystallized is transferred to the outer tube and recrystallized in the usual way. The inner plunger

Crystals

Rough ground
glass surfaces

Centrifuge tube

Crystals
and solvent

Solvent

is added, and a wire hanger is fashioned so that the assembly can be removed from
the centrifuge tube without the plunger falling off. Filtration in this device occurs
through the rough surface that has been ground into the shoulder of the outer tube.

The Wilfilter possesses several advantages: a special recrystallization device
is not needed, no transfers of material are needed, it is not as limited in capacity
(which is 4.5 mL), and its cost is one-fifth that of a Craig tube assembly.

### (D) Filtration into a Reaction Tube on a Hirsch Funnel

If it is desired to have the filtrate in a reaction tube instead of spread on the bottom
of the 25-mL filter flask, then the process described in the previous section can be
carried out in the apparatus shown in Figure 4.17. The vacuum hose is connected to
the side arm using the thermometer adapter and a short length of glass tubing. Evaporate the filtrate in the reaction tube to collect a second crop of crystals.

### (E) Filtration into a Reaction Tube on a Micro Büchner Funnel

If the quantity of material being collected is very small, the bottom of the chromatography column is a micro Büchner funnel, which can be fitted into the top of
the thermometer adapter, as shown in Figure 4.18. Again, it is good practice to
cover the frit with a piece of 6-mm filter paper (cut with a cork borer).

### (F) The Micro Büchner Funnel in an Enclosed Filtration Apparatus

In the apparatus shown in Figure 4.19, recrystallization is carried out in the upper
reaction tube in the normal way. The apparatus is then turned upside down, the
crystals are shaken down onto a micro Büchner funnel, and a vacuum is applied

### ■ FIG. 4.17

A microscale Hirsch filtration assembly. The Hirsch funnel gives a vacuum-tight seal to the 105° adapter.

Hirsch funnel with fritted disk

To aspirator

### ■ FIG. 4.18

Filtration using a microscale Büchner funnel.

Thermometer adapter

Micro Büchner funnel with fritted disk (base of chromatography column)

To aspirator

Vacuum-tight connector also serves as clamp for apparatus

Reaction tube with filtrate

### ■ FIG. 4.19

The Schlenk-type filtration apparatus. The apparatus is inverted to carry out the filtration.

Micro Büchner funnel with fritted disk

To aspirator

### ■ FIG. 4.20

Matching filter assemblies. The 6.0-cm polypropylene Büchner funnel (right) resists breakage and can be disassembled for cleaning.

Büchner funnels

No. 2 7.5 cm

No. 0 4.2 cm

No. 000 5 cm

Hirsch funnel

Neoprene adapter

250 mL

125 mL

50 mL

6 cm

7 cm

4.25 cm

3.3 cm

5.5 cm

through the side arm. In this apparatus, crystals can be collected in an oxygen-free atmosphere (Schlenk conditions).

## *Macroscale Apparatus*

### Filtration on a Hirsch Funnel and a Büchner Funnel

If the quantity of material is small (<2 g), a Hirsch funnel can be used in exactly the way described in a previous section. For larger quantities, a Büchner funnel is used. Properly matched Büchner funnels, filter paper, and flasks are shown in Figure 4.20. The Hirsch funnel shown in the figure has a 5-cm bottom plate to accept 3.3-cm paper.

***Online Study Center***

Video: Microscale Filtration on the Hirsch Funnel

Place a piece of filter paper in the bottom of the Büchner funnel. Wet it with solvent and be sure it lies flat so that crystals cannot escape around the edge and under the filter paper. Then, with the vacuum off, pour the cold slurry of crystals into the center of the filter paper. Apply the vacuum; as soon as the liquid disappears from the crystals, break the vacuum to the flask by disconnecting the hose. Rinse the Erlenmeyer flask with cold solvent. Add this to the crystals and reapply the vacuum just until the liquid disappears from the crystals. Repeat this process as many times as necessary and then leave the vacuum on to dry the crystals.

Clamp the filter flask.

### The Water Aspirator and Trap

The most common way to produce a vacuum for filtration in the organic laboratory is by employing a *water aspirator*. Air is entrained efficiently in the water rushing through the aspirator to produce a vacuum roughly equal to the vapor pressure of the water going through it (17 torr at 20°C, 5 torr at 4°C). A check valve is built into the aspirator, but when the water is turned off, it will often back up into the evacuated system. For this reason a *trap* is always installed in the line (Fig. 4.21). *The water passing through the aspirator should always be turned on full force.*

Opening the screw clamp on the trap can open the system to the atmosphere as well as removing the hose from the small filter flask. Open the system and then turn

### ■ FIG. 4.21

An aspirator, a filter trap, and a funnel. Clamp the small filter flask to prevent it from turning over.

Clamp

Thin-walled tubing

Two-hole rubber stopper

Heavy-walled tubing

Splashgon

500-mL Filter flask

25-mL
Filter flask

off the water to avoid having water sucked back into the filter trap. Thin rubber tubing on the top of the trap will collapse and bend over when a good vacuum is established. You will, in time, learn to hear the differences in the sound of an aspirator when it is pulling a vacuum and when it is working on an open system.

### Collecting a Second Crop of Crystals

Regardless of the method used to collect crystals on either a macroscale or a microscale, the filtrate and washings can be combined and evaporated to the point of saturation to obtain a second crop of crystals—hence this advocates having a clean receptacle for the filtrate. This second crop will increase the overall yield, but the crystals will not usually be as pure as the first crop.

## Step 7. Drying the Product

### *Microscale Procedure*

If possible, dry the product in the reaction tube after removing the solvent with a Pasteur pipette. Simply connecting the tube to the water aspirator can do this. If the tube is clamped in a beaker of hot water, the solvent will evaporate more rapidly under vacuum but make sure not to melt the product (*see* Fig. 4.13 on page 73). Water, which has a high heat of vaporization, is difficult to remove in this way. Scrape the product onto a watch glass and allow it to dry to constant weight, which will indicate that all the solvent has been removed. If the product is collected on a Hirsch funnel or a Wilfilter, the last traces of solvent can be removed by squeezing the crystals between sheets of filter paper before drying them on the watch glass.

### *Macroscale Procedure*

Once the crystals have been washed on a Hirsch funnel or a Büchner funnel, press them down with a clean cork or other flat object and allow air to pass through them until they are substantially dry. Final drying can be done under reduced pressure (Fig. 4.22). The crystals can then be turned out of the funnel and squeezed between sheets of filter paper to remove the last traces of solvent before final drying on a watch glass.

**■ FIG. 4.22**
Drying a solid by reduced air pressure.

## Experiments

## 1. Solubility Tests

To test the solubility of a solid, transfer an amount roughly estimated to be about 10 mg (the amount that forms a symmetrical mound on the end of a stainless steel spatula) into a reaction tube and add about 0.25 mL of solvent from a calibrated dropper or pipette. Stir with a fire-polished stirring rod that is 4 mm in diameter, break up any lumps, and determine if the solid is readily soluble at room temperature (about 22°C). If the substance is readily soluble at about 22°C in methanol, ethanol, acetone, or acetic acid, add a few drops of water from a wash bottle to see if a solid precipitates. If so, heat the mixture, adjust the composition of the solvent pair to

*Test Compounds:*

**Resorcinol**

**Anthracene**

**Benzoic acid**

**4-Amino-1-naphthalenesulfonic acid, sodium salt**

produce a hot solution saturated at the boiling point, let the solution stand undisturbed, and note the character of the crystals that form.

If the substance fails to dissolve in a given solvent at about 22°C, heat the suspension and observe if a solution occurs. If the solvent is flammable, heat the test tube on a steam bath or in a small beaker of water kept warm on a steam bath or a hot plate. If the solid dissolves completely, it can be declared readily soluble in the hot solvent; if some but not all dissolves, it is said to be moderately soluble, and further small amounts of solvent should then be added until solution is complete.

When a substance has been dissolved in hot solvent, cool the solution by holding the flask under the tap and, if necessary, induce recrystallization by rubbing the walls of the tube with a stirring rod to make sure that the concentration permits recrystallization. Then reheat to dissolve the solid, let the solution stand undisturbed, and inspect the character of the ultimate crystals.

Perform solubility tests on the test compounds shown in the margin in each of the the following solvents: water (hydroxylic and ionic), toluene (an aromatic hydrocarbon), and ligroin (a mixture of aliphatic hydrocarbons). Note the degree of solubility in the solvents—cold and hot—and suggest suitable solvents, solvent pairs, or other expedients for the recrystallization of each substance. Record the crystal form, at least to the extent of distinguishing between needles (pointed crystals), plates (flat and thin), and prisms. How do your observations conform to the generalization that like dissolves like?

**Cleaning Up.** Place organic solvents and solutions of the compounds in the organic solvents container. Dilute the aqueous solutions with water and flush down the drain. (For this and all other "Cleaning Up" sections, refer to the complete discussion of waste disposal procedures in Chapter 2.)

## 2. Recrystallization of Pure Phthalic Acid, Naphthalene, and Anthracene

**Phthalic acid**          **Naphthalene**          **Anthracene**

The process of recrystallization can be readily observed using phthalic acid. In the *CRC Handbook of Chemistry and Physics*, in the table "Physical Constants of Organic Compounds," the entry for phthalic acid gives the following solubility data (in grams of solute per 100 mL of solvent). The superscripts refer to temperature in degrees Celsius:

| Water | Alcohol | Ether, etc. |
|---|---|---|
| $0.54^{14}$ | $11.71^{18}$ | $0.69^{15}$ eth., i. chl. |
| $18^{99}$ | | |

The large difference in solubility in water as a function of temperature suggests that water is the solvent of choice. The solubility in alcohol is high at about 22°C. Ether is difficult to use because it is so volatile; the compound is insoluble in chloroform (i. chl.).

## Microscale Procedure for Phthalic Acid

*Online Study Center*

Video: Recrystallization

Set the heater control to about 20% of the maximum.

Recrystallize 60 mg (0.060 g) of phthalic acid from the minimum volume of water, using the previous data to calculate the required volume. First, turn on an electrically heated sand bath. Add the solid to a 10 × 100 mm reaction tube and add water dropwise with a Pasteur pipette. Use the calibration marks found in Chapter 1 (*see* Fig. 1.18 on page 19) to measure the volume of water in the pipette and the reaction tube. Add a boiling stick (a wood applicator stick) to facilitate even boiling and prevent bumping. After a portion of the water has been added, gently heat the solution to boiling on a hot sand bath in the electric heater. The deeper the tube is placed in the sand, the hotter it will be. As soon as boiling begins, continue to add water dropwise until the entire solid just dissolves. Cork the tube, clamp it as it cools, and observe the phenomenon of recrystallization.

Alternate procedure: Dry the crystals under vacuum in a steam bath in the reaction tube.

After the tube reaches about 22°C, cool it in ice, stir the crystals with a Pasteur pipette, and expel the air from the pipette as the tip is pushed to the bottom of the tube. When the tip is firmly and squarely seated in the bottom of the tube, release the bulb and withdraw the water. Rap the tube sharply on a wood surface to compress the crystals and remove as much of the water as possible with the pipette. Then cool the tube in ice and add a few drops of ice-cold ethanol to the tube to remove water from the crystals. Connect the tube to a water aspirator and warm it in a beaker of hot water (*see* Fig. 4.13 on page 73). Once all the solvent is removed, using the stainless steel spatula, scrape the crystals onto a piece of filter paper, fold the paper over the crystals, and squeeze out excess water before allowing the crystals to dry to constant weight. Weigh the dry crystals and calculate the percent recovery of product.

## Microscale Procedure for Naphthalene and Anthracene

These compounds can also be isolated using a Wilfilter.

Following the previous procedure, recrystallize 40 mg of naphthalene from 80% aqueous methanol or 10 mg of anthracene from ethanol. You may find it more convenient to use a hot water bath to heat these low-boiling alcohols. These are more typical of compounds to be recrystallized in later experiments because they are soluble in organic solvents. It will be much easier to remove these solvents from the crystals under vacuum than it is to remove water from phthalic acid. You will seldom encounter the need to recrystallize less than 30 mg of a solid in these experiments.

**Cleaning Up.** Dilute the aqueous filtrate with water and flush the solution down the drain. Phthalic acid is not considered toxic to the environment and can be recycled for future recrystallization experiments. Methanol and ethanol filtrates go into the organic solvents container.

## Macroscale Procedure

**Online Study Center**

Video: Macroscale Crystallization

Using the solubility data for phthalic acid to calculate the required volume, recrystallize 1.0 g of phthalic acid from the minimum volume of water. Add the solid to the smallest practical Erlenmeyer flask and then, using a Pasteur pipette, add water dropwise from a full 10-mL graduated cylinder. A boiling stick (a wood applicator stick) facilitates even boiling and will prevent bumping. After a portion of the water has been added, gently heat the solution to boiling on a hot plate. As soon as boiling begins, continue to add water dropwise until the entire solid dissolves. Place the flask on a cork ring or other insulator and allow it to cool undisturbed to about 22°C, during which time the recrystallization process can be observed. Slow cooling favors large crystals. Then cool the flask in an ice bath, decant (pour off) the mother liquor (the liquid remaining with the crystals), and remove the last traces of liquid with a Pasteur pipette. Scrape the crystals onto a filter paper using a stainless steel spatula, squeeze the crystals between sheets of filter paper to remove traces of moisture, and allow the crystals to dry. Alternatively, the crystals can be collected on a Hirsch funnel. Compare the calculated volume of water to the volume of water actually used to dissolve the acid. Calculate the percent recovery of dry, recrystallized phthalic acid.

**Cleaning Up.** Dilute the filtrate with water and flush the solution down the drain. Phthalic acid is not considered toxic to the environment and can be recycled for future recrystallization experiments.

# 3. Decolorizing a Solution with Decolorizing Charcoal

**Online Study Center**

Video: Decolorization of a Solution with Norit

Decolorizing using pelletized Norit

Into a reaction tube place 1.0 mL of a solution of methylene blue dye that has a concentration of 10 mg per 100 mL of water. Add to the tube about 10 or 12 pieces of decolorizing charcoal, shake, and observe the color over a period of 1–2 min. Heat the contents of the tube to boiling (reflux) and observe the color by holding the tube in front of a piece of white paper from time to time. How rapidly is the color removed? If the color is not removed in a minute or so, add more charcoal pellets.

**Cleaning Up.** Place the Norit in the nonhazardous solid waste container.

# 4. Decolorization of Brown Sugar (Sucrose, $C_{12}H_{22}O_{11}$)

Raw sugar is refined commercially with the aid of decolorizing charcoal. The clarified solution is seeded generously with small sugar crystals, and excess water is removed under vacuum to facilitate recrystallization. The pure white crystalline product is collected by centrifugation. Brown sugar is partially refined sugar and can be easily decolorized using charcoal.

In a 50-mL Erlenmeyer flask, dissolve 15 g of dark brown sugar in 30 mL of water by heating and stirring. Pour half the solution into another 50-mL flask. Heat one of the solutions nearly to its boiling point, allow it to cool slightly, and add 250 mg (0.25 g) of decolorizing charcoal (Norit pellets) to it. Bring the solution back to near the boiling point for 2 min; then filter the hot solution into an Erlenmeyer flask through a fluted filter paper held in a previously heated funnel. Treat the other half of the sugar solution in exactly the same way but use only 50 mg of decolorizing charcoal. In collaboration with a fellow student, try heating the solutions for only 15 s after adding the charcoal. Compare your results.

**Cleaning Up.** Decant (pour off) the aqueous layer. Place the Norit in the nonhazardous solid waste container. The sugar solution can be flushed down the drain.

# 5. Recrystallization of Benzoic Acid from Water and a Solvent Pair

**COOH**

**Benzoic acid**

Recrystallize 50 mg of benzoic acid from water in the same way phthalic acid was recrystallized. Then in a dry reaction tube dissolve another 50-mg sample of benzoic acid in the minimum volume of hot methanol and add water to the hot solution dropwise. When the hot solution becomes cloudy and recrystallization has begun, allow the tube to cool slowly to about 22°C; then cool it in ice and collect the crystals. Compare recrystallization in water to that in the solvent pair.

**Cleaning Up.** The methanol-water solution can be disposed in the organic solvents waste container or, if regulations permit, diluted with water and flushed down the drain.

# 6. Recrystallization of Naphthalene from a Mixed Solvent

**Naphthalene**

Do not try to grasp Erlenmeyer flasks with a test tube holder.

Support the funnel in a ring stand.

Add 2.0 g of impure naphthalene (a mixture of 100 g of naphthalene, 0.3 g of a dye such as Congo Red, and another substance such as magnesium sulfate, or dust) to a 50-mL Erlenmeyer flask along with 3 mL of methanol and a boiling stick to promote even boiling. Heat the mixture to boiling over a steam bath or a hot plate and then add methanol dropwise until the naphthalene just dissolves when the solvent is boiling. The total volume of methanol should be 4 mL. Remove the flask from the heat and cool it rapidly in an ice bath. Note that the contents of the flask set to a solid mass, which would be impossible to handle. Add enough methanol to bring the total volume to 25 mL, heat the solution to its boiling point, remove the flask from the heat, allow it to cool slightly, and add 30 mg of decolorizing charcoal pellets to remove the colored impurity in the solution. Heat the solution to its boiling point for 2 min; if the color is not gone, add more Norit and boil again, and then filter through a fluted filter paper in a previously warmed stemless funnel into a 50-mL Erlenmeyer flask. Sometimes filtration is slow because the funnel fits so snugly into the mouth of the flask that a back pressure develops. If you note that raising the funnel increases the flow of filtrate, fold a small strip of paper two or three times and insert it between the funnel and flask. Wash the used flask with 2 mL of hot methanol and use this liquid to wash the filter paper, transferring the solvent with a Pasteur pipette in a succession of drops around the upper rim of the filter paper. When the filtration is complete, the volume of methanol should be 15 mL. If it is not, evaporate the excess methanol.

Because the filtrate is far from being saturated with naphthalene at this point, it will not yield crystals on cooling; however, the solubility of naphthalene in methanol can be greatly reduced by the addition of water. Heat the solution to its boiling point and add water dropwise from a 10-mL graduated cylinder using a Pasteur pipette (or a precalibrated pipette). After each addition of water, the solution will become cloudy for an instant. Swirl the contents of the flask and heat to redissolve any precipitated naphthalene. After the addition of 3.5 mL of water, the solution will almost be saturated with naphthalene at the boiling point of the solvent. Remove the flask from the heat and place it on a cork ring or other insulating surface to cool, without being disturbed, to about 22°C.

Immerse the flask in an ice bath along with another flask containing a ~~30:7~~ 15:3.5 mixture of methanol and water. This cold solvent will be used for washing the crystals. The cold recrystallization mixture is collected by vacuum filtration on a small 50-mm Büchner funnel (Fig. 4.23). The water flowing through the aspirator should always be turned on full force. In collecting the product by suction filtration, use a spatula to dislodge crystals and ease them out of the flask. If crystals still remain in the flask, some filtrate can be poured back into the recrystallization flask as a rinse for washing as often as desired because it is saturated with solute. To free the crystals from contaminating the mother liquor, break the suction, pour a few milliliters of the fresh cold solvent mixture into the Büchner funnel, and immediately reapply suction. Repeat this process until the crystals and the filtrate are free of color. Press the crystals with a clean cork to eliminate excess solvent, pull air through the filter cake for a few minutes, and then put the large, flat, platelike crystals out on a filter paper to dry. The yield of pure white crystalline naphthalene

■ **FIG. 4.23**
A suction filter assembly clamped to provide firm support. The funnel must be pressed down on the Filtervac to establish reduced pressure in the flask.

Filtervac

should be about 1.6 g. The mother liquor contains about 0.25 g, and about 0.15 g is retained in the charcoal and on the filter paper.

**Cleaning Up.** Place the Norit in the nonhazardous solid waste container. The methanol filtrate and washings are placed in the organic solvents container.

# 7. Purification of an Unknown

Recall the seven-step recrystallization procedure:

1. Choose the solvent.
2. Dissolve the solute.
3. Decolorize the solution (if necessary).
4. Filter suspended solids (if necessary).
5. Recrystallize the solute.
6. Collect and wash the crystals.
7. Dry the product.

You will purify an unknown provided by your instructor, 2.0 g if working on a macroscale and 100 mg if working on a microscale. Conduct tests for solubility and the ability to recrystallize in several organic solvents, solvent pairs, and water. Conserve your unknown by using very small quantities for the solubility tests. If only a drop or two of solvent is used, heating the test tube on a steam bath or a sand bath can evaporate the solvent, and the residue can be used for another test. If decolorization is necessary, dilute the solution before filtration. It is very difficult to filter a hot, saturated solution from decolorizing carbon without recrystallization occurring in the filtration apparatus. Evaporate the decolorized solution to the point of saturation and proceed with the recrystallization. Submit as much pure product as possible with evidence of its purity (i.e., the melting point). From the posted list identify the unknown. If an authentic sample is available, your identification can be verified by a mixed melting point determination (*see* Chapter 3).

**Cleaning Up.** Place decolorizing charcoal, if used, and filter paper in the nonhazardous solid waste container. Put organic solvents in the organic solvents container and flush aqueous solutions down the drain.

## Recrystallization Problems and Their Solutions

## Induction of Crystallization

Occasionally, a sample will not crystallize from solution on cooling, even though the solution is saturated with the solute at elevated temperature. The easiest method for inducing crystallization is to add to the supersaturated solution a seed crystal that has been saved from the crude material (if it was crystalline

Seeding

Scratching

before crystallization was attempted). In a probably apocryphal tale, the great sugar chemist Emil Fischer merely had to wave his beard over a recalcitrant solution, and the appropriate seed crystals would drop out, causing recrystallization to occur. In the absence of seed crystals, scratching the inside of the flask with a stirring rod at the liquid-air interface can often induce recrystallization. One theory holds that part of the freshly scratched glass surface has angles and planes corresponding to the crystal structure, and crystals start growing on these spots. Recrystallization is often very slow to begin. Placing the sample in a refrigerator overnight will bring success. Other expedients are to change the solvent (usually to a less soluble one) and to place the sample in an open container where slow evaporation and dust from the air may help induce recrystallization.

## Oils and "Oiling Out"

*Online Study Center*

Video: Formation of an Oil
Instead of Crystals

When cooled, some saturated solutions—especially those containing water—deposit not crystals but small droplets referred to as oils. "Oiling out" occurs when the temperature of the solution is above the melting point of the crystals. If these droplets solidify and are collected, they will be found to be quite impure. Similarly, the melting point of the desired compound may be depressed to a point such that a low-melting eutectic mixture of the solute and the solvent comes out of solution. The simplest remedy for this problem is to lower the temperature at which the solution becomes saturated with the solute by simply adding more solvent. In extreme cases it may be necessary to lower this temperature well below 22°C by cooling the solution with dry ice. If oiling out persists, use another solvent.

Crystallize at a lower
temperature

## Recrystallization Summary

*Online Study Center*

Video: Picking a Solvent

Video: Recrystallization

Video: Decolorization of a
Solution with Norit

Photo: Preparation of a Filter
Pipette; Video: Microscale
Crystallization

1. **Choosing the solvent.**   "Like dissolves like." Some common solvents are water, methanol, ethanol, ligroin, and toluene. When you use a solvent pair, dissolve the solute in the better solvent and add the poorer solvent to the hot solution until saturation occurs. Some common solvent pairs are ethanol-water, ether-ligroin, and toluene-ligroin.

2. **Dissolving the solute.**   In an Erlenmeyer flask or reaction tube, add solvent to the crushed or ground solute and heat the mixture to boiling. Add more solvent as necessary to obtain a hot, saturated solution.

3. **Decolorizing the solution.**   If it is necessary to remove colored impurities, cool the solution to about 22°C and add more solvent to prevent recrystallization from occurring. Add decolorizing charcoal in the form of pelletized Norit to the cooled solution and then heat it to boiling for a few minutes, making sure to swirl the solution to prevent bumping. Remove the Norit by filtration and then concentrate the filtrate.

4. **Filtering suspended solids.**   If it is necessary to remove suspended solids, dilute the hot solution slightly to prevent recrystallization from occurring during filtration. Filter the hot solution. Add solvent if recrystallization begins in the funnel. Concentrate the filtrate to obtain a saturated solution.

Photo: Recrystallization; Video: Recrystallization

**5. Recrystallizing the solute.**   Let the hot saturated solution cool to about 22°C spontaneously. Do not disturb the solution. Then cool it in ice. If recrystallization does not occur, scratch the inside of the container or add seed crystals.

Photos: Use of the Wilfilter, Filtration Using a Pasteur Pipette; Videos: Microscale Filtration on the Hirsch Funnel, Filtration of Crystals Using the Pasteur Pipette

**6. Collecting and washing the crystals.**   Collect the crystals using the Pasteur pipette method, the Wilfilter, or by vacuum filtration on a Hirsch funnel or a Büchner funnel. If the latter technique is employed, wet the filter paper with solvent, apply vacuum, break vacuum, add crystals and liquid, apply vacuum until solvent just disappears, break vacuum, add cold wash solvent, apply vacuum, and repeat until crystals are clean and filtrate comes through clear.

Photo: Drying Crystals Under Vacuum; Video: Recrystallization

**7. Drying the product.**   Press the product on the filter to remove solvent. Then remove it from the filter, squeeze it between sheets of filter paper to remove more solvent, and spread it on a watch glass to dry.

## QUESTIONS

1. A sample of naphthalene, which should be pure white, was found to have a grayish color after the usual purification procedure. The melting point was correct, and the melting point range was small. Explain the gray color.

2. How many milliliters of boiling water are required to dissolve 25 g of phthalic acid? If the solution were cooled to 14°C, how many grams of phthalic acid would recrystallize out?

3. Why should activated carbon be used during a recrystallization?

4. If a little activated charcoal does a good job removing impurities in a recrystallization, why not use a larger quantity?

5. Under which circumstances is it wise to use a mixture of solvents to carry out a recrystallization?

6. Why is gravity filtration rather than suction filtration used to remove suspended impurities and charcoal from a hot solution?

7. Why is a fluted filter paper used in gravity filtration?

8. Why are stemless funnels used instead of long-stem funnels to filter hot solutions through fluted filter paper?

9. Why is the final product from the recrystallization process isolated by vacuum filtration rather than gravity filtration?

# CHAPTER

# 5

# Distillation

**Online Study Center**

This icon will direct you to
techniques, equipment setups,
and online resources at
http://college.hmco.com/PIC/
williamsonMMOE5e.

Distillation is a common
method for purifying liquids
and can be used to determine
their boiling points.

**PRELAB EXERCISE:** Predict what a plot of temperature versus the
volume of distillate will look like for the simple distillation and
the fractional distillation of (a) a cyclohexane-toluene mixture and
(b) an ethanol-water mixture.

The origins of distillation are lost in antiquity, when humans in their thirst for
more potent beverages found that dilute solutions of fermented alcohol could be
separated into alcohol-rich and water-rich portions by heating the solution to boiling
and condensing the vapors above the boiling liquid—the process of distillation.

Because ethyl alcohol (ethanol) boils at 78°C and water boils at 100°C, one
might naively assume that heating a 50:50 mixture of ethanol and water to 78°C
would cause the ethanol molecules to leave the solution as a vapor that could be
condensed as pure ethanol. However, in such a mixture of ethanol and water, the
water boils at about 87°C, and the vapor above the mixture is not 100% ethanol.

A liquid contains closely packed but mobile molecules of varying energy.
When a molecule of the liquid approaches the vapor-liquid boundary and possesses sufficient energy, it may pass from the liquid phase into the gas phase.
Some of the molecules present in the vapor phase above the liquid may, as they
approach the surface of the liquid, reenter the liquid phase and thus become part
of the condensed phase. In so doing, the molecules relinquish some of their kinetic energy (i.e., their motion is slowed). Heating the liquid causes more molecules to enter the vapor phase; cooling the vapor reverses this process.

When a closed system is at equilibrium, many molecules are escaping into the
vapor phase from the liquid phase, and an equal number are returning from the vapor
phase to the liquid phase. The extent of this equilibrium is measured as the vapor
pressure. Even when energy is increased and more molecules in the liquid phase have
sufficient energy to escape into the vapor phase, equilibrium is maintained because
the number moving from the vapor phase into the liquid phase also increases. However, the number of molecules in the vapor phase increases, which increases the
vapor pressure. The number of molecules in the vapor phase depends primarily on the

volume of the system, the temperature, the combined pressure of all the gaseous components, and the strength of the intermolecular forces exerted in the liquid phase. Review the introduction to Chapter 3 about the types of intermolecular forces.

## Simple Distillation

Simple distillation involves boiling a liquid in a vessel (a distilling flask) and directing the resulting vapors through a condenser, in which the vapors are cooled and converted to a liquid that flows back into a collection vessel (a receiving flask). (*See* Fig. 5.5 on page 95.) Simple distillation is used to purify liquid mixtures by separating one liquid component either from nonvolatile substances or from another liquid that differs in boiling point by at least 75°C. The initial condensate will have essentially the same mole ratio of liquids as the vapor just above the boiling liquid. The closer the boiling points of the components of a liquid mixture, the more difficult they are to completely separate by simple distillation.

## Fractional Distillation

Fractional distillation differs from simple distillation in that a fractionating column is placed between the distilling flask and the condenser. This fractionating column allows for successive condensations and distillations and produces a much better separation between liquids with boiling points closer than 75°C. The column is packed with material that provides a large surface area for heat exchange between the ascending vapor and the descending liquid. As a result, multiple condensations and vaporizations occur as the vapors ascend the column. Condensing of the higher-boiling vapor releases heat, which causes vaporization of the lower-boiling liquid on the packing so that the lower-boiling component moves up while the higher-boiling component moves down. Some of the lower-boiling component will run back into the distilling flask. Each successive condensation-vaporization cycle, also called a *theoretical plate*, produces a vapor that is richer in the more volatile fraction. As the temperature of the liquid mixture is increased, the lower-boiling fractions become enriched in the vapor.

Heat exchange between ascending vapor and descending liquid

A large surface area for the packing material is desirable, but the packing cannot be so dense that pressure changes take place within the column to cause nonequilibrium conditions. Also, if the column packing has a very large surface area, it will absorb (hold up) much of the material being distilled. Several different packings for distilling columns have been tried, including glass beads, glass helices, and carborundum chips. One of the best packings in our experience is a copper or steel sponge (Chore Boy). It is easy to insert into the column; it does not come out of the column as beads do; and it has a large surface area, good heat transfer characteristics, and low holdup. It can be used in both microscale and macroscale apparatus.

Column packing

Holdup: unrecoverable distillate that wets the column packing

The ability of different column packings to separate two materials of differing boiling points is evaluated by calculating the number of theoretical plates. Each theoretical plate corresponds to one condensation-vaporization cycle. Other things being equal, the number of theoretical plates is proportional to the height of

Height equivalent to a theoretical plate (HETP)

the column, so various packings are evaluated according to the *height equivalent to a theoretical plate* (*HETP*); the smaller the HETP, the more plates the column will have and the more efficient it will be. The calculation is made by analyzing the proportion of lower- to higher-boiling material at the top of the column and in the distillation pot.[1]

Although not obvious, the most important variable that contributes to good fractional distillation is the rate at which the distillation is carried out. A series of simple distillations take place within a fractionating column, and it is important that complete equilibrium be attained between the ascending vapors and the descending liquid. This process is not instantaneous. It should be an adiabatic process; that is, heat should be transferred from the ascending vapor to the descending liquid with no gain of heat or net heat loss to the surroundings. Advanced distillation systems use thermally insulated, vacuum-jacketed fractionating columns. They also allow the adjustment of the ratio between the amount of condensate that is directed to the receiving flask and the amount returned to the distillation column. A reflux ratio of 30:1 or 50:1 is not uncommon for a 40-plate column. Although a distillation of this type takes several hours, this is far less time than if one to had to do 40 distillations, one after the other, and yields much better separated compounds.

Perhaps it is easiest to understand the series of redistillations that occur in fractional distillation by examining the bubble plate column used to fractionally distill crude oil (Fig. 5.1). These columns dominate the skyline at oil refineries, with some being 150 ft high and capable of distilling 200,000 barrels of crude oil per day. The crude oil enters the column as a hot vapor. Some of this vapor with high-boiling components condenses on one of the plates. The more volatile substances travel through the bubble cap to the next higher plate, where some of the less-volatile components condense. As high-boiling liquid material accumulates on a plate, it descends through the overflow pipe to the next lower plate, and vapor rises through the bubble cap to the next higher plate. The temperature of the vapor that is rising through a cap is above the boiling point of the liquid on that plate. As bubbling takes place, heat is exchanged, and the less volatile components on that plate vaporize and go on to the next plate. The composition of the liquid on a plate is the same as that of the vapor coming from the plate below. So, on each plate a simple distillation takes place. At equilibrium, vapor containing low-boiling material is ascending through the column, and high-boiling liquid is descending.

As a purification method, distillation, particularly fractional distillation, requires larger amounts of material than recrystallization, liquid/liquid extraction, or chromatography. Performing a fractional distillation on less than 1 g of material is virtually impossible. Fractional distillation can be carried out on a scale of about 3–4 g. As will be seen in Chapters 8, 9, and 10, various types of chromatography are employed for separations of milligram quantities of liquids.

Equilibration is slow.

Good fractional distillation takes a long time.

■ **FIG. 5.1**
**A bubble plate distilling column.**

Overflow pipe

Plate

Riser
for
vapor

1. Weissberger, A., ed. *Techniques of Organic Chemistry*, Vol. IV; Wiley-Interscience: New York, 1951.

## Liquid Mixtures

If two different liquid compounds are mixed, the vapor above the mixture will contain some molecules of each component. Let us consider a mixture of cyclohexane and toluene. The vapor pressures, as a function of temperature, are plotted in Figure 5.2. When the vapor pressure of the liquid equals the applied pressure, the liquid boils. Figure 5.2 shows that, at 760 mm Hg (standard atmospheric pressure), these pure liquids boil at about 81°C and 111°C, respectively. If one of these pure liquids were to be distilled, we would find that the boiling point of the liquid would equal the temperature of the vapor and that the temperature of the vapor would remain constant throughout the distillation.

Figure 5.3 is a boiling point–composition diagram for the cyclohexane-toluene system. If a mixture of 75 mole percent toluene and 25 mole percent cyclohexane is heated, we find that it boils at 100°C (point A). Above a binary mixture of cyclohexane and toluene, the vapor pressure has contributions from each component. **Raoult's law** states that the vapor pressure of the cyclohexane is equal to the product of the vapor pressure of pure cyclohexane and the mole fraction of cyclohexane in the liquid mixture:

$$P_c = P_c^{\circ} N_c$$

where $P_c$ is the partial pressure of cyclohexane, $P_c^{\circ}$ is the vapor pressure of pure cyclohexane at the given temperature, and $N_c$ is the mole fraction of cyclohexane in the mixture. For toluene,

$$P_t = P_t^{\circ} N_t$$

*Raoult's law of partial pressures*

*The mole fraction of cyclohexane is equal to the moles of cyclohexane in the mixture divided by the total number of moles (cyclohexane plus toluene) in the mixture.*

**FIG. 5.2**

Vapor pressure versus temperature plots for cyclohexane and toluene.

**■ FIG. 5.3**
Boiling point–composition curves
for a mixture of cyclohexane and
toluene.

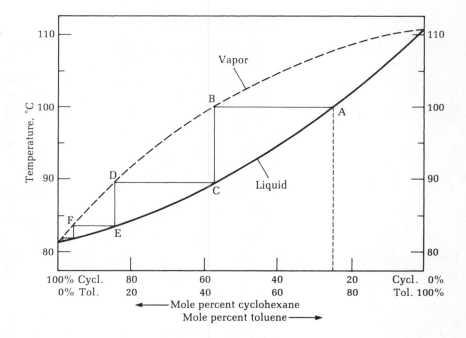

**■ FIG. 5.3**
Boiling point–composition curves for a mixture of cyclohexane and toluene.

The total vapor pressure above the solution ($P_{Tot}$) is given by the sum of the partial pressures due to cyclohexane and toluene:

$$P_{Tot} = P_c + P_t$$

Dalton's law states that the mole fraction of cyclohexane ($X_c$) in the vapor at a given temperature is equal to the partial pressure of the cyclohexane at that temperature divided by the total pressure:

$$X_c = \frac{P_c}{\text{total vapor pressure}}$$

At 100°C cyclohexane has a partial pressure of 433 mm Hg, and toluene has a partial pressure of 327 mm Hg; the sum of the partial pressures is 760 mm Hg, so the liquid boils. If some of the liquid in equilibrium with this boiling mixture were condensed and analyzed, it would be found to be 433/760, or 57 mole percent cyclohexane (point B, Fig. 5.3). This is the best separation that can be achieved on a simple distillation of this mixture. As the simple distillation proceeds, the boiling point of the mixture moves toward 111°C along the line from point A, and the vapor composition becomes richer in toluene as it moves from point B to 110°C. To obtain pure cyclohexane, it would be necessary to condense the liquid at point B and redistill it. When this is done, it is found that the liquid boils at 90°C (point C), and the vapor equilibrium with this liquid is about 85 mole percent cyclohexane (point D). Therefore, to separate a mixture of cyclohexane and toluene, a series of fractions would be collected, and each of these would be partially redistilled. If this fractional distillation were done enough times, the two components could be completely separated.

# Azeotropes

Not all liquids form ideal solutions and conform to Raoult's law. Ethanol and water are two such liquids. Because of molecular interaction, a mixture of 95.5% (by weight) of ethanol and 4.5% of water boils *below* the boiling point of pure ethanol (78.15°C versus 78.3°C). Thus, no matter how efficient the distilling apparatus, 100% ethanol cannot be obtained by distillation of a mixture of, say, 75% water and 25% ethanol. A mixture of liquids of a certain definite composition that distills at a constant temperature without a change in composition is called an *azeotrope*; 95% ethanol is such an azeotrope. The boiling point–composition curve for the ethanol-water mixture is seen in Figure 5.4. To prepare 100% ethanol, the water can be removed chemically (by reaction with calcium oxide) or it can be removed as an azeotrope with still another liquid. An azeotropic mixture of 32.4% ethanol and 67.6% benzene (bp 80.1°C) boils at 68.2°C. A ternary azeotrope containing 74.1% benzene, 18.5% ethanol, and 7.4% water boils at 64.9°C. Absolute alcohol (100% ethanol) is made by adding benzene to 95% ethanol followed by removing the water in the volatile ternary azeotrope of benzene, ethanol, and water.

Ethanol and water form a minimum boiling azeotrope. Substances such as formic acid (bp 100.7°C) and water (bp 100°C) form maximum boiling azeotropes. The boiling point of a formic acid–water azeotrope is 107.3°C.

The ethanol-water azeotrope

**■ FIG. 5.4**

Boiling point–composition curves for a mixture of ethanol and water.

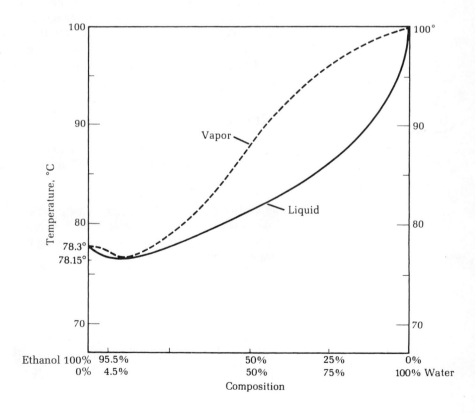

# Boiling Points and Distillation

A constant boiling point on distillation does not guarantee that the distillate is a single pure compound.

A pure liquid has a constant boiling point. A change in boiling point during distillation is an indication of impurity. The converse proposition, however, is not always true; that is, constancy of a boiling point does not necessarily mean that the liquid consists of only one compound. For instance, two miscible liquids of similar chemical structure that boil at the same temperature individually will have nearly the same boiling point as a mixture. And, as noted previously, azeotropes have constant boiling points that can be either above or below the boiling points of the individual components.

Distilling a mixture of sugar and water

When a solution of sugar in water is distilled, the boiling point recorded on a thermometer located in the vapor phase is 100°C (at 760 torr) throughout the distillation, whereas the temperature of the boiling sugar solution itself is initially somewhat above 100°C and continues to rise as the concentration of sugar in the remaining solution increases. The vapor pressure of the solution is dependent on the number of water molecules present in a given volume; hence, with increasing concentration of nonvolatile sugar molecules and decreasing concentration of water, the vapor pressure at a given temperature decreases, and a higher temperature is required for boiling. However, sugar molecules do not leave the solution, and the drop clinging to the thermometer is pure water in equilibrium with pure water vapor.

Boiling point changes with pressure.

When a distillation is carried out in a system open to the air (the boiling point is thus dependent on existing air pressure), the prevailing barometric pressure should be noted and allowance made for appreciable deviations from the accepted boiling point temperature (Table 5.1). Distillation can also be done at the lower pressures that can be achieved using an oil pump or an aspirator, which produces a substantial reduction in the boiling point.

## EXPERIMENTS

Before beginning any distillation, calibrate the thermometer to ensure accurate readings are made; refer to Part 3 of Chapter 3 for calibration instructions.

TABLE 5.1  •  **Variation in Boiling Point with Pressure**

|  | Boiling Point | |
| --- | --- | --- |
| Pressure (mm Hg) | Water (°C) | Benzene (°C) |
| 780 | 100.7 | 81.2 |
| 770 | 100.4 | 80.8 |
| 760 | 100.0 | 80.3 |
| 750 | 99.6 | 79.9 |
| 740 | 99.2 | 79.5 |
| 584* | 92.8 | 71.2 |

*Instituto de Quimica, Mexico City, altitude 7700 ft (2310 m).

# 1. Simple Distillation

Apparatus for simple distillation

**IN THIS EXPERIMENT** the two liquids to be separated are placed in a 5-mL round-bottomed, long-necked flask that is fitted to a distilling head (Fig. 5.5). The flask has a larger surface area exposed to heat than does the reaction tube, so the necessary thermal energy can be put into the system to cause the materials to distill. The hot vapor rises and completely envelops the bulb of the thermometer before passing over it and down toward the receiver. The downward-sloping portion of the distilling head functions as an air condenser. The objective is to observe how the boiling point of the mixture changes during the course of its distillation. (Another simple distillation apparatus is shown in Figure 5.6. Here, the long air condenser will condense even low-boiling liquids, and the receiver is far from the heat.)

■ **FIG. 5.6**
**A simple distillation apparatus.**

Distillation head

Air condenser

Boiling chip

Short-necked 5-mL flask

■ **FIG. 5.5**
A small-scale simple distillation apparatus. This apparatus can be adapted for fractional distillation by packing the long neck with a copper sponge. The temperature is regulated by either scraping sand away from or piling sand up around the flask.

Distillation head with air condenser

Receiver vial

Long-necked 5-mL flask

Boiling chip

The rate of distillation is determined by the heat input to the apparatus. This is most easily and effectively controlled by using a spatula to pile up or scrape away hot sand from around the flask.

Viton is resistant to hot aromatic vapors.

The thermometer bulb must be *completely below* the side arm.

Cyclohexane
bp 81°C
MW 84.16
$n_D^{20}$ 1.4260

Toluene
bp 111°C
MW 92.14
$n_D^{20}$ 1.4960

Throughout this text information regarding the physical properties of substances has been placed with each various structures in the margin. MW is molecular weight, bp is boiling point, den. is density in g/mL, and $n_D^{20}$ is the refractive index.

There are 21 ± 3 drops per milliliter.

## (A) Simple Distillation of a Cyclohexane-Toluene Mixture

To a 5-mL long-necked, round-bottomed flask, add 2.0 mL of dry cyclohexane, 2.0 mL of dry toluene, and a boiling chip (*see* Fig. 5.5). This flask is joined by means of a Viton (black) connector to a distilling head fitted with a thermometer using a rubber connector. The thermometer bulb should be completely below the side arm of the Claisen head so that the mercury reaches the same temperature as the vapor that distills. The end of the distilling head dips well down into a receiving vial, which rests on the bottom of a 30-mL beaker filled with ice. The distillation is started by piling up hot sand to heat the flask. As soon as boiling begins, the vapors can be seen to rise up the neck of the flask. Adjust the rate of heating by piling up or scraping away sand from the flask so that it takes *several minutes* for the vapor to rise to the thermometer. **The rate of distillation should be no faster than 2 drops per minute.**

Record the temperature versus the number of drops during the entire distillation process. If the rate of distillation is as slow as it should be, there will be sufficient time between drops to read and record the temperature. Continue the distillation until only about 0.4 mL remains in the distilling flask. **Never distill to dryness.** On a larger scale explosive peroxides can sometimes accumulate. At the end of the distillation, measure as accurately as possible, perhaps with a syringe, the volume of the distillate and, after it cools, the volume left in the pot; the difference is the holdup of the column if none has been lost by evaporation. Note the barometric pressure, make any thermometer corrections necessary, and make a plot of milliliters (drop number) versus temperature for the distillation.

**Cleaning Up.** The pot residue should be placed in the organic solvents container. The distillate can also be placed there or recycled.

## (B) Simple Distillation of an Ethanol-Water Mixture

In a 5-mL round-bottomed, long-necked flask place 4 mL of a 10% to 20% ethanol-water mixture. Assemble the apparatus as described previously and carry out the distillation until you believe a representative sample of ethanol has collected in the receiver. In the hood place 3 drops of this sample on a Pyrex watch glass and try to ignite it with the blue cone of a microburner flame. Does it burn? Is any unburned residue observed? There was a time when alcohol-water mixtures were mixed with gunpowder and ignited to give proof that the alcohol had not been diluted. One hundred proof alcohol is 50% ethanol by volume.

**Cleaning Up.** The distillate and pot residue can be disposed in the organic solvents waste container or, if regulations permit, diluted with water and flushed down the drain.

# 2. Fractional Distillation

**IN THIS EXPERIMENT**, just as in the last one, you will distill a mixture of two liquids and again record the boiling points as a function of the volume of distillate (drops). The necessity for a very slow rate of distillation cannot be overemphasized.

## *Apparatus*

Assemble the apparatus shown in Figure 5.7. The 10-cm column is packed with 1.5 g of copper sponge and connected to the 5-mL short-necked flask using a black (Viton) connector. The column should be vertical and care should be taken to ensure that the bulb of the thermometer does not touch the side of the distilling head. The column, but not the distilling head, will be insulated with glass wool or cotton at the appropriate time to ensure that the process is adiabatic. Alternatively, the column can be insulated with a cut-off 15-mL polyethylene centrifuge tube.

■ **FIG. 5.7**
A small-scale fractional distillation apparatus. The 10-cm column is packed with 1.5 g of copper sponge (Chore Boy).

Receiver vial

Distilling column packed with copper sponge

Temperature regulated by moving sand

Boiling chip

### (A) Fractional Distillation of a Cyclohexane-Toluene Mixture

To a short-necked flask add 2.0 mL of cyclohexane, 2.0 mL of toluene, and a boiling chip. The distilling column is packed with 1.5 g of copper sponge (Fig. 5.7). The mixture is brought to a boil over a hot sand bath. Observe the ring of condensate that should rise slowly through the column; if you cannot at first see this ring, locate it by touching the column with your fingers. It will be cool above the ring and hot below. Reduce the heat by scraping sand away from the flask and wrap the column, but not the distilling head, with glass wool or cotton if it is not already insulated.

The distilling head and the thermometer function as a small reflux condenser. Again, apply the heat, and as soon as the vapor reaches the thermometer bulb, reduce the heat by scraping away sand. **Distill the mixture at a rate no faster than 2 drops per minute** and record the temperature as a function of the number of drops. If the heat input has been *very* carefully adjusted, the distillation will cease, and the temperature reading will drop after the cyclohexane has distilled. Increase the heat input by piling up the sand around the flask to cause the toluene to distill. Stop the distillation when only about 0.4 mL remains in the flask and measure the volume of distillate and the pot residue as before. Make a plot of the boiling point versus the milliliters of distillate (drops) and compare it to the simple distillation carried out in the same apparatus. Compare your results with those in Figure 5.8.

**Cleaning Up.** The pot residue should be placed in the organic solvents container. The distillate can also be placed there or recycled.

### (B) Fractional Distillation of an Ethanol-Water Mixture

Distill 4 mL of the same ethanol-water mixture used in the simple distillation experiment, following the procedure used for the cyclohexane-toluene mixture

■ **FIG. 5.8**
Simple and fractional distillation curves for cyclohexane and toluene.

with either the short or the long distilling column. Remove what you regard to be the ethanol fraction and repeat the ignition test. Is any difference noted?

**Cleaning Up.** The pot residue and distillate can be disposed in the organic solvents waste container or, if regulations permit, diluted with water and flushed down the drain.

## 3. Instant Microscale Distillation

*Online Study Center*

Video: Instant Microscale
Distillation

Frequently, a very small quantity of freshly distilled material is needed in an experiment. For example, two compounds that need to be distilled freshly are aniline, which turns black because of the formation of oxidation products, and benzaldehyde, a liquid that easily oxides to solid benzoic acid. The impurities that arise in both of these compounds have much higher boiling points than the pure compounds, so a very simple distillation suffices to separate them. This can be accomplished as follows.

Place a few drops of the impure liquid in a reaction tube along with a boiling chip. Clamp the tube in a hot sand bath and adjust the heat so that the liquid refluxes about halfway up the tube. Expel the air from a Pasteur pipette, thrust it down into the hot vapor, and then pull the hot vapor into the cold upper portion of the pipette. The vapor will immediately condense and can then be expelled into another reaction tube that is held adjacent to the hot one (Fig. 5.9). In this way enough pure material can be distilled to determine a boiling point, run a spectrum, make a derivative, or carry out a reaction. Sometimes the first drop or two will be cloudy, which indicates the presence of water. This fraction should be discarded in order to obtain pure dry material.

## 4. Simple Distillation

### Apparatus

In any distillation, the flask should be no more than two-thirds full at the start. Great care should be taken not to distill to dryness because, in some cases, high-boiling explosive peroxides can become concentrated.

Assemble the apparatus for macroscale simple distillation, as shown in Figure 5.10, starting with the support ring followed by the electric flask heater and then the flask. One or two boiling stones are put in the flask to promote even boiling. Each ground joint is greased by putting three or four stripes of grease lengthwise around the male joint and pressing the joint firmly into the other without twisting. The air is thus eliminated, and the joint will appear almost transparent. (Do not use excess grease because it will contaminate the product.) Water enters the condenser at the tublature nearest the receiver. Because of the large heat capacity of water, only a very small stream (3 mm diameter) is needed; too much water pressure will cause the tubing to pop off. A heavy rubber band, or better a Keck clamp, can be used to hold the condenser to the distillation head. Note that the bulb of the thermometer is below the opening into the side arm of the distillation head.

■ **FIG. 5.9**
An apparatus for instant
microscale distillation.

Refluxing
vapor

Boiling liquid

Boiling chip

■ **FIG. 5.10**
An apparatus for macroscale simple
distillation.

Distillation
head

Keck clamp

Clamp

Condenser

Keck clamp

Adapter

Water
out

Water in

Graduated
cylinder

⚠ **CAUTION**: Cyclohexane and
toluene are flammable; make
sure the distilling apparatus is
tight.

Do not add a boiling chip to a hot
liquid. It may boil over.

### (A) Simple Distillation of a Cyclohexane-Toluene Mixture

Place a mixture of 30 mL cyclohexane and 30 mL toluene and a boiling chip in a
dry 100-mL round-bottomed flask and assemble the apparatus for simple distillation. After assuring that all connections are tight, heat the flask strongly until boiling begins. Then adjust the heat until the distillate drops at a regular rate of about
1 drop per second. Record both the temperature and the volume of distillate at

regular intervals. After 50 mL of distillate is collected, discontinue the distillation. Record the barometric pressure, make any thermometer correction necessary, and plot the boiling point versus the volume of distillate. Save the distillate for fractional distillation.

**Cleaning Up.** The pot residue should be placed in the organic solvents container. The distillate can also be placed there or recycled.

Dispose of cyclohexane and toluene in the container provided. Do not pour them down the drain.

### (B) Simple Distillation of an Ethanol-Water Mixture

In a 500-mL round-bottomed flask place 200 mL of a 20% aqueous solution of ethanol. Follow the previous procedure for the distillation of a cyclohexane-toluene mixture. Discontinue the distillation after 50 mL of distillate has been collected. Working in the hood, place 3 drops of distillate on a Pyrex watch glass and try to ignite it with the blue cone of a microburner flame. Does it burn? Is any unburned residue observed?

**Cleaning Up.** The pot residue and distillate can be disposed in the organic solvents container or, if regulations permit, diluted with water and flushed down the drain.

 ## 5. Fractional Distillation

### *Apparatus*

Assemble the apparatus shown in Figures 5.11 and 5.12. The fractionating column is packed with one-fourth to one-third of a metal sponge. The column should be perfectly vertical and be insulated with glass wool covered with aluminum foil (shiny side in). However, insulation is omitted for this experiment so that you can observe what is taking place in the column.

### (A) Fractional Distillation of a Cyclohexane-Toluene Mixture

After the flask from the simple macroscale distillation experiment has cooled, pour the 50 mL of distillate back into the distilling flask, add one or two new boiling chips, and assemble the apparatus for fractional distillation. The stillhead delivers into a short condenser fitted with a bent adapter leading into a 10-mL graduated cylinder. Gradually turn up the heat to the electric flask heater until the mixture of cyclohexane and toluene just begins to boil. As soon as boiling starts, turn down the power. Heat slowly at first. A ring of condensate will rise slowly through the column; if you cannot at first see this ring, locate it by cautiously touching the column with your fingers. The rise should be very gradual so that the column can acquire a uniform temperature gradient. Do not apply more heat until you are sure that the ring of condensate has stopped rising; then increase the heat gradually. In a properly conducted operation, the vapor-condensate mixture reaches the top of the column only after several minutes. Once distillation has commenced, it should continue steadily without any drop in temperature at a rate no greater than 1 mL in 1.5–2 min. Observe the flow and keep it steady by slight increases in heat as required. Protect the column from drafts by wrapping it with

**■ FIG. 5.11**
An apparatus for macroscale fractional distillation.
The position of the thermometer bulb is critical.

**■ FIG. 5.12**
A fractionating column and its
packing. Use one-third of a copper
sponge (Chore Boy).

Keck clamp

Rubber tubing

Keck clamp

Fractionating
column

Stainless steel
sponge
packing

Water out

Water in

aluminum foil, glass wool, or even a towel. This insulation will help prevent
flooding of the column, as will slow and steady distillation.

Record the temperature as each milliliter of distillate collects and take more
frequent readings when the temperature starts to rise abruptly. Each time the graduated cylinder fills, quickly empty it into a series of labeled 25-mL Erlenmeyer
flasks. Stop the distillation when a second constant temperature is reached. Plot a
distillation curve and record what you observed inside the column in the course of
the fractionation. Combine the fractions that you think are pure and turn in the

product in a bottle labeled with your name, desk number, the name of the product, the boiling point range, and the weight.

**Cleaning Up.**  The pot residue should be placed in the organic solvents container. The cyclohexane and toluene fractions can also be placed there or recycled.

### (B) Fractional Distillation of Ethanol-Water Mixture

Place the 50 mL of distillate from the simple distillation experiment in a 100-mL round-bottomed flask, add one or two boiling chips, and assemble the apparatus for fractional distillation. Follow the previous procedure for the fractional distillation of a cyclohexane-toluene mixture. Repeat the ignition test. Is any difference noted? Alternatively, distill 60 mL of the 10%–20% ethanol-water mixture that results from the fermentation of sucrose (*see* Chapter 64).

**Cleaning Up.**  The pot residue and distillate can be disposed in the organic solvents container or, if regulations permit, diluted with water and flushed down the drain.

 # 6. Fractional Distillation of Unknowns

You will be supplied with an unknown, prepared by your instructor, that is a mixture of two solvents listed in Table 5.2, only two of which form azeotropes. The solvents in the mixture will be mutually soluble and differ in boiling point by more than 20°C. The composition of the mixture (in percentages of the two components) will be either 20:80, 30:70, 40:60, 50:50, 60:40, 70:30, or 80:20. Identify the two compounds and determine the percent composition of each. Perform a fractional distillation on 4 mL of the unknown for microscale; use at least 50 mL of unknown for macroscale. Fractionate the unknown and identify the components from the boiling points. Prepare a distillation curve. You may be directed to analyze your distillate by gas chromatography (*see* Chapter 10) or refractive index (*see* Chapter 14).

| TABLE 5.2  •  **Some Properties of Common Solvents** | |
|---|---|
| **Solvent** | **Boiling Point (°C)** |
| Acetone | 56.5 |
| Methanol | 64.7 |
| Hexane | 68.8 |
| 1-Butanol | 117.2 |
| 2-Methyl-2-propanol | 82.2 |
| Water | 100.0 |
| Toluene* | 110.6 |

*Methanol and toluene form an azeotrope with a boiling point of 63.8°C (69% methanol).

**Cleaning Up.** Organic material goes in the organic solvents container. Water and aqueous solutions can be flushed down the drain.

## QUESTIONS

**1.** In either of the simple distillation experiments, can you account for the boiling point of your product in terms of the known boiling points of the pure components of your mixture? If so, how? If not, why not?

**2.** From the plots of the boiling point versus the volume of distillate in the simple distillation experiments, what can you conclude about the purity of your product?

**3.** From the plots of the boiling point versus the volume of distillate in either of the fractional distillations of the cyclohexane-toluene mixture, what conclusion can you draw about the homogeneity of the distillate?

**4.** From the plots of the boiling point versus the volume of distillate in either of the fractional distillations of the ethanol-water mixture, what conclusion can you draw about the homogeneity of the distillate? Does it have a constant boiling point? If constant, is it a pure substance?

**5.** What is the effect on the boiling point of a solution (e.g., water) produced by a soluble nonvolatile substance (e.g., sodium chloride)? What is the effect of an insoluble substance such as sand or charcoal? What is the temperature of the vapor above these two boiling solutions?

**6.** In the distillation of a pure substance (e.g., water), why does all of the water not vaporize at once when the boiling point is reached?

**7.** In fractional distillation, liquid can be seen running from the bottom of the distillation column back into the distilling flask. What effect does this returning condensate have on the fractional distillation?

**8.** Why is it extremely dangerous to attempt to carry out a distillation in a completely closed apparatus (one with no vent to the atmosphere)?

**9.** Why is better separation of two liquids achieved by slow rather than fast distillation?

**10.** Explain why a packed fractionating column is more efficient than an unpacked one.

**11.** In the distillation of the cyclohexane-toluene mixture, the first few drops of distillate may be cloudy. Explain this occurrence.

**12.** What effect does the reduction of atmospheric pressure have on the boiling point? Can cyclohexane and toluene be separated if the external pressure is 350 mm Hg instead of 760 mm Hg?

**13.** When water-cooled condensers are used for distillation or for refluxing a liquid, the water enters the condenser at the lowest point and leaves at the highest. Why?

*Online Study Center*

**General Resources**
Web Links

# CHAPTER 6

# Steam Distillation, Vacuum Distillation, and Sublimation

**Online Study Center**

This icon will direct you to techniques, equipment setups, and online resources at http://college.hmco.com/PIC/williamsonMMOE5e.

**PRELAB EXERCISE:** If a mixture of toluene and water is distilled at 97°C, what weight of water would be necessary to carry over 1 g of toluene? Compare the operation of a barometer, which is used to measure atmospheric pressure, with that of a mercury manometer, which is used to measure the pressure in an evacuated system.

## PART 1: Steam Distillation

When a mixture of cyclohexane and toluene is distilled (*see* Chapter 5), the boiling point of these two *miscible* liquids is *between* the boiling points of each of the pure components. By contrast, if a mixture of benzene and water (immiscible liquids) is distilled, the boiling point of the mixture will be *below* the boiling point of each pure component. Because benzene and water are essentially insoluble in each other, the molecules in a droplet of benzene are not diluted by water molecules from nearby water droplets; thus, the vapor pressure exerted by benzene in the mixture is the same as that of benzene alone at the existing temperature. The same is true for water. These two immiscible liquids exert pressures against the common external pressure independently, and when the sum of the two partial pressures equals the external pressure, boiling occurs. Benzene has a vapor pressure of 760 torr at 80.1°C, and if it is mixed with water, the combined vapor pressure must equal 760 torr at some temperature below 80.1°C. This temperature, the boiling point of the mixture, can be calculated from known values of the vapor pressures of the separate liquids at that temperature. Vapor pressures for water and benzene in the range 50°C–80°C are plotted in Figure 6.1. The dashed line cuts the two curves at points where the sum of the vapor pressures is 760 torr; hence the boiling point of the mixture is 69.3°C.

The boiling point of immiscible liquids is below the boiling point of either pure liquid.

Each liquid exerts its own vapor pressure.

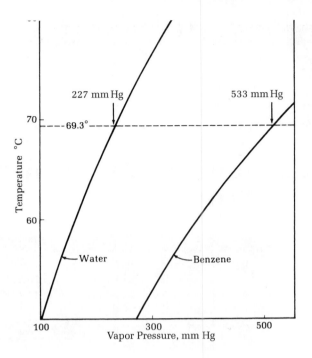

Practical use can sometimes be made of the fact that many water-insoluble
liquids and solids behave as benzene does when mixed with water, volatilizing at
temperatures below their boiling points. Naphthalene, a solid, boils at 218°C but
distills with water at a temperature below 100°C. Because naphthalene is not very
volatile, considerable water is required to entrain it, and the conventional way of
conducting the distillation is to pass steam into a boiling flask containing naphthalene and water, a process called *steam distillation*. With more volatile compounds,
or with a small amount of material, the substance can be heated with water in a
simple distillation flask, with the steam generated in situ.

Some high-boiling substances decompose before the boiling point is reached
and, if impure, cannot be purified by ordinary distillation. However, these substances can be freed from contaminating substances by steam distillation at a
lower temperature, at which they are stable. Steam distillation also offers the advantage of selectivity because some water-insoluble substances are volatile with
steam while others are not; some volatilize so slowly that sharp separation is possible. This technique is useful in processing natural oils and resins, which can be
separated into steam-volatile and non-steam-volatile fractions. It is useful for the
recovery of a non-steam-volatile solid from its solution in a high-boiling solvent
such as nitrobenzene (bp 210°C); all traces of the solvent can be eliminated, and
the temperature can be kept low.

Used for the isolation of
perfume and flavor oils.

The boiling point remains constant during a steam distillation as long as adequate amounts of both water and the organic component are present to saturate the
vapor space. Determination of the boiling point and correction for any deviation

TABLE 6.1 • **The Vapor Pressure of Water at Different Temperatures**

| t (°C) | p (mm Hg) | t (°C) | p (mm Hg) | t (°C) | p (mm Hg) | t (°C) | p (mm Hg) |
|---|---|---|---|---|---|---|---|
| 0 | 4.58 | 20 | 17.41 | 24 | 22.18 | 28 | 28.10 |
| 5 | 6.53 | 21 | 18.50 | 25 | 23.54 | 29 | 29.78 |
| 10 | 9.18 | 22 | 19.66 | 26 | 24.99 | 30 | 31.55 |
| 15 | 12.73 | 23 | 20.88 | 27 | 26.50 | 35 | 41.85 |
| | | | | | | | |
| 60 | 149.3 | 70 | 233.7 | 80 | 355.1 | 90 | 525.8 |
| 61 | 156.4 | 71 | 243.9 | 81 | 369.7 | 91 | 546.0 |
| 62 | 163.8 | 72 | 254.6 | 82 | 384.9 | 92 | 567.0 |
| 63 | 171.4 | 73 | 265.7 | 83 | 400.6 | 93 | 588.6 |
| 64 | 179.3 | 74 | 277.2 | 84 | 416.8 | 94 | 610.9 |
| 65 | 187.5 | 75 | 289.1 | 85 | 433.6 | 95 | 633.9 |
| 66 | 196.1 | 76 | 301.4 | 86 | 450.9 | 96 | 657.6 |
| 67 | 205.0 | 77 | 314.1 | 87 | 468.7 | 97 | 682.1 |
| 68 | 214.2 | 78 | 327.3 | 88 | 487.1 | 98 | 707.3 |
| 69 | 223.7 | 79 | 341.0 | 89 | 506.1 | 99 | 733.2 |

Dalton's law

from normal atmospheric pressure permits calculation of the amount of water required for the distillation of a given amount of organic substance. According to Dalton's law, the molecular proportion of the two components in the distillate is equal to the ratio of their vapor pressures ($p$) in the boiling mixture; the more volatile component contributes the greater number of molecules to the vapor phase. Thus,

$$\frac{\text{moles of water}}{\text{moles of substance}} = \frac{p_{\text{water}}}{p_{\text{substance}}}$$

The vapor pressure of water ($p_{\text{water}}$) at the boiling temperature in question can be found by interpolating the data in Table 6.1. For the organic substance the vapor pressure is equal to $760 - p_{\text{water}}$. Hence, the weight of water required per gram of substance is given by the expression

$$\text{wt of water per g} \atop \text{of substance} = \frac{18 \times p_{\text{water}}}{\text{MW of substance} \times (760 - p_{\text{water}})}$$

From the data given in Figure 6.1 for benzene and water, the fact that the mixture boils at 69.3°C, and the molecular weight of 78.11 for benzene, the water required for the steam distillation of 1 g of benzene is only $227 \times 18/533 \times 1/78 = 0.10$ g. Nitrobenzene (bp 210°C, MW 123.11) steam distills at 99°C and requires 4.0 g of water per gram of nitrobenzene. The low molecular weight of water makes water a favorable liquid for the two-phase distillation of organic compounds.

Steam distillation is employed in several experiments in this text. On a small scale steam is generated in the flask that contains the substance to be steam distilled by simply boiling a mixture of water and the immiscible substance.

## Isolation of Citral

Terpenes

Citral is an example of a very large group of natural products called *terpenes*. They are responsible for the characteristic odors of plants such as eucalyptus, pine, mint, peppermint, and lemon. The odors of camphor, menthol, lavender, rose, and hundreds of other fragrances are due to terpenes, which have 10 carbon atoms with double bonds along with aldehyde, ketone, or alcohol functional groups (Fig. 6.2).

These natural terpenes all arise from a common precursor: isopentenyl pyrophosphate. At one time these were thought to come from the simple diene, isoprene (2-methyl-1,3-butadiene), because the skeletons of terpenes can be dissected into isoprene units, each having five carbon atoms arranged as in 2-methylbutane. These isoprene units are almost always arranged in a "head-to-tail" fashion.

The isoprene rule

Geranial and neral are geometric isomers.

In the present experiment citral, a mixture of geranial and neral, is isolated by steam distillation of lemongrass oil, which is used to make lemongrass tea, a popular drink in Mexico. The distillate contains 90% geranial and 10% neral, with the isomer about the 2,3-bond. Citral is used in a commercial synthesis of vitamin A.

Lemongrass oil contains a number of substances; simple or fractional distillation would not be a practical method for obtaining pure citral. And because lemongrass oil boils at 229°C, it has a tendency to polymerize, oxidize, and decompose during distillation. For example, heating with potassium bisulfate, an acidic compound, converts citral to 1-methyl-4-isopropylbenzene (*p*-cymene).

Steam distillation: the isolation of heat-sensitive compounds

Steam distillation is thus a very gentle method for isolating citral. The distillation takes place below the boiling point of water. The distillate consists of a mixture of citral (and some neral) and water. It is isolated by shaking the mixture with *t*-butyl methyl ether. The citral dissolves in the *t*-butyl methyl ether, which is immiscible with water, and the two layers are separated. The *t*-butyl methyl ether is dried (it dissolves some water) and evaporated to leave citral.

Store the citral in the smallest possible container in the dark for later characterization. The homogeneity of the substance can be investigated using thin-layer chromatography (TLC; *see* Chapter 8) or gas chromatography (*see* Chapter 10). An infrared spectrum (*see* Chapter 11) would confirm the presence of the aldehyde; an ultraviolet spectrum (*see* Chapter 14) would indicate it is a conjugated aldehyde; and a nuclear magnetic resonance (NMR) spectrum (*see* Chapter 12) would clearly show the aldehyde proton, the three methyl groups, and the two olefinic protons. Chemically, the molecule can be characterized by reaction with bromine and also permanganate, which shows it contains double bonds; the Tollens test confirms the presence of an aldehyde (*see* Chapter 36).

Geranial

Isopentenyl pyrophosphate

Isoprene
(2-Methyl-1,3-butadiene)

Neral

Limonene
(lemons)

Myrcene
(bayberry)

α-Pinene
(turpentine)

Menthol
(mint)

Carvone
(caraway seeds)

Camphor

1,8-Cineole
(eucalyptus)

Pulegone
(pennyroyal oil)

Anethole
(anise seed)

---

## EXPERIMENTS

# 1. Isolation of Citral

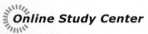
Wrap the vertical part of the
apparatus in cotton or glass
wool to speed the distillation.

Into a 5-mL short-necked, round-bottomed flask place a boiling chip, 0.5 mL of
lemongrass oil (*not* lemon oil)[1], and 3 mL of water. Assemble the apparatus as
depicted in Figure 6.3 and distill as rapidly as possible, taking care that all the
distillate condenses. Using a syringe, inject water dropwise through the septum
to keep the volume in the flask constant. Distill into an ice-cooled vial, as
shown, or better into a 15-mL centrifuge tube, until no more oily drops can be
detected, about 10 to 12 mL. Use this citral and water mixture for the next
experiment.

■ **FIG. 6.3**
A small-scale steam distillation
apparatus. Add water via syringe
at A.

**Citral (mixture of geranial
and neral)**
MW 152.24
bp 229°C, den. 0.888
$n_D^{20}$ 1.4880

---

1. Lemongrass oil can be obtained from Pfaitz and Bauer, 172 E. Aurora St., Waterbury, CT 06708.

**2. Extraction of Citral**

> **IN THIS EXPERIMENT** you will extract the pleasant-smelling liquid citral from water (from the previous experiment) using ether as the solvent. The ether solution is dried, and the ether is evaporated to leave pure citral. Extraction is a common isolation technique (*see* Chapter 7 for information on the theory and practice of extraction).

Add 2.5 mL of *t*-butyl methyl ether to the cold distillate. Shake, let the layers separate, draw off the aqueous layer with a Pasteur pipette, and transfer it to another centrifuge tube. Continue this extraction process by again adding 1.5 mL of ether to the aqueous layer, shaking, and separating to extract the citral from the steam distillate. (*See* Chapter 7 for the many advantages of *t*-butyl methyl ether as an extraction solvent.)

*In Sweden they say, "Add anhydrous drying agent until it begins to snow."*

Place the ether in a reaction tube and then add about 2 mL of saturated sodium chloride solution to the ether. Stopper the tube, flick the tube or shake it, and then remove and discard the aqueous layer. Add anhydrous calcium chloride pellets to the ether until these no longer clump together, stopper the tube, and shake it over a period of 5–10 min. This removes water adhering to the reaction tube and dissolved in the ether. Force a Pasteur pipette to the bottom of the reaction tube, expel the air from the pipette, draw up the ether, and expel it into another tared reaction tube. Add fresh ether, which will serve to wash off the drying agent, and add that ether to the tared reaction tube. Add a boiling stick or boiling chip to the ether in the reaction tube, place it in a beaker of hot water, and evaporate the ether by boiling and drawing the ether vapors into the water aspirator or, better, by blowing a gentle stream of air or nitrogen into the tube (in the hood). The last traces of ether may be removed by connecting the reaction tube directly to the water aspirator. The residue should be a clear, fragrant oil. If it is cloudy, it is wet.

*See Chapter 7 for the removal of water from ether solutions.*

Determine the weight of the citral and calculate the percentage of citral recovered from the lemongrass oil, assuming that the density of the lemongrass oil is the same as that of citral (0.89). Transfer the product to the smallest possible airtight container and store it in the dark for later analysis to confirm the presence of double bonds (alkenes) and the aldehyde group (*see* Chapters 18 and 36). Analyze the infrared spectrum in detail.

**Cleaning Up.** The aqueous layer can be flushed down the drain. Any ether goes into the organic solvents container. Allow ether to evaporate from the calcium chloride in the hood and then place the calcium chloride in the nonhazardous solid waste container. However, be aware of the regulations in your area as they may not allow evaporation of small amounts of organic solvents in a hood. If this is the case, special containers should be available for the disposal of these wet solids.

## 3. Isolation of Citral

Citral, a fragrant terpene aldehyde made up of two isoprene units, is the main component of the steam-volatile fraction of lemongrass oil and is used in a commercial synthesis of vitamin A.

### Steam Distillation Apparatus

In the assembly shown in Figure 6.4, steam is passed into a 250-mL, round-bottomed flask through a section of 6-mm glass tubing fitted into a stillhead with a piece of 5-mm rubber tubing connected to a trap, which in turn is connected to the steam line. The trap serves two purposes: (1) it allows water, which is in the steam line, to be removed before it reaches the round-bottomed flask; and (2) adjustment of the clamp on the hose at the bottom of the trap allows precise control of the steam flow. As shown, the stopper in the trap should be wired on as a precaution. A bent adapter attached to a *long* condenser delivers the condensate into a 250-mL Erlenmeyer flask.

### Alternative Steam Distillation Setup

If a steam line is not available, the steam distillation setup shown in Figure 6.5 may be used. A 100-mL or 250-mL boiling flask containing the mixture to be distilled is equipped with a magnetic stirrer and is heated on a heating mantle. Equip the boiling flask with a distillation adapter to which is attached a thermometer adapter and a separatory funnel, which serves to add water to the boiling flask during the distillation. A condenser is attached to the thermometer adapter and is equipped with a vacuum adapter, which is attached to the receiving flask. The receiving flask can be cooled by placing it in an ice-water bath. Make sure you

■ **FIG. 6.4**
A steam distillation
apparatus.

**CAUTION**: Live steam causes severe burns. If the apparatus leaks steam, immediately turn off the steam valve.

Perform steam distillations within fume hoods.

■ **FIG. 6.5**
An alternative macroscale steam distillation setup. Water is to be added through the separatory funnel.

■ **FIG. 6.5**
An alternative macroscale steam distillation setup. Water is to be added through the separatory funnel.

securely clamp the distillation flask and condenser. Also make sure to use a rubber band or a Keck clip to hold the vacuum adapter on the end of the condenser.

## Steam Distillation

Using a graduated cylinder, measure out 10 mL of lemongrass oil (*not* lemon oil) into a 250-mL boiling flask. Rinse the remaining contents of the graduated cylinder into the flask with a small amount of *t*-butyl methyl ether. Add 100 mL of water, make the connections as in Figure 6.4, heat the flask with a low flame, and pass in the steam.

To start the steam distillation, heat the flask containing the mixture on a Thermowell to prevent water from condensing in the flask to the point where water and product splash over into the receiver. Then turn on the steam valve, being sure that the screw clamp on the bottom of the trap is open. Slowly close the clamp and allow steam to pass into the flask. Try to adjust the rate of steam addition and the rate of heating so that the water level in the flask remains constant. Unlike ordinary distillations, steam distillations are usually run as fast as possible, with proper care to avoid splashing of material into the receiver and the escape of uncondensed steam. Distill as rapidly as the cooling facilities allow and continue until droplets of oil no longer appear at the tip of the condenser (about 250 mL of distillate).

## Extraction

Pour 50 mL of *t*-butyl methyl ether into a 125-mL separatory funnel, cool the distillate, if necessary, and pour a part of it into the funnel. Shake, let the layers separate, discard the lower layer, add another portion of distillate, and repeat. When the last portion of distillate has been added, rinse the flask with a small amount of *t*-butyl methyl ether to recover adhering citral. Use the techniques described in

Chapter 7 for drying, filtering, and evaporating the *t*-butyl methyl ether. Take the tare of a 1-g tincture bottle, transfer the citral to it with a Pasteur pipette, and determine the weight and the yield from the lemongrass oil. Label the bottle and store (in the dark) for later testing for the presence of the double bonds (alkenes) and the aldehyde group (*see* Chapters 18 and 36).

**Cleaning Up.** The aqueous layer can be flushed down the drain. Any *t*-butyl methyl ether goes in the organic solvents container. Allow *t*-butyl methyl ether to evaporate from the calcium chloride pellets in the hood and then place the calcium chloride in the nonhazardous solid waste container. However, be aware of the regulations in your area as they may not allow evaporation of small amounts of organic solvents in a hood. If this is the case, special containers should be available for the disposal of these wet solids.

# 4. Recovery of a Dissolved Substance

**Anthracene**
mp 216°C

**Toluene**
bp 110.8°C
den. 0.866

Measure 50 mL of a 0.2% solution of anthracene in toluene into a 250-mL round-bottomed flask and add 100 mL of water (*see* Fig. 6.4 on page 112). For an initial distillation to determine the boiling point and composition of the toluene-water azeotrope, fit the stillhead with a thermometer instead of the steam-inlet tube (*see* Fig. 5.4 on page 93). Heat the mixture with a hot plate, a sand bath, or an electric flask heater; distill about 50 mL of the azeotrope; and record a value for the boiling point. After removing the heat source, pour the distillate into a graduated cylinder and measure the volumes of toluene and water. Calculate the weight of water per gram of toluene and compare the result to the theoretical value calculated from the vapor pressure of water at the observed boiling point (*see* Table 6.1 on page 107).

Replace the thermometer with the steam-inlet tube. To start the steam distillation, heat the flask containing the mixture with a Thermowell to prevent water from condensing in the flask to the point where water and product splash over into the receiver. Then turn on the steam valve, being sure that the screw clamp on the bottom of the trap is open. Slowly close the clamp and allow steam to pass into the flask. Try to adjust the rate of steam addition and the rate of heating so that the water level in the flask remains constant. Unlike ordinary distillations, steam distillations are usually run as fast as possible, with proper care to avoid splashing of material into the receiver and the escape of uncondensed steam.

Continue the distillation by passing in steam until the distillate is clear and then until fluorescence appearing in the stillhead indicates that a trace of anthracene is beginning to distill. Stop the steam distillation by first opening the clamp at the bottom of the trap and then turning off the steam valve. Grasp the round-bottomed flask with a towel when disconnecting it and, using the clamp to support it, cool it under the tap. The bulk of the anthracene can be dislodged from the flask walls and collected on a small suction filter. To recover any remaining anthracene, add a small amount of acetone to the flask, warm the flask on the steam bath to dissolve the material, add water to precipitate it, and then collect the precipitate on the same suction filter. About 80% of the hydrocarbon in the original toluene solution should be recoverable. When dry, recrystallize the material from about 1 mL of toluene and note that the crystals are more intensely fluorescent than the solution or the amorphous solid. The characteristic fluorescence is quenched by mere traces of impurities.

**Cleaning Up.** Place the toluene and the anthracene in the organic solvents container. The aqueous layer can be flushed down the drain.

# 5. Isolation and Bioassay of Eugenol from Cloves

In the 15th century Philippus Aureolus Theophrastus Bombast von Hohenheim, otherwise known as Paracelsus, urged all chemists to make extractives for medicinal purposes rather than try to transmute base metals into gold. He believed every extractive possessed a quintessence that was the most effective part in effecting cures. There was then an upsurge in the isolation of essential oils from plant materials, some of which are still used for medicinal purposes. Among these are camphor, quinine, oil of cloves, cedarwood, turpentine, cinnamon, gum benzoin, and myrrh. Clove oil, which consists almost entirely of eugenol and its acetate, is a food flavoring agent as well as a dental anesthetic. The Food and Drug Administration has declared clove oil to be the most effective nonprescription remedy for toothache.

The biggest market for essential oils is for perfumes, and prices for these oils reflect their rarity. Recently, worldwide production of orange oil was 1500 tons and sold for $0.75 per lb, whereas 400 tons of clove oil sold for $14 per lb, and 10 tons of jasmine oil sold for $2000 per lb. These three oils represent the most common extractive processes. Orange oil is obtained by expression (squeezing) of the peel in presses; clove oil is obtained by steam distillation, as performed in this experiment; and jasmine oil is obtained by the extraction of the flower petals using ethanol.

Cloves are the flower buds from a tropical tree that can grow to 50 ft in height. The buds are handpicked in the Moluccas (the Spice Islands) of Indonesia, the East Indies, and the islands of the Indian Ocean, where a single tree can yield up to 75 lb of the sun-dried buds we know as cloves.

Cloves contain between 14% and 20% by weight of essential oil, of which only about 50% can be isolated. The principal constituent of clove oil is eugenol and its acetyl derivative. Eugenol boils at 255°C, but because it is insoluble in water, it will form an azeotrope with water and steam distill at a temperature slightly below the boiling point of water. Being a phenol, eugenol will dissolve in aqueous alkali to form the phenolate anion, which forms the basis for its separation from its acetyl derivative. The relative amounts of eugenol and acetyleugenol and the effectiveness of the alkali extraction in separating them can be analyzed by thin-layer chromatography.

**Eugenol**

**Acetyleugenol**

## *Procedure*

**IN THIS EXPERIMENT** the essential oil–a mixture of eugenol and acetyleugenol–is steam distilled from cloves. These compounds are isolated from the aqueous distillate by extraction into dichloromethane. The dichloromethane solution is shaken with aqueous sodium hydroxide, which reacts with the acidic eugenol. The dichloromethane layer is dried and evaporated to give acetyleugenol. The basic layer is acidified, and the eugenol is extracted into dichloromethane, which is in turn dried and evaporated to give pure eugenol. The isolated crude oil, eugenol, and acetyleugenol can then be evaluated for antibacterial properties.

### Steam Distillation

Place 25 g of whole cloves in a 250-mL round-bottomed flask, add 100 mL of water, and set up an apparatus for steam distillation (*see* Fig. 6.4 on page 112 or Fig. 6.5 on page 113) or simple distillation (*see* Fig. 5.10 on page 100). In the latter procedure steam is generated in situ. Heat the flask strongly until boiling begins and then reduce the heat just enough to prevent foam from being carried over into the receiver. Instead of using a graduated cylinder as a receiver, use an Erlenmeyer flask and transfer the distillate periodically to the graduated cylinder; then, if any material does foam over, the entire distillate will not be contaminated. Collect 60 mL of distillate, remove the flame, and add 60 mL of water to the flask. Resume the distillation and collect an additional 60 mL of distillate.

### Extraction

The eugenol is sparingly soluble in water and is extracted easily from the distillate with dichloromethane. Place the 120 mL of distillate in a 250-mL separatory funnel and extract with three 15-mL portions of dichloromethane. In this extraction very gentle shaking will fail to remove all of the product, and long and vigorous shaking will produce an emulsion of the organic layer and water. The separatory funnel will appear to have three layers in it. It is better to err on the side of vigorous shaking and draw off the clear lower layer to the emulsion line for the first two extractions. For the third extraction, shake the mixture less vigorously and allow a longer period of time for the two layers to separate.

Combine the dichloromethane extracts (the aqueous layer can be discarded) and add just enough anhydrous calcium chloride pellets so that the drying agent no longer clumps together but appears to be a dry powder as it settles in the solution. This may require as little as 2 g of drying agent. Swirl the flask for 1–2 min to complete the drying process and then decant the solvents into an Erlenmeyer flask. It is quite easy to decant the dichloromethane from the drying agent, so it is not necessary to set up a filtration apparatus to accomplish this separation.

Measure the volume of the dichloromethane in a dry graduated cylinder, place exactly one-fifth of the solution in a tared Erlenmeyer flask, add a wood boiling stick (easier to remove than a boiling chip), and evaporate the solution on a steam bath or a hot plate in a hood. The residue of crude clove oil will be used in the TLC (thin-layer chromatography) analysis and/or in the bioassay. From its weight, the total yield can be calculated. This residue can then be dissolved in acetone (a 5%–10% solution) for TLC analysis and/or for a bioassay.

*See* Chapter 7 for the theory of acid/base extractions.

To separate eugenol from acetyleugenol, extract the remaining four-fifths of the dichloromethane solution (about 30 mL) with 5% aqueous sodium hydroxide solution. Carry out this extraction three times, using 10-mL portions of sodium hydroxide each time. Dry the dichloromethane layer over anhydrous calcium chloride pellets, decant the solution into a tared Erlenmeyer flask, and evaporate the solvent. The residue should consist of acetyleugenol and other steam-volatile neutral compounds from cloves.

Acidify the combined aqueous extracts to pH 1 with concentrated hydrochloric acid (use Congo Red or litmus paper) and then extract the liberated eugenol with three 8-mL portions of dichloromethane. Dry the combined extracts, decant into a tared Erlenmeyer flask, and evaporate the solution on a steam bath or a hot plate. Calculate the weight percent yields of eugenol and acetyleugenol on the basis of the weight of cloves used.

Analyze your products using infrared spectroscopy and thin-layer chromatography. Obtain an infrared spectrum of either eugenol or acetyleugenol using the thin-film method. Compare your spectrum to the spectrum of a fellow student who has examined the other compound.

**Chapter 11**: Infrared Spectroscopy

Use plastic sheets precoated with silica gel for thin-layer chromatography. One piece the size of a microscope slide should suffice. Spot crude clove oil, eugenol, and acetyleugenol (1% solutions) side by side about 5 mm from one end of the plate. The spots should be very small. Immerse the end of the slide in a dichloromethane–hexane mixture (1:2 or 2:1) about 3 mm deep in a covered beaker. After running the chromatogram, observe the spots under an ultraviolet lamp or by developing them in an iodine chamber.

**Chapter 8**: Thin-Layer Chromatography

Wash hands after handling bacteria.

## Antibiotic Behavior of Spice Essential Oils

You will bioassay your isolated clove oil with *Bacillus cereus* to test for its antibiotic behavior. Many bacterial cultures are available commercially, but *Bacillus cereus* is preferred because it is nonpathogenic, requires no special medium for growth, and grows at room temperature in a short time. Cultures of *E. coli* have to be handled with care because of their potential pathogenicity. Their optimal growth is at 37°C, which means that an incubator is needed to work with these cultures.

Agar culture plates are prepared in sterile Petri dishes or can be purchased.[2] These plates are stored in the refrigerator. Whenever you handle the sterile agar Petri dishes (bottom dish and top cover), try to minimize any exposure to random forms of bacteria or dirt in the laboratory. Always wear gloves when handling the plates and the *Bacillus cereus* because your hands can easily contaminate the agar. Open the cover of the dish a few inches vertically so that you have just enough room to streak the plate with the bacteria or insert a testing disk. Immediately cover the agar again to reduce the risk of outside contamination. You will use a pair of tweezers to handle the test disk and sterile swabs to streak the plates with the *Bacillus cereus*. After preparing your assay, you will observe the growth of bacteria on the plate. If the bacteria do not grow in the area surrounding the disk treated with your essential oil, then one or more of the compounds in this isolated mixture is exhibiting antibiotic behavior.

## Bioassay Using Agar Medium Bacterial Growth Plates

Two students can use one agar plate. Draw a line down the middle of the plate to separate each student's disks from the other's. Label the bottom of the culture plate with your name and each disk's identity so that you will know which disk is which.

_____

2. Agar plates and bacteria can be purchased from Carolina Biological Supply Co. www.carolina.com. Nutrient Agar Plates: part number CE-82-1862 (packages of 10); *Bacillus cereus*: part number CE-15-4869.

Sterilize a copper wire loop (a piece of copper wire with a loop 5-mm wide and 4–5 in. [10–12 cm] long) by heating in a flame. Once cool, use the loop to transfer a small amount of the bacteria from the starter culture plate to a previously prepared sterile agar culture plate.

Using a pair of tweezers cleaned by wiping with an ethanol-soaked tissue, pick up a sterile disk (1.3-cm filter paper) and dip it into the prepared acetone-oil solution [5% to 10% (wt/vol) solution of oil in acetone]. Allow the disk to soak for 10 min, remove the disk, and lean it against a clean glass-stirring rod on a paper towel to air dry. Prepare a control disk by soaking a disk in ethanol and letting it dry. Lift the lid on the agar plate just enough to insert each disk using clean forceps. Tape the cover to the bottom. Place the plate in a quiet area at room temperature. The results of the assay *must* be assessed after 24 h (Fig. 6.6).

Bioassays of the isolated eugenol and acetyleugenol may also be performed and compared.

**Cleaning Up.** Combine all of the aqueous layers, neutralize with sodium carbonate, dilute with water, and flush down the drain. Any solutions containing dichloromethane should be placed in the halogenated organic solvents container. Allow the solvent to evaporate from the calcium chloride pellets in the hood and then place the drying agent in the nonhazardous solid waste container. If local regulations do not allow for the evaporation of solvent from the pellets in a hood, dispose of the wet pellets in a special container. The acetone solution of oil can be placed in the nonhalogenated solvent container. If an autoclave is available, the agar plates and contaminated materials should be autoclaved for at least 1 h. If an autoclave is not available, soak the agar plates and contaminated materials in bleach for several days. The bleach solution can then be flushed down the drain with lots of water.

# PART 2: Vacuum Distillation

**M**any substances cannot be satisfactorily distilled in the ordinary way, either because they boil at such high temperatures that decomposition occurs or because they are sensitive to oxidation. Steam distillation will work only if the compound is volatile. When neither ordinary distillation nor steam distillation is practical, purification can be accomplished by distillation at reduced pressure.

## Vacuum Distillation Assemblies

### Macroscale Vacuum Distillation

A typical macroscale vacuum distillation apparatus is illustrated in Figure 6.7. It is constructed of a round-bottomed flask (often called the *pot*) containing the material to be distilled, a Claisen distilling head fitted with a hair-fine capillary mounted through a rubber tubing sleeve, and a thermometer with the bulb extending below the side-arm opening. The condenser fits into a vacuum adapter that is connected to the receiver and, via heavy-walled rubber tubing, to a mercury manometer and then to the trap and water aspirator.

⚠️
Carry out vacuum distillation behind a safety shield.
Before using, check the apparatus for scratches and cracks.

**■ FIG. 6.7**
A macroscale vacuum distillation apparatus.

Thermometer bulb below side arm of Claisen distilling head

Keck clamp

Water out

Capillary

Claisen distilling head

Pot

Condenser

Keck clamp

Vacuum adapter

Water in

Keck clamp

To trap and aspirator

Keck clamp

Receiver

Manometer

Liquids usually bump vigorously when boiled at reduced pressure, and most boiling stones lose their activity in an evacuated system; it is therefore essential to make special provision to control bumping. This is done by allowing a fine stream of air bubbles to be drawn into the boiling liquid through a glass tube drawn to a hair-fine capillary. The capillary should be so fine that even under vacuum only a few bubbles of air are drawn in each second; smooth boiling will be promoted, and the pressure will remain low. The capillary should extend to the very bottom of the flask and should be slender and flexible so that it will whip back and forth in the boiling liquid. Another method used to prevent bumping when small quantities of material are being distilled is to introduce sufficient glass wool into the flask to fill a part of the space above the liquid. Rapid magnetic stirring will also prevent bumping.

Making a hair-fine capillary

Handle flames in the organic laboratory with great care.

A hair-fine capillary is made as follows. A 6-in. length of 6-mm glass tubing is rotated and heated in a small area over a *very hot* flame to collapse the glass and thicken the side walls, as shown in Figure 6.8a. The tube is removed from the flame, allowed to cool slightly, and then drawn into a thick-walled, coarse-diameter capillary (Fig. 6.8b). This coarse capillary is heated at point X over the wing top of a Bunsen burner turned 90°. When the glass is very soft, but not soft enough to collapse the tube entirely, the coarse capillary is lifted from the flame and without hesitation drawn smoothly and rapidly into a hair-fine capillary by stretching the hands as far as they will reach (about 2 m; Fig. 6.8c). The two capillaries so produced can be snapped off to the desired length. To ascertain that there is indeed a hole in a hair-fine capillary, place the fine end into a low-viscosity liquid such as acetone or *t*-butyl methyl ether and blow in the large end. A stream of very small bubbles should be seen. If the stream of bubbles is not extremely small, make a new capillary. The flow of air through the capillary *cannot* be controlled by attaching a rubber tube and clamp to the top of the capillary tube. Should the right-hand capillary of Figure 6.8c break when in use, it can be fused to a scrap of glass (for use as a handle) and heated again at point Y (Fig. 6.8c). In this way a capillary can be redrawn many times.

Heating baths

The pot is heated via a heating bath rather than a Thermowell to promote even boiling and make accurate determination of the boiling point possible. The

**■ FIG. 6.8**
A capillary for vacuum distillation.

bath is filled with a suitable heat-transfer liquid (water, cottonseed oil, silicone oil, or molten metal) and heated to a temperature about 20°C higher than that at which the substance in the flask distills. The bath temperature is kept constant throughout the distillation. The surface of the liquid in the flask should be below that of the heating medium so as to lessen the tendency to bump. Heating of the flask is begun only after the system has been evacuated to the desired pressure; otherwise, the liquid might boil too suddenly on reduction of the pressure.

Changing fractions

To change fractions, the following must be done in sequence: Remove the source of heat, release the vacuum, change the receiver, restore the vacuum to its former pressure, and resume heating.

A fractional vacuum distillation apparatus is shown in Figure 6.9. The distillation neck of the Claisen adapter is longer than that of other adapters, and it has a series of indentations intruding from four directions such that the points nearly meet in the center. These indentations increase the surface area over which rising vapor can come to equilibrium with descending liquid, and the distillation neck then serves as a fractionating column (a Vigreux column). A column packed with a metal sponge has a high tendency to become filled with liquid (flood) at reduced pressure. The apparatus illustrated in Figure 6.9 also has a fraction collector, which allows the removal of a fraction without disturbing the vacuum in the system. While the receiver is being changed, the distillate collects in the small reservoir A. The clean receiver is evacuated by another aspirator at tube B before being reconnected to the system.

## Semi-microscale and Microscale Vacuum Distillation

If only a few milliliters of a liquid are to be distilled, the apparatus shown in Figure 6.10 has the advantage of low holdup—that is, not much liquid is lost to wetting the surface area of the apparatus. The fraction collector illustrated is

**■ FIG. 6.9**
A vacuum distillation apparatus with Vigreux column and fraction collector.

**■ FIG. 6.10**
A semi-microscale vacuum distillation apparatus.

Condensing water
Out
In
To vacuum
Air in
Capillary
Cow
A
B

known as a *cow*. Rotation of the cow about the lightly greased standard taper joint will allow four fractions to be collected without interrupting the vacuum.

A distillation head of the type shown in Figure 6.11 allows fractions to be removed without disturbing the vacuum and control of the reflux ratio (*see* Chapter 5) by manipulation of the condenser and stopcock A. These can be adjusted to remove all the material that condenses or only a small fraction, with the bulk of the liquid being returned to the distilling column to establish equilibrium between the descending liquid and the ascending vapor. Thus liquids with small boiling-point differences can be separated.

On a truly micro scale (<10 mg), simple distillation is not practical because of mechanical losses. Fractional distillation of small quantities of liquids is also difficult. The necessity for maintaining an equilibrium between ascending vapors and descending liquid means that there will inevitably be large mechanical losses. The best apparatus for the fractional distillation of 0.5 mL to 20 mL of liquid under reduced pressure is the *micro spinning band column*, which has a holdup of about 0.1 mL. In research, preparative-scale gas chromatography (*see* Chapter 10) and high-performance liquid chromatography (HPLC) have supplanted vacuum distillation of very small quantities of material.

It is possible, however, to assemble several pieces of apparatus that can be used for simple vacuum distillation using components from the microscale kit that accompanies this text. One such assembly is shown in Figure 6.12. The material to be distilled is placed in a 5-mL short-necked flask packed loosely with glass wool, which will help prevent bumping. The thermometer bulb extends entirely below the side arm, and the distillate is collected in a reaction tube, which can be cooled in a beaker of ice. The vacuum source is connected to a syringe needle;

■ **FIG. 6.11**
A vacuum distillation head.

■ **FIG. 6.12**
A microscale apparatus for simple vacuum distillation.

To aspirator

Glass wool

**■ FIG. 6.13**
A Kugelrohr bulb-to-bulb distillation apparatus.

**■ FIG. 6.14**
A closed-end mercury manometer.

alternatively, the rubber tubing from the aspirator, if of the proper diameter, can be connected directly to the adapter, dispensing with the septum and needle.

## The Kugelrohr

The Kugelrohr (German, meaning "bulb-tube") is a widely used piece of research apparatus that consists of a series of bulbs that can be rotated and heated under vacuum. Bulb-to-bulb distillation frees the desired compound of very low-boiling and very high-boiling or nonvolatile impurities. The crude mixture is placed in bulb A (Fig. 6.13) in the heated glass chamber. At bulb C is a shutter mechanism that holds in the heat but allows the bulbs to be moved out of the heated chamber one by one as distillation proceeds. The lowest boiling material collects in bulb D; then bulb C is moved out of the heated chamber, the temperature increased, and the next fraction is collected in bulb C. Finally, this process is repeated for bulb B. The bulbs rotate under vacuum via a mechanism similar to the one used on rotary evaporators. The same apparatus is used for bulb-to-bulb fractional sublimation (see Part 3 for sublimation).

## Manometers

The pressure of the system is measured with a closed-end mercury manometer. The manometer (Fig. 6.14) is connected to the system by turning the stopcock until its V-groove is aligned with the side arm. To avoid contamination of the manometer, it should be connected to the system only when a reading is being made. The pressure (in mm Hg) is given by the height (in mm) of the central mercury column

above the reservoir of mercury and represents the difference in pressure between
the nearly perfect vacuum in the center tube (closed at the top, open at the bottom)
and the large volume of the manometer, which is at the pressure of the system.

Another common type of closed-end mercury manometer is shown in
Figure 6.15. At atmospheric pressure the manometer is completely full of mercury,
as shown in Figure 6.15a. When the open end is connected to a vacuum, the
pressure is measured as the distance in millimeters between the mercury in the
right and left legs. Note that the pressure in the left leg is essentially zero (mercury
has a low vapor pressure). If the applied vacuum were perfect (0 mm Hg), then the
two columns would have the same height.

## The Water Aspirator in Vacuum Distillation

A water aspirator in good order produces a vacuum nearly corresponding to the
vapor pressure of the water flowing through it. Polypropylene aspirators perform
well and are not subject to corrosion, which does affect brass aspirators. If a
manometer is not available, the assembly is free of leaks, and the trap and lines are
clean and dry, an approximate estimate of the pressure can be made by measuring
the water temperature and reading the pressure from Table 6.1 (on page 107).

## The Rotary Oil Pump

To obtain pressures below 10 mm Hg, a mechanical vacuum pump of the type
illustrated in Figure 6.16a is used. A pump of this type in good condition can attain
pressures as low as 0.1 mm Hg. These low pressures are measured with a tilting-type
McLeod gauge (Fig. 6.16b). When a reading is being made, the gauge is tilted to the
vertical position shown, and the pressure is read as the difference between the heights
of the two columns of mercury. Between readings the gauge is rotated clockwise 90°.

Never use a mechanical vacuum pump before placing a mixture of dry ice
and isopropyl alcohol in a Dewar flask (Fig. 6.16c) around the trap and never
pump corrosive vapors (e.g., hydrogen chloride gas) into the pump. Should this

The Dewar flask should be
wrapped with tape or otherwise
protected from implosion.

(a)

(b)

Dewar flask

Dry ice/isopropyl alcohol

Trap

To McLeod gauge

(c)

Oil pump

happen, change the pump oil immediately. With care, the pump will give many years of good service. The dry-ice trap condenses organic vapors and water vapor, both of which would otherwise contaminate the vacuum pump oil and exert enough vapor pressure to destroy a good vacuum.

For an exceedingly high vacuum ($5 \times 10^{-8}$ mm Hg) a high-speed, three-stage mercury diffusion pump is used (Fig. 6.17).

## Relationship between Boiling Point and Pressure

It is not possible to calculate the boiling point of a substance at some reduced pressure from a knowledge of the boiling temperature at 760 mm Hg because the relationship between boiling point and pressure varies from compound to

**FIG. 6.17**

A high-speed, three-stage mercury diffusion pump that is capable of producing a vacuum of $5 \times 10^{-8}$ mm Hg. Mercury is boiled in the flask; the vapor rises in the center tube, is deflected downward in the inverted cups, and entrains gas molecules, which diffuse in from the space to be evacuated, A. The mercury condenses to a liquid and is returned to the flask; the gas molecules are removed at B by an ordinary rotary vacuum pump.

■ **FIG. 6.18**

Boiling point curves.

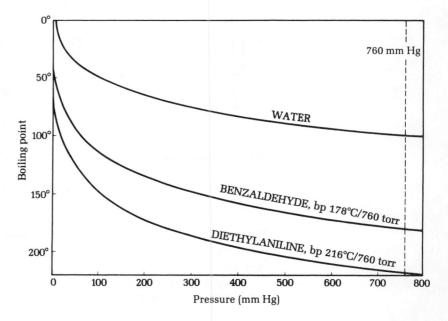

compound and is somewhat unpredictable. It is true, however, that boiling point curves for organic substances have much the same general disposition, as illustrated by the two lower curves in Figure 6.18. These are similar and do not differ greatly from the curve for water. For substances boiling in the range of 150°C to 250°C at 760 mm Hg, the boiling point at 20 mm Hg is 100°C–120°C lower than that at 760 mm Hg. Benzaldehyde, which is very sensitive to air oxidation at the

| TABLE 6.2 Distillation of a (Hypothetical) Substance at Various Pressures | | | |
|---|---|---|---|
| **Method** | | **Pressure (mm Hg)** | **bp (°C)** |
| Ordinary distillation | | 760 | 250 |
| Aspirator | summer | 25 | 144 |
| | winter | 15 | 134 |
| | poor condition | 10 | 124 |
| Rotary oil pump | good condition | 3 | 99 |
| | excellent condition | 1 | 89 |
| Mercury vapor pump (high vacuum distillation) | | 0.01 | 30 |

normal boiling point of 178°C, distills at 76°C at 20 mm Hg, and the concentration of oxygen in the rarefied atmosphere is just 3% of what it would be in an ordinary distillation.

The curves all show a sharp upward inclination in the region of very low pressure. The lowering of the boiling point that attends a reduction in pressure is far more pronounced at low pressures rather than at high pressures. A drop in atmospheric pressure of 10 mm Hg lowers the normal boiling point of an ordinary liquid by less than 1°C, but a reduction of pressure from 20 mm Hg to 10 mm Hg causes a drop of about 15°C in the boiling point. The effect at pressures below 1 mm is still more striking, and with the development of highly efficient oil vapor or mercury vapor diffusion pumps, distillation at a pressure of a few thousandths or ten-thousandths of a millimeter has become a standard operation in many research laboratories. High-vacuum distillation—that is, at a pressure below 1 mm Hg—affords a useful means of purifying extremely sensitive or very slightly volatile substances. Table 6.2 indicates the order of magnitude of the reduction in boiling point attainable by operating in different ways and illustrates the importance of keeping vacuum pumps in good repair.

The boiling point of a substance at varying pressures can be estimated from a pressure-temperature nomograph, such as the one shown in Figure 6.19. If the boiling point of a nonpolar substance at 760 mm Hg is known, for example, 145°C (column B), and the new pressure measured, for example, 28 mm Hg (column C), then a straight line connecting these values on columns B and C when extended intersects column A to give the observed boiling point of 50°C. Conversely, a substance observed to boil at 50°C (column A) at 1.0 mm Hg (column C) will boil at approximately 212°C at atmospheric pressure (column B). If the material being distilled is polar, such as a carboxylic acid, column D is used. Thus a substance that boils at 145°C at atmospheric pressure (column B) will boil at 28 mm Hg (column C) if it is nonpolar and at a pressure of 10 mm Hg (column D) if it is polar. The boiling point will be 50°C (column A).

Vacuum distillation is not confined to the purification of substances that are liquid at ordinary temperatures; it can often be used to advantage for solid substances. This operation is conducted for a different purpose and by a different

Observed
boiling point
at *P* mm Hg

Boiling point
corrected
to 760 mm Hg

Pressure,
*P* mm Hg

technique. A solid is seldom distilled in order to separate constituents of differing degrees of volatility; distillation is intended to purify the solid. Often, one vacuum distillation can remove foreign coloring matter and tar without any appreciable loss of product, whereas several wasteful recrystallizations might be required to attain the same purity. It is often good practice to distill a crude product and then to recrystallize it. Time is saved in the recrystallization process because the hot solution usually requires neither filtration nor clarification. The solid must be dry, and a test should be made to determine if it will distill without decomposition at the pressure of the available pump. A compound lacking the required stability at high temperatures is sometimes indicated by its structure, but a high melting point should not be taken as an indication that distillation will fail. Substances melting as high as 300°C have been distilled with success at the pressure of an ordinary rotary vacuum pump.

## Molecular Distillation

At $1 \times 10^{-3}$ mm Hg the mean free path of nitrogen molecules is 56 mm at 25°C. This means that an average $N_2$ molecule can travel 56 mm before bumping into another $N_2$ molecule. Specially designed apparatus makes it possible to distill almost any volatile molecule by operating at a very low pressure and having the

■ **FIG. 6.20**
A molecular distillation apparatus.

Water→

To high vacuum

Distillate

Material to
be distilled

condensing surface close to the material being distilled. The thermally energized molecule escapes from the liquid, moves about 10 mm without encountering any other molecules, and condenses on a cold surface. A simple apparatus for molecular distillation is illustrated in Figure 6.20. The bottom of the apparatus is placed on a hot surface, and the distillate moves a very short distance before condensing. If it is not too viscous, it will drip from the point on the glass condensing surface and run down to the receiver.

The efficiency of this molecular distillation apparatus is less than one theoretical plate (*see* Chapter 5). It is used to remove very volatile substances such as solvents and to separate volatile, high-boiling substances at low temperatures from nonvolatile impurities.

In more sophisticated apparatus of quite different design, the liquid to be distilled falls down a heated surface, which is again close to the condenser. This movement of a film prevents a buildup of nonvolatile material on the surface of the material to be distilled, which would cause the distillation to cease. Vitamin A is distilled commercially in this manner.

*Online Study Center*

Photo: Sublimation Apparatus

# PART 3: Sublimation

**S**ublimation is the process whereby a solid evaporates from a warm surface and condenses on a cold surface, again as a solid (Fig. 6.21 and Fig. 6.22). This technique is particularly useful for the small-scale purification of solids because there is so little loss of material in transfer. If the substance possesses the correct properties, sublimation is preferred over recrystallization when the amount of material to be purified weighs less than 100 mg.

As demonstrated in the experiment that follows, sublimation can occur readily at atmospheric pressure. For substances with lower vapor pressures, vacuum sublimation is used. At very low pressures sublimation becomes very similar to molecular distillation, where the molecule leaves the warm solid and passes unobstructed to a cold condensing surface and condenses in the form of a solid.

Because sublimation occurs from the surface of the warm solid, impurities can accumulate and slow down or even stop the sublimation, in which case it is

■ **FIG. 6.21**
A small-scale vacuum sublimation apparatus.

15-mL Centrifuge tube, to be filled with ice

Adapter (Pluro stopper)

Pipette bulb

25-mL Filter flask

Material to be sublimed

Sublimate

Heat source

■ **FIG. 6.22**
(a) A Mallory sublimator.
(b) A large vacuum sublimator.

Vacuum

Sublimate

Water in

Rubber O-ring

Material to be sublimed

Heat here

(a)

Vacuum

Rubber O-ring

Ice

Sublimate goes to here
Material to be sublimed
Heat this surface

(b)

necessary to open the apparatus, grind the impure solid to a fine powder, and restart the sublimation.

Sublimation is much easier to carry out on a small scale than on a large one. It would be unusual for a research chemist to sublime more than about 10 g of material because the sublimate tends to fall back to the bottom of the apparatus. In an apparatus such as that illustrated in Figure 6.22a, the solid to be sublimed is placed in the lower flask and connected via a lubricant-free rubber O-ring to the condenser, which in turn is connected to a vacuum pump. The lower flask is immersed in an oil bath at the appropriate temperature, and the product is sublimed and condensed onto the cool walls of the condenser. The parts of the apparatus are gently separated, and the condenser is inverted; the vacuum connection serves as a convenient funnel for product removal. For large-scale work the sublimator in Figure 6.22b is used. The inner well is filled with a coolant (ice or dry ice). The sublimate clings to this cool surface, from which it can be removed by scraping and dissolving in an appropriate solvent.

*A very simple sublimation apparatus: an inverted stemless funnel on a watch glass*

## Lyophilization

Lyophilization, also called *freeze-drying*, is the process of subliming a solvent, usually water, with the object of recovering the solid that remains after the solvent is removed. This technique is used extensively to recover heat- and oxygen-sensitive substances of natural origin, such as proteins, vitamins, nucleic acids, and other biochemicals, from a dilute aqueous solution. The aqueous solution of the substance to be lyophilized is usually frozen to prevent bumping and then subjected to a vacuum of about $1 \times 10^{-3}$ mm Hg. The water sublimes and condenses as ice on the surface of a large and very cold condenser. The sample remains frozen during this entire process, without any external cooling being supplied, because of the very high heat of vaporization of water; thus the temperature of the sample never exceeds 0°C. Freeze-drying on a large scale is employed to make "instant" coffee, tea, soup, and rice and all sorts of dehydrated foods.

## EXPERIMENT

## Sublimation of an Unknown Substance
### *Apparatus and Technique*

The apparatus consists of a 15-mL centrifuge tube thrust through an adapter (use a drop of glycerol to lubricate the adapter) fitted in a 25-mL filter flask. Once the adapter is properly positioned on the centrifuge tube, it should remain there permanently. This tube is used only for sublimation.

The sample is either ground to a fine powder and uniformly distributed on the bottom of the flask or introduced into the flask as a solution and then the solvent evaporated to deposit the substance on the bottom of the filter flask. If the compound sublimes easily, care must be exercised using the latter technique to ensure that the sample does not evaporate as the last of the solvent is being removed.

The 15-mL centrifuge tube and adapter are placed in the flask so that the tip of the centrifuge tube is about 3–8 mm above the bottom of the flask (*see* Fig. 6.21). The flask is clamped with a three-prong clamp. Many substances sublime at

| TABLE 6.3 • Sublimation Unknowns | | | |
|---|---|---|---|
| **Substance** | **mp (°C)** | **Substance** | **mp (°C)** |
| 1,4-Dichlorobenzene | 55 | Benzoic acid | 122 |
| Naphthalene | 82 | Salicylic acid | 159 |
| 1-Naphthol | 96 | Camphor | 177 |
| Acetanilide | 114 | Caffeine | 235 |

atmospheric pressure, including the unknowns listed in Table 6.3. For vacuum sublimation, the side arm of the flask leads to an aspirator or vacuum pump for reduced pressure sublimation. The centrifuge tube is filled with ice and water. The ice is not added before closing the apparatus, so atmospheric moisture will not condense on the tube.

Sublimation: Add sample to flask, close flask openings, add ice, sublime sample, remove ice, add room-temperature water, open apparatus, and scrape off product.

The filter flask is warmed cautiously on a hot sand bath until the product just begins to sublime. The heat is maintained at that temperature until sublimation is complete. Because some product will collect on the cool upper parts of the flask, it should be fitted with a loose cone of aluminum foil to direct heat there and cause the material to collect on the surface of the centrifuge tube. Alternatively, tilt the flask and rotate it in the hot sand, or remove the flask from the sand bath, and then use a heat gun to warm the flask.

Roll the filter flask in the sand bath or heat it with a heat gun to drive material from the upper walls of the flask.

Once sublimation is judged complete, the ice water is removed from the centrifuge tube with a pipette and replaced with water at room temperature. This will prevent moisture from condensing on the product once the vacuum is turned off and the tube removed from the flask. The product is scraped from the centrifuge tube with a metal spatula onto a piece of glazed paper. It is much easier to scrape the product from a centrifuge tube than from a round-bottomed test tube. The last traces can be removed by washing off the tube with a few drops of an appropriate solvent, if that is deemed necessary.

## Procedure

Into the bottom of a 25-mL filter flask, place 50 mg of an impure unknown taken from the list in Table 6.3. These substances can be sublimed at atmospheric pressure, although some will sublime more rapidly at reduced pressure. Close the flask with a rubber pipette bulb and then place ice water in the centrifuge tube. Cautiously warm the flask until sublimation starts and then maintain that temperature throughout the sublimation. Fit the flask with a loose cone of aluminum foil to direct heat up the sides of the flask, which causes the product to collect on the tube. Once sublimation is complete, remove the ice water from the centrifuge tube and replace it with water at room temperature. Collect the product, determine its weight and the percent recovery, and from the melting point identify the unknown. Hand in the product in a neatly labeled vial.

**■ FIG. 6.23**
The $^1H$ NMR spectrum of eugenol (250 MHz).

**Cleaning Up.** Wash material from the apparatus with a minimum quantity of acetone, which, except for the chloro compound, can be placed in the organic solvents container. The 1,4-dichlorobenzene solution must be placed in the halogenated organic waste container.

## For Further Investigation

Room deodorants, for example, the Renuzit Long Last adjustable air freshner, appear to sublime. Place about 150 mg, carefully weighed, of the green solid in the filter flask and sublime it with very little heat applied to the flask. Identify the substance that sublimes. Weigh the residue. As an advanced experiment, run the IR spectrum (*see* Chapter 11) of the residue (*not* of the substance that sublimes). Comment on the economics of air fresheners.

## QUESTIONS

1. Assign the peaks in the $^1H$ NMR spectrum of eugenol (Fig. 6.23) to specific protons in the molecule. The OH peak is at 5.1 ppm.
2. A mixture of ethyl iodide ($C_2H_5I$, bp 72.3°C) and water boils at 63.7°C. What weight of ethyl iodide would be carried over by 1 g of steam during steam distillation?

3. Iodobenzene ($C_6H_5I$, bp 188°C) steam distills at 98.2°C. How many molecules of water are required to carry over 1 molecule of iodobenzene? How many grams of water per gram of iodobenzene?

4. The condensate from a steam distillation contains 8 g of an unknown compound and 18 g of water. The mixture steam distilled at 98°C. What is the molecular weight of the unknown?

5. Bearing in mind that a sealed container filled with steam will develop a vacuum if the container is cooled, explain what would occur if a steam distillation was stopped by turning off the steam valve before opening the screw clamp on the adapter trap.

6. A mixture of toluene (bp 110.8°C) and water is steam distilled. Visual inspection of the distillate reveals that there is a greater volume of toluene than water present, yet water (bp 100°C) has the higher vapor pressure. Explain this observation.

7. Did sublimation aid in the purification of the unknown in part 3? Justify your answer.

8. From your experience and knowledge of the process of sublimation, cite some examples of substances that sublime.

9. How might freeze-dried foods be prepared?

10. Describe the mercury level in a manometer like that shown in Figure 6.15a if the left arm is 1000 cm tall.

**Online Study Center**

**General Resources**
Web Links

11. What criteria would you use to decide whether distillation is an appropriate technique for the purification of a reaction product? What type of distillation method—simple, fractional, vacuum, or steam—should be used?

# CHAPTER 7

# Extraction

**Online Study Center**

This icon will direct you to techniques, equipment setups, and online resources at http://college.hmco.com/PIC/williamsonMMOE5e.

**PRELAB EXERCISE:** Describe how to separate a mixture of 3-toluic acid and 4'-aminoacetophenone using acid-base liquid/liquid extraction. What species will end up in the aqueous layer if you mix a solution of benzoic acid and aniline in ether with a solution of $NaHCO_3$ (aq)? Draw the structure of this species.

**E**xtraction is one of the oldest chemical operations known to humankind. The preparation of a cup of coffee or tea involves the extraction of flavor and odor components from dried vegetable matter with hot water. Aqueous extracts of bay leaves, stick cinnamon, peppercorns, and cloves, along with alcoholic extracts of vanilla and almond, are used as food flavorings. For the past 150 years or so, organic chemists have extracted, isolated, purified, and characterized the myriad compounds produced by plants that for centuries have been used as drugs and perfumes—substances such as quinine from cinchona bark, morphine from the opium poppy, cocaine from coca leaves, and menthol from peppermint oil. The extraction of compounds from these natural products is an example of solid/liquid extraction—the solid being the natural product and the liquid being the solvent into which the compounds are extracted. In research, a Soxhlet extractor (Fig. 7.1) is often used for solid/liquid extraction.

Although solid/liquid extraction is the most common technique for brewing beverages and isolating compounds from natural products, liquid/liquid extraction is a very common method used in the organic laboratory, specifically when isolating reaction products. Reactions are typically homogeneous liquid mixtures and can therefore be extracted with either an organic or aqueous solvent. Organic reactions often yield a number of byproducts—some inorganic and some organic. Also, because some organic reactions do not go to 100% completion, a small amount of starting material is present at the end of the reaction. When a reaction is complete, it is necessary to do a *workup*, that is, separate and purify the desired product from the mixture of byproducts and residual starting material. Liquid/liquid extraction is a common separation step in this workup, which is then followed by

■ **FIG. 7.1**
The Soxhlet extractor for the
extraction of solids such as dried
leaves or seeds. The solid is put in
a filter paper thimble. Solvent
vapor rises in the tube on the right;
condensate drops onto the solid in
the thimble, leaches out soluble
material, and, after initiating an
automatic siphon, carries it to the
flask where nonvolatile extracted
material accumulates. Substances
of low solubility can be extracted
by prolonged operation.

Condenser ——

Paper thimble ——      —— Soxhlet

purification of the product. There are two types of liquid/liquid extraction: neutral and acid/base. The experiments in this chapter demonstrate solid/liquid extraction and the two types of liquid/liquid extraction.

Organic products are often separated from inorganic substances in a reaction mixture by liquid/liquid extraction with an organic solvent. For example, in the synthesis of 1-bromobutane (*see* Chapter 16), 1-butanol, also a liquid, is heated with an aqueous solution of sodium bromide and sulfuric acid to produce the product and sodium sulfate.

$$2\ CH_3CH_2CH_2CH_2OH\ +\ 2\ NaBr\ +\ H_2SO_4\ \rightarrow$$
$$2\ CH_3CH_2CH_2CH_2Br\ +\ 2\ H_2O\ +\ Na_2SO_4$$

The 1-bromobutane is isolated from the reaction mixture by extraction with *t*-butyl methyl ether, an organic solvent in which 1-bromobutane is soluble and in which water and sodium sulfate are insoluble. The extraction is accomplished by simply adding *t*-butyl methyl ether to the aqueous mixture and shaking it. Two layers will result: an organic layer and an aqueous layer. The *t*-butyl methyl ether is less dense than water and floats on top; it is easily removed/drained away from the water layer and evaporated to leave the bromo product free of inorganic substances, which reside in the aqueous layer.

## Partition Coefficient

The extraction of a compound such as 1-butanol, which is slightly soluble in water as well as very soluble in ether, is an equilibrium process governed by the solubilities of the alcohol in the two solvents. The ratio of the solubilities is known as the *distribution coefficient*, also called the *partition coefficient* (*k*), and is an equilibrium constant with a certain value for a given substance, pair of solvents, and temperature.

The *concentration* of the solute in each solvent can be well correlated with the *solubility* of the solute in the pure solvent, a figure that is readily found in solubility tables in reference books. For substance C

$$k = \frac{\text{concentration of C in } t\text{-butyl methyl ether}}{\text{concentration of C in water}}$$

$$> \frac{\text{solubility of C in } t\text{-butyl methyl ether (g/100 mL)}}{\text{solubility of C in water (g/100 mL)}}$$

Consider compound A that dissolves in *t*-butyl methyl ether to the extent of 12 g/100 mL and dissolves in water to the extent of 6 g/100 mL.

$$k = \frac{12 \text{ g/100 mL } t\text{-butyl methyl ether}}{6 \text{ g/100 mL water}} = 2$$

If a solution of 6 g of A in 100 mL of water is shaken with 100 mL of $t$-butyl methyl ether, then

$$k = \frac{x \text{ g of A}/100 \text{ mL } t\text{-butyl methyl ether}}{6 - x \text{ g of A}/100 \text{ mL water}}$$

from which

$$x = 4.0 \text{ g of A in the ether layer}$$
$$6 - x = 2.0 \text{ g of A left in the water layer}$$

It is, however, more efficient to extract the 100 mL of aqueous solution twice with 50-mL portions of $t$-butyl methyl ether rather than once with a 100-mL portion.

$$k = \frac{x \text{ g of A}/50 \text{ mL}}{6 - x \text{ g of A}/100 \text{ mL}} = 2$$

from which

$$x = 3.0 \text{ g of A in the } t\text{-butyl methyl ether layer}$$
$$6 - x = 3.0 \text{ g of A in the water layer}$$

If this 3.0 g/100 mL of water is extracted again with 50 mL of $t$-butyl methyl ether, we can calculate that 1.5 g of A will be in the ether layer, leaving 1.5 g in the water layer. So two extractions with 50-mL portions of ether will extract 3.0 g + 1.5 g = 4.5 g of A, whereas one extraction with a 100-mL portion of $t$-butyl methyl ether removes only 4.0 g of A. Three extractions with 33-mL portions of $t$-butyl methyl ether would extract 4.7 g. Obviously, there is a point at which the increased amount of A extracted does not repay the effort of multiple extractions, but remember that several small-scale extractions are more effective than one large-scale extraction.

## Properties of Extraction Solvents

Liquid/liquid extraction involves two layers: the organic layer and the aqueous layer. The solvent used for extraction should possess many properties, including the following:

- It should readily dissolve the substance to be extracted at room temperature.
- It should have a low boiling point so that it can be removed readily.
- It should not react with the solute or the other solvent.
- It should not be highly flammable or toxic.
- It should be relatively inexpensive.

In addition, it should not be miscible with water (the usual second phase). No solvent meets every criterion, but several come close. Some common liquid/liquid

| TABLE 7.1 • Common Solvents Listed by Density | |
|---|---|
| Solvent | Density (g/mL) |
| Hexane | 0.695 |
| Diethyl ether | 0.708 |
| *t*-Butyl methyl ether | 0.740 |
| Toluene | 0.867 |
| Water | 1.000 |
| Dichloromethane | 1.325 |
| Chloroform | 1.492 |

extraction solvent pairs are water-ether, water-dichloromethane, and water-hexane. Notice that each combination includes water because most organic compounds are immiscible in water and therefore can be separated from inorganic compounds. Organic solvents such as methanol and ethanol are not good extraction solvents because they are soluble in water.

### *Identifying the Layers*

One common mistake when performing an extraction is to misidentify the layers and discard the wrong one. It is good practice to save all layers until the desired product is in hand. The densities of the solvents will predict the identities of the top and bottom layers. In general, the densities of nonhalogenated organic solvents are less than 1.0 g/mL and those of halogenated solvents are greater than 1.0 g/mL. Table 7.1 lists the densities of common solvents used in extraction.

Although density is the physical property that determines which layer is on top or on bottom, a very concentrated amount of solute dissolved in either layer can reverse the order. The best method to avoid a misidentification is to perform a drop test. Add a few drops of water to the layer in question and watch the drop carefully. If the layer is water, then the drop will mix with the solution. If the solvent is the organic layer, then the water drop will create a second layer.

Identify layers by a drop test.

### *Ethereal Extraction Solvents*

$CH_3CH_2$—O—$CH_2CH_3$

**Diethyl ether**
**"Ether"**
MW 74.12, den. 0.708
bp 34.6°C, $n_D^{20}$ 1.3530

In the past, diethyl ether was the most common solvent for extraction in the laboratory. It has high solvent power for hydrocarbons and oxygen-containing compounds. It is highly volatile (bp 34.6°C) and is therefore easily removed from an extract. However, diethyl ether has two big disadvantages: it is highly flammable and poses a great fire threat, and it easily forms peroxides. The reaction of diethyl ether with air is catalyzed by light. The resulting peroxides are higher boiling than the ether and are left as a residue when the ether evaporates. If the residue is heated, it will explode because ether peroxides are treacherously high explosives. In recent years, a new solvent has come on the scene—*tert*-butyl methyl ether.

**tert-Butyl methyl ether**
MW 88.14, den. 0.741
bp 55.2°C, $n_D^{20}$ 1.369

■ **FIG. 7.2**
A separatory funnel with Teflon stopcock.

*tert*-Butyl methyl ether, called methyl *tert*-butyl ether (MTBE) in industry, has many advantages over diethyl ether as an extraction solvent. Most important, it does not easily form peroxides, so it can be stored for much longer periods than diethyl ether. And, in the United States, it is less than two-thirds the price of diethyl ether. It is slightly less volatile (bp 55°C), so it does not pose the same fire threat as diethyl ether, although one must be as careful in handling this solvent as in handling any other highly volatile, flammable substance. The explosion limits for *t*-butyl methyl ether mixed with air are much narrower than for diethyl ether, the toxicity is less (it is not a carcinogen), the solvent power is the same, and the ignition temperature is higher (224°C versus 180°C).

The weight percent solubility of diethyl ether dissolved in water is 7.2%, whereas that of *t*-butyl methyl ether is 4.8%. The solubility of water in diethyl ether is 1.2%, while in *t*-butyl methyl ether it is 1.5%. Unlike diethyl ether, *t*-butyl methyl ether forms an azeotrope with water (4% water) that boils at 52.6°C. This means that evaporation of any *t*-butyl methyl ether solution that is saturated with water should leave no water residue, unlike diethyl ether.

The low price and ready availability of *t*-butyl methyl ether came about because it replaced tetraethyl lead as the antiknock additive for high-octane gasoline and as a fuel oxygenate, which helps reduce air pollution, but its water solubility has allowed it to contaminate drinking water supplies in states where leaking underground fuel storage tanks are not well regulated. Consequently, it is being replaced with the much more expensive ethanol. In this text, *t*-butyl methyl ether is *strongly* suggested wherever diethyl ether formerly would have been used in an extraction. It will not, however, work as the only solvent in the Grignard reaction, probably because of steric hindrance. So whenever the word *ether* appears in this text as an extraction solvent, it is suggested that *t*-butyl methyl ether be used and not diethyl ether.

## Mixing and Separating the Layers

For microscale separations, mixing and separating the layers with a pipette normally incurs very little product loss. Because the two solvents are typically in a reaction tube for microscale extraction, the two layers can be mixed by drawing up and rapidly expelling them with a pipette. Then the layers are allowed to separate, and the bottom layer is separated by drawing it up into a pipette and transferring it to a different container.

For macroscale separations, a separatory funnel (Fig. 7.2) is used to mix and separate the organic and aqueous layers. In macroscale experiments, a frequently used method of working up a reaction mixture is to dilute the mixture with water and extract it with an organic solvent, such as ether, in a separatory funnel. When the stoppered funnel is shaken to distribute the components between the immiscible solvents *t*-butyl methyl ether and water, pressure always develops through volatilization of ether from the heat of the hands, and liberation of a gas ($CO_2$) (in acid/base extractions) can increase the pressure. Consequently, the funnel is grasped so that the stopper is held in place by one hand

■ **FIG. 7.3**
The correct position for holding a separatory funnel when shaking. Point outlet away from yourself and your neighbors.

and the stopcock by the other, as illustrated in Figure 7.3. After a brief shake or two, the funnel is held in the inverted position shown, and the stopcock is opened cautiously (with the funnel stem pointed away from nearby persons) to release pressure. The mixture can then be shaken more vigorously, with pressure released as necessary. When equilibration is judged to be complete, the slight, constant terminal pressure due to ether is released, the stopper is rinsed with a few drops of ether delivered by a Pasteur pipette, and the layers are allowed to separate. The organic reaction product is distributed wholly or largely into the upper ether layer, whereas inorganic salts, acids, and bases pass into the water layer, which can be drawn off. If the reaction was conducted in alcohol or some other water-soluble solvent, the bulk of the solvent is removed in the water layer, and the remainder can be eliminated in two or three washings with 1–2 volumes of water conducted with the techniques used in the first equilibration. The separatory funnel should be supported in a ring stand, as shown in Figure 7.2.

Before adding a liquid to the separatory funnel, check the stopcock. If it is glass, see that it is properly greased, bearing in mind that too much grease will clog the hole in the stopcock and also contaminate the extract. If the stopcock is Teflon, see that it is adjusted to a tight fit in the bore. Store the separatory funnel with the Teflon stopcock loosened to prevent sticking. Because Teflon has a much larger temperature coefficient of expansion than glass, a stuck stopcock can be loosened by cooling the stopcock in ice or dry ice. Do not store liquids in the separatory funnel; they often leak or cause the stopper or stopcock to freeze. To have sufficient room for mixing the layers, fill the separatory funnel no more than three-fourths full. Withdraw the lower layer from the separatory funnel through the stopcock and pour the upper layer out through the neck.

All too often the inexperienced chemist discards the wrong layer when using a separatory funnel. Through incomplete neutralization, a desired component may still remain in the aqueous layer, or the densities of the layers may change. Cautious workers save all layers until the desired product has been isolated. The organic layer is not always the top layer. If in doubt, perform a drop test by adding a few drops of each to water in a test tube.

## *Practical Considerations When Mixing Layers*

### Pressure Buildup

The heat of one's hand or heat from acid/base reactions will cause pressure buildup in an extraction mixture that contains a very volatile solvent such as dichloromethane. The extraction container—whether a test tube or a separatory funnel—must be opened carefully to vent this pressure.

Sodium bicarbonate solution is often used to neutralize acids when carrying out acid/base extractions. The result is the formation of carbon dioxide, which can cause foaming and high pressure buildup. Whenever bicarbonate is used, add it very gradually with thorough mixing and frequent venting of the extraction device. If a large amount of acid is to be neutralized with bicarbonate, the process should be carried out in a beaker.

### Emulsions

Imagine trying to extract a soap solution (e.g., a nonfoaming dishwashing detergent) into an organic solvent. After a few shakes with an organic solvent, you would have an absolutely intractable emulsion. An emulsion is a suspension of one liquid as droplets in another. Detergents stabilize emulsions, and so any time a detergent-like molecule is in the material being extracted, there is the danger that emulsions will form. Substances of this type are commonly found in nature, so one must be particularly wary of emulsion formation when creating organic extracts of aqueous plant material, such as caffeine from tea. Emulsions, once formed, can be quite stable. You would be quite surprised to open your refrigerator one morning and see a layer of clarified butter floating on the top of a perfectly clear aqueous solution that had once been milk, but milk is the classic example of an emulsion.

Prevention is the best cure for emulsions. This means shaking the solution to be extracted *very gently* until you see that the two layers will separate readily. If a bit of emulsion forms, it may break simply on standing for a sufficient length of time. Making the aqueous layer highly ionic will help. Add as much sodium chloride as will dissolve and shake the mixture gently. Vacuum filtration sometimes works and, when the organic layer is the lower layer, filtration through silicone-impregnated filter paper is helpful. Centrifugation works very well for breaking emulsions. This is easy on a small scale, but often the equipment is not available for large-scale centrifugation of organic liquids.

*Shake gently to avoid emulsions.*

## Drying Agents

The organic solvents used for extraction dissolve not only the compound being extracted but also water. Evaporation of the solvent then leaves the desired compound contaminated with water. At room temperature water dissolves 4.8% of *t*-butyl methyl ether by weight, and the ether dissolves 1.5% of water. But ether is virtually insoluble in water saturated with sodium chloride (36.7 g/100 mL). If ether that contains dissolved water is shaken with a saturated aqueous solution of sodium chloride, water will be transferred from the *t*-butyl methyl ether to the aqueous layer. So, strange as it may seem, ethereal extracts are routinely dried by shaking them with an aqueous saturated sodium chloride solution.

Solvents such as dichloromethane do not dissolve nearly as much water and are therefore dried over a chemical drying agent. Many choices of chemical drying agents are available for this purpose, and the choice of which one to use is governed by four factors: (1) the possibility of reaction with the substance being extracted, (2) the speed with which it removes water from the solvent, (3) the efficiency of the process, and (4) the ease of recovery from the drying agent.

Some very good but specialized and reactive drying agents are potassium hydroxide, anhydrous potassium carbonate, sodium metal, calcium hydride, lithium aluminum hydride, and phosphorus pentoxide. Substances that are essentially neutral and unreactive and are widely used as drying agents include anhydrous calcium sulfate (Drierite), magnesium sulfate, molecular sieves, calcium chloride, and sodium sulfate.

*Drierite*, a specially prepared form of calcium sulfate, is a fast and effective drying agent. However, it is difficult to ascertain whether enough has been used.

*Drierite, CaSO$_4$*

An indicating type of Drierite is impregnated with cobalt chloride, which turns from blue to red when it is saturated with water. This works well when gases are being dried, but it should not be used for liquid extractions because the cobalt chloride dissolves in many protic solvents.

**Magnesium sulfate, MgSO₄**

*Magnesium sulfate* is also a fast and fairly effective drying agent, but it is so finely powdered that it always requires careful filtration for removal.

**Molecular sieves, zeolites**

*Molecular sieves* are sodium alumino-silicates (zeolites) that have well-defined pore sizes. The 4 Å size adsorbs water to the exclusion of almost all organic substances, making them a fast and effective drying agent. Like Drierite, however, it is impossible to ascertain by appearance whether enough has been used. Molecular sieves in the form of 1/16-in. pellets are often used to dry solvents by simply adding them to the container.

**Calcium chloride (CaCl₂) pellets are the drying agent of choice for small-scale experiments.**

*Calcium chloride*, recently available in the preferred form of pellets (4 to 80 mesh[1]), is a very fast and effective drying agent. It has the advantage that it clumps together when excess water is present, which makes it possible to know how much to add by observing its behavior. Unlike the older granular form, the pellets do not disintegrate into a fine powder. These pellets are admirably suited to microscale experiments where the solvent is removed from the drying agent with a Pasteur pipette. Calcium chloride is much faster and far more effective than anhydrous sodium sulfate; after much experimentation, we have decided that this is the agent of choice, particularly for microscale experiments. These pellets are used for most of the drying operations in this text. Note, however, that calcium chloride reacts with some alcohols, phenols, amides, and some carbonyl-containing compounds. Advantage is sometimes taken of this property to remove not only water from a solvent but also, for example, a contaminating alcohol (*see* Chapter 16—the synthesis of 1-bromobutane from 1-butanol). Because *t*-butyl methyl ether forms an azeotrope with water, its solutions should, theoretically, not need to be dried; evaporation carries away the water. Drying these ether solutions with calcium chloride pellets removes water droplets that get carried into the ether solution.

**Sodium sulfate, Na₂SO₄**

*Sodium sulfate* is a very poor drying agent. It has a very high capacity for water but is slow and not very efficient in the removal of water. Like calcium chloride pellets, it clumps together when wet, and solutions are easily removed from it using a Pasteur pipette. Sodium sulfate has been used extensively in the past and should still be used for compounds that react with calcium chloride.

## PART 1: The Technique of Neutral Liquid/Liquid Extraction

The workup technique of liquid/liquid extraction has four steps: (1) mixing the layers, (2) separating the layers, (3) drying the organic layer, and (4) removing the solvent. The microscale neutral liquid/liquid extraction technique is described in the following sections.

---

1. These pellets are available from Fisher Scientific, Cat. No. C614–3.

# Step 1. Mixing the Layers

Once the organic and aqueous layers are in contact with one another, mixing is required to ensure that the desired compound(s) get extracted into the desired layer. First, place 1–2 mL of an aqueous solution of the compound to be extracted in a reaction tube. Add about 1 mL of extraction solvent, for example, dichloromethane. Note, as you add the dichloromethane, whether it is the top or the bottom layer. (Since dichloromethane is more dense than water, predict what layer it will be.) An effective way to mix the two layers is to flick the tube with a finger. Grasp the tube firmly at the very top between the thumb and forefinger and flick it vigorously at the bottom (Fig. 7.4). You will find that this violent motion mixes the two layers well, but nothing comes out the top. Another good mixing technique is to pull the contents of the reaction tube into a Pasteur pipette and then expel the mixture back into the tube with force. Doing this several times will effect good mixing of the two layers. A stopper can be placed in the top of the tube, and the contents can be mixed by shaking the tube, but the problem with this technique is that the high vapor pressure of the solvent will often force liquid out around the cork or stopper.

# Step 2. Separating the Layers

After thoroughly mixing the two layers, allow them to separate. Tap the tube if droplets of one layer are in the other layer or on the side of the tube. After the layers separate completely, draw up the lower dichloromethane layer into a Pasteur pipette. Leave behind any middle emulsion layer. The easiest way to do this is to attach the pipette to a pipette pump (Fig. 7.5). This allows very precise control of the liquid being removed. It takes more skill and practice to remove the lower layer cleanly with a 2-mL rubber bulb attached to a pipette because the high vapor pressure of the solvent tends to make it dribble out. To avoid losing any of the solution, it is best to hold a clean, dry, empty tube in the same hand as the full tube to receive the organic layer (Fig. 7.6).

*Always draw out the lower layer and place it in another container.*

From the discussion of the partition coefficient, you know that several small extractions are better than one large one, so repeat the extraction process with two further 1-mL portions of dichloromethane. An experienced chemist might summarize all the preceding with the following notebook entry, "Aqueous layer extracted 3 $\times$ 1-mL portions $CH_2Cl_2$," and in a formal report would write, "The aqueous layer was extracted three times with 1-mL portions of dichloromethane."

If you are working on a larger microscale, a microscale separatory funnel (Fig. 7.7) should be used. A separatory funnel, regardless of size, should be filled to only about two-thirds of its capacity so the layers can be mixed by shaking. The microscale separatory funnel has a capacity of 8.5 mL when full, so it is useful for an extraction with a total volume of about 6 mL.

Use a wood boiling stick to poke out the polyethylene frit from the bottom part of the separatory funnel. Store it for later replacement. Close the valve, add up to 5 mL of the solution to be extracted to the separatory funnel, then add the extraction solvent so that the total volume does not exceed 6 mL.

■ **FIG. 7.4**
Mixing the contents of a reaction tube by flicking it. Grasp the tube firmly at the very top and flick it vigorously at the bottom. The contents will mix without coming out of the tube.

■ **FIG. 7.5**
The removal of a solvent from a reaction tube with a pipette and pipette pump.

2-mL Pipette pump

1-mL Pipette graduated to ± 0.01 mL

Reaction tube

■ **FIG. 7.6**
Grasp both reaction tubes in one hand when transferring material from one tube to another with a Pasteur pipette.

Cap the separatory funnel and mix the contents by inverting the funnel several times. If the two layers separate fairly easily, then the contents can be shaken more thoroughly. If the layers do not separate easily, be careful not to shake the funnel too vigorously because intractable emulsions could form.

Remove the stopper from the funnel, clamp it, and then, grasping the valve with two hands, empty the bottom layer into an Erlenmeyer flask or other container. If the top layer is desired, pour it out through the top of the separatory funnel—don't drain it through the valve, which may have a drop of the lower layer remaining in it.

## Step 3. Drying the Organic Layer

Dichloromethane dissolves a very small quantity of water, and microscopic droplets of water are suspended in the organic layer, often making it cloudy. To remove the water, a drying agent, for example, anhydrous calcium chloride pellets, is added to the dichloromethane solution.

■ **FIG. 7.7**
A microscale separatory funnel.
Remove the polyethylene frit from
the micro Büchner funnel before
using.

Polyethylene
cap

Glass
column

Micro
Büchner
funnel

Record the tare of the final
container.

### *How Much Drying Agent Should Be Used?*

When a small quantity of the drying agent is added, the crystals or pellets become sticky with water, clump together, and fall rapidly as a lump to the bottom
of the reaction tube. There will come a point when a new small quantity of drying agent no longer clumps together, but the individual particles settle slowly
throughout the solution. As they say in Scandinavia, "Add drying agent until it
begins to snow." The drying process takes about 10–15 min, during which time
the tube contents should be mixed occasionally by flicking the tube. The solution should no longer be cloudy but clear (although it may be colored).

Once drying is judged complete, the solvent is removed by forcing a Pasteur
pipette to the bottom of the reaction tube and pulling the solvent in. Air is expelled
from the pipette as it is being pushed through the crystals or pellets so that no drying
agent will enter the pipette. It is very important to wash the drying agent left in the reaction tube with several small quantities of pure solvent to transfer all the extract.

## Step 4. Removing the Solvent

If the quantity of extract is relatively small, say 3 mL or less, then the easiest way
to remove the solvent is to blow a stream of air (or nitrogen) onto the surface of the
solution from a Pasteur pipette (Fig. 7.8). Be sure that the stream of air is very gentle before inserting it into the reaction tube. The heat of vaporization of the solvent
will cause the tube to become rather cold during the evaporation and, of course,
slow down the process. The easiest way to add heat is to hold the tube in your hand.

Another way to remove the solvent is to attach the Pasteur pipette to an aspirator and pull air over the surface of the liquid. This is not quite as fast as blowing
air onto the surface of the liquid and runs the danger of sucking up the liquid into
the aspirator.

If the volume of liquid is more than about 3 mL, put it into a 25-mL filter flask,
put a plastic Hirsch funnel in place, and attach the flask to the aspirator. By placing
your thumb in the Hirsch funnel, the vacuum can be controlled, and heat can be applied by holding the flask in the other hand while swirling the contents (Fig. 7.9).

The reaction tube or filter flask in which the solvent is evaporated should be
tared (weighed empty), and this weight recorded in your notebook. In this way, the
weight of material extracted can be determined by again weighing the container.

**EXPERIMENT**

 ## Partition Coefficient of Benzoic Acid

> **IN THIS EXPERIMENT** you will shake a solution of benzoic acid in water with the im-
> miscible solvent dichloromethane. The benzoic acid will distribute (partition) itself be-
> tween the two layers. By removing the organic layer, drying, and evaporating it, the
> weight of benzoic acid in the dichloromethane can be determined and thus the ratio in
> the two layers. This ratio is a constant known as the partition coefficient.

■ **FIG. 7.8**
An aspirator tube being used to
remove solvent vapors.

To water aspirator or
to nitrogen or air supply

Pasteur pipette

To
aspirator

In a reaction tube, place about 100 mg of benzoic acid (weighed to the nearest milligram) and add exactly equal volumes of water followed by dichloromethane (about 1.6 mL each). While making this addition, note which layer is organic and which is aqueous. Put a septum on the tube and shake the contents vigorously for at least 2 min. Allow the tube to stand undisturbed until the layers separate and then carefully draw off, using a Pasteur pipette, *all* of the aqueous layer without removing any of the organic layer. It may be helpful to draw out the tip of the pipette to a fine point in a flame and, using this, to tilt the reaction tube on its side to make this separation as clean as possible.

Add anhydrous calcium chloride pellets to the dichloromethane in very small quantities until it no longer clumps together. Mix the contents of the tube by flicking it and allow it to stand for about 5 min to complete the drying process. Using a dry Pasteur pipette, transfer the dichloromethane to a tared dry reaction tube or a 10-mL Erlenmeyer flask containing a boiling chip. Complete the transfer by washing the drying agent with two more portions of solvent that are added to the original solution and then evaporate the solvent. This can be done by boiling off the solvent while removing solvent vapors with an aspirator tube or by blowing a stream of air or nitrogen into the container while warming it in one's hand (*see* Fig. 7.8). This operation should be performed in a hood.

From the weight of the benzoic acid in the dichloromethane layer, the weight in the water layer can be obtained by difference. The ratio of the weight in dichloromethane to the weight in water is the distribution coefficient because the volumes of the two solvents were equal. Report the value of the distribution coefficient in your notebook.

**Cleaning Up.** The aqueous layer can be flushed down the drain. Dichloromethane goes into the halogenated organic solvents container. After allowing the solvent to evaporate from the sodium sulfate in the hood, place the sodium sulfate in the non-hazardous solid waste container. If local regulations do not allow for the evaporation of solvents in a hood, dispose of the wet sodium sulfate in a special container.

## Part 2: Acid/Base Liquid/Liquid Extraction

Acid/base liquid/liquid extraction involves carrying out simple acid/base reactions to separate strong organic acids, weak organic acids, neutral organic compounds, and basic organic substances. The chemistry involved is given in the following equations, using benzoic acid, phenol, naphthalene, and aniline as examples of the four types of compounds.

Here is the strategy (refer to the flow sheet in Fig. 7.10): The four organic compounds are dissolved in *t*-butyl methyl ether. The ether solution is shaken with a saturated aqueous solution of sodium bicarbonate, a weak base. This will react only with the strong acid, benzoic acid (**1**), to form the ionic salt, sodium benzoate (**5**), which dissolves in the aqueous layer and is removed. The ether solution now contains just phenol (**2**), naphthalene (**4**), and aniline (**3**). A 3 *M* aqueous solution of sodium hydroxide is added, and the mixture is shaken. The hydroxide, a strong base, will react only with the phenol (**2**), a weak acid, to form sodium phenoxide (**6**), an ionic compound that dissolves in the aqueous layer and is removed. The ether now contains only naphthalene (**4**) and aniline (**3**). Shaking it with dilute hydrochloric acid removes the aniline, a base, as the ionic anilinium chloride (**7**). The aqueous layer is removed. Evaporation of the *t*-butyl methyl ether now leaves naphthalene (**4**), the neutral compound. The other three compounds are recovered by adding acid to the sodium benzoate (**5**) and sodium phenoxide (**6**) and base to the anilinium chloride (**7**) to regenerate the covalent compounds benzoic acid (**1**), phenol (**2**), and aniline (**3**).

The ability to separate strong acids from weak acids depends on the acidity constants of the acids and the basicity constants of the bases. In the first equation consider the ionization of benzoic acid, which has an equilibrium constant ($K_a$) of $6.8 \times 10^{25}$. The conversion of benzoic acid to the benzoate anion in equation 4 is governed by the equilibrium constant, $K$ (equation 5), obtained by combining equations 3 and 4.

The p$K_a$ of carbonic acid, $H_2CO_3$, is 6.35.

$$C_6H_5COOH + H_2O \rightleftharpoons C_6H_5COO^- + H_3O^+ \qquad (1)$$

$$K_a = \frac{[C_6H_5COO^-][H_3O^+]}{[C_6H_5COOH]} = 6.8 \times 10^{-5}, \text{p}K_a = 4.17 \qquad (2)$$

■ **FIG. 7.10**

A flow sheet for the separation of a strong acid, a weak acid, a neutral compound, and a base: benzoic acid, phenol, naphthalene, and aniline (this page). Acid/base reactions of the acidic and basic compounds (opposite page).

In reaction tube A,
1) Dissolve all four compounds in *t*-butyl methyl ether.
2) Mix with $HCO_3^-$, remove lower aqueous layer, and place it in tube B.

1) Add $OH^-$, mix, remove lower aqueous layer, and place it in tube C.

1) Add HCl, mix, remove lower aqueous layer, and place it in tube D.

1) Dry solution over anhydrous calcium chloride pellets, remove solvent, and evaporate it to dryness in a clean, dry, tared reaction tube.

**5**

Add HCl

**1**
**Benzoic acid**
Forms a white, crystalline precipitate that can be removed by filtration

**6**

Add HCl

**2**
**Phenol**

**7**

Add NaOH

**3**
**Aniline**

**4**
**Naphthalene**

Phenol and aniline each form oily layers on top of the aqueous layer. Extract each with *t*-butyl methyl ether: Add ether to the tube, mix, separate layers, dry the ether layer over anhydrous calcium chloride pellets, remove solution from drying agent, and evaporate the solvent.

**1**
**Benzoic acid**
$pK_a = 4.17$
Covalent, sol. in org. solvents

$+ Na^+HCO_3^-$ ⟶

**5**
**Sodium benzoate**
Ionic, sol. in water

$+ H_2O + CO_2$

**5**

$+ H^+Cl^-$ ⟶

**1**

$+ Na^+Cl^-$

**2**
**Phenol**
$pK_a = 10$
Covalent, sol. in org. solvents

$+ Na^+OH^-$ ⟶

**6**
**Sodium phenoxide**
Ionic, sol. in water

$+ H_2O$

**6**

$+ H^+Cl^-$ ⟶

**2**

$+ Na^+Cl^-$

**3**
**Aniline**
$pK_b = 9.30$
Covalent, sol. in org. solvents

$+ H^+Cl^-$ ⟶

**7**
**Anilinium chloride**
Ionic, sol. in water

**7**

$+ Na^+OH^-$ ⟶

**3**

$+ H_2O + Na^+Cl^-$

$$K_w = [H_3O^+][OH^-] = 10^{-14} \tag{3}$$

$$C_6H_5COOH + OH^- \rightleftharpoons C_6H_5COO^- + H_2O \tag{4}$$

$$K = \frac{[C_6H_5COO^-]}{[C_6H_5COOH][OH^-]} = \frac{K_a}{K_w} = \frac{6.8 \times 10^{-5}}{10^{-14}} = 3.2 \times 10^8 \tag{5}$$

If 99% of the benzoic acid is converted to $C_6H_5COO^-$,

$$\frac{[C_6H_5COO^-]}{[C_6H_5COOH]} = \frac{99}{1} \tag{6}$$

then from equation 5 the hydroxide ion concentration would need to be $6.8 \times 10^{-7}$ $M$. Because saturated $NaHCO_3$ has $[OH^- = 3 \times 10^{-4}$ $M$, the hydroxide ion concentration is high enough to convert benzoic acid completely to sodium benzoate.

For phenol, with a $K_a$ of $10^{-10}$, the minimum hydroxide ion concentration that will produce the phenoxide anion in 99% conversion is $10^{-2}$ $M$. The concentration of hydroxide in 10% sodium hydroxide solution is $10^{-1}$ $M$, and so phenol in strong base is entirely converted to the water-soluble salt.

## General Considerations

If acetic acid was used as the reaction solvent, it would also be distributed largely into the aqueous phase; if the reaction product is a neutral substance, however, the residual acetic acid in the ether can be removed by one washing with excess 5% sodium bicarbonate solution. If the reaction product is a higher molecular weight acid, for example, benzoic acid ($C_6H_5COOH$), it will stay in the ether layer, while acetic acid is being removed by repeated washing with water; the benzoic acid can then be separated from neutral byproducts by extraction with sodium bicarbonate or sodium hydroxide solution and acidification of the extract. Acids of high molecular weight are extracted only slowly by sodium bicarbonate, so sodium carbonate is used in its place; however, carbonate is more prone than bicarbonate to produce emulsions. Sometimes an emulsion in the lower layer can be settled by twirling the separatory funnel by its stem. An emulsion in the upper layer can be broken by grasping the funnel by the neck and swirling it. Because the tendency to emulsify increases with the removal of electrolytes and solvents, a little sodium chloride solution is added with each portion of wash water. If the layers are largely clear but an emulsion persists at the interface, the clear part of the water layer can be drawn off, and the emulsion run into a second funnel and shaken with fresh ether.

Liquid/liquid extraction and acid/base extraction are employed in the majority of organic reactions because it is unusual to have the product crystallize from the reaction mixture or to be able to distill the reaction product directly from the reaction mixture. In the research literature, one will often see the statement "the reaction mixture was worked up in the usual way," which implies an extraction process of the type described here. Good laboratory practice dictates, however, that the details of the process be written out.

## EXPERIMENTS

# 1. Separation of a Carboxylic Acid, a Phenol, and a Neutral Substance

**Benzoic acid**
mp 123°C, p$K_a$ 4.17

**1,4-Dimethoxybenzene
(Hydroquinone dimethyl ether)**
mp 57°C

*Online Study Center*

Photo: Extraction with Ether;
Video: Extraction with Ether

**4-*tert*-Butylphenol**
mp 101°C, p$K_a$ 10.17

A mixture of equal parts of a carboxylic acid, a phenol, and a neutral substance is to be separated by extraction from an ether solvent. Note carefully the procedure for this extraction. In the next experiment you are to work out your own extraction procedure. Your unknown will consist of either benzoic acid or 2-chlorobenzoic acid (the carboxylic acid), 4-*t*-butyl phenol or 4-bromophenol, and biphenyl or 1,4-dimethoxybenzene (the neutral substance). The object of this experiment is to identify the three substances in the mixture and to determine the percent recovery of each from the mixture.

## Procedure

> **IN THIS EXPERIMENT** three organic solids are separated by reaction with base followed by extraction. Bicarbonate converts carboxylic acids (but not phenols) to ions. Hydroxide ion converts phenols (as well as carboxylic acids) to ions. Ionic substances are soluble in water. The addition of hydrochloric acid to the aqueous ionic solutions regenerates nonionic substances. At each step you should ask yourself, "Have I converted a nonionic substance to an ionic one?" (or vice versa). The ionic substances will be in the aqueous layer; the nonionic ones will be in the organic layer.

Dissolve about 0.18 g of the mixture (record the exact weight) in 2 mL of *t*-butyl methyl ether or diethyl ether in a reaction tube (tube 1). Then add 1 mL of a saturated aqueous solution of sodium bicarbonate to the tube. Use the graduations on the side of the tube to measure the amounts because they do not need to be exact. Mix the contents of the tube thoroughly by pulling the two layers into a Pasteur pipette and expelling them forcefully into the reaction tube. Do this for about 3 min. Allow the layers to separate completely and then draw off the lower layer into another reaction tube (tube 2). Add another 0.15 mL of sodium bicarbonate solution to the tube, mix the contents as before, and add the lower layer to tube 2. Exactly which chemical species is in tube 2? Add 0.2 mL of ether to tube 2, mix it thoroughly, remove the ether layer, and discard it. This is called *backwashing* and serves to remove any organic material that might contaminate the contents of tube 2.

Add 1.0 mL of 3 *M* aqueous sodium hydroxide to tube 1, shake the mixture thoroughly, allow the layers to separate, draw off the lower layer using a clean Pasteur pipette, and place it in tube 3. Extract tube 1 with two 0.15-mL portions of water, and add these to tube 3. Backwash the contents of tube 3 with 0.15 mL of ether and discard the ether wash just as was done for tube 2. Exactly which chemical species is in tube 3?

To tube 1 add saturated sodium chloride solution, mix, remove the aqueous layer, and then add to the ether anhydrous calcium chloride pellets until the drying

⚠️

Add HCl with care; $CO_2$ is released.

OH

**4-Bromophenol**
mp 66°C, p$K_a$ 10.2

**Biphenyl**
mp 71°C

COOH

**2-Chlorobenzoic acid**
mp 141°C, p$K_a$ 2.92

The best way to remove the solvent: under a gentle stream of air or nitrogen.

agent no longer clumps together. Wash it off with ether after the drying process is finished. Allow 5–10 min for the drying of the ether solution.

Using the concentration information given in the inside back cover of this book, calculate exactly how much concentrated hydrochloric acid is needed to neutralize the contents of tube 2. Then, by dropwise addition of concentrated hydrochloric acid, carry out this neutralization while testing the solution with litmus paper. An excess of hydrochloric acid does no harm. This reaction must be carried out with *extreme care* because much carbon dioxide is released in the neutralization. Add a boiling stick to the tube and very cautiously heat the tube to bring most of the solid carboxylic acid into solution. Allow the tube to cool slowly to room temperature and then cool it in ice. Remove the solvent with a Pasteur pipette and recrystallize the residue from boiling water. Again, allow the tube to cool slowly to room temperature and then cool it in ice. At the appropriate time, stir the crystals and collect them on a Hirsch funnel using the procedures detailed in Chapter 4. The crystals can be transferred and washed on the funnel using a small quantity of ice water. The solubility of benzoic acid in water is 1.9 g/L at 0°C and 68 g/L at 95°C. The solubility of chlorobenzoic acid is similar. Turn the crystals out onto a tared piece of paper, allow them to dry thoroughly, and determine the percent recovery of the acid. Assess the purity of the product by checking its melting point.

In exactly the same way, neutralize the contents of tube 3 with concentrated hydrochloric acid. This time, of course, there will be no carbon dioxide evolution. Again, heat the tube to bring most of the material into solution, allow it to cool slowly, remove the solvent, and recrystallize the phenol from boiling water. At the appropriate time, after the product has cooled slowly to room temperature and then in ice, it is also collected on a Hirsch funnel, washed with a very small quantity of ice water, and allowed to dry. The percent recovery and melting point are determined.

The neutral compound is recovered using the Pasteur pipette to remove the ether from the drying agent and to transfer it to a tared reaction tube. The drying agent is washed two or three times with additional ether to ensure complete transfer of the product.

Evaporate the solvent by placing the tube in a warm water bath and directing a stream of nitrogen or air onto the surface of the ether in the hood (*see* Fig. 7.8 on page 146). An aspirator tube can also be used for this purpose. Determine the weight of the crude product and then recrystallize it from methanol-water if it is the low-melting compound. Reread Chapter 4 for detailed instructions on carrying out the process of recrystallization from a mixed solvent. The product is dissolved in about 0.5–1 mL of methanol, and water is added until the solution gets cloudy, which indicates that the solution is saturated. This process is best carried out while heating the tube in a hot water bath at 50°C. Allow the tube to cool slowly to room temperature and then cool it thoroughly in ice. If you have the high-melting compound, recrystallize it from ethanol (8 mL/g).

The products are best isolated by collection on a Hirsch funnel using an ice-cold alcohol–water mixture to transfer and wash the compounds. Determine the percent recovery and the melting point. Turn in the products in neatly labeled $1\frac{1}{2}$-in. $\times$ $1\frac{1}{2}$-in. (4 cm $\times$ 4 cm) ziplock plastic bags attached to the laboratory report. If the yield on recrystallization is low, concentrate the filtrate (the mother liquor) and obtain a second crop of crystals.

# 2. Separation of Neutral and Basic Substances

**IN THIS EXPERIMENT** remember that hydrochloric acid will convert a nonionic amine to an ionic substance and that base will regenerate the nonionic substance from the ionic form. The nonionic and neutral substances are ether soluble, and the ionic substances will be found in the aqueous layer.

A mixture of equal parts of a neutral substance (naphthalene or 1,4-dichlorobenzene) and a basic substance (4-chloroaniline or ethyl 4-aminobenzoate) is to be separated by extraction from an ether solution. Naphthalene and 1,4-dichlorobenzene are completely insoluble in water. The bases will dissolve in hydrochloric acid, while the neutral compounds will remain in ether solution. The bases are insoluble in cold water but will dissolve to some extent in hot water and are soluble in ethanol. Naphthalene and 1,4-dichlorobenzene can be purified as described in Chapter 4. They also sublime very easily. Keep the samples covered.

*Ether vapors are heavier than air and can travel along bench tops, run down drain troughs, and collect in sinks. Be extremely careful to avoid flames when working with volatile ethers.*

**Naphthalene**
mp 82°C

**1,4-Dichlorobenzene**
mp 56°C

**4-Chloroaniline**
mp 68–71°C, $pK_b$ 4.15

**Ethyl 4-aminobenzoate**
mp 90°C, $pK_b$ 4.92

Plan a step-by-step procedure for separating 200 mg of the mixture into its components and have the plan checked by your instructor before proceeding. A flow sheet is a convenient way to present the plan. Select the correct solvent or mixture of solvents for the recrystallization of the bases on the basis of solubility tests. Determine the weights and melting points of the isolated and purified products and calculate the percent recovery of each. Turn in the products in neatly labeled vials or 1 1/2-in. × 1 1/2-in. ziplock plastic bags attached to the report.

**Cleaning Up.** Combine all aqueous filtrates and solutions, neutralize them, and flush the resulting solution down the drain. Used ether should be placed in the organic solvents container, and the drying agent, once the solvent has evaporated from it, can be placed in the nonhazardous solid waste container. If local regulations do not allow for the evaporation of solvents in a hood, dispose of the wet sodium sulfate in a special container. Any 4-chloroaniline or 1,4-dichlorobenzene should be placed in the halogenated waste container.

# 3. Separation of Acidic and Neutral Substances

$pH = -\log [H^+]$
$pK_a$ = acidity constant
$pK_b$ = basicity constant

A mixture of equal proportions of benzoic acid, 4-*t*-butylphenol, and 1,4-dimethoxybenzene is to be separated by extraction from *t*-butyl methyl ether. Note the detailed directions for extraction carefully. Prepare a flow sheet (*see* Fig. 7.10 on page 148) for this sequence of operations. In the next experiment you will work out your own extraction procedure.

### Procedure

Extinguish all flames when working with *t*-butyl methyl ether! The best method for removing the ether is by simple distillation. Dispose of waste ether in the container provided.

**Benzoic acid**
mp 123°C, $pK_a$ 4.17

**4-*tert*-Butylphenol**
mp 101°C, $pK_a$ 10.17

Dissolve 3 g of the mixture in 30 mL of *t*-butyl methyl ether and transfer the mixture to a 125-mL separatory funnel (*see* Fig. 7.2 on page 139) using a little *t*-butyl methyl ether to complete the transfer. Add 10 mL of water and note which layer is organic and which is aqueous. Add 10 mL of a 3 *M* aqueous solution of sodium bicarbonate to the funnel. Swirl or stir the mixture to allow carbon dioxide to escape. Stopper the funnel and cautiously mix the contents. Vent the liberated carbon dioxide and then shake the mixture thoroughly with frequent venting of the funnel. Repeat the process with another 10 mL of bicarbonate solution. Allow the layers to separate completely and then draw off the lower layer into a 50-mL Erlenmeyer flask (labeled flask 1). What does this layer contain?

Add 10 mL of 1.5 *M* aqueous sodium hydroxide to the separatory funnel, shake the mixture thoroughly, allow the layers to separate, and draw off the lower layer into a 50-mL Erlenmeyer flask (labeled flask 2). Repeat the process with another 10 mL of base. Then add an additional 5 mL of water to the separatory funnel, shake the mixture as before, and add this to flask 2. What does flask 2 contain?

Add 15 mL of a saturated aqueous solution of sodium chloride to the separatory funnel, shake the mixture thoroughly, allow the layers to separate, and draw off the lower layer, which can be discarded. What is the purpose of adding saturated sodium chloride solution? Carefully pour the ether layer into a 50-mL Erlenmeyer flask (labeled flask 3) from the top of the separatory funnel, taking great care not to allow any water droplets to be transferred. Add about 4 g of anhydrous calcium chloride pellets to the ether extract and set it aside.

Acidify the contents of flask 2 by dropwise addition of concentrated hydrochloric acid while testing with litmus paper. Cool the flask in an ice bath.

*Cautiously* add concentrated hydrochloric acid dropwise to flask 1 until the contents are acidic to litmus and then cool the flask in ice.

Decant (pour off) the ether from flask 3 into a tared flask, making sure to leave all of the drying agent behind. Wash the drying agent with additional ether to ensure complete transfer of the product. If decantation is difficult, then remove the drying agent by gravity filtration (*see* Fig. 4.6 on page 69). Put a boiling stick in the flask and evaporate the ether in the hood. An aspirator tube can be used for

**■ FIG. 7.11**
An aspirator tube in use.

Aspirator tube
To water aspirator
Clamp here
Tared flask
Boiling stick
Steam in
Outlet to sink
Steam bath (or sand bath)

this purpose (Fig. 7.11). Determine the weight of the crude *p*-dimethoxybenzene and then recrystallize it from methanol. See Chapter 4 for detailed instructions on how to carry out recrystallization.

Isolate the *t*-butylphenol from flask 2, employing vacuum filtration on a Hirsch funnel, and wash it on the filter with a small quantity of ice water. Determine the weight of the crude product and then recrystallize it from ethanol. Similarly isolate, weigh, and recrystallize from boiling water the benzoic acid in flask 1. The solubility of benzoic acid in water is 1.9 g/L at 0°C and 68 g/L at 95°C.

Dry the purified products, determine their melting points and weights, and calculate the percent recovery of each substance, bearing in mind that the original mixture contained 1 g of each compound. Hand in the three products in neatly labeled vials.

**Cleaning Up.** Combine all aqueous layers, washes, and filtrates. Dilute with water, neutralize using either sodium carbonate or dilute hydrochloric acid. This material can then be flushed down the drain with excess water. Methanol filtrate and any ether go in the organic solvents container. Allow ether to evaporate from the calcium chloride in the hood. Then place the calcium chloride in the nonhazardous solid waste container. If local regulations do not allow for the evaporation of solvents in a hood, dispose of the wet sodium sulfate in a special container.

*Online Study Center*

Video: Microscale Filtration on the Hirsch Funnel

OCH$_3$

OCH$_3$

**1,4-Dimethoxybenzene (Hydroquinone dimethyl ether)**
mp 57°C

**Naphthalene**
mp 82°C

NH$_2$

Cl

**4-Chloroaniline**
mp 68–71°C, p$K_b$ 10.0

# 4. Separation of Neutral and Basic Substances

A mixture of equal parts of a neutral substance (naphthalene) and a basic substance (4-chloroaniline) is to be separated by extraction from *t*-butyl methyl ether solution. The base will dissolve in hydrochloric acid, whereas the neutral naphthalene will remain in the *t*-butyl methyl ether solution. 4-Chloroaniline is insoluble in cold water but will dissolve to some extent in hot water and is soluble in ethanol. Naphthalene can be purified as described in Chapter 4.

Plan a procedure for separating 2.0 g of the mixture into its components and have the plan checked by your instructor before proceeding. A flow sheet is a convenient way to present the plan. Using solubility tests, select the correct solvent or mixture of solvents to recrystallize 4-chloroaniline. Determine the weights and melting points of the isolated and purified products and calculate the percent recovery of each. Turn in the products in neatly labeled vials.

**Cleaning Up.** Combine all aqueous filtrates and solutions, neutralize them, and flush the resulting solution down the drain with a large excess of water. Used *t*-butyl methyl ether should be placed in the organic solvents container, and the drying agent, once the solvent has evaporated from it, can be placed in the nonhazardous solid waste container. If local regulations do not allow for the evaporation of solvents in a hood, dispose of the wet sodium sulfate in a special container. Any 4-chloroaniline should be placed in the chlorinated organic compounds container.

⚠ Handle aromatic amines with care. Most are toxic, and some are carcinogenic. Avoid breathing the dust and vapor from the solid and keep the compounds off the skin, which is best done by wearing nitrile gloves.

# 5. Extraction and Purification of Components in an Analgesic Tablet

> **IN THIS EXPERIMENT** a powdered analgesic tablet, Excedrin, is boiled with dichloromethane and filtered. The solid on the filter is boiled with ethanol, which dissolves everything but the binder. The ethanol is evaporated and from the hot solution acetaminophen recrystallizes. The dichloromethane solution is shaken with base that converts aspirin to the water-soluble carboxylate anion. The dichloromethane is then evaporated to give caffeine that is purified by sublimation. The aqueous carboxylate anion solution is made acidic, which frees aspirin; warming the mixture and allowing it to cool allows aspirin to recrystallize. It is isolated by filtration.

Excedrin contains aspirin, caffeine, and acetaminophen as determined by thin-layer chromatography (TLC; *see* Chapter 8) or high performance liquid chromatography. A tablet is held together with a binder to prevent the components from crumbling when stored or while being swallowed. A close reading of the contents on the package will disclose the nature of the binder. Starch is commonly used, as is microcrystalline cellulose or silica gel. All of these have one property in common: They are insoluble in water and common organic solvents.

**Aspirin**
**(Acetylsalicylic acid)**
mp 135°C

**Acetaminophen**
**(*p*-Hydroxyacetanilide)**
mp 169–170.5°C

**Caffeine**
mp 238°C

**TABLE 7.2 • Solubilities**

| | Water | Ethanol | Chloroform | Diethyl ether | Ligroin |
|---|---|---|---|---|---|
| Aspirin | 0.33 g/100 mL at 25°C; 1 g/100 mL at 37°C | 1 g/5 mL | 1 g/17 mL | 1 g/13 mL | |
| Acetaminophen | v. sl. sol. cold; sol. hot | sol. | ins. | sl. sol. | ins. |
| Caffeine | 1 g/46 mL at 25°C; 1 g/5.5 mL at 80°C; 1 g/1.5 mL at 100°C | 1 g/66 mL at 25°C; 1 g/22 mL at 60°C | 1 g/5.5 mL | 1 g/530 mL | sl. sol. |

Inspection of the structures of caffeine, acetylsalicylic acid, and acetaminophen reveals that one is a base, one is a strong organic acid, and one is a weak organic acid. It might be tempting to separate this mixture using exactly the same procedure employed in separating benzoic acid, 4-*t*-butylphenol, and 1,4-dimethoxybenzene (experiment 3)—that is, dissolve the mixture in dichloromethane; separate the strongly acidic component by reaction with bicarbonate ion, a weak base; then remove the weakly acidic component by reaction with hydroxide, a strong base. This process would leave the neutral compound in the dichloromethane solution.

In the present experiment the solubility data (see Table 7.2) reveal that the weak acid, acetaminophen, is not soluble in ether, chloroform, or dichloromethane, so it cannot be extracted by a strong base. We can take advantage of this lack of solubility by dissolving the other two components, caffeine and aspirin, in dichloromethane and removing the acetaminophen by filtration. The binder is also insoluble in dichloromethane, but treatment of the solid mixture with ethanol will dissolve the acetaminophen and not the binder. These can then be separated by filtration, with the acetaminophen isolated by evaporation of the ethanol.

This experiment is a test of technique. It is not easy to separate and recrystallize a few milligrams of a compound that occurs in a mixture.

### Microscale Procedure

In a mortar, grind an Extra Strength Excedrin tablet to a very fine powder. The label states that this analgesic contains 250 mg of aspirin, 250 mg of acetaminophen, and 65 mg of caffeine per tablet. Place 300 mg of this powder in a reaction tube and add 2 mL of dichloromethane. Warm the mixture briefly and note that a large part of the material does not dissolve. Filter the mixture on a microscale Büchner funnel (the base of a chromatography column; Fig. 7.12) into another reaction tube. This is done by transferring the slurry to the funnel with a Pasteur pipette and completing the transfer with a small portion of dichloromethane. This filtrate is solution 1. A Hirsch funnel (Fig. 7.13) or a Wilfilter (Fig. 7.14) can also be used for this procedure. Pressure filtration is another alternative.

Transfer the powder on the filter to a reaction tube, add 1 mL of ethanol, and heat the mixture to boiling on the sand bath (with a boiling stick). Not all the material will go into solution. That which does not is the binder. Filter the mixture on

⚠️ Handle dichloromethane in the hood. It is a suspected carcinogen.

**Online Study Center**

Photos: Micro Büchner Funnel, Vacuum Filtration into Reaction Tube through Hirsch Funnel, Use of the Wilfilter

The binder can be starch, microcrystalline cellulose, or silica gel.

■ **FIG. 7.12**
A microscale Büchner funnel assembly.

Thermometer adapter

Micro Büchner funnel with fritted disk

To aspirator

Vacuum-tight connector

Reaction tube with filtrate

■ **FIG. 7.13**
A Hirsch funnel with an integral adapter, a polyethylene frit, and a 25-mL filter flask.

Filter paper, 12 mm dia.

Polyethylene filter disk (frit), 10 mm dia.

Hirsch funnel

To aspirator

25-mL Filter flask

■ **FIG. 7.14**
The Wilfilter filtration apparatus. *See* Chapter 4 for usage information.

Crystals

Wilfilter

Filtrate

*Online Study Center*

Photos: Vacuum Filtration into Reaction Tube through Hirsch Funnel, Micro Büchner Funnel

Acetaminophen

*Online Study Center*

Video: Filtration of Crystals Using the Pasteur Pipette

Check product purity by TLC (Chapter 8) using 25:1 ethyl acetate–acetic acid to elute the silica gel plates.

the same microscale Büchner funnel into a tared reaction tube and complete the transfer and washing using a few drops of hot ethanol.

Evaporate about two-thirds of the filtrate by boiling off the ethanol or, better, by warming the solution and blowing a stream of air into the reaction tube. Heat the residue to boiling (add a boiling stick to prevent bumping) and, if necessary, add more ethanol to bring the solid into solution. Allow the saturated solution to cool slowly to room temperature to deposit crystals of acetaminophen, which is reported to melt at 169°C–170.5°C. After the mixture has cooled to room temperature, cool it in ice for several minutes, remove the solvent with a Pasteur pipette, wash the crystals once with 2 drops of ice-cold ethanol, remove the ethanol, and dry the crystals under aspirator vacuum while heating the tube on a steam or sand bath.

Alternatively, the original ethanol solution can be evaporated to dryness, and the residue recrystallized from boiling water. The crystals can be collected on a Hirsch funnel (Fig. 7.13) or by use of a Wilfilter (Fig. 7.14). Once the crystals are dry, determine their weight and melting point. TLC analysis (see Chapter 8) and a determination of the melting points of these crystals and the two other components of this mixture will indicate their purity.

**Online Study Center**

Video: Recrystallization; Photos: Vacuum Filtration into Reaction Tube through Hirsch Funnel, Use of the Wilfilter

Video: Extraction with Dichloromethane

Photos: Sublimation Apparatus, Filtration Using a Pasteur Pipette; Videos: Recrystallization, Filtration Using a Pasteur Pipette

Caffeine

See Figure 4.13 on page 73 for drying crystals under vacuum.

The dichloromethane filtered from the binder and acetaminophen mixture (solution 1) should contain caffeine and aspirin. These can be separated by extraction either with acid (which will remove the caffeine as a water-soluble salt) or with base (which will remove the aspirin as a water-soluble salt). We shall use the latter procedure.

To the dichloromethane solution in a reaction tube, add 1 mL of 3 *M* sodium hydroxide solution and shake the mixture thoroughly. Remove the aqueous layer, add 0.2 mL more water, shake the mixture thoroughly, and again remove the aqueous layer, which is combined with the first aqueous extract.

To the dichloromethane, add calcium chloride pellets until the drying agent no longer clumps together. Shake the mixture over a 5-min to 10-min period to complete the drying process; then remove the solvent, wash the drying agent with more solvent, and evaporate the combined extracts to dryness under a stream of air to leave crude caffeine.

The caffeine can be purified by sublimation (Fig. 7.15) or by recrystallization. Recrystallize the caffeine by dissolving it in the minimum quantity of 30% ethanol in tetrahydrofuran. It also can be recrystallized by dissolving the product in a minimum quantity of hot toluene or acetone and adding to this solution ligroin (hexanes) until the solution is cloudy while at the boiling point. In any case, allow the solution to cool slowly to room temperature; then cool the mixture in ice and remove the solvent from the crystals with a Pasteur pipette. Remove the remainder of the solvent under aspirator vacuum and determine the weight of the caffeine and its melting point.

The aqueous hydroxide extract contains aspirin as the sodium salt of the carboxylic acid. To the aqueous solution, add 3 *M* hydrochloric acid dropwise until the solution tests strongly acid with indicator paper; then add 2 more drops of acid. This will give a suspension of white acetylsalicylic acid in the aqueous solution. It could be filtered off and recrystallized from boiling water, but this would cause transfer losses. An easier procedure is to heat the aqueous solution that contains the precipitated aspirin.

Add a boiling stick and heat the mixture to boiling (Fig. 7.16), at which time the aspirin should dissolve completely. If it does not, add more water. Long boiling will hydrolyze the aspirin to salicylic acid (mp 157°C–159°C). Once completely dissolved, the aspirin should be allowed to recrystallize slowly as the solution cools to room temperature in an insulated container. Once the tube has reached room temperature, it should be cooled in ice for several minutes, and then the solvent is removed with a Pasteur pipette. Wash the crystals with a few drops of ice-cold water and isolate them with a Wilfilter or scrape them out onto a piece of filter paper. Squeezing the crystals between sheets of filter paper will hasten drying. Once these crystals are completely dry, determine the weight of the acetylsalicylic acid and its melting point.

**Online Study Center**

Photos: Filtration Using a Pasteur Pipette, Use of the Wilfilter; Videos: Recrystallization, Filtration of Crystals Using the Pasteur Pipette

Aspirin

Alternative procedure: Use a Hirsch funnel or a Wilfilter to isolate the aspirin.

**Cleaning Up.** Place any dichloromethane-containing solutions in the halogenated organic waste container and the other organic liquids in the organic solvents container. The aqueous layers should be diluted and neutralized with sodium carbonate before being flushed down the drain. After it is free of solvent, the calcium

■ **FIG. 7.15**
A sublimation apparatus.

15-mL Centrifuge
tube, to be filled
with ice

Adapter (Pluro stopper)

Pipette
bulb

25-mL Filter flask

Material to
be sublimed

Sublimate

Heat source

■ **FIG. 7.16**
Rerystallization in a reaction tube.

Boiling stick ——

Cool at this poin

Air condenser –

Boiling solvent –

Temperature
controlled by
depth in sand —

chloride can be placed in the nonhazardous solid waste container. If local regulations do not allow for the evaporation of solvents in a hood, dispose of the wet
sodium sulfate in a special container.

## *Macroscale Procedure*

⚠

Handle dichloromethane
in the hood. It is a suspected
carcinogen.

The binder can be starch,
microcrystalline cellulose, or
silica gel.

Acetaminophen

In a mortar, grind two Extra Strength Excedrin tablets to a very fine powder. The
label states that this analgesic contains 250 mg of aspirin, 250 mg of acetaminophen, and 65 mg of caffeine per tablet. Place this powder in a test tube and add
7.5 mL of dichloromethane. Warm the mixture briefly and note that a large part of
the material does not dissolve. Filter the mixture into another test tube. This can
be done by transferring the slurry to a funnel equipped with a piece of filter paper.
Use a Pasteur pipette and complete the transfer with a small portion of
dichloromethane. This filtrate is solution 1.

Transfer the powder on the filter to a test tube, add 4 mL of ethanol, and heat
the mixture to boiling (with a boiling stick). Not all of the material will go into solution. That which does not is the binder. Filter the mixture into a tared test tube
and complete the transfer and washing by using a few drops of hot ethanol. This
is solution 2.

Evaporate about two-thirds of solution 2 by boiling off the ethanol (with a
boiling stick) or, better, by warming the solution and blowing a stream of air into
the test tube. Heat the residue to boiling (add a boiling stick to prevent bumping)
and add more ethanol, if necessary, to bring the solid into solution. Allow the saturated solution to cool slowly to room temperature to deposit crystals of acetaminophen, which is reported to melt at 169°C–170.5°C. After the mixture has

Check product purity by TLC using 25:1 ethyl acetate–acetic acid to elute the silica gel plates.

An alternative to shaking is pipette mixing or using a vortex stirrer, if available.

Caffeine

**Sodium acetylsalicylate**
(soluble in water)

Aspirin

Alternative procedure: Use a Hirsch funnel (*see* Fig. 4.14 on page 74) to isolate the aspirin.

cooled to room temperature, cool it in ice for several minutes, remove the solvent with a Pasteur pipette, wash the crystals once with 2 drops of ice-cold ethanol, remove the ethanol, and dry the crystals under aspirator vacuum while heating the tube on a steam or sand bath.

Alternatively, the original ethanol solution is evaporated to dryness, and the residue is recrystallized from boiling water. The crystals are best collected and dried on a Hirsch funnel (*see* Fig. 7.13 on page 158). Once the crystals are dry, determine their weight and melting point. TLC analysis (*see* Chapter 8) and a determination of the melting points of these crystals and the two other components of this mixture will indicate their purity.

The dichloromethane filtered from the binder and acetaminophen mixture (solution 1) should contain caffeine and aspirin. These can be separated by extraction either with acid (which will remove the caffeine as a water-soluble salt) or with base (which will remove the aspirin as a water-soluble salt). We shall use the latter procedure.

To the dichloromethane solution in a test tube, add 4 mL of 3 *M* sodium hydroxide solution and shake the mixture thoroughly. Remove the aqueous layer, add 1 mL more water, shake the mixture thoroughly, and again remove the aqueous layer, which is combined with the first aqueous extract.

To the dichloromethane add anhydrous calcium chloride pellets until the drying agent no longer clumps together. Shake the mixture over a 5–10-min period to complete the drying process, then remove the solvent, wash the drying agent with more solvent, and evaporate the combined extracts to dryness under a stream of air to leave crude caffeine.

The caffeine can be purified by sublimation or by recrystallization. Recrystallize the caffeine by dissolving it in the minimum quantity of 30% ethanol in tetrahydrofuran. It can also be recrystallized by dissolving the product in a minimum quantity of hot toluene or acetone and adding hexanes to this solution until the solution is cloudy while at the boiling point. In any case, allow the solution to cool slowly to room temperature; then cool the mixture in ice and remove the solvent from the crystals with a Pasteur pipette. Remove the remainder of the solvent under aspirator vacuum and determine the weight of the caffeine and its melting point.

The aqueous hydroxide extract contains aspirin as the sodium salt of the carboxylic acid. To the aqueous solution add 3 *M* hydrochloric acid dropwise until the solution tests strongly acid with indicator paper; then add 2 more drops of acid. This will give a suspension of white acetylsalicylic acid in the aqueous solution. It could be filtered off and recrystallized from boiling water, but this would cause transfer losses. An easier procedure is to simply heat the aqueous solution that contains the precipitated aspirin and allow it to recrystallize on slow cooling.

Add a boiling stick and heat the mixture to boiling, at which time the aspirin should dissolve completely. If it does not, add more water. Long boiling will hydrolyze the aspirin to salicylic acid (mp 157°C–159°C). Once completely dissolved, the aspirin should be allowed to recrystallize slowly as the solution cools to room temperature in an insulated container. Once the tube has reached room temperature, it should be cooled in ice for several minutes, and then the solvent is

*Online Study Center*

Photo: Filtration Using a
Pasteur Pipette; Video: Filtration
of Crystals Using the Pasteur
Pipette

removed with a Pasteur pipette. The crystals are to be washed with a few drops of ice-cold water and then scraped out onto a piece of filter paper. Squeezing the crystals between sheets of the filter paper will hasten drying. Once these crystals are completely dry, determine the weight of the acetylsalicylic acid and its melting point.

**Cleaning Up.** Place any dichloromethane-containing solutions in the halogenated organic waste container and the other organic liquids in the organic solvents container. The aqueous layers should be diluted and neutralized with sodium carbonate before being flushed down the drain. After it is free of solvent, the calcium chloride can be placed in the nonhazardous solid waste container. If local regulations do not allow for the evaporation of solvents in a hood, dispose of the wet sodium sulfate in a special container.

# Extractions from Common Items

## 6. Extraction of Caffeine from Tea

Tea and coffee have been popular beverages for centuries, primarily because they contain caffeine, a stimulant. Caffeine stimulates respiration, the heart, and the central nervous system, and it is a diuretic (i.e., it promotes urination). It can cause nervousness and insomnia and, like many drugs, can be addictive, making it difficult to reduce the daily dose. A regular coffee drinker who consumes just 4 cups per day can experience headache, insomnia, and even nausea upon withdrawal from the drug. On the other hand, it helps people to pay attention and can sharpen moderately complex mental skills as well as prolong the ability to exercise.

Caffeine may be the most widely abused drug in the United States. During the course of a day, an average person may unwittingly consume up to 1 g of caffeine. The caffeine content of some common foods and drugs is given in Table 7.3.

Caffeine belongs to a large class of compounds known as alkaloids. These are of plant origin, contain basic nitrogen, often have a bitter taste and a complex structure, and usually have physiological activity. Their names usually end in *-ine*; many are quite familiar by name if not chemical structure—for example, nicotine, cocaine, morphine, and strychnine.

Tea leaves contain tannins, which are acidic, as well as a number of colored compounds and a small amount of undecomposed chlorophyll (soluble in dichloromethane). To ensure that the acidic substances remain water soluble and that the caffeine will be present as the free base, sodium carbonate is added to the extraction medium.

The solubility of caffeine in water is 2.2 mg/mL at 25°C, 180 mg/mL at 80°C, and 670 mg/mL at 100°C. It is quite soluble in dichloromethane, the solvent used in this experiment to extract the caffeine from water.

Caffeine can be easily extracted from tea bags. The procedure one would use to make a cup of tea—simply "steeping" the tea with very hot water for about 7 min—extracts most of the caffeine. There is no advantage to boiling the tea leaves with water for 20 min. Because caffeine is a white, slightly bitter, odorless,

| TABLE 7.3  •  Caffeine Content of Common Foods and Drugs | |
| --- | --- |
| Espresso | 120 mg per 2 oz |
| Coffee, regular, brewed | 103 mg per cup |
| Instant coffee | 57 mg per cup |
| Coffee, decaffeinated | 2–4 mg per cup |
| Tea | 30–75 mg per cup |
| Cocoa | 5–40 mg per cup |
| Milk chocolate | 6 mg per oz |
| Baking chocolate | 35 mg per oz |
| Coca-Cola, Classic | 46 mg per 12 oz |
| Jolt Cola | 72 mg per 12 oz |
| Anacin, Bromo-Seltzer, Midol | 32 mg per pill |
| Excedrin, Extra Strength | 65 mg per pill |
| Dexatrim, Dietac, Vivarin | 200 mg per pill |
| Dristan | 16 mg per pill |
| No-Doz | 100 mg per pill |

crystalline solid, it is obvious that water extracts more than just caffeine. When the brown aqueous solution is subsequently extracted with dichloromethane, caffeine primarily dissolves in the organic solvent. Evaporation of the solvent leaves crude caffeine, which on sublimation yields a relatively pure product. When the concentrated tea solution is extracted with dichloromethane, emulsions can form very easily. There are substances in tea that cause small droplets of the organic layer to remain suspended in the aqueous layer. This emulsion formation results from vigorous shaking. To avoid this problem, it might seem that one could boil the tea leaves with dichloromethane first and then extract the caffeine from the dichloromethane solution with water. In fact, this does not work. Boiling 25 g of tea leaves with 50 mL of dichloromethane gives only 0.05 g of residue after evaporation of the solvent. Subsequent extractions yield even less material. Hot water causes the tea leaves to swell and is obviously a far more efficient extraction solvent. An attempt to sublime caffeine directly from tea leaves is also unsuccessful.

### Microscale Procedure

In a 30-mL beaker place 15 mL of water, 2 g of sodium carbonate, and a wooden boiling stick. Bring the water to a boil on the sand bath, remove the boiling stick, and brew a very concentrated tea solution by immersing a tea bag (2.4 g tea) in the very hot water for 5 min. After the tea bag cools enough to handle, and being careful not to break the bag, squeeze as much water from the bag as possible. Again bring the water to a boil and add a new tea bag to the hot solution. After 5 min, remove the tea bag and squeeze out as much water as possible. This can be done easily on a Hirsch funnel. Rinse the bag with a few mL of very hot water but be

*Online Study Center*

Video: Caffeine from Tea

*9 or 10*

sure the total volume of aqueous extract does not exceed ~~12~~ mL. Pour the extract into a 15-mL centrifuge tube and cool the solution in ice to below 40°C (the boiling point of dichloromethane).

Using three 2-mL portions of dichloromethane, extract the caffeine from the tea. Cork the tube and use a gentle rocking motion to carry out the extraction. Vigorous shaking will produce an intractable emulsion, whereas extremely gentle mixing will fail to extract the caffeine. If you have ready access to a centrifuge, the shaking can be very vigorous because any emulsions formed can be broken fairly well by centrifugation for about 90 s. After each extraction, remove the lower organic layer into a reaction tube, leaving any emulsion layer behind. Dry the combined extracts over anhydrous calcium chloride pellets for 5–10 min in an Erlenmeyer flask. Add the drying agent in portions with shaking until it no longer clumps together. Transfer the dry solution to a tared 25-mL filter flask, wash the drying agent twice with 2-mL portions of dichloromethane, and evaporate it to dryness (*see* Fig. 7.9 on page 146). The residue will be crude caffeine (determine its weight), which is to be purified by sublimation.

Fit the filter flask with a Pluro stopper or no. 2 neoprene adapter through which is thrust a 15-mL centrifuge tube. Put a pipette bulb on the side arm. Clamp the flask with a large three-prong clamp, fill the centrifuge tube with ice and water, and heat the flask on a hot sand bath (*see* Fig. 7.15 on page 160). Caffeine is reported to sublime at about 170°C. Tilt the filter flask and rotate it in a hot sand bath to drive more caffeine onto the centrifuge tube. Use a heat gun to heat the upper walls of the filter flask. When sublimation ceases, remove the ice water from the centrifuge tube and allow the flask to cool somewhat before removing the centrifuge tube. Scrape the caffeine onto a tared weighing paper, weigh and, using a plastic funnel, transfer it to a small vial or a plastic bag. At the discretion of your instructor, determine the melting point with a sealed capillary. The melting point of caffeine is 238°C. Using the centrifugation technique to separate the extracts; about 30 mg of crude caffeine can be obtained. This will give you 10–15 mg of sublimed material, depending on the caffeine content of the particular tea being used. The isolated caffeine can be used to prepare caffeine salicylate (experiment 9).

**Cleaning Up.** Discard the tea bags in the nonhazardous solid waste container. Allow the solvent to evaporate from the drying agent and discard in the same container. If local regulations do not allow for the evaporation of solvents in a hood, dispose of the wet sodium sulfate in a special container. Place any unused and unrecovered dichloromethane in the chlorinated organic compounds container. The apparatus can be cleaned with soap and hot water. Caffeine can be flushed down the drain because it is biodegradable.

## Macroscale Procedure

To an Erlenmeyer flask containing 25 g of tea leaves (or 10 tea bags) and 20 g of sodium carbonate, add 225 mL of vigorously boiling water. Allow the mixture to stand for 7 min and then decant into another Erlenmeyer flask. To the hot tea leaves,

⚠️ **CAUTION**: Do not breathe the vapors of dichloromethane and, if possible, work with this solvent in the hood.

Balance the centrifuge tubes.

*Remove tubing before turning water off*

add another 50 mL of hot water and then immediately decant and combine with the first extract. Very little, if any, additional caffeine is extracted by boiling the tea leaves for 20 min. Decantation works nearly as well as vacuum filtration and is much faster.

Cool the aqueous solution to near room temperature and extract it twice with 30-mL portions of dichloromethane. Do not shake the separatory funnel so vigorously as to cause emulsion formation, bearing in mind that if it is not shaken vigorously enough the caffeine will not be extracted into the organic layer. Use a gentle rocking motion of the separatory funnel. Drain off the dichloromethane layer on the first extraction; include the emulsion layer on the second extraction. Dry the combined dichloromethane solutions and any emulsion layer with anhydrous calcium chloride pellets. Add sufficient drying agent until it no longer clumps together on the bottom of the flask. Carefully decant or filter the dichloromethane solution into a tared Erlenmeyer or distilling flask. Silicone-impregnated filter paper passes dichloromethane and retains water. Wash the drying agent with a further portion of solvent and evaporate or distill the solvent. A wood applicator stick is better than a boiling chip to promote smooth boiling because it is easily removed once the solvent is gone. The residue of greenish-white crystalline caffeine should weigh about 0.25 g.

### Recrystallization of Caffeine

**Online Study Center**

Video: Recrystallization

To recrystallize the caffeine, dissolve it in 5 mL of hot acetone, transfer it with a Pasteur pipette to a small Erlenmeyer flask and, while it is hot, add ligroin to the solution until a faint cloudiness appears. Set the flask aside and allow it to cool slowly to room temperature. This mixed-solvent method of recrystallization depends on the fact that caffeine is far more soluble in acetone than ligroin, so a combination of the two solvents can be found where the solution is saturated in caffeine and will appear cloudy. Cool the solution containing the crystals and remove them by vacuum filtration, employing a Hirsch funnel or a very small Büchner funnel. Use a few drops of ligroin to transfer and wash the crystals. If you wish to obtain a second crop of crystals, collect the filtrate in a test tube, concentrate it to the cloud point using an aspirator tube (*see* Fig. 7.11 on page 155), and repeat the recrystallization process.

**Cleaning Up.** The filtrate can be diluted with water and washed down the drain. Any dichloromethane collected goes into the halogenated organic waste container. After the solvent is allowed to evaporate from the drying agent in the hood, the drying agent can be placed in the nonhazardous solid waste container; otherwise it goes in the hazardous waste container. If local regulations do not allow for the evaporation of solvents in a hood, dispose of the wet sodium sulfate in a special container. The tea leaves go in the nonhazardous solid waste container.

## 7. Extraction of Caffeine from Cola Syrup

Coca-Cola was originally flavored with extracts from the leaves of the coca plant and the kola nut. Coca is grown in northern South America; the Indians of Peru and Bolivia have for centuries chewed the leaves to relieve the pangs of hunger and sensitivity to high mountain cold. The cocaine from the leaves causes local anesthesia of the stomach. It has limited use as a local anesthetic for surgery on the eye, nose, and throat. Unfortunately, it is now a widely abused and illicit drug. Kola nuts contain about 3% caffeine as well as a number of other alkaloids. The kola tree is in the same family as the cacao tree from which cocoa and chocolate are obtained. Modern cola drinks do not contain cocaine; however, Coca-Cola contains 46 mg of caffeine per 12-oz serving. The acidic taste of many soft drinks comes from citric, tartaric, phosphoric, and benzoic acids.

**Caffeine**
mp 238°C

**Cocaine**

Automatic soft drink dispensing machines mix a syrup with carbonated water. In the following experiment caffeine is extracted from concentrated cola syrup.

### Microscale Procedure

Add 1 mL of concentrated ammonium hydroxide to a mixture of 5 mL of commercial cola syrup and 5 mL of water in a 15-mL centrifuge tube. Add 1 mL of dichloromethane and tip the tube gently back and forth for 5 min. Do not shake the mixture as in a normal extraction because an emulsion will form, and the layers will not separate. After the layers have separated as much as possible, remove the clear lower layer, leaving the emulsion behind. Using 1.5 mL of dichloromethane, repeat the extraction in the same way two more times. At the final separation, include the emulsion layer with the dichloromethane. If a centrifuge is available, the mixture can be shaken vigorously, and the emulsion broken by centrifugation for 90 s. Combine the extracts in a reaction tube and dry the solution with anhydrous calcium chloride pellets. Add the drying agent with shaking until it no longer clumps together. After 5–10 min, remove the solution with a Pasteur pipette and place it in a tared filter flask. Wash off the drying agent with more dichloromethane and evaporate the mixture to dryness. Determine the crude weight of caffeine; then sublime it as described in the preceding experiment.

**⚠ CAUTION**: Do not breathe the vapor of dichloromethane. Work with this solvent in the hood.

### Macroscale Procedure

Add 10 mL of concentrated ammonium hydroxide to a mixture of 50 mL of commercial cola syrup and 50 mL of water. Place the mixture in a separatory funnel, add 50 mL of dichloromethane, and swirl the mixture and tip the funnel back and forth for at least 5 min. Do not shake the solutions together as in a normal extraction because an emulsion will form, and the layers will not separate. An emulsion is made up of droplets of one phase suspended in the other. (Milk is an emulsion.) Separate the layers. Repeat the extraction with a second 50-mL portion of dichloromethane. From your knowledge of the density of dichloromethane and water, you should be able to predict which is the top layer and which is the bottom layer. If in doubt, add a few drops of each layer to water. The aqueous layer will be soluble; the organic layer will not. Combine the dichloromethane extracts and any emulsion that has formed in a 125-mL Erlenmeyer flask; then add anhydrous calcium chloride pellets to remove water from the solution. Add the drying agent until it no longer clumps together at the bottom of the flask but swirls freely in solution. Swirl the flask with the drying agent from time to time over a 10-min period. Carefully decant (pour off) the dichloromethane or remove it by filtration through a fluted filter paper, add about 5 mL more solvent to the drying agent to wash it, and decant this also. Combine the dried dichloromethane solutions in a tared flask and remove the dichloromethane by distillation or evaporation on a steam or sand bath. Remember to add a wood applicator stick to the solution to promote even boiling. Determine the weight of the crude product.

Chlorinated solvents are toxic, insoluble in water, and expensive and should never be poured down the drain.

### Recrystallization of Caffeine

To recrystallize the caffeine, dissolve it in 5 mL of hot acetone, transfer it with a Pasteur pipette to a small Erlenmeyer flask, and, while it is hot, add ligroin to the solution until a faint cloudiness appears. Set the flask aside and allow it to cool slowly to room temperature. This mixed-solvent method of recrystallization depends on the fact that caffeine is far more soluble in acetone than ligroin, so a combination of the two solvents can be found where the solution is saturated in caffeine (the cloud point). Cool the solution containing the crystals and remove them by vacuum filtration, employing a Hirsch funnel or a very small Büchner funnel. Use a few drops of ligroin to transfer and wash the crystals. If you wish to obtain a second crop of crystals, collect the filtrate in a test tube, concentrate it to the cloud point using an aspirator tube (*see* Fig. 7.11 on page 155), and repeat the recrystallization process.

**Cleaning Up.** Combine all aqueous filtrates and solutions, neutralize them, and flush the resulting solution down the drain. Used dichloromethane should be placed in the halogenated waste container, and the drying agent, once the solvent has evaporated from it, can be placed in the nonhazardous solid waste container. If local regulations do not allow for the evaporation of solvents in a hood, dispose of the wet sodium sulfate in a special container. The ligroin-acetone filtrates should be placed in the organic solvents container.

**Sublimation of Caffeine.** Sublimation is a fast and easy way to purify caffeine. Using the apparatus depicted in Figure 7.15 (on page 160), sublime the crude caffeine at atmospheric pressure following the procedure in part 3 of Chapter 6.

*Online Study Center*
Photos of Extraction Procedure

# 8. Isolation of Caffeine from Instant Coffee

Instant coffee, according to manufacturers, contains between 55 mg and 62 mg of caffeine per 6-oz cup, and a cup is presumably made from a teaspoon of the powder, which weighs 1.3 g; so 2 g of the powder should contain 85–95 mg of caffeine. Unlike tea, however, coffee contains other compounds that are soluble in dichloromethane, so obtaining pure caffeine from coffee is not easy. The objective of this experiment is to extract instant coffee with dichloromethane (the easy part) and then to try to devise a procedure for obtaining pure caffeine from the extract.

From TLC analysis (*see* Chapter 8), you may deduce that certain impurities have a high $R_f$ value in hydrocarbons (in which caffeine is insoluble). Consult reference books (see especially the *Merck Index*[2]) to determine the solubility (and lack of solubility) of caffeine in various solvents. You might try trituration (grinding the crude solid with a solvent) to dissolve impurities preferentially. Column chromatography is another possible means of purifying the product. Or you might convert all of it to the salicylate and then regenerate the caffeine from the salicylate. Experiment! Or you can simply use the following procedure.

## Procedure

> **IN THIS EXPERIMENT** a very concentrated aqueous solution of coffee is prepared and shaken vigorously with an organic solvent to make an intractable emulsion that can be broken (separated into two layers) by centrifugation. Caffeine is isolated by recrystallization.

In a 10-mL Erlenmeyer flask, place 2 g of sodium carbonate and 2 g of instant coffee powder. Add 9 mL of boiling water, stir the mixture well, bring it to a boil again with stirring, cool it to room temperature, and then pour it into a 15-mL plastic centrifuge tube fitted with a screw cap. Add 2 mL of dichloromethane, cap the tube, shake it vigorously for 60 s; then centrifuge it at high speed for 90 s. Remove the clear yellow dichloromethane layer and place it in a 10-mL Erlenmeyer flask. Repeat this process twice more. To the combined extracts add anhydrous calcium chloride pellets until they no longer clump together, allow the solution to dry for a few minutes; then transfer it to a tared 25-mL filter flask and wash the drying agent with more solvent. Remove the solvent as was done in the tea extraction experiment and determine the weight of the crude caffeine. You should obtain about 60 mg of crude product. Sublimation of this orange powder gives an impure orange sublimate that smells strongly of coffee, so sublimation is not a good way to purify this material.

Caffeine has no odor.

Dissolve a very small quantity of the product in a drop of dichloromethane and perform a TLC analysis of the crude material. Dissolve the remainder of the material in 1 mL of boiling 95% ethanol; then dilute the mixture with 1 mL of

---

2. O'Neill, M. J.; Smith, A.; Heckelman, P. E.; Budavari, S., eds. *The Merck Index*, 13th ed.; Merck and Co., Inc.: Rahway, NJ, 2001.

*t*-butyl methyl ether, heat to boiling, and allow to cool slowly to room temperature. Long, needlelike crystals should form in the orange solution. Alternatively, recrystallize the product from a 1:1 mixture of ligroin (hexanes) and 2-propanol, using about 2 mL. Cool the mixture in ice for at least 10 min and then collect the product on a Hirsch funnel. Complete the transfer with the filtrate and then wash the crystals twice with cold 50/50 ethanol/*t*-butyl methyl ether. The yield of white fluffy needles of caffeine should be more than 30 mg.

**Cleaning Up.** Allow the solvent to evaporate from the drying agent and discard it in the nonhazardous waste container. If local regulations do not allow for the evaporation of solvents in a hood, dispose of the wet sodium sulfate in a special container. Place any unused and unrecovered dichloromethane in the chlorinated organic solvents container.

Preparation of a derivative of caffeine.

## 9. Caffeine Salicylate

One way to confirm the identity of an organic compound is to prepare a derivative of it. Caffeine melts and sublimes at 238°C. It is an organic base and can therefore accept a proton from an acid to form a salt. The salt formed when caffeine combines with hydrochloric acid, like many amine salts, does not have a sharp melting point; it merely decomposes when heated. But the salt formed from salicylic acid, even though ionic, has a sharp melting point and can thus be used to help characterize caffeine. Figure 7.17 is the $^1$H NMR spectrum of caffeine.

Caffeine + Salicylic acid ⟶ Caffeine salicylate

CAUTION: Petroleum ether is very flammable. Extinguish all flames.

Recrystallization from mixed solvents

*Online Study Center*

Video: Filtration of Crystals Using the Pasteur Pipette; Photo: Drying Crystals Under Vacuum

### Procedure

The quantities given can be multiplied by 5 or 10, if necessary. To 10 mg of sublimed caffeine in a tared reaction tube, add 7.5 mg of salicylic acid and 0.5 mL of dichloromethane. Heat the mixture to boiling and add petroleum ether (a poor solvent for the product) dropwise until the mixture just turns cloudy, indicating that the solution is saturated. If too much petroleum ether is added, then clarify it by adding a very small quantity of dichloromethane. Insulate the tube to allow it to cool slowly to room temperature; then cool it in ice. The needlelike crystals are isolated by removing the solvent with a Pasteur pipette while the reaction tube is in the ice bath. Evaporate the last traces of solvent under vacuum (Fig. 7.18) and

**FIG. 7.17**
The $^1$H NMR spectrum of caffeine (250 MHz).

**FIG. 7.18**
The drying of crystals under vacuum in beaker of warm water.

determine the weight of the derivative and its melting point. Caffeine salicylate is reported to melt at 137°C.

**Cleaning Up.** Place the filtrate in the halogenated organic solvents container.

## QUESTIONS

1. Suppose a reaction mixture, when diluted with water, afforded 300 mL of an aqueous solution of 30 g of the reaction product malononitrile [$CH_2(CN)_2$], which is to be isolated by extraction with ether. The solubility of malononitrile in ether at room temperature is 20.0 g/100 mL and in water is 13.3 g/100 mL. What weight of malononitrile would be recovered by extraction with (a) three 100-mL portions of ether and (b) one 300-mL portion of ether? *Suggestion*: For each extraction let $x$ equal the weight extracted into the ether layer. In part (a) the concentration in the ether layer is $x/100$ and in the water layer is $(30 - x)/300$; the ratio of these quantities is equal to $k = 20/13.3$.

2. Why is it necessary to remove the stopper from a separatory funnel when liquid is being drained from it through the stopcock?

3. The p$K_a$ of *p*-nitrophenol is 7.15. Would you expect this to dissolve in sodium bicarbonate solution? The p$K_a$ of 2,5-dinitrophenol is 5.15. Will it dissolve in bicarbonate solution?

4. The distribution coefficient, $k$ = conc. in ligroin ÷ conc. in water, between ligroin and water for solute A is 7.5. What weight of A would be removed from a solution of 10 g of A in 100 mL of water by a single extraction with 100 mL of ligroin? What weight of A would be removed by four successive extractions with 25-mL portions of ligroin? How much ligroin would be required to remove 98.5% of A in a single extraction?

5. In experiment 1, how many moles of benzoic acid are present? How many moles of sodium bicarbonate are contained in 1 mL of a 10% aqueous solution? (A 10% solution has 1 g of solute in 9 mL of solvent.) Is the amount of sodium bicarbonate sufficient to react with all of the benzoic acid?

6. To isolate benzoic acid from a bicarbonate solution, it is acidified with concentrated hydrochloric acid, as in experiment 1. What volume of acid is needed to neutralize the bicarbonate? The concentration of hydrochloric acid is expressed in various ways on the inside back cover of this laboratory manual.

7. How many moles of 4-*t*-butylphenol are in the mixture to be separated in experiment 1? How many moles of sodium hydroxide are contained in 1 mL of 5% sodium hydroxide solution? (Assume the density of the solution is 1.0.) What volume of concentrated hydrochloric acid is needed to neutralize this amount of sodium hydroxide solution?

8. Draw a flow sheet to show how you would separate the components of a mixture containing an acid substance, toluic acid, a basic substance, *p*-bromo-aniline, and anthracene, a neutral substance.

9. Write equations showing how caffeine could be extracted from an organic solvent and subsequently isolated.

10. Write equations showing how acetaminophen might be extracted from an organic solvent such as an ether, if it were soluble.

11. Write detailed equations showing the mechanism by which aspirin is hydrolyzed in boiling, slightly acidic water.

**Online Study Center**

**General Resources**
Web Links

# Thin-Layer Chromatography: Analyzing Analgesics and Isolating Lycopene from Tomato Paste

***Online Study Center***

This icon will direct you to techniques, equipment setups, and online resources at http://college.hmco.com/PIC/williamsonMMOE5e.

**PRELAB EXERCISE:** Based on the number and polarity of the functional groups in aspirin, acetaminophen, ibuprofen, and caffeine, whose structures are shown on page 184, predict which of these four compounds has the highest $R_f$ value and which has the lowest.

Chromatography is the separation of two or more compounds or ions caused by their molecular interactions with two phases—one moving and one stationary. These two phases can be a solid and a liquid, a liquid and a liquid, a gas and a solid, or a gas and a liquid. You very likely have seen chromatography carried out on paper towels or coffee filters to separate inks and food dyes. The cellulose paper is the stationary or solid phase, and a propanol-water mixture is the mobile or liquid phase. The samples are spotted near one edge of the paper, and this edge is dipped into the liquid phase. The solvent is drawn through the paper by capillary action, and the molecules are separated based on how they interact with the paper. Although there are several different forms of chromatography, the principles are essentially the same.

*Thin-layer chromatography (TLC)* is a sensitive, fast, simple, and inexpensive analytical technique that you will use repeatedly while carrying out organic experiments. It is a micro technique; as little as $10^{-9}$ g of material can be detected, although the usual sample size is from $1 \times 10^{-6}$ g to $1 \times 10^{-8}$ g. The stationary phase is normally a polar solid adsorbent, and the mobile phase can be a single solvent or a combination of solvents.

TLC requires micrograms of material.

## Uses of Thin-Layer Chromatography

1. **To determine the number of components in a mixture.** TLC affords a quick and easy method for analyzing such things as a crude reaction mixture, an extract from a plant substance, or the ingredients in a pill. Knowing the

number and relative amounts of the components aids in planning further analytical and separation steps.

2. **To determine the identity of two substances.**   If two substances spotted on the same TLC plate give spots in identical locations, they *may* be identical. If the spot positions are not the same, the substances cannot be the same. It is possible for two or more closely related but not identical compounds to have the same positions on a TLC plate. Changing the stationary or mobile phase will usually effect their separation.

3. **To monitor the progress of a reaction.**   By sampling a reaction at regular intervals, it is possible to watch the reactants disappear and the products appear using TLC. Thus, the optimum time to halt the reaction can be determined, and the effect of changing such variables as temperature, concentrations, and solvents can be followed without having to isolate the product.

4. **To determine the effectiveness of a purification.**   The effectiveness of distillation, crystallization, extraction, and other separation and purification methods can be monitored using TLC, with the caveat that a single spot does not guarantee a single substance.

5. **To determine the appropriate conditions for a column chromatographic separation.**   In general, TLC is not satisfactory for purifying and isolating macroscopic quantities of material; however, the adsorbents most commonly used for TLC—silica gel and alumina—are also used for column chromatography, which is discussed in Chapter 9. Column chromatography is used to separate and purify up to 1 g of a solid mixture. The correct adsorbent and solvent to use for column chromatography can be rapidly determined by TLC.

6. **To monitor column chromatography.**   As column chromatography is carried out, the solvent is collected in a number of small flasks. Unless the desired compound is colored, the various fractions must be analyzed in some way to determine which ones have the desired components of the mixture. TLC is a fast and effective method for doing this.

## The Principles of Chromatography

To thoroughly understand the process of TLC (and other types of chromatography), we must examine the process at the molecular level. All forms of chromatography involve a dynamic and rapid equilibrium of molecules between the liquid and the stationary phases. For the chromatographic separation of molecules A and B shown in Figure 8.1, there are two states:

1. **Free**—dissolved in the liquid or gaseous mobile phase
2. **Adsorbed**—sticking to the surface of the solid stationary phase

Molecules A and B are continuously moving back and forth between the dissolved (free) and adsorbed states, with billions of molecules adsorbing and billions of other molecules desorbing from the solid stationary phase each second. The equilibrium between the free and adsorbed states depends on the relative strength of the

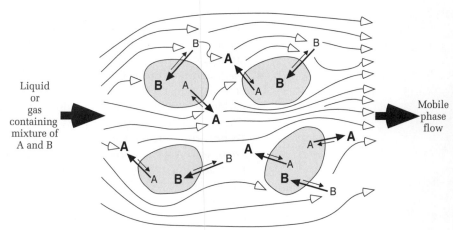

attraction of A and B to the liquid phase molecules *versus* the strength of attraction of A and B to the stationary phase structure. As discussed in the introduction to Chapter 3, the strength of these attractive forces depends on the following factors:

- Size, polarity, and hydrogen bonding ability of molecules A and B
- Polarity and hydrogen bonding ability of the stationary phase
- Polarity and hydrogen bonding ability of the mobile phase solvent

Molecules distribute themselves, or *partition*, between the mobile and stationary phases depending on these attractive forces. As implied by the equilibrium arrows in Figure 8.1, the A molecules are less polar and are thus weakly attracted to a polar stationary phase, spending most of their time in the mobile phase. In contrast, equilibrium for the more polar B molecules lies in the direction of being adsorbed onto the polar stationary phase. The equilibrium constant $k$ (also called the *partition coefficient*) is a measure of the distribution of molecules between the mobile phase and the stationary phase and is similar to the distribution coefficient for liquid/liquid extraction. This constant changes with structure.

Simply adding a mixture to a combination of a liquid phase and a stationary phase will not separate it into its pure components. For separation to happen, the liquid phase must be mobile and be flowing past the stationary phase, as depicted in Figure 8.2. Because the A molecules spend more time in the mobile phase, they will be carried through the stationary phase and be eluted faster and move farther

**■ FIG. 8.3**
A chromatographic separation. Over time, the mobile phase carries the less weakly adsorbed A molecules ahead of the more strongly adsorbed B molecules.

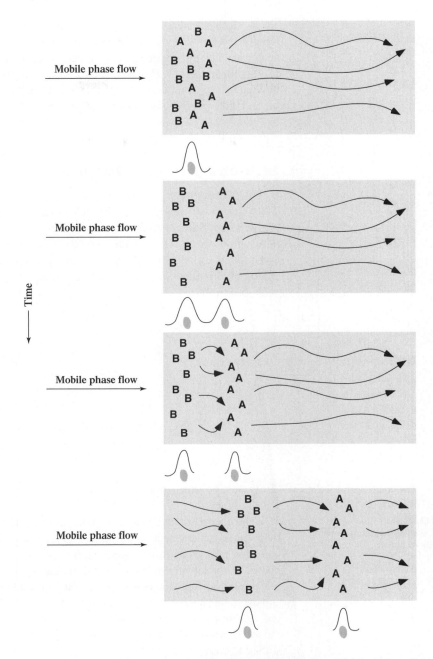

in a given amount of time. Because the B molecules are adsorbed on the stationary phase more than A molecules, the B molecules spend less time in the mobile phase and therefore migrate through the stationary phase more slowly and are eluted later. The B molecules do not migrate as far in the same amount of time. The consequence of this difference is that A is gradually separated from B by moving ahead in the flowing mobile phase as time passes, as shown in Figure 8.3.

A simple analogy may help to illustrate these concepts. Imagine a group of hungry and not-so-hungry people riding a moving sidewalk (the mobile phase in this analogy) that moves beside a long buffet table covered with all sorts of delicious food (the stationary phase). Hungry people, attracted to the food, will step off and on the moving belt many times in order to fill their plates. The not-so-hungry people will step off and on to get food far less often. Consequently, the more strongly attracted hungry people will lag behind, while the not-so-hungry ones will move ahead. The two types of people are thus separated based on the strength of their attraction for the food.

## Stationary Phase Adsorbents

In TLC, the stationary phase is a polar adsorbent, usually finely ground alumina $[(Al_2O_3)_x]$ or silica $[(SiO_2)_x]$ particles, coated as a thin layer on a glass slide or plastic sheet. Silica, commonly called *silica gel* in the laboratory, is simply very pure white sand. The extended covalent network of these adsorbents creates a very polar surface. Partial structures of silica and alumina are shown below. The silicon or aluminum atoms are the smaller, darker spheres:

Silica, $(SiO_2)_x$                    Alumina, $(Al_2O_3)_x$

■ **FIG. 8.4**
A partial silica structure showing polar Si–O bonds.

The electropositive character of the aluminum or silicon and the electronegativity of oxygen create a very polar stationary phase (Fig. 8.4). Therefore, the more polar the molecules to be separated, the stronger the attraction to the stationary phase. Nonpolar molecules will tend to stay in the mobile phase. In general, the more polar the functional group, the stronger the adsorption on the stationary phase and the more slowly the molecules will move. In an extreme situation, the molecules will not move at all. This problem can be overcome by increasing the polarity of the mobile phase so that the equilibrium between the free and adsorbed state is shifted toward the free state.

Although silica is the most common stationary phase used for TLC, many other types are used, ranging from paper to charcoal, nonpolar to polar, and reverse phase to normal phase. Several different types of stationary phases are listed according to polarity in Table 8.1.

Silica gel and alumina are commonly used in column chromatography for the purification of macroscopic quantities of material (*see* Chapter 9). Of the two,

| TABLE 8.1 • Common Stationary Phases Listed by Increasing Polarity |
| --- |
| Polydimethyl siloxane* |
| Methyl- or Phenylsiloxane* |
| Cyanopropylsiloxane* |
| Carbowax [poly(ethyleneglycol)]* |
| Reverse phase (hydrocarbon-coated silica, e.g., $C_{18}$) |
| Paper |
| Cellulose |
| Starch |
| Calcium sulfate |
| Silica (silica gel) |
| Florisil (magnesium silicate) |
| Magnesium oxide |
| Alumina (aluminum oxide; acidic, basic, or neutral) |
| Activated carbon (charcoal or Norit pellets) |

Increasing polarity ↓ (left margin)

*Stationary phase for gas chromatography

alumina, when anhydrous, is the more active; that is, it will adsorb substances more strongly. It is thus the adsorbent of choice for the separation of relatively nonpolar substrates, such as hydrocarbons, alkyl halides, ethers, aldehydes, and ketones. To separate more polar substrates, such as alcohols, carboxylic acids, and amines, the less active adsorbent, silica gel, is often used.

## Molecular Polarity and Elution Sequence

Assuming we are using a polar adsorbent, how can we determine how rapidly the compounds in our particular mixture move, that is, their elution sequence? Because the more polar compounds will adsorb more strongly to the polar stationary phase, they will move the slowest and the shortest distance on a TLC plate. Nonpolar compounds will move rapidly; they will elute first or move the greatest distance on the TLC plate. Table 8.2 lists several common compound classes according to how they move or elute on silica or alumina.

You should be able to look at a molecular structure, identify its functional group(s), and easily determine whether it is more or less polar than another structure with different functional groups. Note that the polarity of a molecule increases as the number of functional groups in that molecule increases. Thus, ethyl acetoacetate, with both ketone and ester groups, is more polar than ethyl pentanoate, which has only an ester group. However, it should be noted that chromatography is not an exact science. The rules discussed here can be used to help predict the order of elution; however, only performing an experiment will give definitive answers.

Elution sequence is the order in which the components of a mixture move during chromatography.

**Ethyl acetoacetate**

**Ethyl pentanoate**

| TABLE 8.2 • Elution Order for Some Common Functional Groups with a Silica or Alumina Stationary Phase |
| --- |

*Highest/fastest (elute with nonpolar mobile phase)*

Alkane hydrocarbons

Alkyl halides (halocarbons)

Alkenes (olefins)

Dienes

Aromatic hydrocarbons

Aromatic halides

Ethers

Esters

Ketones

Aldehydes

Amines

Alcohols

Phenols

Carboxylic acids

Sulfonic acids

*Lowest/slowest (need polar mobile phase to elute)*

*(left margin, vertical:) Increasing polarity of functional group*

## *Mobile Phase Solvent Polarity*

The key to a successful chromatographic separation is the mobile phase. You cannot change the polarities of the compounds in your mixture, and you normally use silica gel or alumina as the stationary phase. In extreme situations very polar substances chromatographed on alumina will not migrate very far from the starting point (i.e., give low $R_f$ values), and nonpolar compounds chromatographed on silica gel will travel with the solvent front (i.e., give high $R_f$ values). These extremes of behavior are markedly affected, however, by the solvents used to carry out the chromatography. A polar solvent will carry along with it polar substrates, and nonpolar solvents will do the same with nonpolar compounds—another example of the generalization "like dissolves like." By using different solvents, either alone or as mixtures, we can adjust the polarity of the mobile phase and affect the equilibria between the free and adsorbed states. Changing the polarity of the mobile phase can optimize the chromatographic separation of mixtures of compounds with a wide variety of polarities.

Table 8.3 lists, according to increasing polarity, some solvents that are commonly used for both TLC and column chromatography. Because the polarities of benzene, carbon tetrachloride, or chloroform can be matched by other, less toxic solvents, these three solvents are seldom used. In general, the solvents for TLC and column chromatography are characterized by having low boiling points that allow them to be easily evaporated and low viscosities that allow them to migrate

Avoid using benzene, carbon tetrachloride, and chloroform. Benzene is known to be a carcinogen when exposure is prolonged; the others are suspected carcinogens.

| TABLE 8.3 • Common Mobile Phases Listed by Increasing Polarity |
| --- |
| Helium |
| Nitrogen |
| Pentanes (petroleum ether) |
| Hexanes (ligroin) |
| Cyclohexane |
| Carbon tetrachloride* |
| Toluene |
| Chloroform* |
| Dichloromethane (methylene chloride) |
| t-Butyl methyl ether |
| Diethyl ether |
| Ethyl acetate |
| Acetone |
| 2-Propanol |
| Pyridine |
| Ethanol |
| Methanol |
| Water |
| Acetic acid |

Increasing polarity →

*Suspected carcinogens

rapidly. A solvent more polar than methanol is seldom needed. Often, two solvents are used in a mixture of varying proportions; the polarity of the mixture is a weighted average of the two. Hexane and ether mixtures are often employed.

Finding a good solvent system is usually the most critical aspect of TLC. If the mobile phase has not been previously determined, start with a nonpolar solvent such as hexane or ligroin and observe the separation. If the mixture's components do not move very far, try adding a polar solvent such as ether or ethyl acetate to the hexane. Compare the separation to the previous plate. In most cases, a combination of two solvents is the best choice. If the spots stay at the bottom of the plate, add more of the polar solvent. If they run with the solvent front (move to the top), increase the proportion of the nonpolar solvent. Unfortunately, some trial and error is usually involved in determining which solvent system is the best. There is a large amount of literature on the solvents and adsorbents used in the separation of a wide variety of substances.

## Spotting the TLC Plate

It is recommended that you use commercially available TLC plates, poly(ethylene terephthalate) (Mylar) sheets coated with silica gel using polyacrylic acid as a binder; these fluoresce under ultraviolet (UV) light.[1] The TLC plates must be

**■ FIG. 8.5**
A marked and spotted TLC plate.

Ac As C I

Aspirin spot

Acetaminophen spot

Caffeine spot

Ibuprofen spot

Depth of solvent

Pencil line 1 cm from bottom edge

**■ FIG. 8.6**
(a) A spotting capillary. (b) Soften the glass by heating at the base of the flame as shown and then stretch to a fine capillary.

(a)

(b)

Too much sample is a frequent problem. Use a 1% solution of the mixture. Apply very small spots.

**CAUTION**: Bunsen burners should be used only in lab areas that are far from flammable organic solvents.

handled gently, or the 100-mm-thick coating of silica gel can be easily scratched off. With a pencil, lightly draw a faint line 1 cm from the end and then three or four short hash marks to guide spotting. Lightly write identifying letters at the top of the plate to keep track of the placement of the compound spots (Fig. 8.5). Note that a pencil is always used to mark TLC plates because the graphite (carbon) is inert. If ink is used to mark the plate, it will chromatograph just as any other organic compound and give flawed results.

You need to dissolve only a few milligrams of material because one can detect a few micrograms of compound on a TLC plate. Choose a volatile solvent. Even if the material is only partially soluble, you will normally be able to observe the compound because only low concentrations are needed. It is extremely important that the spots be as small as possible and that they be applied using a 1% (not more than 2%) solution of the compounds being separated.

Once the sample is prepared, a spotting capillary must be used to add the sample to the plate. Spotting capillaries can be made by drawing out open-end melting point tubes or Pasteur pipette stems in a burner flame (Fig. 8.6).[2] The bore of these capillaries should be so small that once a liquid is drawn into them, it will not flow out to form a drop. Practice spotting just pure solvent onto an unmarked TLC plate. Dip the capillary into the solvent and let a 2–3 cm column of solvent flow into it by

1. Whatman flexible plates for TLC, cat. no. 4410 222 (Fisher cat. no. 05-713-162); cut with scissors to 1″ × 3″ or 2.5 × 7.5 cm Baker-flex, Silica Gel IB-F (J.T. Baker, Phillipsburg, N.J.). Unlike student-prepared plates, these coated sheets give very consistent results. A supply of these plates makes it a simple matter to examine most of the reactions in this book for completeness of reaction, purity of product, and side reactions.

2. Three-inch pieces of old and unusable gas chromatography capillary columns are also effective spotting capillaries.

■ **FIG. 8.7**
Using a wide-mouth bottle
to develop a TLC plate.

Plastic-capped jar

Filter paper

Solvent

Plastic chromatographic sheet

■ **FIG. 8.8**
Using a foil-covered beaker
to develop a TLC plate.

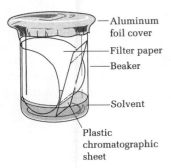

Aluminum foil cover

Filter paper

Beaker

Solvent

Plastic chromatographic sheet

■ **FIG. 8.9**
A UV lamp used to visualize spots.

capillary action, hold this vertically over the *coated* side of the plate, and lower the pipette until the tip just touches the adsorbent. Only then will liquid flow onto the plate; quickly withdraw the capillary when the spot is about 1 mm in diameter. The center of the letter *o* on this page is more than 1 mm in diameter. The distance between the sides of the letter *n* is 1 mm. The solvent should evaporate quickly, leaving your mixture behind on the plate. You may have to spot the plate a couple of times to ensure that sufficient material is present; do not spot too much sample because this will lead to a poor separation. Smearing, smudging, and overlapping of spots will make the identification of separated components difficult. Practice spotting a number of times until you develop good spotting technique. You are now ready to spot the mixture solutions as described in Experiments 1 and 2.

## Development

Once the dilute solution of the mixture has been spotted on the plate, the next step is the actual chromatographic separation, called *plate development*. The marked and spotted TLC plate is inserted into a 4-oz wide-mouth bottle (Fig. 8.7) or beaker (Fig. 8.8) containing 4 mL of an organic solvent or solvent mixture. The bottle is lined with filter paper that is wet with solvent to saturate the atmosphere within the container. Use tweezers to place the plate in the development chamber; oils from your fingers can sometimes smear or ruin a TLC plate. Also make sure that the origin spots are not below the solvent level in the chamber. If the spots are submerged in the solvent, they are washed off the plate and lost. The top of the bottle is put in place and the time noted. (If a beaker is used, the beaker is to be covered with aluminium foil.) The solvent travels up the thin layer by capillary action, and if the substance is a pure colored compound, one soon sees a spot traveling either along with the solvent front or, more commonly, at some distance behind the solvent front. Once the solvent has run within a centimeter of the top of the plate, remove the plate with tweezers. Immediately, before the solvent evaporates, use a pencil to draw a line across the plate where the solvent front can be seen. The proper location of this solvent front line is important for $R_f$ calculations.

## Visualization

If you are fortunate enough to be separating organic molecules that are colored, such as dyes, inks, or indicators, then visualizing the separated spots is easy. However, because most organic compounds are colorless, this is rarely the case.

For most compounds a UV light works well for observing the separated spots. TLC plates normally contain a fluorescent indicator that makes them glow green under UV light of wavelength 254 nm. Compounds that adsorb UV light at this wavelength will quench the green fluorescence, yielding dark purple or bluish spots on the plate. Simply hold the plate by its edges under a UV lamp as shown in Figure 8.9, and the compound spots become visible to the naked eye. Lightly circle the spots with a pencil so that you will have a permanent record of their location for later calculations.

Another useful visualizing technique is to use an iodine ($I_2$) chamber. Certain compounds, such as alkanes, alcohols, and ethers, do not absorb UV light sufficiently

Never look into a UV lamp.

to quench the fluorescence of the TLC plate and therefore will not show up under a UV lamp. However, they will adsorb iodine vapors and can be detected (after any residual solvent has evaporated) by placing the plate for a few minutes in a capped 4-oz bottle containing some crystals of iodine. Iodine vapor is adsorbed by the organic compound to form brown spots. These brown spots should be outlined with a pencil immediately after removing the plate from the iodine bottle because they will soon disappear as the iodine sublimes away; a brief return to the iodine chamber will regenerate the spots. Using both the UV lamp and iodine vapor visualization methods will ensure the location of all spots on the TLC plate.

Many specialized spray reagents have also been developed to give specific colors for certain types of compounds.

## $R_f$ Values

$R_f$ is the ratio of the distance the spot travels from the origin to the distance the solvent travels.

In addition to qualitative results, TLC can also provide a chromatographic parameter known as an $R_f$ value. The $R_f$ value is the retardation factor or the ratio-to-front value expressed as a decimal fraction. The $R_f$ value is the ratio of the distance the spot travels from the point of origin to the distance the solvent travels. The $R_f$ value can be calculated as follows:

$$R_f = \frac{\text{distance spot travels}}{\text{distance solvent travels}}$$

This number should be calculated for each spot observed on a TLC plate. Figure 8.10 shows a diagram of a typical TLC plate and how the distances are measured to calculate the $R_f$ value. The best separations are usually achieved when the $R_f$ values fall between 0.3 and 0.7.

■ **FIG. 8.10**
A developed TLC plate with spots visualized and $R_f$ values determined.

If two spots travel the same distance or have the same $R_f$ value, then it might be concluded that the two components are the same molecule. Just as many organic molecules have the same melting point and color, many can have the same $R_f$ value, so identical $R_f$ values do not necessarily mean identical compounds. For comparisons of $R_f$ values to be valid, TLC plates must be run under the exact same conditions for stationary phase, mobile phase, and temperature. Even then, additional information such as a mixed melting point or an IR spectrum should be obtained before concluding that two substances are identical.

## Comparison of Different Types of Chromatography

Table 8.4 summarizes the terminology used in chromatography and how these apply to different types of chromatography. All of the chromatographic types involve the same principles but vary in the nature of the stationary phase and the mobile phase and the measure of separation.

---

## EXPERIMENTS

## 1. Analgesics

Analgesics are substances that relieve pain. The most common of these is aspirin, a component of more than 100 nonprescription drugs. In Chapter 41, the history of this most popular drug is discussed. In this experiment, analgesic tablets will be analyzed by TLC to determine which analgesics they contain and whether they contain caffeine, which is often added to counteract the sedative effects of the analgesic.

In addition to aspirin and caffeine, the most common components of the currently used analgesics are acetaminophen and ibuprofen. In addition to one or

---

**TABLE 8.4 • Chromatography Terms and Their Definitions with Examples**

| Chromatography Term | Definition | Examples |
|---|---|---|
| Mixture | A collection of different compounds | Aspirin, ibuprofen, caffeine, and fluorene/fluorenone |
| Stationary phase | A fixed material that can adsorb compounds | Alumina, silica gel, and silicone gum |
| Mobile phase | A moving liquid or gas that dissolves compounds and carries them along | Hexane, $CH_2Cl_2$, and ethyl acetate (TLC and column chromatography); helium gas (gas chromatography) |
| Adsorption | The strength of attraction between the compounds and the stationary phase | London forces, hydrogen bonds, and dipole-dipole attractive forces |
| Separation | A measure of the elution or migration rate of compounds | $R_f$ (TLC); elution volume (column chromatography); retention time (gas chromatography) |

more of these substances, each tablet contains a binder—often starch, microcrystalline cellulose, or silica gel. And to counteract the acidic properties of aspirin, an
inorganic buffering agent is added to some analgesics. An inspection of analgesic
labels will reveal that most cold remedies and decongestants contain both aspirin
and caffeine in addition to the primary ingredient.

**Aspirin**
Acetylsalicylic acid

**Acetaminophen**
4-Acetamidophenol

**Ibuprofen**
2-(4-Isobutylphenyl)propionic acid

**Caffeine**

To identify an unknown by TLC, the usual strategy is to run chromatograms
of known substances (the standards) and the unknown at the same time. If the unknown has one or more spots that correspond to spots with the same $R_f$ values as
the standards, then those substances are probably present.

Proprietary drugs that contain one or more of the common analgesics and
sometimes caffeine are sold under some of the following brand names: Bayer
Aspirin, Anacin, Datril, Advil, Excedrin, Extra Strength Excedrin, Tylenol, and
Vanquish. Note that ibuprofen has a chiral carbon atom. The S-(+)-enantiomer is
more effective than the other.

## Procedure

Before proceeding, practice the TLC spotting technique described earlier. Following that procedure, draw a light pencil line about 1 cm from the end of a chromatographic plate. On this line spot aspirin, acetaminophen, ibuprofen, and

caffeine, which are available as reference standards. Use a separate capillary for each standard (or rinse the capillary carefully before reusing). Make each spot as small as possible, preferably less than 0.5 mm in diameter. Examine the plate under UV light to see that enough of each compound has been applied; if not, add more. On a separate plate, run the unknown and one or more of the standards.

The unknown sample is prepared by crushing a part of a tablet, adding this powder to a test tube or small vial along with an appropriate amount of ethanol, and then mixing the suspension. Not all of the crushed tablet will dissolve, but enough will go into solution to spot the plate. The binder—starch or silica—will not dissolve. Weigh out only part of the tablet to try to prepare a 1% solution of the unknown. Typically, ibuprofen tablets contain 200 mg of the active ingredient, aspirin tablets contain 325 mg, and acetaminophen tablets contain 500 mg.

*Because of the insoluble binder, not all of the unknown will dissolve.*

To the developing jar or beaker (*see* Fig. 8.7 or Fig. 8.8 on page 181), add 4 mL of the mobile phase, a mixture of 95% ethyl acetate and 5% acetic acid. Insert the spotted TLC plates with a tweezers. After the solvent has risen nearly to the top of the plate, remove the plate from the developing chamber, mark the solvent front with a pencil, and allow the solvent to dry. Examine the plate under UV light to see the components as dark spots against a bright green-blue background. Outline the spots with a pencil. The spots can also be visualized by putting the plate in an iodine chamber made by placing a few crystals of iodine in the bottom of a capped 4-oz jar. Calculate the $R_f$ values for the spots and identify the components in the unknown.

**Cleaning Up.** Solvents should be placed in the organic solvents container; dry, used chromatographic plates can be discarded in the nonhazardous solid waste container.

## 2. Plant Pigments

The botanist Michael Tswett discovered the technique of chromatography and applied it, as the name implies, to colored plant pigments. The leaves of plants contain, in addition to chlorophyll-a and chlorophyll-b, other pigments that are revealed in the fall when the leaves die and the chlorophyll rapidly decomposes. Among the most abundant of the other pigments are the carotenoids, which include the carotenes and their oxygenated homologs, the xanthophylls. The bright orange β-carotene is the most important of these because it is transformed in the liver to vitamin A, which is required for night vision.

**Chlorophyll-a**

**β-Carotene (C$_{40}$H$_{56}$)**
mp 183°C, $\lambda_{max}^{hexane}$ 451 nm

■ **FIG. 8.11**
An energy-minimized, space-filling model of lycopene. The molecule is flat, but steric hindrance of the methyl groups causes the molecule to bend into an S shape.

**3′-Dehydrolutein (a xanthophyll)**

**Butter Yellow**
**(carcinogenic)**

**Lycopene (C$_{40}$H$_{56}$)**
MW 536.85
mp 173°C, $\lambda_{max}^{hexane}$ 475 nm

Cows eat fresh, green grass that contains carotene, but they do not metabolize the carotene entirely, so it ends up in their milk. Butter made from this milk is therefore yellow. In the winter the silage cows eat does not contain carotene because that compound readily undergoes air oxidation, and the butter made at that time is white. For some time an azo dye called Butter Yellow was added to winter butter to give it the accustomed color, but the dye was found to be a carcinogen. Now winter butter is colored with synthetic carotene, as is all margarine.

**Isoprene**

Lycopene from tomato paste and β-carotene from strained carrots

As an interesting variation, try extracting lycopene from commercial ketchup.

**CAUTION**: Do not breathe the vapors of dichloromethane. Carry out the extraction in the hood.

Lycopene (Fig. 8.11), the red pigment of the tomato, is a $C_{40}$-carotenoid made up of eight isoprene units. β-Carotene, the yellow pigment of the carrot, is an isomer of lycopene in which the double bonds at $C_1$—$C_2$ and $C'_1$—$C'_2$ are replaced by bonds extending from $C_1$ to $C_6$ and from $C'_1$ to $C'_6$ to form rings. The chromophore in each case is a system of 11 all-*trans* conjugated double bonds; the closing of the two rings causes β-carotene to absorb at shorter wavelengths than lycopene does, shifting its color from red to yellow.

In 1911, Richard Willstätter and Heinrich R. Escher isolated 20 mg of lycopene per kilogram of fresh tomatoes, which contain about 96% water.[3] They then found a more convenient source in commercial tomato paste: the seeds and skin were eliminated, and the water content was reduced by evaporation in vacuum to a content of 26% solids. From this they isolated 150 mg of lycopene per kilogram of paste. The expected yield in the following experiment is 0.075 mg, which is not enough to weigh on a balance.

A jar of strained carrots sold as baby food serves as a convenient source of β-carotene. The German investigators isolated 1 g of β-carotene per kilogram of dried, shredded carrots of unstated water content.

The following procedure calls for the dehydration of tomato or carrot paste with ethanol and extraction with dichloromethane, an efficient solvent for lipids.

## Experimental Considerations

Carotenoids are very sensitive to light-catalyzed air oxidation. Perform this experiment as rapidly as possible; keep the solutions as cool and dark as possible. This extraction produces a mixture of products that can be analyzed by both TLC and column chromatography (*see* Chapter 9). If enough material for TLC only is desired, use one-tenth the quantities of starting material and solvents employed in the following procedure. This extraction can also be carried out with hexane if the ventilation is not adequate enough to use dichloromethane. However, hexane is more prone to form emulsions than the chlorinated solvent.

 ## Procedure

**IN THIS EXPERIMENT** some tomato or carrot paste is treated with acetone, which will remove water and lipids but not the highly colored carotenoid hydrocarbons. The carotenoids are extracted by dichloromethane and analyzed by TLC.

A 5-g sample of fresh tomato or carrot paste (baby food) is transferred to the bottom of a 25 × 150 mm test tube, followed by 10 mL of acetone. The mixture is stirred and shaken before being filtered on a Hirsch funnel. Scrape as much of the material from the tube as possible and press it dry on the funnel. Let the tube drain thoroughly. Place the filtrate in a 125-mL Erlenmeyer flask.

3. Willstätter, R.; Escher, H. R. *Z. Physiol. Chem.* **1911**, *64*, 47–61.

Return the solid residue to the test tube, shake it with a 10-mL portion of dichloromethane, and again filter the material on a Hirsch funnel. Add the filtrate to the 125-mL flask. Repeat this process two more times and then pour the combined filtrates into a separatory funnel. Add water and sodium chloride solution (which aids in the breaking of emulsions) and shake the funnel gently. This aqueous extraction will remove the acetone and any water-soluble components from the mixture, leaving the hydrocarbon carotenoids in the dichloromethane. Dry the colored organic layer over anhydrous calcium chloride and filter the solution into a dry flask. Remove about 0.5 mL of this solution and store it under nitrogen in the dark until it can be analyzed by TLC. Evaporate the remainder of the dichloromethane solution to dryness under a stream of nitrogen or under vacuum on a rotary evaporator. This material can be used for column chromatography (*see* Chapter 9). If it is to be stored, fill the flask with nitrogen and store it in a dark place.

*Air can be used for the evaporation, but nitrogen is better because these hydrocarbons air oxidize with great rapidity.*

### Thin-Layer Chromatography

Spot the mixture on a TLC plate about 1 cm from the bottom and 8 mm from the edge. Make one spot concentrated by repeatedly touching the plate, but ensure that the spot is as small as possible, certainly less than 1 mm in diameter. The other spot can be of lower concentration. Develop the plate with an 80:20 hexane-acetone mixture. With other plates you could try cyclohexane and toluene as eluents and also hexane-ethanol mixtures of various compositions.

*Many spots may be seen. There are two common carotene and chlorophyll isomers and four xanthophyll isomers.*

The container in which the chromatography is carried out should be lined with filter paper that is wet with the solvent so that the atmosphere in the container will be saturated with the solvent vapor. After elution is completed, remove the TLC plate and mark the solvent front with a pencil and outline the colored spots. Examine the plate under UV light. Also place the plate in an iodine chamber to visualize the spots.

**Cleaning Up.** The aqueous saline filtrate containing acetone can be flushed down the drain. Recovered and unused dichloromethane should be placed in the halogenated organic waste container; the solvents used for TLC should be placed in the organic solvents container. If local regulations allow, evaporate any residual solvent from the drying agents in the hood and place the dried solid in the nonhazardous waste container. Otherwise, place the wet drying agent in a waste container designated for this purpose. Used plant material and dry TLC plates can be discarded in the nonhazardous waste container.

 ### Procedure

In a small mortar grind 2 g of green or brightly colored fall leaves (do not use ivy or waxy leaves) with 10 mL of ethanol, pour off the ethanol (which serves to break up and dehydrate the plant cells), and grind the leaves successively with three 1-mL portions of dichloromethane that are decanted or withdrawn with a Pasteur pipette and placed in a test tube. The pigments of interest are extracted by the dichloromethane. Alternatively, place 0.5 g of carrot paste (baby food) or tomato paste in a test tube, stir and shake the paste with 3 mL of ethanol until the paste has a somewhat dry or fluffy appearance, remove the ethanol, and extract the dehydrated paste with three 1-mL portions of dichloromethane. Stir and shake the plant material with the solvent to extract as much of the pigments as possible.

Fill the tube containing the dichloromethane extract from leaves or vegetable paste with a saturated sodium chloride solution and shake the mixture. Remove the aqueous layer; add anhydrous calcium chloride pellets to the dichloromethane solution until the drying agent no longer clumps together. Shake the mixture with the drying agent for about 5 min and then withdraw the solvent with a Pasteur pipette and place it in a test tube. Add to the solvent a few pieces of Drierite to complete the drying process. Gently stir the mixture for about 5 min, transfer the solvent to a test tube, wash off the drying agent with more solvent, and then evaporate the combined dichloromethane solutions under a stream of nitrogen while warming the tube in your hand or in a beaker of warm water. Carry out this evaporation in the hood.

These hydrocarbons air oxidize with great rapidity.

Immediately cork the tube filled with nitrogen and then add 1 or 2 drops of dichloromethane to dissolve the pigments for TLC analysis. Carry out the analysis without delay by spotting the mixture on a TLC plate about 1 cm from the bottom and 8 mm from the edge. Make one spot concentrated by repeatedly touching the plate, but ensure that the spot is as small as possible—less than 1.0 mm in diameter. The other spot can be of lower concentration. Develop the plate with a 70:30 hexane-acetone mixture. With other plates try cyclohexane and toluene as eluents and also hexane-ethanol mixtures of various compositions. The container in which the chromatography is carried out should be lined with filter paper that is wet with the solvent so that the atmosphere in the container will be saturated with solvent vapor. After elution is completed, mark the solvent front with a pencil and outline the colored spots. Examine the plate under the UV light. Are any new spots seen? Report colors and $R_f$ values for all of your spots and identify each as lycopene, carotene, chlorophyll, or xanthophyll.

**Cleaning Up.** The ethanol used for the dehydration of the plant material can be flushed down the drain along with the saturated sodium chloride solution. Recovered and unused dichloromethane should be placed in the halogenated organic waste container. The solvents used for TLC should be placed in the organic solvents container. If local regulations allow, evaporate any residual solvent from the drying agents in the hood and place the dried solid in the nonhazardous waste container. Otherwise, place the wet drying agent in a waste container designated for this purpose. Used plant material and dry TLC plates can be discarded in the nonhazardous waste container.

## 3. For Further Investigation

Many of the pigments in plants are made up of compounds called *anthocyanins.* Grind about 4 g of colored plant tissue (flower petals, blueberries, strawberries, cranberries, apple skins, red cabbage, red or purple grapes, etc.) and a small amount of alumina or fine sand with about 4 mL of a mixture of 99% methanol and 1% hydrochloric acid. Spot the extract on cellulose TLC plates and elute with a solvent mixture of 20% concentrated hydrochloric acid, 40% water, and 4% formic acid. Note the number and color of the spots. Look up the structures of the possible anthocyanins, of which many are glycosides of the aglycones delphinidin, peonidin, malvidin, and cyanidin, among others.

# 4. Colorless Compounds

You will now apply the thin-layer technique to a group of colorless compounds. The spots can be visualized under UV light if the plates have been coated with a fluorescent indicator; chromatograms can also be developed in a 4-oz bottle containing crystals of iodine. During development, spots will appear rapidly but remember that they also disappear rapidly. Therefore, outline each spot with a pencil immediately on withdrawal of the plate from the iodine chamber. Some suggested solvents are pure cyclohexane, pure toluene, toluene (3 mL) plus dichloromethane (1 mL), or toluene (4.5 mL) plus methanol (1/2 mL).

The compounds for trial are to be selected from the following list (all 1% solutions in toluene; those compounds with an asterisk are fluorescent under UV light):

1. Anthracene*
2. Cholesterol
3. 2,7-Dimethyl-3,5-octadiyne-2,7-diol
4. Diphenylacetylene
5. *trans,trans*-1,4-Diphenyl-1,3-butadiene*
6. *p*-Di-*t*-butylbenzene
7. 1,4-Di-*t*-butyl-2,5-dimethoxybenzene
8. *trans*-Stilbene
9. 1,2,3,4-Tetraphenylnaphthalene*
10. Tetraphenylthiophene
11. *p*-Terphenyl*
12. Triphenylmethanol
13. Triptycene

Except for tetraphenylthiophene, the structures for all of these molecules will be found in this book.

Make your own selections.

It is up to you to make selections and to plan your own experiments. Do as many as time permits. One plan would be to select a pair of compounds that are estimated to be separable and that have $R_f$ values determinable with the same solvent. One can assume that a hydroxyl compound will travel less rapidly with a hydrocarbon solvent than a hydroxyl-free compound; you will therefore know what to expect if the solvent contains a hydroxylic component. An aliphatic solvent should carry along an aromatic compound with aliphatic substituents better than one without such groups. However, instead of relying on assumptions, you can do brief preliminary experiments on used plates on which previous spots are visible or outlined. If you spot a pair of compounds on such a plate and let the solvent rise about 3 cm from the starting line before development, you might be able to tell if a certain solvent is appropriate for a given sample. Alternatively, make some spots on a plate (new or used) and then touch each spot with a different solvent held in a capillary. In Figure 8.12a, the mixture did not move away from the point of origin; in Figure 8.12b, two concentric rings are seen between the origin and the solvent front. This is how a good solvent behaves. In Figure 8.12c, the mixture of compounds traveled with the solvent front.

Preliminary trials on used plates

■ **FIG. 8.12**
A fast method for determining
the correct solvent for TLC. See
the text for the procedure.

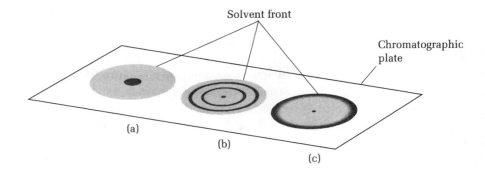

Once a solvent is chosen, run a complete chromatogram on the two compounds
on a fresh plate. If separation of the two compounds seems feasible, put two spots
of one compound on a plate, let the solvent evaporate, and put spots of the second
compound over the first ones. Run a chromatogram and see if you can detect two
spots in either lane (with colorless compounds, it is advisable not to attempt a three-
lane chromatogram until you have acquired considerable practice and skill).

**Cleaning Up.**  Solvents should be placed in the organic solvents container, and dry,
used chromatographic plates can be discarded in the nonhazardous solid waste
container.

## Discussion

If you have investigated hydroxylated compounds, you doubtless have found that
it is reasonably easy to separate a hydroxylated from a nonhydroxylated compound or a diol from a mono-ol. How, by a simple reaction followed by a thin-
layer chromatogram, could you separate cholesterol from triphenylmethanol?
Heating a sample of each with acetic anhydride and a trace of pyridine catalyst for
5 min on a steam bath, followed by chromatography, should do it. A first trial of a
new reaction leaves questions about what has happened and how much, if any,
starting material is present. A comparative chromatogram of the reaction mixture
with starting material may tell the story. How crude is a crude reaction product?
How many components are present? The thin-layer technique may give the answers to these questions and suggest how best to process the product. A preparative column chromatogram may afford a large number of fractions of eluent (say,
1 to 30). Some fractions probably contain nothing and should be discarded, while
others should be combined for evaporation and workup. How can you identify the
good and the useless fractions? Take a few used plates and put numbered circles
on clean places of each; spot samples of each of the fractions; and, without any
chromatography, develop the plates with iodine. Negative fractions for discard
will be obvious, and the pattern alone of positive fractions may allow you to infer
which fractions can be combined. Thin-layer chromatograms of the first and last
fractions of each suspected group would then show whether or not your inferences
are correct.

# Fluorescence

Four of the compounds listed in Experiment 4 are fluorescent under UV light. These compounds give colorless spots that can be picked up on a chromatogram by fluorescence (after removal from the UV-absorbing glass bottle). If a UV light source is available, spot the four compounds on a used plate and observe the fluorescence.

Take this opportunity to examine a white shirt or handkerchief under UV light to see if it contains a brightener, that is, a fluorescent white dye or optical bleach. These substances are added to counteract the yellow color that repeated washing gives to cloth. Brighteners of the type of Calcofluor White MR, a sulfonated *trans*-stilbene derivative, are commonly used in detergent formulations for cotton; the substituted coumarin derivative is typical of brighteners used for nylon, acetate, and wool. Detergents normally contain 0.1%–0.2% of optical bleach. The amount of dye on a freshly laundered shirt is approximately 0.01% of the weight of the fabric.

**Calcofluor White MR**

**7-Diethylamino-4-methylcoumarin**

## QUESTIONS

1. Why might it be very difficult to visualize the separation of *cis*- and *trans*-2-butene by TLC?
2. What error is introduced into the determination of an $R_f$ value if the top is left off the developing chamber?
3. What problem will ensue if the level of the developing liquid is higher than the applied spot in a TLC analysis?

4.  In what order (from top to bottom) would you expect to find naphthalene, butyric acid, and phenyl acetate on a silica gel TLC plate developed with dichloromethane?

5.  In carrying out an analysis of a mixture, what do you expect to see when the TLC plate has been allowed to remain in the developing chamber too long, so that the solvent front has reached the top of the plate?

6.  Arrange the following in order of increasing $R_f$ with TLC: acetic acid, acetaldehyde, 2-octanone, decane, and 1-butanol.

7.  What will be the result of applying too much compound to a TLC plate?

8.  Why is it necessary to run TLC in a closed container and to have the interior vapor saturated with the solvent?

9.  What will be the appearance of a TLC plate if a solvent of too low polarity is used for the development? a solvent of too high polarity?

10.  A TLC plate showed two spots with $R_f$ values of 0.25 and 0.26. The plate was removed from the developing chamber, the residual solvent was allowed to evaporate from the plate, and then the plate was returned to the developing chamber. What would you expect to see after the second development was complete?

11.  One of the analgesics has a chiral center. Which compound is it? One of the two enantiomers is far more effective at reducing pain than the other.

12.  Using a ruler to measure distances, calculate the $R_f$ value for substance B in Figure 8.10.

# Column Chromatography: Fluorenone, Cholesteryl Acetate, Acetylferrocene, and Plant Pigments

**PRELAB EXERCISE:** Compare column chromatography and thin-layer chromatography (TLC) with regard to the (1) quantity of material that can be separated, (2) time needed for the analysis, (3) solvent systems used, and (4) ability to separate compounds.

Column chromatography is one of the most useful methods for the separation and purification of both solids and liquids when carrying out microscale experiments. It becomes expensive and time consuming, however, when more than about 10 g of material must be purified. Column chromatography involves the same chromatographic principles as detailed for TLC in Chapter 8, so be sure that you understand those before doing the experiments in this chapter.

As discussed in Chapter 1, organic chemists obtain new compounds by synthesizing or isolating natural products that have been biosynthesized by microbes, plants, or animals. In most cases, initial reaction products or cell extracts are complex mixtures containing many substances. As you have seen, recrystallization, distillation, liquid/liquid extraction, and sublimation can be used to separate and purify a desired compound from these mixtures. However, these techniques are frequently not adequate for removing impurities that are closely related in structure. In these cases, column chromatography is often used. The broad applicability of this technique becomes obvious if you visit any organic chemistry research lab, where chromatography columns are commonplace.

Three of the five experiments in this chapter involve synthesis and may be your first experience in running an organic reaction. Experiments 1 and 2 involve the synthesis of a ketone. In Experiment 3 an ester of cholesterol is prepared. Experiment 4 demonstrates the separation of colored compounds. Experiment 5 involves the isolation and separation of natural products (plant pigments), which is analogous to Experiment 2 in Chapter 8 but on a larger scale.

The most common adsorbents for column chromatography—silica gel and alumina—are the same stationary phases as used in TLC. The sample is dissolved in a small quantity of solvent (the eluent) and applied to the top of the column. The eluent, instead of rising by capillary action up a thin layer, flows down through the column filled with the adsorbent. Just as in TLC, there is an equilibrium established between the solute adsorbed on the silica gel or alumina and the eluting solvent flowing down through the column, with the less strongly absorbed solutes moving ahead and eluting earlier.

Three mutual interactions must be considered in column chromatography: the activity of the stationary adsorbent phase, the polarity of the eluting mobile solvent phase, and the polarity of the compounds in the mixture being chromatographed.

# Additional Principles of Column Chromatography

## Adsorbents

A large number of adsorbents have been used for column chromatography, including cellulose, sugar, starch, and inorganic carbonates; but most separations employ alumina [$(Al_2O_3)_x$] or silica gel [$(SiO_2)_x$]. Alumina comes in three forms: acidic, neutral, and basic. The neutral form of Brockmann activity grade II or III, 150 mesh, is most commonly employed. The surface area of this alumina is about 150 $m^2$/g. Alumina as purchased will usually be activity grade I, meaning that it will strongly adsorb solutes. It must be deactivated by adding water, shaking, and allowing the mixture to reach equilibrium over an hour or so. The amount of water needed to achieve certain activities is given in Table 9.1. The activity of the alumina on TLC plates is usually about III. Silica gel for column chromatography, 70–230 mesh, has a surface area of about 500 $m^2$/g and comes in only one activity.

**TABLE 9.2 • Elutropic Series for Solvents**

n-Pentane (least polar)

Petroleum ether

Cyclohexane

Hexanes

Carbon disulfide

t-Butyl methyl ether

Dichloromethane

Tetrahydrofuran

Dioxane

Ethyl acetate

2-Propanol

Ethanol

Methanol

Acetic acid (most polar)

| TABLE 9.1 • Alumina Activity | | | | | |
|---|---|---|---|---|---|
| Brockmann activity grade | I | II | III | IV | V |
| Percent by weight of water | 0 | 3 | 6 | 10 | 15 |

## Solvents

Solvent systems for use as mobile phases in column chromatography can be determined from TLC, the scientific literature, or experimentally. Normally, a separation will begin with a nonpolar or low-polarity solvent, allowing the compounds to adsorb to the stationary phase; then the polarity of the solvent is *slowly* switched to desorb the compounds and allow them to move with the mobile phase. The polarity of the solvent should be changed gradually. A sudden change in solvent polarity will cause heat evolution as the alumina or silica gel adsorbs the new solvent. This will vaporize the solvent, causing channels to form in the column that severely reduce its separating power.

Several solvents are listed in Table 9.2, arranged in order of increasing polarity (elutropic series), with n-pentane being the least polar. The order shown in the table reflects the ability of these solvents to dislodge a polar substance adsorbed onto either silica gel or alumina, with *n*-pentane having the lowest solvent power.

As a practical matter, the following sequence of solvents is recommended in an investigation of unknown mixtures: elute first with petroleum ether (pentanes); then hexanes; followed by hexanes containing 1%, 2%, 5%, 10%, 25%, and 50% ether; pure ether; ether and dichloromethane mixtures; followed by dichloromethane and methanol mixtures. Either diethyl ether or *t*-butyl methyl ether can be used, but *t*-butyl methyl ether is recommended. Solvents such as methanol and water are normally not used because they can destroy the integrity of the stationary phase by dissolving some of the silica gel. Some typical solvent combinations are hexanes-dichloromethane, hexanes–ethyl acetate, and hexanes-toluene. An experimentally determined ratio of these solvents can sufficiently separate most compounds.

Petroleum ether: mostly isometric pentanes

## Compound Mobility

The ease with which different classes of compounds elute from a column is indicated in Table 9.3. Molecules with nonpolar functional groups are least adsorbed and elute first, while more polar or hydrogen-bonding molecules are more strongly adsorbed and elute later. The order is similar to that of the eluting solvents—another application of "like dissolves like."

## Sample and Column Size

Chromatography columns can be as thin as a pencil for milligram quantities to as big as a barrel for the industrial-scale separation of kilogram quantities. A microscale column for the chromatography of about 50 mg of material is shown in Figure 9.1; columns with larger diameters, as shown in Figures 9.2 and 9.3, are used for macroscale procedures. The amount of alumina or silica gel used should generally weigh at least 30 times as much as the sample, and the column, when packed, should have a height at least 10 times the diameter. The density of silica gel is 0.4 g/mL, and the density of alumina is 0.9 g/mL, so the optimum size for any column can be calculated.

## Packing the Column

### Microscale Procedure

Before you pack the column, tare several Erlenmeyer flasks, small beakers, or 20-mL vials to use as receivers. Weigh each one carefully and mark it with a number on the etched circle.

Uniform packing of the chromatography column is critical to the success of this technique. Two acceptable methods for packing a column are dry packing and slurry packing, which normally achieve the best results. Assemble the column as depicted in Figure 9.1. To measure the amount of adsorbent, fill the column one-half to two-thirds full; then pour the powder out into a small beaker or flask. Clamp the column in a vertical position and close the valve. Always grasp the

| TABLE 9.3 • Elution Order for Solutes |
| --- |
| Alkanes (first) |
| Alkenes |
| Dienes |
| Aromatic hydrocarbons |
| Ethers |
| Esters |
| Ketones |
| Aldehydes |
| Amines |
| Alcohols |
| Phenols |
| Acids (last) |

*Online Study Center*

Photo: Column Chromatography;
Video: Column Chromatography

**■ FIG. 9.1**
A microscale chromatographic
column.

Funnel

Glass
column

Alumina

Micro
Büchner
funnel

Polyethylene
frit

10-mL
Erlenmeyer
flask

**■ FIG. 9.2**
A macroscale chromatographic
column.

Solvent

Sand

Alumina or
silica gel

Sand
Glass wool

**■ FIG. 9.3**
A chromatographic tube on
ring stand.

valve with one hand while turning it with the other. Fill the column with a non-polar solvent such as hexanes almost to the top.

● **Dry Packing Method.**    This is the simplest method for preparing a microscale column. Slowly add the powedered alumina or silica gel through the funnel while gently tapping the side of the column with a pencil. The solid should "float" to the bottom of the column. Try to pack the column as evenly as possible; cracks, air bubbles, and channels will lead to a poor separation.

● **Slurry Packing Method.**    To slurry pack a column, add about 8 mL of hexanes to the adsorbent in a flask or beaker, stir the mixture to eliminate air bubbles, and then (this is the hard part) swirl the mixture to get the adsorbent suspended in the solvent and immediately pour the entire slurry into the funnel. Open the valve, drain some solvent into the flask that had the adsorbent in it and finish transferring the slurry to the column. Place an empty flask under the column and allow the solvent to drain to about 5 mm above the top surface of the adsorbent. Tap the column with a pencil until the packing settles to a minimum height. Try to pack the column as evenly as possible; cracks, air bubbles, and channels will lead to a poor separation.

The slurry method normally gives the best column packing, but it is also the more difficult technique to master. Whether the dry packing or slurry packing method is chosen, the most important aspect of packing the column is creating an evenly distributed and packed stationary phase. The slurry method is often used for macroscale separations.

Once the column is loaded with solvent and adsorbent, place a flask under it, open the stopcock (use two hands for the microscale column), and allow the solvent level to drop to the *top* of the packing. Avoid allowing the solvent level to go below the stationary phase (known as letting the column "run dry") because this allows air bubbles and channel formation to occur, which leads to a poor separation.

### Macroscale Procedure

Before you pack the column, prepare several Erlenmeyer flasks to use as receivers by taring (weighing) each one carefully and marking each with a number on the etched circle.

*Extinguish all flames; work in laboratory hood.*

### Dry Packing Method

The column can be prepared using a 50-mL burette such as the one shown in Figure 9.2 or using the less expensive and equally satisfactory chromatographic tube shown in Figure 9.3, in which the flow of solvent is controlled by a screw pinchclamp. Weigh the required amount of silica gel (12.5 g in the first experiment), close the pinchclamp on the tube, and fill about half full with a 90:10 mixture of hexanes and ether. With a wooden dowel or glass rod, push a small plug of glass wool through the liquid to the bottom of the tube, dust in through a funnel enough sand to form a 1-cm layer over the glass wool, and level the surface by tapping the tube. Unclamp the tube. With your right hand grasp both the top of the tube and the funnel so that the whole assembly can be shaken to dislodge silica gel that may stick to the walls; with your left hand pour in the silica gel slowly (Fig. 9.4) while tapping the column with a rubber stopper fitted on the end of a pencil. If necessary, use a Pasteur pipette full of a 90:10 mixture of hexanes and ether to wash down any silica gel that adheres to the walls of the column above the liquid. When the silica gel has settled, add a little sand to provide a protective layer at the top. Open the pinchclamp, let the solvent level fall until it is just slightly above the upper layer of sand, and then stop the flow.

■ **FIG. 9.4**
A useful technique for filling a chromatographic tube with silica gel.

### Slurry Packing Method

Alternatively, the silica gel can be added to the column (half filled with hexanes) by slurrying the silica gel with a 90:10 mixture of hexanes and ether in a beaker. The powder is stirred to suspend it in the solvent and immediately poured through a wide-mouth funnel into the chromatographic tube. Rap the column with a rubber stopper to cause the silica gel to settle and to remove bubbles. Add a protective layer of sand to the top. The column is now ready for use.

**Cleaning Up.** After use, the tube is conveniently emptied by pointing the open end into a beaker, opening the pinchclamp, and applying gentle air pressure to the tip. If the plug of glass wool remains in the tube after the alumina leaves, wet it with acetone and reapply air pressure. Allow the adsorbent to dry in the hood and then dispose of it in the nonhazardous waste container.

## Adding the Sample

Dissolve the sample completely in a very minimum volume of dichloromethane (just a few drops) in a small flask or vial. Add to this solution 300 mg of the adsorbent, stir, and evaporate the solvent completely by heating the slurry *very gently* with *constant* stirring to avoid bumping. Remember that dichloromethane boils at 41°C. Pour this dry powder into the funnel of the chromatography column, wash it down onto the column with a few drops of hexane, and then tap the column to remove air bubbles from the layer of adsorbent-solute mixture just added. Open the valve and carefully add new solvent in such a manner that the top surface of the column is not disturbed. A thin layer of fine sand can be added to the column after the sample to avoid disturbance of the column surface when the solvent is being added. Run the solvent down near to the surface several times to apply the sample as a narrow band at the top of the column.

## Eluting the Column

Fill the column with solvent, open the stopcock, and continue to add more solvent while collecting 1–3 mL fractions in small tared containers. Collecting small fractions is important to the success of your column separation. Fractions that are too small can always be pooled together; however, if the collected fractions are too large, you may get more than one compound in any particular fraction. If this occurs, the only way to attain separation is to redo the chromatography. Column chromatography is a lengthy process, so collecting large fractions is discouraged.

## Isolating the Separated Compounds

If the mixture to be separated contains colored compounds, then monitoring the column is very simple. The colored bands will move down the column along with the solvent, and as they approach the end of the column, you can collect the separated colors in individual containers. However, most organic molecules are colorless. In this case, the separation must be monitored by TLC. Spot each fraction on a TLC plate (Fig. 9.5). Four or five fractions can be spotted on a single plate. Before you develop the plate, do a quick examination under UV light to see if there is any compound where you spotted. If not, you can spot the next fraction in that location. Note which fraction is in which lane. Develop the plate and use the observed spot(s) to determine which compound is in each of the collected fractions. Spotting some of the starting material or the product (if available) on the TLC plate as a standard will help in the identification.

The colors of the fractions or the results from analyzing the fractions by TLC will indicate which fractions contain the compound(s) you are interested in isolating. Combine fractions containing the same compound and evaporate the

**■ FIG. 9.5**
Spot each fraction on a TLC plate. Examine under UV light to see which fractions contain the compound.

solvent. Recrystallization may be used to further purify a solid product. However, on a milligram scale, there is usually not enough material to do this.

## Other Types of Chromatography

*Online Study Center*

**General Resources**
Flash Chromatography

Relying only on gravity, liquid flow through a column can be quite slow, especially if the column is tightly packed. One method to speed up the process is *flash chromatography*. This method uses a pressure of about 10 psi of air or nitrogen on top of the column to force the mobile phase through the column. Normally, doing this would give a poorer separation. However, it has been found that with a finer mesh of alumina or silica gel, flash chromatography can increase the speed without lowering the quality of the separation. Go to this book's website for an illustrated set of instructions for packing and using a flash chromatography column.

High-performance liquid chromatography

High-performance liquid chromatography (HPLC) is a high-tech version of column chromatography, which is capable of separating complex mixtures with dozens of components. A high-pressure pump forces solvent at pressures up to 10,000 psi through a stainless steel tube packed tightly with extremely small adsorbent particles. The eluent flows from the column to a detector, such as a tiny UV absorbance cell or a mass spectrometer that is able to detect extremely small amounts of separated components, as little as a picogram ($10^{-12}$ g).

Reverse-phase chromatography

HPLC is used extensively in biochemistry to separate cellular components such as proteins, lipids, and nucleic acids. Mixtures of these types of compounds can be dissolved only in a predominantly aqueous mobile phase such as methanol-water or acetonitrile-water, and normal silica gel or alumina stationary phases do not work well with high concentrations of water. Rather than polar stationary phases, highly nonpolar ones called *reverse-phase packings* are used. These are manufactured by bonding lots of hydrocarbon molecules to the surfaces of silica gel particles, which convert the particles into highly nonpolar grease balls. With this packing, the order of elution is the reverse of that observed for a normal silica gel phase. On a reverse-phase column, the more nonpolar compounds will adhere to the nonpolar stationary phase more strongly, and the polar compounds will elute first.

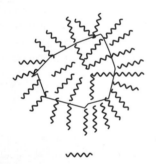

$$CH_3$$
$$|$$
$$—O—Si—(CH_2)_{17}CH_3$$
$$|$$
$$CH_3$$

**C$_{18}$ reverse-phase packing**

Chiral chromatography

Is it possible to separate two enantiomers (optical isomers), both of which have the same intermolecular attractive forces? Chiral stationary phases can be used to separate enantiomers. Giving the stationary phase an asymmetry or handedness allows one enantiomer to be specifically retained on the column. Such columns are quite expensive and are limited to a particular type of separation, but they have led to great achievements in separation science. This separation technique is of great importance in the pharmaceutical industry because the U.S. Food and Drug Administration (FDA) specifies the amounts of impurities, including enantiomers, that can be found in drugs. For example, Thalidomide, a drug prescribed as a sedative and an antidepressant in the 1960s, was found to be a potent teratogen that caused birth defects when pregnant women took the drug. It was quickly pulled from the market. Thalidomide has two enantiomers, and further research demonstrated that only one of the enantiomers caused the birth defects.

## EXPERIMENTS

# 1. Air Oxidation of Fluorene to Fluorenone

The 9-position of fluorene is unusually reactive for a hydrocarbon. The protons on this carbon atom are acidic by virtue of being doubly benzylic, and, consequently, this carbon can be oxidized by several reagents, including oxygen from the air. Here the oxidation is carried out using a phase-transfer catalyst called methyltricapryl ammonium chloride, commonly known as Stark's catalyst or Aliquat 336. In the presence of this catalyst, the hydroxide is carried into the organic layer, where it can remove one of the acidic fluorene protons, creating a carbanion that can react with oxygen in the air. An intermediate hydroperoxide is formed and loses water to give the ketone.

Sufficiently acidic to be removed by strong base

H   H

**Fluorene**
mp 114°C
MW 166.22

$+ O_2$

CH₃ Cl⁻
N⁺

Methyltricapryl ammonium chloride
(Aliquat 336, Stark's catalyst)

10 $M$ NaOH
Toluene
10–60 min

O

**Fluorenone**
mp 83°C
MW 180.21

$+ H_2O$

## Microscale Procedure

**IN THIS EXPERIMENT** fluorene is oxidized to fluorenone by oxygen from the air in a base-catalyzed reaction involving a phase-transfer catalyst. The reaction is monitored by TLC and stopped when about a 50:50 mixture of product and starting material has formed. The toluene layer is separated and washed with 5% hydrochloric acid and saturated sodium chloride. After drying over calcium chloride, the toluene is removed and the fluorene-fluorenone mixture is separated by microscale column chromatography on alumina, using hexanes as the mobile phase. Fractions are analyzed by TLC; common fractions are pooled and weighed after solvent removal to give pure fluorene and fluorenone.

**CAUTION**: Sodium hydroxide is a strong base and is very corrosive. Wash your hands immediately if contact occurs.

This experiment requires a 50-mL to 125-mL separatory funnel. To a 25-mL Erlenmeyer flask clamped to a ring stand above a magnetic stirrer, add 5 mL of 10 $M$ NaOH and 70 mg of fluorene while stirring with a $\frac{1}{2}$-inch stir bar. Add 5 mL of toluene and stir until all of the solid has dissolved. (Observe the color and identify which layer is organic and which is aqueous.) Add approximately 3 drops of Stark's catalyst (Aliquat 336) to the solution. Stir vigorously but without splashing the solution. The reaction can take anywhere from 10 min to 30 min. Follow the rate of the reaction by TLC. Develop the TLC plate by using 20% dichloromethane in hexanes and use a UV lamp to visualize the products. Also spot the plate with a 1% fluorene standard. When approximately half of the fluorene has been converted to fluorenone (as evidenced by the fact that the product spot is about the same size and intensity), pour the reaction mixture into a separatory funnel, rinsing the beaker with an additional 3 mL of toluene, which is also added to the separatory funnel.

Separate the organic layer from the aqueous layer. Wash the organic layer in the separatory funnel with 1.5 $M$ hydrochloric acid (three separate times with 5 mL each time) and then saturated sodium chloride (three separate times with 5 mL each time). After each washing, drain the aqueous layer from the separatory funnel into a waste beaker. Dry the remaining toluene layer over anhydrous calcium chloride pellets in a 125-mL Erlenmeyer flask. Add the pellets until they no longer clump together (3–5 scoops). Allow the product to dry for 5–10 min before decanting the toluene from the solid calcium chloride and transferring the toluene to a 100-mL tared beaker. Wash the solid calcium chloride with 3 mL of toluene, adding this to the main portion of toluene to complete the transfer of product. The crude mixture of fluorene and fluorenone will be separated by column chromatography in the next lab period. The beaker containing the toluene extract can be allowed to stand in your hood (labeled properly) until the next lab period. The toluene will evaporate in the interim. Alternatively, if you have time, reduce the volume of solvent (~1 mL) by heating gently on a sand bath. Insert a boiling stick if you do this.

**Cleaning Up.** Carefully dilute the strongly basic aqueous reaction layer 10- to 20-fold with water in a large beaker, add the hydrochloric acid washes, and then neutralize by slowly adding additional hydrochloric acid. Flush this and the sodium chloride washes down the drain with lots of water. If local regulations allow, evaporate any residual solvent from the drying agents in the hood and place the dried solid in the nonhazardous waste container. Otherwise, place the wet drying agent in a waste container designated for this purpose.

## Column Chromatography of the Fluorene-Fluorenone Mixture

Set up a sand bath at a heating setting of 50 before you do anything else.

Figure 9.1 on page 197 shows the typical setup you will use for the chromatographic separation of fluorene and fluorenone, and a general outline of the procedure is given on page 196. It is essential to have at least 10 clean 10-mL Erlenmeyer flasks, reaction tubes, small beakers, test tubes, or vials available to collect the chromatography fractions as they elute. At least two of these should be weighed, with their tare weight recorded. Once you have this done and the column is assembled, you can pack the column with alumina according to the following instructions.

### Packing the Column

Before you assemble the column, check the small plug that fits into the bottom of the column to make sure that it has a small fritted disk inside. Next, make sure that the plug fits snugly into the glass column and is not easy to pull out. If it is loose, get a new bottom plug from the stockroom. Finish assembling the chromatography column as depicted in Figure 9.1. Be sure to clamp the column securely and vertically.

Grasp the valve with one hand and turn it with the other. Close the valve and fill the column with hexanes to the bottom of the plastic funnel. Weigh out approximately 4.5 g of activity grade III alumina in a small beaker and slowly sprinkle the dry alumina into the hexanes in the column while you tap the column with a pen or pencil. It may be necessary to drain off some of the solvent to keep it from flowing over the top. This amount of alumina should fill the column to a height of about 10 cm. It is extremely important *not* to let the column run dry at any time. This will allow air to enter the column, which will result in uneven bands and poor separation.

After all of the alumina has been added to the column, open the stopcock and continue to tap the column as you allow the solvent to drain slowly until the solvent just barely covers the surface of the alumina, collecting the solvent in an Erlenmeyer flask.

### Adding the Sample

It is important to use a minimum amount of solvent when dissolving the sample. If too much is used, poor separation will result.

The solvent is drained just to the surface of the alumina, which should be perfectly flat. Dissolve the crude mixture of fluorene and fluorenone in 10 drops of dichloromethane and 10 drops of toluene and add this with a pipette to the surface of the alumina. Be sure to add the sample as a solution; should any sample crystallize, add 1 more drop of dichloromethane. (This is done so that the sample to be added to the column is in the most concentrated solution possible.) Drain some liquid from the column until the dichloromethane-toluene solution just barely covers the surface of the alumina. Then add a few drops of hexanes and drain out some solvent until the liquid just covers the alumina. Repeat until the sample is seen as a narrow band at the top of the column. Carefully add a 4–5 mm layer of sand and then fill the column with hexanes.

Collect 3-mL fractions in a combination of small vials, 10-mL Erlenmeyer flasks, 13 × 100 mm test tubes, vials, and small beakers. You will probably collect close to ten 3-mL fractions. While the chromatography is running, you will be determining the amount of fluorene or fluorenone in each fraction by TLC. Once you determine which fractions contain which compound, you will combine the "like" fractions and evaporate the solvent.

After you collect each 3-mL fraction in a flask, apply it to a TLC plate by spotting it 2 or 3 times on the plate in the same location. Four or five fractions can be applied and analyzed at the same time using one TLC plate. Allow the solvent to completely evaporate from the spot and examine the plate under a UV lamp to determine if there is any material present, as evidenced by a dark blue spot. The TLC plates can be developed using 20% dichloromethane in hexanes. You will probably collect a few fractions that contain little or no material; these fractions are likely to be the first or the middle of the series.

After spotting and developing each fraction on the TLC plate, combine like fractions and immediately start to boil off the solvent on the sand bath *in the hood* using a boiling stick broken in half. Tilt the flasks and vials on their side as much as possible to allow the heavy vapors to escape. As soon as all the liquid seems to have boiled off, set the flask *on its side* on the bench top in the hood to allow the last traces of heavy solvent vapors to escape. After the flask has cooled to room temperature, if that fraction contains any material, crystals may appear. If an oily, gooey residue is present, you may have to scratch it with a glass stirring rod to induce crystallization. With good organizational effort, you can do the TLC analysis and evaporate off the solvent at about the same rate at which you collect fractions; thus you can follow the progress of the chromatography simply by noting the amount of material in each flask, vial, or test tube. If, after solvent removal and cooling, the flasks are perfectly clean on careful inspection, they can be used to collect subsequent fractions.

If the yellow band has not moved one third of the way down the column after two fractions have been collected, you can speed up the elution by replacing the hexanes solvent at the top of the column with 20% dichloromethane in hexanes. Once the first component has completely eluted, you can speed up the elution of the second component by using 50% dichloromethane in hexanes. Decide when the product has been completely eluted from the column by using visual cues and TLC. Using a few drops of dichloromethane, wash all the fractions that contain fluorene as determined by TLC analysis into a tared container. Do the same for the fluorenone fractions. Evaporate the dichloromethane and obtain dry weights for the product and the recovered fluorene.

Mark all compound spots on all TLC plates with a pencil. Tape your developed TLC plates in your notebook with wide transparent tape. Calculate the theoretical yield and the percent yield of your pure fluorenone both with and without taking into account the amount of fluorene starting material recovered. Calculate the percent recovery of fluorene.

**Cleaning Up.** When you are done with the column, pour out the excess solvent into the proper waste container, pull out the bottom, and leave the wet column propped in a beaker in your desk hood. The column will dry out by the next lab, and the dry, used alumina can then be easily emptied out into a waste bin.

# 2. Chromium(VI) Oxidation of Fluorene to Fluorenone

Chromium(VI) oxidations are less favored today because of environmental concerns based on chromium's toxicity.

The 9-position of fluorene is unusually reactive for a hydrocarbon. The protons on this carbon atom are acidic by virtue of being doubly benzylic, and, consequently, this carbon can be oxidized by several reagents, including elemental oxygen. In this experiment, the very powerful and versatile oxidizing agent chromium(VI), in the form of chromium trioxide, is used to carry out the oxidation. Chromium(VI) in a variety of other forms is used for about a dozen oxidation reactions in this text. The *dust* of chromium(VI) salts is reported to be a carcinogen, so avoid breathing it.

**Fluorene**
mp 114°C
MW 166.22

$$\xrightarrow[\text{CH}_3\text{COOH}]{\text{Na}_2\text{Cr}_2\text{O}_7}$$

**Fluorenone**
mp 83°C
MW 180.21

## Microscale Procedure

> **IN THIS EXPERIMENT** the hydrocarbon fluorene is oxidized to the ketone fluorenone by sodium dichromate in acetic acid with heating. The mixture is diluted with water, and the crude product is isolated by filtration; then—in a standard procedure—it is dissolved in ether. The ether is dried and evaporated to give a mixture of fluorene and fluorenone, which is separated by column chromatography.

**CAUTION**: Sodium dichromate is toxic. The dust is corrosive to nasal passages and skin and is a suspected carcinogen. Hot acetic acid is very corrosive to skin. Handle sodium dichromate and acetic acid in the hood; always wear gloves.

In a reaction tube dissolve 50 mg of fluorene in 0.25 mL of acetic acid by heating and add this hot solution to a solution of 0.15 g of sodium dichromate dihydrate in 0.5 mL of acetic acid. Heat the reaction mixture to 80°C for 15 min in a hot water bath; then cool it and add 1.5 mL of water. Stir the mixture for 2 min; then filter it on a Hirsch funnel. Wash the product well with water and press out as much water as possible. Return the product to the reaction tube, add 2 mL of ether, and add anhydrous calcium chloride pellets until it no longer clumps together. Cork and shake the tube and allow the product to dry for 5–10 min before evaporating the ether in another tared reaction tube. Use ether to wash off the drying agent and to complete the transfer of product. Use this ether solution to spot a TLC plate. This crude mixture of fluorene and fluorenone will be separated by column chromatography.

### Column Chromatography of the Fluorene-Fluorenone Mixture

**Online Study Center**

Photo: Column Chromatography;
Video: Column Chromatography

The sample can also be applied to the column using the technique described on page 196.

Prepare a microscale chromatographic column exactly as described at the beginning of this chapter (*see* Fig. 9.1 on page 197). Use alumina as the adsorbent. Dissolve the crude mixture of fluorene and fluorenone in a mixture of 10 drops of dichloromethane and 10 drops of toluene and add this to the surface of the alumina. Be sure to add the sample as a solution; should any sample crystallize, add 1 more drop of dichloromethane. Run the hexanes down to the surface of the alumina, add a few drops more of hexanes, and repeat the process until the sample is seen as a narrow band at the top of the column. Carefully add a 3-mm layer of sand, fill the column with hexanes, and collect 5-mL fractions in tared 10-mL Erlenmeyer flasks. Sample each flask for TLC (*see* Fig. 9.5 on page 199) and evaporate each to dryness. Final drying can be done under vacuum using the technique shown in Figure 9.6. You are to decide when all of the product has been eluted from the column. The TLC plates can be developed using 20% dichloromethane in hexanes. Combine fractions that are identical and determine the melting points of the two substances.

**Cleaning Up.** The filtrate probably contains unreacted dichromate. To destroy it, add 3 $M$ sulfuric acid until the pH is 1; then complete the reduction by adding solid sodium thiosulfate until the solution becomes cloudy and blue colored. Neutralize with sodium carbonate and then filter the flocculent precipitate of $Cr(OH)_3$ through Celite in a Büchner funnel. If regulations allow, the filtrate can be diluted with water and flushed down the drain; otherwise, it should be disposed of in a designated waste container. The precipitate and Celite should be placed in the heavy metal hazardous waste container.

## Macroscale Procedure

Handle dichromate and acetic acid in laboratory hood; always wear gloves.

One-half hour unattended heating

In a 250-mL Erlenmeyer flask dissolve 5.0 g of practical grade fluorene in 25 mL of acetic acid by heating on a steam bath with occasional swirling. In a 125-mL Erlenmeyer flask dissolve 15 g of sodium dichromate dihydrate in 50 mL of acetic acid by swirling and heating on a hot plate. Adjust the temperature of the dichromate solution to 80°C, transfer the thermometer, and adjust the fluorene–acetic acid solution to 80°C; then, *in the hood*, pour in the dichromate solution. Note the time and the temperature of the solution and heat on a steam bath for 30 min. Observe the maximum and final temperature; then cool the solution and add 150 mL of water. Swirl the mixture for a full 2 min to coagulate the product and promote rapid filtration; collect the yellow solid in an 8.5-cm Büchner funnel using vacuum filtration (if filtration is slow, empty the funnel and flask into a beaker and stir vigorously for a few minutes). Wash the filter cake well with water and then suck the filter cake as dry as possible. Either let the product dry overnight or dry it quickly as follows: Put the moist solid into a 50-mL Erlenmeyer flask, add ether (20 mL) and swirl to dissolve, and add anhydrous calcium chloride (10 g) to scavenge the water. Decant the ethereal solution through a cone of anhydrous calcium chloride in a funnel into a 125-mL Erlenmeyer flask; then rinse the flask and funnel with ether. Evaporate on a steam bath under an aspirator, heat until all the ether is removed, and pour the hot oil into a 50-mL beaker to cool and solidify. Scrape out the yellow solid. The yield should be about 4.0 g.

### Separation of Fluorene and Fluorenone

Prepare a column of 12.5 g of alumina, run out excess solvent, and pour onto the column a solution of 0.5 g of fluorene-fluorenone mixture. Elute at first with hexanes and use tared 50-mL flasks as receivers. The yellow color of fluorenone provides one index of the course of the fractionation, and the appearance of solid around the delivery tip provides another. Frequently wash the solid on the tip into the receiver with ether. When you believe that one component has been eluted completely, change to another receiver until you judge that the second component is beginning to appear. Then, when you are sure the second component is being eluted, change to a 1:1 hexanes and ether mixture and continue until the column is exhausted. It is possible to collect practically all of the two components in the two receiving flasks, with only a negligible intermediate fraction. After evaporation of the solvent, evacuate each flask under vacuum (Fig. 9.7) and determine the weight and melting point of the products. A convenient method for evaporating fractions is to use a rotary evaporator (Fig. 9.8).

**■ FIG. 9.6**
Drying a solid by reduced air pressure.

**Cleaning Up.** All organic material from this experiment can go in the organic solvents container. If local regulations allow, evaporate any residual solvent from

■ **FIG. 9.7**
An aspirator tube in use. A boiling stick may be necessary to promote even boiling.

Aspirator tube
To water aspirator
Clamp here
Tared flask
Boiling stick

Steam in
Outlet to sink
Steam bath (or sand bath)

■ **FIG. 9.8**
A rotary evaporator. The rate of evaporation with this apparatus is very fast due to the thin film of liquid spread over the entire inner surface of the rotating flask, which is heated under vacuum. Foaming and bumping are also greatly reduced.

To aspirator
Condenser
Liquid to be evaporated
Condensate
Ice

the alumina in the hood and place the dried solid in the nonhazardous waste container. Otherwise, place the wet drying agent in a waste container designated for this purpose.

## 3. Acetylation of Cholesterol

Cholesterol is a solid alcohol; the average human body contains about 200 g distributed in brain, spinal cord, and nerve tissue and occasionally clogs the arteries and the gall bladder.

In the following experiment, cholesterol is dissolved in acetic acid and allowed to react with acetic anhydride to form the ester cholesteryl acetate. The reaction does not take place rapidly and consequently does not go to completion under the conditions of this experiment. Thus, when the reaction is over, both unreacted cholesterol and the product, cholesteryl acetate, are present. Separating these by fractional crystallization would be extremely difficult; but because they differ in polarity (the hydroxyl group of cholesterol is the more strongly adsorbed on alumina), they are easily separated by column chromatography. Both molecules are colorless and hence cannot be detected visually. Each fraction should be sampled by TLC. In that way not only the presence but also the purity of each fraction can be assessed. It is also possible to put 1 drop of each fraction on a watch glass and evaporate it to see if the fraction contains product. Solid will also appear on the tip of the column while a compound is being eluted.

### *Microscale Procedure*

**IN THIS EXPERIMENT** cholesterol is refluxed with acetic acid and acetic anhydride. In the standard procedure the mixture is diluted with water and extracted with ether, and the ether is dried and evaporated. The resulting mixture of cholesterol and cholesteryl acetate is separated by column chromatography on silica gel, eluting with hexanes and then hexanes and ether mixtures. Cholesteryl acetate comes off first, followed by cholesterol.

Cholesterol

Acetic anhydride

Cholesteryl acetate

$+ CH_3COOH$

In a reaction tube, add 0.5 mL of acetic acid to 50 mg of cholesterol. The initial thin slurry may set into a stiff paste of the molecular complex consisting of one molecule of cholesterol and one molecule of acetic acid. Add 0.10 mL of acetic anhydride and a boiling chip and gently reflux the reaction mixture on a hot sand bath for no more than 30 min (*see* Fig. 1.1 on page 2).

While the reaction is taking place, prepare the microscale chromatography column as described previously, using silica gel as the adsorbent. Refer to Figure 9.1 on page 197 and the associated procedure earlier in the chapter. Cool the mixture, add 2 mL of water, and extract the product with three 2-mL portions of ether that are placed in the 15-mL centrifuge tube. Wash the ether extracts in the tube with two 2-mL portions of water and one 2.5-mL portion of 3 *M* sodium hydroxide (these three washes remove the acetic acid) and dry the ether by shaking it with 2.5 mL of saturated sodium chloride solution. Then complete the drying by adding enough anhydrous calcium chloride pellets to the solution so that the drying agent does not clump together.

Shake the ether solution with the drying agent for 10 min; then transfer it in portions to a tared reaction tube and evaporate to dryness. Use 1 drop of this ether solution to spot a TLC plate for later analysis. If the crude material weighs more than the theoretical weight, you will know that it is not dry or that it contains acetic acid, which can be detected by its odor. Dissolve this crude cholesteryl acetate in the minimum quantity of ether and apply it to the top of the chromatography column.

To prevent the solution from dribbling from the pipette, use a pipette pump to make the transfer. It also could be applied as a dry powder adsorbed on silica gel, as in Experiment 1. Elute the column with hexanes, collecting two 5-mL fractions in tared 10-mL Erlenmeyer flasks or other suitable containers. Add a boiling stick to each flask and evaporate the solvent under an aspirator tube on a steam bath or a sand bath (*see* Fig. 9.7). If the flask appears empty, it can be used to collect later fractions. Lower the solvent layer to the top of the sand and elute with 15 mL of a 70:30 mixture of hexanes and ether, collecting five 3-mL fractions. Evaporate the solvent under an aspirator tube (Fig. 9.7). The last traces of solvent can be removed using reduced pressure, as shown in Figure 9.6 on page 206. Follow the 70:30 mixture with 10 mL of a 50:50 mixture of hexanes and ether, collecting four 2-mL fractions. Save any flask that has any visible residue. Analyze the original mixture and each fraction by TLC on silica gel plates using a 1:1 mixture of hexanes and ether to develop the plates and either UV light or iodine vapor to visualize the spots.

Cholesteryl acetate (mp 115°C) and cholesterol (mp 149°C) should appear, respectively, in early and late fractions with a few empty fractions (no residue) in between. If so, combine consecutive fractions of early and late material and determine the weights and melting points. Calculate the percentage of the acetylated material compared to the total recovered and calculate the percentage yield from cholesterol.

**Cleaning Up.** After neutralization, acetic acid, the aqueous layers, and the saturated sodium chloride layers from the extraction can be flushed down the drain with water. Ether, hexanes, and TLC solvents should be placed in the organic solvents container. If local regulations allow, evaporate any residual solvent from the drying agents and the chromatography packing in the hood and place the dried solid in the nonhazardous waste container. Otherwise, place the wet materials in waste containers designated for this purpose.

## *Macroscale Procedure*

Carry out procedure in laboratory hood.

Cover 0.5 g of cholesterol with 5 mL of acetic acid in a small Erlenmeyer flask; swirl and note that the initially thin slurry soon sets to a stiff paste of the molecular compound $C_{27}H_{45}OH \cdot CH_3CO_2H$. Add 1 mL of acetic anhydride and heat the mixture on a steam bath for any convenient period of time from 15 min to 1 h; record the actual heating period. While the reaction takes place, prepare the chromatographic column. Cool the reaction mixture, add 20 mL of water, and extract with two 25-mL portions of ether. Wash the combined ethereal extracts twice with 15-mL portions of water and once with 25 mL of 3 *M* sodium hydroxide and dry by shaking the ether extracts with 25 mL of saturated sodium chloride solution; then dry the ether over anhydrous calcium chloride pellets for 10 min in an Erlenmeyer flask, filter, and evaporate the ether. Save a few crystals of this material for TLC analysis. Dissolve the residue in 3–4 mL of ether, transfer the solution with a Pasteur pipette onto a column of 12.5 g of silica gel, and rinse the flask with another small portion of ether.[1] In order to apply the ether solution to the top of the sand and avoid having it coat the interior of the column, pipette the solution down a 6-mm-diameter glass tube that is resting on the top of the sand. Label a series of

---

1. Ideally, the material to be adsorbed is dissolved in hexanes (ligroin), the solvent of lowest elutant power. The present mixture is not soluble enough in hexanes, and so ether is used, but the volume is kept to a minimum.

■ **FIG. 9.9**
A bubbler for adding solvent
automatically.

50-mL Erlenmeyer flasks as fractions 1 to 10. Open the pinchclamp, run the eluant solution into a 50-mL Erlenmeyer flask, and as soon as the solvent in the column has fallen to the level of the upper layer of sand, fill the column with a part of a measured 125 mL of a 70:30 mixture of hexanes and ether. When about 25 mL of eluant has collected in the flask (fraction 1), change to a fresh flask; add a boiling stone to the first flask and evaporate the solution to dryness on a steam bath under an aspirator tube (*see* Fig. 9.7 on page 207). Evacuation using an aspirator helps to remove last traces of ligroin (*see* Figure 9.6 on page 206). If fraction 1 is negative (no residue), use the flask for collecting further fractions. Continue adding the hexanes and ether mixture until the 125-mL portion is exhausted; then use 100 mL of a 1:1 hexanes and ether mixture. A convenient bubbler (Fig. 9.9) made from a 125-mL Erlenmeyer flask, a short piece of 10-mm-diameter glass tubing, and a cork will automatically add solvent. A separatory funnel with a stopper and a partially open stopcock serves the same purpose. Collect and evaporate successive 25-mL fractions of eluant. Save any flask that has any visible solid residue. The ideal method for the removal of solvents involves the use of a rotary evaporator (*see* Fig. 9.8 on page 207). Analyze the original mixture and each fraction by TLC on silica gel plates using a 1:1 mixture of hexanes-ether to develop the plates and either UV light or iodine vapor to visualize the spots.

Cholesteryl acetate (mp 115°C) and cholesterol (mp 149°C) should appear, respectively, in early and late fractions with a few empty fractions (no residue) in between. If so, combine consecutive fractions of early and of late material and determine the weights and melting points. Calculate the percentage of the acetylated material compared to the total recovered and compare your result with those of others in your class employing different reaction periods.

**Cleaning Up.** After neutralization, acetic acid, the aqueous layers, and the saturated sodium chloride layers from the extraction can be flushed down the drain with water. Ether, hexanes, and TLC solvents should be placed in the organic solvents container. Hexanes and ether from the chromatography go into the organic solvents container. If local regulations allow, evaporate any residual solvent from the drying agents and silica gel in the hood and place the dried solid in the nonhazardous waste container. Otherwise, place the wet drying agent and silica gel in waste containers designated for this purpose.

# 4. Chromatography of a Mixture of Ferrocene and Acetylferrocene

**IN THIS EXPERIMENT** a mixture of two compounds is separated by chromatography on alumina. The first compound to come down the column is ferrocene (a yellow band). The solvent polarity is changed so that acetylferrocene is eluted as an orange band. The solvents are evaporated from the collection flasks and the compounds recrystallized. Both of these compounds are colored (*see* Chapter 49 for their preparation), so it is easy to follow the progress of the chromatographic separation.

Prepare the microscale alumina column exactly as described at the beginning of the chapter. Then add a dry slurry of 90 mg of a 50:50 mixture of acetylferrocene

(a)

**Ferrocene**
MW 186.04
mp 172–174°C

(b)

Fe with CH₃ and C=O group

**Acetylferrocene**
MW 228.08
mp 85–86°C

### ■ FIG. 9.10

Evaporation of a low-boiling liquid under vacuum. Heat is supplied by the hand, the contents of the flask are swirled, and the vacuum is controlled with the thumb.

To aspirator

**Isoprene**

and ferrocene that has been adsorbed onto 300 mg of alumina, following the procedure for preparing and adding the sample given at the beginning of the chapter.

Carefully add hexanes to the column, open the valve (use both hands), and elute the two compounds. The first to be eluted, ferrocene, will be seen as a yellow band. Collect this in a 10-mL flask. Any crystalline material seen at the tip of the valve should be washed into the flask with a drop or two of ether. Without allowing the column to run dry, add a 50:50 mixture of hexanes and ether, and elute the acetylferrocene, which will be seen as an orange band. Collect it in a 10-mL flask. Spot a silica gel TLC plate with these two solutions. Evaporate the solvents from the two flasks and determine the weights of the residues. An easy way to evaporate the solvent is to place it in a tared 25-mL filter flask and heat the flask in your hand under vacuum while swirling the contents (Fig. 9.10).

Recrystallize the products from the minimum quantities of hot hexanes. Isolate the crystals, dry them, and determine their weights and melting points. Calculate the percent recovery of the crude and recrystallized products based on the 45 mg of each in the original mixture.

The TLC plate is eluted with a 30:1 mixture of toluene and absolute ethanol. Do you detect any contamination of one compound by the other?

**Cleaning Up.** If regulations allow, empty the chromatography column onto a piece of aluminum foil in the hood. After the solvent has evaporated, place the alumina and sand in the nonhazardous waste container. Otherwise, place the wet alumina and sand in a designated waste container. Evaporate the crystallization mother liquor to dryness and place the residue in the hazardous waste container.

## 5. Isolation of Lycopene and β-Carotene

Lycopene, the red pigment in tomatoes, is a $C_{40}$-carotenoid made up of eight five-carbon isoprene units. β-Carotene, the yellow pigment of the carrot, is an isomer of lycopene in which the double bonds at $C_1$—$C_2$ and $C'_1$—$C'_2$ are replaced by bonds extending from $C_1$ to $C_6$ and from $C'_1$ to $C'_6$ to form rings. The chromophore in each case is a system of 11 all-*trans* conjugated double bonds; the closing of the two rings renders β-carotene less highly pigmented than lycopene.

These colored hydrocarbons have been encountered in the TLC experiment (*see* Chapter 8). The isolation procedure described here affords sufficient carotene and lycopene to carry out analytical spectroscopy and some isomerization reactions. It might be of interest if some students isolate carotene from strained carrot baby food while others isolate lycopene from tomato paste.

Lycopene is responsible not only for the red color of tomatoes but also of red grapefruit and flamingos. If flamingos do not include foods containing lycopene in their diet, they will be white.

Until recently, lycopene was thought to have no utility until a study showed a lower incidence of prostate cancer among men who consumed such foods as spaghetti and pizza (but not tomato juice). It is theorized that the lycopene, which is insoluble in water, is dissolved in the fat of pasta sauce and pizza and thus absorbed through the intestine. Carotene does not have the same anticancer effect.

**Lycopene (C$_{40}$H$_{56}$)**
MW 536.85
mp 173°C, $\lambda_{max}^{hexane}$ 475 nm

**β-Carotene (C$_{40}$H$_{56}$)**
mp 183°C, $\lambda_{max}^{hexane}$ 451 nm

Lycopene is the predominant carotenoid in blood plasma and prostate tissue. It is not converted into vitamin A as carotene is, but it is a powerful antioxidant and is an efficient scavenger of singlet oxygen.

Carotenoids are highly sensitive to photochemical air oxidation; therefore, protect solutions and solids from undue exposure to light and heat and work as rapidly as possible. Do not heat solutions when evaporating solvents and, if possible, flush apparatus with nitrogen to exclude oxygen. Research workers isolate these compounds in dimly lit rooms and/or wrap all containers and chromatographic columns in aluminum foil and carry out extractions and crystallizations using solvents that have been deoxygenated.

## Dehydration and Extraction of Tomato or Carrot Paste

**IN THIS EXPERIMENT** a vegetable paste is stirred with acetone to remove water, but not the coloring matter, from the paste. The mixture is filtered, the yellow filtrate discarded, and the material on the filter squeezed as dry as possible. This solid is then extracted three times with dichloromethane; the solution is dried over calcium chloride and evaporated at room temperature under vacuum to leave the crude carotenoids.

Add 5 g of tomato or carrot paste to a 15-mL centrifuge tube or 25 × 150-mm test tube; then add about 7 mL of acetone and stir the paste for several minutes until it is no longer gummy. This acetone treatment removes most of the water from the cellular mixture. Filter the mixture on a small Büchner funnel. Scrape out the tube with a spatula, let it drain thoroughly, and squeeze out as much liquid as possible from the solid residue in the funnel with a spatula. Discard the yellow filtrate. Then return the solid residue to the centrifuge tube and add 5 mL of dichloromethane to effect extraction. Cap the tube and shake the mixture vigorously. Filter the mixture on a Büchner funnel once more, repeat the extraction and filtration with two or three further 5-mL portions of dichloromethane, clean the tube thoroughly, and place the filtrates in it. Dry the solution over anhydrous calcium chloride pellets, filter the solution into a small flask, and evaporate the solution to dryness with a stream of nitrogen or under vacuum using a rotary evaporator (Fig. 9.8 on page 207) or the apparatus shown in Figure 9.10 on page 211, never heating the sample above 50°C.

Determine the weight of the crude material. It will be very small. If the residue is dry, as it should be, add just enough dichloromethane to dissolve the residue. Save 1 drop of this solution to carry out a TLC analysis (using dichloromethane as the eluent on silica gel plates; *see* Chapter 8). Then add 200 mg of alumina to the remaining dichloromethane solution and evaporate the mixture to dryness, again without heat.

## Column Chromatography

*Online Study Center*

Video: Column Chromatography;
Photo: Column Chromatography

The crude carotenoid is to be chromatographed on an 8-cm column of basic or neutral alumina, prepared with hexanes as solvent (see the detailed procedure at the beginning of this chapter). Run out excess solvent or remove it from the top of the chromatography column with a Pasteur pipette. Using the dry sample loading method described at the beginning of this chapter, add the 300 mg of alumina that has the crude carotenoids absorbed on it. Add a few drops of hexanes to wash down the inside of the chromatography column and to consolidate the carotenoid mixture at the top of the column. Elute the column with hexanes, discard the initial colorless eluate, and collect all yellow or orange eluates together. Place a drop of solution on a microscope slide and evaporate the remainder to dryness using a stream of nitrogen or a rotary evaporator (*see* Fig. 9.8 on page 207). Examination of the material spotted on the slide may reveal crystallinity. If you are using tomato paste, a small amount of yellow β-carotene will come off first, followed by lycopene. Collect the red lycopene separately by eluting with a mixture of 10% acetone in hexane and also evaporate that solution to dryness.

Finally, dissolve the samples obtained by evaporating the solvent in the least possible amount of dichloromethane and carry out TLC of the two products in order to ascertain their purity (*see* Chapter 8, Experiment 2). You may want to combine your purified products with those of several other students, evaporate the solution to dryness, dissolve the residue in deuterochloroform, $CDCl_3$, and determine the $^1H$ NMR spectrum (*see* Chapter 12). Also obtain an infrared spectrum and a visible spectrum (in hexane).

Note that β-carotene is in demand as a source of vitamin A and is manufactured by an efficient synthesis. Until very recently no use for lycopene had been found.

**Cleaning Up.** Place recovered and unused dichloromethane in the halogenated organic waste container; the solvents used for TLC in the organic solvents container. If local regulations allow, evaporate any residual solvent from the drying agents in the hood and place the dried solid in the nonhazardous waste container. Otherwise, place the wet drying agent in a waste container designated for this purpose. Used plant material and dry TLC plates can be discarded in the nonhazardous waste container.

## For Further Investigation

The carotenoids of any leaf can be isolated in the manner described in this experiment. Grind the leaf material (about 10 g) in a mortar with some sand; then follow the above procedure. Waxy leaves do not work well. The carotenoids are present in the leaf during its entire life span, so a green leaf from a maple tree or euonymus shrub, also known as burning bush, known to turn bright red in the fall will show lycopene even when the leaf is green. In the fall the chlorophyll decomposes before the carotenoids, so the leaves appear in a variety of orange and red hues.

It is of interest to investigate the carotenoids of the tomato, of which there are some 80 varieties. The orange-colored tangerine tomato contains an isomer of lycopene. If a hexane solution of the prolycopene from this tomato is treated with a drop of a very dilute solution of iodine in hexane and then exposed to bright light, the solution will turn deep-orange in color, indicating that a *cis*-double bond has isomerized to the *trans* form. The product is, however, still not identical to natural lycopene.

## Isomerization

Prepare a hexane solution of either carotene or lycopene and save a drop for TLC. Treat the solution with a very dilute solution of iodine in hexane, expose the resulting mixture to strong light for a few minutes, and then carry out TLC on the resulting solution. Also compare the visible spectrum before and after isomerization.

Iodine serves as a catalyst for the light-catalyzed isomerization of some of the *trans*-double bonds to an equilibrium mixture containing *cis*-isomers.

## QUESTIONS

1. Predict the order of elution of a mixture of triphenylmethanol, biphenyl, benzoic acid, and methyl benzoate from an alumina column.
2. What would be the effect of collecting larger fractions when carrying out the experiments in this chapter?
3. What would have been the result if a large quantity of petroleum ether alone were used as the eluent in either of the experiments described?
4. Once the chromatographic column has been prepared, why is it important to allow the level of the liquid in the column to drop to the level of the alumina before applying the solution of the compound to be separated?
5. A chemist started to carry out column chromatography on a Friday afternoon, reached the point at which the two compounds being separated were about three-fourths of the way down the column, and then returned on Monday to find that the compounds came off the column as a mixture. Speculate the reason for this. The column had not run dry over the weekend.
6. Write a detailed mechanism for the formation of fluorenone from fluorene. Explain the purpose of the phase-transfer catalyst.
7. The primary cause of low yields in the isolation of lycopene is oxidation of the product during the procedure. Once crystalline, it is reasonably stable. Speculate on the primary products formed by photochemical air oxidation of lycopene.

**Online Study Center**

**General Resources**
Web Links

# CHAPTER

# 10

# Gas Chromatography: Analyzing Alkene Isomers

**Online Study Center**

This icon will direct you to techniques, equipment setups, and online resources at http://college.hmco.com/PIC/ williamsonMMOE5e.

**PRELAB EXERCISE:** If the dehydration of 2-methyl-2-butanol occurred on a purely statistical basis, what would be the relative proportions of 2-methyl-1-butene and 2-methyl-2-butene?

Gas chromatography (GC) is a rapid and sensitive method of separating and analyzing mixtures of gaseous or liquid compounds. The information it provides can tell you

1. if you have successfully synthesized your product;
2. whether your product contains unreacted starting material or other impurities;
3. whether your product is a mixture of isomers; and
4. the relative amounts of different materials or isomers in a mixture.

Gas chromatography is one of several *instrumental* analysis methods used by organic chemists. Unlike the apparatus for thin-layer (TLC) or column chromatography, the apparatus used for gas chromatography—a modern gas chromatograph like that shown in Figure 10.1—costs thousands of dollars and is therefore shared among many users. There are two major reasons for spending this much money for this instrument. First, very complex mixtures containing hundreds of components can be separated. Second, separated components can be detected in very small amounts, $10^{-6}$ to $10^{-15}$ g. The chromatogram in Figure 10.2 is a good example of the power of gas chromatography; it shows the separation of 1 mg of crude oil into the hundreds of compounds that are present: alkanes, alkenes, and aromatics, including many isomers. The exceptional sensitivity of GC instruments is the major reason they are often used in environmental and forensic chemistry labs, where the detection of trace amounts is necessary.

**■ FIG. 10.1**
A typical modern gas chromatograph with a data acquisition and analysis computer.

**■ FIG. 10.2**
A chromatogram from the GC analysis of 1 mg of crude petroleum. The larger peaks represent the series of *n*-alkanes from $C_5$ to >$C_{30}$. The column was temperature programmed from 40°C to 300°C at 20°C/min.

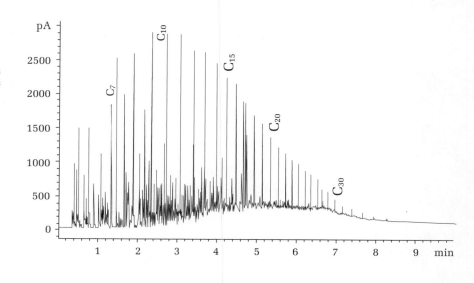

**Polydimethylsiloxane**

**Carbowax**

Gas chromatography involves the same principles that apply to all forms of chromatography, which were covered in Chapter 8. The mobile phase is a gas, usually helium. The stationary phase is often a fairly nonpolar polymer such as polydimethylsiloxane or poly(ethylene glycol) (Carbowax) that is stable at temperatures as high as 350°C. The most commonly used phase is polydimethylsiloxane in which 5% of the methyl groups have been replaced by phenyl groups.

There are two types of GC columns:

1. *Packed columns* in which the stationary phase consists of solid particles similar to the alumina or silica gel used in column chromatography but coated with a nonpolar polymer (e.g., polydimethylsiloxane or Carbowax) and packed into 3–6 mm diameter metal or glass tubes that are 1–3 m long and rolled into a compact coil.

2. *Capillary columns* are long, thin, flexible quartz tubes with a very thin coating of a nonpolar stationary phase polymer on the inside wall. They can be 15–60 m long with internal diameters of 0.1–0.5 mm. Their successful development came about because of the invention of fiber optics and the discovery that coating the outside of hollow quartz fibers with a polyimide polymer prevented them from being easily scratched and broken.

Packed columns can separate milligram quantities of materials, while capillary columns work best with microgram quantities or less. On the other hand, capillary columns have much better separating power than packed columns.

The GC instrument has the following components (Fig. 10.3). Helium at a pressure of 10–30 psi from a compressed gas cylinder flows through the heated injector port, the column (located in a temperature-controlled oven), and the heated detector at a flow rate of 10–60 mL/min for a packed column and about 1 mL/min for a capillary column. A microliter syringe is used to inject 1–25 $\mu$L of sample through a rubber septum into the hot (~250°C) injector, where any liquid compounds are instantly vaporized to the gaseous state. As the vaporized molecules are swept through the column, they interact with the stationary phase and are adsorbed and desorbed many times each second in a dynamic equilibrium. As discussed in Chapter 8, the equilibria between adsorbed and free states depends on each molecule's size (London forces), polarity, and ability to hydrogen

■ **FIG. 10.3**
A diagram of a gas chromatograph.

Syringe for
sample injection

Injector

Data acquisition and
analysis system

Detector

Carrier gas (He)

Column

Oven

■ **FIG. 10.4**

■ **FIG. 10.4**
The differential adsorption of
two compounds in a flowing gas
leads to the separation of the
compounds over time.

$t_1$

Flow →

$t_2$

Flow →

$t_3$

Flow →

bond. Molecules spending more time absorbed to the stationary phase will not be carried through the column as rapidly as those that are weakly bound and free. A molecular view of this differential adsorption and flow is depicted in Figure 10.4.

    The separated components pass one after the other into the detector. In column chromatography (*see* Chapter 9), human eyes are the detectors, either seeing colored compounds flow out of the column or visualizing the presence of colorless compounds on a fluorescent TLC plate. Gas chromatographs have electronic detectors that produce a signal voltage proportional to the number of molecules passing through them at any instant in time. A record of this voltage versus time is a gas chromatogram, with peak areas representing the amount of each individual component passing through the detector. Figure 10.5 is the chromatogram produced by the injection of a mixture of two compounds that were separated on the

**■ FIG. 10.5**

A gas chromatogram of a mixture of two compounds, A and B.

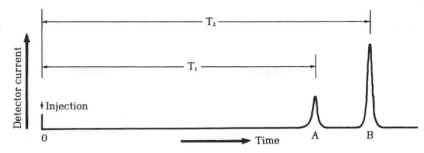

column and detected to produce peaks A and B. The amount of each is proportional to the area under its peak, so we can say that there appears to be about twice as much B as A. If an integrating recorder or computer data acquisition system is used, the exact areas under all peaks are automatically determined and can be used to quantify the amount of each component in a mixture.

Most GC instruments use one of the following types of detectors:

- The *thermal conductivity detector* (TCD) consists of an electrically heated wire or thermistor. The temperature of the sensing element depends on the thermal conductivity of the gas flowing around it. Changes in thermal conductivity, such as when organic molecules displace some of the carrier gas, cause a temperature rise in the element that is sensed as a change in resistance. The TCD is the least sensitive (detecting micrograms per second, $10^{-6}$ g/s, of material) of the four detectors described here, but it is quite rugged. It will detect all types of molecules, not just those containing C—H bonds, and compounds passing through it are unchanged and can be collected.

- The *flame ionization detector* (FID) is at least a thousand times more sensitive than the TCD, easily detecting nanograms per second ($10^{-9}$ g/s) of material. The molecules, however, must contain C—H bonds because the sample is actually burned to form ions. These ions carry a tiny current between two electrodes, which is greatly amplified to produce the signal output.

- The *electron capture detector* (ECD) contains a tiny amount of a radioactive substance, such as $^{63}$Ni, that emits high-energy electrons ($\beta^-$ particles). The sample molecules can capture these electrons, become charged, and carry a current between two electrodes, as in the FID detector. Halocarbons have exceptionally good cross sections for electron capture because of the electronegativity of the halogen atoms, and these types of compounds can be detected at picogram per second ($10^{-12}$ g/s) levels. For this reason, ECD is used to measure levels of chlorocarbon pollutants, such as chloroform, DDT, and dioxin in the environment.

- A *mass spectrometer* detector is the only one of these four detectors that provides information about the structure of the molecules that pass into it. For this reason, combination gas chromatography–mass spectrometry (GC–MS) systems are often used in organic research, forensic science, and environmental analysis in spite of their high cost ($40,000 to $250,000). Sensitivity is at the nanogram to picogram per second level. Chapter 13 discusses mass spectrometry in detail.

The instant you inject your sample into the GC instrument, you normally press a button that starts a clock in the data recorder, either an electronic integrator or a computer data system. Then, as each eluting component produces a peak in the detector, the top of the peak is detected, and the elapsed time since injection, called the *retention time*, is recorded and stored. In Figure 10.5, the retention time of compound A is $T_1$ and of B is $T_2$. Which compound is more strongly absorbed by the stationary phase?

One powerful variable in gas chromatography that is not available in TLC or column chromatography is temperature. Because the GC column is in an oven whose temperature is programmable—that is, the temperature can be raised from near room temperature to as high as 350°C at a constant and reproducible rate—we are able to separate much more complex mixtures than if the oven could be set to only one temperature. At room temperature, a mixture containing lower-boiling, weakly adsorbed components (such as ether and dichloromethane) and higher-boiling, strongly adsorbed components (such as butanol and toluene) might take hours to separate because the latter would move significantly more slowly through the column at low temperature. If the column were kept at a high temperature, the higher-boiling compounds would come out in a shorter time, but the lower-boiling ones would not be retained at all and therefore would not be sep-

*Temperature programming*

arated. By programming the oven temperature, we can inject such a mixture at a low initial temperature so that weakly adsorbed components are separated. Then, as the temperature slowly rises, the more strongly retained components will move through the column faster and also have reasonable retention times. Temperature programming allows the separation of very complex mixtures containing both low-boiling small molecules and high-boiling large ones, as demonstrated by the GC analysis of crude petroleum in Figure 10.2 on page 216.

## The General GC Analysis Procedure

Most GC separations are of gaseous and liquid mixtures. Although some solids will pass through a gas chromatograph at higher temperatures, it is best if all of the samples are distilled or vacuum transferred before injection into the GC instrument to ensure that only volatile compounds are present. If samples contain materials that cannot be vaporized and swept through the GC column, even at high temperatures, these will remain on the column and can ultimately ruin its performance. GC columns are expensive and can cost up to $400 to replace.

The guidelines given here are for GC analysis using a capillary column and an FID detector, which require very dilute solutions to avoid overwhelming the column, even if 90%–95% of the sample is split away so that it does not enter the column, a common practice in capillary chromatography. Samples analyzed with a packed column and a TCD detector can be much more concentrated or even undiluted (neat).

## Sample Preparation

There are different ways of preparing a GC sample, depending on whether your product is a solid, a liquid (usually obtained by distillation), or a gaseous mixture; on the type of column (capillary or packed); and on the type of detector (FID, ECD, or TCD) being used.

### Solid Samples

Low-melting, sublimable solids can be analyzed at higher column temperatures. Run a GC analysis on solids only if your experimental procedure explicitly says that you can do so. To prepare a solid sample for capillary column analysis, put a *small* crystal of the solid into a small vial and dissolve in 1 mL of dichloromethane.

### Liquid Samples

Dilute solutions of liquid organic mixtures in dichloromethane are the most common form of GC samples. Place 1 or 2 drops of your distilled product into a small vial and add 1 mL of dichloromethane. Because GC columns can be damaged by moisture, it is suggested that enough *anhydrous* sodium sulfate be added to cover the bottom of the vial to a depth of 1–2 mm to ensure that all traces of water are removed. The sodium sulfate can be left in the sample because it usually will not go into the fine needle of the microliter syringe used to inject the sample. Injection of 1 µL of a sample prepared in this manner usually provides strong signals in an FID detector. In special cases, undiluted neat or pure liquids can be injected directly without dilution as noted, for example, the methylbutenes in Experiment 1. Also, if preparative gas chromatography is used to separate and collect the components of a mixture, neat liquids are usually injected.

### Gaseous Samples

Gases are normally collected over water, as in Experiment 2. Do not remove the septum-capped collection tube from the beaker of water, but bring it and the beaker of water together to the gas chromatograph. A gas syringe is inserted through the rubber septum and 10–500 µL (depending on the type of the gas chromatograph) of the gas sample is withdrawn and injected into the unit.

## Sample Analysis

Make sure that the required gases are flowing and that the instrument has been on at least 1 h so that all zones, the injector, the oven, and the detector have come to the required temperatures. Enter a temperature program: initial temperature, heating rate (in degrees per min), and final temperature, as required by the particular experiment. Fill the microliter syringe with the sample solution, normally 1 µL (Fig. 10.6a). A 10-µL capacity syringe with a plunger guide is recommended. Handle this expensive syringe carefully; it has a sharp needle, and the glass barrel is easily dropped and broken. Make sure that there are no bubbles in the syringe barrel when you draw up the sample. If bubbles are present, fill the syringe about

■ FIG. 10.6

The steps in sample injection: (a) Fill the microliter syringe. (b) Insert the needle through the injector septum. (c) Depress the plunger. (d) Withdraw the syringe.

(a)                          (b)                          (c)                          (d)

Do not touch the metal injector cap. It is very HOT.

halfway and depress the plunger quickly. Draw up the sample again slowly. Repeat as necessary to remove all bubbles. Inject the sample as shown in Figure 10.6c. To obtain a quality chromatogram, these steps should be done rapidly, and the syringe should be removed as soon as the plunger has been depressed and the sample injected (Fig. 10.6d). Then quickly press the start analysis button on the gas chromatograph, computer, or integrator to start data acquisition or mark the paper on the pen recorder.

Once all the compounds have eluted and the oven temperature program has finished, the run should be stopped. Plot your chromatogram using the computer or remove the chromatogram chart from the integrator or flat bed recorder. Record the GC analysis parameters: column diameter and length, column packing type, carrier gas and its flow rate, column temperature (initial, final, and heating rate if programmed), detector and injector temperatures, sample injection amount and solvent (if used), signal attenuation, and chart speed; keep these with your gas chromatogram. Write the structures of all identifiable peaks on the chromatogram. If you used dichloromethane to dilute your sample, the largest peak will be due to this substance and will have a short retention time.

Remember that retention times are not constants. If the programmed oven temperature rate is higher or lower than for a prior analysis, then the retention times for all of the peaks will be consistently shorter or consistently longer. Changes in helium flow and the aging of the column also affect retention times. Chemists primarily look for similarities in the pattern of eluting peaks, not for perfect matches in retention times. In general, for simple mixtures chromatographed on nonpolar stationary phases, the retention time order is the same as the order of increasing boiling points. For example, you would expect cyclohexene (bp 83°C) to have a shorter retention time than toluene (bp 110°C). You can often ascertain the identity of a certain peak by adding a small amount of a known standard to the sample and rerunning the GC analysis. The peak corresponding to the standard will increase significantly and can be identified. Often the chromatogram will show a number of small peaks that cannot be assigned to the product, the starting material, or the solvents; these are likely due to byproducts. As long as these impurities are minor, say, less than 10% of the total area (not including the solvent), they can probably be ignored. The ultimate method of identifying every peak is to analyze the sample on a GC–MS instrument.

**Cleaning Up.** Rinse out the syringe by filling it with dichloromethane and squirting it onto a tissue or paper towel at least four times. Clean and return any sample collection tubes, if used.

## Collecting a Sample for an Infrared Spectrum

The small amount of sample analyzed in gas chromatography is an advantage in many cases, but it precludes isolating the separated components. Some specialized chromatographs can separate samples as large as 0.5 mL per injection and automatically collect each fraction in a separate container. At the other extreme, gas chromatographs equipped with FIDs can detect micrograms of substances, such as traces of pesticides in food or drugs in blood and urine. Clearly, a gas chromatogram gives little information about the chemical nature of the sample being detected. However, certain *preparative* chromatographs, like that shown in Figure 10.7, allow the collection of enough sample at the exit port to obtain an infrared spectrum. About 10–15 µL of a mixture (not diluted in solvent) is injected, and as the peak for the compound of interest appears, a 2-mm-diameter glass tube, 3 in. long and packed with glass wool, is inserted into the rubber septum at the exit port. The sample, if it is not too volatile, will condense in the cold glass tube. Subsequently, the sample is washed out with 1 or 2 drops of solvent, and an infrared spectrum is obtained. This process can be repeated to collect enough sample for obtaining an NMR spectrum (*see* Chapter 12), using a few drops of deuterochloroform ($CDCl_3$) to wash out the tube each time. See Figure 10.8 for another collection device.

**■ FIG. 10.7**

Some gas chromatographs with packed columns allow the collection of small amounts of separated compounds.

**■ FIG. 10.8**

A gas chromatographic collection device. Fill the container with ice or a dry ice–acetone mixture and attach to the outlet port of the gas chromatograph.[1]

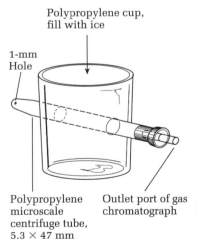

Polypropylene cup, fill with ice

1-mm Hole

Polypropylene microscale centrifuge tube, 5.3 × 47 mm

Outlet port of gas chromatograph

---

1. This apparatus is available from Kimble Kontes (Vineland, NJ).

---

**EXPERIMENTS**

 ## 1. 2-Methyl-1-Butene and 2-Methyl-2-Butene by Dehydration of an Alcohol[2]

The dilute sulfuric acid–catalyzed dehydration of 2-methyl-2-butanol (*t*-amyl alcohol) proceeds readily to give a mixture of alkenes that can be analyzed by gas chromatography. The mechanism of this reaction involves the intermediate formation of the relatively stable tertiary carbocation followed by the loss of a proton either from a primary carbon atom to give the terminal olefin, 2-methyl-1-butene, or from a secondary carbon to give 2-methyl-2-butene.

$$CH_3CH_2\underset{\underset{CH_3}{|}}{\overset{\overset{OH}{|}}{C}}CH_3 + H_2SO_4 \underset{}{\overset{fast}{\rightleftharpoons}} CH_3CH_2\underset{\underset{CH_3}{|}}{\overset{\overset{\overset{+}{O}H_2}{|}}{C}}CH_3 + HSO_4^-$$

**2-Methyl-2-butanol**
bp 102°C
den. 0.805
MW 88.15

$$CH_3CH_2\underset{\underset{CH_3}{|}}{\overset{\overset{\overset{+}{O}H_2}{|}}{C}}CH_3 \overset{slow}{\rightleftharpoons} CH_3CH_2\underset{\underset{CH_3}{|}}{\overset{+}{C}}CH_3 + H_2O$$

$$CH_3CH_2\underset{\underset{CH_3}{|}}{\overset{+}{C}}-CH_2 \rightleftharpoons CH_3CH_2\underset{\underset{CH_3}{|}}{C}=CH_2$$

**2-Methyl-1-butene**
bp 31.16°C
den. 0.662
MW 70.14

---

2. Schimelpfenig, C. W. *J. Chem. Educ.* **1962**, *39*, 310.

$$CH_3CH-\overset{+}{C}CH_3 \overset{|}{\underset{CH_3}{}} \rightleftharpoons CH_3CH=CCH_3 \overset{|}{\underset{CH_3}{}}$$

**2-Methyl-2-butene**
bp 38.57°C
den. 0.662
MW 70.14

2-Methyl-2-butanol can also be dehydrated in high yield using iodine as a catalyst:

$$CH_3CH_2\overset{OH}{\underset{CH_3}{C}}CH_3 \xrightarrow{I_2 \text{ (trace)}} CH_3CH_2\overset{OI}{\underset{CH_3}{C}}CH_3 \xrightarrow{-HOI}$$

$$CH_3CH_2\overset{}{\underset{CH_3}{C}}=CH_2 + CH_3CH=CCH_3\overset{}{\underset{CH_3}{}} + HI + HOI$$

$$HI + HOI \longrightarrow H_2O + I_2$$

Each step of this $E_1$ elimination reaction is reversible, and thus the reaction is driven to completion by removing one of the products, the alkene. Often these reactions produce several alkene isomers. The Saytzeff rule states that the more substituted alkene is the more stable and thus the one formed in larger amount. The *trans*-isomer is more stable than the *cis*-isomer. With this information it should be possible to deduce which peak on the gas chromatogram corresponds to a given alkene and to predict the ratios of the products.

In analyzing your results from this experiment, consider the fact that the carbocation can lose any of six primary hydrogen atoms but only two secondary hydrogen atoms to give the product olefins.

## Microscale Procedure

**IN THIS EXPERIMENT** a tertiary alcohol is dehydrated to a mixture of alkenes in a reaction catalyzed by sulfuric acid. The products are collected by simple distillation into an ice-cold vial. Cold aqueous base is added to remove the acid, and the organic layer is dried with calcium chloride. The products are again distilled and then analyzed by gas chromatography.

■ **FIG. 10.9**
An apparatus for the dehydration of 2-methyl-2-butanol. The beaker can be clamped with a large three-prong clamp.

Air condenser

Ice

Receiver vial

Boiling chip

Handle sulfuric acid with care.

Keep the side arm down in the cold vial to avoid losing product.

Into a 5-mL long-necked, round-bottomed flask place 1 mL of water and then add dropwise with thorough mixing 0.5 mL of concentrated sulfuric acid. Cool the hot solution in an ice bath; then add to the cold solution 1.0 mL (0.80 g) of 2-methyl-2-butanol. Mix the reactants thoroughly, add a boiling stone, and set up the apparatus for simple distillation, as shown in Figure 10.9. Hold the flask in a towel while connecting the apparatus to guard against spills or flask breakage. Cool the receiver (a vial) in ice to avoid losing the highly volatile products. Warm the flask on a sand bath to start the reaction and distill the products over the temperature range 30°C–45°C. After all the products have distilled, the rate of distillation will decrease markedly. Cease distillation at this point. After this, the temperature registered on the thermometer will rise rapidly as water and sulfuric acid begin to distill.

To the distillate add 0.3 mL cold 3 *M* sodium hydroxide solution to neutralize sulfurous acid, mix well, draw off the aqueous layer, and dry the product over anhydrous calcium chloride pellets, adding the drying agent in small quantities until it no longer clumps together. Keep the vial cold. While the product is drying, rinse out the distillation apparatus with water, ethanol, and then a small amount of

Products are easily lost by evaporation. Store the mixture in a spark-proof refrigerator if GC analysis must be postponed.

acetone and dry it thoroughly by drawing air through it using an aspirator. If this is not done carefully, the products will become contaminated with acetone. Carefully transfer the dry mixture of butenes to the distilling apparatus using a Pasteur pipette, add a boiling chip to the now dry products, and distill into a tared, cold receiver. Collect the portion boiling up to 43°C. Weigh the product and calculate the yield, which is normally around 50% on this small scale compared to 84% on macroscale. Inject a few microliters of product into a gas chromatograph maintained at room temperature (packed column suggestion: 6-mm-diameter × 3-m column 10% SE-30 silicone rubber on Chromosorb-W).

For capillary column analysis, inject 1–10 μL of just the gaseous *vapors* above the liquid. Oven temperatures should be as low as room temperature (oven door open) to obtain baseline separation of the methylbutene isomers. In a few minutes two peaks should appear. Measure the relative areas under the two peaks. From your knowledge of the mechanisms of the dehydration of secondary alcohols, which olefin should predominate? Does this agree with the boiling points? (In general, the compound with the shorter retention time has the lower boiling point.)

**Cleaning Up.** The residue in the reaction tube should be diluted with water, neutralized with sodium carbonate, and then flushed down the drain.

## *Macroscale Procedure*

Handle sulfuric acid with care.

Pour 18 mL of water into a 100-mL round-bottomed flask and cool in an ice-water bath while slowly pouring in 9 mL of concentrated sulfuric acid. Cool this 1:2 acid further with swirling while slowly pouring in 18 mL (15 g) of 2-methyl-2-butanol. Shake the mixture thoroughly and then mount the flask for fractional distillation over a flask heater, with the arrangement for ice cooling of the distillate seen in Figure 19.2 (on page 355). Cooling is needed because the olefin is volatile. Use a long condenser and an ample stream of cooling water. Heat the flask slowly with an electric flask heater until distillation of the hydrocarbon is complete. If the ice-cooled test tube will not hold 15 mL, be prepared to collect half of the distillate in another ice-cooled test tube. Transfer the distillate to a separatory funnel and shake it with about 5 mL of a 3 *M* sodium hydroxide solution to remove any traces of sulfurous acid. The aqueous solution sinks to the bottom and is drawn off. Dry the hydrocarbon layer by adding sufficient anhydrous calcium chloride pellets until the drying agent no longer clumps together. After about 5 min remove the drying agent by gravity filtration or careful decantation into a dry 25-mL round-bottomed flask and distill the dried product through a fractionating column (*see* Fig. 19.2 on page 355), taking the same precautions as before to avoid evaporation losses. Rinse the fractionating column with acetone and dry it with a stream of air to remove water from the first distillation. Collect in a tared bottle the portion boiling at 30°C–43°C. The yield reported in the literature is 84%; the average student yield is about 50%. Analyze the sample by injecting a few microliters of product into a gas chromatograph maintained at room temperature and equipped with a 6-mm-diameter × 3-m column packed with 10% SE-30 silicone rubber on Chromosorb-W or a similar inert packing. In a few minutes two peaks should appear. From your knowledge of the mechanisms of dehydration of secondary alcohols, which olefin should predominate? Does this agree with the boiling points? (In general, the compound with the shorter retention time has the lower boiling point.) Measure the relative areas under the two peaks.

**Cleaning Up.** The pot residue from the reaction is combined with the sodium hydroxide wash and neutralized with sodium carbonate. The pot residue from the distillation of the product is combined with the acetone used to wash out the apparatus and placed in the organic solvents container. If the calcium chloride pellets are dry, they can be placed in the nonhazardous solid waste container. If they are wet with organic solvents, they must be placed in the hazardous solid waste container for solvent-contaminated drying agent.

## 2. Computational Chemistry

The steric energies of the two isomeric butenes produced in Experiment 1 can be calculated using a molecular mechanics program, but the results are not valid because the bonding type is not the same. It is possible to compare the steric energies of *cis*- and *trans*-isomers of an alkene, but not a 1,1-disubstituted alkene with a 1,2-disubstituted alkene.

Compare your yields of the two isomers to their heats of formation, calculated using AM1 or a similar semiempirical calculation as described in Chapter 15. Do the calculated heats of formation correlate with the relative percentages of the isomers? What does this correlation or lack thereof tell you about the mechanism of the reaction?

## 3. 1-Butene and *cis*- and *trans*-2-Butene by Dehydration of an Alcohol[3]

| 2-Butanol | 1-Butene | *cis*-2-Butene | *trans*-2-Butene |
|---|---|---|---|
| MW 74.12 | MW 56.11 | MW 56.11 | MW 56.11 |
| bp 98°C | bp −6.3°C | bp 3.7°C | bp 0.9°C |

**IN THIS EXPERIMENT** 2-butanol is dehydrated with sulfuric acid to give a mixture of gaseous alkenes that are collected by the downward displacement of water. The gaseous mixture is analyzed by gas chromatography.

3. Helmkamp, G. K.; Johnson, H. W. Jr. *Selected Experiments in Organic Chemistry*, 3rd ed., Freeman: New York, 1983; 99.

■ **FIG. 10.10**

To thread a polyethylene tube through a septum, make a hole in the septum with a needle and then push a toothpick through the hole. Push the polyethylene tube firmly onto the toothpick; then pull and push on the toothpick. The tube will slide through the septum. Finally, pull the tube from the toothpick. A blunt syringe needle can also be used instead of a toothpick.

■ **FIG. 10.11**

An apparatus for the dehydration of 2-butanol. Butenes are collected by the downward displacement of water.

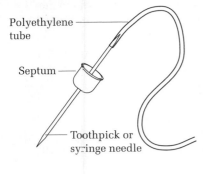

**Online Study Center**

Video: Butenes from Butanol; Photo: Placing a Polyethylene Tube through a Septum

Bend the polyethylene tubing to shape in hot sand or steam.

In a $10 \times 100$ mm reaction tube place 0.10 mL (81 mg) of 2-butanol and 0.05 mL (1 drop) of concentrated sulfuric acid. It is not necessary to measure the quantities of alcohol and acid exactly. Mix the reactants well and add a boiling chip. Stirring the reaction mixture is not necessary because it is homogeneous. Insert a polyethylene tube through a septum (Fig. 10.10), place this septum on the reaction tube, and lead the tubing into the mouth of the distilling column/collection tube, which is capped with a septum. The distilling column/collection tube is filled with water, capped with a finger, and inverted in a beaker of water and clamped for the collection of butene by the downward displacement of water (Fig. 10.11).

Lower the reaction tube into a hot ($\sim$100°C) sand bath and increase the heat slowly to complete the reaction. Collect a few milliliters of butene and *remove the polyethylene tube from the water bath*. If the polyethylene tube is not removed from the water bath, water will be sucked back into the reaction tube when it cools. Do not remove the septum-capped collection tube from the beaker of water, but bring it and the beaker of water together to the gas chromatograph. Use about 0.5 mL of the gaseous butene mixture to carry out the GC analysis with a TCD detector or a smaller amount if an FID detector is used.

**Cleaning Up.** The residue in the reaction tube may be diluted with water and flushed down the drain.

**Online Study Center**

**General Resources**
Additional Experiments

## QUESTIONS

1. Write the structures of the three olefins produced by the dehydration of 3-methyl-3-pentanol.

2. When 2-methylpropene is bubbled into dilute sulfuric acid at room temperature, it appears to dissolve. What new substance has been formed?

3. A student wished to prepare ethylene gas by the dehydration of ethanol at 140°C, using sulfuric acid as the dehydrating agent. A low-boiling liquid was obtained instead of ethylene. What was the liquid, and how might the reaction conditions be changed to give ethylene?

4. What would be the effect on retention time of increasing the carrier gas flow rate?

5. What would be the effect on retention time of raising the column temperature?

6. What would be the effect of raising the temperature or increasing the carrier gas flow rate on the ability to resolve two closely spaced peaks?

7. If you were to use a column one-half the length of the one you actually used, how do you think the retention times of the butenes would be affected? How do you think the separation of the peaks would be affected? How do you think the width of each peak would be affected?

8. From your knowledge of the dehydration of tertiary alcohols, which olefin should predominate in the product of the dehydration of 2-methyl-2-butanol? Why?

9. What is the maximum volume (at standard temperature and pressure, STP) of the butene mixture that could be obtained by the dehydration of 81 mg of 2-butanol?

10. What other gases are in the collection tube besides the three butenes at the end of the reaction in Experiment 3?

11. Using a computational chemistry program (*see* Chapter 15), calculate the steric energies or heats of formation of the three isomeric butenes produced in Experiment 3. Do these energies correlate with the product distributions found? What does a correlation or lack thereof tell you about the mechanism of the reaction?

# CHAPTER

# 11

# Infrared Spectroscopy

**PRELAB EXERCISE:** When an infrared (IR) spectrum is run, it is possible that the chart paper is not properly placed or that the spectrometer is not mechanically adjusted. Describe how you could calibrate an IR spectrum.

The types and molecular environment of functional groups in organic molecules can be identified by infrared (IR) spectroscopy. Like nuclear magnetic resonance (NMR) and ultraviolet (UV) spectroscopy, IR spectroscopy is nondestructive. Moreover, the small quantity of sample needed, the speed with which a spectrum can be obtained, the relatively low cost of the spectrometer, and the wide applicability of the method combine to make IR spectroscopy one of the most common structural elucidation tools used by organic chemists.

IR radiation consists of wavelengths that are longer than those of visible light. It is detected not with the eyes but by a feeling of warmth on the skin. When absorbed by molecules, radiation of these wavelengths (typically 2.5–5 μm) increases the amplitude of vibrations of the chemical bonds joining atoms.

IR spectra are measured in units of frequency or wavelength. The wavelength is measured in micrometers[1] or microns, μm ($1 \; \mu m = 1 \times 10^{-6}$ m). The positions of absorption bands are measured in frequency units called wavenumbers $\nu$, which are expressed in reciprocal centimeters, $cm^{-1}$, corresponding to the number of cycles of the wave in each centimeter.

2.5–25 μm equals 4000–400 $cm^{-1}$. Wavenumber($cm^{-1}$) is proportional to frequency (c is the speed of light).

$$\bar{\nu}(cm^{-1}) = \frac{\nu}{c}$$

$$cm^{-1} = \frac{10,000}{\mu m}$$

Examine the scale carefully.

Unlike UV and NMR spectra, IR spectra are inverted, with the strongest absorptions at the bottom, and are not always presented on the same scale. Some

---

1. Although micrometers are also known as microns, the micron is not the official SI unit.

231

IR spectroscopy easily detects:

| Hydroxyl groups | —OH |
|---|---|
| Amines | —NH$_2$ |
| Nitriles | —C≡N |
| Nitro groups | —NO$_2$ |

IR spectroscopy is especially useful for detecting and distinguishing among all carbonyl-containing compounds:

*Acids*

$$-\overset{\overset{\textstyle O}{\|}}{C}-OH$$

*Amides*

$$-\overset{\overset{\textstyle O}{\|}}{C}-NH_2$$

*Anhydrides*

$$-\overset{\overset{\textstyle O}{\|}}{C}-O-\overset{\overset{\textstyle O}{\|}}{C}-$$

*Aldehydes*

$$-\overset{\overset{\textstyle O}{\|}}{C}-H$$

*Ketones*

$$-\overset{\overset{\textstyle O}{\|}}{C}-$$

*Esters*

$$-\overset{\overset{\textstyle O}{\|}}{C}-O-$$

*Lactones*

$$-\overset{\overset{\textstyle O}{\|}}{C}-O-$$

spectrometers record the spectra on an ordinate linear in microns, but this compresses the low-wavelength region. Other spectrometers present the spectra on a scale linear in reciprocal centimeters, but linear on two different scales: one between 4000 and 2000 cm$^{-1}$, which spreads out the low-wavelength region; and the other a smaller one between 2000 and 667 cm$^{-1}$. Consequently, spectra of the same compound run on two different spectrometers will not always look the same.

To picture the molecular vibrations that interact with IR light, imagine a molecule as being made up of balls (atoms) connected by springs (bonds). The vibration can be described by Hooke's law from classical mechanics, which says that the frequency of a stretching vibration is directly proportional to the strength of the spring (bond) and inversely proportional to the masses connected by the spring. Thus we find C—H, N—H, and O—H bond-stretching vibrations are high frequency (short wavelength) compared to those of C—C and C—O because of the low mass of hydrogen compared to that of carbon or oxygen. The bonds connecting carbon to bromine and iodine, atoms of large mass, vibrate so slowly that they are beyond the range of most common IR spectrometers. A double bond can be regarded as a stiffer, stronger spring, so we find C=C and C=O vibrations at higher frequency than C—C and C—O stretching vibrations. And C≡C and C≡N stretch at even higher frequencies than C=C and C=O (but at lower frequencies than C—H, N—H, and O—H). These frequencies are in keeping with the bond strengths of single (~100 kcal/mol), double (~160 kcal/mol), and triple bonds (~220 kcal/mol).

These stretching vibrations are intense and particularly easy to analyze. A nonlinear molecule of $n$ atoms can undergo $3n - 6$ possible modes of vibration, which means cyclohexane with 18 atoms can undergo 48 possible modes of vibration. A single CH$_2$ group can vibrate in six different ways because it is connected to two other atoms. Each vibrational mode produces a peak in the spectrum because it corresponds to the absorption of energy at a discrete frequency. These many modes of vibration create a complex spectrum that defies simple analysis, but even in very complex molecules, certain functional groups have characteristic frequencies that can easily be recognized. Within these functional groups are the above-mentioned atoms and bonds, C—H, N—H, O—H, C=C, C=O, C≡C, and C≡N. Their absorption frequencies are given in Table 11.1.

When the frequency of IR light is the same as the natural vibrational frequency of an interatomic bond, light will be absorbed by the molecule, and the amplitude of the bond vibration will increase. The intensity of IR absorption bands is proportional to the change in dipole moment that a bond undergoes when it stretches. Thus, the most intense bands (peaks) in an IR spectrum are often from C=O and C—O stretching vibrations, whereas the C=C stretching band for a symmetrical acetylene is almost nonexistent because the molecule undergoes no net change of dipole moment when it stretches:

$$\overset{\overset{\textstyle +}{\longrightarrow}}{\diagdown C=O} \longleftrightarrow \overset{\overset{\textstyle +}{\longrightarrow}}{\diagdown C=O} \qquad H_3C-C\equiv C-CH_3 \longleftrightarrow H_2C-C\equiv C-CH_3$$

Change in dipole moment           No change in dipole moment

| TABLE 11.1  •  Characteristic IR Absorption Wavenumbers | |
| --- | --- |
| **Functional Group** | **Wavenumber (cm$^{-1}$)** |
| O—H | 3600–3400 |
| N—H | 3400–3200 |
| C—H | 3080–2760 |
| C≡N | 2260–2215 |
| C≡C | 2150–2100 |
| C=O | 1815–1650 |
| C=C | 1660–1600 |
| C—O | 1200–1050 |

The intensity of absorption is proportional to the change in dipole moment.

Unlike proton NMR spectroscopy, where the area of the peaks is strictly proportional to the number of hydrogen atoms causing the peaks, the intensities of IR peaks are not proportional to the numbers of atoms causing them. Given the chemical shifts and coupling constants, it is not too difficult to calculate a theoretical NMR spectrum that is an exact match to the experimental one. The calculation of all possible stretching and bending frequencies, for larger molecules, requires large amounts of time on a fast computer. Every peak or group of peaks in an NMR spectrum can be assigned to specific hydrogens in a molecule, but the assignment of the majority of peaks in an IR spectrum is usually not possible. Peaks to the right (longer wavelength) of 1250 cm$^{-1}$ are the result of combinations of vibrations that are characteristic not of individual functional groups but of the molecule as a whole. This part of the spectrum is often referred to as the *fingerprint region* because it is uniquely characteristic of each molecule. Although two organic compounds can have the same melting points or boiling points and can have identical UV and NMR spectra, they cannot have identical IR spectra (except, as usual, for enantiomers). IR spectroscopy is thus the final arbiter in deciding whether two compounds are identical.

## Analysis of IR Spectra

Only a few simple rules or equations govern IR spectroscopy. Because it is not practical to calculate theoretical spectra, the analysis is done almost entirely by correlation with other spectra. In printed form, these comparisons take the form of lengthy discussions, so detailed analysis of a spectrum is best done with a good reference book at hand.

In a modern analytical or research laboratory, a collection of many thousands of spectra is maintained on a computer. When the spectrum of an unknown compound is run, the analyst picks out five or six of the strongest peaks and asks the computer to list all the compounds that have peaks within a few reciprocal centimeters of the listed peaks. From the printout of a dozen or so compounds, it is often possible to pinpoint all the functional groups in the molecule being analyzed. There may be a perfect match of all peaks, in which case the unknown will have been identified.

For relatively simple molecules, a computer search is hardly necessary. Much information can be gained about the functional groups in a molecule from relatively few correlations.

To carry out an analysis, (1) pay most attention to the strongest absorptions; (2) pay more attention to peaks to the left (shorter wavelength) of 1250 cm$^{-1}$; and (3) pay as much attention to the absence of certain peaks as to the presence of others. The absence of characteristic peaks will definitely exclude certain functional groups. Be wary of weak O—H peaks because water is a common contaminant of many samples. Because potassium bromide is hygroscopic, water is often found in the spectra of KBr pellets.

## The Step-by-Step Analysis of IR Spectra

IR spectra are analyzed as follows:

1. Is there a peak between 1820 cm$^{-1}$ and 1625 cm$^{-1}$? If not, go to Step 2.
   (a) Is there a strong, wide O—H peak between 3200 cm$^{-1}$ and 2500 cm$^{-1}$? If so, the compound is a carboxylic acid (Fig. 11.1, oleic acid). If not . . .
   (b) Is there a medium-to-weak N—H band between 3520 cm$^{-1}$ and 3070 cm$^{-1}$? If there are two peaks in this region, the compound is a primary amide; if not, it is a secondary amide. If there is no peak in this region . . .
   (c) Are there two strong peaks, one in the region 1870 cm$^{-1}$ to 1800 cm$^{-1}$ and the other in the region 1800 cm$^{-1}$ to 1740 cm$^{-1}$? If so, an acid anhydride is present. If not . . .
   (d) Is there a peak in the region of 2720 cm$^{-1}$? If so, is the carbonyl peak in the region 1715 cm$^{-1}$ to 1680 cm$^{-1}$? If so, the compound is a conjugated aldehyde; if not, it is an isolated aldehyde (Fig. 11.2, benzaldehyde). However, if there is no peak near 2720 cm$^{-1}$ . . .
   (e) Does the strong carbonyl peak fall in the region 1815 cm$^{-1}$ to 1770 cm$^{-1}$ and the compound give a positive Beilstein test? If so, it is an acid halide. If not . . .

■ **FIG. 11.1**

The IR spectrum of oleic acid (thin film).

**■ FIG. 11.2**
The IR spectrum of benzaldehyde (thin film).

**■ FIG. 11.3**
The IR spectrum of *n*-butyl acetate (thin film).

(f) Does the strong carbonyl peak fall in the range 1690 cm$^{-1}$ to 1675 cm$^{-1}$? If so, the compound is a conjugated ketone. If not . . .

(g) Does the strong carbonyl peak fall in the range 1670 cm$^{-1}$ to 1630 cm$^{-1}$? If so, the compound is a tertiary amide. If not . . .

(h) Does the spectrum have a strong, wide peak in the range 1310 cm$^{-1}$ to 1100 cm$^{-1}$? If so, does the carbonyl peak fall in the range 1730 cm$^{-1}$ to 1715 cm$^{-1}$? If so, the compound is a conjugated ester; if not, the ester is not conjugated (Fig. 11.3, *n*-butyl acetate). If there is no strong, wide peak in the range 1310 to 1100 cm$^{-1}$, then . . .

(i) The compound is an ordinary nonconjugated ketone.

■ **FIG. 11.4**
The IR spectrum of cyclohexanol
(thin film).

■ **FIG. 11.5**
The IR spectrum of benzonitrile
(thin film).

2. If the spectrum lacks a carbonyl peak in the range 1820 cm$^{-1}$ to 1625 cm$^{-1}$, does it have a broad band in the region 3650 cm$^{-1}$ to 3200 cm$^{-1}$? If so, does it also have a peak at about 1200 cm$^{-1}$, a C—H stretching peak to the left of 3000 cm$^{-1}$, and a peak in the region 1600 cm$^{-1}$ to 1470 cm$^{-1}$? If so, the compound is a phenol. If the spectrum does not meet these latter three criteria, the compound is an alcohol (Fig. 11.4, cyclohexanol). However, if there is no broad band in the region 3650 cm$^{-1}$ to 3200 cm$^{-1}$, then . . .

(a) Is there a broad band in the region 3500 cm$^{-1}$ to 3300 cm$^{-1}$, does the compound smell like an amine, or does it contain nitrogen? If so, are there two peaks in this region? If so, the compound is a primary amine; if not, it is a secondary amine. However, if there is no broad band in the region 3500 cm$^{-1}$ to 3300 cm$^{-1}$, then . . .

**■ FIG. 11.6**
The IR spectrum of
3-methylpentane (thin film).

**■ FIG. 11.7**
The IR spectrum of *t*-butylbenzene
(thin film).

(b) Is there a sharp peak of medium-to-weak intensity at 2260 cm$^{-1}$
to 2100 cm$^{-1}$? If so, is there also a peak at 3320 cm$^{-1}$ to 3310 cm$^{-1}$?
If so, then the compound is a terminal acetylene. If not, the compound
is most likely a nitrile (Fig. 11.5, benzonitrile), although it might be an
asymmetrically substituted acetylene. If there is no sharp peak of
medium-to-weak intensity at 2260 cm$^{-1}$ to 2100 cm$^{-1}$, then . . .

(c) Are there strong peaks in the region 1600 cm$^{-1}$ to 1540 cm$^{-1}$ and
1380 cm$^{-1}$ to 1300 cm$^{-1}$? If so, the molecule contains a nitro group.
If not . . .

(d) Is there a strong peak in the region 1270 cm$^{-1}$ to 1060 cm$^{-1}$? If so, the
compound is an ether. If not . . .

(e) The compound is either a tertiary amine (odor?), a halogenated hydrocarbon (Beilstein test?), or just an ordinary hydrocarbon (Fig. 11.6,
3-methylpentane, and Fig. 11.7, *t*-butylbenzene).

The effect of ring size on the carbonyl frequencies of lactones and esters:

1727 cm$^{-1}$

1745 cm$^{-1}$

1740 cm$^{-1}$

1775 cm$^{-1}$

1832 cm$^{-1}$

Many comments can be added to this bare outline. For example, dilute solutions of alcohols will show a sharp peak at about 3600 cm$^{-1}$ for a nonhydrogen-bonded O—H in addition to the usual broad hydrogen-bonded O—H peak.

Aromatic hydrogens give peaks just to the left of 3000 cm$^{-1}$, whereas aliphatic hydrogens appear just to the right of 3000 cm$^{-1}$. However, NMR spectroscopy is the best method for identifying aromatic hydrogens.

The carbonyl frequencies listed earlier refer to an open chain or an unstrained functional group in a nonconjugated system. If the carbonyl group is conjugated with a double bond or an aromatic ring, the peak will be displaced to the right by 30 cm$^{-1}$. When the carbonyl group is in a ring smaller than six members or if there is oxygen substitution on the carbon adjacent to an aldehyde or ketone carbonyl, the peak will be moved to the left (refer to the margin notes about the effect of ring size and Table 11.2).

| TABLE 11.2 • Characteristic IR Carbonyl Stretching Peaks (Chloroform Solutions) | | |
|---|---|---|
| | Carbonyl-Containing Compounds | Wavenumber (cm$^{-1}$) |
| RCR (O) | Aliphatic ketones | 1725–1705 |
| RCCl (O) | Acid chlorides | 1815–1785 |
| R—C=C—C—R (O) | α,β-Unsaturated ketones | 1685–1666 |
| ArCR (O) | Aryl ketones | 1700–1680 |
| (cyclohexanone)=O | Cyclohexanones | 1725–1705 |
| —C—CH$_2$—C— (O)(O) | β-Diketones | 1640–1540 |
| RCH (O) | Aliphatic aldehydes | 1740–1720 |
| R—C=C—CH (O) | α,β-Unsaturated aldehydes | 1705–1685 |
| ArCH (O) | Aryl aldehydes | 1715–1695 |

The effect of ring size on carbonyl frequency of ketones:

$1705 \text{ cm}^{-1}$

$1715 \text{ cm}^{-1}$

$1715 \text{ cm}^{-1}$

$1745 \text{ cm}^{-1}$

$1780 \text{ cm}^{-1}$

$1815 \text{ cm}^{-1}$

| TABLE 11.2 • (Continued) | | |
|---|---|---|
| | Carbonyl-Containing Compounds | Wavenumber $(\text{cm}^{-1})$ |
| RCOOH | Aliphatic acids | 1725–1700 |
| R—C=C—COOH | $\alpha,\beta$-Unsaturated acids | 1700–1680 |
| ArCOOH | Aryl acids | 1700–1680 |
| RCOR' | Aliphatic esters | 1740 |
| R—C=C—COR' | $\alpha,\beta$-Unsaturated esters | 1730–1715 |
| Ar COR | Aryl esters | 1730–1715 |
| HCOR | Formate esters | 1730–1715 |
| $CH_2$=CHOCCH$_3$ $C_6H_5$OCCH$_3$ | Vinyl and phenyl acetate | 1776 |
| R—C—O R—C—O | Acyclic anhydrides (two peaks) | 1840–1800 1780–1740 |
| RCNH$_2$ | Primary amides | 1694–1650 |
| RCNHR' | Secondary amides | 1700–1670 |
| RCNR'$_2$ | Tertiary amides | 1670–1630 |

Methyl groups often give a peak near $1375 \text{ cm}^{-1}$, but NMR is a better method for detecting this group.

The pattern of substitution on an aromatic ring (mono-, *ortho*-, *meta*-, and *para*, di-, tri-, tetra-, and penta-) can be determined from C—H out-of-plane bending

The band patterns of toluene and *o-*, *m-*, and *p*-xylene. These peaks are *very weak*. They are characteristic of aromatic substitution patterns in general, not just for these four molecules. Try to find these patterns in other spectra throughout this text.

vibrations in the region 670–900 cm$^{-1}$. Much weaker peaks between 1650 cm$^{-1}$ and 2000 cm$^{-1}$ are illustrated in Figure 11.8.

Extensive correlation tables and discussions of characteristic group frequencies can be found in the specialized references listed at the end of this chapter.

# The Fourier Transform Infrared (FTIR) Spectrometer

Like their NMR counterparts, Fourier transform IR instruments are computer based, can sum a number of scans to increase the signal-to-noise ratio, and allow enormous flexibility in the ways spectra can be analyzed and displayed because they are digitally oriented. In addition, they are faster and more accurate than double-beam spectrometers.

A schematic diagram of a single-beam FTIR spectrometer is given in Figure 11.9. IR radiation goes to a Michelson interferometer in which half the light beam passes through a partially coated :nirror to a fixed mirror and half goes to a moving mirror. The combined beams pass through the sample and are then focused on the detector. The motion of the mirror gives a signal that varies sinusoidally. Depending on the mirror position, the different frequencies of light either reinforce or cancel each other, resulting in an interferogram. The mirror, the only moving part of the spectrometer, thus affects the light frequencies that encode the very high optical frequencies, transforming them to low-frequency signals that change as the mirror moves back and forth. The Fourier transform convert the digitized signal into a signal that is a function of frequency—creating a spectrum.

Like Fourier transform NMR spectroscopy, all IR frequencies are sampled simultaneously instead of being scanned successively, so a spectrum is obtained in seconds. Because there is no slit to sort out the wavelengths, high resolution is possible without losing signal strength. Extremely small samples or very dilute solutions can

The optical path diagram of an FTIR spectrometer.

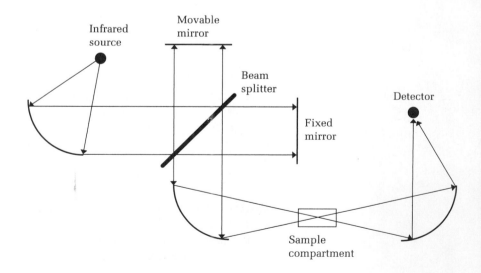

be examined because it is easy to sum hundreds of scans in the computer. It is even possible to take a spectrum of an object seen under a microscope. The spectrum of one compound can be subtracted from, say, a binary mixture spectrum to reveal the nature of the other component. Data can be smoothed, added, resized, and so on because of the digital nature of the output from the spectrometer.

## Experimental Considerations

The sample, solvents, and equipment must be dry.

Glass, quartz, and plastics are opaque to IR radiation and absorb it. Therefore, sample holders must be made of other transparent materials. Metal halide salts such as sodium or silver chloride, calcium or barium fluoride, and potassium bromide are transparent to IR radiation; these are prepared as large polished crystals for use in cell holders, with NaCl and KBr being the most common. They are fragile and are easily attacked by moisture, so they must be handled gently by the edges and kept away from aqueous solutions. IR spectra can be determined on neat (undiluted) liquids, on solutions with an appropriate solvent, and on solids using mulls and KBr pellets, and by diffuse reflectance of the solid mixed with KBr. All IR cells or KBr powder must be stored in a desiccator when not in use.

## The Spectra of Neat Liquids

To run a spectrum of a neat (free of water!) liquid, remove a demountable cell (Fig. 11.10) from the desiccator, place a drop of the liquid between the salt plates, press the plates together gently to remove any air bubbles, and add the top rubber gasket and metal top plate. Next, put on all four of the nuts and *gently* tighten them to apply an even pressure to the top plate. Place the cell in the sample compartment (nearest the front of the spectrometer) and run the spectrum.

Although running a spectrum on a neat liquid is convenient and results in no extraneous bands to interpret, it is not possible to control the path length of the light through the liquid in a demountable cell. A low-viscosity liquid when squeezed between the salt plates may be so thin that the short path length gives peaks that are too weak. A viscous liquid, on the other hand, may give peaks that are too intense. A properly run spectrum will have the most intense peak with a transmittance of about 10%.

A fast, simple alternative: Place a drop of the compound on a round salt plate, add a top plate, and mount the two on the holder pictured in Figure 11.11.

Another demountable cell is pictured in Figure 11.11. The plates are thin wafers of silver chloride, which is transparent to IR radiation. This cell has advantages over the salt cell in that the silver chloride disks are more resistant to breakage than NaCl plates, less expensive, and not affected by water. Because silver chloride is photosensitive, the wafers must be stored in a dark place to prevent them from turning black. Because one side of each wafer is recessed, the thickness of the sample can be varied according to the manner in which the cell is assembled. In general, the spectra of pure liquids are run as the thinnest possible films. Most of the spectra in this chapter have been obtained in this way. The disks are cleaned by rinsing them with an organic solvent such as acetone or ethanol and wiping dry with an absorbent paper towel.

An exploded view of a demountable salt cell for analyzing the IR spectra of neat liquids. The salt plates are fragile and expensive. Do not touch the surfaces. Use only dry solvents and samples.

■ **FIG. 11.11**
A demountable silver chloride or sodium chloride cell.

Thin film of liquid to be analyzed

Cross-section of wafers or disks

O-rings

AgCl (or NaCl) disks

A very simple cell consists of two circular NaCl disks (1-in. diameter × 3/16-in. thick, 25 mm × 4 mm). The sample, 1 drop of a pure liquid or solution, is applied to the center of a disk with a polyethylene pipette (to avoid scratching the NaCl disk). The other disk is added, the sample is squeezed to a thin film, and the two disks are placed on the V-shaped holder shown in Figure 11.11.

## The Spectra of Solutions

The most widely applicable method of running the spectra of solutions involves dissolving an amount of the liquid or solid sample in an appropriate solvent to give a 10% solution. Just as in NMR spectroscopy, the best solvents to use are carbon disulfide and carbon tetrachloride; because these compounds are not polar enough to dissolve many substances, chloroform is used as a compromise. Unlike NMR solvents, no solvent suitable in IR spectroscopy is entirely free of absorption bands in the frequency range of interest (Figs. 11.12, 11.13, and 11.14). In chloroform, for instance, no light passes through the cell between 650 cm$^{-1}$ and 800 cm$^{-1}$. As can be seen from the figures, spectra obtained using carbon disulfide and chloroform cover the entire IR frequency range. Carbon tetrachloride would appear to be a good choice because it has few interfering peaks, but it is a poor solvent for polar compounds. In practice, a background spectrum is run with the same solvent in the same cell used to run the sample solution, and the data system subtracts the solvent peaks in background spectrum from the solution spectrum to yield an IR spectrum of the sample only.

Solvents: $CHCl_3$, $CS_2$, $CCl_4$

**CAUTION**: Chloroform, carbon tetrachloride, and carbon disulfide are toxic. The first two are carcinogens. Use a laboratory hood to minimize exposure to vapors.

■ FIG. 11.12
The IR spectrum of chloroform (thin film).

■ FIG. 11.13
The IR spectrum of carbon disulfide (thin film).

■ FIG. 11.14
The IR spectrum of carbon tetrachloride (thin film).

■ **FIG. 11.15**
A sealed IR sample cell.

Hypodermic syringe port
Top plate
NaCl plates
Lead gasket 0.1 mm thick
Bottom plate

■ **FIG. 11.16**
Flushing an IR sample cell. The solvent used to dissolve the sample is used in this process.

Three drops are needed to fill the cell.

Three large drops of solution will fill the usual sealed IR cell (Fig. 11.15). A 10% solution of a liquid sample can be approximated by diluting 1 drop of the liquid sample with 9 drops of the solvent. Because weights are more difficult to estimate, solid samples should be weighed to obtain a 10% solution.

The solvent and the sample must be dry. Do not touch or breathe on NaCl plates. Cells are very expensive.

First, obtain a background spectrum of the solvent. The IR cell is filled by inclining it slightly and placing about 3 drops of the solvent in the lower hypodermic port with a capillary dropper. The cell must be completely dry because the new sample will not enter if the cell contains solvent. The liquid can be seen rising between the salt plates by looking into the window. In the most common sealed cell, the salt plates are spaced 0.1 mm apart. Be sure that the cell is filled past the window and that no air bubbles are present. Then place a Teflon stopper lightly but firmly in the hypodermic port. Be particularly careful not to spill any of the sample on the outside of the cell windows. Place the cell in the instrument cell holder and acquire a background spectrum. Blow dry nitrogen or draw dry air through the cell to remove the solvent and dry the cell, fill with the sample solution prepared above, and acquire the spectrum.

After running the sample, force clean solvent through the cell using a syringe attached to the top port of the cell (Fig. 11.16). Dry using dry nitrogen or air and store in a desiccator.

Dispose of waste solvents in the container provided.

**Cleaning Up.** Discard halogenated liquids in the halogenated organic waste container. Other solutions should be placed in the organic solvents container.

## Mulls

The sample must be finely ground.

Solids insoluble in the usual solvents can be run as mulls. In preparing a mull, the sample is ground to a particle size less than that of the wavelength of light going through the sample (2.5 μm) to avoid scattering light. About 15–20 mg of the

■ **FIG. 11.17**
The IR spectrum of Nujol
(paraffin oil).

sample is ground for 3–10 min in an agate mortar until it is spread over the entire inner surface of the mortar and has a caked and glassy appearance. Then to make a mull, add 1 or 2 drops of paraffin oil (Nujol; Fig. 11.17) and grind the sample for 2–5 more minutes. The mull, which should have the consistency of thin margarine, is transferred to the bottom salt plate of a demountable cell (*see* Fig. 11.10 on page 242) using a rubber policeman, the top plate is added and rotated back and forth to distribute the sample evenly to eliminate all air pockets, and the spectrum is run. Because the bands from Nujol obscure certain frequency regions, running another mull using Fluorolube as the mulling agent will allow the entire IR spectral region to be covered. If the sample has not been ground sufficiently fine, there will be marked loss of transmittance at the short-wavelength end of the spectrum. After running the spectrum, the salt plates are wiped clean with a cloth saturated with an appropriate solvent.

A fast, simple alternative: Grind a few milligrams of the solid with 1 drop of tetrachloroethylene between two round salt plates. Mount on the holder as shown in Figure 11.11 on page 242.

## Potassium Bromide Disk

The spectrum of a solid sample can also be run by incorporating the sample in a potassium bromide (KBr) disk. This procedure needs only one disk to cover the entire spectral range because KBr is completely transparent to IR radiation. Although very little sample is required, making the disk calls for special equipment and time to prepare it. Because KBr is hygroscopic, water is a problem.

Into a stainless steel capsule containing a ball bearing are weighed 1.5–2 mg of the compound and 200 mg of spectroscopic-grade KBr (previously dried and stored in a desiccator). The capsule is shaken for 2 min on a Wig-L-Bug (the device used by dentists to mix silver amalgam). The sample is evenly distributed over the face of a 13-mm die and subjected to a pressure of 14,000–16,000 psi for 3–6 min while under vacuum in a hydraulic press. A transparent disk is produced, which is removed from the die with tweezers and placed in a holder like that shown in Figure 11.11 on page 242 prior to running the spectrum.

■ FIG. 11.18
A simplified diagram of a diffuse
reflectance system.

■ FIG. 11.18
A simplified diagram of a diffuse
reflectance system.

Sample
cup

IR source          Detector

# Diffuse Reflectance

Reasonable IR spectra of solids can be obtained without pressing a KBr pellet if the IR instrument is equipped with diffuse reflectance optics. A mixture of the sample and KBr is put in a small sample cup, and this is placed in the diffuse reflectance optics system (Fig. 11.18). When the IR beam enters the sample, it can either be reflected off the surface of a particle or be transmitted through a particle. The IR beam that passes through a particle can either reflect off the next particle or be transmitted through the next particle. This transmission-reflectance event can occur many times in the sample, which increases the path length. Finally, any scattered IR energy that is not absorbed by the sample is collected by a spherical mirror that is focused onto a detector. A background spectrum of only pure KBr should be run before the sample to subtract out any absorbance that is not from the sample.

Weigh out 0.160–0.170 g of KBr quickly (this compound readily absorbs atmospheric humidity) and place it in a vial. Weigh out between 4 mg and 7 mg of your solid and add this to the vial. Mix and dump into a small, dry mortar and pestle. Grind for 1–2 min until a fine powder is obtained. Transfer to weighing paper and then to the cup of the diffuse reflectance sample holder. Place the holder in the reflectance attachment and acquire the spectrum.

# Gas Phase IR Spectroscopy:
# The Williamson Gas Phase IR Cell

In the past it was difficult to obtain the IR spectra of gases. A commercial gas cell costs hundreds of dollars; the cell must be evacuated, and the dry gas carefully introduced—not a routine process. But we have found that a uniquely simple gas phase IR cell can be assembled at virtually no cost from a 105° microscale connecting adapter.[2]

■ FIG. 11.19
The Williamson Gas Phase IR Cell
utilizing the 105° connecting
adapter from the microscale kit
that accompanies this text.

Wad of cotton

Fine wire
or thread

Polyethylene
film

2. The microscale connecting adapter is a available from Kimble Kontes (catalog no. 748001-1000).

**■ FIG. 11.20**
The Williamson Gas Phase IR Cell mounted in a cell holder.

**■ FIG. 11.21**
The IR spectrum of polyethylene film.

720 cm⁻¹

1463 cm⁻¹

2926 cm⁻¹    2852 cm⁻¹

Wavenumber (cm⁻¹)

To make the Williamson gas phase IR cell, attach a thin piece of polyethyene film from, for example, a sandwich bag to each end of the adapter with a small rubber band, fine wire, or thread. Place a loose wad of cotton in the side arm and mount the IR cell in the beam of the spectrometer (Fig. 11.19 and Fig. 11.20).

The adapter can be supported on a V-block as shown in Figure 11.20 or, temporarily, on a block of modeling clay. Be sure the beam, usually indicated by a red laser, strikes both ends of the cell. Run a background spectrum on the cell. The spectrum of polyethylene (Fig. 11.21) is very simple, so interfering bands will be few. This spectrum is automatically subtracted when you put a sample in the cell and run a spectrum.

A more permanent cell can be constructed by attaching two thin plates of silver chloride to the ends of the adapter with epoxy glue. These silver chloride plates can be made in the press used for making KBr discs, or they can be purchased.[3]

## Obtaining a Spectrum

To run a gas phase spectrum, remove the cell from the spectrometer and spray a very short burst of a gas from an aerosol can onto the cotton. The nonvolatile components of the spray will stick to the cotton, and the propellant gas will diffuse into the cell. Run the spectrum in the usual way. The concentration of the gas is about the same as a thin film of liquid run between salt plates. If the concentration of the gas is too high, remove the cell from the spectrometer, remove the cotton, and "pour out" half the invisible gas by holding the cell in a vertical position.

You may be fortunate in having in your school's library a copy of Volume 3 of *The Aldrich Library of FT-IR Spectra: Vapor Phase* or an online database of IR spectra on your IR instrument data system. Either resource contains hundreds of reference spectra that will help you analyze the IR spectra of propellants in aerosol cans and any other gases you may encounter. Otherwise you may want to compare your spectra to the few in this text and available on the website for this text and assemble your own library of gas phase spectra. It is possible to make approximate calculations of IR vibrational frequencies using a semi-empirical calculation at the AM1 level (*see* Chapter 15).

Throughout the remainder of this book, representative IR spectra of starting materials and products are presented, and the important bands in each spectrum are identified.

■ **FIG. 11.22**
The IR spectrum of carbon dioxide (gas phase).

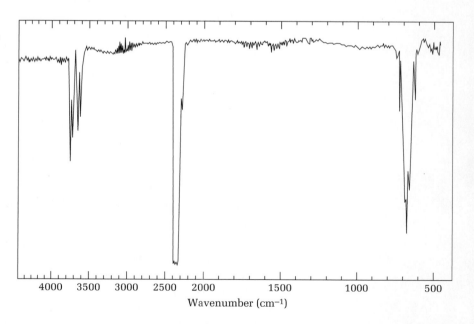

Wavenumber (cm$^{-1}$)

---

3. The silver chloride disks can be purchased from Fisher Scientific (catalog no. 14-385-860).

■ **FIG. 11.23**
The IR spectrum of methane
(gas phase).

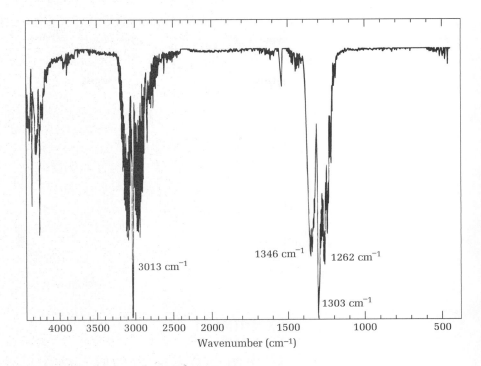

**■ FIG. 11.23**
The IR spectrum of methane (gas phase).

## For Further Investigation

### *The Gas Phase IR Spectra of Aerosol Propellants*

Acquire gas phase spectra from a number of substances from aerosol cans. Some cans will be labeled with the name of the propellant gas; others will not. Analyze the spectra, bearing in mind that the gas will be a small molecule such as carbon dioxide (Fig. 11.22), methane (Fig. 11.23), nitrous oxide (Fig. 11.24), propane, butane, isobutane, or a fluorocarbon such as 1,1,1,2-tetrafluoroethane. A very old aerosol can may contain an illegal chlorofluorocarbon, now outlawed because these substances destroy the ozone layer. Some examples of aerosol cans include hairsprays, lubricants (WD-40, Teflon), Freeze-It, inhalants for asthma, cigarette lighters, butane torches, and natural gas from the lab.

### *Other Gas Phase Spectral Research*

You can deliberately generate hydrogen chloride (or deuterium chloride) and sulfur dioxide and then analyze the gas that is evolved from a thionyl chloride reaction. Can you detect carbon monoxide in automobile exhaust? How about marsh gas that you see bubbling out of ponds? With perhaps dozens of scans, can you record the spectra of vanillin over vanilla beans, eugenol vapor over cloves, anisaldehyde over anise seeds? What is the nature of the volatile components of chili peppers, orange peel, nail polish, and nail polish remover? Could you use this method to determine if someone has been drinking alcohol?

■ **FIG. 11.24**
The IR spectrum of nitrous oxide
(gas phase).

Wavenumber (cm⁻¹)

---

## EXPERIMENT

## Unknown Carbonyl Compound

The sample, solvents, and
apparatus must be dry.

Run the IR spectrum of an unknown carbonyl compound obtained from your laboratory instructor. Be particularly careful that all apparatus and solvents are completely free of water, which will damage the NaCl plates. Determine the frequency of the carbonyl peak in the unknown *and* list the possible types of compounds that could correspond to this frequency (Table 11.2).

---

## REFERENCES

1.  Colthrup, Norman B., Lawrence H. Daly, and Stephen E. Wiberley, *Introduction to Infrared and Raman Spectroscopy*, 3rd ed. New York: Academic Press, 1990.

2.  Cooper, James W., *Spectroscopic Techniques for Organic Chemists*. New York: John Wiley, 1980.

3.  Lin-Vein, Daimay, Norma B. Colthrup, William G. Fateley, and Jeannette G. Grasselli, *The Handbook of Infrared and Raman Characteristic Frequencies of Organic Molecules*. Boston: Academic Press, 1991.

4.  Pouchert, Charles J. *The Aldrich Library of FT-IR Spectra*, 2nd ed. Milwaukee: Aldrich Chemical Co., 1997.

5.  Roeges, Noel P. G., *A Guide to the Complete Interpretation of Infrared Spectra of Organic Structures*. New York: John Wiley, 1994.

6.  Silverstein, Robert M., Francis X. Webster, and David Kiemle, *Spectrometric Identification of Organic Compounds*, 7th ed.[4] New York: Wiley, 2005.

---

4. This book includes IR, UV, and NMR spectra.

**CHAPTER**

# 12

# Nuclear Magnetic Resonance Spectroscopy

⚙ *Online Study Center*

This icon will direct you to techniques, equipment setups, and online resources at http://college.hmco.com/PIC/williamson MMOE5e.

[1]H NMR: Determination of the number, kind, and relative locations of hydrogen atoms (protons) in a molecule

Chemical shift, δ (ppm)

**PRELAB EXERCISE:** Outline the preliminary solubility experiments you would carry out using inexpensive solvents before preparing a solution of an unknown compound for nuclear magnetic resonance (NMR) spectroscopy using expensive deuterated solvents.

**M**ost organic chemists would agree that the most powerful instrumental method for revealing the structure of organic molecules is nuclear magnetic resonance (NMR) spectroscopy. The 2002 Nobel Prize in Chemistry was awarded, in part, to Kurt Wüthrich for advances in NMR spectroscopy that allowed the determination of the three-dimensional structure of biological macromolecules in solution. Because of such capabilities, introductory organic chemistry courses devote considerable time to the study of NMR. This concise chapter assumes that you have had some prior exposure to the concepts discussed; it focuses on the practical aspects of sample preparation, data acquisition, and interpretation of NMR spectra to elucidate and confirm organic structures.

Let us briefly review the theory of NMR. Certain nuclei such as [1]H, [13]C, [15]N, [18]O, [19]F, and [31]P are tiny magnets that can assume two energy states when placed in a magnetic field: aligned (lower energy) and opposed (higher energy). Like the energy states in electronic spectra, the difference in energy between these states is quantized, and energy is absorbed and emitted only at certain radio frequencies. These frequencies depend on the type of nucleus and the magnetic field strength, which is constant for a given instrument. More importantly, however, is the fact that slight differences in electronegativity and bonding state ($sp$, $sp^2$, and $sp^3$) of surrounding atoms cause small—parts per million (ppm)—variations in the magnetic field felt by each nucleus, causing them to have absorption signals at slightly different frequencies. These small variations, called *chemical shifts*, are plotted versus signal intensity to produce the NMR spectrum. The interpretation of these signals and other spectral features such as splitting patterns and peak areas, as described in the following sections, facilitates organic structure elucidation.

# Interpretation of $^1$H NMR Spectra

There are two approaches to interpreting proton NMR spectra:

1. The *structure from the spectrum* approach is the strategy of using the information in the NMR spectrum to draw the structure of the molecule based on reference tables and rules. This approach is used if the compound's structure is unknown. In this case, the NMR spectrum alone is often insufficient to "solve" the complete structure and must be combined with knowledge about the compound's source (synthetic reaction or natural product) and complementary spectral data (infrared, ultraviolet, and/or mass spectrometric).

2. The *spectrum from the structure* approach might be thought of as the reverse of the first approach and is commonly used when verifying products of known reactions using known starting materials. A hypothetical NMR spectrum is created, based on the known compound's structure (or related structural possibilities), using the same reference tables and rules as in the first approach. Computer programs are available that allow the calculation of a hypothetical NMR spectrum for any molecule simply by entering its structure. (Refer to the supplemental information for this chapter on this book's website.) Even better, there may be a published spectrum of the compound available in a collection of reference spectra. Either way, the sample spectrum should be compared to the hypothetical or reference spectrum, and if the two are closely matched, one can be confident the correct compound has been obtained.

$H_3C$   I
$CH_2$
**Ethyl iodide**

NMR spectra contain considerably more structural information than do infrared spectra, and this additional information should be used in interpreting them. The following four most informative features are the principal ones to look for when interpreting spectra. We will use ethyl iodide as a specific illustration of each feature.

## 1. The Number of Signals Due to Equivalent Hydrogen Nuclei or Protons

The three hydrogen atoms of the methyl group are said to be chemically and magnetically equivalent. Replacing any of these hydrogens with another atom will give the same compound. Similarly, the two methylene protons are magnetically and chemically equivalent. This is because there is very fast rotation about the carbon-carbon bond. Individual hydrogens or each set of *equivalent* hydrogens will experience slightly different magnetic fields depending on their chemical environment and, therefore, will produce a signal at different places in the spectrum. Thus, we would predict that ethyl iodide would produce two signals, one for the three equivalent $CH_3$ protons and the other from the equivalent protons on the $CH_2$ bonded to the electronegative iodine. The NMR spectra of ethyl iodide, regardless of whether it is obtained at low field strength (Fig. 12.1) or high field strength (Fig. 12.2), shows three signals, or peaks, at 0.00 ppm, 1.83 ppm, and 3.20 ppm, with the latter two split into patterns of lines. The peak at 0.00 ppm is due to the addition of a reference compound, tetramethylsilane (TMS), used as the zero reference point to assure that the instrument frequency assignments are accurate. Thus, there are only two sets of signal peaks from ethyl iodide as predicted.

$$\begin{array}{c} CH_3 \quad CH_3 \\ \diagdown \quad \diagup \\ Si \\ \diagup \quad \diagdown \\ CH_3 \quad CH_3 \end{array}$$

**Tetramethylsilane**

■ **FIG. 12.1**

The $^1H$ NMR spectrum of ethyl iodide (60 MHz). The stair-step like line is the integral. In the integral mode of operation, the recorder pen moves from left to right and moves vertically a distance proportional to the areas of the peaks over which it passes. Hence the relative areas of the quartet of peaks at 3.20 ppm and the triplet of peaks at 1.83 ppm are given by the relative heights of the integrals (4 cm is to 6 cm as 2 is to 3). The relative numbers of hydrogen atoms are proportional to the peak areas (2 H and 3 H).

■ **FIG. 12.2**

The $^1H$ NMR spectrum of ethyl iodide (250 MHz). Compare this spectrum to Figure 12.1, run at 60 MHz. The peaks at 1.83 ppm and 3.20 ppm have been expanded and plotted on the left side of the spectrum.

## 2. The Position of Each Signal on the Horizontal Axis

Each signal's position on the horizontal axis is called the *chemical shift*, and it indicates the effective magnetic field around a single proton or group of equivalent protons as affected by nearby bonds and atoms. This *chemical environment* depends on structure, that is, the bond types ($sp^3$, $sp^2$, $sp$, or aromatic) and the electronegativities of the atoms one, two, and three bonds away from the particular protons. Chemical shifts, symbolized by $\delta$, are measured in dimensionless parts per million (ppm) to the left of, or *downfield* from, the reference TMS peak, according to the equation

$$\delta(ppm) = \frac{\text{shift of peak downfield from TMS (in Hz)}}{\text{spectrometer frequency (in MHz)}}$$

CH$_3$—

**Methyl**

—CH$_2$—

**Methylene**

—CH—
|

**Methine**

Table 12.1 gives the chemical shifts for protons in different chemical environments, spanning the region from 0 ppm to 12.5 ppm. Each signal in the spectrum of ethyl iodide can be assigned to a particular structural feature using this and similar tables. According to Table 12.1, the hydrogens on CH$_2$ groups attached to both an alkyl group and iodine in RCH$_2$I (where R can be methyl, methylene, or methine) should yield a signal between ~2.3 ppm and 3.2 ppm. We see a signal at 3.20 ppm in the spectrum of ethyl iodide, which is in this range. The table shows a range of 0.7 ppm to 1.1 ppm for a methyl group attached to a methylene or methine R group (RCH$_3$); R is –CH$_2$I for ethyl iodide. In actuality, the other signal in the spectrum of methyl iodide is not in this range; it is at 1.83 ppm, which is further downfield than predicted. This type of inconsistency occurs occasionally and demonstrates that reference tables are only generally representative of the majority of molecules that have been examined. It appears that in this particular case, the electronegative iodine, even though it is three bonds away, still strongly affects the magnetic field and shifts the peak for the methyl protons downfield. An examination of the other features of the spectrum will show that this peak assignment is correct.

## 3. Splitting Patterns and Coupling Constants

Patterns and coupling constants in the spectrum help define which groups are next to each other in a molecule's carbon skeleton. In an open-chain molecule with free rotation about the bonds and no chiral centers, protons couple with each other over three chemical bonds to give characteristic patterns of lines. If one or more equivalent protons couple to *one* adjacent proton, then the coupling hydrogen(s) appears as a *doublet* of equal intensity lines separated by the *coupling constant* ($J$) measured in Hertz (Hz). If the coupling proton or protons couple equally to *two* protons three chemical bonds away, they appear as a *triplet* of lines with relative intensities of 1:2:1, again separated by $J$. Finally, a *quartet* of lines in the ratio of 1:3:3:1 arises when one or more protons couple to the *three* protons on a methyl group. In general, chemically equivalent protons give a pattern of lines containing one more line than the number of protons being coupled to, and the intensities of the peaks follow the binomial expansion, conveniently represented by Pascal's triangle. Thus the methylene group of ethyl iodide appears as a quartet of lines at 3.20 ppm with relative intensities of 1:3:3:1 because the two methylene protons

Pascal's triangle

| | | | | | | | |
|---|---|---|---|---|---|---|---|
| | | | 1 | | | | singlet, s |
| | | 1 | | 1 | | | doublet, d |
| | 1 | | 2 | | 1 | | triplet, t |
| | 1 | 3 | | 3 | 1 | | quartet, q |
| 1 | 4 | | 6 | | 4 | 1 | pentet |
| 1 | 5 | 10 | | 10 | 5 | 1 | sextet |

TABLE 12.1 • **Proton Chemical Shifts**

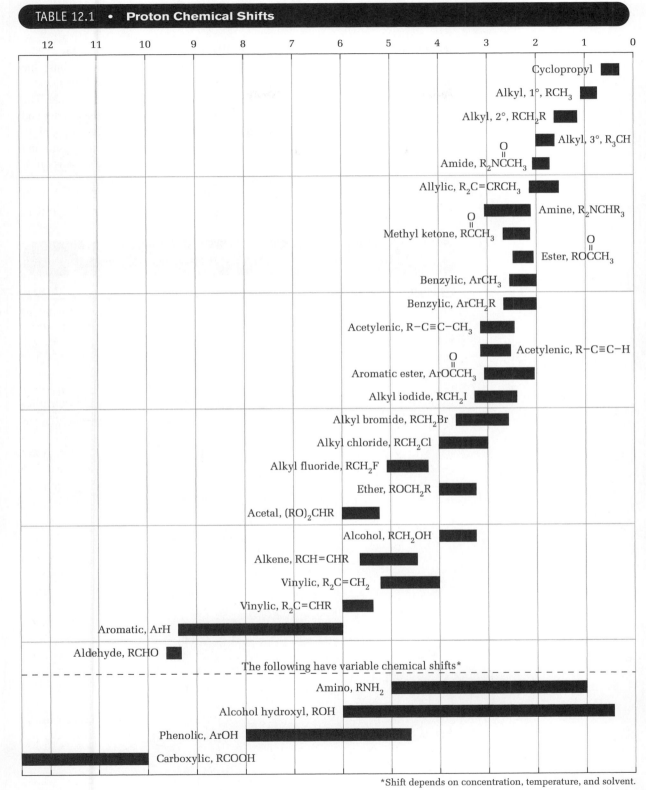

*Shift depends on concentration, temperature, and solvent.

are coupled to the three equivalent methyl protons (*see* Fig. 12.1 and Fig. 12.2). The methyl peak is split into a 1:2:1 triplet centered at 1.83 ppm because the methyl protons are coupled to the two adjacent methylene protons. This quartet-triplet combination is a strong indication of an ethyl group in any molecule. The *J* indicated in Figure 12.1 is the same 7.6 Hz between all lines in the quartet and triplet because the two groups are adjacent; therefore, the protons are coupled. In the 250 MHz spectrum, Figure 12.2, the splitting patterns are compressed and harder to distinguish because there are more Hz per ppm, but the expansion of these peaks on the left of the spectrum clearly shows a quartet and triplet.

Other characteristic proton coupling constants are given in Table 12.2. In alkenes *trans* coupling is larger than *cis* coupling, and both are much larger than

**TABLE 12.2 • Spin-Spin Coupling Constants for Various Geometries**

| Fragment | J (Hz) | Fragment | J (Hz) |
|---|---|---|---|
| | 7–12 | | 12–15 |
| | 13–18 | | 0–10 |
| | 0.5–3 | | 6–9 |
| | | | 1–3 |
| | | | 0–1 |
| | 0.5–2.5 | | 1–3 |
| | 4–10 | $CH_3 - CH_2 -$ | 6.5–7.5 |
| | 0–3 | | 5.5–7 |
| | 9–13 | | 5–9 |
| | | | 2–4 |

geminal coupling (coupling that takes place between two groups on the same carbon). *Ortho*, *meta*, and *para* couplings in aromatic rings range from 0 Hz to 9 Hz. The couplings in a rigid system of saturated bonds are strongly dependent on the dihedral angle between the coupling protons, as seen in cyclohexane.

## 4. The Integral

The relative numbers of distinctive hydrogen atoms (protons) in the molecule of ethyl iodide are determined from the *integral*, the stair-step line over the peaks shown in Figures 12.1 and 12.2. The height of the step is proportional to the area under the NMR signal for each group of equivalent protons. In NMR spectroscopy (contrasted with infrared spectroscopy, for instance) the area for each signal, including all splitting peaks, is directly proportional to the number of hydrogen atoms causing that signal. The methyl protons give a triplet of peaks, and the methylene protons give a quartet of peaks. Both ethyl iodide spectra in Figures 12.1 and 12.2 have a ratio of 3 to 2 for the area of the methyl triplet relative to the area of the methylene quartet, which is further evidence for the presence of an ethyl group. Integrators are part of all NMR spectrometers; running the integral takes little more time than running the spectrum. Most spectrometers print out a numerical value for the integral (with many more significant figures than are justified!).

In actuality, when chemists interpret NMR spectra, they use a combination of the *structure from the spectrum* and the *spectrum from the structure* approaches. A typical mental analysis of the NMR spectrum of 4-iodotoluene (Fig. 12.3) might go like this:

> This molecule should produce three signals: a singlet for the methyl protons, a doublet for the two equivalent aromatic protons *ortho* to the methyl group, and a doublet for the other two equivalent aromatic protons *ortho* to the iodo group. The methyl group is attached to an aromatic

**4-Iodotoluene**

■ **FIG. 12.3**
The $^1$H NMR spectrum of 4-iodotoluene (90 MHz in CDCl$_3$).

PPM (δ)

ring, and according to the table of proton chemical shifts, this should be detected as a singlet peak between 1.8 ppm and 2.8 ppm. The NMR spectrum of my product shows a singlet at 2.28 ppm, which is consistent with this. The aromatic protons show up as doublets in the right range, 6 ppm to 9 ppm according to the table. The spectrum has doublets at 6.95 ppm and 7.60 ppm. It also shows a singlet at 5.2 ppm and, according to the table, this implies that my product contains hydrogen on an alkene or in an acetal. Bad news! From other information I know, my product has neither of these functional groups. But wait! If I look at spectra of common sample impurities, I see that dichloromethane, the solvent I extracted the product with, appears at 5.25 ppm as a singlet. So this peak is not from my product and can be ignored . . .

This back-and-forth analysis correlating structure and NMR data, including coupling constants and integrals, is continued until all NMR peaks, splitting patterns, and integrals can be accounted for, and the identity of the product is verified.

When NMR data are reported in the literature, it is usually in a concise numerical form. For example, the NMR data for ethyl iodide derived from its spectrum (Fig. 12.2) would be reported as $^{1}$H NMR (CDCl$_3$): 1.83 (3H, t, $J$ = 7.6 Hz), 3.20 (2H, q, $J$ = 7.6 Hz), where CDCl$_3$ (deuterochloroform) is the solvent, 1.83 and 3.20 are the chemical shifts in ppm, 3H and 2H are integrals, t = triplet, q = quartet, and $J$ is the coupling constant in Hz.

A more complex spectrum

Some NMR spectra are not as easily analyzed as the spectrum for ethyl iodide. Consider the spectra shown in Figure 12.4. The proton spectrum of this unsaturated chloroester has been run at 500 MHz. Each chemically and magnetically nonequivalent proton is well resolved so that all of the couplings can be seen. Only the quartet of peaks at 4.1 ppm (relative area 2) and the triplet at 1.2 ppm (relative area 3) follow the simple first-order coupling rules outlined earlier. This quartet/triplet pattern is very characteristic of the commonly encountered ethyl group.

Because of the chiral carbon (marked with an asterisk), the protons on carbons 5 and 7 are diastereotopic, have different chemical shifts, and couple with each other and with adjacent protons to give the patterns seen in the spectrum. Many of the peaks can be assigned to specific hydrogens based simply on their chemical shifts. The coupling patterns then confirm these assignments.

## Carbon-13 Spectroscopy

The element carbon consists of 98.9% carbon atoms with mass 12 ($^{12}$C) and spin 0 (NMR inactive) and only 1.1% carbon atoms with mass 13 ($^{13}$C) and spin 1/2 (NMR active). Carbon, with such a low concentration of spin 1/2 nuclei, gives a very small signal when run under the same conditions as used for a proton spectrum. Carbon resonates at 75 MHz in a spectrometer where protons resonate at 300 MHz. Because only 1 in 100 carbon atoms has mass 13, the chances of a molecule having two $^{13}$C atoms adjacent to one another are small. Consequently, coupling of one carbon with another is not observed.

Each $^{13}$C atom couples to hydrogen atoms over one, two, and three bonds. Because the coupling constants are large, there is a high probability of peak overlap. To simplify the spectra as well as to increase the signal-to-noise ratio, a

The 500-MHz $^1$H and 75-MHz $^{13}$C NMR spectra of ethyl (3-chloromethyl)-4-pentenoate.[1] *I. $^{13}$C DEPT[2] spectrum.* The CH$_3$ and CH peaks are upright, and the CH$_2$ peaks are inverted. The quaternary carbon (the carbonyl carbon) does not appear. *II. The normal 75-MHz noise-decoupled $^{13}$C spectrum.* Note the small size of the carbonyl peak. *III. Expansions of each group of proton NMR peaks.* Protons E and F as well as protons H and I are not equivalent to each other. These pairs of protons are diastereotopic because they are on a carbon adjacent to a chiral carbon atom. The frequencies of all peaks are found on this book's website. *IV. The integral.* The height of the integral is proportional to the number of protons under it. *V. The 500-MHz $^1$H spectrum.*

TABLE 12.3 • **Carbon Chemical Shifts**

**PPM**

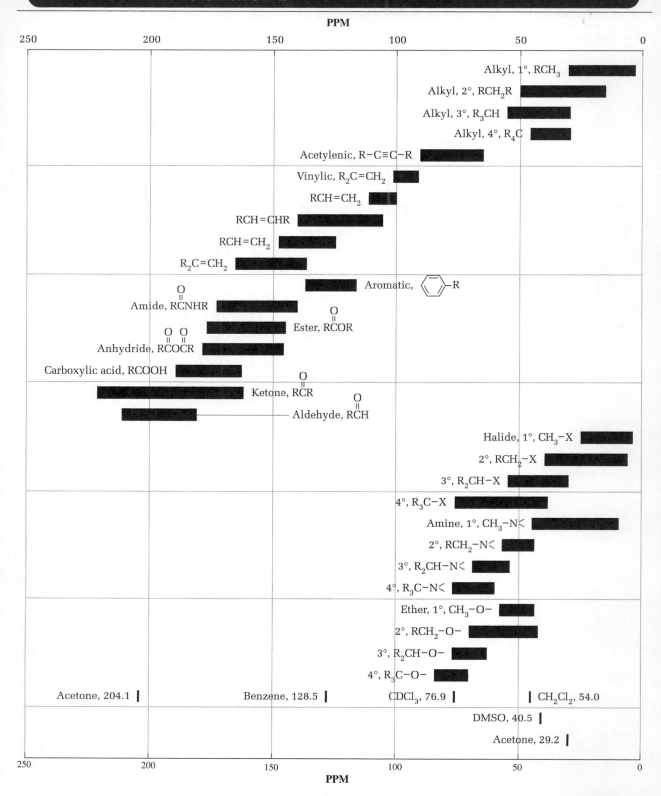

Alkyl, 1°, $RCH_3$

Alkyl, 2°, $RCH_2R$

Alkyl, 3°, $R_3CH$

Alkyl, 4°, $R_4C$

Acetylenic, $R-C{\equiv}C-R$

Vinylic, $R_2C=CH_2$

$RCH=CH_2$

$RCH=CHR$

$RCH=CH_2$

$R_2C=CH_2$

Aromatic, $\langle\!\!\!\bigcirc\!\!\!\rangle-R$

Amide, $\overset{O}{\overset{\|}{R}CNHR}$

Ester, $R\overset{O}{\overset{\|}{C}}OR$

Anhydride, $R\overset{O}{\overset{\|}{C}}O\overset{O}{\overset{\|}{C}}R$

Carboxylic acid, RCOOH

Ketone, $R\overset{O}{\overset{\|}{C}}R$

Aldehyde, $R\overset{O}{\overset{\|}{C}}H$

Halide, 1°, $CH_3-X$

2°, $RCH_2-X$

3°, $R_2CH-X$

4°, $R_3C-X$

Amine, 1°, $CH_3-N{<}$

2°, $RCH_2-N{<}$

3°, $R_2CH-N{<}$

4°, $R_3C-N{<}$

Ether, 1°, $CH_3-O-$

2°, $RCH_2-O-$

3°, $R_2CH-O-$

4°, $R_3C-O-$

Acetone, 204.1 |     Benzene, 128.5 |     $CDCl_3$, 76.9 |     | $CH_2Cl_2$, 54.0

DMSO, 40.5 |

Acetone, 29.2 |

**PPM**

$^{13}$C spectra: Broadband noise
decoupling gives a single line for
each carbon.

special technique is routinely used in obtaining $^{13}$C spectra: *broadband noise decoupling*. Decoupling has the effect of collapsing all multiplets (quartets, triplets, etc.) into a single peak. Further, the energy put into decoupling the protons can be looked on as appearing in the carbon spectrum in the form of an enhanced peak. This nuclear Overhauser enhancement (NOE) effect makes the peak appear three times larger than it would otherwise be. The result of decoupling is that every chemically and magnetically distinct carbon atom will appear as a single sharp line in the spectrum. Because the NOE effect is somewhat variable and does not affect carbons bearing no protons, one cannot do carbon counting from peak integrals the way hydrogen counting is done from proton NMR spectra.

### Carbon Chemical Shifts

The range of carbon chemical shifts is 200 ppm compared to the 10-ppm range for protons. It is not common to have an accidental overlap of carbon peaks. In the spectrum for the unsaturated chloroester in Figure 12.4 (spectrum II), there are eight sharp peaks corresponding to the eight carbon atoms in the molecule.

The generalities governing carbon chemical shifts are very similar to those governing proton shifts, as seen by comparing Table 12.1 to Table 12.3. Most of the downfield peaks are due to those carbon atoms near electron-withdrawing groups. In Fig. 12.4 (spectrum II) the furthest downfield peak is that from the carbonyl carbon of the ester. The attached electronegative oxygens make the peak appear at about 172 ppm. The peak is smaller than any other in the spectrum because it does not have an attached proton and thus does not benefit from a NOE effect.

The $^{13}$C spectrum of sucrose in Figure 12.5 displays a single line for each carbon atom; in Figure 20.3 (on page 364) a single line is seen for each of the 27 carbon atoms in cholesterol.

**■ FIG. 12.5**
The $^{13}$C NMR spectrum of sucrose
(22.6 MHz). Not all lines have been
assigned to individual carbon
atoms.

# Fourier Transform Spectroscopy

The Fourier transform
spectrometer

Even with broadband decoupling, the signal from $^{13}C$ spectra can get lost in the random noise produced by the spectrometer. Similarly, dilute solutions of samples give very noisy proton spectra. To increase the signal-to-noise ratio, the data for a number of spectra are averaged in the spectrometer computer. Because noise is random and the signal coherent, the signal will increase in size, and the noise will decrease as many spectra are averaged together. The improvement in the signal-to-noise ratio is proportional to the square root of the number of scans.

Spectra are accumulated rapidly by applying a very short pulse of radio-frequency energy to the sample and then storing the resulting free induction decay signal in digital form in the computer. The free induction decay signal, which takes seconds to acquire, contains frequency information about all the signals in the spectrum. In a few minutes several hundred free induction decay signals can be obtained and averaged in the computer. The free induction decay signal is converted to a spectrum of conventional appearance by carrying out a Fourier transform (FT) computation on the signal using the spectrometer's computer.

Because the spectral information is in digital form, a large number of operations can be performed on it. The spectrum can be smoothed at the expense of resolution, or the resolution can be enhanced at the expense of noise. As indicated earlier, the spectrum can be numerically integrated and the position of each peak, in hertz or ppm, printed above the peak. The spectra can be expanded both vertically and horizontally to clarify complex couplings, and all the parameters used to acquire the spectrum can be printed as well.

## Protons with Variable Chemical Shifts

| Labile proton(s) | δ (ppm) |
| --- | --- |
| ROH | 0.5–6 |
| ArOH | 4.5–8 |
| RCOOH | 10–12 |
| Enols | 10–17 |
| RNH$_2$ | 1–5 |
| Amides | 5–6.5 |

Molecules that have protons attached to oxygen and nitrogen form hydrogen bonds with each other and with protic solvents. As a result, the chemical shifts of these protons depend on solvent, temperature, and concentration and can appear almost anywhere in a spectrum. They will exchange with each other, a process catalyzed by small concentrations of acid or base. If this exchange is rapid (a common occurrence in alcohols, for example), only a single sharp line for the hydroxyl proton will be seen. At intermediate rates of exchange, the line becomes broad; at slow rates of exchange, it couples to other protons within three chemical bonds.

Protons bound to nitrogen often give a broad line because of the nonuniform distribution of charge on the nucleus of $^{14}N$. Obviously, NMR spectroscopy is a poor means for detecting hydroxyl, amine, and amide protons; infrared spectroscopy, on the other hand, is an excellent means of doing this. Because labile protons can undergo rapid exchange with each other, they can also exchange with deuterium. If 1 drop of $D_2O$ is added to a $CDCl_3$ solution of an NMR sample of a compound that contains such protons and gently mixed, the protons will exchange for deuterium atoms. The water will, of course, float on top of the denser chloroform. The peak for the labile proton will disappear, thus simplifying the spectrum and allowing for assignment of the peak.

Deuterium exchange

## ■ FIG. 12.6

The 60-MHz $^1$H NMR spectrum of 2-methyl-3-pentanol (0.4 $M$ in CS$_2$) with various amounts of shift reagent present. (A) No shift reagent present. All methyl peaks are superimposed. The peak for the proton adjacent to the hydroxyl group is downfield from the others because it is adjacent to the electonegative oxygen atom. (B) 2-Methyl-3-pentanol + Eu(dpm)$_3$. The mole ratio of Eu(dpm)$_3$ to alcohol is 0.05. The hydroxyl proton peak at 1.6 ppm in spectrum A appears at 6.2 ppm in spectrum B because it is closest to the Eu in the complex formed between Eu(dpm)$_3$ and the alcohol. The next closest proton, the one on the hydroxyl-bearing carbon atom, gives a peak at 4.4 ppm. Peaks due to the three different methyl groups at 1.1 ppm to 1.5 ppm begin to differentiate. (C) The mole ratio of Eu(dpm)$_3$ to alcohol is 0.1. The hydroxyl proton does not appear in this spectrum because its chemical shift is greater than 8.6 ppm with this much shift reagent present. (D) The mole ratio of Eu(dpm)$_3$ to alcohol is 0.25. Further differentiation of methyl peaks (2.3 ppm to 3.0 ppm) is evident. (E) The mole ratio of Eu(dpm)$_3$ to alcohol is 0.5. Separate groups of peaks begin to appear in the region 4.2 ppm to 6.2 ppm. (F) The mole ratio of Eu(dpm)$_3$ to alcohol is 0.7. Three groups of peaks (at 6.8 ppm, 7.7 ppm, and 8.1 ppm) due to the protons on C-2 and C-4 are evident, and three different methyls are now apparent. (G) The mole ratio of Eu(dpm)$_3$ to alcohol is 0.9. Only the methyl peaks appear on the spectrum. The two doublets come from the methyls attached to C-2, and the triplet comes from the terminal methyl at C-5.

Tris(dipivaloylmethanato)europium(III), Eu(dpm)$_3$
MW 701.78, mp 188–189°C

## Shift Reagents

Some spectra can be simplified by adding shift reagents to the sample. The resulting spectrum is similar to a spectrum one might obtain in a high-field spectrometer.

Adding a few milligrams of a hexacoordinate complex of the lanthanide europium called tris(dipivaloylmethanato)europium(III), or Eu(dpm)$_3$, to an NMR sample that contains a Lewis base center (an amine or basic oxygen, such as a hydroxyl group) has a dramatic effect on the spectrum. This reagent (soluble in CDCl$_3$) causes large shifts in the positions of peaks arising from the protons near the metal atom in this molecule and is therefore referred to as a *shift reagent*. It produces the shifts by complexing with the unshared electrons of the hydroxyl oxygen, the amine nitrogen, or other Lewis base center.

With no shift reagent present, the $^1$H NMR spectrum of 2-methyl-3-pentanol (Fig. 12.6a) is not readily analyzed. Adding 10-mg portions of the shift reagent to the sample causes very large downfield shifts of peaks, owing to protons near the coordination site (Figs. 12.6b–12.6g).[3] The two protons on C-4 and the two methyls on C-2 are diastereotopic and thus are magnetically nonequivalent because they are adjacent to a chiral center, C-3; each gives a separate set of peaks. When enough shift reagent is added, this spectrum can be analyzed by inspection (Fig. 12.7). Quantitative information about molecular geometry can be obtained from such shifted spectra. The shift induced by the shift reagent is related to the distance and the angle of the proton with respect to the europium atom.

**Chiral shift reagents**

Chiral shift reagents (derivatives of camphor) will cause differential shifts of the protons or carbon atoms in enantiomers. The separated peaks can be integrated to determine enantiomeric purity. A chiral shift reagent is used in Chapter 65 to determine the enantiomeric purity of a chiral alcohol made by the enzymatic reduction of a ketone.

■ **FIG. 12.7**
The $^1$H NMR spectrum of 2-methyl-3-pentanol containing Eu(dpm)$_3$ at 60 MHz. The mole ratio of Eu(dpm)$_3$ to alcohol is 1.0. Compare this spectrum with those shown in Figure 12.6. Protons nearest the hydroxyl group are shifted most. Methyl groups are recorded at reduced spectrum amplitude. Note the large chemical shift difference between the two diastereotopic protons on C-4.

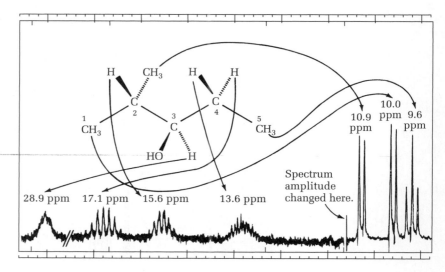

3. Williamson, K. L.; Clutter, D. R.; Emch, R.; Alexander, M.; Burroughs, A. E.; Chua, C.; Bogel, M. E. *J. Am. Chem. Soc.* **1974**, *96*, 1471.

## EXPERIMENTS

# 1. Running an NMR Spectrum

## Sample Preparation

A typical $^1$H NMR sample consists of about 10–50 mg of sample dissolved in about 1.0 mL of CDCl$_3$ that contains a very small percentage of TMS, the reference compound. Because the tiniest particles of dust, especially metal particles, can adversely affect resolution, it is best to filter every sample routinely through a wad of cotton or glass wool in a Pasteur pipette (Fig. 12.8).

If very high resolution spectra (all lines very sharp) are desired, oxygen, a paramagnetic impurity, must be removed by bubbling a fine stream of pure nitrogen through the sample for 60 s. Routine samples do not require this treatment.

It is most convenient if all the tubes in a laboratory are filled to exactly the same height with the CDCl$_3$ solution; this will greatly facilitate tuning of the spectrometer. High-quality NMR tubes give the best spectra, which are free of spinning side bands. A good tube should roll at a slow, even rate down a very slightly inclined piece of plate glass.

Chemical shifts of protons and carbon are measured relative to the sharp peak in the TMS (taken as 0.0 ppm). Because CDCl$_3$ is not 100% pure, there will always be a very small peak at 7.27 ppm in the proton spectrum that arises from the tiny amount of residual ordinary chloroform, CHCl$_3$, in the solvent (Fig. 12.10). In the carbon spectrum three lines of equal intensity appear at 77 ppm due to coupling of the deuterium atom with the carbon in CDCl$_3$.

If the sample does not dissolve in CHCl$_3$, a number of other deuterated solvents are available, including deuteroacetone (CD$_3$COCD$_3$), deuterodimethylsulfoxide (DMSO-$d_6$), and deuterobenzene (C$_6$D$_6$). All of these solvents are expensive, so the solubility of the sample should first be checked in nondeuterated solvents. For highly polar samples, a mixture of the more expensive DMSO-$d_6$ with the less expensive CDCl$_3$ is often satisfactory. If it is necessary to use D$_2$O as the solvent, then a special water-soluble reference, sodium 2,2-dimethyl-2-silapentane-5-sulfonate [(CH$_3$)$_3$Si(CH$_2$)$_3$SO$_3$$^-$Na$^+$ (DSS)], must be used. The protons on the three methyl groups bound to the silicon in this salt absorb at 0.0 ppm.

The usual NMR sample has a volume of about 1.0 mL, even though the volume sensed by the spectrometer receiver coils (referred to as the *active volume*) is much smaller. To average the magnetic fields produced by the spectrometer within the sample, the tube is spun by an air turbine at about 30 revolutions per second while taking the spectrum. Spinning that is too rapid or an insufficient amount of solution will cause the vortex produced by the spinning to penetrate the active volume (Fig. 12.9), giving erratic and nonreproducible spectra.

**Cleaning Up.** Place halogenated solvents and compounds in the halogenated organic waste container. All other substances go into the organic solvents container.

**■ FIG. 12.8**
Microfiltration for NMR samples. The solution to be filtered is placed in the top of the Pasteur pipette, the rubber bulb is put in place, and pressure is applied to force the sample through the cotton or glass wool into an NMR sample tube.

— Pasteur pipette

— Solution to be filtered

— Cotton or glass wool

— NMR tube

■ FIG. 12.9
The effect of too rapid spinning or insufficient sample. The active volume is the only part of the sample detected by the spectrometer.

## Adjusting the Spectrometer

To be certain the spectrometer is correctly adjusted and working properly, record the spectrum of the standard sample of chloroform and TMS usually found with the instrument. A 5% ethylbenzene in *d*-chloroform (CDCl$_3$) is also a useful performance check sample.

Small peaks symmetrically placed on each side of a principal peak are artifacts called *spinning side bands*. They are recognized as such by changing the spin speed, which causes the spinning side bands to change positions. They can be minimized by proper adjustment of the homogeneity controls.

Because Fourier transform spectrometers lock on the resonance of deuterium to achieve field/frequency stabilization, all samples must be dissolved in a solvent containing deuterium. Confirmation that the lock is obtained is registered on a meter or oscilloscope. The $Z$ and $Z^2$ controls are used to maximize the field homogeneity and achieve the highest resolution. This adjustment is needed for virtually every sample, but, as noted earlier, if all sample tubes in a laboratory are filled to exactly the same height, these adjustments will be minimized. Many modern instruments have automatic tuning capabilities.

## Common Contaminant Peaks

Almost every NMR spectrum contains spurious peaks that arise from a number of sources: residual protium ($^1$H) in deuterated ($^2$H) NMR solvents, dirty NMR tubes, and impure or inadequately dried products. NMR spectra of common contaminants are shown in Figure 12.10. The deuterated NMR solvents used are usually only about 99% deuterated and the residual protium present will produce detectable signals, particularly if sample signals are weak relative to the solvent. This type of residual peak is shown for CDCl$_3$, DMSO-$d_6$, and D$_2$O in Figure 12.10. Due to exchange with H$_2$O from glass surfaces and the atmosphere, D$_2$O is especially prone to contain a lot of HOD. If a product is crystallized from a solvent or isolated by removal of a solvent such as water, ethanol, dichloromethane, or ether, any solvent that remains will produce interfering signals, as shown in this figure. Atmospheric moisture absorbed by samples or solutions also gives rise to water peaks of varying intensities. NMR tubes are often rinsed out with acetone and even after prolonged drying, traces of acetone will remain and show up at about 2.2 ppm. All chemical shifts vary over a considerable range, depending on the solvent. For example, dichloromethane appears at 4.5 ppm in *d*-benzene, and 5.8 ppm in DMSO-$d_6$. The H$_2$O peak varies over an even wider range.

## Two-Dimensional NMR Spectroscopy

The fast computers associated with FT spectrometers allow for a series of precisely timed pulses and data accumulations to give a large data matrix that can be subjected to Fourier transformation in two dimensions to produce a two-dimensional (2D) NMR spectrum.

■ **FIG. 12.10**

The ¹H NMR spectra of common sample impurities and deuterated solvents. Note how the chemical shift due to water as an impurity varies over a wide range.

One of the most common and useful of these is the COSY (correlated spectroscopy) spectrum (Fig. 12.11). In this spectrum, two ordinary one-dimensional (1D) spectra are correlated with each other through spin-spin coupling. The 2D spectrum is a topographic representation, where spots represent peaks. The 1D spectra at the top and side are projections of these peaks. Along the diagonal of the 2D spectrum is a spot for each group of peaks in the molecule.

Figure 12.11 shows the 2D spectrum of citronellol. (Fig. 12.12 is the 1D spectrum of this compound.) Consider spot A on the diagonal of the 2D spectrum. From a table of chemical shifts, we know that this is a vinyl proton, 7, the single proton on the double bond. From the structure of citronellol, we can expect this proton to have a small coupling over four bonds to the protons on methyls 8 and 9, and a stronger coupling to the protons on carbon-6. In the absence of a 2D spectrum, it is not obvious which group of peaks belongs to carbon-6, but the spot B correlates with spot C on the diagonal, which is directly below the peak labeled 6 on the 2D spectrum.

The diagonal spot C, which we have just assigned to carbon-6, correlates through the off-diagonal spot D with the diagonal spot E, which lies just below the group of peaks labeled 5. Spot A on the diagonal also correlates through spots F and G with spots H and I on the diagonal, which lie directly below methyl peaks 8 and 9. In this way we can determine the complete connectivity pattern of the molecule, seeing which protons are coupled to other protons.

Other 2D experiments allow proton spectra on one axis to be correlated with carbon spectra on the other axis. NOESY (nuclear Overhauser effect spectroscopy) experiments give cross-peaks for protons that are near to each other in space but not spin coupled to each other.

## 2. Identification of an Unknown Alcohol or Amine by $^1$H NMR

*Shift reagents are expensive.*

Prepare a solution of the unknown in $CDCl_3$ and run a spectrum in the normal way. To the unknown solution add about 5 mg of $Eu(dpm)_3$ (the shift reagent), shake thoroughly to dissolve, and run another spectrum. Continue adding $Eu(dpm)_3$ in 5-mg to 10-mg portions until the spectrum is shifted enough for easy analysis. Integrate peaks and groups of peaks if in doubt about their relative areas. To protect the $Eu(dpm)_3$ from moisture, store it in a desiccator.

**Cleaning Up.** All samples containing shift reagents go into a hazardous waste container for heavy metals.

## QUESTIONS

1. Propose a structure(s) consistent with the proton NMR spectrum in Figure 12.13. Numbers adjacent to groups of peaks refer to relative peak areas. Account for missing lines.
2. Propose a structure(s) consistent with the proton NMR spectrum in Figure 12.14. Numbers adjacent to groups of peaks refer to relative peak areas.
3. Propose a structure(s) consistent with the proton NMR spectrum in Figure 12.15.
4. Propose structures for (a), (b), and (c), consistent with the $^1H$ and $^{13}C$ NMR spectra in Figures 12.16, 12.17, and 12.18. These are isomeric alcohols with the empirical formula $C_4H_{10}O$.

■ **FIG. 12.13**
The $^1H$ NMR spectrum (60 MHz) for question 1.

■ **FIG. 12.14**
The $^1H$ NMR spectrum (60 MHz) for question 2.

■ **FIG. 12.15**
The $^1H$ NMR spectrum (60 MHz)
for question 3.

$C_7H_7ClO$

■ **FIG. 12.16**
The $^1H$ NMR spectrum (60 MHz)
of compound (a), $C_4H_{10}O$, for
question 4.

■ **FIG. 12.17**
The $^{13}C$ NMR spectrum
(22.6 MHz) of compound (b),
$C_4H_{10}O$, for question 4.

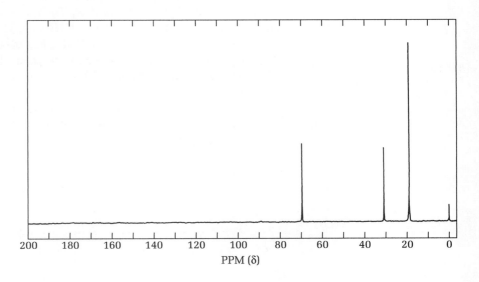

**■ FIG. 12.18**
The $^{13}C$ NMR spectrum (22.6 MHz) of compound (c), $C_4H_{10}O$, for question 4.

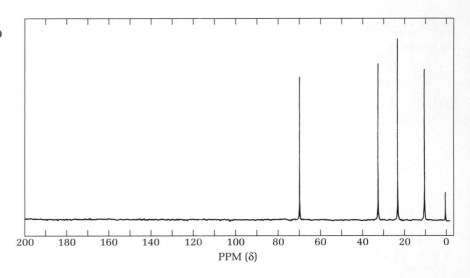

PPM ($\delta$)

---

## REFERENCES

**Online Study Center**

**General Resources**
Additional Experiments, Web Links

1. Abraham, Raymond J., J. Fisher, and P. Loftus, *Introduction to NMR Spectroscopy*. Chichester, UK: John Wiley & Sons, 1988.

2. Croasmun, William R., and Robert M. K. Carlson, *Two-Dimensional NMR Spectroscopy: Applications for Chemists and Biochemists*, 2nd ed. New York: VCH Publishers, 1994.

3. Friebolin, Horst, *Basic One- and Two-Dimensional NMR Spectroscopy*, 3d ed. New York: VCH, 1998.

4. Gadian, David G., *NMR and Its Applications to Living Systems*, Oxford, UK: Oxford University Press, 1996.

5. Pavia, Donald L., Gary M. Lampman, and George S. Kriz, *Introduction to Spectroscopy*, 3rd ed.[4] Brooks/Cole Thomson Learning, 2001.

6. Sanders, Jeremy K. M., and Brian K. Hunter, *Modern NMR Spectroscopy: A Guide for Chemists*, 2d ed. Oxford, UK: Oxford University Press, 1993.

7. Silverstein, Robert M., Francis X. Webster, and David Kiemle, *Spectrometric Identification of Organic Compounds*, 7th ed.[4] New York: Wiley, 2005.

8. Young, Ian R., Dave M. Grant, and Ryan K. Harris, eds., *Methods in Biomedical Magnetic Resonance Imaging and Spectroscopy*, 2 vol. New York: John Wiley & Sons, 2000.

---

4. This book includes IR, UV, and NMR spectra.

# Mass Spectrometry

**Online Study Center**

This icon will direct you to
techniques, equipment setups,
and online resources at http://
college.hmco.ccm/PIC/williamson
MMOE5e.

**PRELAB EXERCISE:** Calculate the molecular weight of bromobutane (in grams per mole) and the exact masses of its molecular
ions (in daltons) as measured by mass spectrometry.

Although its origins date back to the early part of the 20th century, mass spectrometry (MS) has experienced a renaissance in the past 2 decades due to
major improvements in ion formation and analysis. Because mass spectrometry is
capable of providing composition and structure information for a broad range of
compounds at sensitivities much greater than those of other techniques, more
money is invested worldwide in mass spectrometers than in any other type of in-
strumentation. In the chemical and life sciences, mass spectrometry is used to
identify and quantify compounds present in complex organic mixtures; to identify
structures of biomolecules such as carbohydrates, nucleic acids, and steroids; to
sequence proteins and oligosaccharides; to determine how drugs are used by the
body; to monitor the breath of patients anesthetized during surgery; and to check
fermentation processes for the biotechnology industry. In environmental science,
mass spectrometry is used to detect environmental pollutants such as dioxins in
fish and humans and to determine gene damage from environmental causes. In
forensic science, mass spectrometry is used to confirm and quantify drugs of
abuse and steroid use by athletes and to identify accelerants used in arson. Mass
spectrometry is used to determine the age and origins of specimens in geochem-
istry and archaeology; to locate oil deposits by measuring petroleum precursors in
rock; to report the composition of molecular species found in space; to perform
ultrasensitive, multielement inorganic analyses; and to establish the elemental
composition of semiconductor materials.

Notice that the infrared (IR), nuclear magnetic resonance (NMR), and ultraviolet (UV) methods are called *spectroscopy*, but MS is mass *spectrometry*. This
difference helps remind us that mass spectrometry does not involve the absorption
of specific electromagnetic energies resulting in spectra of frequency versus intensity; mass spectrometry involves the production and detection of different
masses to yield spectra recorded as ion mass versus ion abundance.

# The Mass Spectrometer

Mass spectrometers have the following components, connected as shown in Figure 13.1.

## 1. Ion Source

Neutral molecules in the gas state move randomly, and it is impossible to move them in any particular direction without using a pump to create a pressure gradient. However, if molecules can be positively or negatively charged to form ions, these ions can be attracted and repelled by charged metal surfaces so that they can be moved or accelerated in any chosen direction. The ionization of molecules is done in numerous ways, but the most common way is shown in Figure 13.2. Sample molecules in the gaseous state are introduced into an *ion source*, a small metal enclosure containing a hot filament, like that in a light bulb, on one side. Electrons are emitted from the filament and attracted to a small metal collector electrode on the other side by a slight positive voltage, usually +70 eV, relative to the filament. These energetic electrons can collide with the gaseous sample molecules and ionize them by knocking an electron out of their electron clouds, producing *molecular ions*, most having a single positive charge, but a few with two positive charges. Some negative ions are formed by the capture of an electron, but these are much less abundant. Therefore, most electron ionization spectra are of positive ions.

**■ FIG. 13.1**
The components of a mass spectrometer.

**■ FIG. 13.2**
A view into the electron ionization source enclosure. Sample molecules, $CO_2$ in this illustration, are ionized by high-energy electron impacts and together with fragment ions are accelerated into the mass analyzer as a continuous beam of pulsed ion packets.

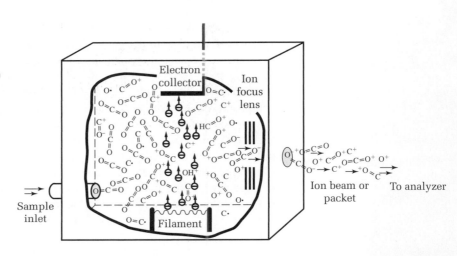

The ionization process is illustrated in Figure 13.2 and in the following text using a sample of $CO_2$. Electron ionization produces $O{=}C{=}O^{\dot{+}}$ molecular ions written using the cation radical symbol $\dot{+}$ because they are odd electron species.

*Ionization:*

$$O{=}C{=}O \;+\; e^{\ominus} \;\longrightarrow\; O{=}C{=}O^{\dot{+}} \;+\; 2e^{\ominus}$$

*Fragmentation:*

$$O{=}C{=}O^{\dot{+}} \;\longrightarrow\; O{\cdot} \;+\; C{=}O^{+} \quad\text{or}\quad O^{+} \;+\; C{=}O{\cdot}$$

$$C{=}O^{+} \;\longrightarrow\; C^{\dot{+}} \;+\; O^{\cdot} \quad\text{or}\quad C{\cdot} \;+\; O^{\dot{+}}$$

The energy of the bombarding electrons (about 70 eV) is generally much greater than that of the bonds holding the molecule together. Thus, when high-energy electrons interact with the molecule, not only does ionization occur, but bonds are broken and *fragmentation* occurs, giving rise to *fragment ions* and *fragment neutrals*. The $C{=}O$ bonds of $CO_2$ can be broken to produce either a $C{=}O^{+}$ ion and a neutral O radical or an $O^{+}$ ion and a neutral $C{=}O$ radical. If both $C{=}O$ bonds are broken, a $C^{+}$ ion can be produced.

The electron ionization source, as well as the other parts of a mass spectrometer, must be maintained at very low pressure (high vacuum; $\sim10^{-8}$ atm), and only very small amounts of sample can be introduced; if the pressure is too high, the ions would collide with unionized molecules or other ions instead of being accelerated from the ion source and into the ion mass analyzer. Also, the electron filament would burn out if an atmosphere of air were present. Therefore, sample inlets are typically a fine capillary or a tiny pinhole to restrict sample flow into the source chamber. Solids are introduced on a metal rod that slides through a sealed vacuum lock. Molecules that are not ionized and neutral fragments are pumped away.

There are several other ionization methods that have the advantage of producing intense molecular ions or protonated molecular ions, which are just as informative. For samples that are gaseous or that can be vaporized in a vacuum, chemical ionization and field ionization are often used. These will not be discussed here, but the references at the end of this chapter provide details. High molecular weight nonvolatile compounds can be ionized by electrospray ionization (ESI) or matrix-assisted laser desorption ionization (MALDI), both of which are discussed in a later section on bioanalytical mass spectrometry.

## 2. Mass Analyzer

Positive ions are propelled into the analyzer by maintaining the ion source at a more positive electrical potential relative to the analyzer and by focusing the ions with voltages applied to an electronic lens system located between the source and the analyzer. The beam of ions passes through the *mass analyzer* where the ions are separated according to mass. Actually, ions are separated based on the mass to charge ratio ($m/z$). For example, for $CO_2{}^{+}$ ions $z = 1$, so $m/z = 44/1$, or 44; for $CO_2{}^{++}$ ions

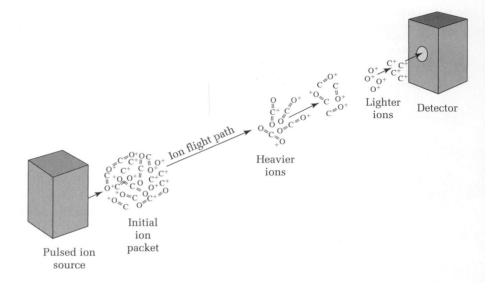

$z = 2$, so $m/z = 44/2$, or 22. Doubly charged ions are rare, so $z = 1$ in almost all cases, and $m/1 = m$; therefore, for practical purposes, ions are separated according to their mass, and the ensuing discussion will mainly use mass instead of $m/z$.

In the earliest mass spectrometers, the ions of different masses were separated by passing the ion beam through a magnetic field. Today, however, two types of mass analyzers are commonly used. The simplest in theory is the *time-of-flight (TOF) analyzer* (Fig. 13.3). The ion beam is pulsed through an electronic gate in little packets, and the start time of each packet of ions is recorded. The packet of ions travels down a long tube, typically 1–2 m in length. Lower mass, lighter ions travel faster and arrive at the detector to produce signals with short times, while heavier ions produce signals with longer times. The development of high-speed electronics now allows the detection of ions that differ in mass by only 1 part in 10,000, which corresponds to a difference in flight times of a few microseconds ($10^{-6}$ s) or less.

*Quadrupole mass analyzers* are commonly used in GC–MS systems (discussed in a later section of this chapter). The continuous beam of ions from the source passes between four parallel metal rods or poles (Fig. 13.4). Depending on the magnitude and frequency of the direct current (DC) and alternating current (AC) voltages on the rods, only ions of a certain *resonant* mass will pass between the rods and into the detector. Ions of other masses are nonresonant and will be deflected to the side and miss the detector. The mass range is scanned by increasing the rod voltage and AC frequency so that ions of increasing masses hit the detector.

### 3. Detector

After separation by mass, the ions pass into the *detector* where they impact a metal electrode called a dynode that has a high negative voltage. This releases a number of electrons (~5–10) that are drawn toward a slightly more positive

■ **FIG. 13.4**

A quadrupole mass spectrometer. Ions oscillate as they move through the electrical fields on the four rods. Changing these fields allows a specific resonant ion to pass to the detector. Ions of other masses are deflected.

■ **FIG. 13.5**

A diagram of an electron multiplier detector.

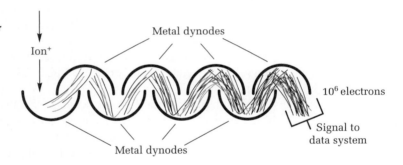

dynode; each of these electrons releases a number of electrons, which accelerate toward a third dynode, releasing even more electrons (Fig. 13.5). These *electron multipliers* have enough dynode stages that a single positive ion impact produces a cascade of a million or more electrons at the final electrode. Single ions can be detected, thus making mass spectrometry a very sensitive method.

The signal output voltage is proportional to the number of ions hitting the detector at a given time (TOF analysis) or AC and DC voltages (quadrupole analysis) and is continuously digitized and stored in a computer. The computer has files of mass versus calibrated TOFs or quadrupole voltages and can therefore plot the mass spectrum as the *percent relative abundance* versus mass or more correctly $m/z$ (see earlier discussion). All signal peaks are normalized with the most abundant ion as 100%.

The electron ionization spectrum for a sample of $CO_2$ is shown in Figure 13.6. The ionized $CO_2$ molecule (or molecular ion) appears at mass 44. The ion is singly

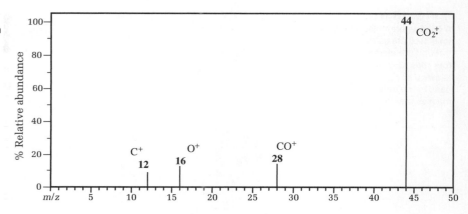

A dalton (Da) is a mass unit used
in mass spectrometry and is
defined as 1/12 of the mass of
$^{12}C$, which has the atomic weight
12.00000.

charged, and the *nominal ion mass* is 44 atomic mass units or daltons (Da): carbon =
12 Da and oxygen = 16 Da (in calculating nominal ion mass, atomic masses are
rounded to the nearest integer). Because the ionization process breaks up or fragments some of the $CO_2$ molecules, a fraction of the ions appear in the spectrum at
mass values less than the molecular ion. Cleavage of a carbon-oxygen bond in
the molecular ion, which produces ionized carbon monoxide ($CO^+$) and a neutral
oxygen radical (O·) or ionized atomic oxygen (an $O^+$ ion) and a neutral carbon
monoxide radical (C=O·), results in the fragment ions at masses 28 and 16; loss
of two neutral oxygen atoms results in an additional fragment at mass 12 for carbon (a $C^+$ ion). Note that only charged fragments, not neutral fragments, appear in
the spectrum.

Most mass spectrometers are set up so that they do not scan below $m/z = 35$
because there are very large background signals at masses 32 ($O_2$), 28 ($N_2$), and
18 ($H_2O$) due to residual air and moisture in the instrument, in spite of the high
vacuum used. Therefore, unlike Figure 13.6, the remaining spectra in this chapter
start at mass 35. Weak background signals at masses 40 and 44 from atmospheric
argon and carbon dioxide are also seen in many spectra but are usually ignored.

## Interpretation of Electron Ionization Spectra

### *Molecular Ions*

The mass of a molecular ion $M^{+}$, as measured by a mass spectrometer, is not the
same as the molecular weight conventionally used to weigh out mole quantities on
a balance. The analyzer of a mass spectrometer separates individual molecular ions
based on the masses of the isotopes in those ions. The important isotopes for organic mass spectrometry are listed in Table 13.1. Every organic compound contains both $^{12}C$ isotopes with mass 12.0000 Da and $^{13}C$ isotopes with mass 13.0034
Da, with a 1.1% probability of a given carbon being a $^{13}C$. (Radioactive $^{14}C$ is at

**TABLE 13.1 • The Natural Abundance of Isotopes (atoms per 10,000 atoms and percent abundance) and the Atomic Mass of Each Isotope**

| | Abundance and Mass of the Most Abundant Isotope (given 10,000 atoms) | | | Abundance and Mass of Other Isotopes (given 10,000 atoms) | | | | |
|---|---|---|---|---|---|---|---|---|
| | Isotope | No. of Atoms | Mass (Da) | % Abundance | Isotope | No. of Atoms | Mass (Da) | % Abundance | Average Mass (g/mol) |
| C | $^{12}C$ | 9890 | 12.0000 | 98.9 | $^{13}C$ | 110 | 13.0034 | 1.1 | 12.0107 |
| H | $^{1}H$ | 9999 | 1.0078 | 99.99 | $^{2}H$ | 1 | 2.0141 | 0.01 | 1.00794 |
| O | $^{16}O$ | 9985 | 15.9949 | 99.76 | $^{17}O$ | 4 | 16.9991 | 0.04 | 15.9994 |
| | | | | | $^{18}O$ | 20 | 17.9992 | 0.2 | |
| N | $^{14}N$ | 9963 | 14.0031 | 99.63 | $^{15}N$ | 37 | 15.0001 | 0.37 | 14.0067 |
| Cl | $^{35}Cl$ | 7577 | 34.9689 | 75.8 | $^{37}Cl$ | 2423 | 36.9659 | 24.2 | 35.453 |
| Br | $^{79}Br$ | 5069 | 78.9183 | 50.7 | $^{81}Br$ | 4931 | 80.9163 | 49.3 | 79.904 |
| S | $^{32}S$ | 9493 | 31.9721 | 94.93 | $^{33}S$ | 76 | 32.9715 | 0.76 | 32.065 |
| | | | | | $^{34}S$ | 429 | 33.9679 | 4.29 | |

The elements $^{19}F$, $^{31}P$, and $^{127}I$ are monoisotopic.

**■ FIG. 13.7**

The mass spectrum of decane showing isotopic molecular ions, fragment ions, and neutral losses due to bond cleavages at a, b, c, and d.

a much lower natural abundance.) About 89% of molecules in a sample of decane contain only $^{12}C$ and have the composition $^{12}C_{10}H_{22}$, and 11% of the molecules have one $^{13}C$ (10 carbons times 1.1% probability of each one being a $^{13}C$) and the composition $^{12}C_9{}^{13}C_1H_{22}$. When 1 mol of decane is weighed out, huge numbers of carbon atoms ($\sim 60 \times 10^{23}$) are present, and therefore it is reasonable to use an average mass of the two isotopes, 12.0107 g/mol, to calculate the molecular weight of decane as 142.285 g/mol. But in a mass spectrometer, $^{12}C_{10}H_{22}{}^+$ ions are separated from $^{12}C_9{}^{13}C_1H_{22}{}^+$ ions, as seen in the mass spectrum of decane shown in Figure 13.7.

■ **FIG. 13.8**

The mass spectrum of 4-chloro-3-methyl phenol.

■ **FIG. 13.9**

Chlorine and bromine isotope patterns.

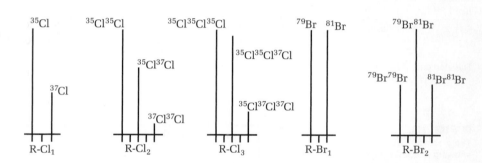

To calculate the exact masses of these ions, isotopic masses for $^{12}C$ (12.0000 Da) and $^{13}C$ (13.0034 Da) must be used. The $^{12}C_{10}H_{22}{}^{+}$ molecular ion has a mass of 142.1716 Da and that of $^{12}C_9{}^{13}C_1H_{22}{}^{+}$ has a mass of 143.1750 Da. The heavier isotopes of hydrogen ($^2H$ and $^3H$) and oxygen ($^{17}O$ and $^{18}O$) have abundances that are so low that they are difficult to detect and are usually ignored. For decane, the *nominal* mass of the most abundant molecular ion containing just $^{12}C$, obtained by rounding the exact mass to the closest integer, is 142 Da, the same as the integer molecular weight of decane, 142 g/mol. For other isotopes, this may not be true. In the mass spectrum of 4-chloro-3-methylphenol (Fig. 13.8), there are four molecular ions with isotopic compositions shown due to the presence of not only $^{13}C$ but also the two isotopes of chlorine, $^{35}Cl$ and $^{37}Cl$. The exact mass of the largest molecular ion, $^{12}C_7H_7O^{35}Cl$ is 142.0184 Da. This is quite different than the molecular weight of this compound, 142.5774 g/mol. After rounding, the nominal mass of the ion is 142 Da, but the integer molecular weight is 143 g/mol. This difference is the primary reason that daltons are used in mass spectrometry—to help differentiate exact masses measured by mass spectrometry from the conventional molecular weight in grams/mole used in lab weighing.

Both chlorine and bromine have isotopes separated by 2 Da, creating unique and easily recognized patterns in a mass spectrum. The ratio of peaks depends on

| TABLE 13.2  •  Six Compounds with Molecular Ions of Nominal Mass 142 That Can Be Differentiated by Accurate Mass Measurement |||

| Compound | Molecular Formula | Exact Mass of Molecular Ion |
|---|---|---|
| 3-Fluoro-4-methoxyphenol | $C_7H_7FO_2$ | 142.0430 |
| Octenoic acid | $C_8H_{14}O_2$ | 142.0993 |
| Cyclohexylurea | $C_7H_{14}N_2O$ | 142.1106 |
| 3-Nonanone | $C_9H_{18}O$ | 142.1357 |
| 1,2-Diaminocyclooctane | $C_8H_{18}N_2$ | 142.1469 |
| Decane | $C_{10}H_{22}$ | 142.1721 |

the number of chlorines or bromines in the molecule, as shown in Figure 13.9. The $Cl_1$ pattern is seen in the mass spectrum of the chlorophenol in Figure 13.8.

Many modern instruments are able to measure ion masses to within a few parts per million of accuracy. For example, even though all of the compounds listed in Table 13.2 (as well as hundreds of other possible compounds) have molecular ions at nominal mass 142, exact mass measurement would easily differentiate among them.

Conversely, given an exact mass measured by mass spectrometry, a computer can compare this with all possible exact masses calculated for all reasonable combinations of $^{12}C$, $^{1}H$, $^{14}N$, $^{16}O$, $^{35}Cl$, and so forth and produce a list of compositions that come within a few parts per million of the measured mass. Compositions determined in this manner are required whenever a new compound is reported in a publication. (Note, however, that the composition $C_9H_2O_2$, for example, is not reasonable even though it has a mass of 142.)

Compounds containing C, H, O, F, Cl, Br, I, and S or an even number of nitrogens will have an even mass molecular ion. *The nitrogen rule* states that all molecules with an odd number of nitrogen atoms will have an odd mass molecular ion. Compare the molecular ions in the mass spectrum of a nitrogen-containing compound (Fig. 13.10) with the other spectra in this chapter. If we look at the simplest compounds of C, O, and N, the reason becomes apparent. C and O have even masses and form an even number of bonds, so $H_2O$ has a molecular mass of 18 and $CH_4$ has a molecular mass of 16, both even. Nitrogen ($^{14}N$) has an even mass but forms an odd number of bonds, so $NH_3$ has an odd mass, 17. Organic compounds are simply derivatives of these simple molecules. The halogens form only single bonds (odd) but have odd masses, so replacement of a hydrogen by a halogen does not change the molecular mass from even to odd.

## Fragmentation

With electron ionization, the percent relative abundance, or intensity, of the molecular ion varies dramatically with compound structure. Carbon dioxide (*see* Fig. 13.6 on page 278) and aromatic compounds (*see* Fig. 13.8) contain multiple

■ **FIG. 13.10**

The mass spectrum of methyl 3-nitrobenzoate.

■ **FIG. 13.11**

The mass spectrum of 4,4-dimethyloctane, an isomer of decane.

double bonds that are stronger than single bonds, and the $\pi$ bonding systems can help stabilize the positive charge so that less bond fragmentation occurs and the molecular ion abundance is high. Decane has all single bonds and fragments more readily. The most abundant ions are fragments, and a weak molecular ion is observed. Although the mass of the molecular ion is considered the most important piece of information to be derived from a mass spectrum, the sad truth is that many compounds, like hydrocarbons, alcohols, and alkyl halides, do not show any molecular ion signal. For example, the mass spectrum of 4,4-dimethyloctane (Fig. 13.11) shows no signal at mass 142, which indicates that it fragments even more readily than decane. Nevertheless, the masses and mass differences of abundant fragment ions can provide useful structural information.

We have already discussed the fragment ions (detected) and fragment neutrals (not detected) generated by bond cleavage in $CO_2$. It is clear that these peak masses are calculated by the simple arithmetic of adding up the atomic isotopic masses of the atoms in the fragments formed by the cleavage of bonds in the molecule. Table 13.3 gives the masses and possible structures for several commonly observed fragment

| TABLE 13.3 • Common Fragment Ion Masses and Structures | |
|---|---|
| **Fragment Ion Mass (Da)** | **Structure(s)** |
| . . . 113, 99, 85, 71, 57, 43 <br> Alkyl series | . . . $C_8H_{17}^+$, $C_7H_{15}^+$, $C_6H_{13}^+$, $C_5H_{11}^+$, $C_4H_9^+$, $C_3H_7^+$ <br> Differ by $CH_2$ or 14 Da |
| . . . 111, 97, 83, 69, 55, 41 <br> Alkenyl series | . . . $C_8H_{15}^+$, $C_7H_{13}^+$, $C_6H_{11}^+$, $C_5H_7^+$, $C_4H_7^+$, $C_3H_5^+$ <br> Differ by $CH_2$ or 14 Da |
| . . . 113, 99, 85, 71, 57, 43 <br> Alkanoyl series | (a)  $C_6H_{13}\overset{\overset{\displaystyle O}{\|}}{C}{}^+$, $C_5H_{11}\overset{\overset{\displaystyle O}{\|}}{C}{}^+$, $C_4H_9\overset{\overset{\displaystyle O}{\|}}{C}{}^+$, $C_3H_7\overset{\overset{\displaystyle O}{\|}}{C}{}^+$, $C_2H_5\overset{\overset{\displaystyle O}{\|}}{C}{}^+$ <br> Differ by $CH_2$ or 14 Da |
| 105 plus mass of any group on ring | (b)  (benzoyl) |
| 91 plus mass of any group on ring | (c)  (benzyl) |
| 77 plus mass of any group on ring | (d)  (phenyl) |

ions. The abundance of a particular fragment ion depends on two factors: the ease with which bonds to it can be broken and the stability of the resulting ion. In the mass spectrum of decane (*see* Fig. 13.7 on page 279), we can see a series of fragment ions at masses 41, 57, 71, 85, 99, and 113, differing by increments of 14 Da, which is the mass of a $CH_2$ unit. Table 13.3 lists this alkyl series of ions. These arise from the cleavage of bonds at a, b, c, and d, as marked on the decane molecular ion, to form a charged fragment and an uncharged or neutral fragment. From the ion abundances, it is clear that the positive charge prefers to reside on fragments containing three or four carbons, with masses 43 and 57, so the neutral fragments formed by bond cleavage contain the other seven or six carbons.

This alkyl series of fragment ions is common in all alkanes or molecules with large alkyl groups and is observed in the mass spectrum 4,4-dimethyloctane (Fig. 13.11). However, there are several important differences in the percentage ion abundances in its spectrum when compared to that of decane (Fig. 13.7). These can be explained in terms of the increasing stability of carbocations as we go from primary to secondary to tertiary carbocations, the same stability observed in solution organic reactions. The structure of dimethyloctane promotes fragmentation at b and c to give the stable tertiary cations shown at masses 85 and 99. Both of these ions are much more abundant in this spectrum than in the spectrum of decane. Notice also the decreased intensity at mass 71, for the $C_6H_{11}$ alkyl ion. Because there are no $C_6$ alkyl groups in 4,4-dimethyloctane, the formation of an ion of mass 71 requires the breaking of two C—C bonds and a hydrogen transfer, a much less favorable process. Finally, unlike decane, no molecular ion signal is detected, which is more evidence of the facile fragmentation of this molecule.

Strong peak intensities at masses 77, 91, and 105 are uniquely diagnostic for the presence of an aromatic ring (*see* Table 13.3). These ions are increased in mass by any groups attached to the aromatic ring. Chapter 28 describes the synthesis of methyl 3-nitrobenzoate. The mass spectrum of this product is shown in Figure 13.10. The most abundant ion at mass 150 corresponds to the benzoyl ion of mass 105 shown in Table 13.3, plus the weight of a nitro group, $NO_2$, mass 46, minus 1, the mass of a hydrogen that the $NO_2$ replaces.

Notice that the fragment ion masses of the alkanoyl series are the same as that of the alkyl series; therefore, it might seem impossible to distinguish carbonyl compounds like ketones, acids, and esters from alkanes. However, the presence of a C=O leads to some diagnostic fragmentations. For example, the mass spectrum of 3-nonanone in Figure 13.12 shows a very diagnostic fragment ion at mass 72, which is even. This ion arises by a fragmentation of the molecular ion involving a rearrangement called the McLafferty rearrangement:

$$M^{+}_{\cdot} = 142$$

**Mass 72**          **Neutral loss**
**fragment ion**          **fragment**

Electron pair bonds shift in the six-member intermediate, and a hydrogen is transferred from carbon 6 of the ketone to the carbonyl oxygen. The bond rearrangement releases the neutral alkene, and the positive charge remains on the oxygen. Fragmentation with a hydrogen rearrangement is designated with ~H on the spectrum. Such fragmentations are observed quite frequently.

## Neutral Loss Masses

When a molecule breaks into two fragments, only one of them carries the positive charge of the parent molecular ion and can be detected. The uncharged piece, normally a radical or in certain cases a small molecule, is not detected. However, its

■ FIG. 13.12
The mass spectrum of
3-nonanone.

■ FIG. 13.12
The mass spectrum of
3-nonanone.

mass can be easily calculated by subtracting the mass of the charged fragment from the molecular ion mass. Like fragment ion masses, the neutral fragment difference masses provide structural information. And they are often easier to understand. Common neutral loss masses are shown in Table 13.4.

Fragment ions can be specified in either of two ways: the fragment structure with a positive charge attached or the $M^{+}$ minus the neutral fragment. Both ways are used in the spectra discussed here. The latter method is much more concise and readily understood for the higher mass fragments close to the $M^{+}$. For example, the fragment at mass 113 in the decane spectrum (Fig. 13.7) is much easier to designate as $M^{+}$ minus ethyl ($M^{+} - C_2H_5$) than as $CH_3CH_2CH_2CH_2CH_2CH_2CH_2CH_2^{+}$. The same is true for fragments at masses 99 and 85. In the spectrum of the chlorophenol in Figure 13.8, the difference in mass between the largest $M^{+}$ at mass 142 and the major fragment ion at mass 117 is 35. From Table 13.4, we see this can be attributed to a neutral loss of chlorine and the fragment ion is specified as $M^{+} - Cl$, where Cl is either $^{35}Cl$ or $^{37}Cl$ depending on the isotopes in the fragmenting molecular ion.

One of the major differences between the mass spectra of decane and 4,4-dimethyloctane (Fig. 13.7 and Fig. 13.11) lies in the neutral loss of methyl. This fragmentation, specified here as $M^{+} - CH_3$, is seen for the dimethyloctane because it leads to a stable tertiary carbocation fragment. This type of fragment is not possible with the linear alkane decane, and a neutral loss of methyl is not observed. The major fragmentation of the ester shown in Figure 13.10 involves a neutral loss with a mass difference of 31, which is attributable to cleavage of neutral $OCH_3$ radical. In Figure 13.12, the ion at mass 113 is specified both as its fragment ion structure and neutral loss designation.

Many neutral losses involve the elimination of an intact neutral molecule such as water, carbon monoxide, carbon dioxide, or acetic acid. Figure 13.13 shows the mass spectrum of the alcohol 2-nonanol, and, as is typical of alcohols, no molecular ion is seen because of the facile neutral loss of $H_2O$. The mass of the molecular ion can be deduced, however, by examining other ions observed at

| TABLE 13.4 • Some Common Neutral Loss Masses and Possible Fragment Structures | |
| --- | --- |
| **Mass Difference (Da)** | **Neutral Radical or Molecule Lost** |
| 1 | ·H (hydrogen) |
| 15 | ·CH₃ (methyl) |
| 17 | ·OH (hydroxyl) |
| 18 | H₂O (water) |
| 28 | C≡O (carbon monoxide) or CH₂=CH₂ (ethene) |
| 29 | ·CH₂CH₃ (ethyl) or $\overset{O}{\overset{\|}{·C}}$—H (formyl) |
| 31 | ·OCH₃ (methoxyl) |
| 35 or 37 | ³⁵Cl· or ³⁷Cl· (chloro) |
| 42 | CH₂=C=O (from acetyl group) |
| 43 | $CH_3$—$\overset{O}{\overset{\|}{C}}$· (acetyl)  or ·C₃H₇ (propyl) |
| 44 | CO₂ (carbon dioxide) |
| 45 | ·OCH₂CH₃ (ethoxyl) or $·\overset{O}{\overset{\|}{C}}OH$  (carboxyl) |
| 60 | $CH_3-\overset{O}{\overset{\|}{C}}-OH$ (acetic acid) |
| 79 or 81 | ⁷⁹Br· or ⁸¹Br· (bromo) |

higher molecular weight. In particular, the presence of two ions 3 Da apart (masses 126 and 129 in Fig. 13.13) indicates neutral losses of water and a methyl group from a molecular ion. For 2-nonanol, adding 15 to mass 129 for a methyl radical equals 144 and adding 18 to mass 126 for a water molecule also equals mass 144. This is strong evidence that the molecular ion mass is indeed 144. The fact that the alkene fragment ion formed by the loss of water has an even mass (except for a compound containing nitrogen) is another indicator of the loss of a molecule rather than a radical.

Each of the bond fragmentations designated a, b, and c in Figure 13.13 is called an *alpha cleavage* because each involves the cleavage of the bond

■ **FIG. 13.13**
The mass spectrum of 2-nonanol.

adjacent, or alpha, to the C—OH bond. The electron movement for each of the fragmentations to give ions at masses 143, 129 and 45 are shown below (R = CH₂CH₂CH₂CH₂CH₂CH₂CH₃):

## Computer-Aided Mass Spectral Identification

Today, there are computer searchable databases of digitized mass spectra of over 200,000 different organic compounds. With less than 1 mg of sample, a mass spectrum can be acquired by a mass spectrometer and stored in a computer. This digitized spectrum can be compared with all the spectra in these collections, and those reference spectra that have ion masses and intensities similar to the sample spectrum can be displayed. If the sample and a reference library spectrum match closely enough, the chemist can feel confident that the sample has been correctly identified. This sort of analytical power is critical to the success of forensic and environmental chemistry laboratories. Be aware, however, that mass spectra of many structural isomers, for example, the three possible dimethylbenzenes, will be so similar that a secure identification is not possible. In this case, other spectral data such as IR and NMR must be acquired or known standards must be run using a GC–MS system to compare retention times as well as mass spectra.

## Gas Chromatography–Mass Spectrometry (GC–MS)

The development of capillary column gas chromatography greatly simplified the direct coupling of a gas chromatograph with a mass spectrometer. The combination is one of the most powerful organic analysis techniques available, and thousands of GC–MS instruments are found in research, analytical, and teaching labs around the world. As described in Chapter 10, gas chromatography has the ability to separate complex mixtures containing hundreds of compounds (*see* Fig. 10.2 on page 216). However, many detectors, such as the flame ionization detector or the electron capture detector, while being very sensitive, do not provide any structural information. Replacing these with a mass spectrometer maintains the sensitivity but provides the rich structural information available from this instrument.

The typical mass analyzer is of the quadrupole type (*see* Fig. 13.4 on page 277), but TOF analyzers and other types are also used. The advantages of the TOF analyzer are rapid scanning (10 ms per spectrum or less) and accurate mass measurement. Quadrupole instruments acquire spectra at the rate of about one spectrum per second for a mass range of 35–600 Da. A typical GC–MS analysis using temperature programming of the GC oven to elute all the compounds in a mixture requires about 30 min. Every spectrum acquired by the mass spectrometer is digitized and stored by the data system; almost 2000 digitized spectra are acquired and stored in 30 min. A typical capillary gas chromatography peak is 3–10 s wide, so from three to six spectra are recorded as each compound enters the mass spectrometer ion source from the end of the GC column. The computer constructs a gas chromatogram of the separation by plotting the sum of all the ion signal intensities in each scan, called the total ion current (TIC), versus time. The instrument operator can click on any point on the chromatogram, typically at the top of each peak, to view the mass spectrum at that time in a separate window. These can be compared to a digital library of reference spectra as discussed in the previous section and identified if there is a good match. In this way complex mixtures of hundreds of compounds can be separated and structurally characterized.

## Bioanalytical Mass Spectrometry

In 2002, John Fenn of the United States and Koichi Tanaka of Japan shared the Nobel Prize in Chemistry for their development of *soft ionization* methods for the analysis and structural identification of biological macromolecules such as proteins. Because the electron ionization method requires that the sample be in the gaseous vapor state in order to be ionized, most of the high molecular weight

**■ FIG. 13.14**

The MALDI ionization process.

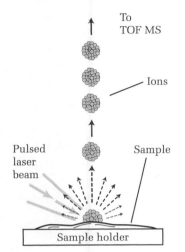

**■ FIG. 13.15**

The MALDI mass spectrum of polypropylene glycol (structure shown). Each peak differs by the mass of one monomer unit.

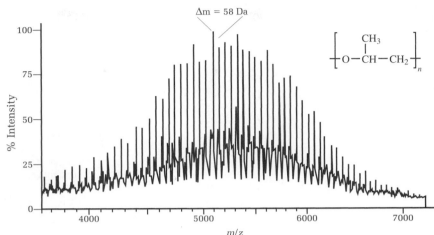

polar materials, such as polysacharrides, nucleic acids, and proteins, would decompose long before they could be heated to a high enough temperature to produce appreciable gaseous molecules. Therefore, for most of its history, mass spectrometry was limited to molecules of molecular weight 2000 Da or less, mostly less than 1000 Da. Then in the 1990s, two new methods were developed that allowed "soft" ionization, without heating, and thus the mass analysis of materials with molecular weights of over 1 million Da. Both of these new techniques—MALDI and ESI—have the advantage that fragmentation is minimal and primarily molecular ions are observed. Actually, the ions formed are *pseudomolecular ions* produced by the attachment of one or more protons or other positive ions such as sodium or potassium to the neutral molecule.

MALDI (*matrix-assisted laser desorption ionization*) uses laser pulses to sputter ions from a mixture of a high molecular weight sample embedded in a crystalline matrix of a low molecular weight compound, such as a cyanobenzoic acid (Fig. 13.14). The matrix facilitates the lifting of the large macromolecules from the surface into the gas phase, where they are mass analyzed, typically using TOF mass spectrometry. A MALDI mass spectrum of polypropylene glycol is shown in Figure 13.15.

ESI (electrospray ionization) is illustrated in Figures 13.16a and 13.16b. When a liquid solution of sample molecules flows from the tip of a fine metal capillary maintained at a high voltage toward a flat metal plate with a potential of thousands of volts between them (Fig. 13.16a), a fine spray of charged droplets forms. Flowing nitrogen gas aids the evaporation of the solvent, and the charged droplets shrink (Fig. 13.16b). At some point the charge becomes too concentrated, and repulsive forces break the droplet into finer droplets. The evaporation, shrinking, and droplet fracture continues until only charged sample molecules remain. Macromolecules typically pick up many charges; the larger the molecule, the

■ **FIG. 13.16**

(a) The electrospray ionization source. (b) The electrospray ionization process.

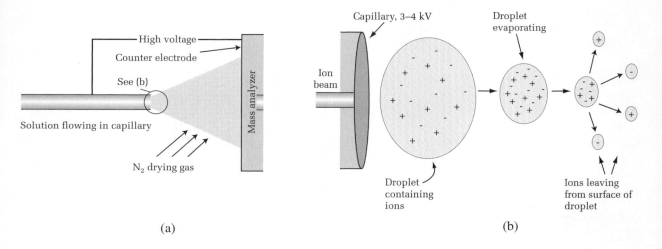

(a)                     (b)

■ **FIG. 13.17**

The electrospray mass spectrum of hen egg lysozyme, showing z and m/z for seven charge states observed. A is the adduct ion of the molecule with 7 to 13 protons attached to give the multiply charged ion masses shown.

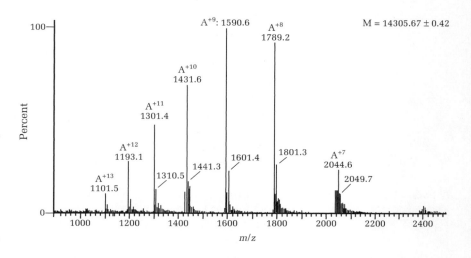

more the charges it can hold. A pin size hole in the plate leads into the high vacuum of a mass analyzer of the quadrupole, TOF, or some other type. The fact that molecules have multiple charges complicates the mass spectrum because each different charge state appears at a different m/z. The electrospray spectrum of the hen egg lysozyme is shown in Figure 13.17. Ions with seven different charge states are detected. Computer-aided calculations allow the determination of these charge states, and from this a molecular weight of 14,305.67 Da is calculated.

As in the case of GC–MS, the development of ESI has allowed the coupling of high performance liquid chromatography (HPLC) with mass spectrometry to give HPLC–MS. This has become a powerful tool for the analysis of complex biological mixtures, such as protein digests.

## QUESTIONS

1. Show by calculation that if you weighed 10,000 molecules of decane consisting of the expected percentages of $^{12}C_{10}H_{22}$ and $^{12}C_9{}^{13}C_1H_{22}$ molecules, you would obtain an average molecular weight that is the same as that listed in reference books.

2. Give both the isotopic composition formula and the relative intensity for the two peaks marked with a question mark in the mass spectra in Figures 13.18 and 13.19.

■ **FIG. 13.18**
A mass spectrum for question 2.

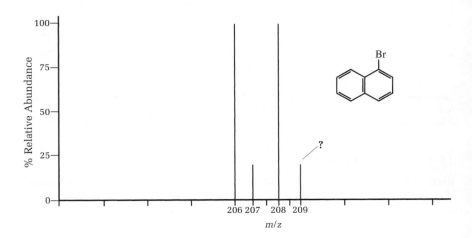

■ **FIG. 13.19**
A mass spectrum for question 2.

3. Give a structure for the compound with the empirical formula $C_8H_8O_2$ and the following spectral data:

   NMR: Singlet at 3.7 ppm (3 H); multiplet at 7.2 ppm (5 H)

   IR: 2850 cm$^{-1}$ (sharp), 1720 cm$^{-1}$, 1600 cm$^{-1}$, 1503 cm$^{-1}$

   MS: as shown in Figure 13.20

## REFERENCES

1. Lee, Terrence A., *A Beginner's Guide to Mass Spectral Interpretation.* New York: John Wiley and Sons, 1998.

2. McLafferty, Fred W., and František Tureček, *Interpretation of Mass Spectra,* 4th ed. Sausalito, CA: University Science Books, 1993.

3. McMaster, Marvin, and Christopher McMaster, *GC/MS: A Practical User's Guide.* New York: Wiley-VCH, 1998.

4. Silverstein, Robert M., Francis X. Webster, and David Kiemle, *Spectrometric Identification of Organic Compounds,* 7th ed.[1] New York: Wiley, 2005.

5. Sparkman, O. David, *Mass Spectrometry Desk Reference.* Pittsburgh: Global View Publishing, 2000.

---

1. This book includes IR, UV, and NMR spectra.

# CHAPTER 14

# Ultraviolet Spectroscopy, Refractive Indices, and Qualitative Instrumental Organic Analysis

**PRELAB EXERCISE:** In the identification of an unknown organic compound, certain procedures are more valuable than others. For example, far more information is obtained from an IR or NMR spectrum than from a UV spectrum or a refractive index measurement. Outline, in the order of priority, the steps you will employ in identifying an organic unknown.

## Ultraviolet Spectroscopy

UV: Electronic transitions within molecules

Ultraviolet (UV) spectroscopy produces data about electronic transitions within molecules. Whereas absorption of low-energy infrared (IR) radiation causes bonds in a molecule to stretch and bend, the absorption of short-wavelength, high-energy ultraviolet (UV) radiation causes electrons to move from one energy level to another with energies that are capable of breaking chemical bonds. Many molecules produce only a major absorption signal (*see* Figs. 14.2 to 14.7 on pages 294–298) due to a single, extended structural feature involving $\pi$ bonded systems: carbon-carbon single bonds, carbon-oxygen or carbon-nitrogen double bonds, triple bonds, and aromatic rings. Because many different combinations of these $\pi$ systems can absorb at nearly the same wavelength, the structure is usually elucidated by other techniques, and its UV spectrum is interpreted based on this structure.

We are most concerned with transitions of $\pi$ electrons in conjugated and aromatic ring systems. These transitions occur in the wavelength region of 200–800 nm (nanometers, $10^{-9}$ meters). Most common UV spectrometers cover the region of 200–400 nm as well as the visible spectral region of 400–800 nm. Below 200 nm air (oxygen) absorbs UV radiation; spectra in that region must therefore be obtained in a vacuum or in an atmosphere of pure nitrogen.

Consider ethylene, even though it absorbs UV radiation in the normally inaccessible region at 163 nm. The double bond in ethylene has two *s* electrons in a π molecular orbital and two, less tightly held, *p* electrons in a π molecular orbital. Two unoccupied, high-energy-level, antibonding orbitals are associated with these orbitals. When ethylene absorbs UV radiation, one electron moves up from the bonding π molecular orbital to the antibonding π* molecular orbital (Fig. 14.1). As Figure 14.1 indicates, this change requires less energy than the excitation of an electron from the σ to the σ* molecular orbital.

By comparison with IR spectra and nuclear magnetic resonance (NMR) spectra, UV spectra are mostly featureless (Fig. 14.2). This condition results as molecules in a number of different vibrational states undergo the same electronic transition to produce a band spectrum instead of a line spectrum.

**Band spectra**

Unlike IR spectroscopy, UV spectroscopy lends itself to precise quantitative analysis of substances. The intensity of an absorption band is usually given by the molar extinction coefficient, ε, which, according to the Beer-Lambert law, is equal to the absorbance (*A*) divided by the product of the molar concentration (*c*) and the path length (*l*) in centimeters.

**Beer-Lambert law**

$$\varepsilon = \frac{A}{cl}$$

$\lambda_{max}$ = wavelength of maximum absorption

The wavelength of maximum absorption (the tip of the peak) is given by $\lambda_{max}$. Because UV spectra are so featureless, it is common practice to describe a

**■ FIG. 14.1**
The electronic energy levels of ethylene.

**■ FIG. 14.2**
The UV spectrum of cholesta-3,5-diene in ethanol.

spectrum like that of cholesta-3,5-diene (Fig. 14.2) as $\lambda_{max}$ = 234 nm ($\varepsilon$ = 20,000) and not bother to reproduce the actual spectrum.

The extinction coefficients of conjugated dienes and enones are in the range of 10,000–20,000, so only very dilute solutions are needed for spectra. In the example in Figure 14.2, the absorbance at the tip of the peak is 1.2, and the path length is the usual 1 cm; so the molar concentration, $c$, needed for this spectrum is $6 \times 10^{-5}$ mol/L,

ε, extinction coefficient

The path length (*l*) is the distance (in centimeters) that light travels through a sample.

$$c = \frac{A}{l\varepsilon} = \frac{1.2}{20,000} = 6 \times 10^{-5} \text{ mol/L}$$

which is 0.221 mg/10 mL of solvent. The usual laboratory balance cannot accurately weigh such small quantities; therefore sample preparation usually requires the quantitative serial dilution of more concentrated solutions.

The usual solvents for UV spectroscopy are 95% ethanol; methanol; water; and saturated hydrocarbons such as hexane, trimethylpentane, and isooctane. The three hydrocarbons are often especially purified to remove impurities that absorb in the UV region. Any transparent solvent can be used for spectra in the visible region.

*Spectro* grade solvents

Sample cells for spectra in the visible region are made of glass or clear plastic, but UV cells must be composed of the more expensive fused quartz because glass absorbs UV radiation. The cells and solvents must be clean and pure because very little of a substance produces a UV spectrum. A single fingerprint will give a spectrum!

UV cells are expensive; handle with care.

Ethylene has $\lambda_{max}$ = 163 nm ($\varepsilon$ = 15,000), and butadiene has $\lambda_{max}$ = 217 nm ($\varepsilon$ = 20,900). As the conjugated system is extended, the wavelength of maximum absorption moves to longer wavelengths (toward the visible region). For example, lycopene, with 11 conjugated double bonds, has $\lambda_{max}$ = 470 nm ($\varepsilon$ = 185,000; Fig. 14.3). Because lycopene absorbs blue visible light at 470 nm, the substance appears bright red. It is responsible for the color of tomatoes; its isolation and analysis are described in Chapters 8 and 9.

**■ FIG. 14.3**
The UV/Vis spectrum of lycopene in isooctane.

**TABLE 14.1 • Rules for Predicting the $\lambda_{max}$ for Conjugated Dienes and Polyenes**

| | Increment (nm) |
|---|---|
| Parent acyclic diene (butadiene) | 217 |
| Parent heteroannular diene | 214 |
| Double bond extending the conjugation | 30 |
| Alkyl substituent or ring residue | 5 |
| Exocyclic location of double bond to any ring | 5 |
| Groups: OAc, OR | 0 |
| Solvent correction (see Table 14.3) | ( ) |
| $\lambda_{max}^{EtOH} =$ | Total |

**TABLE 14.2 • Rules for Predicting $\lambda_{max}$ for Conjugated Enones and Dienones:**

$$\overset{\beta}{\underset{|}{\beta}}-\overset{\alpha}{\underset{|}{C}}=\overset{R}{\underset{|}{C}}-C=O \quad \text{and} \quad \overset{\delta}{\underset{|}{\delta}}-\overset{}{\underset{|}{C}}=\overset{\gamma}{\underset{|}{\gamma}}-\overset{\beta}{\underset{|}{C}}=\overset{\alpha}{\underset{|}{C}}-\overset{R}{\underset{|}{C}}=O$$

| | Increment (nm) |
|---|---|
| Parent $\alpha,\beta$-unsaturated system | 215 |
| Double bond extending the conjugation | 30 |
| R (alkyl or ring residue), OR, $OCOCH_3$ | |
| $\quad \alpha$ | 10 |
| $\quad \beta$ | 12 |
| $\quad \gamma, \delta$ and higher | 18 |
| $\alpha$-Hydroxyl, enolic | 35 |
| $\alpha$-Cl | 15 |
| $\alpha$-Br | 23 |
| exo-Location of double bond to any ring | 5 |
| Homoannular diene component | 39 |
| Solvent correction (see Table 14.3) | ( ) |
| $\lambda_{max}^{EtOH} =$ | Total |

**■ FIG. 14.4**

The UV spectra of (1) pulegone and (2) carvone in hexane.

## TABLE 14.3 • Solvent Correction

| Solvent | Factor for Correction to Ethanol |
|---|---|
| Hexane | +11 |
| Ether | +7 |
| Dioxane | +5 |
| Chloroform | +1 |
| Methanol | 0 |
| Ethanol | 0 |
| Water | −8 |

## TABLE 14.4 • Calculation of $\lambda_{max}$ for Pulegone (See Fig. 14.4)

| | |
|---|---|
| Parent $\alpha,\beta$-unsaturated system | 215 nm |
| $\alpha$-Ring residue, R | 10 |
| $\beta$-Alkyl group (two methyls) | 24 |
| Exocyclic double bond | 5 |
| Solvent correction (hexane) | −11 |
| Calculated $\lambda_{max}$ = | 245 nm; found = 244 nm |

## TABLE 14.5 • Calculation of $\lambda_{max}$ for Carvone (See Fig. 14.4)

| | |
|---|---|
| Parent $\alpha,\beta$-unsaturated system | 215 nm |
| $\alpha$-Alkyl group (two methyls) | 10 |
| $\beta$-Ring residue | 12 |
| Solvent correction (hexane) | −11 |
| Calculated $\lambda_{max}$ = | 226 nm; found = 229 nm |

Woodward and Fieser rules for dienes and dienones

The wavelengths of maximum absorption of conjugated dienes and polyenes and conjugated enones and dienones are given by the Woodward and Fieser rules (Tables 14.1 and 14.2). The application of the rules is demonstrated by the spectra of pulegone (1) and carvone (2) in Figure 14.4. The solvent correction is given in Table 14.3. The calculations are given in Tables 14.4 and 14.5.

No simple rules exist for the calculation of aromatic ring spectra, but several generalizations can be made. From Figure 14.5 it is obvious that as polynuclear aromatic rings are extended linearly, $\lambda_{max}$ shifts to longer wavelengths.

Effect of acid and base on $\lambda_{max}$

As alkyl groups are added to benzene, $\lambda_{max}$ shifts from 255 nm for benzene to 261 nm for toluene to 272 nm for hexamethylbenzene. Substituents bearing nonbonding electrons also cause shifts of $\lambda_{max}$ to longer wavelengths—for example, from 255 nm for benzene to 257 nm for chlorobenzene, 270 nm for phenol, and 280 nm for aniline ($\varepsilon$ = 6200–8600). That these effects are the result of the

■ **FIG. 14.5**

The UV spectra of (1) naphthalene, (2) anthracene, and (3) tetracene.

■ **FIG. 14.6**

The UV spectra of (1) neutral vanillin and (2) the anion of vanillin.

■ **FIG. 14.7**

The UV spectra of (1) aniline and (2) aniline hydrochloride.

interaction of the $\pi$-electron system with nonbonded electrons is seen dramatically in the spectra of vanillin and the anion derived by deprotonation of its phenolic OH (Fig. 14.6). The two additional nonbonding electrons in the anion cause $\lambda_{max}$ to shift from 279 nm to 351 nm and $\varepsilon$ to increase. Protonation of the nonbonding electrons on the nitrogen of aniline to give the anilinium cation causes $\lambda_{max}$ to decrease from 280 nm to 254 nm (Fig. 14.7). These changes of $\lambda_{max}$ as a function of pH have obvious analytical applications.

Intense bands result from $\pi$-$\pi$ conjugation of double bonds and carbonyl groups with the aromatic ring. Styrene, for example, has $\lambda_{max} = 244$ nm ($\varepsilon = 12{,}000$), and benzaldehyde has $\lambda_{max} = 244$ nm ($\varepsilon = 15{,}000$).

## EXPERIMENT

# UV Spectrum of an Unknown Acid, Base, or Neutral Compound

Determine whether an unknown compound obtained from your instructor is acidic, basic, or neutral from the UV spectra in a neutral solvent such as pure ethanol or methanol as well as under acidic conditions (add 1 drop of 5% HCl to the solution in the cuvette and mix) and basic conditions (add 1 drop of 1 $M$ NaOH to the acidic solution in the cuvette, mix, and check that the pH is basic).

**Cleaning Up.** Because UV samples are extremely dilute solutions in ethanol or methanol, they can normally be flushed down the drain.

# Refractive Indices

The *refractive index*, symbolized by $n$, is a physical constant that, like the boiling point, can be used to characterize liquids. It is the ratio of the velocity of light traveling in air to the velocity of light moving in the liquid (Fig. 14.8). It is also equal to the ratio of the sine of the angle of incidence $\phi$ to the sine of the angle of refraction $\phi'$:

$$n = \frac{\text{velocity of light in air}}{\text{velocity of light in liquid}} = \frac{\sin \phi}{\sin \phi'}$$

The angle of refraction is also a function of temperature and the wavelength of light (consider the dispersion of white light by a prism). Because the velocity of light in air (strictly speaking, a vacuum) is always greater than that through a liquid, the refractive index is a number greater than 1, for example, hexane, $n_D^{20} = 1.3751$; diiodobenzene, $n_D^{20} = 1.7179$. The superscript 20 indicates that the refractive index was measured at 20°C, and the subscript D refers to the yellow D-line from a sodium vapor lamp, which produces light with a wavelength of 589 nm.

The measurement is made on a refractometer using a few drops of liquid. Compensation is made within the instrument for the fact that white light, not sodium vapor light, is used, and a temperature correction must be applied to the observed reading by adding 0.00045 for each degree above 20°C:

$$n_D^{20} = n_D^{t} + 0.00045(t - 20°C)$$

The refractive index can be determined to 1 part in 10,000, but because the value is quite sensitive to impurities, full agreement in the literature regarding the last figure does not always exist. For this reason, the refractive indices in this book have been rounded to the nearest 1 part in 1000, as have the refractive indices reported in the Aldrich catalog of chemicals. To master the technique of using the refractometer, measure the refractive indices of several known, pure liquids before measuring an unknown.

Specialized hand-held refractometers that read over a narrow range are used to determine the concentration of sugar, salt, or alcohol in water.

■ **FIG. 14.8**
The refraction of light.

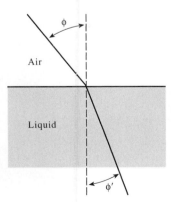

# Using a Refractometer

Refractometers come in many designs. In the most common, the Abbé design (Fig. 14.9), two or three drops of the sample are placed on the measuring prism using a polyethylene Beral pipette (to avoid scratching the prism face). The illuminating prism is closed, and the lamp is turned on and positioned for maximum brightness as seen through the eyepiece. If the refractometer is set to a nearly correct value, then a partially gray image will be seen, as shown in Figure 14.10a. Turn the index knob so that the line separating the dark and light areas is at the crosshairs, as shown in Figure 14.10b. Sometimes the line separating the dark and light areas is fuzzy and colored (Fig. 14.10c). Turn the chromatic adjustment until the

■ **FIG. 14.9**
An Abbé refractometer.
The sample block can be
thermostatted.

Chromatic
adjustment

Thermometer

Measuring prism;
place sample here

Index knob

Illuminating
prism

Lamp

■ **FIG. 14.10**
(a) The view into a refractometer
when the index knob is out of
adjustment. (b) The view into a
refractometer when properly
adjusted. (c) The view when the
chromatic adjustment is incorrect.

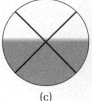

(a)     (b)     (c)

For most organic liquids, the index of refraction decreases approximately $0.00095 \pm 0.0001$ for every °C increase in temperature.

demarcation line is sharp and colorless. Then read the refractive index. On a newer instrument, press a button or hold down the on/off switch to light up the scale in the field of vision or give the digital readout. On older models, read the refractive index through a separate eyepiece. Read the temperature on the thermometer attached to the refractometer and make the appropriate temperature correction to the observed index of refraction. When the measurement is completed, open the prism and wipe off the sample with lens paper, using ethanol, acetone, or hexane only as necessary.

# Qualitative Instrumental Organic Analysis

As indicated in many of the previous chapters, physical characterization and structural elucidation are major activities in organic chemistry. In many areas, such as drug metabolism studies and forensic or environmental chemistry, only milligram or microgram quantities of organic compounds are available. Fortunately, today's instrumental methods have the requisite sensitivity to meet this challenge. It may be possible to establish the structure of a compound on the basis of spectra alone (IR, NMR, MS [mass spectrometry], and/or UV), but often these spectra must be supplemented with other information about the unknown: physical state, solubility, and confirmatory tests for functional groups. Before spectra are run, other information about the unknown must be obtained. Is it pure or a mixture (test by thin-layer [TLC], gas, or liquid chromatography)? Once a substance is known to be homogeneous, it can often be identified by spectroscopy alone. What are its physical properties: melting point, boiling point, and color and solubility in various solvents, such as those commonly used in NMR? A mass spectrum can determine a compound's molecular weight and, if measured with sufficient accuracy, identify the elements present and the molecular formula.

For the millions of organic compounds that have been synthesized or isolated from nature, spectral data are included when they are reported in the chemical literature. Thousands of the more common chemicals that can be purchased or easily synthesized have had their IR, NMR, MS, and UV spectra printed in multivolume collections that can be searched by compound class, formula, or spectral features. Today, these collections have been converted into digital form that can be searched by computer and compared with the spectra obtained for an unknown, yielding a list of all closely matching compounds. If two substances have identical IR spectra, they can be regarded as identical. Such is not always true of other spectra. Many substances can have identical or nearly identical UV spectra, and it is possible for the MS or NMR spectra for two different substances to be almost identical. When new substances are encountered in research laboratories, their spectra can be compared with those in commercial databases. Even though the particular new compound is not represented in those databases, a list of very similar substances can be generated, which will guide the determination of the structure of the new substance.

# Identifying an Unknown Compound

Most students consider the identification of an unknown organic compound to be one of the most enjoyable and challenging organic lab activities. In this laboratory course you may receive an "unknown" substance, which, of necessity, is usually a commercially available compound. At least initially, you may not be given access to a commercial spectral database for searching and comparison, but you will use the skills you have learned to interpret the spectral data you acquire. After you have done the best you can with your IR, NMR, and UV spectra and have arrived at a short list of possible compounds, you will be provided with the mass spectrum of your unknown. Interpretation of this spectrum usually eliminates most of the candidate structures and should allow you to complete the identification of the unknown.

## *Physical State*

### Check for Sample Purity

Distill or recrystallize as necessary. A constant boiling point and sharp melting point are indicators of purity but beware of azeotropes and eutectic substances. Check homogeneity by TLC, GC, or HPLC.

### Note the Color

Common colored compounds include nitro and nitroso compounds (yellow), diketones (yellow), quinones (yellow to red), azo compounds (yellow to red), and polyconjugated olefins and ketones (yellow to red). Phenols and amines are often brown to dark purple because of traces of air oxidation products.

### Note the Odor

Some liquid and solid amines are recognizable by their fishy odors; esters are often pleasantly fragrant. Alcohols, ketones, aromatic hydrocarbons, and aliphatic olefins have characteristic odors. On the unpleasant side in terms of odor are thiols, isonitriles, and low-molecular-weight carboxylic acids.

## *Ignition Test*

Heat a small sample on a spatula; first hold the sample near the side of a microburner to see if it melts normally and then burns. Then heat directly in the flame. If a large ashy residue is left after ignition, the unknown is probably a metal salt. Aromatic compounds often burn with a smoky flame.

## *Beilstein Test for Halogens*

Although the presence of a halogen can usually be determined by mass spectrometry, this test is so simple that it can easily be run to confirm the MS data. Heat the tip of a copper wire in a burner flame until no further coloration of the flame is noticed. Allow the wire to cool slightly, then dip it into the unknown (solid or liquid), and again heat it in the flame. A green flash is indicative of chlorine, blue-green of bromine, and blue of iodine; fluorine is not detected because copper fluoride is not volatile. The Beilstein test is very sensitive; halogen-containing impurities may give misleading results. Run the test on a compound known to contain a halogen for comparison to your unknown.

## Spectra

Obtain IR and NMR spectra following the procedures in Chapters 11 and 12. If these spectra indicate the possible presence of conjugated double bonds, aromatic rings, or conjugated carbonyl compounds, obtain the UV spectrum following the procedures in this chapter. Interpret the spectra as fully as possible by investigating the reference sources cited at the end of the spectroscopy chapters. Once you have interpreted these spectra to the best of your ability, you may obtain or be provided with a mass spectrum (Chapter 13), which should make a secure identification of the unknown possible.

## Solubility Tests

There is a logical sequence for determining the solubility of an organic compound in order, for example, to dissolve it for spectroscopic analysis. In the process of determining the solubility of the unknown, much information about the nature of the compound can be obtained.

Like dissolves like.

*Like dissolves like; a substance is most soluble in that solvent to which it is most closely related in structure.* This statement serves as a useful classification scheme for all organic molecules. The solubility measurements are done at room temperature with 1 drop of a liquid or 5 mg of a solid (finely crushed) and 0.2 mL of solvent. The mixture should be rubbed with a rounded stirring rod and agitated vigorously. Lower members of a homologous series are easily classified; higher members become more like the hydrocarbons from which they are derived. If a very small amount of the sample fails to dissolve when added to some of the solvent, it can be considered insoluble; conversely, if several portions dissolve readily in a small amount of the solvent, the substance is obviously soluble. If an unknown seems to be more soluble in dilute acid or base than in water, the observation can be confirmed by neutralizing the solution; the original material will precipitate if it is less soluble in a neutral medium. If both acidic and basic groups are present, the substance may be amphoteric and therefore soluble in both acid and base. Aromatic aminocarboxylic acids are amphoteric, like aliphatic ones, but they do not exist as zwitterions. They are soluble in both dilute hydrochloric acid and sodium hydroxide but not in bicarbonate solution. Aminosulfonic acids exist as zwitterions; they are soluble in alkali but not in acid. The solubility tests are not infallible, and many borderline cases are known. Carry out the tests according to the scheme in Figure 14.11 and the following Notes to Solubility Tests and tentatively assign the unknown to one of the groups I–X.

**Group I** Monofunctional polar compounds
(up to about 5 carbons)
(a) Acids, (b) Alcohols, (c) Aldehydes,
(d) Amines, (e) Anhydrides, (f) Esters,
(g) Ketones, (h) Nitriles, (i) Phenols

**Group II**
(a) Polycarboxylic acids
(b) Amine salts

**Group III**
(a) Metal salts of carboxylic acids (infusible)
(b) Ammonium salts (fusible)
Di- and polyfunctional compounds including:
(c) Sugars, (d) Amino acids, (e) Polyhydric
     alcohols

**Group IV**
(a) Carboxylic acids
(b) Certain phenols (3)

**Group V**
(a) Phenols
(b) Enols
(c) 1° and 2° nitro compounds
(d) Sulfonamides of 1° amines

**Group VI**
(a) Amines, 1°, 2°, 3°

**Group VII** (5)
(a) Aromatic nitro compounds (yellow)

**Group VIII** (6)
(a) Amides
(b) Nitriles
(c) Sulfonamides of 2° amines

**Group IX** Neutral polar compounds (most
containing oxygen)
(a) Alcohols, (b) Aldehydes, (c) Anhydrides,
(d) Esters, (e) Ethers, (f) Ketones,
(g) Unsaturated hydrocarbons

**Group X** Inert compounds
(a) Alkyl and Aryl halides
(b) Aromatic hydrocarbons
(c) Saturated hydrocarbons
(d) Diaryl ethers

■ **FIG. 14.11**
A solubility classification scheme.
The numbers following the tests
refer to the Notes to Solubility
Tests.

**Notes to Solubility Tests**

1. Groups I, II, III (soluble in water). Test the solution with pH paper. If the compound is not easily soluble in cold water, treat it as water insoluble but test with indicator paper.

2. If the substance is insoluble in water but dissolves partially in 5% sodium hydroxide, add more water; the sodium salts of some phenols are less soluble in alkali than in water. If the unknown is colored, be careful to distinguish between the *dissolving* and the *reacting* of the sample. Some quinones (colored) *react* with alkali and give highly colored solutions. Some phenols (colorless) *dissolve and then* become oxidized to give colored solutions. Some compounds (e.g., benzamide) are hydrolyzed with such ease that careful observation is required to distinguish them from acidic substances.

3. Nitrophenols (yellow), aldehydophenols, and polyhalophenols are sufficiently strongly acidic to react with sodium bicarbonate.

4. Oxygen- and nitrogen-containing compounds form oxonium and ammonium ions in concentrated sulfuric acid and dissolve.

5. On reduction in the presence of hydrochloric acid, nitro compounds form water-soluble amine hydrochlorides. Dissolve 250 mg of tin(II) chloride in 0.5 mL of concentrated hydrochloric acid, add 50 mg of the unknown, and warm. The material should dissolve with the disappearance of the color and give a clear solution when diluted with water.

6. Most amides can be hydrolyzed by short boiling with a 10% sodium hydroxide solution; the acid dissolves with evolution of ammonia. Reflux 100 mg of the sample and a 10% sodium hydroxide solution for 15–20 min. Test for the evolution of ammonia, which confirms the elementary analysis for nitrogen and establishes the presence of a nitrile or amide.

**Cleaning Up.** Because the quantities of material used in these tests are extremely small, and because no hazardous substances are handed out as unknowns, it is possible to dilute the material with a large quantity of water and flush down the drain, unless it is very smelly or is dissolved in water-insoluble nonhalogenated or halogenated solvents, in which case it should be disposed of in the appropriate containers.

# QUESTIONS

1. Calculate the UV absorption maximum for 2-cyclohexene-1-one.
2. Calculate the UV absorption maximum for 3,4,4-trimethyl-2-cyclohexene-1-one.
3. Calculate the UV absorption maximum for
4. What concentration, in g/mL, of a substance with MW 200 should be prepared to give an absorbance value equal to 0.8 if the substance has $\varepsilon = 16{,}000$ and a cell with a path length of 1 cm is employed?

5. When borosilicate glass (Kimax, Pyrex), $n_D^{20}$ 1.474, is immersed in a solution having the same refractive index, it is almost invisible. Soft glass with $n^{20}$ 1.52 is quite visible. This is an easy way to distinguish between the two types of glass. Calculate the mole percents of toluene and heptane that will have a refractive index of 1.474, assuming a linear relationship between the refractive indices of the two.

## REFERENCES

**Online Study Center**

**General Resources**
Web Links

1. Gillam, Albert. E., Edward S. Stern, and Christopher J. Timmons, *Gillam and Stern's Introduction to Electronic Absorption Spectroscopy in Organic Chemistry*, 3rd. ed. London: Edward Arnold, 1970.
2. Jaffe, Hans H., and Milton Orchin, *Theory and Applications of Ultraviolet Spectroscopy*. New York: John Wiley, 1962.
3. Lambert, Joseph B., Herbert F. Shurvell, David A. Lightner, and Robert G. Cooks, *Organic Structural Spectroscopy*. Upper Saddle River, NJ: Prentice-Hall, 1998.
4. Rao, Chintamani N. R., *Ultraviolet and Visible Spectroscopy: Chemical Applications*, 3rd ed. London: Butterworths, 1975.
5. Silverstein, Robert M., Francis X. Webster, and David Kiemle, *Spectrometric Identification of Organic Compounds*, 7th ed.[1] New York: Wiley, 2005.
6. Williams, Dudley H., and Ian Fleming, *Spectroscopic Methods in Organic Chemistry*, 5th ed. New York: McGraw-Hill, 1995.

---

1. This book includes IR, UV, and NMR spectra.

# Computational Chemistry

**Online Study Center**

This icon will direct you to techniques, equipment setups, and online resources at http://college.hmco.com/PIC/williamsonMMOE5e.

The high-speed digital computer has, as in so many other aspects of modern life, changed the way we do and visualize organic chemistry. The ability to predict the outcome, in a quantitative sense, of a chemical reaction and to visualize organic molecules in three dimensions was, until fairly recently, the province of the specialist. Calculations that once taxed the largest computer facilities at a university can now be carried out on a desktop computer. This computational approach to chemistry is being used extensively in drug design, protein mutagenesis, biomimetics, catalysis, studies of DNA-protein interactions, and the determination of structures of molecules using nuclear magnetic resonance (NMR) spectroscopy, to name just a few applications.

## Graphical and Mathematical Models

Physical properties and chemical reactivity are a direct consequence of molecular structure, commonly represented by graphical models showing atomic geometry and electronic distribution. Models can be simple or complex; although all models are simplified representations of reality, it is not absolutely necessary that a particular model correspond exactly to reality. The level of complexity depends on how much information is needed to understand and predict chemical behavior. For example, all of the models for methanol shown in Figure 15.1 are correct, but certain models provide more information than others. What is essential is that models allow chemists to predict some measurable quantity with sufficient accuracy. For example, the simplest model of methanol in Figure 15.1, the formula $CH_4O$, provides all the information necessary to calculate the amount of $CO_2$ that will be released if one burns a certain quantity of methanol. However, this simple model is not adequate for predicting the hydrogen bonding that occurs between methanol molecules. To do this, a Lewis structure that shows the nonbonding pairs of electrons on the oxygen is a more appropriate model. Chemists often use Lewis structures and think about these representations in terms of the valence bond theory. A Lewis structure shows a molecule as a skeleton of atoms with their accompanying *valence shell electrons*. Valence shell electrons are shown as either

CH₄O

Elemental
formula

Lewis structure,
hand drawn

3-D
VSEPR,
hand drawn

3-D spoke and ball,
molecular mechanics

3-D space-filling,
molecular mechanics

3-D electron density,
semiempirical (AM1)

3-D LUMO/LUMO+,
semiempirical (AM1)

electron pair bonds connecting atoms or as nonbonded pairs localized on atoms. Lewis structures are simple to draw, and the information conveyed is usually sufficient, within the rules of *valence shell electron pair repulsion* (*VSEPR*) *theory*, to obtain an approximate prediction of the three-dimensional shape of a molecule. Lewis structures are also usually adequate to give a general picture of the electronic changes that occur during the course of a reaction and are used to write mechanisms involving electron "pushing," which is represented by arrows. Simple hand-drawn structures provide enough information and insight that most of the organic structure and reactivity can be understood using just these. However, hand-drawn structures do not provide detailed quantitative information about bond lengths, angles, and strengths; ionization potentials; electron affinities; or dipoles and molecular energies. There are many times when such information is useful in solving a problem.

By way of analogy, if only the skeletal structure of an animal were presented to an untrained person, that person would probably make mistakes when trying to predict the animal's complete shape and behavior. A trained anatomist, however, will more likely be able to make reasonable predictions based on past experience. Hand-drawn structures present a similar situation. Novices may have more difficulty making correct predictions, whereas a trained chemist is more likely to derive useful information about the structure and properties of a molecule based on its Lewis structure. When drawing accurate structures and predicting physical properties and chemical behavior as determined by molecular energies, however, both the expert and the novice alike encounter problems. Clearly, the development of more numerically accurate, mathematical models would help. They yield tables of numbers such as bond lengths, molecular energies, and orbital information that can be used directly or converted to graphical representations. These more accurate and detailed graphical structures can be rotated in any direction on a computer screen and would be hard to draw by hand. They help to visualize complex numerical data, providing much greater insight into molecular and reaction chemistry.

There are two major types of mathematical models. In the *molecular mechanics* model, molecules are made up of atoms and bonds (as opposed to nuclei and electrons), and the model uses the ball and spring equations developed in physics for classical mechanics. Atom positions are adjusted to best match known empirical parameters, such as bond lengths and angles derived from experimental measurement. In the *quantum chemical* model, molecules are considered to be positively charged atomic nuclei in a cloud of negative electrons. This model relies on the pure theory of quantum physics and mathematical equations to describe the attractive and repulsive interactions between the charges that hold the molecule together. It uses no experimental parameters, working only from theoretical principles, and is therefore also called the *ab initio* ("from the beginning") model.

There is also a third model, called the *semiempirical* model, which is a hybrid of the pure theory-based quantum chemical model and the experimentally parameterized molecular mechanics model. It helps overcome certain limitations associated with these two models and can be effectively applied to a broad range of problems.

You do not need to be a highly trained theoretician to run these calculations, but you should be critical and even skeptical of results produced by computational chemistry. One of the major limitations of all these computations is that they assume noninteracting molecules, in other words, gaseous molecules in a vacuum. Obviously, reactions run in solvents involve solvent-molecule interactions that can significantly change the experimentally measured energetics from those predicted by calculation. (There are models that can estimate solvent effects.) It is quite possible to run a series of calculations that produce results that are absolutely meaningless. Therefore, it is imperative that you use your knowledge of organic chemistry and repeatedly ask, "Are these results consistent with what is observed experimentally? Do they make sense using the traditional hand-drawn, arrow-pushing chemistry taught in textbooks?" If they don't, it probably means that you have chosen the wrong computational model to do the calculation. Computers just generate numbers; they cannot tell you whether the chosen method is valid for the problem being studied. People have to make these decisions to correctly solve real problems.

The biggest challenge in computational chemistry lies in understanding the capabilities and limitations of each model. These will be emphasized in the following sections. One also has to choose which computational method will provide useful information in a minimum amount of time. Ab initio calculations can take hours and even days to complete. If the same information can be derived from a semiempirical calculation in one-tenth the time, the fancier method is unnecessary and wasteful of computer resources.

# Molecular Mechanics Models

The simplest, fastest, and easiest to understand calculations involve molecular mechanics, and the results from this approach are usually adequate for many of the modeling exercises in this text. Molecular mechanics is used to calculate the structures of molecules based not on a complete solution of the Schrödinger equation but on a mechanical model for molecules. This model regards molecules as being masses (atoms) connected by forces somewhat like springs (bonds) at certain preferred lengths and angles.

Every organic chemist routinely uses a set of models to examine the three-dimensional structures of molecules because it is difficult to visualize these from a two-dimensional drawing. Molecular mechanics programs allow one to determine quickly the best (lowest-energy) structure for a molecule. Experienced chemists have been doing this intuitively for years, but simple plastic or metal models can lead to incorrect structures in many cases, and, of course, one obtains no quantitative data regarding the stability of one structure compared to another.

Modern computational chemistry software makes it quite easy to draw an approximate structure of a molecule on the computer screen. After the rough structure is drawn and converted to a digital format, a molecular mechanics program such as SYBYL or MMFF automatically steps through a series of iterative calculations; such programs adjust every bond angle, bond length, dihedral angle, and van der Waals interaction to produce a new structure, often called the *equilibrium geometry*, having the lowest possible *strain energy*. The absolute value of the strain energy often differs from one program to another, and so it can be used only in a relative sense to determine the difference in energies between two possible structural geometries or conformations. Molecular mechanics assumes that all bond properties are the same in all molecules. For exceptional molecules such as cyclopropane and cyclobutane, it merely substitutes new parameters for these unusual rings. The experimental data on which many of the parameters for molecular mechanics rest come from x-ray diffraction studies of crystals, electron diffraction studies of gases, and microwave spectroscopy. Even if an ab initio quantum chemical calculation is ultimately planned, the starting structure should always be energy minimized using the faster molecular mechanics program before going to the next level.

Molecular mechanics calculations do *not* provide any information about the locations of electrons and other electronic properties such as ionization potentials, and they *cannot* be used to explore processes that involve bond breaking or bond making. This computational method cannot evaluate the energy of a transition state, and thus the activation energy from which reaction rates can be derived, nor can it calculate the heat of formation or of combustion. These sorts of calculations are the province of the quantum chemical computational models; unlike them, molecular mechanics calculations are empirical in nature and are not derived from first principles.

If a molecule has a single minimum-energy conformation and that conformation has been found, then the shape of the final molecule can be used to predict or confirm approximate vicinal (three-bond) proton-proton NMR coupling constants, which are dependent on the dihedral angle between the protons.

(Couplings are also dependent to a smaller degree on electronegativity of substituents, bond angles, etc.) Optical rotatory dispersion curves can also be predicted when molecular conformations are known.

In a typical empirical force field used in molecular mechanics, the steric energy of a molecule can be represented as the sum of five energies given by the following equation:

$$E_{steric} = E_{str} + E_{bnd} + E_{tor} + E_{oop} + E_{vdW}$$

where $E_{str}$ is the energy needed to stretch or compress a bond. Plastic student molecular models are realistic in that there is no way to adjust this value; in molecular mechanics calculations one finds that there is a big force constant (more correctly, potential constant) for this parameter. Most sets of molecular models are constructed so that modest bending of bond angles is possible, and similarly one finds that the force constant for angle bending ($E_{bnd}$) has an intermediate value. The force constant for torsional motion ($E_{tor}$) is very small as seen in the free rotation about bonds in most mechanical models. The out-of-plane bending term ($E_{oop}$) arises in molecules such as methylene cyclobutene where, without this term, the methylene group would want to be bent out of the plane to relieve angle strain. The energy due to van der Waals interactions ($E_{vdW}$) is very important. It takes two forms, either attractive or repulsive, depending on internuclear distance.

Molecular mechanics allows the calculation of the equilibrium geometry of a molecule (Fig. 15.2). The individual mathematical equations describing the mechanics of a molecule are not terribly complex. For a typical organic molecule containing 30–40 atoms, only a few seconds at most are needed by a modern desktop computer to carry out the millions of calculations and summations for the pair-wise interactions between atoms described by these equations.

Let us examine in detail the five most important factors entering into molecular mechanics calculations. The bond between two atoms can be regarded just like a spring. It can be stretched, compressed, or bent and, provided these distortions are small, will follow Hooke's law from classical physics.

For bond stretching, the energy is given by

$$E_{str} = (1/2)k_r(r - r_0)^2$$

**Methylene cyclobutene**

All of these energies are empirical in nature, not calculated from first principles.

**■ FIG. 15.2**

Cyclodecapentaene. From this energy-minimized structure, it is easy to see that the molecule cannot be flat and thus cannot benefit from aromatic stabilization.

where $E_{str}$ is the bond stretching energy in kJ/mol, $k_r$ is the bond stretching force constant in kJ/mol·Å², $r$ is the bond length, and $r_0$ is the equilibrium bond length.

Similarly, the energy for bond angle bending (in-plane bending) is given by

$$E_{bnd} = (1/2)k_\theta(\theta - \theta_0)^2$$

where $E_{bnd}$ is the bond bending energy in kJ/mol, $k_\theta$ is the angle bending force constant in kJ/mol·deg², $\theta$ is the angle between two adjacent bonds in degrees, and $\theta_0$ is the equilibrium value for the angle between the two bonds in degrees.

The energy for twisting a bond, $E_{tor}$, is given by

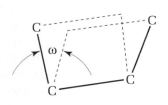

$$E_{tor} = (1/2)k_\omega[1 - \cos(j\omega)]$$

where $E_{tor}$ is the torsional energy in kJ/mol, $k_\omega$ is the torsional force constant in kJ/mol·deg, $\omega$ is the torsion angle in degrees, and $j$ is usually 2 or 3 depending on the symmetry of the bond (e.g., twofold in ethylene or threefold in ethane).

The energy term for out-of-plane bending ($E_{oop}$) is not used very often, but it is a necessary term. It arises in cases such as cyclobutanone, where the C—C—O bond angle is not the $sp^2$ bond angle of 120° but 133°. To relieve this bond angle strain, the oxygen could tip up out of the plane, but this would require a large amount of energy because of the distortion of the $\pi$ bonding of the carbonyl. So the force constant for out-of-plane bending is different from that for in-plane bending (bond angle bending; $E_{bnd}$) discussed earlier.

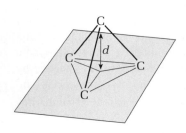

$$E_{oop} = (1/2)k_\delta(d^2)$$

where $E_{oop}$ is the out-of-plane bending energy in kJ/mol, $k_\delta$ is the out-of-plane bending constant in kJ/mol·Å², and $d$ is the height of the central atom in angstroms above the plane of its substituents.

Very important is the van der Waals energy term ($E_{vdW}$) given by

$$E_{vdW} = \varepsilon[(r_0/r)^{12} - 2(r_0/r)^6]$$

where $E_{vdW}$ is the van der Waals energy in kJ/mol, $\varepsilon$ is the van der Waals potential of the two interacting atoms in kJ/mol, $r_0$ is the equilibrium distance between the two atoms, and $r$ is the interatomic distance in the molecule in angstroms.

This important function is best understood by referring to Figure 15.3, which shows a curve that describes the energy of two atoms as a function of their distance ($r$) from each other. At very short distances, the two atoms repel each other very strongly. This interaction raises the energy of the system as the 12th power of the interatomic distance. At a slightly longer distance, a weak attraction is described by an $r^6$ term, and then at large separation the interaction energy goes asymptotically to zero. The value of $\varepsilon$ determines the depth of the potential well. At distance $r_0$, when the attractive forces overwhelm the repulsive forces, the system has minimum energy. It is $r_0$ that we call the *bond length*. The molecular mechanics van der Waals term is just an approximation of the ideal curve. It comes close enough and is easy to compute.

■ **FIG. 15.3**
The van der Waals energy function.

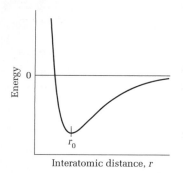

Carrying out molecular mechanics calculations on a personal computer or a small workstation is not difficult. One first draws the structure of the molecule on a computer screen in the same way or even using the same program that is used to draw organic structures for publication in reports and papers. This results in a wire model such as the hexagon we associate with cyclohexane with the proper $x$, $y$, $z$ coordinates for each carbon. The program can be directed to add hydrogens at all the unfilled valences. Then the molecular mechanics program is told to minimize the structure, and in a few seconds a minimum energy structure is calculated. Usually, the molecule appears to wiggle on the screen as the minimum energy geometry is calculated.

Consider, for example, the molecule butane. When you initially draw the molecule, you will undoubtedly not draw a perfectly correct structure in terms of bond lengths, bond angles, and dihedral angles. The computer will calculate the energy of the structure you draw and then make a slight change in the coordinates of the atoms and recalculate the energy of the molecule. If the newly calculated energy is lower than the previously calculated one, it will continue to change the coordinates until a minimum energy is reached. This iterative process is continued until it is found that no change in any bond angle, bond length, torsion angle, or van der Waals interaction can be made without raising the energy of the molecule. At this point the calculation stops.

But there is one possible pitfall in this process: The computer program usually has no way of knowing whether it has calculated the structure with the lowest possible energy for the molecule (the *global minimum*) or just a *local minimum*. A geographical analogy may make this statement more clear. It is as if you were instructing the computer to roll a ball to the lowest place in California (Death Valley). The computer might find the lowest place in the Sacramento Valley, but the algorithm on which it operates has no way of knowing whether that is the global minimum or if there is any way to climb the mountains that separate the local minimum from the global minimum.

Now consider once more the molecule butane. If the initial structure put into the computer has a conformation with a torsion angle (dihedral angle) of between $0°$ and just a bit less than $120°$ (margin figure a), the molecular mechanics program will calculate the *gauche* conformation (margin figure b) with a $60°$ torsion angle as the lowest-energy form. In fact, the lowest-energy form (margin figure c) is *anti* and has a torsion angle of $180°$. The global minimum is the *anti*conformer, but the calculation can easily fall into one of the two local minima, the *gauche* conformations, which are not the lowest-energy conformations. The full potential energy diagram for all rotational conformers of butane generated by molecular mechanics, shown in Figure 15.4, illustrates this concept clearly.

There are several approaches to the solution to this problem, but no one approach is completely satisfactory. If one has enough experience and intuition, it is often possible to spot the fact that the minimum-energy conformation calculated by the computer is not the global minimum. A better solution, however, is to calculate energies for an arbitrary set of dihedral angles about one of the bonds. In butane, one could ask for the energies to be calculated at dihedral angles of $0°$, $30°$, $60°$, $90°$, $120°$, $150°$, and $180°$ using a part of the program called a *dihedral driver*. In this way you would be sure to catch the lowest-energy form.

(a)

(b)

(c)

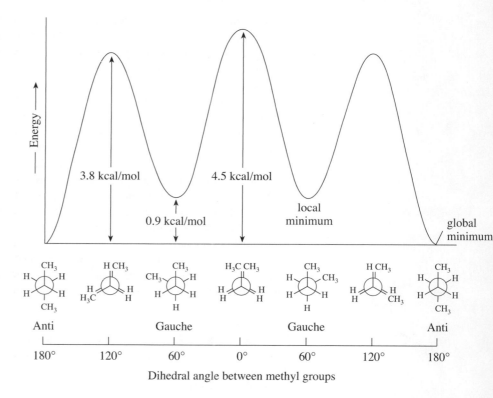

Now consider a larger molecule. If one were to search for the best conformation at angles of 0°, 30°, 60°, 90°, 120°, 150°, and 180° around each carbon-carbon bond, it would be necessary to make $7^n$ calculations, where $n$ is the number of carbon-carbon bonds. In 2-bromononane, for example, this would be $7^8 = 5,764,801$ calculations. The problem gets out of hand very rapidly.

It is not possible to use a dihedral driver within a ring, so the lowest-energy form is found by other means (ring flip and flap, molecular dynamics, and Monte Carlo methods).

An elaboration of this computational method, called *molecular dynamics*, is often used to study the motions of large molecules such as proteins and other large polymers. Molecular dynamics is basically a series of molecular mechanics calculations, with each conformation being a frame in a movie animation of the movements of a molecule in the process of folding into a preferred shape.

# Quantum Chemical or ab initio Models

In the 1920s, a more sophisticated theory of atomic and molecular structure was developed; it improved on Lewis structures and yielded a more accurate representation of molecular properties. This new theory relied on quantum mechanics and the Schrödinger equation for predicting the behavior of an electron in the presence of a positively charged nucleus. However, the mathematics for the exact solution of the equations involved was so complex that a full calculation could only be performed on the one-electron hydrogen atom. However, by assuming that the nuclei don't move (Born-Oppenheimer approximation) and that electrons move independently of one another [Hartree-Fock (HF) approximation; ab initio models are often called HF models], the Schrödinger equation could be simplified so that it was practical to perform useful calculations on multiatom systems. Many scientists have worked on further refinements of the equations and computer programs that have helped to gradually improve the quality of the results. Because of the intensive nature of the calculations, the availability of fast desktop computers and efficient computational programs has also been very important. Today, calculations involving more than 25 nonhydrogen atoms are feasible. Two commonly used HF model programs are HF/3-21G and HF/6-31G*.[1] If further accuracy is required, one can enhance the HF models by using perturbation theory to compensate for electron-electron repulsion within a molecular orbital. Approximations that add perturbation theory were developed by Möller and Plesset, and a commonly used MP-level model is MP2/6-31G*. Note that different approximations often lead to different results. The mathematical complexity of quantum calculations puts a full explanation of the principles beyond the scope of this text.

Unlike molecular mechanics, ab initio models use no empirical data. The calculation is truly from first principles, using approximate Schrödinger equations to calculate the locations of the atomic nuclei and *all* electrons: valence, unpaired, and inner shell. The equilibrium (minimum-energy) geometries are usually improved relative to those from molecular mechanics, better reflecting the true structure. Many properties that are defined by electron distribution, such as dipole moments, ionization potentials, and the shape and sign of molecular orbitals, are calculated. It is also possible to calculate vibrational energies to produce an approximate infrared (IR) spectrum. More importantly, the energies of molecules, transition states, and bond-making or bond-breaking reactions can be computed and compared. This information cannot be derived from molecular mechanics, but it can be calculated using the semiempirical models to be discussed in the following section.

These capabilities are best demonstrated by the following example. In the presence of a free radical initiator (AIBN, azoisobutylnitrile; *see* Chapter 18),

---

1. The asterisk means that the program has been improved.

The four possible energy–
reaction coordinate diagrams
for the intramolecular cyclization
of the hexenyl radical to either
the cyclopentylmethyl radical
or the cyclohexyl radical.

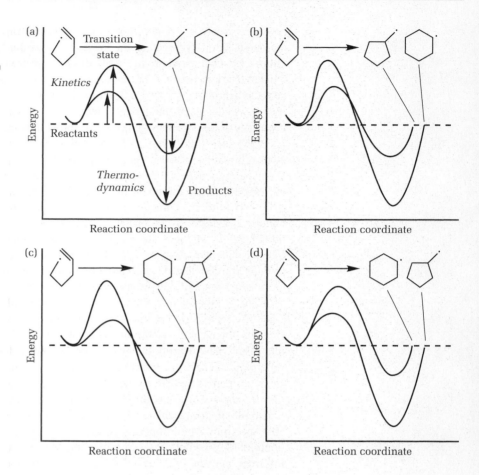

tributyltin hydride cleaves carbon-bromine bonds to give free radicals. In the case
of 6-bromo-1-hexene, the initially formed radical can either abstract a hydrogen
from surrounding molecules to form 1-hexene or undergo intramolecular radical
cyclization to form either a cyclopentylmethyl or cyclohexyl radical, either of
which also picks up hydrogen to give the products shown.

Organic chemistry fundamentals would predict that cyclohexane should be
the major cyclic product because the formation of six-membered rings and secondary radicals in the intermediate cyclohexyl radical would be favored. Contrary
to this prediction, the major product (81%) is derived from the five-membered cyclopentylmethyl radical, with the less stable primary radical. Let us see whether an
analysis of the energetics of this reaction using ab initio methods provides insight
into this apparent contradiction.

As in the energy analysis of the conformers of butane (Fig. 15.4), it is useful
to use energy–reaction coordinate diagrams to understand the possible reaction
pathways. Figure 15.5 shows four diagrams depicting the energy needed to cause
the radical and π bond to react (the transition state) and the energy released as a

new bond is formed. The former energy determines the kinetics or rate of the reaction, and the latter determines the thermodynamic stability of the products relative to the reactant and to each other. There are four different energy pathways possible, as shown in the four diagrams.

Diagrams (a) and (b) in Figure 15.5 show the cyclohexyl radical to be more thermodynamically stable than the cyclopentylmethyl radical, consistent with the chemical fundamentals discussed previously. Diagrams (c) and (d) make the cyclopentylmethyl radical the more stable one, which would be inconsistent. When the energy minima of these two radicals are calculated using a 6-31G* ab initio computation, the cyclohexyl radical is roughly 5 kcal/mol more stable than the cyclopentylmethyl radical, consistent with our chemical knowledge. Energy diagrams (c) and (d) are therefore incorrect depictions of the reaction.

If the cyclohexyl radical is the more stable, as calculated and depicted in diagrams (a) and (b), there must be another explanation of why the cyclopentylmethyl product is formed 40 times more readily. A possible explanation is that the transition state energy necessary to get the initial hexenyl radical to react at carbon 2, forming a five-membered ring, is less than the energy needed to react at carbon 6. Because ab initio methods also allow calculation of the energy maxima of transition states, this question can also be examined computationally. The calculated difference in transition state energies is 2.7 kcal/mol lower in favor of ring closure to the cyclopentylmethyl radical, making diagram (b) the only correct representation of the energetics. Computational chemistry shows that the final product ratio is determined by kinetics, the rate of cyclopentylmethyl closure being faster than the rate of cyclohexyl closure due to the lower transition state energy.

Note that the addition of the initial hexenyl radical to the double bond is for this case *irreversible* because the reverse transition state energy is quite high. If the energy difference were low enough, a dynamic equilibrium between product and reactant radicals might eventually lead to the formation of the more thermodynamically stable product (cyclohexyl radical), given sufficient time and energy. Most *reversible* reaction product ratios are determined by thermodynamics, not kinetics.

Product ratios and reactant:product ratios, whether determined by thermodynamics or transition state energy differences, are roughly proportional to the energy differences, as shown in Table 15.1. Notice that even small energy differences lead to large differences in these ratios.

| TABLE 15.1 • The Relationship of Product Ratio to Energy Differences | | |
|---|---|---|
| **Energy Difference (kcal/mol)** | **Energy Difference (kJ/mol)** | **Product Ratio, Major:Minor** |
| 0.5 | 2 | 80:20 |
| 1 | 4 | 90:10 |
| 2 | 8 | 95:5 |
| 3 | 12 | 99:1 |

# Semiempirical Models

Most molecular computations done by organic chemists, especially those examining minimum energy geometries, are done using a semiempirical model because such a model provides the best compromise between speed and accuracy. This kind of model can be thought of as a hybrid of molecular mechanics models based on experimentally measured *empirical* data and pure quantum chemical theory or ab initio models, thus the name *semiempirical*. Semiempirical models use Schrödinger equation approximations like those described previously but, in order to make the calculations less time consuming, *only* the locations of *valence electrons* are calculated, *not all electrons*. For the inner shell electrons, empirical data from typical organic molecules is used to estimate their locations. This and other approximations allow semiempirical models to be used for molecules containing up to 100 or more nonhydrogen atoms, much larger than those that can be modeled by the more rigorous ab initio models. However, semiempirical models are often less accurate, especially for calculations on novel molecules with unusual bonding or reactivity. Two commonly used programs are AM1 and PM3.

Most computational chemistry texts not only provide explanations of the different mathematical model equations but also do a large number of comparisons of calculated properties (bond lengths, angles, dipoles, various energies, etc.) from the different mathematical models versus actual properties measured experimentally in the lab. The data can be quite sobering, with differences between calculated and experimental values averaging 7%–10% and with many examples having much larger errors. As seen from Table 15.1, these types of errors in energy calculations would lead to even larger discrepancies in calculated and experimental product ratios. As stated earlier, every computational result has to be examined critically to make sure it fits with both chemical logic and experimental results. Nevertheless, being able to model transient intermediates such as free radicals or transition states, species that are virtually impossible to isolate and study in the lab, provides some exciting ways to extend our understanding of chemistry.

# Types of Calculations

The model used to perform the calculation of any particular property or graphical representation determines the amount and accuracy of information generated in the output. The first two types of calculations, geometry optimization and single point calculations, can be performed with any of the three computational models described previously. The last three types of calculations, involving electron locations and energies, are not solvable by molecular mechanics.

## 1. Geometry Optimization

*Geometry optimization* is a standard computational chemistry calculation to find the lowest-energy, or most relaxed, conformation for a molecule. The approach is the same for all levels of calculation, involving an iterative jiggling process like that described for molecular mechanics. At each step, the molecular geometry is modified slightly, and the energy of the molecule is compared with the last cycle.

A reaction coordinate diagram for cyclohexane, showing local skew boat and global chair minimum-energy geometries.

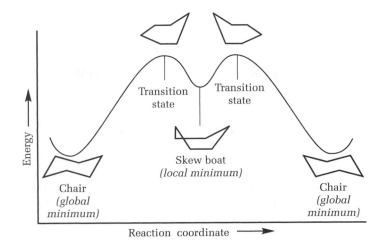

The computer moves the atoms in the molecule a little, calculates the energy, moves them a little more, and keeps going until it finds the lowest energy. This is the energy minimum of the molecule and is obtained at the optimized geometry. Recall that the energies from molecular mechanics can only be used in a relative sense, while those from quantum electronic structure models can be compared in an absolute sense, like heats of formation.

Be wary of structures that may be stuck in a local, not global, energy minimum. As you may know, the chair conformation of cyclohexane is lower in energy than the boat conformation (Fig. 15.6). If you entered the boat structure, geometry optimization should find the chair as the structure with the lowest energy. Remember that sometimes, however, the computer gets "stuck" trying to find a minimum due to the presence of a local minimum. For example, if you start with a boat structure of cyclohexane, the computer may get to a "skew-boat" local energy minimum but may not flip the molecule completely into the chair conformation that corresponds to the *global minimum*. This can happen when the energy jump from skew boat to chair is too large. A knowledge of fundamental organic chemistry as presented in introductory textbooks helps to recognize these types of problems.

## 2. Single Point Calculations

*Single point calculations* are often used in combination with geometry optimization to investigate steric hindrance. In this case the method only performs one computational cycle to calculate the energy of a particular fixed geometry. In a thermodynamically controlled reaction, the energetic difference between two conformations is often due to steric hindrance. If the product molecule optimizes in one conformation, you can use single point calculations to determine how much more energy is needed to form the nonpreferred conformation. The structure drawing and manipulation part of the software will allow you to move only that part of the molecule that changes in the higher energy form, leaving the rest of the molecule optimized. The single point calculation performed on this modified molecule will give an energy that you can directly compare with the optimized energy to find the energy difference between the lower and higher energy conformers. For example, the energetic difference between having a constituent in the axial or equatorial position on a cyclohexane ring can be determined.

### 3. Transition State Calculations

*Transition state calculations* can be thought of as the reverse of geometry optimizations. In this case, the method searches for a structure of *maximum* energy, a transient intermediate that cannot be isolated experimentally. For example, this type of calculation allows one to examine transition state energies and geometries of intermediates involved in carbocation rearrangements. The literature contains standard models that should used as the starting point for these calculations. It takes an appreciable amount of effort and experience to properly analyze transition state structures and energies.

### 4. Vibrational Frequency Calculations

*Vibrational frequency calculations* allow the calculation of IR stretching and bending absorption frequencies, and it is a lot of fun to view animations of these types of motions in molecules. The vibrational frequency of a two-atom system is proportional to the square root of the force constant (the second derivative of the energy with respect to the interatomic distance) divided by the reduced mass of the system (which depends on the masses of the two atoms). In fact, the analysis of transition states (as well as other essential quantities such as entropy) really starts by exploring slight movements of atoms that can be considered as vibrations.

The frequencies calculated using semiempirical models are usually higher than the frequencies measured on an IR instrument by about 10%, and the accuracy depends on the computational model used. The best way to check the calculated values is see whether the vibration animations make sense; for example, if an animation shows a $C{=}O$ stretching vibration, the calculated frequency can be matched with the measured frequency of that type, which is usually easy to assign in the acquired IR spectrum.

### 5. Electronic Density and Spin Calculations, Graphical Models, and Property Maps

*Electronic density and spin calculations, graphical models, and property maps* allow the visualization of electronic properties such as electron densities, electrostatic potentials, spin densities, and the shapes and signs of molecular orbitals. The values for a particular property at each point in the three-dimensional space around a molecule are displayed on a two-dimensional computer screen as a surface of constant numerical value, often called an *isosurface*, which can be rotated in any direction to study it. Alternately, numerical variations of a given property (such as electron density) at a defined distance from the molecule can be displayed as a property map using color as a key. Carrying out surface calculations and viewing their graphical representations are major activities in computational chemistry and can provide useful insight into the mechanisms of organic reactions. A few examples of isosurfaces, orbitals, and property maps are shown in Figure 15.1 (on page 308). More examples and explanations, including color images, are available on the website for this book.

*Online Study Center*

**General Resources**
Molecular Property Maps

## Examples of Calculations

Try to reduce computational time by understanding the various computational methods and answering the question "What kind of information do I need?" If only geometry or ring strain information is needed, using the more intensive ab initio calculation to produce high-precision molecular orbitals is not necessary. Even if you need the high accuracy of an ab initio calculation, do not start there. First, approximate the molecular geometry using a molecular mechanics calculation. Then use this structure to run an AM1 semiempirical calculation for a good approximation of the valence electron properties. Finally, if even more accuracy is required, use the AM1 output for the 3-21G input to see how the inner shell electrons affect the electronic properties.

An instructive exercise is to calculate the energy of butane at dihedral angles of 0°, 60°, 120°, and 180°, using single point energy calculations for each. You should be able to reproduce the *form* of the curve shown in Figure 15.4 (on page 314), but you usually will not obtain the experimentally derived energies shown.

Remember that the calculated steric energy has no physical significance. Differences in steric energies can be useful, however, if the molecules being compared have the same bonding pattern. The molecule with the lower energy will be the more stable isomer (geometric, conformational, or stereo). Other types of comparisons cannot be made. A molecular mechanics calculation cannot be used to compare the stabilities of unrelated molecules.

If the program you are using breaks down the steric energy into the six terms $E_{steric}$, $E_{str}$, $E_{bnd}$, $E_{tor}$, $E_{oop}$, and $E_{vdW}$, do not give the individual terms much credence. Differing parameter sets in other programs may weight these differently. You should pay the most attention to the total steric energy.

Molecular mechanics calculations, like IR and NMR spectroscopy, are a tool available to the chemist. As such, they are presented in this text as adjuncts to experimental work in the laboratory, not as an end in themselves. In most of the computational problems presented in the following chapters, you are asked to carry out computations of the steric energies or the heats of formations of isomers.

One must be careful when using molecular mechanics computations to draw conclusions. The computation can be made on any molecule, but the resulting steric energy cannot always be compared to the steric energy of another molecule, even though they are isomers. For example, *cis*- and *trans*-butene and other similar *cis*- and *trans*-isomers can be compared to each other because they are both 1,2-disubstituted alkenes, but a valid comparison cannot be made between 1,2- and 1,1-disubstituted alkenes.

Although not intending to mislead you deliberately, we have included computations on reactions and products that for one reason or another are not expected to give valid comparisons or answers. It is for you to decide whether the computations are in agreement with your experiment.

Higher levels of computation, the semiempirical molecular orbital methods and the even higher ab initio methods, will give heats of formation that can be compared to one another. The problem with these higher levels of computation is that they require more time and more computer power. But often an AM1 semiempirical molecular orbital calculation on a relatively small molecule can be done in a few seconds. The cost (in computer time) of ab initio calculations goes up more than the fourth power of the number of atoms, whereas molecular mechanics calculations go up as the square of the number of atoms in the molecule.

The radical chlorination of 1-chlorobutane gives four isomeric dichlorobutanes. In Chapter 18 these chlorobutanes are synthesized, and their relative amounts are measured by gas chromatography with the object of determining the relative reactivity of the various hydrogens in the starting material. Using a semiempirical program, you can calculate the energies of the four products to see whether or not these energies correlate with the product distributions. A correlation or lack thereof may give information regarding the mechanism of the reaction.

Similarly, the dehydration of 2-butanol (*see* Chapter 10) gives three isomeric butenes that are separated and analyzed by gas chromatography. Calculation of their heats of formation or steric energies may or may not correlate with the percentage of each isomer formed in the reaction. Again, a correlation or lack thereof may give information about the mechanism of the reaction or about the applicability of computational chemistry to problems of this type.

Treatment of *cis*-norbomene-5,6-endo-dicarboxylic acid with concentrated sulfuric acid gives the isomeric compound X (Chapter 48). The energies of any proposed structures for X can be calculated and compared with that of the starting material. If the reaction conditions (hot concentrated sulfuric acid) are regarded as favoring an equilibrium between product and starting material, then X would be expected to have the lower energy if the correct structure is proposed.

In Chapter 50 you are asked to calculate the most stable conformation of the product *p*-terphenyl. Hexaphenylethane is a molecule that does not exist, but this does not preclude calculation of its steric energy and bond lengths (Chapter 31). One can calculate the energy of the isomeric dimer that does form. This calculation gives a clear picture of why the triphenylmethyl radical does not simply dimerize, even though the steric energy may not be correct.

The oxidation of citronellol can give four possible isomeric isopulegols (*see* Chapter 23). Molecular mechanics calculations can indicate which of these is the most stable. Similarly, oxidation of these alcohols can give two possible isopulegones that theoretically can equilibrate. Molecular mechanics calculations can again indicate which is the more stable. These rearrange to pulegone. Again, calculation may disclose whether this product is more stable than the starting material, throwing light on the mechanism of the isomerization reaction.

The increase in time needed for a calculation on a molecule having $n$ atoms:

molecular mechanics, $n^2$

semiempirical, $n^3$

ab initio, $>n^4$

The Wittig reaction usually gives a mixture of *cis-* and *trans-*isomers. In Chapter 39 two Wittig reactions are carried out, in which one isomer predominates in each reaction. A molecular mechanics calculation discloses not only what these isomers look like but also their energies.

A very easy reaction to carry out is the aldol condensation of benzaldehyde with acetone to give in high yield dibenzalacetone (mp 110°C–111°C). It is interesting to explore the products of this reaction with molecular mechanics because three geometric isomers can be formed, and seven single-bond *cis-* or *trans-*isomers of the geometric isomers lie at the energy minima. The question is, which of these 10 isomers is formed in the reaction? Computational aspects of this question are dealt with at length in Chapter 37.

In all of these calculations, not only will you calculate the energies of the molecules, you will also be able to see what the lowest-energy conformation of the molecule looks like. In this regard it is interesting to look at the calculated low-energy conformation of benzophenone (Chapter 38), tetraphenylcyclopentadienone (Chapter 51), and pseudopellitierene (Chapter 66). *Z-* and *E-*stilbene are synthesized in Chapters 58 and 60, and in Chapter 59 the Perkin reaction is used to make a mixture of *Z-* and *E-*phenylcinnamic acids. Energy calculations, as well as a picture of the molecular conformations, help rationalize the relative amounts of the isomers formed in these synthetic reactions as well as their ultraviolet (UV) spectra.

Other examples of synthetic experiments with computational chemistry supplements are available on this book's website.

## REFERENCES

Burkert, Ulrich, and Norman L. Allinger, *Molecular Mechanics* (ACS Monograph 177). Washington, DC: American Chemical Society, 1982.

Clark, Tim, *A Handbook of Computational Chemistry: A Practical Guide to Chemical Structure and Energy Calculations*. New York: John Wiley & Sons, 1985.

Earl, Boyd L., David W. Emerson, Brian J. Johnson, and Richard L. Titus, "Teaching Practical Computer Skills to Chemistry Majors," *J. Chem. Educ.* 71 (1994): 1065.

Freeman, Fillmore, Zufan M.Tsegai, K. Marc Lasner, Warren J. Hehre, "A Comparison of the ab Initio Calculated and Experimental Conformational Energies of Alkylcyclohexanes," *J. Chem. Educ.* 77 (2000): 661.

Grant, Guy H., and W. Graham Richards, *Computational Chemistry*. Oxford, UK: Oxford University Press, 1995.

Hehre, Warren J., *A Guide to Molecular Mechanics and Quantum Chemical Calculations*. Irvine, CA: Wavefunction Press, 2003.

Hehre, Warren J., *A Laboratory Book of Computational Organic Chemistry*. Irvine, CA: Wavefunction Press, 1998.

Hehre, Warren J., Lonnie D. Burke, Alan J. Shusterman, and William J. Pietro, *Experiments in Computational Organic Chemistry*. Irvine, CA: Wavefunction Press, 1993.

Hehre, Warren J., *The Molecular Modeling Workbook for Organic Chemistry*, 2nd ed. Irvine, CA: Wavefunction Press, 2005.

Jarret, Ronald M., and Ny Sin, "Molecular Mechanics as an Organic Chemistry Laboratory Exercise," *J. Chem. Educ.* 67 (1990): 153–155.

Jensen, Frank, *Introduction to Computational Chemistry*. New York: John Wiley & Sons, 1999.

Lipkowitz, Kenny B., Raima Larter, and Tom Cundari, eds., *Reviews in Computational Chemistry*, Vol. 20. New York: John Wiley & Sons, 2004.

Lipkowitz, Kenny B., and Daniel Robertson, "Conformer Hunting: An Open-Ended Computational Chemistry Exercise That Expresses Real-World Complexity and Student Forethought," *J. Chem. Educ.* 77 (2000): 206.

The *Journal of Chemical Education*. Since 1989, many articles on molecular mechanics calculations and computational chemistry in general have been published. Use electronic journal searches to find these articles.

# CHAPTER
# 16

# The S$_N$2 Reaction: 1-Bromobutane

**Online Study Center**

This icon will direct you
to techniques, equipment
setups, and online resources at
http://college.hmco.com/PIC/
williamsonMMCE5e.

**PRELAB EXERCISE:** Prepare a detailed flow sheet for the isolation
and purification of *n*-butyl bromide. Indicate how each reaction
byproduct is removed and which layer is expected to contain the
product in each separation step.

Choice of reagents

In this experiment 1-butanol is converted to 1-bromobutane by an S$_N$2 reaction. In general, a primary alkyl bromide can be prepared by heating the corresponding alcohol with (1) constant-boiling hydrobromic acid (47% HBr); (2) an aqueous solution of sodium bromide and excess sulfuric acid, which is an equilibrium mixture containing hydrobromic acid; or (3) a solution of hydrobromic acid produced by bubbling sulfur dioxide into a suspension of bromine in water. Reagents 2 and 3 contain sulfuric acid at a concentration high enough to dehydrate secondary and tertiary alcohols to undesirable byproducts (alkenes and ethers); hence the HBr method is preferred for preparing halides of the types R$_2$CHBr and R$_3$CBr. Primary alcohols are more resistant to dehydration and can be converted efficiently to the bromides by the more economical methods (2 and 3), unless these are of such high molecular weight that they lack adequate solubility in the aqueous mixtures. The NaBr-H$_2$SO$_4$ method is preferred to the Br$_2$-SO$_2$ method because of the unpleasant, choking property of sulfur dioxide. Here is the overall equation, along with key properties of the starting material and principal product.

$$CH_3CH_2CH_2CH_2OH \xrightarrow{\text{NaBr, H}_2\text{SO}_4} CH_3CH_2CH_2CH_2Br + NaHSO_4 + H_2O$$

**1-Butanol**
bp 118°C
den. 0.810
MW 74.12
$n_D^{20}$ 1.399

**1-Bromobutane**
bp 101.6°C
den. 1.275
MW 137.03
$n_D^{20}$ 1.439

The procedure that follows specifies a certain proportion of 1-butanol, sodium bromide, sulfuric acid, and water; defines the reaction temperature and

## TABLE 16.1 • Properties of Reagents, Products, and Byproducts

### Reagents

| Reagent | MW | Den. | Bp (°C) | Wt used (g) | Moles Theory | Used |
|---------|-----|------|---------|-------------|--------|------|
| n-C$_4$H$_9$OH | 74.12 | 0.810 | 118 | 8.0 | 0.11 | 0.11 |
| NaBr | 102.91 | — | — | 13.3 | 0.11 | 0.13 |
| H$_2$SO$_4$ | 98.08 | 1.84 | — | 20.0 | 0.11 | 0.20 |

### Products and Byproducts

| Compound | MW | Den. | Bp (°C) Given | Found | Theoretical Yield Moles | g | Found g | % |
|----------|-----|------|-------|-------|-------|-----|-----|-----|
| n-C$_4$H$_9$Br | 137.03 | 1.275 | 101.6 | _____ | 0.11 | 14.8 | _____ g | _____% |
| CH$_3$CH$_2$CH=CH$_2$ | | | −6.3 | _____ | | | | |
| C$_4$H$_9$OC$_4$H$_9$ | | | 141 | _____ | | | | |

***Online Study Center***

**General Resources**
Chemical Data Table Template

The laboratory notebook

time; and describes operations to be performed in working up the reaction mixture. The prescription of quantities is based on considerations of stoichiometry as modified by the results of experimentation. Before undertaking a preparative experiment, you should analyze the procedure and calculate the molecular proportions of the reagents. A table of the properties of starting material, reagents, products, and byproducts provides guidance in regulating the temperature and in separating and purifying the product (*see* Table 16.1, for example) and should be entered in your laboratory notebook. Construct a flow sheet of the operations to be performed.

$$NaBr + H_2SO_4 \rightleftharpoons HBr + NaHSO_4$$

One mole of 1-butanol theoretically requires 1 mol each of sodium bromide and sulfuric acid, but this procedure calls for using a slight excess of bromide and twice the theoretical amount of acid. Excess acid is used to shift the equilibrium in favor of a high concentration of hydrobromic acid. The amount of sodium bromide taken, arbitrarily set at 1.2 times the theory as an insurance measure, is calculated as follows:

$$\frac{8 \text{ (g of C}_4\text{H}_9\text{OH)}}{74.12 \text{ (MW of C}_4\text{H}_9\text{OH)}} \times 102.91 \text{ (MW of NaBr)} \times 1.2 = 13.3 \text{ g of NaBr}$$

## Theoretical Yield Calculation

The theoretical yield is 0.11 mol of product, corresponding to the 0.11 mol of butyl alcohol taken. The maximal weight of product is calculated thus:

0.11 (mole of alcohol) × 137.03 (MW of product)
                     = 14.8 g 1-bromobutane (theoretical yield of product)

The probable byproducts are 1-butene, dibutyl ether, and the starting alcohol. The alkene is easily separated by distillation, but the other substances are in the same boiling point range as the product. However, all three possible byproducts can be eliminated by extraction with concentrated sulfuric acid. See Chapter 19 for another discussion of yield calculations.

## EXPERIMENTS

 # 1. Synthesis of 1-Bromobutane

$$CH_3CH_2CH_2CH_2OH \xrightarrow{\text{NaBr, H}_2\text{SO}_4} CH_3CH_2CH_2CH_2Br + NaHSO_4 + H_2O$$

**1-Butanol**
bp 118°C
den. 0.810
MW 74.12
$n_D^{20}$ 1.399

**1-Bromobutane**
bp 101.6°C
den. 1.275
MW 137.03
$n_D^{20}$ 1.439

**CAUTION:** Handle sulfuric acid with care because it is corrosive to tissue.

On a microscale greater accuracy is achieved by weighing than by volumetric measurement. It is much easier to weigh 0.80 g of a liquid than to measure 0.80 mL with a pipette.

Control heating by piling sand around the flask or by scraping it away.

**IN THIS EXPERIMENT** sodium bromide and 1-butanol are dissolved in water. Sulfuric acid is added cautiously, which generates hydrobromic acid, which in turn reacts with the alcohol upon heating to make 1-bromobutane. This alkyl bromide is then codistilled with water (steam distillation). The bromobutane is removed from the water and then treated with sulfuric acid to remove impurities. The sulfuric acid layer is removed, the product is treated with aqueous base, the base is removed, and the product is dried over calcium chloride. A small quantity of xylene is added to the product, and the mixture is fractionally distilled to give pure 1-bromobutane. The xylene acts as a chaser in this distillation. This is a fairly complex experiment.

In a 5-mL round-bottomed, long-necked flask, dissolve 1.33 g of sodium bromide in 1.5 mL of water and 0.80 g of 1-butanol. Cautiously, while constantly swirling, add 1.1 mL (2.0 g) of concentrated sulfuric acid dropwise to the solution. The NaBr will dissolve during heating. Fit the flask with a distillation head (Fig. 16.1) and reflux the reaction mixture on a sand bath for 45 min, taking care that none of the reactants distill during the reaction period. Wrap the upper end of the apparatus with a damp pipe cleaner if escaping vapor is a problem. The upper layer that soon separates in the reaction flask is the alkyl bromide because the aqueous solution of inorganic salts has the greater density. Remove the pipe cleaner, insulate the flask with cotton or glass wool, and distill the product into a collection vial until no more water-insoluble droplets come over, by which time the temperature of the distillate should have reached 115°C. If in doubt about whether the product has completely distilled, collect some of the distillate in a small tube and examine it carefully. The sample collected in the receiver is an azeotrope of 1-bromobutane and water containing some sulfuric acid, 1-butene, unreacted 1-butanol, and di-*n*-butyl ether. Rinse the distillation head with acetone

■ **FIG. 16.1**
An apparatus for refluxing and
distilling a reaction mixture, which
is shown in the distillation mode.

—Air condenser

—Receiver vial

—Boiling chip

so that it will be dry for use later in the experiment. Also, to prepare for the final step, take time to clean the round-bottomed flask by rinsing it with 1 mL of ethanol followed by 1 mL of acetone and then drawing air through it using a water aspirator. It is very important when drying an apparatus of this type to remove all the wash solvent, acetone in this case; otherwise, it will contaminate the final product.

Identifying the organic layer

Transfer the distillate to a reaction tube, rinsing the vial with about 1 mL of water, which is then mixed with the sample in the reaction tube. Note that the 1-bromobutane now forms the lower layer. Remove the 1-bromobutane with a Pasteur pipette and place it in a dry reaction tube. Add 1 mL of concentrated sulfuric acid and mix the contents well by flicking the tube. The acid removes any unreacted starting material as well as any alkenyl or ethereal byproducts. Allow the two layers to separate completely and then remove the sulfuric acid layer. The relative densities given previously will help identify the two layers. An empirical method of distinguishing the layers is to remove a drop of the lower layer into a test tube of water to see whether the material is soluble ($H_2SO_4$) or not (1-bromobutane). Separate the layers and wash the 1-bromobutane layer with 1 mL of 3 $M$ sodium hydroxide solution (den. 1.11) to remove traces of acid, separate, and be careful to save the proper layer. In experiments of this type, it is good practice to save all layers until the product is in hand.

**p-Xylene**
MW 106.17
bp 137–138°C

Using a chaser solvent.

Viton is very resistant to
aromatic solvents.

Dry the cloudy 1-bromobutane by adding anhydrous calcium chloride pellets and mixing until the liquid clears and the calcium chloride no longer clumps together. After 5 min, decant the dried liquid into the 5-mL round-bottomed flask that has been dried. Rinse the drying agent with two 1-mL portions of *p*-xylene (bp 137°C–138°C), which is then transferred to the round-bottomed flask. The high-boiling xylene is a "chaser" solvent; it chases all the bromobutane from the distilling flask. Otherwise, about 0.3 mL would remain behind.

Add a boiling stone, pack the neck of the flask with a stainless steel sponge for fractional distillation, and fit it with a dry distilling head and thermometer. Use a Viton (black) connector between the flask and the distilling head. Wrap the column with cotton or glass wool and distill, collecting material boiling in the range 99°C–103°C. Stop collecting the moment the temperature begins to rise above 103°C. Most of the product will boil at 102°C. A typical yield is in the range 1–1.2 g.

Put the sample in a vial of appropriate size and make a neatly printed label stating the name and formula of the product and your name.

**Cleaning Up.** Carefully dilute all nonorganic material with water (the pot residue, the sulfuric acid wash, the sodium hydroxide wash) and combine and neutralize with sodium carbonate before flushing down the drain. Because all the organic material has been contaminated with small quantities of product, it must be placed in the halogenated organic waste container. If local regulations allow, evaporate any residual solvent from the drying agents in the hood and place the dried solid in the nonhazardous waste container. Otherwise, place the wet drying agent in a waste container designated for this purpose.

## 2. Synthesis of 1-Bromobutane

Put 13.3 g of sodium bromide, 15 mL of water, and 10 mL of *n*-butyl alcohol in a 100-mL round-bottomed flask, cool the mixture in an ice-water bath, and slowly add 11.5 mL of concentrated sulfuric acid while swirling and cooling. Place the flask in an electric flask heater, clamp it securely, and fit it with a short condenser for reflux condensation (Fig. 16.2). Heat to the boiling point, note the time, and adjust the heat for brisk and steady refluxing. The upper layer that soon separates is the alkyl bromide because the aqueous solution of inorganic salts has a greater density. Reflux for 45 min, remove the heat, and let the condenser drain for a few minutes (extension of the reaction period to 1 h increases the yield by only 1%–2%). Remove the condenser, mount a stillhead in the flask, and set the condenser for simple distillation (*see* Fig. 5.10 on page 100) through a bent or vacuum adapter into a 50-mL Erlenmeyer flask. Distill the mixture, make frequent readings of the temperature, and distill until no more water-insoluble droplets come over, by which time the temperature should have reached 115°C (collect a few drops of distillate in a test tube and see if it is water soluble). The increasing boiling point is due to azeotropic distillation of *n*-butyl bromide with water containing increasing amounts of sulfuric acid, which raises the boiling point.

Pour the distillate into a separatory funnel, shake with about 10 mL of water, and note that *n*-butyl bromide (1-bromobutane) now forms the lower layer. A pink coloration in this layer due to a trace of bromine can be discharged by adding a

**■ FIG. 16.2**
Refluxing a reaction mixture.

Water out

Water in

pinch of sodium bisulfite and shaking again. Drain the lower layer of 1-bromobutane into a clean flask, clean and dry the separatory funnel, and return the 1-bromobutane to it. Then cool 10 mL of concentrated sulfuric acid thoroughly in an ice bath, add the acid to the funnel, shake well, and allow 5 min for separation of the layers. Use care in handling concentrated sulfuric acid. Check to see that the stopcock and stopper do not leak. The relative densities given in Table 16.1 identify the two layers; an empirical method of telling the layers apart is to draw off a few drops of the lower layer into a test tube and see whether the material is soluble in water ($H_2SO_4$) or insoluble in water (bromobutane). Separate the layers, allow 5 min for further drainage, and separate again. Then wash the 1-bromobutane with 10 mL of 3 $M$ sodium hydroxide (den. 1.11) solution to remove traces of acid, separate, and be careful to save the proper layer.

Calcium chloride removes both water and alcohol from a solution.

Dry the cloudy 1-bromobutane by adding 1 g of anhydrous calcium chloride pellets while swirling until the liquid clears. After 5 min decant the dried liquid into a 25-mL flask or filter it through a fluted filter paper, add a boiling stone, distill, and collect material boiling in the range 99°C–103°C. A typical student yield is in the range 10–12 g. Note the approximate volumes of forerun[1] and residue.

A proper label is important.

Put the sample in a narrow-mouth bottle of appropriate size; make a neatly printed label stating the name and formula of the product and your name. Press the label onto the bottle under a piece of filter paper and be sure that it is secure. After all the time spent on the preparation, the final product should be worthy of a carefully executed and secured label.

Check purity by infrared, refractive index, or gas chromatographic analysis.

**Cleaning Up.** Carefully dilute all nonorganic material with water (the reaction pot residue, the sulfuric acid wash, and the sodium hydroxide wash) and combine and neutralize with sodium carbonate before flushing down the drain with excess water. The residue from the distillation of 1-bromobutane goes in the container for halogenated organic solvents. If local regulations allow, evaporate any residual solvent from the drying agents in the hood and place the dried solid in the nonhazardous waste container. Otherwise, place the wet drying agent in a waste container designated for this purpose.

The $^{13}$C NMR spectra of 1-bromobutane (Fig. 16.3) and 1-butanol (Fig. 16.4), as well as the $^1$H NMR spectrum of 1-butanol (Fig. 16.5) are shown as a verification for your work.

## QUESTIONS

1. Which experimental method would you recommend for the preparation of 1-bromooctane? of *t*-butyl bromide?
2. Explain why the crude product is apt to contain certain definite organic impurities.

---

1. The forerun is the material that distills before the temperature levels off between 99°C and 103°C, which is the boiling point of the product.

4. How should the reaction conditions in this experiment be changed to try to produce 1-chlorobutane?
5. Write a balanced equation for the reaction of sodium bisulfite with bromine.
6. What is the purpose of refluxing the reaction mixture for 45 min? Why not simply boil the mixture in an Erlenmeyer flask?
7. Why is it necessary to remove all water from 1-bromobutane before distilling it?
8. Write reaction mechanisms showing how 1-butene and di-*n*-butyl ether are formed.
9. Why is the resonance of the bromine-bearing carbon atom and the hydroxyl-bearing carbon atom farthest downfield in Figures 16.3 and 16.4?

**FIG. 16.3**
The $^{13}$C NMR spectrum of *n*-butyl bromide (22.6 MHz).

**■ FIG. 16.4**
The $^{13}$C NMR spectrum of 1-butanol (22.6 MHz).

■ **FIG. 16.5**
The $^1H$ NMR spectrum of
1-butanol (250 MHz).

## CHAPTER

# 17

# Nucleophilic Substitution
# Reactions of Alkyl Halides

*Online Study Center*

This icon will direct you to
techniques, equipment setups,
and online resources at
http://college.hmco.com/PIC/
williamsonMMOE5e.

**PRELAB EXERCISE:** Predict the outcomes of the two sets of
experiments to be carried out with the 11 halides used in this
chapter.

The alkyl halides, R—X (where X = Cl, Br, I, and sometimes F), play a
central role in organic synthesis. These can easily be prepared from, among
others, alcohols, alkenes, and, industrially, alkanes. In turn, they are the starting
materials for the synthesis of a large number of new functional groups. The syntheses are often carried out by nucleophilic substitution reactions in which the
halide is replaced by another group such as cyano, hydroxyl, ether, ester, alkyl—
the list is long. As a consequence of the importance of this substitution reaction, it
has been studied carefully by employing reactions such as the two used in this
chapter. Some of the questions that can be asked are as follows:

- How does the structure of the alkyl part of the alkyl halide affect the
  reaction?
- What is the effect of changing the . . .

  nature of the halide?
  nature of the solvent?
  relative concentrations of the reactants?
  temperature of the reaction?
  nature of the nucleophile?

In this chapter we explore the answers to some of these questions.

In free radical reactions the covalent bond undergoes homolysis when it
breaks:

$$R\!:\!\ddot{\underset{\cdot\cdot}{Cl}}\!: \longrightarrow R\!\cdot\; +\; \cdot\ddot{\underset{\cdot\cdot}{Cl}}\!:$$

A carbocation is planar and has an empty *p*-orbital.

Order of carbocation stability

Nucleophilic substitution

One step

Two steps

Reaction kinetics

whereas in ionic reactions it undergoes heterolysis:

$$R{:}\ddot{\underset{..}{Cl}}{:} \longrightarrow R^+ + {:}\ddot{\underset{..}{Cl}}{:}^-$$

A carbocation is often formed as a reactive intermediate in these reactions. This carbocation is $sp^2$ hybridized and trigonal-planar in structure with a vacant *p*-orbital. Much experimental evidence of the type obtained in this chapter indicates that the order of stability of carbocations is

$$
\underset{\overset{|}{R}}{\overset{\overset{R}{|}}{R{-}C^+}} >
\underset{\overset{|}{H}}{\overset{\overset{R}{|}}{R{-}C^+}} >
\underset{\overset{|}{H}}{\overset{\overset{H}{|}}{R{-}C^+}} >
\underset{\overset{|}{H}}{\overset{\overset{H}{|}}{H{-}C^+}}
$$

The alkyl (R) groups stabilize the positive charge of the carbocation by displacing or releasing electrons toward the positive charge. As the number of alkyl groups attached to the carbocation increases, the stabilization increases.

Many organic reactions occur when a nucleophile (a species with an unshared pair of electrons) reacts with an alkyl halide to replace the halogen with the nucleophile.

$$Nu{:}^- + R{:}\ddot{\underset{..}{X}}{:} \longrightarrow Nu{:}R + {:}\ddot{\underset{..}{X}}{:}^-$$

This substitution reaction can occur in one smooth step:

$$Nu{:}^- + R{:}\ddot{\underset{..}{X}}{:} \longrightarrow \left[ \overset{\delta-}{Nu} \cdots R \cdots \overset{\delta-}{\underset{..}{X}}{:} \right] \longrightarrow Nu{:}R + {:}\ddot{\underset{..}{X}}{:}^-$$

or it can occur in two discrete steps:

$$R{:}\ddot{\underset{..}{X}}{:} \longrightarrow R^+ + {:}\ddot{\underset{..}{X}}{:}^-$$

$$Nu{:}^- + R^+ \longrightarrow Nu{:}R$$

depending primarily on the structure of the R group. The nucleophile ($Nu{:}^-$) can be a substance with a full negative charge, such as $:\ddot{\underset{..}{I}}:^-$ or $H{:}\ddot{\underset{..}{O}}{:}^-$, or an uncharged molecule with an unshared pair of electrons such as exists on the oxygen atom in water, $H{-}\ddot{\underset{..}{O}}{-}H$. Not all of the halides, $:\ddot{\underset{..}{X}}{:}^-$, depart with equal ease in nucleophilic substitution reactions. In this chapter we investigate the ease with which the different halogens leave in one of the substitution reactions.

To distinguish between the reaction that occurs as one smooth step and the reaction that occurs as two discrete steps, it is necessary to study the kinetics of the reaction. If the reaction were carried out with several different concentrations of $R{:}\ddot{\underset{..}{X}}{:}^-$ and $Nu{:}^-$, we could determine if the reaction is bimolecular or unimolecular.

In the case of the smooth, one-step reaction, the nucleophile must collide with the alkyl halide. The kinetics of the reaction

$$\text{Nu:}^- + \text{R:}\ddot{\text{X}}\text{:} \longrightarrow \left[ \overset{\delta-}{\text{Nu}} \cdots \text{R} \cdots \overset{\delta-}{\ddot{\text{X}}\text{:}} \right] \longrightarrow \text{Nu:R} + :\ddot{\text{X}}\text{:}^-$$

$$\text{Rate} = k \left[ \text{Nu:}^- \right] \left[ \text{R} :\ddot{\text{X}}\text{:} \right]$$

are found to depend on the concentration of both the nucleophile and the halide. Such a reaction is said to be a bimolecular nucleophilic substitution reaction, $S_N2$. If the reaction occurs as a two-step process,

$$\text{R:}\ddot{\text{X}}\text{:} \xrightarrow{\text{slow}} \text{R}^+ + :\ddot{\text{X}}\text{:}^-$$

$$\text{Nu:}^- + \text{R}^+ \xrightarrow{\text{fast}} \text{Nu:R}$$

$$\text{Rate} = k \left[ \text{R:}\ddot{\text{X}}\text{:} \right]$$

the rate of the first step, the slow step, depends only on the concentration of the halide, and it is said to be a unimolecular nucleophilic substitution reaction, $S_N1$.

The $S_N1$ reaction proceeds through a planar carbocation. Even if the starting material were chiral, the product would be a 50:50 mixture of enantiomers because the planar intermediate can form a bond with the nuceophile on either face.

*Racemization: $S_N1$*

**Chiral alkyl chloride**   **Planar carbocation**

(1)

(2)

**Enantiomers**

Inversion: $S_N2$

The $S_N2$ reaction occurs with inversion of configuration to give a product of the opposite chirality from the starting material.

$$\text{Nu:}^- + \quad \underset{\underset{\text{D}}{\text{R}}}{\overset{\text{H}}{\text{C}}}\text{—Cl} \quad \longrightarrow \quad \overset{\text{H}}{\underset{\underset{\text{D}}{\text{R}}}{\text{Nu------}\overset{\delta^-}{\text{C}}\text{------}\overset{\delta^-}{\text{Cl}}}} \quad \longrightarrow \quad \text{Nu—}\underset{\underset{\text{D}}{\text{R}}}{\overset{\text{H}}{\text{C}}} \quad + \text{ :}\ddot{\text{Cl}}\text{:}^-$$

The order of reactivity for *simple* alkyl halides in the $S_N2$ reaction is

Order of $S_N2$ reactivity

$$CH_3\text{—X} > \text{R—}CH_2\text{—X} > \text{R—}\underset{\underset{\text{R}}{|}}{\overset{|}{\text{CH}}}\text{—X} > \text{R—}\underset{\underset{\text{R}}{|}}{\overset{\overset{\text{R}}{|}}{\text{C}}}\text{—X}$$

The tertiary halide is in parentheses because it usually does not react by an $S_N2$ mechanism. The primary factor in this order of reactivity is steric hindrance, that is, the ease with which the nucleophile can come within bonding distance of the alkyl halide; 2,2-dimethyl-1-bromopropane, even though it is a primary halide, reacts 100,000 times slower than ethyl bromide ($CH_3CH_2Br$) because of steric hindrance to attack on the bromine atom in the dimethyl compound.

Steric hindrance toward $S_N2$

$$CH_3\text{—}\underset{\underset{CH_3}{|}}{\overset{\overset{CH_3}{|}}{\text{C}}}\text{—}CH_2\text{—Br} \qquad\qquad CH_3CH_2Br$$

$$\text{Rate} = 1 \qquad\qquad\qquad \text{Rate} = 100,000$$

The primary factor in $S_N1$ reactivity is the relative stability of the carbocation that is formed. For simple alkyl halides, this means that only tertiary halides react by this mechanism. The tertiary halide must be able to form a planar carbocation. Only slightly less reactive are the allyl carbocations, which derive their great stability from the delocalization of the charge on the carbon by resonance.

The allyl carbocation

$$CH_2\text{=CH—}CH_2\text{—}\ddot{\text{Br}}\text{:} \longrightarrow CH_2\text{=CH—}\overset{+}{\text{CH}}_2 \longleftrightarrow \overset{+}{\text{CH}}_2\text{—CH=}CH_2 + \text{ :}\ddot{\text{Br}}\text{:}^-$$

The nature of the solvent has a large effect on the rates of $S_N2$ reactions. In a solvent with a hydrogen atom attached to an electronegative atom such as oxygen, the protic solvent forms hydrogen bonds to the nucleophile. These solvent molecules get in the way during an $S_N2$ reaction.

Aprotic: no ionizable protons

If the solvent is polar and aprotic, solvation of the nucleophile cannot occur, and the $S_N2$ reaction can occur up to a million times faster. Some common polar, aprotic solvents are N,N-dimethylformamide and dimethylsulfoxide:

**N,N-Dimethylformamide**     **Dimethylsulfoxide**

In the $S_N1$ reaction a polar protic solvent such as water stabilizes the transition state more than it does the reactants, lowering the energy of activation for the reaction and thus increasing the rate, relative to the rate in a nonpolar solvent. Acetic acid, ethanol, and acetone are relatively nonpolar solvents and have lower dielectric constants than the polar solvents water, dimethylsulfoxide, and N,N-dimethylformamide.

The leaving group

The rate of $S_N1$ and $S_N2$ reactions depends on the nature of the leaving group, the best leaving groups being the ones that form stable ions. Among the halogens we find that the iodide ion ($I^-$) is the best leaving group as well as the best nucleophile in the $S_N2$ reaction.

Vinylic and aryl halides

do not normally react by $S_N1$ reactions because the resulting carbocations

| TABLE 17.1 • Summary of S$_N$1 and S$_N$2 Reactions | | |
|---|---|---|
| | **Unimolecular Nucleophilic Substitution (S$_N$1)** | **Second-Order Nucleophilic Substitution (S$_N$2)** |
| Kinetics | First order | Second order |
| Mechanism | Two steps; unimolecular in the rate-determining step via a carbocation | One-step, bimolecular |
| Stereochemistry | Racemization predominates | Inversion of configuration |
| Reactivity of structure | 3° > 2° > 1° > CH$_3$ > vinyl | 3° < 2° < 1° < CH$_3$ < vinyl |
| Rearrangements | May occur | No rearrangements because no carbocation intermediate |
| Effect of leaving group | —I > —Br > —Cl>> —F | —I > —Br > —Cl >> —F |
| Effect of nucleophile | Not important because it is not in the rate-determining step | I$^-$ > Br$^-$ > Cl$^-$ > F$^-$ |
| Concentration of nucleophile | S$_N$1 is favored by low concentration | S$_N$2 is favored by high concentration |
| Solvent polarity | High, favors S$_N$1 | Low, favors S$_N$2 |

**Temperature dependence**

are relatively difficult to form. For S$_N$2 reactions, electrons in the nearby double bonds repel the nucleophile, which is either an ion or a polarized neutral species.

The rates of both S$_N$1 and S$_N$2 reactions depend on the temperature of the reaction. As the temperature increases, the kinetic energy of the molecules increases, leading to a greater rate of reaction. The rate of many organic reactions will approximately double when the temperature increases about 10°C. This information is summarized in Table 17.1.

## EXPERIMENTS

In the experiments that follow, 11 representative alkyl halides are treated with sodium iodide in acetone and with an ethanolic solution of silver nitrate.

## 1. Sodium Iodide in Acetone

Acetone, with a dielectric constant of 21, is a relatively nonpolar solvent that will readily dissolve sodium iodide. The iodide ion is an excellent nucleophile, and the nonpolar solvent (acetone) favors the S$_N$2 reaction; it does not favor ionization of the alkyl halide. The extent of reaction can be observed because sodium bromide and sodium chloride are not soluble in acetone and precipitate from solution if a reaction occurs.

Organic halides that can react by an S$_N$2 mechanism give a precipitate of NaX with sodium iodide in acetone.

$$Na^+I^- + R—Cl \longrightarrow R—I + NaCl\downarrow$$
$$Na^+I^- + R—Br \longrightarrow R—I + NaBr\downarrow$$

## 2. Ethanolic Silver Nitrate Solution

When an alkyl halide is treated with an ethanolic solution of silver nitrate, the silver ion coordinates with an electron pair of the halogen. This weakens the carbon-halogen bond because a molecule of insoluble silver halide is formed, thus promoting an $S_N1$ reaction of the alkyl halide. The solvent, ethanol, favors ionization of the halide, and the nitrate ion is a very poor nucleophile, so alkyl nitrates do not form by an $S_N2$ reaction.

Organic halides that can react by an $S_N1$ mechanism give a precipitate of AgX with ethanolic silver nitrate solution.

$$R\text{—}\ddot{\underset{\cdot\cdot}{X}}: \;\overset{Ag^+}{\rightleftharpoons}\; R\overset{\delta^+}{\ddot{\underset{\cdot\cdot}{X}}}\overset{\delta^+}{Ag} \longrightarrow R^+ + AgX\downarrow$$

On the basis of the foregoing discussion, tertiary halides would be expected to react with silver nitrate most rapidly and primary halides least rapidly. The $R^+$ ion can react with the solvent to give either an alkene or an ether.

## Microscale Procedure

Label 11 small containers (reaction tubes, 3-mL centrifuge tubes, 10 × 75-mm test tubes, or 1-mL vials), and place 0.1 mL or 100 mg of each of the following halides in the tubes.

CH₃CH₂CH₂CH₂Cl

**1-Chlorobutane**
bp 77–78°C

CH₃CH₂CH₂CH₂Br

**1-Bromobutane**
bp 100–104°C

Cl
|
CH₃CH₂CHCH₃

**2-Chlorobutane**
bp 68–70°C

CH₃
|
CH₃—C—Cl
|
CH₃

**2-Chloro-2-methylpropane**
bp 51–52°C

Br (Bromobenzene)

**Bromobenzene**
bp 156°C

CH₃CH=CHCH₂Cl

**1-Chloro-2-butene mixture
of *cis* and *trans* isomers**
bp 635°C (*cis*)
bp 68°C (*trans*)

*allyl chloride*

CH₃
|
CH₃CHCH₂Cl

**1-Chloro-2-methylpropane**
bp 68–69°C

Br
|
CH₃CH₂CHCH₃

**2-Bromobutane**
bp 91°C

CH₂CH=C(Cl)CH₃

**2-Chloro-2-butene
(mixture of isomers)**
bp 62–67°C

*1Chloro-2methyl
propene*

CH₃CH₂I

**Iodoethane**
bp 72°C

*iodobutane
(using this one)*

Cl (1-Chloroadamantane)

**1-Chloroadamantane**
mp 165–166°C

*solid
measure
100mg*

To each tube then rapidly add 1 mL of an 18% solution of sodium iodide in acetone, stopper each tube, mix the contents thoroughly, and note the time. Note the time of first appearance of any precipitate. If no reaction occurs within about 5 min, place those tubes in a 50°C water bath and watch for any reaction over the next 5 or 6 min.

Empty the tubes, rinse them with ethanol, place the same amount of each of the alkyl halides in each tube as in the first part of the experiment, add 1 mL of 1% ethanolic silver nitrate solution to each tube, mix the contents well, and note the time of addition as well as the time of appearance of the first traces of any precipitate. If a precipitate does not appear in 5 min, heat the tubes containing these unreactive halides in a 50°C water bath for 5 to 6 min and watch for any reaction.

To test the effect of solvent on the rate of $S_N1$ reactivity, compare the time needed for a precipitate to appear when 2-chlorobutane is treated with 1% ethanolic silver nitrate solution and when treated with 1% silver nitrate in a mixture of 50% ethanol and 50% water.

In your analysis of the results from these experiments, consider the following for both $S_N1$ and $S_N2$ conditions:

- the nature of the leaving group (Cl vs. Br) in the 1-halobutanes
- the effect of structure, that is, compare simple primary, secondary, and tertiary halides, unhindered primary vs. hindered primary halides, a simple tertiary halide vs. a complex tertiary halide, and an allylic halide vs. a tertiary halide
- the effect of solvent polarity on the $S_N1$ reaction
- the effect of temperature on the reaction

**Cleaning Up.** Because all of the test solutions contain halogenated material, all test solutions and washes as well as unused starting materials should be placed in the halogenated organic waste container.

## QUESTIONS

1. What would be the effect of carrying out the sodium iodide in acetone reaction with the alkyl halides using an iodide solution half as concentrated?
2. The addition of sodium or potassium iodide catalyzes many $S_N2$ reactions of alkyl chlorides or bromides. Explain.
3. In $S_N1$ reactions, the intermediate carbocations can eliminate a proton to yield alkenes or react with the solvent to yield ethers. Draw the structures of the byproducts of this type that would be derived from the reaction of the carbocation derived from 2-bromo-2-methylbutane in ethanol.

# Radical Initiated Chlorination
# of 1-Chlorobutane

**Online Study Center**

This icon will direct you to
techniques, equipment setups,
and online resources at
http://college.hmco.com/PIC/
williamsonMMOE5e.

**PRELAB EXERCISE:** Write two chain propagation steps for the
reaction of butane with sulfuryl chloride.

Most of the alkanes from petroleum are used to produce energy by combustion, but a small percentage are converted to industrially useful compounds by controlled reaction with oxygen or chlorine. The alkanes are inert to attack by most chemical reagents but will react with oxygen and halogens under the special conditions of radical initiated reactions.

At room temperature an alkane such as butane will not react with chlorine. In order for the following reaction to occur,

$$CH_3CH_2CH_2CH_3 + Cl_2 \longrightarrow CH_3CH_2CH_2CH_2Cl + HCl$$

a precisely oriented four-center collision of the $Cl_2$ molecule with the butane molecule must occur, with two bonds broken and two bonds formed simultaneously:

It is unlikely that the necessarily precise orientation of the two reacting molecules will be found. They must, of course, be within bonding distance of each other as well, so steric factors play a part.

If a concerted four-center reaction does not work, then an alternative possibility is a stepwise mechanism:

$$:\ddot{C}l:\ddot{C}l: \xrightarrow{\text{slow}} 2 :\ddot{C}l\cdot \qquad (1)$$

$$CH_3CH_2CH_2CH_3 \xrightarrow{\text{slow}} CH_3CH_2CH_2CH_2\cdot + H\cdot \qquad (2)$$

$$:\overset{\cdot\cdot}{\underset{\cdot\cdot}{Cl}}\cdot + CH_3CH_2CH_2CH_2\cdot \xrightarrow{\text{fast}} CH_3CH_2CH_2CH_2Cl \qquad (3)$$

$$:\overset{\cdot\cdot}{\underset{\cdot\cdot}{Cl}}\cdot + H\cdot \xrightarrow{\text{fast}} H:\overset{\cdot\cdot}{\underset{\cdot\cdot}{Cl}}: \qquad (4)$$

For this series of reactions to occur, the chlorine molecule must dissociate into two chlorine atoms. Because chlorine has a bond energy of 58 kcal/mole, we would not expect any significant numbers of molecules to dissociate at room temperature. Thermal motion at 25°C can only break bonds having energies less than 30–35 kcal/mole. Although thermal dissociation would require a high temperature, the dissociation of chlorine into atoms (chlorine radicals) can be caused by violet and ultraviolet light:

$$Cl_2 \xrightarrow{\text{h}\nu} 2 :\overset{\cdot\cdot}{\underset{\cdot\cdot}{Cl}}\cdot$$

The photon energy of red light is 48 kcal/mole, whereas light of 300 nm (ultraviolet) has a photon energy of 96 kcal/mole.

The chlorine radical can react with butane by abstraction of a hydrogen atom:

$$CH_3CH_2CH_2CH_3 + :\overset{\cdot\cdot}{\underset{\cdot\cdot}{Cl}}\cdot \rightleftharpoons CH_3CH_2CH_2CH_2\cdot + HCl$$

a reaction that is very slightly exothermic (and thus written as a reversible reaction). The butyl radical can react with chlorine:

$$CH_3CH_2CH_2CH_2\cdot + Cl_2 \longrightarrow CH_3CH_2CH_2CH_2Cl + :\overset{\cdot\cdot}{\underset{\cdot\cdot}{Cl}}\cdot$$

a reaction that evolves 26 kcal/mole of energy. The net result of these two reactions is a reaction of $Cl_2$ with butane catalyzed by $:\overset{\cdot\cdot}{\underset{\cdot\cdot}{Cl}}\cdot$, because a chlorine radical is produced for every chlorine radical consumed. The whole process can be terminated by the reaction of radicals with each other:

$$CH_3CH_2CH_2CH_2\cdot + :\overset{\cdot\cdot}{\underset{\cdot\cdot}{Cl}}\cdot \longrightarrow CH_3CH_2CH_2CH_2Cl$$

$$2\ CH_3CH_2CH_2CH_2\cdot \longrightarrow CH_3(CH_2)_6CH_3$$

$$2 :\overset{\cdot\cdot}{\underset{\cdot\cdot}{Cl}}\cdot \longrightarrow Cl_2$$

To summarize, the process of light-induced radical chlorination involves three steps: chain initiation in which chlorine radicals are produced, chain propagation that involves no net consumption of chlorine radicals, and chain termination that destroys radicals.

$$Cl_2 \longrightarrow 2 \: \ddot{\underset{..}{C}}l \cdot \qquad \textit{Chain initiation}$$

$$CH_3CH_2CH_2CH_3 \: + \: \ddot{\underset{..}{C}}l \cdot \rightleftharpoons CH_3CH_2CH_2CH_2 \cdot \: + \: HCl \: \Big\}$$
$$CH_3CH_2CH_2CH_2 \cdot \: + \: Cl_2 \longrightarrow CH_3CH_2CH_2CH_2Cl \: + \: \ddot{\underset{..}{C}}l \cdot \Big\}$$

$\textit{Chain propagation}$

$$CH_3CH_2CH_2CH_2 \cdot \: + \: \ddot{\underset{..}{C}}l \cdot \longrightarrow CH_3CH_2CH_2CH_2Cl$$
$$2 \: CH_3CH_2CH_2CH_2 \cdot \longrightarrow CH_3(CH_2)_6CH_3$$
$$2 \: \ddot{\underset{..}{C}}l \cdot \longrightarrow Cl_2$$

$\textit{Chain termination}$

## Relative Reactivities of Hydrogen

In the preceding example the product is shown to be 1-chlorobutane, but in fact the reaction produces a mixture of 1-chlorobutane and 2-chlorobutane. If the reaction were to occur purely by chance, we would expect the ratio of products to be 6:4 because there are six primary hydrogens in the two $CH_3$ groups and four secondary hydrogens in the two $CH_2$ groups of butane. But because a secondary C—H bond is weaker than a primary C—H bond (95 kcal/mole vs. 98 kcal/mole), we might expect more 2-chlorobutane than chance would dictate.

The chlorination of 2-methylbutane is summarized in Table 18.1. As can be seen from the table, the relatively weak tertiary C—H bond (92 kcal/mole) gives rise to 22% of product in contrast to the 8% expected on the basis of a random attack of $\: \ddot{\underset{..}{C}}l \cdot$ on the starting material.

The relative reactivities of the various hydrogens of 2-methylbutane on a per-hydrogen basis (referred to the primary hydrogens of C-4 as 1.0) can be calculated. The molecule has nine primary hydrogens, chlorination of which accounts for 45% of the products. This means that each primary hydrogen is responsible for $45\% \div 9 = 5.0\%$ of the product. Similarly, the two secondary hydrogens account for 33% of the product, so each secondary hydrogen is responsible for 16.5% of the product. Finally, the one tertiary hydrogen is responsible for 22% of the product. These results mean that tertiary hydrogens are $22 \div 5.0 = 4.4$ times as reactive as primary hydrogens toward chlorination, and secondary hydrogens are $16.5 \div 5.0 = 3.3$ times as reactive toward chlorination as primary hydrogens. Results very similar to these have been found for a large number of hydrocarbons. The relative reactivity toward chlorination is as follows:

$$R_3CH \: > \: R_2CH_2 \: > \: RCH_3$$
$$4.4 \qquad\quad 3.3 \qquad\quad 1$$

A reaction of this type is of little use unless you happen to need the four products in the ratios found and can manage to separate them (their boiling

**TABLE 18.1 • Product Distribution for the Chlorination of 2-Methylbutane**

| Reactant | | Products | Ratio | Statistical Expectation Percent | Found (%) | Relative Reactivity |
|---|---|---|---|---|---|---|
| CH₃—C(CH₃)(H)—CH₂CH₃ | Cl₂ / 300° → | CH₃—C(CH₃)(H)—CHCl—CH₃ | 2/12 | 17 | 33 | 3.3 |
| | | ClCH₂—C(CH₃)(H)—CH₂—CH₃ | 6/12 | 50 | 30 | 1 |
| | | CH₃—C(CH₃)(Cl)—CH₂—CH₃ | 1/12 | 8 | 22 | 4.4 |
| | | CH₃—C(CH₃)(H)—CH₂—CH₂Cl | 3/12 | 25 | 15 | 1 |

points are very similar). Industrially, however, the radical chlorination of methane and ethane is important, and the products can be separated easily:

$$CH_4 \xrightarrow[\Delta]{Cl_2} \quad CH_3Cl \quad + \quad CH_2Cl_2 \quad + \quad CHCl_3 \quad + \quad CCl_4$$

| | **Chloromethane** | **Dichloromethane** | **Trichloromethane** | **Tetrachloromethane** |
|---|---|---|---|---|
| *Common names:* | Methyl chloride | Methylene chloride | Chloroform | Carbon tetrachloride |
| *Boiling points:* | −24°C | 40°C | 62°C | 77°C |

In the first experiment we will chlorinate 1-chlorobutane because it is easier to handle in the laboratory than gaseous butane, and we will use sulfuryl chloride as our source of chlorine radicals because it is easier to handle than gaseous chlorine. Instead of using light to initiate the reaction, we will use a chemical initiator, 2,2'-azobis-(2-methylpropionitrile). This azo compound (R—N=N—R) decomposes at moderate temperatures (80°C–100°C) to give two relatively stable radicals and nitrogen gas:

The common name for this catalyst is AIBN, which stands for azoisobutylnitrile.

$$
\underset{\underset{\displaystyle CH_3}{|}}{\overset{\overset{\displaystyle CN}{|}}{CH_3-C}}-N=N-\underset{\underset{\displaystyle CH_3}{|}}{\overset{\overset{\displaystyle CN}{|}}{C}}-CH_3 \xrightarrow{80-100°} 2\ \underset{\underset{\displaystyle CH_3}{|}}{\overset{\overset{\displaystyle CN}{|}}{CH_3-C\cdot}} + N_2 \uparrow
$$

**2,2'-Azobis-(2-methylpropionitrile)**
MW 164.21, mp 102–103° (dec.)

$$
\underset{\underset{\displaystyle CH_3}{|}}{\overset{\overset{\displaystyle CN}{|}}{CH_3-C\cdot}} + \underset{\underset{\displaystyle O}{\|}}{\overset{\overset{\displaystyle O}{\|}}{Cl-S}}-Cl \longrightarrow \underset{\underset{\displaystyle CH_3}{|}}{\overset{\overset{\displaystyle CN}{|}}{CH_3-C}}-Cl + \underset{\underset{\displaystyle O}{\|}}{\overset{\overset{\displaystyle O}{\|}}{\cdot S}}-Cl
$$

**Sulfuryl chloride**

$$
\underset{\underset{\displaystyle O}{\|}}{\overset{\overset{\displaystyle O}{\|}}{\cdot S}}-Cl \longrightarrow \underset{\underset{\displaystyle O}{\|}}{\overset{\overset{\displaystyle O}{\|}}{S}} + :\ddot{\underset{}{Cl}}\cdot
$$

Before conducting the experiment, try to predict the ratios of products.

The radical monochlorination of 1-chlorobutane can give four products: 1,1-, 1,2-, 1,3-, and 1,4-dichlorobutane. If the reaction occurred completely at random, we would expect products in the ratios of the number of hydrogen atoms on each carbon, that is, 2:2:2:3, respectively (22%, 22%, 22%, 33%).

Will these ratios correlate with the calculated heats of formation of the products?

The object of the following experiments is to carry out the radical chlorination of 1-chlorobutane and then to determine the ratios of products using gas chromatography. From these ratios the relative reactivities of the hydrogens can be calculated.

$$
CH_3CH_2CH_2CH_2Cl + SO_2Cl_2 \xrightarrow{R\cdot} CH_3CH_2CH_2CHCl_2 + CH_3CH_2CHClCH_2Cl
$$
$$
\text{bp } 114°C \qquad\qquad \text{bp } 124°C
$$

**1-Chlorobutane**
MW 92.57
den. 0.886
bp 77–78°C

**Sulfuryl chloride**
MW 134.97
den. 1.67
bp 69°C

$$
+ CH_3CHClCH_2CH_2Cl + CH_2ClCH_2CH_2CH_2Cl
$$
$$
\text{bp } 134°C \qquad\qquad \text{bp } 162°C
$$

$$
+ SO_2 + HCl
$$

## EXPERIMENTS

 # 1. Radical Chlorination of 1-Chlorobutane

> **IN THIS EXPERIMENT** chlorine radicals (from sulfuryl chloride and a catalyst) react with 1-chlorobutane to form four different dichlorobutanes. The amounts of each product depend on the number and kinds of hydrogens in the starting material. The product is analyzed by gas chromatography. Because the reaction gives off both hydrogen chloride and sulfur dioxide, these gases are trapped. The crude product is diluted with water, the organic phase is washed with dilute base and water and then dried before carrying out the gas chromatography.

*Online Study Center*

Photos: Placing a Polyethylene
Tube through a Septum,
Gas Trap

To a 10 ×100 mm reaction tube add 1-chlorobutane (0.50 mL, 0.432 g, 4.6 mmol), sulfuryl chloride (0.16 mL, 0.27 g, 2.0 mmol), 2,2′-azobis-(2-methylpropionitrile) (4 mg, 0.025 mmol), and a boiling chip. Use a 1-mL graduated pipette to measure the 1-chlorobutane and use a dispenser or a 0.5-mL syringe to measure the sulfuryl chloride (in the hood). Rinse the syringe immediately after use and leave it disassembled. Weigh the azo compound on a balance. In this experiment none of the reagents need to be measured with great care; the 1-chlorobutane is in large excess, and the azo compound is present in catalytic amounts.

Fit the reaction tube with a rubber septum and a piece of polyethylene tubing that leads down into another reaction tube, the mouth of which has a piece of damp cotton placed in it (Fig. 18.1). During this reaction, sulfur dioxide and hydrogen chloride gas are evolved, and the damp cotton will absorb the gas. The technique for threading a polyethylene tube through a septum is shown in Figure 18.2. The amount of HCl evolved in this experiment (2 mmol) is equal to 0.15 mL of concentrated hydrochloric acid. Be sure that the end of the polyethylene tube does not touch any water because it would be sucked back into the reaction tube. Clamp the reaction tube with the reactants in a beaker of hot water maintained at 80°C–85°C so that just the tip of the tube is immersed in the water. This will cause the contents to boil gently and the vapors to condense on the cool upper walls of the reaction tube.

At the end of the reaction period, remove the tube from the beaker of water, allow it to cool, and then carefully add 0.5 mL of water dropwise to the tube from a Pasteur pipette. Note which layer is the aqueous one. Mix the contents thoroughly and then draw off and discard the water layer. Wash the organic phase in the same way with a 0.5-mL portion of 5% sodium bicarbonate solution and once with 0.5 mL water. Carefully remove all the water with a Pasteur pipette and then add anhydrous calcium chloride pellets to dry the product. Transfer the dry product (it should be perfectly clear, not cloudy) to a small, tared, screw-capped sample vial or corked reaction tube. Do not store the product in a polyethylene or rubber-capped container because it will be readily absorbed by these materials. It is not a good idea to store the product for a prolonged period before analysis because the composition will change depending on which components evaporate or are

■ **FIG. 18.1**

The microscale apparatus for the chlorination of 1-chlorobutane with HCl and an $SO_2$ trap. Heat the polyethylene tubing carefully in a steam bath to bend it permanently. The tubes are to be clamped appropriately.

■ **FIG. 18.2**

To thread a polyethylene tube through a septum, make a hole through the septum with a needle and then push a toothpick through the hole. Push the polyethylene tube firmly onto the toothpick, then pull and push on the toothpick. The tube will slide through the septum. Finally, pull the tube from the toothpick. A blunt syringe needle can be used instead of a toothpick.

absorbed by the cap on the container. Determine the weight of the product and then analyze it by gas chromatography.

**Cleaning Up.** Rinse the damp cotton with water and combine the rinse with all the aqueous layers and washes. Neutralize the aqueous solution with sodium carbonate and flush the solution down the drain. The drying agent will be coated with chlorinated product and must therefore be disposed of in the hazardous waste container. Any unused starting material and product must be placed in the halogenated organic waste container.

## 2. Free-Radical Chlorination of 1-Chlorobutane

To a 25-mL round-bottomed flask add 1-chlorobutane (5 mL, 4.32 g, 0.046 mole), sulfuryl chloride (1.6 mL, 2.7 g, 0.02 mole), 2,2′-azobis-(2-methylpropionitrile)

■ **FIG. 18.3**
The apparatus for the chlorination
of 1-chlorobutane with $SO_2$ and an
HCl gas trap.

Steam bath

Steam in

To sink

500-mL Filter flask

Glass tube *which does
not dip below water
surface*

Cold water

⚠️

Conduct this experiment
in a laboratory hood.

(0.03 g), and a boiling chip. Equip the flask with a condenser and gas trap
(Fig. 18.3). Heat the mixture to gently reflux on a steam bath for 20 min. Remove the flask from the steam bath, allow it to cool somewhat, and quickly—
to minimize the escape of sulfur dioxide and hydrochloric acid—lift the condenser from the flask and add a second 0.03-g portion of the initiator. Heat the
reaction mixture for an additional 10 min, remove the flask and condenser from
the steam bath, and cool the flask in a beaker of water. Pour the contents of the
flask through a funnel into about 10 mL of water in a small separatory funnel,
shake the mixture, and separate the two phases. Wash the organic phase with
two 4-mL portions of 0.5 $M$ sodium bicarbonate solution and once with a 4-mL
portion of water; then dry the organic layer over anhydrous calcium chloride
pellets (about 1 g) in a dry Erlenmeyer flask. The mixture can be analyzed by
gas chromatography at this point, or the unreacted 1-chlorobutane can be removed by fractional distillation (up to bp 85°C) and the pot residue analyzed
by gas chromatography.

**Cleaning Up.** Empty the gas trap and combine the contents with all the aqueous layers and washes. Neutralize the aqueous solution with sodium carbonate and flush the solution down the drain with a large excess of water. The drying agent will be coated with chlorinated product and must therefore be disposed of in the hazardous waste container for solvent-contaminated drying agent. Any unused starting material and product must be placed in the halogenated organic waste container.

## Gas Chromatography

See Chapter 10 for information about gas chromatography. A 30 m × 0.32 mm i.d. 95% methyl/5% phenyl capillary column programmed from 35°C to 150°C at 6°/min should separate all of the dichlorobutanes as well as any trichlorobutane isomers. Alternately, a Carbowax packed column works well, although any other nonpolar phase such as methyl silicone should work as well. A typical set of packed column operating conditions would be column temperature of 100°C, a helium flow rate of 35 mL/min, a column size of 5-mm dia × 2 m, a sample size of 5 μL, and attenuation of 16. With a nonpolar column, the products are expected to come out in the order of their boiling points.

The molar amounts of each compound present in the reaction mixture are proportional to the areas under the peaks in the chromatogram. Because 1-chlorobutane is present in large excess, let this peak run off the paper but be sure to keep the four product peaks on the paper. If the chromatograph is not equipped with an integrator, determine the relative peak areas by simply cutting out the peaks with a pair of scissors and weighing them. This method of peak integration works very well and depends on the uniform thickness of paper, which results in its weight being proportional to its area. Calculate the relative percent of each product molecule and the partial rate factors relative to the primary hydrogens on carbon-4. Compare your results to those for the chlorination of 2-methylbutane, the data for which were given in Table 18.1.

**Online Study Center**

**General Resources**
Additional Experiments

## Computational Chemistry

Construct the four dichlorobutanes and minimize their energies using a molecular mechanics program. Compare the steric energies for the different conformers to determine if the global minimum for each one has been found. Record the four steric energies and then submit each energy-minimized molecule to an AM1 semi-empirical molecular orbital calculation to calculate the heat of formation. Because one would not expect a correlation to exist between the steric energies and the heats of formation, only the heats of formation can be used to compare the stabilities of the different isomers. Do the heats of formation correlate with the product distributions of the four dichlorobutanes? Does this correlation or lack thereof give information regarding the mechanism of the reaction? Remember that the isomers would need to be in equilibrium with each other so that the most stable is also the one formed in highest yield.

## QUESTIONS

1. Using the relative reactivities of the primary and secondary hydrogens in butane as given in Table 18.1 and given that the reactivity of the hydrogens of the $-CH_2Cl$ group is only 2 times that of the primary hydrogens on the methyl group, calculate the ratio of all products based on the number of each type of hydrogen and the relative reactivity.

2. Arrange the following radicals in order of increasing stability, low to high:

# Alkenes from Alcohols: Cyclohexene from Cyclohexanol

*Online Study Center*

This icon will direct you to techniques, equipment setups, and online resources at http://college.hmco.com/PIC/ williamsonMMOE5e.

**PRELAB EXERCISE:** Prepare a detailed flow sheet for the preparation of cyclohexene, indicating at each step which layer contains the desired product.

OH

$\xrightarrow{\text{H}_3\text{PO}_4}$

+ H$_2$O

**Cyclohexanol**
mp 25°C, bp 161°C
den. 0.96, MW 100.16

**Cyclohexene**
bp 83°C
den. 0.81, MW 82.14

The dehydration of cyclohexanol to cyclohexene can be accomplished by pyrolysis of the cyclic secondary alcohol with an acid catalyst at a moderate temperature or by distillation over alumina or silica gel. The procedure selected for these experiments involves catalysis by phosphoric acid; sulfuric acid is not more efficient, causes charring, and gives rise to sulfur dioxide. When a mixture of cyclohexanol and phosphoric acid is heated in a flask equipped with a fractionating column, the formation of water soon becomes evident. On further heating, the water and the cyclohexene that form distill together by the principle of steam distillation, and any higher-boiling cyclohexanol that may volatilize is returned to the flask. However, after dehydration is complete and the bulk of the product has distilled, the fractionating column remains saturated with a water-cyclohexene mixture that merely refluxes and does not distill. Hence, for the recovery of otherwise lost reaction product, a chaser solvent is added, and distillation is continued. A suitable chaser solvent is the water-immiscible, aromatic solvent toluene (bp 110°C); as it steam distills, it carries over the more volatile cyclohexene. When the total water-insoluble layer is separated, dried, and redistilled through the dried column, the chaser again drives the cyclohexene from the column; the difference in boiling points is such that a sharp separation is possible. The holdup in a metal sponge-packed column is so great (about 1.0 mL) that if a chaser solvent is not used in the procedure, the yield will be much lower.

The mechanism of this reaction involves initial rapid protonation of the hydroxyl group by the phosphoric acid:

$$
\text{(cyclohexanol)} + H_3PO_4 \rightleftharpoons \text{(protonated cyclohexanol)} + H_2PO_4^-
$$

This is followed by loss of water to give the unstable secondary carbocation, which quickly loses a proton to water to give the alkene:

$$
\text{(protonated)} \longrightarrow \text{(carbocation)} \longrightarrow \text{(cyclohexene)} + H_3O^+
$$

## EXPERIMENTS

# 1. The Preparation of Cyclohexene

**IN THIS EXPERIMENT** cyclohexanol undergoes acid-catalyzed dehydration to give cyclohexene. The product is distilled from the reaction flask along with the water generated. Toluene is used as a chaser solvent. The distillate is washed with salt solution, dried, and fractionally distilled.

**CAUTION**: Handle phosphoric acid with care because it is corrosive to tissue.

Use a black Viton connector.

Introduce 2.0 g of cyclohexanol followed by 0.5 mL of 85% phosphoric acid and a boiling chip into a 5-mL round-bottomed, long-necked flask. Add a copper sponge to the neck of the flask and shake to mix the layers. Heat will evolve. Use the arrangement for fractional distillation as shown in Figure 19.1. Note that the bulb of the thermometer must be completely below the side arm of the distilling head. Wrap the fractionating column and distilling head with glass wool or cotton. (Refer to the fractional distillation experiments in Chapter 5 for details of this technique.)

Heat the mixture gently on a sand bath and then distill until the residue in the flask has a volume of about 0.5 mL to 1.0 mL and very little distillate is being formed; note the temperature range. Let the assembly cool slightly after removing

■ FIG. 19.1
The apparatus for synthesizing
cyclohexene from cyclohexanol.
The column is packed with a
copper sponge.

Vial

Ice

30-mL
Beaker

Copper sponge

Hot sand bath

Using a chaser solvent

it from the sand, remove the thermometer briefly, and add 2 mL of toluene (the chaser solvent) into the top of the column using a Pasteur pipette. Note the amount of the upper layer in the boiling flask and distill again until the volume of this layer has been reduced by about half. Transfer the contents of the vial into a reaction tube and rinse out the vial with a little toluene. Use toluene for rinsing in subsequent operations. Wash the mixture with an equal volume of saturated sodium chloride solution, remove the aqueous layer, and then add sufficient anhydrous calcium chloride to the reaction tube so that it does not clump together. Shake the solution with the drying agent and let it dry for at least 5 min. While this is taking place, clean the distilling apparatus first with water, then ethanol, and finally a little acetone. It is absolutely essential that the apparatus be completely dry; otherwise, the product will become contaminated with whatever solvent remains in the apparatus. Before starting the distillation, note the barometric pressure and determine the reading expected given that the boiling point is 83°C at 1 atm. Transfer the dry cyclohexene solution to the distilling flask, add a boiling chip, and distill the product. At the moment the temperature starts to rise above the plateau at which the product distills, stop the distillation to avoid contamination of the

⚠
CAUTION: Cyclohexene is
extremely flammable; keep away
from open flame.

cyclohexene with toluene. A typical yield of this volatile alkene is about 1 g. Report your yield in grams and your percentage yield. Run the infrared (IR) spectrum and interpret it for purity. Look for peaks due to starting material and toluene. Gas chromatography is especially useful in the analysis of this compound because the expected impurities differ markedly in boiling point from the product.

**Cleaning Up.** The aqueous solutions (pot residues and washes) should be diluted with water and neutralized before flushing down the drain. The ethanol wash and sodium chloride solution also can be flushed down the drain, whereas the acetone wash and all toluene-containing solutions should be placed in the organic solvents container. Once dried and free of solvent, the calcium chloride can be placed in the nonhazardous solid waste container, if local regulations permit. Otherwise, the wet drying agent should be placed in the designated waste container.

# 2. The Preparation of Cyclohexene

**CAUTION**: Handle phosphoric acid with care because it is corrosive to tissue.

Using a chaser solvent

Introduce 20.0 g of cyclohexanol (technical grade), 5 mL of 85% phosphoric acid, and a boiling stone into a 100-mL round-bottomed flask and shake to mix the layers. Note the evolution of heat. Use the typical arrangement for fractional distillation shown in Figure 5.11 (on page 102) but modified by use of a bent adapter delivering into an ice-cooled test tube in a 125-mL Erlenmeyer receiver, as shown in Figure 19.2.

Note the initial effect of heating the mixture and then distill until the residue in the flask has a volume of 5–10 mL and very little distillate is being formed; note the temperature range. Then let the assembly cool slightly, remove the thermometer briefly, and pour 20 mL of toluene (the chaser solvent) into the top of the column through a long-stemmed funnel. Note the amount of the upper layer in the boiling flask and distill again until the volume of the layer has been reduced by about half. Pour the contents of the test tube into a small separatory funnel and rinse with a little chaser solvent; use this solvent for rinsing in subsequent operations. Wash the mixture with an equal volume of saturated sodium chloride solution, separate the water layer, run the upper layer into a clean flask, and add 5 g of anhydrous calcium chloride (10 × 75-mm test tube-full) to dry it. Before the final distillation, note the barometric pressure, apply any necessary thermometer corrections, and determine the reading expected for a boiling point of 83°C. Dry the boiling flask, column, and condenser; decant the dried liquid into the flask through a stemless funnel plugged with a bit of cotton; and fractionally distill, with all precautions against evaporation losses. The ring of condensate should rise very slowly as it approaches the top of the fractionating column so that the thermometer can record the true boiling point soon after distillation begins. Record both the corrected boiling point of the bulk of the cyclohexene fraction and the temperature range, which should be no more than 2°C. If the cyclohexene has been dried thoroughly, it will be clear and colorless; if wet, it will be cloudy. A typical student yield is 13.2 g. Report your yield in grams and your percent yield.

**CAUTION**: Cyclohexene is extremely flammable; keep away from open flame.

**Cleaning Up.** The aqueous solutions (pot residues and washes) should be diluted with water and neutralized before flushing down the drain with a large excess of

water. The ethanol wash and sodium chloride solution can also be flushed down the
drain; the acetone wash and all toluene-containing solutions should be placed in the
organic solvents container. The calcium chloride should be placed in the contaminated drying agents waste container or, if local regulations allow, once free of solvent, the calcium chloride can be placed in the nonhazardous solid waste container.

## Yield Calculations

See Chapter 16 for an example table of the properties of starting materials,
reagents, products, byproducts, and the limiting reagent.

Rarely do organic reactions give 100% yields of one pure product; an important objective of every experiment is to obtain the highest yield of the desired product. The microscale experiment uses 2 g of starting cyclohexanol. This corresponds to 0.02 mol because the molecular weight of cyclohexanol is 100:

$$\frac{2.0 \text{ g}}{100.16 \text{ g/mol}} = 0.020 \text{ mol} = 20 \text{ mmol}$$

If the reaction gave a 100% yield of cyclohexene, then 0.02 mol of the alkene would be produced. The weight of 0.020 mol of cyclohexene is

$$0.020 \text{ mol} \times 82.14 \text{ g/mol} = 1.64 \text{ g}$$

See Chapter 16 for another example of yield calculations.

We call the 1.64 g the theoretical yield; it could be obtained if the reaction proceeded perfectly. If the actual yield is only 0.82 g, then the reaction would be said to give a 50% yield:

$$\frac{0.82 \text{ g}}{1.64 \text{ g}} \times 100 = 50\%$$

**Online Study Center**

**General Resources**
Additional Experiments

Typical student yields are included throughout this text. These are not theoretical yields; they suggest what an average or above-average student can expect to obtain for the experiment.

## QUESTIONS

1. Assign the peaks in the $^1$H NMR spectrum of cyclohexene (Fig. 19.3) to specific groups of protons on the molecule.
2. Which product(s) would be obtained by the dehydration of 2-heptanol? of 2-methyl-1-cyclohexanol?
3. Mixing cyclohexanol with phosphoric acid is an exothermic process, whereas the production of cyclohexene is endothermic. Referring to the two chemical reactions on page 352, construct an energy diagram showing the course of this reaction. Label the diagram with the starting alcohol, the oxonium ion (the protonated alcohol), the carbocation, and the product.
4. Is it reasonable that the $^1$H spectrum of cyclohexene (Fig. 19.3) should closely resemble the $^{13}$C spectrum (Fig. 19.4)? Compare these spectra with the spectrum of cyclohexanol (Fig. 19.5).

■ **FIG. 19.3**
The $^1H$ NMR spectrum of
cyclohexene (250 MHz).

■ **FIG. 19.4**
The $^{13}C$ NMR spectrum
of cyclohexene.

■ **FIG. 19.5**
The $^{13}C$ NMR spectrum
of cyclohexanol.

# Bromination and Debromination: Purification of Cholesterol

**PRELAB EXERCISE:** Make a molecular model of 5α,6β-dibromo-
cholestan-β-ol and convince yourself that the bromine atoms in
this molecule are trans and diaxial.

The bromination of a double bond is an important and well-understood organic reaction. In the experiments in this chapter, it is employed for the very
practical purpose of purifying crude cholesterol through the process of bromination, crystallization, and then zinc dust debromination.

**Bromine**
MW 159.81
bp 59.5°C
den. 3.10

$+$ $Br_2$

CH$_3$COOH

**Zn**
**Zinc**
At Wt 65.38

**Cholesterol (1)**
**(5-Cholesten-3β-ol)**
MW 386.66, bp 149–150°C

**5α,6β-Dibromocholestan-3β-ol (3)**

The reaction involves nucleophilic attack by the alkene on bromine with the formation of a tertiary carbocation that probably has some bromonium ion character resulting from the sharing of the nonbonding electrons on bromine with the electron-deficient C-5 carbon. This ion is attacked from the backside by the bromide ion to form dibromocholesterol, with the bromine atoms in the trans and diaxial configuration, the usual result when brominating a cyclohexene.

Cholesterol isolated from natural sources contains small amounts (0.1%–3%) of 3β-cholestanol, 7-cholesten-3β-ol, and 5,7-cholestadien-3β-ol.[1] These substances are so very similar to cholesterol in solubility that their removal by crystallization is not feasible. However, complete purification can be accomplished through the sparingly soluble dibromo derivative 5α,6β-dibromocholestan-3β-ol. 3β-Cholestanol is saturated and does not react with bromine; thus, it remains in the mother liquor. 7-Cholesten-3β-ol and 5,7-cholestadien-3β-ol are dehydrogenated by bromine to dienes and trienes, respectively, that likewise remain in the mother liquor and are eliminated along with colored byproducts.

**3β-Cholestanol**

**7-Cholesten-3β-ol**

---

1. A fourth companion, cerebrosterol, or 25-hydroxycholesterol, is easily eliminated by crystallization from alcohol.

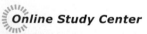

**5,7-Cholestadien-3β-ol**

The cholesterol dibromide that crystallizes from the reaction solution is collected, washed free of impurities or their dehydrogenation products, and debrominated with zinc dust, resulting in the regeneration of very pure cholesterol. Specific color tests can differentiate between pure cholesterol and tissue cholesterol purified by ordinary methods.

## EXPERIMENTS

## 1. Microscale Bromination of Cholesterol

> **IN THIS EXPERIMENT** an ether solution of cholesterol is brominated with bromine in acetic acid. The product crystallizes and is isolated by filtration. Impurities in the starting material do not crystallize and are removed in the washing process.

⚠️ **CAUTION**: Be extremely careful to avoid getting the solution of bromine and acetic acid on the skin. Carry out the reaction in the hood and wear disposable gloves for maximum safety.

**Online Study Center**

Video: Microscale Filtration on the Hirsch Funnel

In a 10 × 75 mm test tube, dissolve 100 mg of gallstone cholesterol or of commercial cholesterol (content of 7-cholesten-3β-ol about 0.6%) in 0.7 mL of ether by gentle warming. Then add 0.5 mL of a solution of bromine and sodium acetate in acetic acid.[2] Cholesterol dibromide begins to crystallize in 1–2 min. Cool in an ice bath and stir the crystalline paste with a stirring rod for about 10 min to ensure complete crystallization; at the same time, cool a mixture of 0.3 mL of ether and 0.7 mL of acetic acid in ice. Then collect the crystals on a Hirsch funnel and wash with the iced ether–acetic acid solution to remove the yellow mother liquor. A short test tube is used to facilitate the transfer of the crystalline paste. This material is not easy to filter on a very small scale because the crystals are tiny. Alternatively, carry out the bromination in a reaction tube and isolate the product with a Wilfilter (Fig. 20.1). Finally, wash the crystals on the filter with a few drops of ethanol, while continuing to apply suction, and transfer the white solid without drying it (dry weight 120 mg) to a reaction tube.

**Cleaning Up.** The filtrate from this reaction contains halogenated material and must be placed in the halogenated organic waste container.

---

2. Weigh a 125-mL Erlenmeyer flask on a balance placed in the hood, add 4.5 g bromine using a Pasteur pipette (avoid breathing the vapor), and add 50 mL acetic acid and 0.4 g sodium acetate (anhydrous).

**■ FIG. 20.1**
A Wilfilter filtration apparatus.

Crystals

Wilfilter

Filtrate

## 2. Macroscale Bromination of Cholesterol

In a 25-mL Erlenmeyer flask dissolve 1 g of commercial cholesterol (content of 7-cholesten-3β-ol about 0.6%) in 7 mL of ether by gentle warming and, with a pipette fitted with a pipetter or a plastic syringe, or from a burette in the hood, add 5 mL of a solution of bromine and sodium acetate in acetic acid.[3] Cholesterol dibromide begins to crystallize in 1–2 min. Cool in an ice bath and stir the crystalline paste with a stirring rod for about 10 min to ensure complete crystallization; at the same time, cool a mixture of 3 mL of ether and 7 mL of acetic acid in ice. Then collect the crystals on a Hirsch funnel and wash with the iced ether–acetic acid solution to remove the yellow mother liquor. Finally, wash with a little methanol, while continuing to apply suction,

**CAUTION:** Be extremely careful to avoid getting the solution of bromine and acetic acid on the skin. Carry out the reaction in the hood and wear disposable gloves for maximum safety.

---

3. *See* footnote 2.

and transfer the white solid without drying it (dry weight 1.2 g) to a 50-mL Erlenmeyer flask.

**Cleaning Up.** The filtrate from this reaction contains halogenated material and so must be placed in the halogenated organic waste container.

## 3. Microscale Zinc Dust Debromination

> **IN THIS EXPERIMENT** the dibromo compound from Experiment 1 is dissolved in ether and acetic acid and debrominated with zinc dust to regenerate the cholesterol. The ether solution, after separation from any zinc and zinc acetate, is washed with water, base, and salt solution. The solution is dried over calcium chloride and then separated from the drying agent. Methanol is added, and the solution is evaporated until the pure product begins to crystallize.

Add 2 mL of ether, 0.5 mL of acetic acid, and 40 mg of zinc dust[4] to the dibromocholesterol from Experiment 1 and mix the contents. In about 3 min, the dibromide dissolves; after 5–10 min of mixing (a magnetic stirrer could be used), zinc acetate usually separates to form a white precipitate (the dilution sometimes is such that no separation occurs). Stir for an additional 5 min and then add water (no more than 1 drop) until any solid present (zinc acetate) dissolves to make a clear solution. Decant the solution from the zinc into a reaction tube, wash the ethereal solution twice with 2 mL of water and then with 1 mL of 3 $M$ sodium hydroxide and 1 mL of saturated sodium chloride solution. Then shake the ether solution with anhydrous calcium chloride pellets, pipette off the dry solution, wash the drying agent with a little ether, add 1 mL methanol (and a boiling stick), and evaporate the solution to the point where most of the ether is removed and the purified cholesterol begins to crystallize. Remove the solution from the heat; let crystallization proceed at room temperature and then in an ice bath. Collect the crystals on a Hirsch funnel or use a Wilfilter (Fig. 20.1). Wash the crystals with cool methanol; you should obtain 60 to 70 mg of cholesterol (mp 149°C–150°C).

**Cleaning Up.** The unreacted zinc dust can be discarded in the nonhazardous solid waste container after it has been allowed to dry and then sit exposed to the air on a watch glass for about 30 min. Sometimes the zinc dust at the end of this reaction will get quite hot as it air oxidizes. The aqueous and acetic acid solutions after neutralization can be flushed down the drain. The organic filtrates containing ether and methanol should be placed in the organic solvents container. If local regulations allow, evaporate any residual solvent from the drying agents in the hood and place the dried solid in the nonhazardous waste container. Otherwise, place the wet drying agent in a waste container designated for this purpose.

*Online Study Center*

Video: Microscale Filtration on the Hirsch Funnel; Photo: Use of the Wilfilter

Collect the product on a Wilfilter

---

4. If the reaction is slow, add more zinc dust or place the tube or flask in an ultrasonic cleaning bath, which will activate the zinc. The amount specified is adequate if material is taken from a freshly opened bottle, but zinc dust deteriorates on exposure to air.

# 4. Macroscale Zinc Dust Debromination

To the flask containing dibromocholesterol, add 20 mL of ether, 5 mL of acetic acid, and 0.2 g of zinc dust and swirl.[5] In about 3 min the dibromide dissolves; after swirling for 5–10 min, zinc acetate usually separates to form a white precipitate (the dilution sometimes is such that no separation occurs). Stir for an additional 5 min and then add water by drops (no more than 0.5 mL) until any solid present (zinc acetate) dissolves to make a clear solution. Decant the solution from the zinc into a separatory funnel and wash the ethereal solution twice with water and then with 10% sodium hydroxide (to remove traces of acetic acid). Then shake the ether solution with an equal volume of saturated sodium chloride solution to reduce the water content, dry the ether with anhydrous calcium chloride, remove the drying agent, add 10 mL of methanol (and a boiling stone), and evaporate the solution on a steam bath to the point where most of the ether is removed and the purified cholesterol begins to crystallize. Remove the solution from the steam bath; let crystallization proceed at room temperature and then in an ice bath, collect the crystals, and then wash them with cold methanol; you should obtain 0.6–0.7 g of cholesterol (mp 149°C–150°C). Figures 20.2 and 20.3 show, respectively, proton and carbon NMR spectra of cholesterol.

**Cleaning Up.** The unreacted zinc dust can be discarded in the nonhazardous solid waste container after it has been allowed to dry and then sit exposed to the air on a watch glass for about 30 min. Sometimes the zinc dust at the end of this reaction will get quite hot as it air oxidizes. The aqueous and acetic acid solutions after neutralization can be flushed down the drain. The organic filtrates containing ether and methanol should be placed in the organic solvents container. If local regulations allow, evaporate any residual solvent from the drying agents in the hood and place the dried solid in the nonhazardous waste container. Otherwise, place the wet drying agent in a waste container designated for this purpose.

**■ FIG. 20.2**
The $^1$H NMR spectrum of cholesterol (60 MHz).

PPM (δ)

5. *See* footnote 4.

■ **FIG. 20.3**
The ¹³C NMR spectrum of
cholesterol (22.6 MHz).

■ **FIG. 20.4**
The IR spectrum of cholesterol
(KBr disk).

**Online Study Center**

**General Resources**
Additional Experiments

## QUESTIONS

1. What is the purpose of the acetic acid in this reaction?
2. Why might old zinc dust not react in the debromination reaction?
3. Why does acetate ion not attack the intermediate bromonium ion to give the 5-acetoxy-6-bromo compound in the bromination of cholesterol?
4. Assign the 3400 cm$^{-1}$ peak in the infrared (IR) spectrum of cholesterol (Fig. 20.4).

# Dichlorocarbene

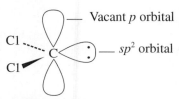

Cl ---- C ⟨ — *sp²* orbital
Cl

— Vacant *p* orbital

**Dichlorocarbene**

**Like a carbanion—electrons in a hybrid orbital. Like a carbocation—vacant *p*-orbital.**

$CH_3OCH_2CH_2OCH_2CH_2OCH_3$
**2-Methoxyethyl ether
Diglyme**
MW 134.17, bp 161°C
miscible with water

**PRELAB EXERCISE:** Propose a synthesis of dichlorocarbene using the principles of phase-transfer catalysis. Why is phase-transfer catalysis particularly suited to this reaction?

Dichlorocarbene is a highly reactive intermediate of bivalent carbon with only six valence electrons around the carbon. It is electrically neutral and a powerful electrophile. As such, it reacts with alkenes, forming cyclopropane derivatives by *cis*-addition to the double bond.

There are a dozen or so ways by which dichlorocarbene can be generated. In this experiment the thermal decomposition of anhydrous sodium trichloroacetate in an aprotic solvent (diglyme) in the presence of *cis, cis*-1,5-cyclooctadiene generates dichlorocarbene to give 5,5,10,10-tetrachlorotricyclo[7.1.0.0⁴,⁶]decane (see page 366).

## The Thermal Decomposition of Sodium Trichloroacetate; Reaction of Dichlorocarbene with 1,5-Cyclooctadiene

The thermal decomposition of sodium trichloroacetate initially gives the trichloromethyl anion, which, in the absence of a proton-donating solvent, gives dichlorocarbene by loss of a chloride ion (equation 1). In the presence of a proton-donating solvent (or moisture), this anion gives chloroform (equation 2).

The conventional method for carrying out the reaction is to add the salt portionwise, over 1–2 h, to a magnetically stirred solution of the olefin in diethylene glycol dimethyl ether (2-methoxyethyl ether; diglyme) at a temperature maintained in a bath at 120°C. Under these conditions the reaction mixture becomes almost black, isolation of a pure product is tedious, and the yield is low.

**Sodium trichloroacetate**
MW 185.39

**Trichloro-methyl anion**

**Dichlorocarbene** (1)

**Chloroform**
MW 119.39, bp 61°C (2)

*cis,cis*-**1,5-Cyclo-octadiene**
MW 108.14
bp 149–150°C

**1**
*cis*, mp 176°C

**2**
*trans*, mp 230°C dec

**5,5,10,10-Tetrachlorotricyclo[7.1.0.0$^{4,6}$]decane**
MW 274.03

(3)

**Tetrachloroethylene**
bp 121°C, den. 1.623

Tetrachloroethylene, a nonflammable solvent used in the dry-cleaning industry, boils at 121°C and is relatively inert toward electrophilic dichlorocarbene. On generation of dichlorocarbene from either chloroform or sodium trichloroacetate in the presence of tetrachloroethylene, the yield of hexachlorocyclopropane (mp 104°C) is only 0.2%–10%. The first idea for simplifying the procedure[1] was to use tetrachloroethylene to control the temperature to the desired range, but sodium trichloroacetate is insoluble in this solvent, and no reaction occurs. Diglyme, or an equivalent, is required to provide some solubility. The reaction proceeds better in a 7:10 mixture of diglyme to tetrachloroethylene than in diglyme alone, but because the salt dissolves rapidly in this mixture, it has to be added in several small portions; the reaction mixture becomes very dark. The situation is vastly improved by the simple expedient of decreasing the amount of diglyme to a 2.5:10 ratio. The salt is so sparingly soluble in this mixture that it can be added at the start of the experiment; it will dissolve slowly as the reaction proceeds. The boiling and evolution of carbon dioxide provide adequate stirring, the mixture can be left unattended, and the little color that develops is eliminated by washing the crude product with methanol.

The main reaction product, 5,5,10,10-tetrachlorotricyclo[7.1.0.0$^{4,6}$]decane, crystallizes from ethyl acetate in beautiful prismatic needles (mp 174°C–175°C), and this was the sole product encountered in runs made in the solvent

1. Fieser, L. F.; Sachs, D. H. *J. Org. Chem.* **1964**, *29*, 1113.

mixture recommended. In earlier runs made in diglyme with the manual control of temperature, the ethyl acetate mother liquor material on repeated crystallization from toluene afforded small amounts of a second isomer (mp 230°C, dec). Elemental analyses indicated a pair of *cis-trans* isomers (**1** and **2** on page 366) had been formed, and both gave negative permanganate tests. To distinguish between them, the junior author of the paper undertook an X-ray analysis that showed that the lower melting isomer is *cis* and the higher melting isomer is *trans*.

**X-ray analysis**

## 1. Sodium Trichloroacetate

> **IN THIS EXPERIMENT** sodium trichloroacetate is prepared by exactly neutralizing the acid with sodium hydroxide. The resulting solution is evaporated to complete dryness.

Use caution in handling the corrosive trichloroacetic acid. Carry out the procedure in the hood.

Sodium trichloroacetate will decompose completely to sodium chloride in a drying oven at 150°C.

Use commercial sodium trichloroacetate (dry) or prepare it as follows. Place 6.4 g of trichloroacetic acid in a 125-mL Erlenmeyer flask, dissolve 1.6 g of sodium hydroxide pellets in 6 mL of water in a 25-mL Erlenmeyer flask, cool the solution thoroughly in an ice bath, and swirl the flask containing the acid in the ice bath while slowly dropping in about nine-tenths of the alkali solution. Then add 1 drop of 0.04% Bromocresol green solution to produce a faint yellow color, visible when the flask is dried and placed on white paper. With a Pasteur pipette, titrate the solution to an endpoint where a single drop produces a change from yellow to blue. If the endpoint is overshot, add a few crystals of acid and titrate more carefully. Close the flask with a rubber stopper, connect to an aspirator, place the flask within the rings of the steam bath, and wrap a towel around both for maximum heat. Turn on the water at full force for maximum suction. The evaporation requires no further attention and should be complete in 15–20 min. When you have an apparently dry white solid, scrape it out with a spatula and break up the large lumps. If you see any evidence of moisture, or if the weight exceeds the calculated value, dry it to constant weight under vacuum.

Be sure that the salt is completely dry.

## 2. Dichlorocarbene Reaction

> **IN THIS EXPERIMENT** absolutely dry sodium trichloroacetate is refluxed in a special solvent mixture with a diene. The trichloroacetate decomposes to give sodium chloride (a precipitate), dichlorocarbene (which reacts with the diene), and carbon dioxide (the evolution of which is used to follow the progress of the reaction). At the end of the reaction, the solvent is removed by steam distillation, and the solid product is extracted into dichloromethane, which is dried and evaporated. Crude product is washed with methanol and recrystallized from ethyl acetate.

■ **FIG. 21.1**
A carbon dioxide bubbler, reflux
air condenser, and reaction flask
for dichlorocarbene synthesis.
See Figure 18.2 (on page 347)
for the technique of threading
a polyethylene tube through
a septum.

Polyethylene
tubing

Rubber
septum

0.5 mL
Diglyme

Hot sand bath

*Online Study Center*

Photo: Placing a Polyethylene
Tube through a Septum

**CAUTION:** Tetrachloroethylene
is a suspected cancer-causing
agent and mutagen. Handle in
the hood.

Test the diglyme for peroxides as
described in Chapter 2.

Place 0.91 g of dry sodium trichloroacetate in a 5-mL round-bottomed
short-necked flask mounted over a hot sand bath. Add a magnetic stirring bar,
1 mL of tetrachloroethylene, 0.25 mL of diglyme, and 0.25 mL (220 mg) of
*cis,cis*-1,5-cyclooctadiene.[2] Attach the empty distilling column as an air condenser and cap it with a septum bearing polyethylene tubing, which in turn dips
below the surface of 0.5 mL of diglyme contained in a reaction tube (Fig. 21.1).
This bubbler will show when the evolution of carbon dioxide ceases. The solvent should be the same as that of the reaction mixture, should there be a suckback. Heat the reaction mixture to boiling, note the time, and reflux gently until
the reaction is complete. You will notice foaming, due to liberated carbon dioxide, and separation of finely divided sodium chloride. Inspection of the bottom of the flask will show lumps of undissolved sodium trichloroacetate, which
gradually disappear.

When the reaction is complete, add 3.5 mL of water to the hot mixture,
connect an adapter and a distilling head to the flask (Fig. 21.2), add a boiling chip,
heat the mixture to boiling, and steam distill until the tetrachloroethylene is
eliminated—the product will separate in the flask as an oil or semisolid. If necessary, inject more water through the septum to complete the steam distillation.

---

2. Diglyme, tetrachloroethylene, and 1,5-cyclooctadiene should be stored over Linde 5A molecular sieves or
anhydrous magnesium sulfate to guarantee that the reagents are dry. The reaction will not work if any reagent is
wet. For safety reasons, dispense tetrachloroethylene in the hood. Do the same with the diene; it has a bad odor.

**■ FIG. 21.2**
A microscale steam distillation
apparatus. Add water via a syringe.

*Online Study Center*

Video: Steam Distillation
Apparatus

Ice

Hot sand bath

Cool the flask to room temperature, transfer the contents to a reaction tube using dichloromethane, and remove the aqueous layer. Dry the solution over anhydrous calcium chloride pellets, remove the solvent, wash the drying agent with more dichloromethane, and evaporate the solvent in a reaction tube. The weight of the crude product should be about 400 mg.

Cover the crude product with methanol, break up the cake with a flattened stirring rod, and crush the lumps. Cool in ice, collect the product crystals, and wash the product with methanol. The yield of colorless, or nearly colorless, 5,5,10,10-tetrachlorotricyclo[7.1.0.0$^{4,6}$]decane is about 300 mg. This material (mp 174°C–175°C) consists almost entirely of the *cis* isomer. Dissolve it in ethyl acetate (1.5–2 mL) and let the solution stand undisturbed for crystallization at room temperature. The pure *cis* isomer separates in large, prismatic needles (mp 175°C–176°C).

*Online Study Center*

Video: Filtration of Crystals
Using the Pasteur Pipette

**Cleaning Up.** The aqueous layer should be neutralized with hydrochloric acid, diluted with water, and flushed down the drain. Organic layers that contain halogenated material must be placed in the halogenated organic solvents container. If local regulations allow, evaporate any residual solvent from the drying agents in the hood and place the dried solid in the nonhazardous waste container. Otherwise, place the wet drying agent in a waste container designated for this purpose. Methanol and ethyl acetate used in crystallization can be placed in the organic solvents container.

# 3. Dichlorocarbene Reaction

Place 7.1 g of dry sodium trichloroacetate in a 100-mL round-bottomed flask mounted over a flask heater and add 10 mL of tetrachloroethylene, 2.5 mL of diglyme, and 2.5 mL (2.2 g) of *cis,cis*-1,5-cyclooctadiene.[3] Attach a reflux condenser and, in the top opening of the condenser, insert a rubber stopper carrying a glass tube connected by rubber tubing to a short section of glass tube, which can be inserted below the surface of 2–3 mL of tetrachloroethylene in a 20 × 150-mm test tube mounted at a suitable level. This bubbler will show when the evolution of carbon dioxide ceases; the solvent should be the same as that of the reaction mixture, should there be a suckback. Heat to boiling, note the time, and reflux gently until the reaction is complete. You will notice foaming, due to liberated carbon dioxide, and separation of finely divided sodium chloride. Inspection of the bottom of the flask will show lumps of undissolved sodium trichloroacetate, which gradually disappear. A large flask is specified because it will serve later for the removal of tetrachloroethylene by steam distillation. Make advance preparations for this operation.

When the reaction is complete, add 40 mL of water to the hot mixture, heat over a hot electric flask heater, and steam distill until the tetrachloroethylene is eliminated—the product separates as an oil or semisolid. Cool the flask to room temperature, decant the supernatant liquid into a separatory funnel, and extract with dichloromethane. Run the extract into the reaction flask and use enough additional dichloromethane to dissolve the product; use a Pasteur pipette to rinse down material adhering to the adapter. Run the dichloromethane solution of the product into an Erlenmeyer flask through a cone of anhydrous calcium chloride pellets in a funnel and evaporate the solvent (bp 41°C). The residue is a tan or brown solid (14 g). Cover it with methanol, break up the cake with a flattened stirring rod, and crush the lumps. Cool in ice, collect the product, and wash with methanol. The yield of colorless, or nearly colorless, 5,5,10,10-tetrachlorotricyclo[7.1.0.0$^{4,6}$]decane is 3.3 g. This material (mp 174°C–175°C) consists almost entirely of the *cis* isomer. Dissolve it in ethyl acetate (15–20 mL) and let the solution stand undisturbed for crystallization at room temperature. The pure *cis* isomer separates in large,

Reflux time: about 75 min

Easy workup

---

3. Diglyme, tetrachloroethylene, and 1,5-cyclooctadiene should be stored over Linde 5A molecular sieves or anhydrous magnesium sulfate to guarantee that the reagents are dry. The reaction will not work if any reagent is wet. For safety reasons, dispense tetrachloroethylene in the hood. Do the same with the diene; it has a bad odor.

■ **FIG. 21.3**
The $^1$H NMR spectrum
of 5,5,10,10-tetrachloro-
tricyclo[7.1.0.0$^{4,6}$]decane
(60 MHz).

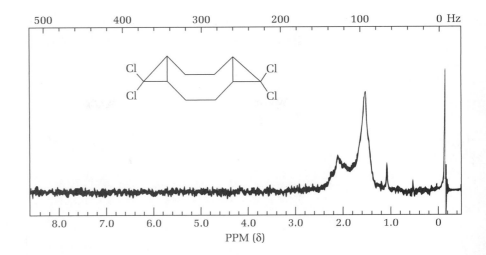

prismatic needles (mp 175°C–176°C). The proton NMR spectrum is shown in Figure 21.3.

**Cleaning Up.** The aqueous layer can be diluted with water and flushed down the drain. Organic layers, which contain halogenated material, must be placed in the halogenated organic solvents container. If local regulations allow, evaporate any residual solvent from the drying agents in the hood and place the dried solid in the nonhazardous waste container. Otherwise, place the wet drying agent in a waste container designated for this purpose. Methanol and ethyl acetate used in crystallization can be placed in the organic solvents container.

# 4. Phase-Transfer Catalysis–Reaction of Dichlorocarbene with Cyclohexene

Ordinarily, water must be scrupulously excluded from carbene-generating reactions, as mentioned earlier. Both the intermediate trichloromethyl anion (**3**) and dichlorocarbene (**4**) react with water. But in the presence of a phase-transfer catalyst (a quaternary ammonium salt such as benzyltriethylammonium chloride), it is possible to carry out the reaction in aqueous medium. In the presence of a catalytic amount of the quaternary ammonium salt, chloroform, 50% aqueous sodium hydroxide, and the olefin are stirred for a few minutes to produce an emulsion. After the exothermic reaction is complete (about 30 min), the product is isolated.

Both benzyltriethylammonium chloride (**1**) and benzyltriethylammonium hydroxide (**2**) partition between the aqueous and organic phases. In the aqueous phase, the quaternary ammonium chloride (**1**) reacts with concentrated hydroxide to give the quaternary ammonium hydroxide (**2**). In the chloroform phase, **2** reacts with chloroform to give the trichloromethyl anion (**3**), which eliminates chloride ion to give dichlorocarbene, :CCl$_2$ (**4**), and **1**. The carbene (**4**) reacts with the alkene to give the product (**5**).

*Aqueous phase*
$$\overset{Cl^-}{(C_2H_5)_3\overset{+}{N}CH_2C_6H_5} + OH^- \rightleftharpoons \overset{OH^-}{(C_2H_5)_3\overset{+}{N}CH_2C_6H_5} + Cl^-$$
**1**                                **2**

*CHCl₃ phase*
$$\overset{Cl^-}{(C_2H_5)_3\overset{+}{N}CH_2C_6H_5} \qquad\qquad \overset{OH^-}{(C_2H_5)_3\overset{+}{N}CH_2C_6H_5}$$

$$:CCl_2 \longrightarrow CCl_3^- \quad \mathbf{3} \longleftarrow + \; CHCl_3$$
**4**

$$\text{cyclohexene} + :CCl_2 \longrightarrow \text{product}$$
**4**                        **5**

## 🧪 *Macroscale Procedure*

> **IN THIS EXPERIMENT** a mixture of cyclohexene, chloroform, some cyclohexanol (see "Emulsion Formation"), and a catalytic amount of a quaternary ammonium chloride ($R_4N^+Cl^-$) are stirred to produce a thick emulsion and an exothermic reaction. In a phase-transfer reaction dichlorocarbene is formed and reacts with the alkene to form the product. The product is extracted into ether, washed, dried, and distilled to give the pure product (**5**).

**CAUTION**: Chloroform is a carcinogen and should be handled in the hood. Avoid all contact with the skin.

⚠

Fifty percent sodium hydroxide solution is very corrosive. Handle with care.

To a mixture of 8.2 g of cyclohexene, 12.0 g of chloroform and 20 mL of 50% aqueous sodium hydroxide in a 125-mL Erlenmeyer flask containing a thermometer, add 0.2 g of benzyltriethylammonium chloride.[4] Swirl the mixture to produce a thick emulsion. The temperature of the reaction will rise gradually at first and then markedly accelerate. As it approaches 60°C, prepare to immerse the flask in an ice bath. With the flask alternately in and out of the ice bath, stir the thick paste and maintain the temperature between 50°C and 60°C. After the exothermic reaction is complete (about 10 min), allow the mixture to cool spontaneously to 35°C and then dilute with 50 mL of water. Separate the layers (test to determine which is the organic layer), extract the aqueous layer once with 10 mL of ether, and wash the combined organic layer and ether extract once with 20 mL of water. Dry the cloudy organic layer by shaking it with anhydrous calcium chloride pellets until the liquid is clear. Decant into a 50-mL round-bottomed flask and distill the mixture (with a boiling chip), first from a steam bath to remove the ether and some unreacted

4. Many other quaternary ammonium salts, commercially available, work equally well (e.g., cetyltrimethyl-ammonium bromide). Fabric softeners also work. *See* Lee, A. W. M.; Yip, W. C. *J. Chem. Educ.* **1991**, *68* (1), 69–70.

cyclohexene and chloroform, then over a very hot sand bath or a free flame. Collect 7,7-dichlorobicyclo[4.1.0]heptane over the range 195°C–200°C. The yield is about 9 g. A purer product will result if the distillation is carried out under reduced pressure.

## Emulsion Formation

The success of this reaction is critically dependent on forming a thick emulsion, for only then will the surface between the two phases be large enough for complete reaction to occur. James Wilbur at Southwest Missouri State University found that when freshly distilled, pure reagents were used, no exothermic reaction occurred, and very little or no dichloronorcarane was formed. We have subsequently found that old, impure cyclohexene gives very good yields of product. Cyclohexene is notorious for forming peroxides at the allylic position. Under the reaction conditions this peroxide will be converted to the ketone:

$$\text{cyclohexene} \xrightarrow{O_2,\ h\nu} \text{(OOH)} \xrightarrow{OH^-} \text{(O)}$$

Apparently, small amounts of this ketone are responsible for the formation of stable emulsions. This appears to be one of the rare cases when impure reagents lead to successful completion of a reaction. If only pure reagents are available, then the reaction mixture must be stirred vigorously overnight using a magnetic stirrer. Alternatively, a small quantity of cyclohexene hydroperoxide can be prepared by pulling air through the alkene for a few hours, employing an aspirator, and using this impure reagent, or—as Wilbur found—by adding 1 mL of cyclohexanol to the reaction mixture. Detergents such as sodium lauryl sulfate do not increase the rate of reaction or emulsion formation.

**Cleaning Up.** Organic layers go in the organic solvents container. The aqueous layer should be diluted with water and flushed down the drain. If local regulations allow, evaporate any residual solvent from the drying agents in the hood and place the dried solid in the nonhazardous waste container. Otherwise, place the wet drying agent in a waste container designated for this purpose. The pot residue from the distillation goes in the halogenated organic solvents container.

## Computational Chemistry

Construct models of the *cis-* and *trans-*dichlorocarbene adducts of cyclooctadiene, **1** and **2**, using a molecular mechanics program. Carry out energy minimization and note the steric energies of each isomer. Then carry out AM1 semiempirical molecular orbital calculations on each isomer to determine their heats of formation. Do these calculations confirm that the predominant product, the *cis-*isomer, is also the most stable? Would you have predicted that the *cis-*isomer would be the most stable by simple inspection of the energy-minimized conformers?

***Online Study Center***

**General Resources**
Additional Experiments

**■ FIG. 21.4**
The $^1$H NMR spectrum of
7,7-dichlorobicyclo[4.1.0]heptane
(250 MHz).

## QUESTIONS

**1.** If water is not rigorously excluded from sodium trichloroacetate dichlorocarbene synthesis, what side reaction occurs?

**2.** Why is vigorous stirring or emulsion formation necessary in a phase-transfer reaction?

**3.** Assign the three groups of peaks in the NMR spectrum of 7,7-dichlorobicyclo[4.1.0]heptane (Fig. 21.4).

# Oxidation: Cyclohexanol to Cyclohexanone; Cyclohexanone to Adipic Acid

**Online Study Center**

This icon will direct you to techniques, equipment setups, and online resources at http://college.hmco.com/PIC/ williamsonMMOE5e.

**PRELAB EXERCISE:** Write balanced equations for the dichromate and hypochlorite oxidations of cyclohexanol to cyclohexanone and for the permanganate oxidation of cyclohexanone to adipic acid.

**Cyclohexanol**
bp 161.5°C, den. 0.96
MW 100.16

**Cyclohexanone**
bp 157°C, den. 0.95
MW 98.14
solubility 1.5 g/100 mL $H_2O^{10°}$

**Adipic acid**
mp 153°C, MW 146.14
solubility 1.4 g/100 g $H_2O^{15°}$

PCC or $Na_2Cr_2O_7$/ HOAc or NaOCl

$OH^-$

$HNO_3$ or $KMnO_4$

The oxidation of a secondary alcohol to a ketone is accomplished by many different oxidizing agents, including sodium dichromate, pyridinium chlorochromate, and sodium hypochlorite (household bleach). The ketone can be oxidized further to the dicarboxylic acid, producing adipic acid. Both of these oxidations can be carried out by the permanganate ion to give the diacid. Nitric acid is a powerful oxidizing agent that can oxidize cyclohexane, cyclohexene, cyclohexanol, or cyclohexanone to adipic acid.

Dichromate oxidation

## The Oxidation of an Alcohol to a Ketone

The dichromate mechanism of oxidizing an alcohol to a ketone appears to be the following:

$$H_2O + Cr_2O_7^{2-} \rightleftharpoons 2\ HCrO_4^-$$

A number of intermediate valence states of chromium are involved in this reaction—the orange $Cr^{6+}$ ion is ultimately reduced to the green $Cr^{3+}$ ion. The course of the oxidation can be followed by these color changes.

In microscale Experiment 1, cyclohexanol is oxidized to cyclohexanone using pyridinium chlorochromate in dichloromethane. The progress of the reaction can be followed by thin-layer chromatography (TLC). In Experiment 2, the macroscale version of this reaction is carried out using sodium dichromate in acetic acid because the reagents are less expensive, the reaction is faster, and much less solvent is required.

Chromium(VI) is probably the most widely used and versatile laboratory oxidizing agent; it is used in a number of different forms to carry out selective oxidations in this text. From an environmental standpoint, however, it is far from ideal. The inhalation of the dust from insoluble Cr(VI) compounds may cause cancer of the respiratory system. The product of the reaction [Cr(III)] should not be flushed down the drain because it is toxic to aquatic life at extremely low concentrations. Therefore, as stated in the "Cleaning Up" section of Experiments 1 and 2, the Cr(III) must be precipitated as insoluble $Cr(OH)_3$, and this material is considered a hazardous waste.

Sodium hypochlorite oxidation

Experiments 3 and 4 use an alternative oxidant for secondary alcohols that is just as efficient and much safer from an environmental standpoint: 5.25% (0.75 *M*) sodium hypochlorite solution, which is available in the grocery store as household bleach.[1] The mechanism of the reaction is not clear. It is not a free radical reaction,

1. Mohrig, J.; Nienhuis, D. M.; Linck, C. F.; C. Van Zoeren, C.; Fox, B. G. *J. Chem. Educ.* **1985**, *62*, 519.

the reaction is much faster in acid than in base, elemental chlorine is presumably the oxidant, and hypochlorous acid must be present. It may form an intermediate alkyl hypochlorite ester, which by an $E_2$ elimination gives the ketone and chloride ion.

Excess hypochlorite is easily destroyed with bisulfite; the final product is the chloride ion, which is far less toxic to the environment than Cr(III).

## The Oxidation of a Ketone to a Carboxylic Acid

Nitric acid oxidation

In microscale Experiment 5, nitric acid is the oxidant. The balanced equation for the oxidation of cyclohexanone to adipic acid is as follows:

In this reaction nitric acid is reduced to nitric oxide.

Permanganate oxidation

Experiment 6 involves the permanganate oxidation of a ketone to a dicarboxylic acid. The reaction can be followed as the bright purple permanganate solution reacts to give a brown precipitate of manganese dioxide. A possible mechanism for this oxidation starts by the reaction of $MnO_4^-$ with the enol form of the ketone and continues as shown here:

$$3 \text{ HMnO}_4^{2-} + \text{H}_2\text{O} \longrightarrow 2 \text{ MnO}_2 + \text{MnO}_4^- + 5 \text{ OH}^-$$

## EXPERIMENTS

# 1. Cyclohexanone from Cyclohexanol

> **IN THIS EXPERIMENT** an alcohol is dissolved in dichloromethane and oxidized at room temperature to a ketone with a $Cr^{6+}$ reagent: pyridinium chlorochromate (PCC). The ketone is isolated by simply evaporating the solvent from the filtered reaction mixture.

Pyridinium chlorochromate,
PCC
MW 215.56

Cyclohexanol
bp 161.5°C, den. 0.96
MW 100.16

Cyclohexanone
bp 157°C, den. 0.95
MW 98.14
solubility $1.5 \, g/100 \, mL \, H_2O^{10°}$

**⚠ CAUTION:** Handle open containers of dichloromethane in the hood; it is a suspected cancer-causing agent.

**Online Study Center**

Videos: Microscale Filtration on the Hirsch Funnel, Filtration of Crystals Using the Pasteur Pipette; Photo: Filtration Using a Pasteur Pipette

Using Celite will speed the filtration.

To a 10-mL Erlenmeyer flask add 0.62 g of finely powdered pyridinium chlorochromate (*see* Chapter 23 for its preparation), 4 mL of dichloromethane, and 250 mg of cyclohexanol. The mixture is stirred magnetically or shaken over a period of days (a week does no harm) until TLC on silica gel indicates that the reaction is complete. Elute the TLC plates with dichloromethane and visualize with iodine.

At the end of the reaction period, remove the chromium salts by filtration on a Hirsch funnel or, if the salts are coarse enough, by removing the solvent with a Pasteur pipette. Wash the chromium salts with a few drops of dichloromethane and evaporate the solvent under a gentle stream of nitrogen or air to leave the product—cyclohexanone. Confirm the structure of this product by obtaining an infrared (IR) spectrum of the pure liquid between sodium chloride or silver chloride plates. Look for a band at 3600 to 3400 $cm^{-1}$, which is indicative of unoxidized alcohol (or water if you are not careful).

**Cleaning Up.** Place used dichloromethane in the halogenated organic waste container. The chromium salts should be dissolved in hydrochloric acid and diluted to <5%. The solution, which should have a pH < 3, should then be treated with a 50% excess of sodium bisulfite to reduce the orange dichromate ion to the green chromic ion. Make the solution basic with ammonium hydroxide to precipitate chromium as the hydroxide. This precipitate is collected on filter paper, which is placed in the hazardous solid waste container for heavy metals. The filtrate from this latter treatment can be flushed down the drain.

## 2. Cyclohexanone from Cyclohexanol

In a 50-mL Erlenmeyer flask, dissolve 7.5 g of sodium dichromate dihydrate in 12.5 mL of acetic acid by swirling the mixture on a hot plate; then cool the solution with ice to 15°C. In a second Erlenmeyer flask chill a mixture of 7.5 g of cyclohexanol and 5 mL of acetic acid in ice. After the first solution is cooled to 15°C, transfer the thermometer and adjust the temperature in the second flask to 15°C. Wipe the flask containing the dichromate solution, pour the solution into the cyclohexanol–acetic acid mixture, rinse the flask with a little acetic acid, note the time, and remove the initially light orange solution from the ice bath. Keep the ice bath ready for use when required.[2] The exothermic reaction that is soon evident can get out of hand unless controlled. When the temperature rises to 60°C, cool in ice just enough to prevent a further rise and then, by intermittent brief cooling, keep the temperature close to 60°C for 15 min. No further cooling is needed, but the flask should be swirled occasionally and the temperature monitored. The usual maximal temperature is 65°C (25–30 min). When the temperature begins to drop and the solution becomes pure green, the reaction is complete. Allow 5–10 min more reaction time; then pour the green solution into a 100-mL round-bottomed flask, rinse the Erlenmeyer flask with

**CAUTION:** The dust from $Cr^{6+}$ salts is toxic and a suspected carcinogen; weigh in the hood.

Reaction time: 45 min

**■ FIG. 22.1**
A steam distillation apparatus.

Wire

Steam

Boiling chip

Adapter

Keck clamp

Thermowell

Keck clamp

To drain

Screw clamp

---

2. When the acetic acid solutions of cyclohexanol and dichromate were mixed at 25°C rather than at 15°C, the yield of crude cyclohexanone was only 3.5 g. A clue to the evident importance of the initial temperature is suggested by an experiment in which the cyclohexanol was dissolved in 6 mL of benzene instead of 5 mL of acetic acid, and with the two solutions being mixed at 15°C. Within a few minutes orange-yellow crystals separated and soon filled the flask; the substance probably is the chromate ester $(C_6H_{11}O)_2CrO_2$. When the crystal magma was allowed to stand at room temperature, the crystals soon dissolved, exothermic oxidation proceeded, and cyclohexanone was formed in high yield. Perhaps a low initial temperature ensures complete conversion of the alcohol into the chromate ester before side reactions set in.

50 mL of water, and add the solution to the flask for steam distillation (Chapter 6) of the product (Fig. 22.1). Distill as long as any oil passes over with the water and, because cyclohexanone is appreciably soluble in water, continue somewhat beyond this point (about 40 mL will be collected).

Alternatively, instead of setting up the apparatus for steam distillation, simply add a boiling chip to the 100-mL flask and distill 20 mL of liquid, cool the flask slightly, add 20 mL of water to it, and distill 20 mL more. Note the temperature during the distillation. This is a steam distillation in which steam is generated in situ rather than from an outside source.

## The Isolation of Cyclohexanone from the Steam Distillate

Cyclohexanone is fairly soluble in water. Dissolving inorganic salts such as potassium carbonate or sodium chloride in the aqueous layer will decrease the solubility of cyclohexanone such that it can be completely extracted with ether. This process is known as *salting out*.

To salt out the cyclohexanone, add to the distillate 0.2 g of sodium chloride per milliliter of water present and swirl to dissolve the salt. Next pour the mixture into a separatory funnel, rinse the flask with ether, add more ether to a total volume of 10–15 mL, shake, and draw off the water layer. Then wash the ether layer with 10 mL of 3 $M$ sodium hydroxide solution to remove acetic acid, test a drop of the wash liquor to be sure it contains excess alkali, and draw off the aqueous layer.

To dry the ether containing dissolved water, shake the ether layer with an equal volume of saturated aqueous sodium chloride solution. Draw off the aqueous layer, pour the ether out of the neck of the separatory funnel into an Erlenmeyer flask, add about 2.5 g of anhydrous calcium chloride pellets, and complete final drying of the ether solution by occasional swirling of the solution over a 5-min period. Remove the drying agent by decantation or gravity filtration into a tared Erlenmeyer flask and rinse the flask that contained the drying agent (the calcium chloride) and the funnel with ether. Add a boiling chip to the ether solution and evaporate the ether on a steam bath under an aspirator tube (*see* Figure 7.11 on page 155). Cool the contents of the flask to room temperature, evacuate the crude cyclohexanone under aspirator vacuum to remove the final traces of ether (*see* Fig. 9.7 on page 207 and Fig. 9.10 on page 211), and weigh the product. The yield is 15–16 g.

The crude cyclohexanone can be purified by simple distillation or used directly to make adipic acid (Experiment 6).

**Cleaning Up.** To the residue from the steam distillation, add sodium bisulfite to destroy any excess dichromate ion. Collect the precipitate of chromium hydroxide that forms when the solution is made just slightly basic with ammonium hydroxide and dispose of the filter paper and precipitate in the hazardous waste container designated for heavy metals. The filtrate can be flushed down the drain. Combine the water layers from the extraction process, neutralize with dilute hydrochloric acid, and flush the solution down the drain. Any ether should be placed in the organic solvents container. If local regulations allow, evaporate any residual solvent from the drying agents in the hood and place the dried solid in the nonhazardous waste container. Otherwise, place the wet drying agent in a waste container designated for this purpose.

Using Celite will speed up the filtration.

# 3. Cyclohexanone from Cyclohexanol by Hypochlorite Oxidation

**Cyclohexanol**
bp 161°C, den. 0.96
MW 100.2

**Cyclohexanone**
bp 157°C, den. 0.97
MW 98.14
solubility 1.5 g/100 mL $H_2O^{10°}$

**IN THIS EXPERIMENT** an alcohol is oxidized to a ketone with household bleach (sodium hypochlorite). The product is isolated by steam distillation and is extracted into the distillate with ether that is dried in a column of drying agents. Removal of the ether leaves the product—cyclohexanone. If a 6% hypochlorite solution is used, you might find it more convenient to carry out the experiment on twice the scale indicated. Using a $\frac{1}{2}$-inch stir bar with a magnetic stirrer is suggested.

To a 5-mL long-necked, round-bottomed flask fitted with a connector with support rod, add 150 mg (1.5 mmol) of cyclohexanol followed by a mixture of 80 mg of acetic acid and 2.3 mL of a 5.25% solution of sodium hypochlorite (e.g., Clorox). Do not use scented bleach and note the concentration. Ultra Clorox is a 6% solution, so use 2 mL and 70 mg of acetic acid. This mixture should be acidic. Test with pH paper and if it is not acidic, add more acetic acid. Add a thermometer. The temperature will rise to 45°C. Swirl the contents of the flask; as soon as the temperature begins to drop, remove the thermometer, cap the flask with a septum, and shake it vigorously for about 3 min. Then place the flask in a beaker of water at 45°C for 15 min to complete the reaction.

Destroy excess oxidizing agent by adding saturated sodium bisulfite solution (about 1 or 2 drops). At this point the reaction mixture no longer gives a positive test with starch-iodide paper (a positive test for an oxidant is a blue-purple color from the starch-triiodide complex).

Add a drop of thymol blue indicator to the solution and then neutralize the solution with 0.3 or 0.4 mL 6 M sodium hydroxide solution. Add a boiling chip to the solution; attach to the flask a distilling head, a thermometer adapter, and a thermometer; and then distill about 0.8 mL of a mixture of water and cyclohexanone (Fig. 22.2). This is a steam distillation in which the steam is generated in situ (*see* Chapter 6).

## *Isolation of Cyclohexanone from the Steam Distillate*

Cyclohexanone is fairly soluble in water. Dissolving inorganic salts such as potassium carbonate or sodium chloride in the aqueous layer will decrease the solubility of cyclohexanone such that it can be completely extracted with ether. This process is known as *salting out.*

■ **FIG. 22.2**
A microscale simple distillation
apparatus. The temperature is
regulated by piling up or removing
sand from near the flask.

Air condenser

Receiver vial

Boiling chip

To salt out the cyclohexanone, transfer the distillate to a reaction tube, add 0.1 g of sodium chloride to the distillate, stir, and shake the tube until most of the sodium chloride dissolves. Then extract the solution with three 0.4-mL portions of ether. As a reminder, this process consists of adding 0.4 mL of the ether to the reaction tube, shaking the tube, allowing the layers to separate, drawing off the top layer with a Pasteur pipette, and placing this top layer (the ether extract) in a clean, dry reaction tube.

Because ether dissolves water to a small extent, the ether extract is wet. A convenient way to dry the extract is to pass it through a short column of drying agent and alumina (a powerful drying agent). To the chromatography column add about 1-in.-high column of alumina and on top of this a 1-in.-high column of anhydrous sodium sulfate. Transfer the ether solution of cyclohexanone to the column, let it sit for 1–2 min, and then collect the dry ether in a tared 5-mL round-bottomed flask. Wash out the column with about 2 mL of ether, which is collected in the same flask.

Add a boiling chip to the dry ether solution and fit the flask for simple distillation. Distill all the ether using the short-path microscale distillation apparatus shown in Figure 22.3. The perfectly water-clear residue is cyclohexanone, which can be analyzed by IR and nuclear magnetic resonance (NMR) spectroscopy. If it

■ **FIG. 22.3**
A short-path microscale simple
distillation apparatus.

Boiling chip

is cloudy, it is wet. Determine the weight of the product and calculate the percentage yield.

**Cleaning Up.** The aqueous reaction mixture after extraction of the product contains primarily sodium, chloride, and acetate ions and can therefore be flushed down the drain. Any unused ether goes in the organic solvents container. If local regulations allow, evaporate any residual solvent from the drying agents and alumina in the hood and place the dried solid in the nonhazardous waste container. Otherwise, place the wet materials in a waste container designated for this purpose.

## 4. Cyclohexanone from Cyclohexanol by Hypochlorite Oxidation

*The addition should take 15–20 min.*

Into a 250-mL Erlenmeyer flask in the hood, place 8 mL (0.075 mol) of cyclohexanol. Introduce a thermometer and slowly add to the flask with swirling a mixture of 4 mL of acetic acid and 115 mL of a commercial household bleach such as Clorox (usually 5.25% by weight sodium hypochlorite, which is 0.75 molar). Do not use scented bleach and note the concentration. Ultra Clorox is a 6% solution, so use 100.6 mL and 3.5 mL of acetic acid. This can be added from a separatory funnel clamped to a ring stand or from another Erlenmeyer flask. Take care not to come in contact with the reagent. It should be acidic. Test it with indicator paper and, if necessary, add more acetic acid. During the addition, keep the temperature in the range of 40°C–50°C. Have an ice bath available in case the temperature goes above 50°C but do not allow the temperature to go below 40°C because the oxidation will be incomplete. The addition should take about 15–20 min. Swirl the reaction mixture periodically for the next 20 min to complete the reaction.

Because the exact concentration of the hypochlorite in the bleach depends on its age, it is necessary to analyze the reaction mixture for unreacted hypochlorite ion and to reduce any excess to chloride ion with bisulfite. Add a drop of the reaction mixture to a piece of starch-iodide paper. Any unreacted hypochlorite will cause the appearance of the blue starch-triiodide complex. Add 1-mL portions of saturated sodium bisulfite solution to the reaction mixture until the starch-iodide test is negative.

Add a few drops of thymol blue indicator solution to the mixture and then slowly add with swirling 15–20 mL of 6 $M$ sodium hydroxide solution until the mixture is neutral, which is indicated by its blue color.

Transfer the reaction mixture to a 250-mL round-bottomed flask, add a boiling chip, and set up the apparatus for simple distillation. Distillation under these conditions is a steam distillation (*see* Chapter 6) in which the steam is generated in situ. Continue the distillation until no more cyclohexanone comes over with the water (40 mL of distillate should be collected).

Refer to the section "Isolation of Cyclohexanone from Steam Distillate" in Experiment 2 to complete the reaction sequence.

**Cleaning Up.** The residue from the steam distillation contains only chloride, sodium, and acetate ions and can therefore be flushed down the drain.

*Note the temperature of the vapor during this distillation.*

# 5. Adipic Acid from Cyclohexanone

**Cyclohexanone**
bp 157°C, den. 0.95
MW 98.14
solubility 1.5 g/100 mL $H_2O^{10°}$

HNO$_3$
or
KMnO$_4$

**Adipic acid**
mp 153°C, MW 146.14
solubility 1.4 g/100 g $H_2O^{15°}$

**IN THIS EXPERIMENT** 150 mg of cyclohexanone (no more) is carefully oxidized with concentrated nitric acid in a highly exothermic reaction. The adipic acid crystallizes from the reaction mixture. It must be very carefully washed free of excess nitric acid because it is quite soluble in water.

■ **FIG. 22.4**
**A Wilfilter filtration apparatus.**

Crystals

Wilfilter

Filtrate

Because of the highly exothermic nature of this reaction with nitric acid, it is not advisable to scale up this procedure. In a $10 \times 100$ mm reaction tube place 1.0 mL of concentrated nitric acid and a boiling chip. Clamp the tube and also a Pasteur pipette, which dips into the tube and is connected to a water aspirator pulling a gentle vacuum to remove nitrogen oxides. Add to the tube 1 small drop of cyclohexanone from a vial containing 150 mg of the ketone. Warm the tube on a sand bath until brown oxides of nitrogen are seen emanating from the nitric acid and an exothermic reaction begins. Do not add more cyclohexanone until it is quite clear (evolution of brown oxides of nitrogen, generation of heat) that the reaction has started. Remove the tube from the heat. Add the remaining cyclohexanone to the hot nitric acid over a period of about 3 min. The reaction is extremely exothermic; no external heating is necessary, but add the ketone rapidly enough to keep the reaction going. After completing the cyclohexanone addition, heat the tube to boiling for 1 min.

As the tube cools to room temperature, fine crystals of adipic acid should appear. If they do not, scratch the inside of the tube at the liquid-air interface with a small glass rod to initiate crystallization. Cool the tube in a mixture of ice and water for at least 3 min, then stir the crystals with the tip of a glass Pasteur pipette, and remove the solvent by forcing the pipette to the bottom of the tube. Rap the tube on a hard surface to remove more solvent from the crystals, taking care to keep the tube cold. Add 0.2 mL of ice water to the crystals, cool thoroughly in ice, and remove as much of the wash liquid as possible with a Pasteur pipette, bearing in mind that the product is very soluble in water. Collect the damp crystals on a Wilfilter by centrifugation (Fig. 22.4) or dry the crystals under vacuum in the reaction tube while heating it on a steam bath. Determine the melting point, the IR spectrum, and the purity of the product by TLC. For TLC, the eluent should be 3 parts of 95% ethanol and 1 part of ammonium hydroxide.

**Cleaning Up.** The nitric acid filtrate and aqueous wash should be neutralized with sodium carbonate before flushing down the drain. The TLC solvent should be neutralized with dilute hydrochloric acid and then flushed down the drain.

# 6. Adipic Acid from Cyclohexanone

**IN THIS EXPERIMENT** cyclohexanone is oxidized by potassium permanganate under basic conditions. The precipitated manganese dioxide is filtered off; the filtrate is then concentrated, acidified with hydrochloric acid, and cooled to yield crystalline adipic acid.

The reaction to prepare adipic acid is conducted with 5.0 g of cyclohexanone, 15 g of potassium permanganate, and amounts of water and alkali that can be adjusted to provide an attended exothermic reaction period of 1/2 h or an unattended overnight reaction.

## Short-Term Procedure

For the short-term reaction, mix the cyclohexanone and permanganate with 125 mL of water in a 250-mL Erlenmeyer flask, adjust the temperature to 30°C, note that there is no spontaneous temperature rise, and then add 1 mL of 3 $M$ sodium hydroxide solution. A temperature rise is soon registered by the thermometer. It may be of interest to determine the temperature at which you can just detect warmth, by holding the flask in the palm of your hand or by touching the flask to your cheek. When the temperature reaches 45°C (15 min), slow the oxidation process by brief ice-cooling, and keep the temperature at 45°C for 20 min. Wait for a slight further rise (47°C) and an eventual drop in temperature (about 25 min later) and then heat the mixture by swirling it on a hot plate to complete the oxidation and to coagulate the precipitated manganese dioxide. Make a spot test by withdrawing some reaction mixture on the tip of a stirring rod and touching it to filter paper; permanganate, if present, will appear in a purple ring around the spot of manganese dioxide. If permanganate is still present, add small amounts of sodium bisulfite until the spot test is negative. Then filter the mixture by suction on an 11-cm Büchner funnel, wash the brown precipitate well with water, add a boiling chip, and evaporate the filtrate on a hot plate or over a flame from a large beaker to a volume of 35 mL. If the solution is not clear and colorless, clarify it with decolorizing charcoal and evaporate again to 35 mL. Acidify the hot solution with concentrated hydrochloric acid to pH 1–2, add 5 mL acid in excess, and let the solution stand to crystallize. Collect the crystals on a small Büchner funnel, wash them with a very small quantity of cold water, press the crystals between sheets of filter paper to remove excess water, and set them aside to dry. A typical yield of adipic acid (mp 152°C–153°C) is 3.5 g.

## Long-Term Procedure

In the alternative procedure, the weights of cyclohexanone and permanganate are the same, but the amount of water is doubled (250 mL) to moderate the reaction and make temperature control unnecessary. All the permanganate must be dissolved before the reaction is begun. Heat the flask and swirl the contents vigorously. Test for undissolved permanganate with a glass rod (as done in the short-term procedure). After adjusting the temperature of the solution to 30°C, 5 mL of 3 $M$ sodium hydroxide is added, and the mixture is swirled briefly and left to stand overnight (the maximum temperature should be 45°C–46°C). The workup is the same as in the short-term procedure; a typical yield of adipic acid is 4.1 g.

**Cleaning Up.** Place the manganese dioxide precipitate in the hazardous waste container for heavy metals. Neutralize the aqueous solution with sodium carbonate and flush it down the drain.

The proton and $^{13}$C NMR spectra of cyclohexanone are given in Figures 22.5 and 22.6. The IR spectrum of cyclohexanol is presented in Figure 22.7 and for cyclohexanone in Figure 22.8. The $^{13}$C spectrum of adipic acid is given in Figure 22.9. Compare any spectra you acquired with these reference spectra.

■ **FIG. 22.5**
The $^1$H NMR spectrum of
cyclohexanone (250 MHz).

■ **FIG. 22.6**
The $^{13}$C NMR spectrum of
cyclohexanone (22.6 MHz).

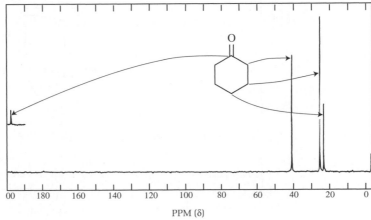

■ **FIG. 22.7**
The IR spectrum of cyclohexanol
(thin film).

■ **FIG. 22.8**
The IR spectrum of cyclohexanone
(thin film).

■ **FIG. 22.9**
The $^{13}C$ NMR spectrum of adipic
acid (22.6 MHz).

---

## QUESTIONS

1. In the hypochlorite oxidation of cyclohexanol to cyclohexanone, what purpose does the acetic acid serve?

2. Explain the order of the chemical shifts of the carbon atoms in the $^{13}C$ spectra of cyclohexanone (Fig. 22.6) and adipic acid (Fig. 22.9).

# CHAPTER 23

# Pulegone from Citronellol: Oxidation with Pyridinium Chlorochromate

**Online Study Center**

This icon will direct you to techniques, equipment setups, and online resources at http://college.hmco.com/PIC/williamsonMMOE5e.

Oxidizing agents:

$Cr_2O_7^{2-}$ in HOAc

$KMnO_4$

$CrO_3$ in aqueous $H_2SO_4$ (Jones reagent)

$CrO_3$ and pyridine (Collins reagent)

Corey reagents (PCC and PDC)

E. J. Corey of Harvard was the winner of the 1991 Nobel Prize in Chemistry.

**PRELAB EXERCISE:** Write a balanced equation for the oxidation of citronellol to citronellal.

A variety of reagents and conditions can be employed to oxidize alcohols to carbonyl compounds. The choice of which reagents and set of conditions to use depends on such factors as the scale of the reaction, the speed of the reaction, the anticipated yield, and the ease of isolating the products. Cyclohexanol can be oxidized to cyclohexanone by dichromate in acetic acid. This is much faster than permanganate oxidation of the same alcohol. Cholesterol can be oxidized to 5-cholesten-3-one by either the Jones reagent (chromium trioxide in sulfuric acid and water) or the Collins reagent—$(C_5H_5N)_2CrO_3$—which is prepared from chromium trioxide in anhydrous pyridine.

E. J. Corey has added two new $Cr^{6+}$ reagents to those that can be used in the oxidation process. Although these reagents might seem similar to the other dichromate reagents, they each have unique advantages. Although not nearly as fast as the Collins or Jones reagents, the Corey reagents are characterized by extraordinary ease of product isolation because one merely removes the reagent by filtration and evaporates the solvent to obtain the product. Both reagents are prepared easily or can be purchased commercially. The reagent used in this experiment, pyridinium chlorochromate (PCC), $C_5H_5NH^+ClCrO_3^-$, will oxidize

$$+ \quad CrO_3 + HCl \longrightarrow$$

**Pyridine**

**Pyridinium chlorochromate, PCC**
MW 215.56

**389**

100%

92%

primary alcohols to aldehydes in high yield without oxidizing the aldehyde further to a carboxylic acid.[1] It will oxidize secondary alcohols to ketones, but, being somewhat acidic, it will cause rearrangement of unconjugated ketones to conjugated ketones.

The other Corey reagent, pyridinium dichromate (PDC), $(C_5H_5NH^+)_2$ $Cr_2O_7^{2-}$, dissolved in dimethylformamide (DMF) will oxidize allylic alcohols to $\alpha,\beta$-unsaturated aldehydes without oxidizing the aldehyde to the carboxylic acid.[2]

**Pyridinium dichromate, PDC**

86%

$$C_6H_5CH = CH - CH_2OH \xrightarrow[\text{4-5 h, 0°C}]{\text{PDC-DMF}} C_6H_5CH = CH - \overset{\overset{\displaystyle O}{\|}}{C}H$$

97%

If the primary alcohol is not conjugated, then it is oxidized to the carboxylic acid:

84%

1. Corey, E. J.; Suggs, J. W. *Tetrahedron Lett.* **1975**, *16*, 2647.

2. Corey, E. J.; Schmidt, G. *Tetrahedron Lett.* **1979**, *20*, 399.

90%

When suspended in $CH_2Cl_2$, PDC will oxidize primary alcohols to the corresponding aldehydes and no further:

83%

## Citronellol–A Terpene

Citronellol is a member of the large class of natural products known as terpenes. Biosynthesized from five-carbon isopentenyl pyrophosphate, these molecules contain 10, 15, 20, . . . atoms; are made up of repeating isoprene units; and have pleasant odors. Many of the ten-carbon molecules are very familiar to you by odor if not by name, for example, pinene from turpentine, limonene from citrus oil, camphor, and menthol.

Geraniol is a sex pheromone released by female bees to attract males. Certain orchids mimic the female bee by also releasing geraniol. The male bee is attracted to the orchid and will even try to copulate with it, in the course of which, as it wanders from orchid to orchid, it collects pollen and pollinates the flowers. See the text website for additional information on chemical communication in insects.

**Isoprene**

# EXPERIMENTS

$$RCH_2OH \xrightarrow[\text{alumina}]{\text{PCC on}} RCHO$$

In the experiments in this chapter, PCC is used to oxidize citronellol, a constituent of rose and geranium oil, to the corresponding aldehyde. If PDC in $CH_2Cl_2$ were used, the reaction would stop at this point, and citronellal could be isolated at 92% yield. If sodium acetate (a buffer) were added to the PCC reaction, the oxidation would also stop at the aldehyde stage, with 82% yield. It has been found that PCC deposited on alumina will effect the same oxidation in less time and even higher yield.[3]

If the buffer is omitted, the PCC is acidic enough to cause the intermediate citronellal to cyclize to isopulegols (four possible isomers). These secondary alcohols are oxidized to two diastereomeric isopulegones. On treatment with base, the double bond will migrate to give pulegone, as shown in the following chemical structures.

**Citronellol**
MW 156.27
bp 222°C

**Citronellal**
MW 154.24

**Isopulegols**
(four possible isomers)

**Isopulegones**
(two possible isomers)

**Pulegone**
MW 152.23
bp 224°C

The course of the reaction can easily be followed by thin-layer chromatography (TLC).[4] The isolation of the oxidation products, citronellal and isopulegone, could not be simpler: The reduced reagent is removed by filtration, and the solvent

3. Cheng, Yu-Shia; Liu, Wen-Liang; Chen, Shu-hsia. *Synthesis* **1980**, *223*.

4. Corey, E. J.; Suggs, J. W. *J. Org. Chem.* **1978**, *41*, 380. TLC $R_f$ values: citronellol, 0.17; citronellal, 0.65; isopulegols, 0.27 and 0.35; isopulegone, 0.41; and pulegone, 0.36; the developer is $CH_2Cl_2$.

is evaporated. In the case of the second experiment, the reaction can be stopped once isopulegone is formed, or the isopulegone can be isomerized to pulegone as a separate reaction.

# 1. Preparation of Pyridinium Chlorochromate (PCC)

Pyridine + $CrO_3$ + HCl ⟶ Pyridinium chlorochromate, PCC

$ClCrO_3^-$

Pyridinium chlorochromate, PCC
MW 215.56

**CAUTION**: The dust (but not a solution of) $Cr^{6+}$ is a suspected carcinogen.

Dissolve 12 g of chromium trioxide in 22 mL of 6 *N* hydrochloric acid and add 9.5 g of pyridine during a 10-min period while maintaining the temperature at 45°C. Cool the mixture to 0°C, collect the crystalline yellow-orange pyridinium chlorochromate on a sintered glass funnel, and dry it for 1 h in vacuum. The compound is not hygroscopic and is stable at room temperature.

**Cleaning Up.** The filtrate should have pH $< 3$ and can then be treated with a 50% excess of sodium bisulfite to reduce the orange dichromate ion to the green chromic ion. This solution should then be made basic with ammonium hydroxide to precipitate chromium as the hydroxide. This precipitate is collected on filter paper, which is placed in the hazardous solid waste container for heavy metals. The filtrate from this latter treatment can be flushed down the drain.

# 2. Preparation of the PCC–Alumina Reagent

To prepare PCC on alumina, cool the reaction mixture described in Experiment 1 to 10°C until the product crystallizes; then reheat the solution to 40°C to dissolve the solid. To the resulting solution add 100 g of alumina with stirring at 40°C. The solvent is removed on a rotary evaporator, and the orange solid is dried in vacuum for 2 h at room temperature. If stored under vacuum in the dark, the reagent will retain its activity for several weeks.

**Cleaning Up.** Because it contains adsorbed hexavalent chromium ion, any used or unused PCC-alumina should be placed in the hazardous solid waste container. Place dichloromethane in the halogenated waste container.

# 3. Citronellal from Citronellol Using PCC on Alumina

| Citronellol | Citronellal |
|---|---|
| MW 156.27 | MW 145.25 |
| bp 222°C | bp 207°C |
| $n_D^{20}$ 1.4560 | $n_D^{20}$ 1.4480 |

## Microscale Procedure

To a 10-mL Erlenmeyer flask, add 1.5 g (1.22 mmol) of PCC on alumina (from Experiment 2), 120 mg of citronellol, and 2 mL of hexane. Stir the mixture using a magnetic stirrer or by shaking for up to 3 h (follow the course of the reaction by TLC), remove the stirring bar, and then remove the solid by filtration on a Hirsch funnel under vacuum. Wash the alumina on the filter with three 2-mL portions of ether and remove the solvents from the filtrate by evaporation under a gentle stream of air or nitrogen while warming the flask in a beaker of water. The last traces of solvent can be removed under vacuum. The residue should be pure citronellal (bp 90°C at 14 mm Hg). Check its purity by TLC on silica gel using dichloromethane as eluent. Using infrared (IR) spectroscopy as a thin film between sodium chloride plates, look for the presence of unoxidized hydroxyl at 3600 to 3400 cm$^{-1}$. See "Cleaning Up" section following the macroscale procedure for disposal procedures.

## Macroscale Procedure

**Pyridinium chlorochromate**
MW 215.56

PCC on alumina (7.5 g, 6.1 mmol; from Experiment 2) is added to a flask containing a solution of citronellol (0.60 g, 3.8 mmol) in 10 mL of *n*-hexane. After stirring or shaking for up to 3 h (follow the course of the reaction by TLC), remove the solid by filtration, wash it with three 10-mL portions of ether, and remove the solvents from the filtrate by distillation or evaporation. The last traces of solvent can be removed under vacuum (*see* Fig. 4.22 on page 79). The residue should be pure citronellal (bp 90°C at 14 mm Hg). Check its purity by TLC and IR spectroscopy. A large number of other primary and secondary alcohols can be oxidized to aldehydes and ketones using this same procedure.

**Cleaning Up.** Because it contains adsorbed hexavalent chromium ion, the used PCC-alumina should be placed in the hazardous heavy metal waste container. Organic material goes in the organic solvents container.

# 4. Isopulegone from Citronellol Using PCC in Dichloromethane

## Microscale Procedure

To 1 g of PCC suspended in 6.25 mL of dichloromethane in a 10-mL Erlenmeyer flask, add 0.25 g of citronellol. The best yields are obtained if the mixture is stirred at room temperature for 36 h or more using a magnetic stirrer. Alternatively, let the mixture sit at room temperature for a week with occasional shaking. The chromium salts are removed by vacuum filtration and washed on the filter with 1 mL of dichloromethane. Remove the solvent by evaporation with a gentle stream of nitrogen or air to give isopulegone. Follow the course of the reaction and analyze the product by TLC on silica gel plastic sheets, developing with dichloromethane. Save 1 mg of the product for the ultraviolet (UV) spectrum (*see* Experiment 5). Refer to the "Cleaning Up" section following the macroscale procedure for disposal procedures.

**Online Study Center**

Video: Microscale Filtration on the Hirsch Funnel

## Macroscale Procedure

To 16 g of PCC suspended in 100 mL of dichloromethane, add 4.0 g citronellol. The best yields are obtained if the mixture is stirred at room temperature for 36 h or more using a magnetic stirrer. Alternatively, let the mixture sit at room temperature for a week with occasional shaking. The chromium salts are removed by vacuum filtration and washed on the filter with 15 mL of dichloromethane. Remove the solvent by evaporation to give isopulegone. Follow the course of the reaction and analyze the product by TLC on silica gel plastic sheets, developing with dichloromethane.

**Cleaning Up.** Dichloromethane should be placed in the halogenated organic waste container. The chromium salts should be dissolved in hydrochloric acid, diluted fourfold, reduced, and precipitated as described under the "Cleaning Up" section of Experiment 1.

**Isopulegone**
(two diastereomers)

# 5. Pulegone from Isopulegone

## Microscale Procedure

Add 250 mg of crude isopulegone isomers to a reaction tube followed by 2 mL of a 2% solution of sodium hydroxide in ethanol. Warm the mixture on a hot water bath or on a steam bath for 1 h. Evaporate the ethanol under a stream of air or nitrogen and then dissolve the residue in 1 mL of dichloromethane. Prepare a chromatographic column from 2.5 g of silica gel in dichloromethane and pour the solution of pulegone through the column followed by 2 mL of dichloromethane. This very crude chromatography is designed to remove the very small quantity of sodium hydroxide present in the pulegone. Evaporate the solvent to leave pure pulegone.

The pulegone can also be isolated from the reaction mixture by evaporating the ethanol, adding 1 mL of water, and extracting the product with two 1-mL portions of ether. The ether extract is washed with 0.5 mL of water, followed by 0.5 mL of saturated sodium chloride solution, and is then dried over calcium

**Pulegone**

***Online Study Center***

Video: Extraction with Ether;
Photos: Extraction with Ether,
Filtration Using a Pasteur
Pipette

Ultraviolet spectra of an
α,β-unsaturated ketone

chloride pellets. The ether is removed from the drying agent with a Pasteur pipette, the drying agent is washed with more ether, and the ether is evaporated under a stream of air or nitrogen to leave pure pulegone.

Pulegone occurs naturally in pennyroyal oil and has a pleasant odor between those of peppermint and camphor. It is an α,β-unsaturated ketone and should therefore have $\lambda_{max}$ of 254 nm as calculated by application of the Fieser and Woodward rules (*see* Chapter 14). The actual value is 253.3 nm. The starting material is not a conjugated ketone and so should not have intense UV absorption. Compare the two compounds by dissolving 1 mg of each in 50 mL ethanol and determining their UV spectra.

**Cleaning Up.** Place dichloromethane in the halogenated organic waste container, ether and ethanol in the organic solvents container. If local regulations allow. evaporate any residual solvent from the drying agent and silica gel in the hood and place the dried solids in the nonhazardous waste container. Otherwise, place the wet solids in waste containers designated for this purpose.

## Macroscale Procedure

Crush and then dissolve one sodium hydroxide pellet under 30 mL of ethanol in a 50-mL Erlenmeyer flask, add 4 g of isopulegone, and warm the mixture on a steam bath for 1 h. Evaporate most of the ethanol on the steam bath; then add 10 mL of cold water and extract the product with 20 mL of ether. Wash the ether extract with about 5 mL of water followed by 5 mL of saturated sodium chloride solution; then dry the ether over anhydrous calcium chloride. Decant the ether into an Erlenmeyer flask, wash the calcium chloride with a few milliliters more ether, and evaporate the ether to leave crude pulegone. Pulegone can be distilled at atmospheric pressure (bp 224°C), but it is best distilled at reduced pressure (bp 103°C at 17 mm Hg). Pulegone occurs naturally in pennyroyal oil and has a pleasant odor somewhere between those of peppermint and camphor. If the final product is not purified by vacuum distillation, this entire experiment can be scaled down by a factor of four or more.

**Cleaning Up.** Combine the aqueous layers, neutralize with dilute hydrochloric acid, and flush down the drain. If local regulations allow, evaporate any residual solvent from the calcium chloride in the hood and place the dried solid in the nonhazardous waste container. Otherwise, place the wet drying agent in a waste container designated for this purpose. Combine all organic material and place it in the organic solvents container.

## Computational Chemistry

Using a molecular mechanics or semi-empirical methods, calculate the steric energies of the two isopulegone isomers to determine which is more stable. Similarly, calculate the steric energies of the four isomeric isopulegols. In each of these two cases, valid comparisons can be made on the basis of steric energies using molecular mechanics because the bonding types are identical. Note that finding the global minimum, especially for the isopulegols, is not easy. You have considerable latitude when picking just how the isopropyl and hydroxyl groups are arranged. Compare your results with those of other students to determine who has the lowest steric energy and why.

Isopulegone isomerizes to pulegone. To compare these, molecular heats of formation must be calculated using AM1 calculations. Is pulegone more stable than each of the isopulegone isomers?

## QUESTIONS

1. Draw the chair conformations of the four possible isopulegols.
2. Assign as many peaks as possible to specific protons in the $^1H$ NMR spectra of citronellol (Fig. 23.1) and citronellal (Fig. 23.2).
3. Give the mechanism for the base-catalyzed isomerization of isopulegone to pulegone.

■ **FIG. 23.1**
The $^1H$ NMR spectrum of citronellol (250 MHz).

■ **FIG. 23.2**
The ${}^1$H NMR spectrum of
citronellal (90 MHz).

# Oxidative Coupling of Alkynes:
# 2,7-Dimethyl-3,5-octadiyn-2,7-diol

**Online Study Center**

This icon will direct you to
techniques, equipment setups,
and online resources at
http://college.hmco.com/PIC/
williamsonMMOE5e.

**PRELAB EXERCISE:** Show the reactions for a two-step method that might be used to convert 2-methyl-3-butyn-2-ol to isoprene.

$$2\ CH_3\underset{\underset{OH}{|}}{\overset{\overset{CH_3}{|}}{C}}-C\equiv CH \xrightarrow[\text{CuCl—Pyridine}]{O_2} CH_3\underset{\underset{OH}{|}}{\overset{\overset{CH_3}{|}}{C}}-C\equiv C-C\equiv C-\underset{\underset{OH}{|}}{\overset{\overset{CH_3}{|}}{C}}CH_3$$

**2-Methyl-3-butyn-2-ol**
MW 84.11, den. 0.868, bp 103°C

**2,7-Dimethyl-3,5-octadiyn-2,7-diol**
MW 166.21, mp 130°C

**Isoprene**

The starting material, 2-methyl-3-butyn-2-ol, is made commercially from acetone and acetylene and is convertible into isoprene. This experiment illustrates the oxidative coupling of a terminal acetylene to produce a diacetylene, commonly known as the Glaser reaction.

Although Glaser reported the coupling of acetylenes using cuprous ion in 1869, the details of the mechanism of this reaction are still obscure. It appears to involve both $Cu^+$ and $Cu^{2+}$ ions. As noted by Glaser, the cuprous acetylide is oxidized by oxygen,

$$C_6H_5C\equiv CH \xrightarrow[NH_4OH]{CuCl} C_6H_5C\equiv CCu \xrightarrow[O_2]{air} C_6H_5C\equiv C-C\equiv CC_6H_5$$

but $Cu^{2+}$ is the actual oxidizing agent. The ammonium hydroxide seems to be needed to keep the acetylide in solution.

The reaction is very useful synthetically in the synthesis of polyenes, vitamins, fatty acids, and the annulenes. Johann Baeyer used the reaction in his historic synthesis of indigo back in 1882. The reaction allowed the unequivocal establishment of the carbon skeleton of this dye:

# EXPERIMENT

## Preparation of 2,7-Dimethyl-3,5-Octadiyn-2,7-Diol
### *Microscale Procedure*

> **IN THIS EXPERIMENT** the reaction time for this coupling reaction is shortened using an excess of catalyst and by supplying one of the reactants, oxygen, under pressure. Pressurized oxygen is obtained by inflating a white rubber pipette bulb. The progress of the reaction is followed by observing the decrease in diameter of the balloon. Swirling or stirring the reaction mixture for about 20 min completes the reaction. Acid is added to neutralize the basic pyridine, and sodium chloride solution is added to precipitate the product. The crystalline product is wet and so is dissolved in ether which, in the usual way, is washed, dried, and evaporated. The residue is crystallized from toluene to give the diol product.

*Measure pyridine in the hood.*

The reaction vessel is a 25-mL filter flask equipped with a stirring bar and a rubber bulb secured to the side arm with copper wire, as shown in Figure 24.1. Add 1.0 mL of 2-methyl-3-butyn-2-ol, 1.0 mL of methanol, 0.3 mL of pyridine, and 0.1 g of copper(I) chloride. Before going to the oxygen station,[1] practice capping the flask with a rubber septum until you can do this quickly. You are to flush out the flask with

---

1. The valves of a cylinder of oxygen should be set to deliver gas at a pressure of 10 lb/in.[2] when the needle valve is opened. The barrel of a 2.5-mL plastic syringe is cut off and thrust into the end of a $\frac{1}{4}$-$\times$ $\frac{3}{16}$-in. rubber delivery tube. Read about the handling of compressed gas cylinders in Chapter 2. Be sure the cylinder is secured to a bench or wall.

**■ FIG. 24.1**
The balloon technique of oxygenation. About 10 lb/in.$^2$ of oxygen pressure is needed to inflate the pipette bulb.

**■ FIG. 24.2**
Open the main tank valve, adjust the gas pressure to 10 lb/in.$^2$ with the regulator valve—clockwise for higher pressure—and finally open the needle valve slowly to obtain oxygen gas.

Outlet pressure gauge — Cylinder pressure gauge — Cylinder valve — Needle valve — Outlet — Gas hose — Clockwise increases pressure

**■ FIG. 24.3**
The appearance of the pipette bulb after oxygenation.

oxygen and cap it before air can diffuse in; the reaction is about twice as fast in an atmosphere of oxygen as in air. Cut the top off a 1-mL plastic syringe barrel, insert it into the oxygen tank supply hose, and put a needle on the syringe barrel. Remove the rubber septum from the flask, insert the needle into the reaction liquid, open the needle valve on the oxygen gas regulator (Fig. 24.2), and let oxygen bubble through the solution in a brisk stream for 2 min. Close the valve and quickly cap the flask and wire the cap. Next, thrust the needle of the delivery syringe through the center of the rubber septum, open the valve, and run in oxygen until you have produced a sizable inflated bulb (Fig. 24.1). Close the valve and withdraw the needle (the hole is self-sealing), note the time, and start swirling the reaction mixture. The rate of oxygen uptake depends on the efficiency of mixing of the liquid and gas phases. Swirl or stir the reaction mixture continuously for 20–25 min.[2] The reaction mixture warms up and becomes deep green. Introduce a second charge of oxygen the same size as the first, note the time, and swirl. In 25–30 min, the balloon reaches a constant volume, indicating that the reaction is complete. A pair of calipers is helpful in recognizing the constant size of the balloon and thus the end of the reaction (Fig. 24.3).

Open the reaction flask, cool if warm, add 0.5 mL of concentrated hydrochloric acid to neutralize the pyridine and keep copper compounds in solution, and cool again; the color changes from green to yellow. Use a spatula to dislodge copper salts adhering to the walls, leaving the salts in the flask. Then add 2.5 mL of saturated

---

2. A magnetic stirrer does not materially shorten the reaction time.

sodium chloride solution to precipitate the diol and stir and cool the thick paste that results. Scrape out the paste onto a Hirsch funnel with the spatula; press down the material to an even cake. Rinse out the flask with water and wash the filter cake with enough additional water to remove all the color from the solid and leave a colorless product. Because the moist product dries slowly, drying is accomplished in ether solution, and the operation is combined with the recovery of diol retained in the mother liquor. Transfer the moist product to a 10-mL Erlenmeyer flask, extract the mother liquor in the filter flask with one 1.5-mL portion of ether, wash the extract once with water, and add the ethereal solution to the flask containing the solid product. Add enough additional ether to dissolve the product, transfer the solution to a test tube, remove the water layer, wash with saturated sodium chloride solution, and dry over anhydrous calcium chloride. The solvent is evaporated on a steam bath, and the solid residue is heated and evacuated until the weight is constant. Recrystallization from toluene then gives colorless needles of the diol (mp 129°C–130°C). The yield of crude product is about 0.5 g. In the recrystallization, the recovery is practically quantitative.

Analyze the product by thin-layer chromatography (TLC) and infrared (IR) spectroscopy. Compare the diyne with the starting material to determine whether the product is pure.

**Cleaning Up.** Combine all aqueous and organic layers from the reaction, add sodium carbonate to make the aqueous layer neutral, and shake the mixture gently to extract pyridine into the organic layer. Place the organic layer in the organic solvents container. Remove any solid (copper salts) from the aqueous layer by filtration. The solid can be placed in the nonhazardous solid waste container, and the aqueous layer can be diluted with water and flushed down the drain. If local regulations allow, evaporate any residual ether from the drying agent in the hood and place the dried solid in the nonhazardous waste container. Otherwise, place the wet drying agent in a waste container designated for this purpose.

 ## Macroscale Procedure

The reaction vessel is a 125-mL or 50-mL filter flask with a rubber bulb secured to the side arm with copper wire, as shown in Figures 24.1 and 24.3. Add 5 mL of 2-methyl-3-butyn-2-ol, 5 mL of methanol, 1.5 mL of pyridine, and 0.25 g of copper(I) chloride. Before going to the oxygen station,[3] practice capping the flask with a rubber septum until you can do this quickly. You are to flush out the flask with oxygen and cap it before air can diffuse in; the reaction is about twice as fast in an atmosphere of oxygen as in air. Insert the oxygen delivery syringe into the flask with the needle under the surface of the liquid, open the valve, and let oxygen bubble through the solution in a brisk stream for 2 min (Fig. 24.2). Close the valve and quickly cap the flask and wire the cap. Next thrust the needle of the delivery syringe through the center of the rubber septum, open the valve, and run in oxygen until you have produced a sizable inflated bulb (Fig. 24.1). Close the valve and withdraw the

Photo: Placing a Polyethylene Tube through a Septum

*Online Study Center*

Measure pyridine in the hood.

---

3. The valves of a cylinder of oxygen should be set to deliver gas at a pressure of 10 lb/in.$^2$ when the needle valve is opened. The barrel of a 2.5-mL plastic syringe is cut off and thrust into the end of a $\frac{1}{4}$-$\times \frac{3}{16}$-in. rubber delivery tube. Read about the handling of compressed gas cylinders in Chapter 2. Be sure the cylinder is secured to a bench or wall.

needle (the hole is self-sealing), note the time, and start swirling the reaction mixture. The rate of oxygen uptake depends on the efficiency of mixing of the liquid and gas phases. By vigorous and continuous swirling it is possible to effect deflation of the balloon to a constant size (*see* Fig. 24.3 on page 401) in 20–25 min.[4] The reaction mixture warms up and becomes deep green. Introduce a second charge of oxygen of the same size as the first, note the time, and swirl. In 25–30 min the balloon reaches a constant volume (e.g., a 5-cm sphere) and the reaction is complete. A pair of calipers is helpful in recognizing the constant size of the balloon and thus the end of the reaction.

Open the reaction flask, cool if warm, add 2.5 mL of concentrated hydrochloric acid to neutralize the pyridine and keep copper compounds in solution, and cool again; the color changes from green to yellow. Use a spatula to dislodge copper salts adhering to the walls. Then add 13 mL of saturated sodium chloride solution to dissolve the copper salts and precipitate the diol. Stir and cool the thick paste that results and scrape it out onto a small Büchner funnel with a spatula; press down the material to an even cake. Rinse out the flask with water and wash the filter cake with enough additional water to remove all the color from the cake and leave a colorless product. Because the moist product dries slowly, drying is accomplished in ether solution, and the operation is combined with the recovery of diol retained in the mother liquor. Transfer the moist product to a 50-mL Erlenmeyer flask, extract the mother liquor in the filter flask with one portion of ether, wash the extract once with water, and run the ethereal solution into the flask containing the solid product. Add enough additional ether to dissolve the product, transfer the solution to a separatory funnel, drain off the water layer, wash with saturated sodium chloride solution, and filter through anhydrous calcium chloride pellets into a tared flask. The solvent is evaporated on a steam bath, and the solid residue is heated and evacuated until the weight is constant. Recrystallization from toluene then gives colorless needles of the diol (mp 129°C–130°C). The yield of crude product is 3–4 g. In the recrystallization, the recovery is practically quantitative.

*Online Study Center*

Video: Macroscale Crystallization

**Cleaning Up.** Combine all aqueous and organic layers from the reaction, add sodium carbonate to make the aqueous layer neutral, and shake the mixture gently to extract pyridine into the organic layer. Place the organic layer in the organic solvents container. Remove any solid (copper salts) from the aqueous layer by filtration. The solid can be placed in the nonhazardous solid waste container, and the aqueous layer can be diluted with water and flushed down the drain. If local regulations allow, evaporate any residual solvent from the drying agent in the hood and place the dried solid in the nonhazardous waste container. Otherwise, place the wet drying agent in a waste container designated for this purpose.

## QUESTIONS

*Online Study Center*

**General Resources**
Web Links

1. Give two oxidation states of copper in this reaction and tell which state forms the copper acetylide. What is oxygen reduced to?
2. What volume of oxygen at standard temperature and pressure is consumed in this reaction?

---

4. A magnetic stirrer does not materially shorten the reaction time.

# Catalytic Hydrogenation

**PRELAB EXERCISE:** Calculate the volume of hydrogen gas generated when 3 mL of 1 *M* sodium borohydride reacts with concentrated hydrochloric acid. Write a balanced equation for the reaction of sodium borohydride with platinum chloride. Calculate the volume of hydrogen that can be liberated by reacting 1 g of zinc with acid.

Catalytic reduction is a very important and widely used industrial process; usually no harmful wastes are produced in the process. Catalytic hydrogenation and dehydrogenation are carried out on an enormous scale, for example, in the catalytic cracking and reforming of crude oil to make gasoline.

Nitrobenzene can be reduced catalytically to aniline with water as the only byproduct:

$$3\ H_2\ +\ \underset{\textbf{Nitrobenzene}}{\overset{NO_2}{\bigcirc}}\ \xrightarrow{\text{catalyst}}\ \underset{\textbf{Aniline}}{\overset{NH_2}{\bigcirc}}\ +\ 2\ H_2O$$

Styrene is made by the catalytic dehydrogenation of ethylbenzene at very high temperatures, but styrene can also be very easily hydrogenated back to ethylbenzene. Palladium, as a catalyst, lowers the energy barrier for the reaction in both directions.

$$\underset{\textbf{Ethylbenzene}}{\overset{CH_2CH_3}{\bigcirc}}\ \underset{\text{Pd}}{\overset{400\text{--}500°C}{\rightleftharpoons}}\ \underset{\textbf{Styrene}}{\overset{CH=CH_2}{\bigcirc}}\ +\ H_2$$

The addition of hydrogen to alkenes is one of the most common reactions. The alkene is more reactive toward this process than is the aromatic ring or functional groups such as esters or ketones.

$$CH{=}CH - COOCH_3 \qquad CH_2{-}CH_2 - COOCH_3$$

$$\text{(benzene ring)} + H_2 \xrightarrow{\text{Pt or Pd}} \text{(benzene ring)}$$

Hydrogenation is stereospecific, so alkynes are reduced to *cis*-alkenes. The metal is usually supported on a high-surface-area material such as charcoal. The alkene and the hydrogen probably are both adsorbed onto the surface of the catalyst before the transfer occurs. This heterogeneous reaction, a reaction that involves reactants in the liquid or gas phase and a catalyst in the solid phase, is difficult to study.

In the experiments in this chapter, catalyic hydrogenation is carried out in several different ways. Experiment 1 is a puzzle for you to solve. In Experiment 2, hydrogen gas from an external supply is used to hydrogenate a long-chain unsaturated alcohol to the corresponding saturated alcohol. In Experiment 3, the hydrogen is generated in situ using the Brown hydrogenation technique. Experiment 4 utilizes a process called *transfer hydrogenation* to produce a saturated fat from an unsaturated one (olive oil). In Experiment 5 olive oil is catalytically reduced by using hydrogen gas.

## EXPERIMENTS

## 1. A Small Research Project

$$\text{(cyclohexene)} \xrightarrow{10\% \text{ Pd/C}} \quad ?$$

The purpose of this experiment is to determine whether a chemical reaction has occurred when cyclohexene is boiled for a short period of time with a small quantity of a catalyst, 10% palladium on carbon, and, if so, to determine the structure of the product(s).

### *Microscale Procedure*

In a reaction tube place about 50 mg of 10% palladium on carbon. To this add about 0.8 mL of cyclohexene and a boiling chip. Reflux this mixture on a warm sand bath (low setting on controller) for about 10–15 min. Be careful not to boil the alkene out of the reaction tube. If necessary, place a coil of pipe cleaner moistened with water around the top of the tube. This will keep the top of the tube cool so that the cyclohexene will not escape. Further insurance against boiling

■ **FIG. 25.1**
An apparatus for refluxing a reaction mixture with maximum cooling.

Air condenser

Viton connector

Damp pipe cleaner if needed for more condensing

Boiling solvent

Temperature controlled by depth in sand

■ **FIG. 25.2**
The filtration of a solution in a Pasteur pipette. This process is usually used to remove charcoal.

Solution to be filtered

Cotton

Celite (optional)

away of the reactant is afforded by adding an empty distilling column to the reaction tube, although with careful adjustment of the heat input (the depth of the tube in the sand), these measures will not be necessary (Fig. 25.1).

At the end of the reflux period, cool the tube to room temperature and filter the solution to remove the catalyst. This is done by stuffing a small ball of cotton into a Pasteur pipette and using another Pasteur pipette to transfer the reaction mixture containing charcoal. Filter the solution into a clean, dry reaction tube. If one filtration does not remove all the charcoal, recycle the filtrate through the filter (Fig. 25.2).

*Online Study Center*

Photo: Preparation of a Filter Pipette

## *Macroscale Procedure*

In a 15-mL tapered 14/20 standard taper flask, place about 100 mg of 10% palladium on carbon. To this add about 2 mL of cyclohexene and a boiling chip. Reflux this mixture on a warm sand bath (low setting on controller) for at least 15 min (Fig. 25.3).

■ **FIG. 25.3**

An apparatus for refluxing a small volume of a liquid.

Water-jacketed condenser

→ Out

← In

15-mL Flask

Sand

■ **FIG. 25.4**

The filtration of a reaction mixture in a Hirsch funnel to remove charcoal. Use fine porosity filter paper.

Filter paper, 12 mm dia.

Polyethylene filter disk (frit), 10 mm dia.

Hirsch funnel

To aspirator

25-mL Filter flask

At the end of the reflux period, cool the flask to room temperature and filter the solution to remove the catalyst. This is done by stuffing a small ball of cotton into a Pasteur pipette and using another Pasteur pipette to transfer the reaction mixture containing palladium on charcoal. Filter the solution into a clean, dry reaction tube or culture tube. Alternatively, the reaction mixture can be filtered in a Hirsch funnel (Fig. 25.4). Use fine porosity filter paper in the funnel.

## *Analysis of the Product*

Run a test for the presence of an alkene on your starting material, cyclohexene, and your product. The decolorization of bromine is probably the best test to run. Count the number of drops of a 3% solution of bromine in dichloromethane or carbon tetrachloride that are decolorized by 0.10 mL of cyclohexene and by 0.10 mL of your filtered reaction mixture.

Analyze your product by gas chromotography or (or a gas chromatography–mass spectrometry system). Review Chapter 10 for the theory of gas chromatography and the procedures for analysis.

Run and interpret nuclear magnetic resonance (NMR), infrared (IR), and ultraviolet (UV) spectra of the starting material and product. The most useful of these is the NMR spectrum, followed by the UV spectrum. On the basis of these data, interpret the results of this experiment.

# 2. Catalytic Hydrogenation of Oleyl Alcohol using Palladium on Charcoal

$$CH_3(CH_2)_7CH{=}CH(CH_2)_8OH + H_2 \xrightarrow{\text{10\% Pd/C}} CH_3(CH_2)_{17}OH$$

| **Oleyl alcohol** | **Octadecanol** |
|---|---|
| MW 268.49, den. 0.849 | MW 270.50 |
| bp 207°C/13 mm | mp 60°C |
| $n_D^{20}$ 1.4600 | |

**IN THIS EXPERIMENT** the 18-carbon unsaturated alcohol oleyl alcohol is reduced to the corresponding saturated alcohol using one of the most widely employed techniques for catalytic hydrogenation. The catalyst is 10% palladium on carbon, which currently sells for $6 per gram; because this reaction is being carried out on a microscale, we will use only a few cents worth.

The hydrogen could come from a compressed gas cylinder, but it is just as convenient to prepare a small quantity by the action of acid on zinc.

The apparatus for this experiment can be set up in a number of ways. If you merely want to hydrogenate a double bond and are not concerned about the quantitative aspects of the reaction, fit the apparatus with a rubber balloon that holds the hydrogen. As soon as the balloon has reached a constant size, hydrogen uptake is complete, and the product can be isolated.

In a modification of the apparatus employed in this experiment, the hydrogen uptake can be followed as water rises in the hydrogen reservoir (a graduated cylinder). With this apparatus, the volume of hydrogen absorbed as a function of time can be measured.

In research-grade apparatus, the hydrogen is stored in a gas burette over mercury. The volume of hydrogen can be determined very accurately so that quantitative estimates of the numbers of double bonds in unknowns can be made or so that selective hydrogenation of the more reactive of several double bonds can be carried out. The reaction is stopped when the theoretical amount of hydrogen has been absorbed.

## Experimental Considerations

**Online Study Center**

Video: Catalytic Hydrogenation

The apparatus in Figure 25.5 will be assembled. The graduated cylinder is filled with hydrogen, and the reaction flask containing the solvent, the catalyst, and a magnetic stirring bar is flushed with hydrogen. The stirrer is started, and as soon as the water level in the reservoir reaches a constant level, the compound to be hydrogenated is injected through the septum of the reaction flask. It can be either a pure liquid or a solid dissolved in alcohol.

**■ FIG. 25.5**
The catalytic hydrogenation
apparatus.

Reservoir filled
with hydrogen

Polyethylene
tube

10-mL
Graduated
cylinder

Stirring
bar

Magnetic
stirrer

A millimole (0.001 mole) of hydrogen at standard temperature and pressure has a volume of 22.4 mL. This will hydrogenate a millimole of a compound that has one double bond. The apparatus in Figure 25.5 holds about 9 mL of gas (0.4 mmol).

For highly accurate measurement of the hydrogen uptake, we should take into account the temperature, the atmospheric pressure, and the vapor pressure of the methanol in the reaction flask. We should also be careful to equalize the internal and external pressures so that precise volume measurements could be made (Fig. 25.6). The experiments in this chapter can be run satisfactorily without this level of accuracy.

The apparatus for hydrogen generation is in Figure 25.7. The 5-mL long-necked flask contains 1 g of zinc. As hydrogen is needed, sulfuric acid is injected through the septum.

**Online Study Center**

Photo: Placing a Polyethylene
Tube through a Septum

■ **FIG. 25.6**

An apparatus for quantitative hydrogenation. Air in the apparatus is removed with an aspirator, and hydrogen is admitted and brought to atmospheric pressure by raising or lowering the leveling bulb. The stirrer is started, and the amount of hydrogen taken up is measured in a gas burette after again bringing the pressure inside to that of the atmosphere.

■ **FIG. 25.7**

A hydrogen generator. *See* Figure 18.2 on page 347 for threading a polyethylene tube through a septum.

## *Generation and Storage of Hydrogen*

**IN THIS EXPERIMENT** hydrogen is generated by the addition of acid to zinc. It is stored in a graduated cylinder by the downward displacement of water.

Assemble the apparatus shown in Figure 25.5. In a 5-mL short-necked, round-bottomed flask place 20 mg of 10% palladium on carbon, a small magnetic stirring bar, and 1.5 mL of methanol. Fill a 250-mL or 400-mL beaker with water. Fill a 10-mL or 25-mL graduated cylinder with water, seal the top with your finger or a loose cork, and invert it in the beaker of water. Clamp it in this position in the beaker of water, removing the cork if used. In the 5-mL long-necked flask place 1 g of granulated or mossy zinc and close it with a septum bearing a polyethylene tube, folding the rubber flange over the outside of the flask neck (Fig. 25.7). Put a kink in the tube by exposing a short section of it to a heat source. This will make insertion of the tube into the cylinder easier. Inject a few drops of 6 *M* (33%) sulfuric acid to generate hydrogen and displace the air from the 5-mL short-necked flask. Thread the polyethylene tubing into the bottom of the graduated cylinder clamped over a beaker of water (*see* Fig. 25.5). Inject more acid onto the zinc and fill the graduated cylinder reservoir with about 9 mL of hydrogen. Adjust the height of the graduated cylinder so that the water level inside is the same as that outside. Remove the septum and the tube from the gas-generating flask and connect them to the short-necked flask as in Figure 25.5.

Turn on the stirrer briefly to saturate the methanol with hydrogen. Note the level of the water in the cylinder. If it comes to the same level as the water on the outside when the cylinder is moved up and down, there is a leak in the system.

## *Hydrogenation*

**IN THIS EXPERIMENT** a liquid alkene is hydrogenated to give a solid product. Uptake of hydrogen is easily observed. The catalyst is removed by filtration, and the solvent is evaporated to give pure product.

Once the water level in the graduated cylinder has stabilized with the stirrer off, inject 0.4 mmol of a compound that has one olefinic double bond to be hydrogenated. In this particular experiment, this will be 107 mg of oleyl alcohol dissolved in 0.6 mL of methanol. If a larger reservoir of hydrogen has been used, for example, a 25-mL graduated cylinder, a proportionately larger quantity of alkene can be hydrogenated. Turn on the stirrer and note the height of the hydrogen in the reservoir as a function of time. Be sure the end of the tube is always above the water level in the reservoir. When the water level no longer changes, the reaction is complete.

Filter the reaction mixture into a tared 25-mL filter flask using a Pasteur pipette; push a piece of cotton firmly into a Pasteur pipette followed by a 5-mm layer of Celite filter aid. With a second pipette, transfer the reaction mixture to the filter pipette and allow the solution to filter through the cotton to remove the charcoal. Force the solution through the cotton with a rubber bulb (*see* Fig. 25.2 on page 406). Rinse the reaction flask, the stirring bar, and the transfer pipette with a few drops of methanol; then filter this solution as well. If the filtrate is gray, refilter the solution in the same pipette without changing the filter.

Remove the methanol from the filter flask under aspirator vacuum (*see* Fig. 25.9 on page 416). Heat the filter flask on a steam bath or a sand bath under vacuum to remove the last traces of methanol. The residue should be pure 1-octadecanol (mp 60°C), although the melting point will depend on the quality of the starting alcohol. Test a small portion of the product for unsaturation using bromine solution.

**Cleaning Up.** Decant the sulfuric acid from the zinc, dilute the solution, neutralize with sodium carbonate, and remove zinc hydroxide by filtration. The filtrate can be flushed down the drain. Zinc and zinc hydroxide are placed in the nonhazardous solid waste container. Wash out the syringe and needle carefully.

***Online Study Center***

Photo: Preparation of a Filter Pipette

***Online Study Center***

Photo: Evaporation of Solvent in a Filter Flask

# 3. Brown Hydrogenation[1]

$$\text{(structure)} \xrightarrow[\text{PtCl}_4]{\text{NaBH}_4-\text{HCl}} \text{(structure)}$$

*cis*-**Norbornene-5,6-*endo*-
dicarboxylic acid**
mp 180–190°C, dec
MW 182.17

*cis*-**Norbornane-5,6-*endo*-
dicarboxylic acid**
mp 170–175°C, dec
MW 184.19

**IN THIS EXPERIMENT** platinum chloride is reduced with borohydride to metallic platinum that is deposited on charcoal. An aqueous solution of an alkene carboxylic acid and a borohydride solution is added to the sealed flask. The hydrogen gas is rapidly absorbed while swirling the solution. The aqueous solution is extracted with ether, and in the usual way the ether is washed, dried, and evaporated. The solid product is recrystallized from aqueous hydrochloric acid.

*See* Figure 25.6 for a quantitative hydrogenation apparatus. However, in this experiment we will measure the uptake of hydrogen in a semiquantitative manner. The reaction vessel is a 125-mL filter flask with a white rubber pipette bulb wired onto the side arm (Fig. 25.8). Introduce 10 mL of water, 1 mL of platinum(IV) chloride solution,[2] and 0.5 g of decolorizing charcoal, and swirl during the addition of 3 mL of stabilized 1 *M* sodium borohydride solution.[3] While allowing 5 min for the formation of the catalyst, dissolve 1 g of *cis*-norbornene-5,6-*endo*-dicarboxylic acid in 10 mL of hot water. Pour 4 mL of concentrated hydrochloric acid into the reaction flask, followed by the hot solution of the unsaturated acid. Cap the flask with a large rubber septum and wire it on. Draw 1.5 mL of the stabilized sodium borohydride solution into the barrel of a plastic syringe, thrust the needle through the center of the rubber septum, and add the solution dropwise with swirling. The initial uptake of hydrogen is so rapid that the balloon may not inflate until you start injecting a second 1.5 mL of borohydride solution through the septum. When the addition is complete and the reaction appears to be reaching an endpoint (about 5 min), heat the flask on a steam bath while swirling and try to estimate the time at which the balloon is deflated to a constant size (about 5 min). When balloon size is constant, heat and swirl for an additional 5 min; then release the pressure by pushing the needle of an open syringe through the rubber septum.

*Handle platinum chloride
carefully to prevent any waste.*

*Ideal reaction time: about
15 min*

1. Brown, H. C.; Brown, C. A. *J. Am. Chem. Soc.* **1962**, *84*, 1495.

2. A solution of 200 mg of PtCl₄ in 4 mL of water.

3. Dissolve 1.6 g of sodium borohydride and 0.30 g of sodium hydroxide pellets (stabilizer) in 40 mL of water. When not in use, the solution should be stored in a refrigerator. If left for some time at room temperature in a tightly stoppered container, enough gas pressure may develop to explode the container.

*Online Study Center*

Video: Microscale Filtration on the Hirsch Funnel; Photo: Extraction with Ether

Common ion effect

Filter the hot solution by suction on a Hirsch funnel and place the catalyst in a jar marked "Catalyst Recovery." Cool the filtrate and extract it with three 15-mL portions of ether. The combined extracts are to be washed with saturated sodium chloride solution and dried over anhydrous calcium chloride pellets. Evaporation of the ether gives about 0.8–0.9 g of white solid.

Acidified water appears to be the best solvent for the crystallization of the saturated *cis*-diacid. The diacid is very soluble in water and crystallizes extremely slowly with poor recovery. However, the situation is materially improved by adding a little hydrochloric acid to decrease the solubility of the diacid.

Scrape out the bulk of the solid product and transfer it to a 25-mL Erlenmeyer flask. Add 1–2 mL of water to the 125-mL filter flask, heat to boiling to dissolve residual solid, and pour the solution into the 25-mL flask. Bring the material into solution at the boiling point with a total of no more than 3 mL of water (as a guide, measure 3 mL of water into a second 25-mL Erlenmeyer). With a Pasteur pipette add 3 drops of concentrated hydrochloric acid and let the solution stand for crystallization. Clusters of heavy prismatic needles soon separate; the recovery is about 90%. The product should give a negative test for unsaturation with acidified permanganate solution (1% potassium permanganate in 10% sulfuric acid).

Observe what happens when a sample of the product is heated in a melting-point capillary to about 170°C. Account for the result. You may be able to confirm your inference by letting the oil bath cool until the sample solidifies and then noting the temperature and behavior on remelting.

**Cleaning Up.** The catalyst removed by filtration may be pyrophoric (i.e., spontaneously flammable in air). Immediately remove it from the Hirsch funnel, wet it with water, and place it either in the hazardous waste container or in the catalyst recovery container. It should be kept wet with water at all times. The combined aqueous filtrates, after neutralization, are diluted with water and flushed down the drain. If local regulations permit, the ether is allowed to evaporate from the calcium chloride pellets in the hood, and they can be placed in the nonhazardous solid waste container. Otherwise, place the wet pellets in the designated container.

# 4. Transfer Hydrogenation of Olive Oil[4,5]

**IN THIS EXPERIMENT** olive oil is treated with a palladium-on-carbon catalyst and the hydrogen donor cyclohexene to produce a solid fat by transfer hydrogenation. In Experiment 5 olive oil is catalytically reduced using hydrogen gas. If both experiments are performed, the results can be compared. Analyze the two products by titration with bromine in carbon tetrachloride (or dichloromethane) and also by NMR spectroscopy. Alternatively, you can carry out just one reduction and compare your results with a classmate who used the other procedure.

---

4. Discussions with Gottfried Brieger regarding transfer hydrogenation are gratefully acknowledged.

5. Barry B. Snider of Brandeis University points out some interesting problems with this experiment. *See* the text website.

Olive oil, like other vegetable oils containing unsaturated fatty acids, when consumed in food, favors the production of high-density lipoproteins (HDL) that do not deposit as much cholesterol in the arteries as the low-density lipoproteins (LDL); LDLs contribute to the disease arteriosclerosis (hardening of the arteries).

Olive oil is a triester consisting of a trihydric alcohol, glycerol, and three long chain fatty (carboxylic) acids. The fatty acid in olive oil is primarily oleic acid, an 18-carbon monounsaturated compound. So olive oil can be regarded as primarily glycerol trioleate, although it contains about 15% saturated fat and an equal quantity of polyunsaturated fat. The double bond in oleic acid has the *cis* configuration; the molecule is bent in the center and does not pack well into a crystal lattice; thus, olive oil is a liquid at room temperature. If the double bonds are saturated, then the triester, glycerol tristearate, is a solid, and melts at almost 70°C. Similarly, a molecule having *trans* double bonds is a solid.

$$\underset{\substack{\text{Glycerol trioleate}\\ \text{mp } -5.5°\text{C}\\ \text{MW } 885.47}}{\begin{array}{l} \text{CH}_2\text{O}\overset{\displaystyle\overset{O}{\|}}{\text{C}}(\text{CH}_2)_7\text{CH}\!=\!\text{CH}(\text{CH}_2)_7\text{CH}_3 \\ \qquad\qquad\qquad\quad cis \\[2pt] \text{CHO}\overset{\displaystyle\overset{O}{\|}}{\text{C}}(\text{CH}_2)_7\text{CH}\!=\!\text{CH}(\text{CH}_2)_7\text{CH}_3 \\ \qquad\qquad\qquad\quad cis \\[2pt] \text{CH}_2\text{O}\overset{\displaystyle\overset{O}{\|}}{\text{C}}(\text{CH}_2)_7\text{CH}\!=\!\text{CH}(\text{CH}_2)_7\text{CH}_3 \\ \qquad\qquad\qquad\quad cis \end{array}} \quad\xrightarrow[\text{10\% Pd/C}]{}\quad \underset{\substack{\text{Glycerol tristearate}\\ \text{mp } 69.9°\text{C}\\ \text{MW } 891.52}}{\begin{array}{l} \text{CH}_2\text{O}\overset{\displaystyle\overset{O}{\|}}{\text{C}}(\text{CH}_2)_{16}\text{CH}_3 \\[2pt] \text{CHO}\overset{\displaystyle\overset{O}{\|}}{\text{C}}(\text{CH}_2)_{16}\text{CH}_3 \\[2pt] \text{CH}_2\text{O}\overset{\displaystyle\overset{O}{\|}}{\text{C}}(\text{CH}_2)_{16}\text{CH}_3 \end{array}}$$

To manufacture margarine, a vegetable oil such as olive oil or, more commonly, corn or soybean oil is partially hydrogenated. This gives a mixture of liquid and solid fats that possesses the consistency of butter. In the process some of the *cis* double bonds are also isomerized to the more stable *trans* isomers. Like saturated fatty acids, the *trans* isomers also contribute to the formation of LDL. When oils such as these are completely hydrogenated, the resulting fat is a solid. Many saturated fats are found in animals. The hard fat (tallow) on a raw steak is made up of saturated fats.

In this procedure we use 100% extra-virgin olive oil from Italy. The label says that of each 14 g of oil, 10 are monounsaturated (mostly from oleic acid), 2 are polyunsaturated (mostly linoleic acid, an 18-carbon acid with two double bonds), and 2 are saturated (this is a mixture of predominantly 18-carbon saturated stearic acid and a smaller amount of 16-carbon palmitic acid).

From this rough analysis, it can be concluded that complete hydrogenation of all the double bonds in olive oil will give a product containing about 90%–95% stearic acid (C-18) and 5%–10% palmitic acid (C-18) esterified to glycerol. Although the product will not be pure glycerol tristearate, it should be a solid with a melting point above 50°C.

The technique of transfer hydrogenation is employed. In the presence of a catalyst, hydrogen is lost from the cyclohexene donor molecules and transferred to the double bonds in the olive oil. Although it cannot be used in all cases, this technique of hydrogenation is quite convenient because it consists of simply re-fluxing for a few minutes the substance to be hydrogenated with the hydrogen donor and the catalyst.

A variety of hydrogen donor molecules can be used, among which are cyclo-hexene, hydrazine, formic acid, cyclohexadiene, and ammonium formate. When ammonium formate decomposes in the presence of a catalyst, it produces hydro-gen, ammonia, and carbon dioxide. Not only does reduction of double bonds take place, but the ammonia can react with esters to form amides. Cyclohexene and cyclohexadiene both lose hydrogen to form the same stable end product. In this experiment, cyclohexene is used as the hydrogen donor.

### Microscale Procedure

In a reaction tube, place 400 mg of olive oil, 1 mL of cyclohexene, 50 mg of 10% palladium-on-carbon catalyst, and a boiling chip. Reflux the mixture for at least 30 min. Be careful not to boil the alkene out of the reaction tube. If necessary, place a coil of damp pipe cleaner around the top of the tube. This will keep the top of the tube cool so the cyclohexene will not escape. The empty distilling column could also be mounted on top of the reaction tube to provide even more condens-ing area (*see* Fig. 25.1 on page 406); however, with careful adjustment of the heat input (the depth of the tube in the sand), this will not be necessary. The isolation procedure is given in Experiment 5.

### Macroscale Procedure

In a standard taper 14/20 flask equipped with a water-cooled condenser, place 0.80 g of olive oil, 2 mL of cyclohexene, 100 mg of 10% palladium-on-carbon catalyst, and a boiling chip (*see* Fig. 25.3 on page 407). Reflux the mixture for at least 15 min. The isolation procedure is given in Experiment 5.

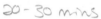

20 – 30 mins

## 5. Catalytic Hydrogenation of Olive Oil

### Microscale Procedure

Follow the procedure of Experiment 2, using the apparatus illustrated in Fig-ure 25.5 on page 409. If 9 mL of hydrogen is generated, this will be sufficient to com-pletely hydrogenate 119 mg of olive oil (assuming that the olive oil is 100 percent glycerol trioleate). Demonstrate in your laboratory notebook that this calculation is correct.

Dissolve the olive oil in 0.6 mL of methanol and inject it into a flask contain-ing 1.5 mL of methanol, 20 mg of 10 percent palladium-on-carbon catalyst, and a small magnetic stirring bar. Stir the reaction mixture until the uptake of hydrogen ceases and then isolate the product.

## Macroscale Procedure

Follow the procedure of Experiment 2. Using an apparatus similar to that in Figure 25.5 on page 409 but with a larger beaker and graduated cylinder, generate 45 mL of hydrogen and hydrogenate 590 mg of olive oil. Dissolve the olive oil in 3 mL of methanol and inject it into a 25-mL flask containing 7.5 mL of methanol, 100 mg of 10 percent palladium-on-carbon catalyst, and a small magnetic stirring bar. Stir the reaction mixture until the uptake of hydrogen ceases and then isolate the product.

## Isolation of Products

Prepare a Pasteur pipette as a micro filter by forcing a small piece of cotton firmly down to the constriction (*see* Fig. 25.2 on page 406). Add the reaction mixture to the filter pipette and then force the solution through the cotton into a tared, 25-mL filter flask. Rinse the reaction tube, filter with a few drops of hexane, and then evaporate the solution to dryness (Fig. 25.9). To remove the last traces of volatile liquid from the product, heat the filter flask in a hot sand bath under vacuum. After cooling on ice, the product should solidify to a hard, white fat. Analyze a portion of the product. The bulk of the product can be used for the synthesis of soap (*see* Experiment 6 in Chapter 40).

Vacuum evaporation while heating is necessary to ensure that the fat will solidify when cooled on ice.

## Analysis of Products

If you performed both Experiments 4 and 5, determine and compare the weights of the products and their physical properties. Count the number of drops of 3% bromine in carbon tetrachloride (or dichloromethane) that can be decolorized by equivalent quantities of each product and of olive oil.

Analyze the NMR spectra of olive oil and of the two products. In olive oil a complex set of peaks centered at 5.38 ppm are produced by protons on a *cis* double bond. If the catalyst has isomerized the *cis* to the *trans*, double-bond peaks for the *trans* isomer will be found centered at 5.35 ppm. Completely hydrogenated olive oil will, of course, have no peaks in this region of the spectrum. A set of 5 peaks centered at 5.26 ppm will always be present and can be used to calibrate the integration. These peaks arise from the hydrogen on the central carbon of the glycerol part of the triester.

■ **FIG. 25.9**
An apparatus for removing a solvent under vacuum.

The IR spectra of olive oil (Fig. 25.10) and of the two products show very small differences. A weak peak in the 3013–3011 cm$^{-1}$ region is characteristic of the *cis* double bond, and a weak peak in the 970–967 cm$^{-1}$ region is characteristic of the *trans* double bond.

## Computational Chemistry

Calculate, using a semiempirical molecular orbital program such as AM1, the heats of formation of cyclohexene, cyclohexane, and benzene. Use this information to rationalize the results when using cyclohexene as the hydrogen donor.

To aspirator

■ **FIG. 25.10**
The IR spectrum (thin film) of olive oil (Bertoli, extra virgin), which consists primarily of glycerol trioleate.

Molecular mechanics steric energies for these molecules do not allow valid comparisons to be made among the three.

### For Further Investigation

Reflux olive oil with 10% palladium-on-carbon catalyst in an inert solvent such as cyclohexane. Analyze the product by NMR spectroscopy. Has the catalyst alone caused the *cis* double bond to isomerize to the *trans*? Analyze samples of commercial margarine by NMR. Are *trans* double bonds present?

## QUESTIONS

1. How would you convert the reduced norbornane diacid product to the corresponding bicyclic anhydride?
2. Why does the addition of hydrochloric acid in Experiment 2 cause the solubility of the bicyclic diacid in water to decrease?
3. Why is decolorizing charcoal added to the hydrogenation reaction mixture in the experiment involving Brown hydrogenation?
4. In the transfer hydrogenation of olive oil, what compound is cyclohexene ultimately converted to?

**Online Study Center**

**General Resources**
Web Links

5. What is the driving force for the transfer hydrogenation of olive oil when using cyclohexene?

6. Write balanced equations for the transfer hydrogenation of pure glycerol trioleate with cyclohexene and ammonium formate as the hydrogen sources. Calculate the number of milligrams of each donor needed to hydrogenate 0.4 g of glycerol trioleate using transfer hydrogenation.

# Sodium Borohydride Reduction of 2-Methylcyclohexanone: A Problem in Conformational Analysis

*Online Study Center*

This icon will direct you to techniques, equipment setups, and online resources at http://college.hmco.com/PIC/ williamsonMMOE5e.

**PRELAB EXERCISE:** Using carefully prepared drawings, try to predict whether *cis*- or *trans*-2-methylcyclohexanol will predominate in this reduction. Then use molecular models to check your conclusion. Finally, if you have access to the appropriate software, see if semiempirical molecular orbital calculations will change your conclusions.

The objective of this experiment is to determine the structures and relative percentage yields of the products formed when 2-methylcyclohexanone is reduced with sodium borohydride. We also compare these results to predictions made using molecular mechanics and semiempirical molecular orbital calculations.

In the reduction of 2-methylcyclohexanone, both the *cis* and *trans* isomers of 2-methylcyclohexanol can be formed:

$$O \quad \xrightarrow{NaBH_4} \quad OH \quad + \quad OH$$

*cis*       *trans*

2-Methylcyclohexanone       2-Methylcyclohexanol

An examination of the models reveals that the two cyclohexanols can, in principle, exist in four chair conformations:

*trans*, diequatorial       *trans*, diaxial

cis, equatorial methyl,
axial hydroxyl

cis, axial methyl,
equatorial hydroxyl

A molecular mechanics program can be used to calculate the relative energies of these four isomers and help you to predict the most stable *trans* and *cis* conformations. Even without calculation, you should be able to predict which of the two *trans* conformations is the more stable.

By carrying out semiempirical molecular orbital calculations on the starting material (*see* Chapter 15), you may be able to decide, based on the shape and location of the lowest unoccupied molecular orbital (LUMO), whether the borohydride anion will attack the carbonyl group from the top, to give predominantly the *trans* isomer, or from the bottom, to give mostly the *cis* isomer. You may also be able to make this prediction by studying a molecular model or even a drawing of 2-methylcyclohexanone:

You can determine the structures and relative percentage yields of the products in this reaction using nuclear magnetic resonance (NMR) spectroscopy. The 250-MHz $^1$H NMR spectrum of a 50:50 mixture of methylcyclohexanols is given in Figure 26.1. The two lowest field groups of peaks are from the hydrogen atom on the hydroxyl-bearing carbon atom.

The reduction of 2-methylcyclohexanone with sodium borohydride will give a mixture of products but not necessarily a 50:50 mixture. Integration of the two low-field groups of peaks will allow determination of the actual percentages of products in this reaction. First, however, the peaks must be assigned to the *cis* and *trans* isomers. The groups of peaks have been expanded and numbered on the spectrum in Figure 26.1, and the frequencies of the 11 peaks are listed in the caption.

The NMR coupling constant of the proton on the hydroxyl-bearing carbon is a function of the dihedral angle between that proton and an adjacent vicinal proton. This is seen most easily by examining molecular models and Newman projections. For the *trans* isomer of 2-methylcyclohexanol, the predominant conformer is the one in which both the methyl group and hydroxyl group are equatorial, and $H_A$ and $H_B$ are diaxial with a dihedral angle of 180° between them. The coupling constant for dihedral hydrogens is normally in the 9–12 Hz range. For the less favorable conformer, $H_A$ and $H_B$ are both equatorial, and the dihedral angle between them is 60°. This results in a coupling constant in the 2–5 Hz range.

**■ FIG. 26.1**

The 250-MHz NMR spectrum of a mixture of cis- and trans-2-methylcyclohexanol. The frequencies of peaks 1 to 5 are 950.3, 947.5, 944.8, 942.2, and 939.5 Hz, respectively. The frequencies of peaks 6 to 11 are 791.2, 786.9, 781.6, 777.1, 771.6, and 767.4 Hz, respectively.

For *cis*-2-methylcyclohexanol, the two possible conformers are about equal in energy. In either conformer, $H_A$ and $H_B$ are in a 60° axial-equatorial stereochemical relationship, which results in a coupling constant in the 4–7 Hz range.

From this information, it should be possible to assign the groups of peaks at 3.1 and 3.8 ppm to either *cis*- or *trans*-2-methylcyclohexanol. From the areas of the two sets of peaks, the relative percentages of the isomers can be determined.

# EXPERIMENT

# Borohydride Reduction of 2-Methylcyclohexanone

This reaction can be run on a scale four times larger in a 25-mL Erlenmeyer flask.

**IN THIS EXPERIMENT** a liquid ketone in methanol is reduced with solid sodium borohydride. Base is added, and the product is extracted into dichloromethane. In the usual way this organic layer is dried and evaporated to leave the liquid alcohol. The same procedure could be used to reduce most ketones to their corresponding alcohols.

To a reaction tube add 1.25 mL of methanol and 300 mg of 2-methylcyclohexanone. Cool this solution in an ice bath contained in a small beaker. While the reaction tube is in the ice bath, carefully add 50 mg of sodium borohydride to the solution. After the vigorous reaction has ceased, remove the tube from the ice bath and allow it to stand at room temperature for 10 min, at which time the reaction should appear to be finished. To decompose the borate ester, add 1.25 mL of 3 *M* sodium hydroxide solution. To the resulting cloudy

**■ FIG. 26.2**
The IR spectrum of a
mixture of *cis*- and *trans*-
2-methylcyclohexanol (thin film).

solution, add 1 mL of water. The product will separate as a small, clear upper layer. Remove as much of this as possible, place it in a reaction tube, and then extract the remainder of the product from the reaction mixture with two 0.5-mL portions of dichloromethane. Add these dichloromethane extracts to the small product layer and dry the combined extracts over anhydrous sodium sulfate (not calcium chloride). After a few minutes, transfer the solution to a dry reaction tube containing a boiling chip. In the hood, boil off the dichloromethane (and any accompanying methanol) and use the residue to run an NMR spectrum in the usual way.

Integrate the two low-field sets of peaks at 3.1 and 3.8 ppm, analyze the coupling constant patterns to assign the two sets of peaks, and report the percentage distribution of *cis*- and *trans*-2-methylcyclohexanol formed in this reaction. The relative amounts of the products can, of course, also be determined by gas chromatography. How do these results compare to your predictions and calculations? Which *cis* and which *trans* conformation is the more stable?

Run an infrared (IR) spectrum of the product as a thin film to determine whether the reduction of the starting material has been completed (Fig. 26.2).

**Cleaning Up.** The reaction mixture is neutralized with acetic acid (to react with sodium borohydride) and flushed down the drain with water.

## *Computational Chemistry*

Using a molecular mechanics program, calculate the steric energies of *cis*- and *trans*-2-methylcyclohexanol. Each of these isomers has two principal conformations. Does reduction of 2-methylcyclohexanone give the more stable isomer? Explain.

***Online Study Center***

**General Resources**
Additional Experiments

## QUESTIONS

**1.** Draw the NMR peak expected for $H_A$ when $H_A$ couples with $H_B$ in the following structure.

**2.** Draw the NMR peak expected when $H_A$ couples with both $H_B$ and $H_C$. Remember that the coupling constants are not equal.

**3.** What is the approximate frequency of the most important peak in the starting material that should be absent in the IR spectrum of the product?

# CHAPTER 27

# Epoxidation of Cholesterol

**Online Study Center**

This icon will direct you to techniques, equipment setups, and online resources at http://college.hmco.com/PIC/williamsonMMOE5e.

**PRELAB EXERCISE:** When milk is irradiated with ultraviolet (UV) light, the vitamin D content increases. What type of reaction is taking place?

**3-Chloroperoxybenzoic acid**
MW 172.57
mp 92–94°C (dec)

**Cholesterol**
MW 386.66
mp 149°C

**5α, 6α-Epoxycholestan-3β-ol**
MW 402.66
mp 110–112°C
mp 139°C

**3-Chlorobenzoic acid**
MW 156.57
mp 157°C

CH₂Cl₂

This experiment carries out an epoxidation reaction on cholesterol, which is representative of a very important group of molecules—the steroids. The rigid cholesterol molecule gives products of well-defined stereochemistry. The epoxidation reaction is stereospecific, and the product can be used to carry out further stereospecific reactions.

Cholesterol is the principal constituent of gallstones and can be readily isolated from them. The average person contains about 200 g of cholesterol, which is primarily found in brain and nerve tissue. The blockage of arteries by cholesterol also leads to the disease arteriosclerosis (hardening of the arteries).

Certain naturally occurring and synthetic steroids have powerful physiological effects. Progesterone and estrone are the female sex hormones, and testosterone is

**Online Study Center**

Video: Cholesterol from Human Gallstones

the male sex hormone; these hormones are responsible for the development of secondary sex characteristics. The closely related synthetic steroid, norethisterone, is an oral contraceptive, and the addition of four hydrogen atoms (reduction of the ethynyl group to the ethyl group) and a methyl group gives an anabolic steroid, ethyltestosterone. This muscle-building steroid is now outlawed for use by Olympic athletes. RU-486, mifepristone, is an abortifacient. Fluorocortisone is used to treat inflammations such as arthritis, and ergosterol on irradiation with UV light is converted to vitamin $D_2$.

**Progesterone**
**Female sex hormone**

**Estrone**
**Female sex hormone**

**Norethisterone**
**Oral contraceptive**

**Ethyltestosterone**
**Anabolic steroid**

**RU-486**
**Mifepristone**
**Abortifacient**

    Much of our present knowledge about the stereochemistry of reactions was developed from steroid chemistry. In the epoxidation of cholesterol, the double bond of cholesterol is stereospecifically converted to the 5α, 6β epoxide. The α designation indicates the epoxide is on the back side of the molecule as it is usually printed. A substituent on the top side (above the plane of the paper) is designated β. A study of molecular models reveals that the angular methyl group prevents top-side attack on the double bond by a perbenzoic acid; thus, the epoxide forms exclusively on the back or α side of the molecule.

**Testosterone**
**Male sex hormone**

**Fluorocortisone**
**Anti-inflammatory**

**Ergosterol**

*hv*

**Vitamin D₂**

Epoxides are most commonly formed by the reaction of a peroxycarboxylic acid with an olefin at room temperature. It is a one-step cycloaddition reaction:

Some peroxyacids are explosive; the reagent used in Experiments 1 and 2 is a particularly stable and convenient peroxycarboxylic acid.

The reaction is carried out in an inert solvent, dichloromethane, and the product is isolated by chromatography. No great care is required in the chromatography to collect fractions because the 3-chlorobenzoic acid, being polar, is adsorbed strongly onto the alumina, while the relatively nonpolar product is eluted easily by ether. After the removal of ether, the product is easily recrystallized from a mixture of acetone and water.

## Steroid Biosynthesis through an Epoxide Intermediate

The biosynthesis of cholesterol and the other naturally occurring steroids shown in this chapter proceeds through an epoxide intermediate. The $C_{30}$ triterpene hydrocarbon squalene (Latin *squalus*, whale) is epoxidized to give squalene oxide. This epoxide is protonated and then cyclizes to form the characteristic four-ring steroid skeleton and, after several steps, cholesterol, in a series of very carefully studied enzyme-catalyzed reactions.

Squalene

Squalene oxide

Several steps

Cholesterol

# EXPERIMENTS

## 1. Cholesterol Epoxide

> **IN THIS EXPERIMENT** a steroid alkene is converted to an epoxide with meta-chloroperbenzoic acid (MCPB) in dichloromethane. The MCPB produced is removed by what amounts to a simple rapid filtration through alumina. Evaporation of the solvent gives the epoxide, which is recrystallized from a mixture of acetone and water.

Because dichloromethane is a suspected carcinogen, handle it in a hood.

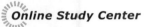

**Online Study Center**

Photo: Column Chromatography;
Video: Column Chromatography

**Online Study Center**

Video: Microscale Filtration on the Hirsch Funnel

Dissolve 194 mg of cholesterol in 0.8 mL of dichloromethane by gentle warming in a 10 × 100 mm reaction tube. In another reaction tube, dissolve 117 mg of 80% 3-chloroperoxybenzoic acid (or 187 mg of 50% material) in 0.8 mL of dichloromethane by gentle warming. Cool the two solutions and then mix them together. They must be cool before mixing because the reaction is exothermic. Stopper the reaction tube and place it in a beaker of water at 40°C for 30 min to complete the reaction. The progress of the reaction can be followed by thin-layer chromatography (TLC) on silica gel plates and using *t*-butyl methyl ether as the eluent.

The reaction mixture is pipetted into a chromatography column (Fig. 27.1) prepared from 3 g of alumina. Follow the procedure described in Chapter 9, except use ether to fill the column and to prepare the alumina slurry. The 3-chlorobenzoic acid will be strongly adsorbed by the alumina. The product is eluted with 30 mL of ether collected in a tared 50-mL round-bottomed flask. The ether can flow through the column by gravity or can be forced out using pressure from a rubber bulb (flash chromatography).

Most of the ether is removed on a rotary evaporator; the last traces are removed using the apparatus depicted in Figure 9.6 (on page 206) for drying a solid under reduced pressure or that depicted in Figure 9.10 (on page 211). The residue should weigh more than 150 mg. If it does not, pass more ether through the column and collect the product as before. Dissolve the product in 1.5 mL of warm acetone and, using a Pasteur pipette, transfer it to a reaction tube. Add 0.2 mL of water to the solution, warm the mixture to bring the solid into solution, and then let the tube and contents cool slowly to room temperature. Cool the mixture in ice and collect the product on a Hirsch funnel. Press the solid down on the filter to squeeze solvent from the crystals; then wash the product with 0.25 mL of ice-cold 90% acetone. Spread the product out on a watch glass to dry. Determine the weight and melting point of the product and calculate the percentage yield.

**Cleaning Up.** Place dichloromethane solutions in the halogenated organic solvents container and organic solvents in the organic solvents container. The alumina should be placed in the hazardous waste container. If it should be necessary to destroy 3-chloroperoxybenzoic acid, add it to an excess of an ice-cold solution of saturated sodium bisulfite in the hood. A peracid will give a positive starch-iodide test (blue-purple color).

**FIG. 27.1**
A chromatography column.

Funnel

Glass
column

Alumina

Micro
Büchner
funnel

Polyethylene
frit

10-mL
Erlenmeyer
flask

## 2. Cholesterol Epoxide

Dissolve 1 g of cholesterol in 4 mL of dichloromethane (*see* caution statement on previous page) in a 25-mL Erlenmeyer flask by gentle warming. In another 25-mL flask dissolve 0.6 g of 80% 3-chloroperoxybenzoic acid in 4 mL of dichloromethane by gentle warming. Cool the two solutions and mix together. The two solutions must be cool before mixing because the reaction is exothermic. Clamp the flask in a beaker of water at 40°C for 30 min to complete the reaction. The progress of the reaction can be followed by TLC on silica gel plates using ether as the eluent.

The reaction mixture is pipetted into a chromatography column prepared from 15 g of alumina. Follow the procedure described in Chapter 9, except use ether to fill the column and to prepare the alumina slurry. The 3-chlorobenzoic acid will be strongly adsorbed by the alumina. The product is eluted with 150 mL of ether collected in a tared 250-mL round-bottomed flask.

Most of the ether is removed on a rotary evaporator, and the last traces are removed using the apparatus depicted in Figure 9.6 (on page 206) for drying a solid under reduced pressure. The residue should weigh more than 0.75 g. If it does not, pass more ether through the column and collect the product as before. Dissolve the product in 7.5 mL of warm acetone and, using a Pasteur pipette, transfer it to a 25-mL Erlenmeyer flask. Add 1 mL of water to the solution, warm the mixture to bring the solid into solution, and then let the flask and contents cool slowly to room temperature. Cool the mixture in ice and collect the product on a Hirsch funnel. Press the solid down on the filter to squeeze solvent from the crystals; then wash the product with 1 mL of ice-cold 90% acetone. Spread the product out on a watch glass to dry. Determine the weight and melting point of the product and calculate the percent yield.

**Cleaning Up.** Place dichloromethane solutions in the halogenated organic solvents container and organic solvents in the organic solvents container. The alumina should be placed in the hazardous waste container. If it should be necessary to destroy 3-chloroperoxybenzoic acid, add it to an excess of an ice-cold solution of saturated sodium bisulfite in the hood. A peracid will give a positive starch-iodide test (blue-purple color).

## QUESTIONS

1. How many moles of each reactant are used in this experiment? Assume the 3-chloroperoxybenzene acid is 80% pure.
2. What simple test could you perform to show that 3-chlorobenzoic acid is not eluted from the chromatography column?

## REFERENCE

Carl Djerassi, *This Man's Pill: Reflections on the 50th Birthday of the Pill.* Oxford, UK: Oxford University Press, 2001.

# CHAPTER

# 28

# Nitration of Methyl Benzoate

**PRELAB EXERCISE:** Draw the complete mechanism for the nitration of chlorobenzene. Chlorine is an *ortho-para* director and deactivator of the benzene ring.

The nitration of methyl benzoate is a typical electrophilic aromatic substitution reaction. The electrophile is the nitronium ion generated by the interaction of concentrated nitric and sulfuric acids:

$$HNO_3 + 2H_2SO_4 \rightleftharpoons NO_2^+ + 2HSO_4^- + H_3O^+$$

**Nitronium ion**

Sulfuric acid protonates the methyl benzoate:

The nitronium ion then reacts with this protonated intermediate at the *meta* position, where the electron density is highest, that is, where there is no positively charged resonance form and yields the intermediate arenium ion which has the four resonance forms shown here:

431

The arenium ion intermediate then transfers a proton to the basic bisulfate ion to give methyl 3-nitrobenzoate:

The ester group is a *meta* director and deactivator of the benzene ring. It is much easier to nitrate a molecule such as anisole, where the methoxyl group is an *ortho-para* director and an activator of the benzene ring, as the following resonance structures indicate:

---

## EXPERIMENTS

## 1. Microscale Nitration of Methyl Benzoate

$$HNO_3 + 2H_2SO_4 \rightleftharpoons NO_2^+ + 2HSO_4^- + H_3O^+$$

**Nitronium ion**          **Hydronium ion**

**Methyl benzoate**
MW 136.16
bp 199.6°C
den. 1.09
$n_D^{20}$ 1.5170

**Methyl 3-nitrobenzoate**
MW 181.15
mp 78°C

> **IN THIS EXPERIMENT** a cold solution of an aromatic ester that has been dissolved in sulfuric acid is reacted with nitric acid. This highly exothermic reaction is kept under control by cooling; then the mixture is poured onto ice. The solid product is isolated by filtration and recrystallized from methanol, in which it is very soluble.

Do not use a plastic syringe and needle to measure either sulfuric acid or nitric acid. The acids react with the metal needle.

Use care in handling concentrated sulfuric and nitric acids.

**Online Study Center**

Videos: Microscale Filtration on the Hirsch Funnel, Recrystallization

To 0.6 mL of concentrated sulfuric acid in a $10 \times 100$ mm reaction tube, add 0.30 g of methyl benzoate. Flick the tube or use magnetic stirring. Cool the mixture to 0°C; add dropwise, using a Pasteur pipette, a mixture of 0.2 mL of concentrated sulfuric acid and 0.2 mL of concentrated nitric acid. Keep the reaction mixture in ice. Using a stirring rod, keep the reaction well mixed during the addition of the acids and do not allow the temperature of the mixture to rise above about 15°C, as judged by touching the reaction tube.

After all the nitric acid has been added, warm the mixture to room temperature and, after 15 min, pour it onto 2.5 g of ice in a small beaker. Isolate the solid product by suction filtration using a Hirsch funnel and a 25-mL filter flask. Wash the product well with water and then with one 0.2-mL portion of ice-cold methanol. If the methanol is not ice-cold, some product can be lost in this washing step. Save a small sample for a melting-point determination and analysis by thin-layer chromatography (TLC) and infrared (IR) spectroscopy.

The remainder is weighed and recrystallized from an equal weight of methanol in a reaction tube. Alternatively, the sample can be dissolved in a slightly larger quantity of methanol, and water is added dropwise to make the hot solution saturated with the product. Slow cooling should produce large crystals with a melting point of 78°C. The crude material can be obtained in about 80% yield with a melting point of 74°C–76°C. If the yield is not as large as expected, concentrate the filtrate and collect a second crop of product.

**Cleaning Up.** Dilute the filtrate from the reaction with water, neutralize with sodium carbonate, and flush down the drain. The methanol from the crystallization should be placed in the organic solvents container.

## 2. Macroscale Nitration of Methyl Benzoate

Use care in handling concentrated sulfuric and nitric acids.

**Online Study Center**

Video: Macroscale Crystallization

In a 125-mL Erlenmeyer flask, cool 12 mL of concentrated sulfuric acid to 0°C and then add 6.1 g of methyl benzoate. Again, cool the mixture to 0°C–10°C. Now add dropwise, using a Pasteur pipette, a cooled mixture of 4 mL of concentrated sulfuric acid and 4 mL of concentrated nitric acid. During the addition of the acids, swirl the mixture frequently (or use magnetic stirring) and maintain the temperature of the reaction mixture in the range of 5°C–15°C.

When all of the nitric acid has been added, warm the mixture to room temperature and, after 15 min, pour it on 50 g of cracked ice in a 250-mL beaker. Isolate the solid product by suction filtration using a small Büchner funnel, wash well with water, and then wash with two 10-mL portions of ice-cold methanol. A small sample is saved for a melting-point determination. The remainder is weighed and crystallized from an equal weight of methanol. The crude product should be

■ **FIG. 28.1**
The $^1H$ NMR spectrum of methyl 3-nitrobenzoate (250 MHz).

■ **FIG. 28.2**
The $^{13}C$ NMR spectrum of methyl 3-nitrobenzoate (22.6 MHz).

obtained in about 80% yield with a melting point of 74°C–76°C. The recrystallized product should have a melting point of 78°C. For your reference, the carbon and proton nuclear magnetic resonance (NMR) spectra of the product are presented in Figures 28.1 and 28.2.

**Cleaning Up.** Dilute the filtrate from the reaction with water, neutralize with sodium carbonate, and flush down the drain. The methanol from the crystallization should be placed in the organic solvents container.

## QUESTIONS

1. Hydrocarbons do not dissolve in concentrated sulfuric acid, but methyl benzoate does. Explain this difference and write an equation showing the ions that are produced.
2. What would you expect the structure of the dinitro ester to be? Consider the directing effects of the ester and the first nitro group on the addition of the second nitro group.
3. Draw resonance structures to show in which position nitrobenzene will nitrate to form dinitrobenzene.
4. Assign the peaks at 3101 cm$^{-1}$, 1709 cm$^{-1}$, and 1390 cm$^{-1}$ in the IR spectrum of methyl 3-nitrobenzoate (Fig. 28.3).

■ **FIG. 28.3**
The IR spectrum of methyl 3-nitrobenzoate. The broad peak at 3400 cm$^{-1}$ comes from water in the KBr disk.

# Friedel–Crafts Alkylation of Benzene and Dimethoxybenzene; Host-Guest Chemistry

**Online Study Center**

This icon will direct you to techniques, equipment setups, and online resources at http://college.hmco.com/PIC/williamsonMMOE5e.

**PRELAB EXERCISE:** Prepare a flow sheet for the alkylation of benzene and the alkylation of dimethoxybenzene, indicating how the catalysts and unreacted starting materials are removed from the reaction mixture.

The Friedel–Crafts[1] alkylation of aromatic rings most often employs an alkyl halide and a strong Lewis acid catalyst. Some of the catalysts that can be used, in order of decreasing activity, are the halides of aluminum, antimony, iron, titanium, tin, bismuth, and zinc. Although useful, the reaction has several limitations. The aromatic ring must be unsubstituted or bear activating groups; because the product—an alkylated aromatic molecule—is more reactive than the starting material, multiple substitution usually occurs. Furthermore, primary halides will rearrange under the reaction conditions:

$+ \ CH_3CH_2CH_2Cl \quad \xrightarrow{AlCl_3}$

CH$_2$CH$_2$CH$_3$

−6°C: 60%
+35°C: 40%

$+$

H$_3$C $-$ C(H) $-$ CH$_3$

40%
60%

---

1. Charles Friedel and James Crafts (who later became the president of MIT) discovered this reaction in 1879.

In this reaction a tertiary halide and the most powerful Friedel–Crafts catalyst, $AlCl_3$, are allowed to react with benzene. (If you prefer not to work with benzene, you can carry out alkylations of dimethoxybenzene or *m*-xylene.) The initially formed *t*-butylbenzene is a liquid, whereas the product, 1,4-di-*t*-butylbenzene, which has a symmetrical structure, is a beautiful crystalline solid. The alkylation reaction probably proceeds through the carbocation under the conditions of the experiments in this chapter:

**Benzene**
MW 78.11, den. 0.88
bp 80°C

**2-Chloro-2-methylpropane**
**(*t*-Butyl chloride)**
MW 92.57, den. 0.85
bp 51°C

**1,4-Di-*t*-butylbenzene**
MW 190.32, mp 77–79°C
bp 167°C

The reaction is reversible. If 1,4-di-*t*-butylbenzene is allowed to react with *t*-butyl chloride and 1.3 mol of aluminum chloride at 0°C–5°C, 1,3-di-*t*-butylbenzene, 1,3,5-tri-*t*-butylbenzene, and unchanged starting material are found in the reaction mixture. Thus, the mother liquor from crystallization of 1,4-di-*t*-butylbenzene probably contains *t*-butylbenzene, the desired 1,4-di-product, the 1,3-di-isomer, and 1,3,5-tri-*t*-butylbenzene.

## Inclusion Complexes: Host-Guest Chemistry

Although the mother liquor probably contains a mixture of several components, the 1,4-di-*t*-butylbenzene can be isolated easily as an inclusion complex. Inclusion complexes are examples of host-guest chemistry. The host molecule thiourea, $NH_2CSNH_2$, has the interesting property of crystallizing in a helical crystal lattice that has a cylindrical hole in it. The guest molecule can reside in this hole if it is the correct size. It is not bound to the host; nuclear magnetic resonance (NMR) studies indicate the guest molecule can rotate longitudinally within the helical crystal lattice. There are often nonintegral numbers (on the average) of host molecules per guest. The inclusion complex of thiourea and 1,4-di-*t*-butylbenzene crystallizes quite nicely from a mixture of the other hydrocarbons; thus more of the product can be obtained. Because thiourea is very soluble in water, the product is recovered from the complex by shaking it with a mixture of ether and water. The complex immediately decomposes, and the product dissolves in the ether layer, from which it can be recovered.

Compare the length of the 1,4-di-*t*-butylbenzene molecule to the length of various *n*-alkanes and predict the host-guest ratio for a given alkane. You can then check your prediction experimentally. *n*-Hexane can be isolated from the mixture of isomers sold under the name *hexanes*.

---

## EXPERIMENTS

# 1. 1,4-Di-*t*-Butylbenzene

**CAUTION**: Benzene is a mild carcinogen. Handle it in the hood; do not breathe its vapors or allow the liquid to come in contact with the skin.

*Online Study Center*

Photos: Removing a Reagent from a Septum-Capped Bottle, Polypropylene Syringe Containing Ether

Aluminum chloride dust is extremely hygroscopic and irritating. It hydrolyzes to hydrogen chloride on contact with moisture. Clean up any spilled material immediately.

*Online Study Center*

Photos: Gas Trap, Placing a Poly-ethylene Tube through a Septum, Extraction with Ether

**IN THIS EXPERIMENT** a mixture of benzene and an alkyl choride are treated with aluminum chloride, a Lewis acid catalyst. During the alkylation reaction hydrogen chloride is evolved and must be trapped. The dialkyl product is isolated by adding water to the reaction mixture and extracting it with ether. In the usual way the ether solution is washed, dried, and evaporated to give the product. This crude material is recrystallized from methanol to give the pure dialkyl product.

Using a 1.0-mL plastic syringe, measure 0.40 mL of dry 2-chloro-2-methyl-propane (*t*-butyl chloride) and 0.20 mL of dry benzene into a dry 10 × 100 mm reaction tube equipped with a septum and tubing (Fig. 29.1). The benzene and the alkyl chloride will usually be found in septum-stoppered containers. Cool the tube in ice and then add 20 mg of aluminum chloride. Weighing and transferring this small quantity is difficult because aluminum chloride reacts with great rapidity with moist air. Keep the reagent bottle closed as much of the time as possible while weighing the reagent into a very small, dry, capped vial. Because the aluminum chloride is a catalyst, the amount need not be exactly 20 mg.

Mix the contents of the reaction tube by flicking the tube with a finger. After an induction period of about 2 min, a vigorous reaction sets in, with bubbling and liberation of hydrogen chloride. The hydrogen chloride is trapped using the apparatus depicted in Figure 29.1. The wet cotton in the empty reaction tube will dissolve and trap the hydrogen chloride. Figure 29.2 illustrates how to thread a polyethylene tube through a septum. Near the end of the reaction, the product separates as a white solid. When this occurs, remove the tube from the ice and let it stand at room temperature for 5 min.

Add about 1.0 mL of ice water to the reaction mixture, mix the contents thoroughly, and extract the product with three 0.8-mL portions of ether. Wash the combined ether extracts with about 1.5 mL of saturated sodium chloride solution and dry the ether over anhydrous calcium chloride pellets. Add sufficient drying agent so that it does not clump together. After 5 min, transfer the ether solution to a dry, tared reaction tube, using more ether to wash the drying agent, and evaporate the ether under a stream of air in the hood. Remove the last traces of ether under water aspirator vacuum (Fig. 29.3). The oily product should solidify on cooling and weigh about 300 mg.

Add anhydrous calcium chloride until it no longer clumps together.

### ■ FIG. 29.1

A hydrogen chloride gas trap for the Friedel–Crafts reaction.

Rubber septum

Polyethylene tubing

Reaction tubes

Damp cotton

Ice bath

### ■ FIG. 29.2

To thread a polyethylene tube through a septum, make a hole through the septum with a needle, then push a toothpick through the hole. Push the polyethylene tube firmly onto the toothpick, then pull and push on the toothpick. The tube will slide through the septum. Finally, pull the tube from the toothpick. A blunt syringe needle can be used instead of a toothpick.

Polyethylene tube

Septum

Toothpick or syringe needle

To aspirator

Glass tube

Thermometer adapter

Reaction tube

Beaker of warm water

Crystals to be dried

### ■ FIG. 29.3

Drying crystals under reduced pressure.

Spontaneous crystallization gives beautiful needles or plates.

**Online Study Center**

Video: Filtration of Crystals Using the Pasteur Pipette

For crystallization, dissolve the product in 0.40 mL of methanol and allow the solution to come to room temperature without disturbance. After thorough cooling at 0°C, remove the methanol with a Pasteur pipette and rinse the crystals with a drop of ice-cold methanol while keeping the reaction tube in ice. Save this methanol solution for analysis by thin-layer chromatography (TLC). The yield of recrystallized material after drying under aspirator vacuum should be about 160 mg. Remove a sample of crystals for analysis by infrared (IR) or NMR spectroscopy, TLC, and melting-point determination (Fig. 29.4 and Fig. 29.5). Using TLC, compare the pure crystalline product to the residue left after evaporation of the methanol.

**Cleaning Up.** Place any unused *t*-butyl chloride in the halogenated organic waste container and any unused benzene in the hazardous waste container for benzene. Any unused aluminum chloride should be mixed thoroughly with a large excess of sodium carbonate, and the solid mixture should be added to a large volume of water

■ **FIG. 29.4**
The IR spectrum of 1,4-di-*t*-butylbenzene.

■ **FIG. 29.5**
The $^1$H NMR spectrum of 1,4-di-*t*-butylbenzene (90 MHz).

before being flushed down the drain. The combined aqueous layers from the reaction should be neutralized with sodium carbonate and then flushed down the drain. Methanol from the crystallization is to be placed in the organic solvents container.

## 2. Preparation of the Thiourea Inclusion Complex

**Thiourea**
MW 76.12

**IN THIS EXPERIMENT** an inclusion complex is prepared by simply cooling a solution of thiourea and di-*t*-butylbenzene. The complex is isolated, weighed, and decomposed with water; then the dialkyl benzene is extracted into ether that is dried and evaporated. The residue is weighed to calculate how many molecules of thiourea complex with a molecule of the di-*t*-butylbenzene.

The inclusion complex starts to crystallize in 10 min.

In a tared reaction tube, dissolve 200 mg of thiourea and 120 mg of 1,4-di-*t*-butylbenzene in 2.0 mL of methanol at room temperature; then cool the mixture in ice, at which time the inclusion complex will crystallize. Using a Pasteur pipette, remove the solvent and wash the product twice with just enough methanol to cover the crystals while keeping the tube on ice. Connect the reaction tube to a water aspirator and, using the heat of your hand, evaporate the remaining methanol under reduced pressure until the weight of the tube is constant. The yield should be about 200 mg.

Remove a small sample and set it aside; carefully determine the weight of the remaining complex and then add about 1.2 mL of water and 1.2 mL of ether to the tube. Shake the mixture until the crystals disappear. This causes the complex to break up, with the thiourea remaining in the aqueous layer and the 1,4-di-*t*-butylbenzene passing into the ether layer. Draw off the aqueous layer and dry the ether layer with anhydrous calcium chloride pellets. Add sufficient drying agent so that it does not clump together. More ether can be added if necessary. Transfer the ether to a tared reaction tube and wash the drying agent twice with fresh portions of ether. The objective is to make a quantitative transfer of the butylbenzene. Evaporate the ether and remove the last traces under aspirator vacuum. After the weight of the tube is constant, record the weight of the hydrocarbon. Calculate the number of molecules of thiourea per molecule of hydrocarbon (probably *not* an integral number).

**Cleaning Up.** Place any unused thiourea and the 1.2 mL of the aqueous solution containing thiourea in the hazardous waste container for thiourea. Alternatively, treat the thiourea with excess aqueous 5.25% sodium hypochlorite solution (household bleach), dilute the mixture with a large amount of water, and flush it down the drain. If local regulations allow, evaporate any residual solvent from the drying agent in the hood and place the dried solid in the nonhazardous waste container. Otherwise, place the wet drying agent in a waste container designated for this purpose.

##  3. 1,4-Di-*t*-Butylbenzene

Measure in the hood 20 mL of 2-chloro-2-methylpropane (*t*-butyl chloride) and 10 mL of benzene in a 125-mL filter flask equipped with a one-holed rubber stopper fitted with a thermometer. Place the flask in an ice-water bath to cool. Weigh 1 g of fresh aluminum chloride onto a creased paper and scrape it with a small spatula into a 10 × 75-mm test tube; close the tube at once with a cork.[2] Connect the side arm of the flask to an aspirator (preferably one made of plastic) and operate it at a rate sufficient to carry away hydrogen chloride formed in the reaction; alternatively, make a trap for the hydrogen chloride similar to that shown in Figure 18.3 (on page 348). Cool the liquid to 0°C–3°C, add about one-quarter of the aluminum chloride, replace the thermometer, and swirl the flask vigorously in the ice bath. After an induction period of about 2 min, a vigorous reaction sets in, with bubbling and liberation of hydrogen chloride. Add the remainder of the catalyst in

Reaction time: about 15 min

---

2. Alternative scheme: Put a wax pencil mark on the test tube 37 mm from the bottom and fill the tube with aluminum chloride to this mark.

three portions at intervals of about 2 min. Toward the end, the reaction product begins to separate as a white solid. When this occurs, remove the flask from the bath and let stand at room temperature for 5 min. Add ice and water to the reaction mixture, allow most of the ice to melt, and then add ether for extraction of the product, stirring with a rod or spatula to help bring the solid into solution. Transfer the solution to a separatory funnel and shake; draw off the lower layer and wash the upper ether layer first with water then with a saturated sodium chloride solution. Dry the ether solution over anhydrous calcium chloride pellets for 5 min, filter the solution to remove the drying agent, remove the ether by evaporation on a steam bath, and evacuate the flask using an aspirator to remove traces of solvent until the weight is constant; the yield of crude product should be 15 g.

The oily product should solidify on cooling. For crystallization, dissolve the product in 20 mL of hot methanol and let the solution come to room temperature without disturbance. If you are in a hurry, with minimal agitation, place it gently in an ice-water bath and observe the result. After thorough cooling at 0°C, collect the product and rinse the flask and product with a little ice-cold methanol. The yield of 1,4-di-*t*-butylbenzene from the first crop is 8.2–8.6 g of satisfactory material. Save the product for the next step as well as the mother liquor, in case you later wish to work it up for a second crop.

## Inclusion Complex Formation

In a 25-mL Erlenmeyer flask dissolve 5 g of thiourea and 3 g of 1,4-di-*t*-butylbenzene in 50 mL of warm methanol (break up lumps with a flattened stirring rod) and let the solution stand for crystallization of the complex, which occurs with ice cooling. Collect the crystals, rinse with a little methanol, and dry to constant weight; the yield is 5.8 g. Remove a small sample and set it aside; carefully determine the weight of the remaining complex and place the material in a separatory funnel along with about 25 mL each of water and ether. Shake until the crystals disappear, draw off the aqueous layer containing thiourea, wash the ether layer with saturated sodium chloride, and dry the ether layer over anhydrous calcium chloride pellets. Remove the drying agent by filtration and collect the filtrate in a tared 125-mL Erlenmeyer flask. Evaporate and evacuate as before, being sure the weight of hydrocarbon is constant before you record it. Calculate the number of molecules of thiourea per molecule of hydrocarbon (probably *not* an integral number).

To work up the mother liquor from the crystallization, first evaporate the methanol. Note that the residual oil does not solidify on ice cooling. Next, dissolve the oil, together with 5 g of thiourea, in 50 mL of methanol, collect the inclusion complex that crystallizes (3.2 g), and recover 1,4-di-*t*-butylbenzene from the complex as before (0.8 g before crystallization). The IR and $^1$H NMR spectra of the product are seen in Figures 29.4 and 29.5.

**Cleaning Up.** Place any unused *t*-butyl chloride in the halogenated organic waste container and any unused benzene in the hazardous waste container for benzene. Any unused aluminum chloride should be mixed thoroughly with a large excess

---

Add anhydrous calcium chloride pellets until they no longer clump together.

Spontaneous crystallization gives beautiful needles or plates.

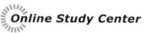

*Online Study Center*

Video: Macroscale Crystallization

⚠ **CAUTION:** Thiourea is a carcinogen. Handle the solid in a hood. Do not breathe its dust.

$$\underset{\text{Thiourea}}{\overset{\displaystyle \overset{S}{\underset{\|}{}}}{H_2NCNH_2}}$$

**Thiourea**
MW 76.12

The inclusion complex starts to crystallize in 10 min.

Workup of mother liquor

of sodium carbonate, and the solid mixture should be added to a large volume of water before being flushed down the drain. The combined aqueous layers from the reaction should be neutralized with sodium carbonate and then flushed down the drain. Methanol from the crystallization is to be placed in the organic solvents container.

# 4. 1,4-Di-*t*-Butyl-2,5-Dimethoxybenzene

**1,4-Dimethoxybenzene (Hydroquinone dimethyl ether)** MW 138.16, mp 57°C

**2-Methyl-2-propanol (*t*-Butyl alcohol)** MW 74.12, den. 0.79 mp 25.5°C, bp 82.8°C $n_D^{20}$ 1.3820

**1,4-Di-*t*-butyl-2,5-dimethoxybenzene** MW 250.37, mp 104–105°C

**Trimethylcarbocation**

This experiment illustrates the Friedel–Crafts alkylation of an activated benzene molecule with a tertiary alcohol in the presence of sulfuric acid as the Lewis acid catalyst. Like the reaction of benzene and *t*-butyl chloride, the substitution involves attack by the electrophilic trimethylcarbocation.

## Microscale Procedure

**⚠ CAUTION:** Handle concentrated sulfuric acid with care.

**IN THIS EXPERIMENT** a mixture of dimethoxybenzene and *t*-butyl alcohol is dissolved in acetic acid and treated with sulfuric acid, a Lewis acid catalyst. Water is added to the reaction mixture, and the solid product is isolated by removing the aqueous material with a Pasteur pipette. The dialkylated product is recrystallized from methanol.

Stir **thoroughly** after each drop of water is added.

In a 10 × 100 mm reaction tube dissolve 120 mg of 1,4-dimethoxybenzene (hydroquinone dimethyl ether) in 0.4 mL of acetic acid with gentle warming and add 0.2 mL of *t*-butyl alcohol (it may be necessary to melt this alcohol). Cool the mixture in ice and then add to it 0.4 mL of concentrated sulfuric acid dropwise from a Pasteur pipette. After each drop of acid is added, mix the solution thoroughly. At the end of this addition, considerable solid reaction product should have separated. Stir the mixture thoroughly with a glass stirring rod, remove the reaction tube from the ice, and allow it to warm to room temperature and remain at 20°C–25°C for at least 10 min to complete the reaction. Next, cool the mixture in ice to cause crystallization to occur. Measure 2.5 mL of water into a container. Very carefully add 1 drop of water to the reaction mixture, stir with the glass rod, and continue to add the remainder of the water dropwise with cooling and stirring. Remove the solvent from the cold solution with a Pasteur pipette and wash the crystals thoroughly with water. Recrystallize the product from methanol. After allowing the mixture to cool to room temperature and then to 0°C in ice, remove the solvent using a Pasteur pipette. The last traces of methanol can be removed under aspirator vacuum while warming the tube in your hand or in a beaker of warm water (Fig. 29.3). The yield of large plates of 1,4-di-*t*-butyl-2,5-dimethoxybenzene should be about 80–100 mg. Analyze the product by IR spectroscopy and TLC, using ligroin as the eluent. Determine the melting point and the percentage yield.

**Cleaning Up.** Combine the aqueous layer, the methanol washes, and the crystallization mother liquor; dilute with water; neutralize with sodium carbonate; and flush down the drain. Any spilled sulfuric acid should be covered with a large excess of solid sodium carbonate, and the mixture should be added to water before being flushed down the drain.

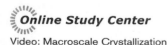

## *Macroscale Procedure*

Place 3 g of 1,4-dimethoxybenzene (hydroquinone dimethyl ether) in a 125-mL Erlenmeyer flask, add 5 mL of *t*-butyl alcohol and 10 mL of acetic acid, and place the flask in an ice-water bath to cool. Measure 15 mL of concentrated sulfuric acid into a 25-mL Erlenmeyer flask and place the flask—properly supported—in an ice bath to cool. For good thermal contact the ice bath should be an ice-water mixture. Put a thermometer in the larger flask, swirl in the ice bath until the temperature is in the range 0°C–3°C, and remove the thermometer (solid, if present, will dissolve later). Do not use the thermometer as a stirring rod. Clamp a small separatory funnel in a position to deliver into the 125-mL Erlenmeyer flask so that the flask can remain in the ice-water bath, wipe the smaller flask dry, and pour the chilled sulfuric acid solution into the funnel. While swirling the 125-mL flask in the ice bath, run in the chilled sulfuric acid by rapid drops during the course of 4–7 min.

By this time considerable solid reaction product should have separated, and insertion of a thermometer should show that the temperature is in the 15°C–20°C range. Swirl the mixture while maintaining the temperature at about 20°C–25°C for an additional 5 min and then cool in ice. Add ice to the mixture to dilute the sulfuric acid; then add water to nearly fill the flask, cool, and collect the product on a Büchner funnel with suction. It is good practice to clamp the filter flask so it does not tip over. Apply only very gentle suction at first to avoid breaking the filter paper, which is weakened by the strong sulfuric acid solution. Wash liberally with water and then turn on the suction to full force. Press down the filter cake with a spatula and let it drain well. Meanwhile, cool a 15-mL portion of methanol for washing to remove a little oil and a yellow impurity. Release the suction, cover the filter cake with one-third of the chilled methanol, and then apply suction. Repeat the washing two more times.

**CAUTION**: Handle concentrated sulfuric acid with care.

Total reaction time: about 12 min

Because air-drying of the crude reaction product takes time, the following short procedure is suggested: Place the moist material in a 50-mL Erlenmeyer flask, add a little dichloromethane (5–8 mL) to dissolve the organic material, and note the appearance of aqueous droplets. Add enough anhydrous calcium chloride pellets to the flask so that the drying agent no longer clumps together, let drying proceed for 10 min, and then remove the drying agent by gravity filtration or careful decantation into another 50-mL Erlenmeyer flask. Add 15 mL of methanol (bp 65°C) to the solution. Remove the dichloromethane (bp 41°C) using a rotary evaporator or by evaporation on a steam bath in a hood. When the volume is estimated to be about 15 mL, let the solution stand for crystallization. When crystallization is complete, cool in ice and collect.

From an environmental standpoint, it would be better to eliminate the solvents by simple distillation using the apparatus depicted in Figure 5.10 (on page 100). Leave 15 mL in the flask and allow the mixture to cool slowly. Large crystals will form. Collect the product on a small Büchner funnel. The yield of large plates of pure 1,4-di-*t*-butyl-2,5-dimethoxybenzene is 2–2.5 g.

## *Antics of Growing Crystals*

Robert Stolow of Tufts University reported[3] that growing crystals of the di-*t*-butyldimethoxy compound change shape in a dramatic manner: Thin plates curl and roll up and then uncurl so suddenly that they propel themselves for a distance of several centimeters. If you do not observe this phenomenon during crystallization of a small sample, you may be interested in consulting the papers cited and pooling your sample with others for trial on a large scale. The solvent mixture recommended by the Tufts workers for observing the phenomenon is 9.7 mL of acetic acid and 1.4 mL of water per gram of product.

■ **FIG. 29.6**
The IR spectrum of 1,4-dimethoxybenzene (KBr disk). Note water contaminant at 3437 cm⁻¹.

3. Stolow, R. D.; Larsen, J. W. *Chem. Ind.* **1963**, *449. See also* Blatchly, J. M.; Hartshorne, N. H. *Trans. Faraday Soc.* **1966**, *62*, 512.

■ **FIG. 29.7**
The $^1$H NMR spectrum
of 1,4-dimethoxybenzene
(60 MHz).

Figures 29.6 and 29.7 present the IR and NMR spectra of the starting hydro-quinone dimethyl ether. Can you predict the appearance of the NMR spectrum of the product?

**Cleaning Up.** Combine the aqueous layer, and the methanol washes, and the crystallization mother liquor; dilute with water; neutralize with sodium carbonate; and flush down the drain. Any spilled sulfuric acid should be covered with a large excess of solid sodium carbonate, and the mixture should be added to water before being flushed down the drain. Dichloromethane mother liquor from the crystallization is placed in the halogenated organic waste container. If local regulations allow, evaporate any residual solvent from the drying agent in the hood and place the dried solid in the nonhazardous waste container. Otherwise, place the wet drying agent in a waste container designated for this purpose.

## 5. For Further Investigation

### *Alkylation of m-Xylene*

*m*-Xylene
MW 106.17
den. 0.868
bp 139°C

2-Chloro-2-methylpropane
MW 92.57
den. 0.851
bp 51°C

> **IN THIS EXPERIMENT** an excess of *m*-xylene (1,3-dimethylbenzene) and *t*-butyl chloride (2-chloro-2-methylpropane) is treated with iron(III) chloride, a Lewis acid catalyst. Hydrogen chloride gas is trapped. When the reaction is complete, water is added; the organic layer is washed, dried, and subjected to instant microscale distillation to remove excess 1,3-dimethylbenzene. The residual alkylated product is analyzed by IR spectroscopy to determine the structure of the product.

The objective of this experiment is to determine the structure of the product formed when *m*-xylene (not benzene, as in Experiment 1) is alkylated with *t*-butyl chloride. Although aluminum chloride is the catalyst most often used for the Friedel–Crafts reaction, it is difficult to store and weigh out because it reacts very rapidly with moisture in the air. Therefore, in this experiment, another Lewis acid catalyst, iron(III) chloride is used. The methyl groups in *m*-xylene are *ortho-para* directors and activators of the benzene ring, but other factors may intervene in this particular experiment. Read about the factors affecting alkylations in the first part of this chapter.

An excess of *m*-xylene is used in this reaction to ensure that the product will be monoalkylated by the *t*-butyl cation. Because the boiling points of the reactants and the product differ highly, a crude but very rapid and efficient distillation is done to remove unreacted xylene and *t*-butyl chloride, leaving only the product. This instant microscale distillation is carried out by boiling the mixture in a reaction tube and then pulling the hot vapors into a Pasteur pipette where they condense. After about half the material has been distilled in this way, the high-boiling residue will consist of almost pure product.

IR spectroscopy can be used to determine unequivocally the structure of the product. It has been found that the hydrogen atoms on a benzene ring give rise to one or two intense, characteristic peaks in the region of 730 to 885 $cm^{-1}$ regardless of the nature of the substituents on the benzene ring. For instance, monosubstituted benzenes have two peaks: one in the range of 770 to 730 $cm^{-1}$, and the other in the range of 710 to 690 $cm^{-1}$ (*see* Table 29.1). NMR spectroscopy can also be used to determine the structure of the product.

Molecular mechanics calculations on the three possible monosubstitution products from this reaction may help you to predict the outcome or help confirm conclusions drawn from experimental evidence.

## Iron (III) Chloride Catalyzed Reaction

**Online Study Center**

Photos: Gas Trap, Placing a Polyethylene Tube through a Septum

Place 0.6 mL of *m*-xylene and 0.5 mL of *t*-butyl chloride in a dry reaction tube equipped with a septum and a polyethylene tube that leads to a tube containing a small wad of damp cotton. See Figures 29.1 and 29.2 (on page 439) for threading a polyethylene tube through a septum. The cotton, dampened with 2 or 3 drops of water, serves to trap the hydrogen chloride evolved during the reaction. Cool the mixture in an ice bath (Fig. 29.1), add 30 mg of iron(III) chloride (purple, free-flowing crystals when pure and dry), and replace the septum and polyethylene

TABLE 29.1 • Infrared C—H Out-of-Plane Bending Vibrations of Substituted Benzenes

| Substituted Benzene | Peak 1 ($cm^{-1}$) | Peak 2 ($cm^{-1}$) |
|---|---|---|
| Benzene | 671 | — |
| Monosubstituted benzenes | 770–730 | 710–690 |
| 1,2-Disubstituted | 770–735 | — |
| 1,3-Disubstituted | 810–750 | 710–690 |
| 1,4-Disubstituted | 835–810 | — |
| 1,2,3-Trisubstituted | 780–760 | 745–705 |
| 1,2,4-Trisubstituted | 825–805 | 885–870 |
| 1,3,5-Trisubstituted | 865–810 | 730–675 |
| 1,2,3,4-Tetrasubstituted | 810–800 | — |
| 1,2,3,5-Tetrasubstituted | 850–840 | — |
| 1,2,4,5-Tetrasubstituted | 870–855 | — |
| Pentasubstituted | 870 | — |

tube. After a short induction period, the reaction will begin a vigorous evolution of hydrogen chloride.

When the vigorous part of this reaction is over, remove the ice bath and allow the reaction mixture to warm to room temperature. When bubbles of hydrogen chloride cease to be evolved or after 15 min, add 1 mL of water to the reaction mixture, mix well, and then remove the water layer with a Pasteur pipette. Repeat this process using about 1 mL of saturated aqueous sodium bicarbonate solution followed by 1 mL of saturated sodium chloride solution. Transfer the organic layer to another reaction tube and dry it with anhydrous calcium chloride pellets.

Transfer the organic layer once again to a dry reaction tube, add a boiling chip, and heat the mixture to boiling. Allow the refluxing vapors to rise about 3 cm in the tube and then draw them into a Pasteur pipette. Squirt the condensate into another reaction tube held in the same hand (Fig. 29.8). Repeat this until about half the mixture has been distilled.

Analyze the residue by IR spectroscopy, which is most easily done as a thin film between sodium chloride or silver chloride plates. The IR spectrum of *m*-xylene is shown in Figure 29.9. Refer to Table 29.1 for the IR absorption frequencies assigned to the different benzene substitution patterns. Note the two strong peaks in Figure 29.9 for *m*-xylene and the frequencies expected for 1,3-disubstitution from Table 29.1. From this correlation chart for the aromatic C—H out-of-plane bending modes, deduce the structure of your product. In your laboratory report, write a mechanism for this reaction that illustrates all the details of your conclusions. Discuss the structure of the product in terms of directive effects of substituents on the benzene ring, the steric effects of substituents, and thermodynamic versus kinetic control of the reaction.

*Online Study Center*

Video: Instant Microscale Distillation

## ■ FIG. 29.8
An apparatus for instant microscale distillation.

Refluxing vapor

Boiling liquid

Boiling chip

## ■ FIG. 29.9
The IR spectrum of *m*-xylene (thin film) from 500 cm$^{-1}$ to 1000 cm$^{-1}$ to show C—H out-of-plane bending vibrations.

Carry out molecular mechanics calculations on the three possible products to determine their relative steric energies or heats of formation. Do these calculations corroborate your conclusions?

**Cleaning Up.** Place any unused *t*-butyl chloride in the halogenated organic waste container and any unused *m*-xylene in the organic solvents container. Unused iron(III) chloride and wash solutions should be combined and neutralized with sodium bicarbonate solution, diluted with water, and flushed down the drain. If local regulations allow, evaporate any residual solvent from the drying agent in the hood and place the dried solid in the nonhazardous waste container. Otherwise, place the wet drying agent in a waste container designated for this purpose.

**Online Study Center**

**General Resources**
Additional Experiments

## QUESTIONS

1. Write equations to explain why the reaction of 1,4-di-*t*-butylbenzene with *t*-butyl chloride and aluminum chloride gives 1,3,5-tri-*t*-butylbenzene.

2. Why must aluminum chloride be protected from exposure to the air?

3. Would you expect to find two strong peaks at $\sim 690\ \text{cm}^{-1}$ and $\sim 770\ \text{cm}^{-1}$ (Fig. 29.9) in your product from the alkylation of *m*-xylene? Why or why not?

4. Draw a detailed mechanism for the formation of *t*-butyl-2,5-dimethoxybenzene.

5. Why is the 1,4 isomer, 1,4-di-*t*-butyl-2,5-dimethoxybenzene, the major product in the alkylation of dimethoxybenzene? Would you expect either of the following compounds to be formed as side products: 1,3-di-*t*-butyl-2,5-dimethoxybenzene or 1,4-dimethoxy-2,3-di-*t*-butylbenzene? Why or why not?

6. Suggest two other compounds that might be used in place of *t*-butyl alcohol to form 1,4-di-*t*-butyl-2,5-dimethoxybenzene.

7. Can you locate the two peaks in Figure 28.3 (on page 435) that show that methyl 3-nitrobenzoate is indeed 1,3-disubstituted? (See Experiment 5 and Table 29.1.)

# Alkylation of Mesitylene

**PRELAB EXERCISE: How much formic acid is consumed in this reaction?**

The reaction of an alkyl halide with an aromatic molecule in the presence of aluminum chloride, the Friedel–Crafts reaction, proceeds through the formation of an intermediate carbocation:

Charles Friedel and James Crafts (who later became the president of MIT) discovered this reaction in 1879, but seven years before, Johann Baeyer and his colleagues carried out very similar reactions using aldehydes as the alkylating agent and strong acids as catalysts. These reactions, like the Friedel–Crafts reaction, proceed through carbocation intermediates. Consider the synthesis of DDT (1,1,1-trichloro-2,2-di(*p*-chlorophenyl)ethane):

**DDT**

Trichloroacetaldehyde (chloral) forms a carbocation on reaction with concentrated sulfuric acid. This reacts primarily at the *para*-position of chlorobenzene; the intermediate alcohol, being benzylic, in the presence of acid readily forms a new carbocation, which in turn attacks another molecule of chlorobenzene. Even though synthesized in 1872, the remarkable insecticidal properties of DDT were not recognized until about 1940. It took another 25 years to realize that DDT, which is resistant to normal biochemical degradation, was building up in rivers, lakes, and streams and causing long-term environmental damage to wildlife. It is now outlawed in many parts of the world.

In the present experiment, discovered by Baeyer in 1872, formaldehyde is allowed to react with mesitylene in the presence of formic acid. The sequence of reactions is very similar to those that form DDT:

**Formaldehyde**   **Formic acid**

**Benzylic carbocation**

**Dimesitylmethane**

When aluminum chloride, a much more powerful catalyst, is used in this reaction, the methyl groups on the mesitylene rearrange and disproportionate to form a number of products, including polymeric material. The strongly activated ring of phenol reacts with formaldehyde at the *ortho-* and *para-* positions to form a polymer called Bakelite (*see* Chapter 67).

A convenient form of formaldehyde to use in a reaction of this type is paraformaldehyde, a polymer that readily decomposes to formaldehyde:

$$ + \ddot{O} - CH_2 - \ddot{O} - CH_2 - \ddot{O} +_n \xrightarrow{\text{heat}} CH_2 = \ddot{O}: $$

**Paraldehyde**                    **Formaldehyde**

| $+CH_2-O+_n$ | $\longrightarrow$ | $CH_2=O$ | + | $\overset{O}{\underset{||}{HCOH}}$ | + | |
|---|---|---|---|---|---|---|
| **Paraldehyde** | | **Formaldehyde** | | **95%** | | **Mesitylene** |
| **(Paraformaldehyde)** | | MW 30.03 | | **Formic acid** | | MW 120.20 |
| mp 163–165°C (dec) | | | | MW 46.03 | | bp 162–164°C |
| | | | | mp 8.5°C | | |

$$ \equiv \quad CH_3 - \cdots - CH_2 - \cdots - CH_3 $$

**Dimesitylmethane**
MW 252.41

## EXPERIMENT

# Dimesitylmethane

## *Microscale Procedure*

**IN THIS EXPERIMENT** two molecules of 1,3,5-trimethylbenzene undergo a Friedel–Crafts alkylation with formaldehyde catalyzed by formic acid. On cooling the product crystallizes from the reaction mixture. It is washed, dried, and crystallized from ligroin.

To 10 mg of paraformaldehyde in a $10 \times 100$ mm reaction tube in the hood, add 0.06 mL (73 mg) of 95% formic acid and a carborundum (black) or Teflon boiling chip. Dissolve the paraformaldehyde in the formic acid by boiling on a hot sand bath; then add 0.133 mL (0.115 g) of mesitylene and another carborundum or Teflon boiling chip. Add the distilling column as an air condenser (Fig. 30.1) and reflux the reaction mixture for 2 h. Cool the mixture first to room temperature and then in ice. Remove the excess formic acid using a Pasteur pipette; wash the crystals with water, aqueous sodium carbonate solution, and again with water; then scrape the crystals out onto a piece of filter paper and pick out the boiling chips. Squeeze the crystals between sheets of filter paper to dry.

Determine the weight of the crude product and save a few crystals for a melting-point determination. Recrystallize the product by dissolving it in a minimum quantity of boiling ligroin. Allow the solution to cool to room temperature, add a seed crystal if necessary, and then cool the mixture for at least 15 min in ice before removing the solvent using a Pasteur pipette. Another solvent for recrystallization is a mixture of 0.75 mL of toluene and 0.1 mL of methanol, which is adequate for 0.5 g of product. Obtain the IR (infrared) and NMR (nuclear magnetic resonance) spectra of the pure material and the melting points of the crude and recrystallized product. Turn in the pure product along with a card giving relevant data on the substance, including the number of moles of each of the starting materials. From the spectra and any tests you may wish to run, confirm the structure of the product.

**Cleaning Up.** At the end of this reaction, there should be no formaldehyde remaining in the reaction mixture. Should it be necessary to destroy formaldehyde, it should be diluted with water and 7 mL of household bleach added to oxidize every 100 mg of paraformaldehyde. After 20 min it can be flushed down the drain. The formic acid solvent from the reaction and aqueous washings should be combined, neutralized with sodium carbonate, and flushed down the drain. Mother liquor from recrystallization is placed in the organic solvents container.

## *Macroscale Procedure*

To 0.75 g of paraformaldehyde in a 50-mL round-bottomed flask in the hood, add 4.5 mL of 95% formic acid. Fit the flask with a reflux condenser (Fig. 30.2) and bring the mixture to a boil on an electrically heated sand bath. After about 5 min,

■ **FIG. 30.1**
An apparatus for refluxing a reaction mixture with maximum cooling.

Air condenser

Viton connector

Damp pipe cleaner, if needed, for more condensing

Boiling solvent

Temperature controlled by depth in sand

■ **FIG. 30.2**
A macroscale reflux apparatus.

Water out

Water in

■ **FIG. 30.3**
The IR of dimesitylmethane.

**CAUTION**: Formic acid is corrosive to tissue. Avoid all contact with the liquid or its vapors. Should any come in contact with your skin, wash it off immediately with a large quantity of water. Carrying out this reaction in the hood protects not only against formic acid but also against formaldehyde, a suspected carcinogen formed transitorily in the reaction.

*Online Study Center*

Video: Macroscale Crystallization

**General Resources**
Additional Experiments

most of the paraformaldehyde will have dissolved. Add through the condenser 10.0 mL of mesitylene. Reflux the reaction mixture for 2 h, cool the mixture to room temperature, and then cool it in ice. Remove the product by vacuum filtration in the hood, taking great care not to come into contact with the residual formic acid. Wash the crystals with water, aqueous sodium carbonate solution, and then again with water and squeeze the crystals between sheets of filter paper to dry. Determine the weight of the crude product and save a few crystals for a melting-point determination. Recrystallize the product. A suggested solvent is a mixture of 7.5 mL of toluene and 1 mL of methanol, which is adequate for 5 g of product. Obtain the melting points of the crude and recrystallized product. Calculate the percent yield and turn in the pure product to your instructor. IR and NMR spectra of the product are found in Figures 30.3, 30.4, and 30.5.

**Cleaning Up.** At the end of this reaction, there should be no formaldehyde remaining in the reaction mixture. Should it be necessary to destroy formaldehyde, it should be diluted with water and 7 mL of household bleach added to oxidize every 100 mg of paraformaldehyde. After 20 min it can be flushed down the drain. The formic acid solvent from the reaction and aqueous washings should be combined, neutralized with sodium carbonate, and flushed down the drain. Mother liquor from recrystallization is placed in the organic solvents container.

■ **FIG. 30.4**
The $^1$H NMR spectrum of dimesitylmethane (90 MHz).

## QUESTIONS

1. The intermediate benzylic carbocation is stabilized by resonance. Draw the contributing resonance structures.
2. Draw the structure of dimesitylmethane, designate on its NMR spectrum all the hydrogens that are equivalent, and predict the relative intensities of all the peaks corresponding to these sets of equivalent hydrogens. The two types of methyls differ from each other by 0.15 ppm.

# CHAPTER 31

# The Friedel–Crafts Reaction: Anthraquinone and Anthracene

**Online Study Center**

This icon will direct you to techniques, equipment setups, and online resources at http://college.hmco.com/PIC/williamsonMMOE5e.

**PRELAB EXERCISE:** Draw the mechanism for the cyclization of 2-benzoylbenzoic acid to anthraquinone using concentrated sulfuric acid.

**Phthalic anhydride**
MW 148.11, mp 132°C

(1) AlCl$_3$ + benzene
(2) H$_2$O

**2-Benzoylbenzoic acid**
MW 226.22, mp 127°C
COOH

H$_2$SO$_4$

**Anthraquinone**
MW 208.20, mp 286°C

SnCl$_2$
or Na$_2$S$_2$O$_4$

**Anthrone**
MW 194.22, mp 156°C
H   H

Zn dust

**Anthracene**
MW 178.22, mp 216°C

O · AlCl$_3$
C
COO$^-$
AlCl$_2$$^+$

**Complex salt**

The Friedel–Crafts reaction of phthalic anhydride with excess benzene as solvent and two equivalents of aluminum chloride proceeds rapidly and gives a complex salt of 2-benzoylbenzoic acid in which 1 mole of aluminum chloride has reacted with the acid function to form the salt RCO$_2$$^-$ AlCl$_2$$^+$; a second mole of aluminum chloride is bound to the carbonyl group. On addition of ice and hydrochloric acid, the complex is decomposed, and the basic aluminum salts are brought into solution (Experiments 1 and 2).

The treatment of 2-benzoylbenzoic acid with concentrated sulfuric acid effects cyclodehydration to anthraquinone, a pale-yellow, high-melting compound of great stability (Experiments 3 and 4). Because anthraquinone can be sulfonated only under forcing conditions, a high temperature can be used to shorten the reaction time without loss in yield of product; the conditions are adjusted so that anthraquinone separates from the hot solution in crystalline form to favor rapid drying (Experiments 3 and 4).

$Na_2S_2O_4$
**Sodium hydrosulfite**

Reduction of anthraquinone to anthrone (Experiments 5 and 6) can be accomplished rapidly on a small scale with tin(II) chloride in acetic acid solution. A second method, which involves refluxing anthraquinone with an aqueous solution of sodium hydroxide and sodium hydrosulfite, is interesting to observe because of the sequence of color changes: Anthraquinone is reduced first to a deep red solution containing anthrahydroquinone diradical dianion; the red color then gives way to a yellow color characteristic of anthranol radical anion; as the alkali is neutralized by the conversion of $Na_2S_2O_4$ to $2$ $NaHSO_3$, anthranol ketonizes to the more stable anthrone. The second method is preferred in industry because (1) sodium hydrosulfite costs less than half as much as tin(II) chloride and (2) water is cheaper than acetic acid, and no solvent recovery problem is involved.

Reduction of anthrone to anthracene (Experiments 7 and 8) is accomplished by refluxing in aqueous sodium hydroxide solution with activated zinc dust. The method has the merit of affording pure, beautifully fluorescent anthracene.

## Fieser's Solution, an Oxygen Scavenger

A solution of 2 g of sodium anthraquinone-2-sulfonate and 15 g of sodium hydrosulfite in 100 mL of a 20% aqueous solution of potassium hydroxide produces a blood-red solution of the diradical dianion:

This solution has a remarkable affinity for oxygen. It is used to remove traces of oxygen from gases such as nitrogen or argon when it is desirable to render them absolutely oxygen free. This solution has a capacity of about 800 mL of oxygen. The color fades, and the solution turns brown when it is exhausted.

## EXPERIMENTS

## 1. 2-Benzoylbenzoic Acid

**Phthalic anhydride**
MW 148.11, mp 132°C

**2-Benzoylbenzoic acid**
MW 226.22, mp 127°C

**IN THIS EXPERIMENT** an aromatic anhydride reacts with benzene and aluminum chloride to give 2-benzoylbenzoic acid and hydrogen chloride gas, which is trapped. Careful temperature control is needed for this exothermic reaction. At the end of the reaction, first ice then water and hydrochloric acid are added, and the product is extracted into ether. The ether is dried and evaporated to a small volume (some benzene is still present), at which point ligroin is added. The product crystallizes and is isolated by filtration.

⚠ **CAUTION**: Benzene is a carcinogen. Carry out this experiment in a hood. Avoid skin contact with the benzene by wearing gloves. Do not carry out this reaction if adequate hood facilities are unavailable.

**Online Study Center**

Photo: Gas Trap; Video: The Grignard Reaction: Mixing Reaction Mixture by Flicking Reaction Tube

Into a dry 10 × 100 mm reaction tube place 150 mg of phthalic anhydride and 0.75 mL of dry benzene. Cool the mixture in an ice bath and then add 300 mg of anhydrous aluminum chloride.[1] Cap the tube with a septum connected to polyethylene tubing leading to another reaction tube, which contains a piece of damp cotton to act as a trap for the hydrogen chloride liberated in the reaction (Fig. 31.1). Figure 31.2 shows how to thread a polyethylene tube through a septum. Mix the contents of the tube thoroughly by flicking the tube. Warm the tube by the heat of your hand. If the reaction does not start, warm the tube *very gently* in a beaker of hot water or hold it over a sand bath or a steam bath for a few seconds. At the first sign of vigorous boiling or evolution of hydrogen chloride, hold the tube over the ice bath in readiness to cool it if the reaction becomes too vigorous. Continue this gentle, cautious heating until the reaction proceeds smoothly enough to reflux it on a hot sand bath. This will take about 5 min.

Heat the reaction mixture on the sand bath until evolution of hydrogen chloride almost ceases; then cool it in ice and add 1 g of ice in small pieces. Allow each little piece to react before adding the next. Using a glass rod, mix the reaction mixture well during this hydrolysis. After the reaction subsides, add 0.3 mL

---

1. Aluminum chloride should be weighed in a small, dry, stoppered vial or reaction tube. The aluminum chloride should be from a freshly opened bottle and should be weighed and transferred in the hood. It is very hygroscopic and releases hydrogen chloride upon reaction with water (including humidity in the air). Weigh the reagent rapidly to avoid exposure of the compound to air. The quality of the aluminum chloride determines the success of this experiment.

## ■ FIG. 31.1

A hydrogen chloride trap for the Friedel–Crafts reaction. The tubing can be bent permanently by heating it in a steam bath.

Rubber septum

Polyethylene tubing

Reaction tubes

Damp cotton

Water bath or ice bath

## ■ FIG. 31.2

To thread a polyethylene tube through a septum, make a hole through the septum with a needle and then push a toothpick through the hole. Push the polyethylene tube firmly onto the toothpick and then pull and push on the toothpick. The tube will slide through the septum. Finally pull the tube from the toothpick.

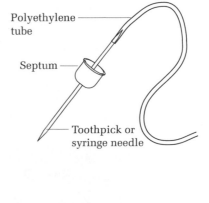

Polyethylene tube

Septum

Toothpick or syringe needle

*Online Study Center*

Photos: Extraction with Ether, Use of the Wilfilter; Video: Microscale Filtration on the Hirsch Funnel

of concentrated hydrochloric acid and 0.5 mL of water. Mix the contents of the tube thoroughly and be sure the mixture is at room temperature. Add 0.5 mL more water, mix the solution well, ascertain that it is at room temperature, and then add 1.5 mL of ether and break up any lumps in the tube with a stirring rod. Mix thoroughly by drawing up the liquid into a Pasteur pipette and rapidly expelling it several times or by stoppering the tube and shaking it vigorously to complete the hydrolysis and extraction. Allow the layers to separate and then remove the aqueous layer with a Pasteur pipette. Add 0.1 mL of concentrated hydrochloric acid and 0.25 mL of water. Pipette mix or stopper and shake the mixture vigorously again and remove the aqueous layers. Transfer the organic layers to a small test tube containing calcium chloride pellets. Allow the solution to dry for 5 min; then transfer it back to the clean dry reaction tube, rinsing the drying agent with a small quantity of ether. Add a boiling chip and boil off the solvents until the volume in the tube is 0.5 mL; then add ligroin until the solution is slightly turbid, indicating that the product is beginning to crystallize. Allow the product to crystallize at room temperature and then at 0°C. Collect the crystals on a Hirsch funnel or on a Wilfilter by centrifugation. This product may be the monohydrate. Heat it at 100°C to drive off any water. Pure 2-benzoylbenzoic acid melts at 127°C–128°C; the yield should be about 0.2 g.

Celite may speed up the filtration of the aluminum hydroxide.

⚠ **CAUTION**: Benzene is a carcinogen. Carry out this experiment in a hood. Avoid skin contact with the benzene by wearing gloves. Do not carry out this reaction if adequate hood facilities are unavailable.

**Cleaning Up.** Combine all aqueous layers, neutralize with sodium carbonate, remove the aluminum hydroxide by filtration, and flush the filtrate down the drain. The aluminum hydroxide and calcium chloride pellets, after being freed of organic solvent (evaporation in the hood), can be placed in the nonhazardous solid waste container. If local regulations do not allow for the evaporation of solvents in a hood, dispose of the wet pellets in a special container. Benzene, ether, and ligroin go in the organic solvents waste container.

## 2. 2-Benzoylbenzoic Acid

This Friedel–Crafts reaction is conducted in a 100-mL round-bottomed flask equipped with a short condenser. A trap for collecting liberated hydrogen chloride is connected to the top of the condenser by rubber tubing of sufficient length to make it possible either to heat the flask on a steam bath or to plunge it into an ice bath. The trap is a suction flask half-filled with water and fitted with a delivery tube inserted to within 1 cm of the surface of the water (*see* Fig. 18.3 on page 348).

Three grams of phthalic anhydride and 15 mL of reagent grade benzene are placed in the flask, and this solution is cooled in an ice bath until the benzene begins to crystallize. Ice cooling serves to moderate the vigorous reaction, which otherwise might be difficult to control. Six grams of anhydrous aluminum chloride[2] are added, the condenser and trap are connected, and the flask is shaken well and warmed for a few minutes by the heat of your hand. If the reaction does not start, the flask is warmed *very gently* by holding it for a few seconds over a steam bath. At the first sign of vigorous boiling, or the evolution of hydrogen chloride, the flask is held over an ice bath in readiness to cool it if the reaction becomes too vigorous. This gentle, cautious heating is continued until the reaction is proceeding smoothly enough to be refluxed on a steam bath. This point is reached in about 5 min.

Continue heating on a steam bath, swirl the mixture, and watch it carefully for sudden separation of the addition compound; the heat of crystallization is such that it may be necessary to plunge the flask into an ice bath to moderate the process. Once the addition compound has separated as a thick paste, heat the mixture for an additional 10 min on a steam bath, remove the condenser, and swirl the flask in an ice bath until cold. (Should no complex separate, heat for an additional 10 min, and then proceed as directed.) Take the flask and ice bath to the hood, weigh out 20 g of ice, add a few small pieces of ice to the mixture, swirl and cool as necessary, and wait until the ice has reacted before adding more. After the 20 g of ice have been added and the decomposition reaction has subsided, add 4 mL of concentrated hydrochloric acid and 20 mL of water, swirl vigorously, and be sure that the mixture is at room temperature. Then add 10 mL of water, swirl vigorously, and again be sure the mixture is at room temperature. Add 10 mL of ether and, with a flattened stirring rod, dislodge solid from the neck and walls of the flask and break up lumps

Short reaction period, high yield

---

2. The aluminum chloride is best weighed in a stoppered test tube. The chloride should be from a freshly opened bottle and should be weighed and transferred in the hood. It is very hygroscopic; work rapidly to avoid exposure of the compound to air. It releases hydrogen chloride on exposure to water (including humidity in the air). The quality of the aluminum chloride determines the success of this experiment.

at the bottom. To further promote hydrolysis of the addition compound, extraction of the organic product, and solution of basic aluminum halides, stopper the flask with a cork and shake vigorously for several minutes.

When most of the solid has disappeared, pour the mixture through a funnel into a separatory funnel until the separatory funnel is nearly filled. Discard the lower aqueous layer. Pour the rest of the mixture into the separatory funnel, rinse the reaction flask with fresh ether, and again drain off the aqueous layer. To reduce the fluffy, dirty precipitate that appears at the interface, add 2 mL of concentrated hydrochloric acid and 5 mL of water, shake vigorously for 2–3 min, and drain off the aqueous layer. If some interfacial dirty emulsion still persists, decant the benzene-ether solution through the mouth of the funnel into filter paper flute-folded for gravity filtration and use fresh ether to rinse the funnel. Clean the funnel and pour in the filtered benzene-ether solution. Shake the solution with a portion of dilute hydrochloric acid; then isolate the reaction product by either of the following procedures:

Alternative procedures

Discard benzene-ether filtrates in the container provided.

Drying time: about 1 h

1. Add 10 mL of 3 $M$ sodium hydroxide solution, shake thoroughly, and separate the aqueous layer.[3] Extract with a further 5-mL portion of aqueous alkali and combine the extracts. Wash with 2 mL of water and add this aqueous solution to the 15 mL of aqueous extract already collected. Discard the benzene-ether solution. Acidify the 17 mL of combined alkaline extract with concentrated hydrochloric acid to pH 1–2 and, if the $o$-benzoylbenzoic acid separates as an oil, cool in ice and rub the walls of the flask with a stirring rod to induce crystallization of the hydrate; collect the product and wash it well with water. This material is the monohydrate $C_6H_5COC_6H_4CO_2H \cdot H_2O$. To convert it into anhydrous $o$-benzoylbenzoic acid, put it in a tared, 50-mL round-bottomed flask, evacuate the flask at the full force of the aspirator, and heat it in the open rings of a steam bath covering the flask with a towel. Check the weight of the flask and contents for constancy after 45 min, 1 h, and 1.25 h. The yield is usually about 4 g (mp 126°C–127°C).

2. Filter the benzene-ether solution through anhydrous calcium chloride pellets for superficial drying, put it into a 50-mL round-bottomed flask, and use a rotary evaporator or distill over a steam bath through a condenser into an ice-cooled receiver until the volume in the distilling flask is reduced to about 11 mL. Add ligroin slowly until the solution is slightly turbid and let the product crystallize first at 25°C and then at 5°C. The yield of anhydrous, colorless, well-formed crystals (mp 127°C–128°C) is about 4 g.

**Cleaning Up.** Combine all aqueous layers, neutralize with sodium carbonate, remove the aluminum hydroxide by filtration, and flush the filtrate down the drain. The aluminum hydroxide and calcium chloride pellets, after being freed of organic solvent (evaporation in the hood), can be placed in the nonhazardous solid

---

3. The nature of the yellow pigment that appears in the first alkaline extract is unknown; the impurity is apparently transient because the final product dissolves in alkali to give a colorless solution.

waste container. If local regulations do not allow for the evaporation of solvents in a hood, dispose of the wet pellets in a special container. Benzene, ether, and ligroin go in the organic solvents waste container.

# 3. Anthraquinone

Anthraquinone can be purified by sublimation. Vapor pressure of anthraquinone:

| Temperature (°C) | Pressure (mm Hg) |
|:---:|:---:|
| 200 | 1.8 |
| 220 | 4.4 |
| 240 | 12.6 |
| 260 | 33.0 |

**2-Benzoylbenzoic acid**
MW 226.22
mp 127°C

**Anthraquinone**
MW 208.20
mp 286°C

**■ FIG. 31.3**
A Wilfilter filtration apparatus. Refer to Chapter 4 for instructions on use.

Crystals

Wilfilter

Filtrate

> **IN THIS EXPERIMENT** 2-benzoylbenzoic acid is dissolved in sulfuric acid and heated to cyclize the material to anthraquinone. With great caution water is added in microdrops to the mixture, which causes the product to crystallize. It is isolated by filtration and then heated with base to remove the unreacted carboxylic acid as a water-soluble ammonium salt. The pure material is collected by filtration, washed, and dried.

In a reaction tube dissolve 100 mg of 2-benzoylbenzoic acid[4] in 0.5 mL of concentrated sulfuric acid by gently heating and stirring. Immerse a thermometer in the reaction mixture and heat it at 150°C–155°C for 45 min. Allow the tube to cool to below 100°C and then, using extreme caution, add a very small drop of water to the mixture. Mix the contents of the tube and continue adding water in minute drops to the mixture. This will cause the product to crystallize and, if done slowly enough, the crystals will be large enough to collect easily by filtration. Fill the tube with water and collect the product by filtration on a Hirsch funnel or a Wilfilter (Fig. 31.3). Return the damp product to the reaction tube and boil the product with 0.5 mL of concentrated ammonium hydroxide to remove unreacted starting material. Filter, wash the product well with water and then with acetone, dry, determine the weight, and then calculate the percent yield. Do not determine the melting point because it is so high (286°C).

**Cleaning Up.** The aqueous filtrate, after neutralization with sodium carbonate, is diluted with water and flushed down the drain.

---

4. Available from the Aldrich Chemical Company.

Reaction time: 5 min

## 4. Anthraquinone

Place 4.0 g of 2-benzoylbenzoic acid[5] (anhydrous) in a 125-mL round-bottomed flask, add 20 mL of concentrated sulfuric acid, and heat on a hot water or steam bath with swirling until the solid is dissolved. Then clamp the flask over a microburner, insert a thermometer, raise the temperature to 150°C, and heat to maintain a temperature of 150°C–155°C for 5 min. Let the solution cool to 100°C, remove the thermometer after letting it drain, and, with a Pasteur pipette, add 4 mL of water by drops with swirling to keep the precipitated material dissolved as long as possible so that it will separate as small, easily filtered crystals. Let the mixture cool further, dilute with water until the flask is full, let cool again, collect the product by suction filtration, and wash well with water. Then remove the filtrate, wash the filter flask, return the funnel to the filter flask without applying suction, and test the filter cake for unreacted starting material as follows: Dilute 8 mL of concentrated ammonia solution with 40 mL of water, pour the solution onto the filter, and loosen the cake so that it is well leached. Then apply suction, wash the cake with water, and acidify a few milliliters of the filtrate. If there is no precipitate upon acidifying, the yield of anthraquinone should be close to the theoretical because it is insoluble in water. Dry the product to constant weight but do not determine the melting point because it is so high (mp 286°C).

**Cleaning Up.** The aqueous filtrate, after neutralization with sodium carbonate, is diluted with water and flushed down the drain.

*Online Study Center*

Video: Microscale Filtration
on the Hirsch Funnel

## 5. Anthrone

**Anthraquinone**
MW 208.20
mp 286°C

SnCl₂
or Na₂S₂O₄

**Anthrone**
MW 194.22
mp 156°C

**IN THIS EXPERIMENT** anthraquinone can be reduced with two different reagents. In the first reaction acetic acid and hydrochloric acid are the solvents; in the second reaction the solvent is water. In both reactions the product crystallizes from the reaction mixture. Compare the two methods of reduction in the microscale experiment.

5. Available from the Aldrich Chemical Company.

### Procedure 1: Tin(II) Chloride Reduction

SnCl₂ · 2H₂O
**Tin(II) chloride dihydrate**

In a reaction tube put 50 mg of crude anthraquinone from the previous experiment, 0.40 mL of acetic acid, and a solution made by warming 0.13 g of tin(II) chloride dihydrate with 0.13 mL of concentrated hydrochloric acid. Add a boiling stone, note the time, and reflux gently until crystals of anthraquinone have completely disappeared (8–10 min); reflux for an additional 15 min and record the total time. Then add water (about 0.12 mL) dropwise until the solution is saturated. Let the solution stand for crystallization. Collect the product using a Wilfilter (Fig. 31.3), dry it, and take the melting point (reported at 156°C). The yield of pale-yellow crystals is about 40 mg.

**Online Study Center**

Photo: Use of the Wilfilter

**Cleaning Up.** Dilute the filtrate with a large volume of water and flush the solution down the drain.

### Procedure 2: Hydrosulfite Reduction

In a 5-mL long-necked, round-bottomed flask put 50 mg of anthraquinone, 60 mg of sodium hydroxide, 150 mg of fresh sodium hydrosulfite, and 1.3 mL of water. Heat over a hot sand bath and swirl for a few minutes to convert the anthraquinone to the deep-red anthrahydroquinone anion. Note that particles of different appearance begin to separate even before the anthraquinone has all dissolved. Add an empty distilling column to the flask and reflux for 45 min; cool, filter the product using a Hirsch funnel or a Wilfilter (Fig. 31.3), wash it well with water, and let dry. Determine the weight and melting point of the crude material and crystallize it from 95% ethanol. Record the approximate volume of solvent used and, if the first crop of crystals recovered is not satisfactory, concentrate the mother liquor and secure a second crop. The usual yield is about 40 mg.

Na₂S₂O₄
**Sodium hydrosulfite**

Decomposes on storage

**Online Study Center**

Video: Microscale Filtration on the Hirsch Funnel; Photo: Use of the Wilfilter

**Cleaning Up.** The aqueous filtrate is treated with household bleach (sodium hypochlorite) until the reaction is complete. Remove the solid by filtration. It may be placed in the nonhazardous solid waste container. The filtrate can be diluted with water and flushed down the drain.

## 6. Anthrone

**Anthraquinone**
MW 208.20
mp 286°C

$\xrightarrow{\text{SnCl}_2 \text{ or Na}_2\text{S}_2\text{O}_4}$

**Anthrone**
MW 194.22
mp 156°C

### Procedure 1: Tin(II) Chloride Reduction

SnCl$_2$ · 2H$_2$O
**Tin(II) chloride dihydrate**

*Online Study Center*

Video: Macroscale Crystallization

In a 50-mL round-bottomed flask provided with a reflux condenser put 2.0 g of an-thraquinone, 16 mL of acetic acid, and a solution made by warming 5.3 g of tin(II) chloride dihydrate with 5.3 mL of concentrated hydrochloric acid. Add a boiling stone to the reaction mixture, note the time, and reflux gently until crystals of an-thraquinone have completely disappeared (8–10 min); then reflux for an addi-tional 15 min and record the total time. Disconnect the flask, heat it on a steam bath, and add water (about 5 mL) in 0.5-mL portions until the solution is satu-rated. Let the solution stand for crystallization. Collect and dry the product and take the melting point (reported at 156°C). The yield of pale-yellow crystals is about 1.7 g.

**Cleaning Up.** Dilute the filtrate with a large volume of water and flush the solution down the drain.

### Procedure 2: Hydrosulfite Reduction

Na$_2$S$_2$O$_4$
**Sodium hydrosulfite**
Decomposes on storage

*Online Study Center*

Video: Microscale Filtration on the Hirsch Funnel

In a 250-mL round-bottomed flask, which can be heated under reflux, put 2.0 g of anthraquinone, 2.4 g of sodium hydroxide, 6 g of fresh sodium hydrosulfite, and 50 mL of water. Heat over a free flame and swirl for a few minutes to con-vert the anthraquinone into the deep-red anthrahydroquinone anion. Note that particles of different appearance begin to separate even before all of the an-thraquinone has dissolved. Arrange for refluxing and reflux for 45 min; cool, filter the product, wash it well with water, and let dry. Note the weight of the product and melting point of the crude material and crystallize it from 95% ethanol; the solution may require filtration to remove insoluble impurities. Record the approximate volume of solvent used and, if the first crop of crystals recovered is not satisfactory, concentrate the mother liquor and secure a second crop.

**Cleaning Up.** The aqueous filtrate is treated with household bleach (sodium hypochlorite) until the reaction is complete. Remove the solid by filtration. It may be placed in the nonhazardous solid waste container. The filtrate can be diluted with water and flushed down the drain.

### Comparison of Results

Compare your results with those obtained by other students using an alternative procedure with respect to yield, quality of product, and working time. Which is the better laboratory procedure? Then consider the cost of the three solvents con-cerned, the cost of the two reducing agents (prices can be found in a current cata-log), the relative ease of recovery of the organic solvents, and the prudent disposal of byproducts and decide which method would be preferred as a manufacturing process.

# 7. Fluorescent Anthracene

**Anthrone**
MW 194.22, mp 156°C

$\xrightarrow[\text{(Cu)}]{\text{Zn dust}}$

**Anthracene**
MW 178.22, mp 216°C

> **IN THIS EXPERIMENT** the ketone anthrone is reduced to the hydrocarbon anthracene using activated zinc dust and sodium hydroxide. The product crystallizes from the reaction mixture, but the excess zinc must be destroyed with hydrochloric acid before isolation. After filtration, the product is recrystallized from toluene.

Put 100 mg of zinc dust into a 5-mL round-bottomed flask and activate the dust by adding 0.6 mL of water and 0.1 mL of a 5% aqueous copper(II) sulfate solution and swirl for a minute or two (or use a magnetic stirrer and stirring bar). Add 40 mg of anthrone, 100 mg of sodium hydroxide, and 1 mL of water, heat to boiling on a sand bath, note the time, and start refluxing the mixture (Fig. 31.4).

**■ FIG. 31.4**
An apparatus for refluxing the reaction mixture in a reaction flask.

Empty distilling column

5-mL Short-necked, round-bottomed flask

Hot sand bath

Hydrogen liberated

Methanol removes water without dissolving much anthracene.

Anthracene can also be recrystallized from ethanol. *See* Chapter 4, Experiment 2.

Anthrone at first dissolves as the yellow anion of anthranol, but anthracene soon begins to separate as a white precipitate.

In about 15 min, the yellow color initially observed on the walls disappears, but refluxing should be continued for a full 30 min. Then remove the heat, rinse down any anthracene in the condenser with a few drops of water, and cool the flask in ice. Remove the solvent with a Pasteur pipette and very carefully add 0.2 mL of concentrated hydrochloric acid dropwise to dissolve the remaining zinc; the hydrogen gas given off causes foaming. Again, cool the mixture and add water to wash down any unreacted zinc; after all the zinc has reacted, remove the acid and wash the product liberally with water. Collect the product on a Hirsch funnel and then wash once with ice-cold methanol to remove as much of the water as possible from the product.

The product need not be dried before recrystallization from about 0.5 mL of boiling toluene. Slow cooling of the toluene solution will give thin, colorless, beautiful fluorescent plates. The yield should be about 20 mg. Collect the product on a Wilfilter (Fig. 31.3). In washing the equipment with acetone, you should be able to observe the striking fluorescence of very dilute solutions. The fluorescence is quenched by a bare trace of impurity.

**Cleaning Up.** The aqueous solution is diluted with water, neutralized with sodium carbonate, and filtered under vacuum to remove zinc hydroxide; the filtrate is diluted with water and flushed down the drain. The solid can be placed in the nonhazardous solid waste container. Toluene and acetone are placed in the organic solvents container.

## 8. Fluorescent Anthracene

Put 3.8 g of zinc dust into a 125-mL round-bottomed flask and activate the dust by adding 23 mL of water and 0.4 mL of copper(II) sulfate solution (Fehling's solution I) and swirling for a minute or two. Add 1.5 g of anthrone, 3.8 g of sodium hydroxide, and 40 mL of water; attach a reflux condenser, heat to boiling, note the time, and start refluxing the mixture. Anthrone at first dissolves as the yellow anion of anthranol, but anthracene soon begins to separate as a white precipitate. In about 15 min the yellow color initially observed on the walls disappears, but refluxing should be continued for a full 30 min. Then stop the heating, use a water wash bottle to rinse down anthracene that has lodged in the condenser, and filter the still-hot mixture on a large Büchner funnel. It usually is possible to decant from, and so remove, a mass of zinc.

Reflux time: 30 min

After a liberal washing with water, blow or shake out the gray cake into a 250-mL beaker and rinse the funnel and paper with water. To remove most of the zinc metal and zinc oxide, add 8 mL of concentrated hydrochloric acid and heat on a steam bath with stirring for 20–25 min, when initial frothing due to liberated hydrogen should have ceased. Collect the now nearly white precipitate on a Büchner funnel, liberally wash with water, release the suction, rinse the walls of the funnel with methanol, use enough methanol to cover the cake, and then apply suction again. Wash again with enough methanol to cover the cake and then remove the solvent thoroughly by suction.

Remove all sources of ignition. Hydrogen is liberated.

Methanol removes water without dissolving much anthracene.

**CAUTION**: Toluene is a highly flammable solvent.

The product need not be dried before recrystallization from toluene. Transfer the methanol-moist material to a 50-mL Erlenmeyer flask and add 22 mL of toluene; a liberal excess of solvent is used to avoid crystallization in the funnel. Be sure there are no flames nearby. Heat the mixture on a hot plate to bring the anthracene into solution, with a small residue of zinc remaining. Filter by gravity using the usual technique (as previously presented) to remove zinc. From the filtrate anthracene is obtained as thin, colorless, beautiful fluorescent plates. The yield should be about 1 g. In washing the equipment with acetone, you should be able to observe the striking fluorescence of very dilute solutions. The fluorescence is quenched by a bare trace of impurity.

**Cleaning Up.** The aqueous solution is diluted with water, neutralized with sodium carbonate, and filtered under vacuum to remove zinc hydroxide. The filtrate is diluted with water and flushed down the drain. The solid waste can be placed in the nonhazardous solid waste container. Toluene and acetone are placed in the organic solvents container.

## QUESTIONS

1. Calculate the number of moles of hydrogen chloride liberated in the synthesis of 2-(4-benzoyl)benzoic acid. If this gaseous acid were dissolved in water, hydrochloric acid would be formed. How many milliliters of concentrated hydrochloric acid would be formed in this reaction? The concentrated acid is 12 $M$ HCl.

2. Write a mechanism for the formation of anthraquinone from 2-benzoyl-benzoic acid, clearly indicating the role of sulfuric acid. What is the name commonly given to this type of reaction?

# Friedel–Crafts Acylation of Ferrocene: Acetylferrocene

*Online Study Center*

This icon will direct you to techniques, equipment setups, and online resources at http://college.hmco.com/PIC/williamsonMMOE5e.

**PRELAB EXERCISE:** How many possible isomers could exist for diacetylferrocene? Explain. Calculate the volume of 3 *M* aqueous sodium hydroxide needed to neutralize the acetic and phosphoric acids in the synthesis of acetylferrocene.

**Ferrocene**
**[Bis(cyclopentadienyl)iron]**
MW 186.04

**Acetic anhydride**
bp 139.5°C, den. 1.08
MW 102.09

85% H₃PO₄

**Acetylferrocene**
**(Acetylcyclopentadienyl)cyclopentadienyliron**
mp 85–86°C
MW 228.08

The Friedel–Crafts acylation of benzene requires aluminum chloride as the catalyst. However, ferrocene, with a very high π-electron density that has been referred to as a superaromatic compound, can be acylated under much milder conditions with phosphoric acid as catalyst. Because the acetyl group is a deactivating substituent, the addition of a second acetyl group, which requires more vigorous conditions, will occur in the nonacetylated cyclopentadienyl ring to give 1,1′-diacetylferrocene. Because ferrocene gives just one monoacetyl derivative and just one diacetyl derivative, it was assigned an unusual sandwich structure.

Acetylferrocene and ferrocene (both highly colored) are easily separated by column chromatography (*see* Chapter 9).

## EXPERIMENT

# Acetylferrocene
## *Microscale Procedure*

> **IN THIS EXPERIMENT** ferrocene reacts with acetic anhydride in an acid-catalyzed Friedel–Crafts reaction. First water is added, which reacts with the anhydride; then base is added to react with the acetic acid. The crude product is isolated by filtration and then chromatographed on alumina. This is an excellent system to observe separation on a chromatography column because the unreacted ferrocene is yellow, and the acetyl product is orange.

Handle acetic anhydride and phosphoric acid with care. Wipe up spills immediately.

To 93 mg of dry sublimed ferrocene (*see* Chapter 49) in a 10 × 100 mm reaction tube, add 0.35 mL (0.38 g) of acetic anhydride followed by 0.1 mL (170 mg) of 85% phosphoric acid. Cap the tube with a septum bearing an empty syringe needle and warm it on a steam bath or in a beaker of boiling water while agitating the mixture to dissolve the ferrocene. Heat the mixture for an additional 10 min and then cool the tube thoroughly in ice. Carefully add to the solution 0.5 mL of ice-water dropwise with thorough mixing, followed by the dropwise addition of 3 *M* aqueous sodium hydroxide solution until the mixture is neutral (test with indicator paper and avoid an excess of base). Collect the product on a Hirsch funnel, wash it thoroughly with water, and press it as dry as possible between sheets of filter paper. Save a sample of this material for melting-point and thin-layer chromatographic (TLC) analyses and purify the remainder by column chromatography.

*Online Study Center*

Photo: Capping a Reaction Tube with a Septum; Video: Microscale Filtration on the Hirsch Funnel

## Column Chromatography

Follow the procedure as given in Chapter 9 for packing a microscale chromatography column and adding the sample (about 90 mg of a ferrocene/acetylferrocene mixture). Use Brockman activity III alumina as the adsorbent.

*Online Study Center*

Photo: Column Chromatography; Video: Column Chromatography

### Elution

Hexanes are extremely flammable. No flames!

Carefully add hexanes to the column and begin to elute the product from the column. Unreacted ferrocene will move down the column as an orange-yellow band. Collect this in a 10-mL flask. Wash any crystalline material that collects on the tip of the valve into the flask with a few drops of ether. Then elute the column with a 50:50 mixture of hexanes and ether. This will move down the acetylferrocene as an orange-red band. Collect this in a separate 10-mL Erlenmeyer flask. Ferrocene is highly symmetrical and nonpolar, while the acetyl compound is the opposite. These properties govern the order of elution from the chromatography column. Spot a TLC plate with these two solutions as well as the crude acetylferrocene and the ferrocene starting material, and analyze as described in the next section. Any diacetylferrocene will be seen as a dark band at the top of the column. Evaporate

**■ FIG. 32.1**
The evaporation of a solvent under reduced pressure. Heat and swirling motion is supplied by one hand; the vacuum is controlled by the thumb of the other hand.

To aspirator

the solvents under reduced pressure and determine the weights of the residues (Fig. 32.1).

## Flash Chromatography

Equilibrium between the solutes adsorbed on the alumina and the eluting solvent is established rapidly, so it does no harm to increase the flow rate of the solvent through the column. This can be done by applying pressure with a rubber bulb above the solvent. There is, however, a possibility that air will be sucked into the bottom of the column when the bulb is removed, which will destroy the uniform packing of the column. Practice this technique on a column that has been eluted in the usual way using gravity flow of the solvent. On a macroscale, special fittings are available for applying air pressure to larger columns.

Recrystallize the acetylferrocene from hot hexanes. Dissolve the residue in a filter flask in the minimum quantity of boiling solvent (about 1 mL) and transfer this hot solution with a Pasteur pipette to a reaction tube. Allow the tube to cool slowly to room temperature and then cool it in ice for at least 10 min. Acetylferrocene crystallizes as dark-red rosettes of needles. Remove the solvent with a Pasteur pipette and then, under vacuum, scrape out the product onto filter paper; after it is dry, determine the weight, calculate the percent yield, and determine the melting point.

Perform a TLC analysis on your product. Dissolve very small samples of pure ferrocene, the crude reaction mixture, and recrystallized acetylferrocene, each in a few drops of toluene; spot the three solutions with microcapillaries on silica gel plates; and develop the chromatogram with 30:1 toluene and absolute ethanol mixture. Visualize the spots under an ultraviolet (UV) lamp if the silica gel has a fluorescent indicator or by adsorption of iodine vapor. Do you detect unreacted ferrocene in the reaction mixture and/or a spot that might be attributed to diacetylferrocene?

**Cleaning Up.** The reaction mixture filtrate can be flushed down the drain. Unused chromatography, recrystallization, and TLC solvents should be placed in the organic solvents container. The alumina from the column should be placed in the

**Online Study Center**

Photos: Crystallization, Filtration Using a Pasteur Pipette, Flash Chromatography; Videos: Recrystallization; Filtration of Crystals Using the Pasteur Pipette hood to allow the hexanes and ether to evaporate from it. Once free of organic solvents, the alumina can be placed in the nonhazardous solid waste container. If local regulations do not allow for the evaporation of solvents, dispose the wet alumina in a special container.

 ## *Macroscale Procedure*

> **IN THIS EXPERIMENT** ferrocene reacts with acetic anhydride in an acid-catalyzed Friedel–Crafts reaction. First water is added, which reacts with the anhydride, then base is added to react with the acetic acid. The crude product is isolated by filtration and then purified by crystallization from hexane. This is an excellent system to observe separation by chromatography because the unreacted ferrocene is yellow, and the acetyl product is orange.

Both acetic anhydride and phosphoric acid are corrosive to tissue. Handle both with care and wipe up any spills immediately.

Hexanes are extremely flammable. No flames!

*Online Study Center*

Video: Macroscale Crystallization

In a 25-mL round-bottomed flask place 3.0 g of ferrocene, 10.0 mL of acetic anhydride, and 2.0 mL of 85% phosphoric acid. Equip the flask with a reflux condenser and a calcium chloride drying tube. Warm the flask gently on a hot water or steam bath while swirling to dissolve the ferrocene; then heat strongly for an additional 10 min. Pour the reaction mixture onto 50 g of crushed ice in a 400-mL beaker and rinse the flask with 10 mL of ice water. Stir the mixture for a few minutes with a glass rod, add 75 mL of 3 *M* sodium hydroxide solution (the solution should still be acidic), and then add solid sodium bicarbonate (be careful of foaming) until the remaining acid has been neutralized. Stir and crush all lumps, allow the mixture to stand for 20 min, and then collect the product by suction filtration. Press the crude material as dry as possible between sheets of filter paper, save a few crystals for TLC analysis, transfer the remainder to an Erlenmeyer flask, and add 40 mL of hexanes to the flask. Boil the solvent on a hot water or steam bath for a few minutes and then decant the dark-orange solution into another Erlenmeyer flask, leaving a gummy residue of polymeric material. Treat the solution with decolorizing charcoal and filter it through a fluted filter paper placed in a warm stemless funnel into an appropriately sized Erlenmeyer flask. Evaporate the solvent (use an aspirator tube; *see* Fig. 8.11 on page 186) until the volume is about 20 mL. Set the flask aside to cool slowly to room temperature. Beautiful rosettes of dark orange-red needles of acetylferrocene will form. After the product has been cooled in ice, collect it on a Büchner funnel and wash the crystals with a small quantity of cold solvent. Pure acetylferrocene has a melting point of 84°C–85°C. Your yield should be about 1.8 g.

Perform a TLC analysis on your product. Dissolve very small samples of pure ferrocene, the crude reaction mixture, and recrystallized acetylferrocene, each in a few drops of toluene; spot the three solutions with microcapillaries on silica gel plates; and develop the chromatogram with 3:1 toluene and absolute ethanol mixture. Visualize the spots under a UV lamp, if the silica gel has a fluorescent indicator, or by adsorption of iodine vapor. Do you detect unreacted ferrocene in the reaction mixture and/or a spot that might be attributed to diacetylferrocene?

**Online Study Center**

**General Resources**
Additional Experiments

**Cleaning Up.** The reaction mixture filtrate can be flushed down the drain. Unused recrystallization and TLC solvents should be placed in the organic solvents container. The decolorizing charcoal, once free of solvent, can be placed in the nonhazardous solid waste container. If local regulations do not allow for the evaporation of solvents, dispose of the wet alumina in a special container.

## QUESTIONS

1. What is the structure of the intermediate species that attacks ferrocene to form acetylferrocene? What other organic molecule is formed?
2. Why does the second acetyl group enter the unoccupied ring to form diacetylferrocene?

# CHAPTER 33

# Reactions of Triphenylmethyl Carbocation, Carbanion, and Radical

Resonance forms contribute to stability.

**PRELAB EXERCISE:** Draw all of the resonance structures of the triphenylmethyl carbocation.

Triphenylmethanol, which is prepared in Chapter 38, has played an interesting role in the history of organic chemistry. It was converted to the first stable carbocation and the first stable free radical. In the experiments in this chapter, triphenylmethanol is easily converted to the triphenylmethyl (trityl) carbocation, carbanion, and radical. Each of these is stabilized by 10 contributing resonance forms and, consequently, is unusually stable. Because of their long conjugated systems, these forms absorb radiation in the visible region of the spectrum and thus can be detected visually.

## EXPERIMENTS

### The Triphenylmethyl (Trityl) Carbocation

The reactions of triphenylmethanol are dominated by the ease with which it dissociates to form the relatively stable triphenylmethyl carbocation. When colorless triphenylmethanol is dissolved in concentrated sulfuric acid, an orange-yellow solution results that gives a fourfold depression of the melting point of sulfuric acid, meaning that 4 moles of ions are produced per mole of triphenylmethanol. If the triphenylmethanol were simply protonated, only 2 moles of ions would result.

$$(C_6H_5)_3COH + 2 H_2SO_4 \rightleftharpoons (C_6H_5)_3C^+ + H_3O^+ + 2 HSO_4^-$$

$$(C_6H_5)_3COH + H_2SO_4 \not\rightleftharpoons (C_6H_5)_3COH_2^+ + HSO_4^-$$

The central carbon atom in the carbocation is $sp^2$ hybridized, and thus the three carbons attached to it are coplanar and disposed at 120° angles:

However, the three phenyl groups, because of steric hindrance, cannot lie on one plane. Therefore, the carbocation is propeller shaped:

## 1. Trityl Methyl Ether

Triphenylmethanol is a tertiary alcohol and undergoes, as expected, $S_N1$ reactions. The intermediate cation is, however, stable enough to be seen in sulfuric acid solution as a red-brown to yellow solution. After dissolution in concentrated sulfuric acid, the hydroxyl anion is protonated; then the $OH_2^+$ portion is lost as $H_2O$ (which is then itself protonated), leaving the carbocation. The bisulfate ion is a very weak nucleophile and does not compete with methanol in the formation of the product, trityl methyl ether.

**Triphenylmethanol**
MW 260.34, mp 163°C

Yellow

**Triphenylmethyl methyl ether**
**Trityl methyl ether**
MW 274.37, mp 96°C

## Microscale Procedure

> **IN THIS EXPERIMENT** triphenylmethanol is dissolved in sulfuric acid, and the resulting ion is allowed to react with methanol to give a crystalline ether. This material is recrystallized from methanol.

In a 10 × 100 mm reaction tube place 50 mg of triphenylmethanol, grind the crystals to a fine powder with a glass stirring rod, and add about 0.4–0.5 mL of concentrated sulfuric acid (use the tube calibration to judge the amount). Continue to stir and grind the solid to dissolve all of the alcohol. Note and explain the color. Using a Pasteur pipette, transfer the sulfuric acid solution to 3 mL of ice-cold methanol in another reaction tube. Use some of the cold methanol to rinse out the first tube. If necessary, induce crystallization by cooling the solution, adding a seed crystal, or scratching the tube with a glass stirring rod at the liquid-air interface. Collect the product by filtration on a Hirsch funnel. Do not use filter paper on top of the polyethylene frit because this strongly acidic solution may attack the paper. Wash the crystals well with water and squeeze them between sheets of filter paper to aid drying. Determine the weight of the crude material, calculate the percent yield, and save a sample for a melting-point determination. Recrystallize the product from boiling methanol and determine the melting point of the purified trityl methyl ether. *See* Figure 33.1 for the $^1$H NMR spectrum.

**Online Study Center**

Video: Microscale Filtration on the Hirsch Funnel

**■ FIG. 33.1**
The $^1$H NMR spectrum of triphenylmethyl methyl ether, also known as trityl methyl ether (250 MHz).

PPM (δ)

**Cleaning Up.** Dilute the filtrate with water, neutralize with sodium carbonate, and flush down the drain.

### Macroscale Procedure

In a 10-mL Erlenmeyer flask place 0.5 g of triphenylmethanol, grind the crystals to a fine powder with a glass stirring rod, and add 5 mL of concentrated sulfuric acid. Continue to stir the mixture to dissolve all of the alcohol. Using a Pasteur pipette, transfer the sulfuric acid solution to 30 mL of ice-cold methanol in a 50-mL Erlenmeyer flask. Use some of the cold methanol to rinse out the first flask. Induce crystallization (if necessary) by cooling the solution, adding a seed crystal, or scratching the solution with a glass stirring rod at the liquid-air interface. Collect the product by filtration on a Hirsch funnel. Do not use filter paper on top of the polyethylene frit because this strongly acidic solution may attack the paper. Wash the crystals well with water and squeeze them between sheets of filter paper to aid drying. Determine the weight of the crude material, calculate the percent yield, and save a sample for a melting-point determination. Recrystallize the product from boiling methanol and determine the melting point of the purified trityl methyl ether. *See* Figure 33.1 for the $^1H$ NMR spectrum.

**Cleaning Up.** Dilute the filtrate with water, neutralize with sodium carbonate, and flush down the drain.

## 2. Triphenylmethyl (Trityl) Bromide

> **IN THIS EXPERIMENT** an acetic acid solution of triphenylmethanol reacts with hydrobromic acid to give the corresponding bromide that is recrystallized from hexane.

Reactions that proceed through
the carbocation

The triphenylmethanol is dissolved in a good ionizing solvent (acetic acid) and allowed to react with a strong acid and nucleophile (hydrobromic acid). The intermediate carbocation reacts immediately with bromide ion. Acetate ion, to the extent it is present, is not a good nucleophile. Experiments 2–9 can be run on five times the quantities specified here.

**Triphenylmethyl bromide**
**Trityl bromide**
MW 323.24, mp 154°C

Measure the acid in a 1-mL graduated pipette fitted with a pipette pump.

## Microscale Procedure

Dissolve 100 mg of triphenylmethanol in 2 mL of warm acetic acid on a steam or hot water bath, add 0.2 mL of 47% hydrobromic acid, and heat the mixture for 5 min on a steam bath or in a beaker of boiling water. Cool it in ice, collect the product on a Hirsch funnel, wash the product with water and hexanes, allow it to dry, determine the weight, and calculate the percent yield. Recrystallize the product from hexanes and then compare the melting points of the recrystallized and crude materials. The compound crystallizes slowly; allow adequate time for crystals to form. Use the Beilstein test[1] to test for halogen in both the product and starting materials.

**Cleaning Up.** Dilute the filtrate from the reaction with water, neutralize with sodium carbonate, and flush down the drain. The hexanes mother liquor from the crystallization goes in the organic solvents container.

## 3. For Further Investigation: Triphenylmethyl (Trityl) Iodide?

> **IN THIS EXPERIMENT** an acetic acid solution of triphenylmethanol is reacted with hydriodic acid to give, by analogy with the previous reactions, a product expected to be the corresponding iodide. Analysis by the Beilstein test will confirm or disprove this.

In a reaction very similar to the preparation of the bromide, the iodide might be prepared. Bisulfite is added to react with any iodine formed.

$$\left( \bigcirc \right)_3 C-OH \quad \xrightarrow[\text{H}_2\text{O, H}^+]{\text{HI, } -\text{H}_2\text{O}} \quad \left( \bigcirc \right)_3 C-I$$

**Triphenylmethyl iodide**
**Trityl iodide**
MW 370.22, mp 183°C

## Microscale Procedure

Dissolve 100 mg of triphenylmethanol in 2 mL of warm acetic acid, add 0.2 mL of 47% hydriodic acid, heat the mixture for 1 h on a steam bath, cool it, and add it to a solution of 0.1 g of sodium bisulfite dissolved in 2 mL water in a 10-mL Erlenmeyer flask. Collect the product on a Hirsch funnel, wash it with

---

1. To carry out the Beilstein test for halogens, heat the tip of a copper wire in a burner flame until no further coloration of the flame is noticed. Allow the wire to cool slightly; then dip the wire into the unknown (solid or liquid) and again heat it in the flame. A green flash is indicative of chlorine, blue-green of bromine, and blue of iodine; fluorine is not detected because copper fluoride is not volatile. The test is very sensitive; halogen-containing impurities may give misleading results. Run the test on a compound known to contain a halogen for comparison to your unknown.

water, press out as much water as possible, and recrystallize the crude and moist product from methanol (about 3–4 mL). Determine the weight of the dry product, calculate the percent yield, and determine the melting point. Run the Beilstein test.[2] What has been produced and why?

**Cleaning Up.** Dilute the filtrate from the reaction with water, neutralize with sodium carbonate, and flush down the drain. The methanol from the crystallization should be placed in the halogenated organic solvents container.

## 4. Triphenylmethyl Fluoborate and the Tropilium Ion

Triphenylmethanol
MW 260.34
mp 163°C

Fluoboric acid
MW 87.83
bp 130°C

Triphenylmethyl carbocation

$BF_4^-, CH_3\overset{O}{\overset{\|}{C}}O\overset{O}{\overset{\|}{C}}CH_3$

Tropilium iodide
MW 218.04

NaI
MW 149.89

Tropilium
fluoborate
MW 177.94

Triphenyl-
methane
MW 244.34
mp 94°C

Cycloheptatriene
MW 92.14
bp 117°C

Triphenylmethyl
fluoborate

$+ 2\ CH_3COOH$

Tropilium ion—an aromatic ion

Reaction of triphenylmethanol with fluoboric acid in the presence of acetic anhydride generates trityl fluoborate, a stable salt. The fluoboric acid protonates the triphenylmethanol, which loses the elements of water in an equilibrium reaction. The water reacts with the acetic anhydride to form acetic acid and thus drives the reaction to completion.

The salt so formed is the fluoborate of the triphenyl carbocation; it is a powerful base. It can be isolated, but in this case it will be used in situ to react with the hydrocarbon cycloheptatriene.

Cycloheptatriene is more basic than most hydrocarbons and will lose a proton to the triphenylmethyl carbocation to give triphenylmethane and the cyclo-heptatrienide carbocation, the tropilium ion. To demonstrate that hydride ion transfer has taken place, isolate triphenylmethane and determine its melting point.

---

2. *See* footnote1.

The Hückel Rule:
$4n + 2$ π electrons

The tropilium ion has a planar structure, each carbon bears a single proton, and the ion contains 6 π electrons; it is aromatic, a characteristic that can be confirmed by nuclear magnetic resonance (NMR) spectroscopy.

The fluoborate group can be displaced by the iodide ion to prepare tropilium iodide. You can see that the iodide is ionic by watching the reaction of the aqueous solution of this ion with silver ion.

### Microscale Procedure

> **IN THIS EXPERIMENT** triphenylmethanol and fluoboric acid react to form a fluoborate salt that is a powerful base. It removes a proton from cycloheptatriene to give tropilium ion, a carbocation that is aromatic because it contains 6 π electrons. This ion is isolated by filtration as the fluoborate salt. The filtrate is made basic and extracted with ether. This ether solution is washed, dried, and evaporated to give triphenylmethane. The fluoborate salt is easily converted to the iodide, which can be isolated and crystallized from methanol.

⚠️ **CAUTION**: Handle fluoboric acid and acetic anhydride with great care. These reagents are toxic and corrosive. Avoid breathing the vapors or any contact with the skin. In case of skin contact, rinse the affected part under running water for at least 10 min. Carry out this experiment in the hood.

*Online Study Center*

Video: Microscale Filtration on the Hirsch Funnel

*Online Study Center*

Photo: Extraction with Ether

*Online Study Center*

Photo: Crystallization; Videos: Picking a Solvent; Recrystallization

Structure proof: NMR

In a 10 × 100-mm reaction tube place 1.75 mL of acetic anhydride, cool the tube, and add 88 mg of fluoboric acid. Add 195 mg of triphenylmethanol with thorough stirring. Warm the mixture to give a homogeneous dark solution of the triphenylmethyl fluoborate and then add 78 mg cycloheptatriene. The color of the trityl cation should fade during this reaction, and the tropilium fluoborate should begin to precipitate. Add 2 mL of anhydrous ether to the reaction tube, stir the contents well while cooling on ice, and collect the product by filtration on a Hirsch funnel. Wash the product with 2 mL of dry ether and then dry the product between sheets of filter paper.

To the filtrate add 3 *M* sodium hydroxide solution and shake the flask to allow all the acetic anhydride and fluoboric acid to react with the base. Test the aqueous layer with indicator paper to ascertain that neutralization is complete; then transfer the mixture to a reaction tube and draw off the aqueous layer. Wash the ether once with 2 mL of water and dry the ether over calcium chloride pellets, adding the drying agent until it no longer clumps together.

Transfer the ether to a tared reaction tube and evaporate the solvent to leave crude triphenylmethane. Remove a sample for a melting-point determination and recrystallize the residue from an appropriate solvent, determined by experimentation. Prove to yourself that the compound isolated is indeed triphenylmethane. Obtain an infrared (IR) spectrum and an NMR spectrum.

To determine the NMR spectrum of the tropilium fluoborate collected on a Hirsch funnel, dissolve about 50 mg of the product in 0.3 mL of deuterated dimethyl sulfoxide that contains 1% tetramethylsilane as a reference. Compare the NMR spectrum obtained to that of the starting material, cycloheptatriene. The spectrum of the latter can be obtained in deuterochloroform, again using tetramethylsilane as the reference compound.

*Online Study Center*

Photo: Filtration Using a
Pasteur Pipette; Video: Filtration
of Crystals Using the Pasteur
Pipette

The tropilium fluoborate can be converted to tropilium iodide by dissolving the fluoborate in a few drops of hot, but not boiling, water and adding to this solution 0.25 mL of a saturated solution of sodium iodide. Cool the mixture in ice, remove the solvent with a pipette, and wash the crystals with 0.5 mL of ice-cold methanol. Scrape most of the crystals onto a piece of filter paper and allow them to dry before determining the weight. To the crystalline residue in the reaction tube, add a few drops of water, warm the tube if necessary to dissolve the tropilium iodide, add a drop of 2% aqueous silver nitrate solution, and note the result.

**Cleaning Up.** Solutions that contain the fluoborate ion (the neutralized filtrate, the NMR sample, the filtrate from the iodide preparation) should be treated with aqueous calcium chloride to precipitate insoluble calcium fluoride, which is removed by filtration and placed in the nonhazardous solid waste container. The filtrate can be flushed down the drain. Under the hood, allow the ether to evaporate from the calcium chloride and then place the drying agent in the nonhazardous solid waste container. If local regulations do not allow for the evaporation of solvents in a hood, dispose of the wet pellets in a special container. The recrystallization solvent goes in the organic solvents container.

## The Trityl Carbanion

The triphenylmethyl carbanion, the trityl anion, can be generated by reacting triphenylmethane with *n*-butyllithium, a very strong base. The reaction generates the blood-red lithium triphenylmethide and butane. The triphenylmethyl anion reacts much as a Grignard reagent does (*see* Chapter 38). In the present experiment it reacts with carbon dioxide to give triphenylacetic acid after acidification. Avoid an excess of *n*-butyllithium; on reaction with carbon dioxide, it gives the vile-smelling pentanoic acid.

**Triphenylmethane**
MW 244.34
mp 94°C

***n*-Butyllithium**
MW 64.06

**Trityl carbanion**
**Lithium triphenylmethide**

$CO_2$, $H^+$

**Triphenylacetic acid**
MW 288.35, mp 270–273°C

# 5. Synthesis of the Trityl Carbanion and Triphenylacetic Acid

**IN THIS EXPERIMENT** triphenylmethane in ether is reacted with the strong base *n*-butyllithium under nitrogen to give the trityl carbanion. When the solution of this carbanion is poured onto dry ice, it is converted to triphenylacetic acid. This acid is made basic with aqueous sodium hydroxide, and the solution is washed with ether. The triphenylacetic acid crystallizes when the aqueous layer is made more acidic.

**Online Study Center**

Photos: Capping a Reaction Tube with a Septum, Removing a Reagent from a Septum-Capped Bottle

Or use 0.25 mL of a 1.6 *M* solution of *n*-butyllithium.

**CAUTION**: Use great care in handling the *n*-butyllithium. It reacts avidly with air and violently with water.

### Microscale Procedure

To a reaction tube (Fig. 33.2) that has been dried for at least 30 min in a 110°C oven and capped with a rubber septum, add 100 mg of triphenylmethane and 3 mL of anhydrous diethyl or *t*-butyl methyl ether. Add an empty syringe needle to the septum; flush out the air in the tube by passing a slow current of nitrogen into the tube for about 30 s. Remove the nitrogen inlet needle and, with thorough mixing of the solution, add to the reaction mixture 0.4 mL of a 1.0 *M* solution of *n*-butyllithium in hexane (a commercial product).

*n*-Butyllithium in hexane

22-Gauge blunt needles

Rubber septum

Triphenylmethane in anhydrous ether

**■ FIG. 33.2**
An apparatus for synthesizing the trityl carbanion.

Reaction with dry ice

In a 30-mL beaker place one or two small pieces of dry ice (solid carbon dioxide) that have been wiped free of any adhering frost. Handle the dry ice with a towel or gloves. Contact with the skin can cause frostbite because the sublimation temperature of dry ice is $-78.5°C$. Calculate how much carbon dioxide is needed to react with the anion prepared from 100 mg of triphenylmethane; use at least a tenfold excess. Using a Pasteur pipette, transfer the solution of the anion to the dry ice; the reaction is immediate. Allow the unreacted dry ice to sublime; then add 3 mL of 1.5 $M$ aqueous sodium hydroxide solution, dissolve as much solid as possible, and transfer the liquid to a reaction tube. Shake the mixture, draw off the ether layer, and wash the aqueous layer with two 2.5-mL portions of ether, which are also discarded. Using 3 $M$ hydrochloric acid, acidify the aqueous layer to pH < 4. Collect the precipitate, wash it with water, and allow it to air-dry. Take the melting point. When pure, triphenylacetic acid melts at 267°C. The purity and identity of the product can be assessed using thin-layer chromatography (TLC) and IR spectroscopy (using a potassium bromide disk).

The sodium hydroxide converts the acid to its water-soluble anion.

**Online Study Center**

Photo: Extraction with Ether

**Cleaning Up.** There should be no excess $n$-butyllithium, but a few drops of the solution in hexane can be destroyed by reaction with $t$-butyl alcohol. Place ether in the organic solvents container. Dilute the aqueous layer with water, neutralize with sodium carbonate, and then flush down the drain.

## Triphenylmethyl: A Stable Free Radical

In the early part of the 19th century, many attempts were made to prepare methyl, ethyl, and similar radicals in a free state, as sodium had been prepared from sodium chloride. Many well-known chemists tried: Gay-Lussac's CN turned out to be cyanogen $(CN)_2$; Bunsen's cacodyl from $(CH_3)_2AsCl$ proved to be $(CH_3)_2As—As(CH_3)_2$; and the Kolbe electrolysis gave $CH_3CH_2CH_2CH_3$ instead of $CH_3CH_2$. Moses Gomberg at the University of Michigan prepared the first free radical in 1900. He had prepared triphenylmethane and chlorinated it to give triphenylmethyl chloride, which he hoped to couple to form hexaphenylethane:

Gomberg: the first free radical

$$2 \left( \bigcirc \right)_3 C-Cl + 2\ Ag \xrightarrow{\quad\times\quad} \left( \bigcirc \right)_3 C-C \left( \bigcirc \right)_3 + 2\ AgCl$$

**Triphenylmethyl chloride**                    **Hexaphenylethane**

The solid that Gomberg obtained was a high-melting, sparingly soluble, white solid, which turned out on analysis to be not a hydrocarbon but instead an oxygen-containing compound $(C_{38}H_{30}O_2)$. Repeating the experiment in a vacuum, he obtained a yellow solution that, on evaporation, deposited crystals of a colorless hydrocarbon that was remarkably reactive. It readily reacted with oxygen, bromine, chlorine, and iodine. On dissolution the white hydrocarbon gave a yellow solution in a vacuum; the hydrocarbon was deposited when he evaporated the solution once more. Gomberg interpreted these results as follows:

> The experimental evidence . . . forces me to the conclusion that we have to deal here with a free radical, triphenylmethyl, $(C_6H_5)_3C\cdot$. The action of zinc results, as it seems to me, in a mere abstraction of the halogen:

$$2\ (C_6H_5)_3C—Cl + Zn \longrightarrow 2\ (C_6H_5)_3C\cdot + ZnCl_2$$

> Now as a result of the removal of the halogen atom from triphenylchloromethane, the fourth valence of the methane is bound either to take up the complicated group $(C_6H_5)_3C\cdot$ or remain as such, with carbon as trivalent. Apparently the latter is what happens.[3]

For a long time after Gomberg first carried out this reaction, it was assumed the radical dimerized to form hexaphenylethane:

$$2\ (C_6H_5)_3C\cdot \longrightarrow (C_6H_5)_3C—C(C_6H_5)_3$$

$$2\ (C_6H_5)_3C\cdot \longrightarrow$$

But in 1968 NMR and UV (ultraviolet) evidence showed that the radical is in equilibrium with a different substance:

Hexaphenylethane has not yet been synthesized, presumably because of steric hindrance.

In Experiment 7 the trityl radical is prepared in much the same fashion as Gomberg used—by reacting trityl bromide with zinc in the absence of oxygen. The yellow solution is then deliberately exposed to air to give the peroxide.

3. Gomberg, M. *J. Am. Chem. Soc.* **1900**, *22*, 757.

### 6. Computational Chemistry

With a molecular mechanics program, construct a model of hexaphenylethane; then carry out an energy minimization of the molecule. Note the steric energy. Construct a model of the compound actually formed when the trityl radical dimerizes. Note the steric energy. What is the length of the central C—C bond in your model of hexaphenylethane? What is the C—C bond length in 1,2-diphenylethane? These numbers are probably not correct because valid parameters do not exist for severely sterically hindered molecules of this type, but the space-filling models are worth studying. Do these calculations give any insight as to why hexaphenylethane has not yet been synthesized?

### 7. Synthesis of Triphenylmethyl Peroxide

$$\left(\bigcirc\right)_3 C-Br + Zn \longrightarrow \left(\bigcirc\right)_3 C\cdot + ZnBr_2$$

**Trityl radical**

$$\Big\downarrow O_2$$

$$\left(\bigcirc\right)_3 C-O-O-C\left(\bigcirc\right)_3$$

**Triphenylmethyl peroxide**
MW 518.67, mp 186°C

The trityl radical

In a small test tube dissolve 100 mg of trityl bromide or chloride in 0.5 mL of toluene. Material prepared in Experiment 2 may be used; it need not be recrystallized. Cap the tube with a septum, insert an empty needle, and flush the tube with nitrogen while shaking the contents. Add 0.2 g of fresh zinc dust as quickly as possible and then flush the tube with nitrogen once more. Shake the tube vigorously for about 10 min and note the appearance of the reaction mixture. Using a Pasteur pipette, transfer the solution, but not the zinc, to a 30-mL beaker. Roll the solution around the inside of the beaker to give it maximum exposure to the air and, after a few minutes, collect the peroxide on a Hirsch funnel. Wash the solid with a little cold toluene, allow it to air dry, and determine the melting point. It is reported to melt at 186°C.

**Cleaning Up.** Transfer the mixture of zinc and zinc bromide to the hazardous waste container.

## *For Further Investigation*

Carry out both of the following reactions. Compare the products formed. Run the necessary solubility and simple qualitative tests. Interpret your results and propose structures for the compounds produced in each of the reactions.

Perform tests on compounds 1 and 2 to deduce their identities. Use 1:10 ether/hexanes for TLC.

HOOCCH$_2$COOH
**Malonic acid**

*Online Study Center*
Photo: Filtration Using a Pasteur Pipette; Video: Filtration of Crystals Using the Pasteur Pipette

### 8. Synthesis of Compound 1

**Microscale Procedure.** In a reaction tube, place 100 mg of triphenylmethanol and 200 mg of malonic acid. Grind the two solids together with a glass stirring rod and then heat the reaction tube at 150°C for 7 min. Use an aspirator tube mounted at the mouth of the tube to carry away undesirable fumes. Allow the tube to cool and then dissolve the contents in about 0.2 mL toluene. Dilute the solution with 1 mL of 60°C–80°C hexanes; after cooling the tube in ice, isolate the crystals, wash them with a little cold hexanes, and determine the weight and melting point of the dry crystals.

**Cleaning Up.** Place the toluene/hexanes mother liquor in the organic solvents container.

CICH$_2$COOH
**Chloroacetic acid**

### 9. Synthesis of Compound 2

**Microscale Procedure.** In a reaction tube dissolve 100 mg of triphenylmethanol in 2 mL of acetic acid; then add to this 0.4 mL of an acetic acid solution containing 5% of chloroacetic acid and 1% of sulfuric acid. Heat the mixture for 5 min, add about 1 mL of water to produce a hot saturated solution, and let the tube cool slowly to room temperature. Collect the product by filtration and wash the crystals with 1:1 methanol-water mixture. Once the crystals are dry, determine the weight and melting point.

**Cleaning Up.** Dilute the filtrate with water, neutralize with sodium carbonate, and flush down the drain.

## QUESTIONS

**1.** Give the mechanism for the free radical chlorination of triphenylmethane.

**2.** Is the propeller-shaped triphenylmethyl carbocation a chiral species?

**3.** Without carrying out the experiments, speculate on the structures of Compounds 1 and 2 synthesized in Experiments 8 and 9.

**4.** Is the production of triphenylmethyl peroxide a chain reaction?

**5.** What product would you expect from the reaction of carbon tetrachloride, benzene, and aluminum chloride?

# CHAPTER 34

# 1,2,3,4-Tetraphenylnaphthalene via Benzyne

**Online Study Center**

This icon will direct you to techniques, equipment setups, and online resources at http://college.hmco.com/ PIC/williamsonMMOE5e.

**PRELAB EXERCISE:** Speculate on the reasons for the stability of the iodonium ion. What type of strain exists in the benzyne molecule?

**1**
**2-Iodobenzoic acid**
MW 248.03, mp 162–163°C

$\dfrac{(OSO_2OK)_2}{H_2SO_4}$

**2**

$C_6H_6$

**3**

$NH_4OH$

**5**
**Benzyne**

$\dfrac{\Delta}{-\ \text{(phenyl-I)}\ ,\ -CO_2}$

**4**
**Diphenyliodonium-2-carboxylate monohydrate**
MW 341.13

· $H_2O$

**5**
**Benzyne**

+

**6**
**Tetraphenylcyclopentadienone**
MW 384.45, mp 219°C

→

→

**7**
**1,2,3,4-Tetraphenylnaphthalene**
MW 432.53, mp 196–199°C and
203–204°C

Synthetic use of an intermediate not known as such

This synthesis of 1,2,3,4-tetraphenylnaphthalene (**7**) demonstrates the transient existence of benzyne (**5**), a hydrocarbon that has not been isolated as such. The precursor, diphenyliodonium-2-carboxylate (**4**), is heated in an inert solvent to a temperature at which it decomposes to benzyne, iodobenzene, and carbon dioxide in the presence of tetraphenylcyclopentadienone (**6**) as the trapping agent. The preparation of the precursor (**4**) illustrates oxidation of a derivative of iodobenzene to an iodonium salt (**2**) and the Friedel–Crafts-like reaction of the substance with benzene to form the diphenyliodonium salt (**3**). Neutralization with ammonium hydroxide then liberates the precursor, the inner salt (**4**), which, when obtained by crystallization from water, is the monohydrate.

## EXPERIMENTS

# 1. Diphenyliodonium-2-Carboxylate Monohydrate (4)

> **IN THIS EXPERIMENT** iodobenzoic acid forms a salt with a peroxyacid in sulfuric acid. This salt reacts with benzene to form another salt that on reaction with ammonium hydroxide forms still another salt, which is extracted into dichloromethane. The dichloromethane is dried and evaporated to give the pure salt, diphenyliodonium-2-carboxylate, which is recrystallized from water.

⚠️ **CAUTION**: Benzene is a mild carcinogen. Carry out this experiment in the hood. Handle concentrated sulfuric acid with care.

$K_2S_2O_8$
**Potassium persulfate
(Potassium peroxydisulfate)**
MW 270.33
*Strong oxidant!*

Measure 0.85 mL of concentrated sulfuric acid into a reaction tube and place it in an ice bath to cool. In a 10-mL Erlenmeyer flask that is clamped in an ice bath, place 200 mg of 2-iodobenzoic acid, prepared using the Sandmeyer reaction (*see* Chapter 44), and 260 mg of potassium persulfate.[1] Remove the tube containing the cold sulfuric acid from the ice bath, wipe it dry, and add the acid to the two solids in the flask. Stir the mixture in the ice bath for 4–5 min to produce an even suspension; then remove it and note the time. The reaction mixture foams somewhat and acquires a succession of colors. After it has stood at room temperature for 20 min, swirl the flask vigorously in an ice bath for 2–3 min, add 0.2 mL of benzene, and swirl and cool until the benzene freezes. Then remove and wipe the flask and note the time at which the benzene melts. Warm the flask in the palm of your hand and swirl frequently at room temperature for 20 min to promote completion of reaction in the two-phase mixture. If available, magnetic stirring of the reaction mixture would be convenient but is not required.

While the reaction is going to completion, place three reaction tubes in an ice bath to chill: one containing 1.9 mL of distilled water, another containing 2.3 mL of concentrated ammonium hydroxide solution, and another containing 4 mL of

---

1. The fine granular material supplied by Fisher Scientific Co. is satisfactory. Persulfate in the form of large prisms should be finely ground prior to use.

dichloromethane. At the end of the 20-min reaction period, chill the reaction mixture thoroughly in an ice bath and add to it dropwise, with vigorous swirling and mixing, the 1.9 mL of ice-cold water that has been chilling. The solid that separates is the potassium bisulfate salt of diphenyliodonium-2-carboxylic acid (**3**).

Pour the chilled dichloromethane into the flask for extracting the product liberated when the ammonia is added. While swirling the flask vigorously, add the ammonia dropwise over a 10-min period. The mixture must be alkaline (pH 9). Test with indicator paper and add more ammonia if necessary. Using a Pasteur pipette, remove the dichloromethane layer, transfer the aqueous layer to a 4-in. test tube, and extract it with two 1.5-mL portions of dichloromethane that are combined with the original dichloromethane extract. Dry the dichloromethane extracts over anhydrous calcium chloride pellets and decant into a tared 10-mL flask. Rinse the drying agent with more solvent; then evaporate the dichloromethane under a stream of air in the hood. Connect the flask to an aspirator and heat it on the steam bath until the weight is constant. The yield is about 240 mg.

For recrystallization, dislodge the solid with a spatula and dissolve it in 2.8 mL of boiling water. Use a wood boiling stick to prevent bumping of the solution. If desired, the tan solution may be decolorized by adding about 15 pieces of pelletized Norit. On cooling the solution after removal of the Norit, diphenyliodonium-2-carboxylate monohydrate (**4**), the benzyne precursor, separates in colorless, rectangular prisms (mp 219°C–220°C). The yield should be about 200 mg.

**Cleaning Up.** Neutralize the aqueous layer with dilute hydrochloric acid. It should then be diluted with water and flushed down the drain. Allow the dichloromethane to evaporate from the drying agent in the hood and then place the drying agent in the nonhazardous solid waste container along with the Norit. If local regulations do not allow for solvent evaporation in the hood, place the wet drying agent in a special container.

**Online Study Center**

Video: Extraction with Dichloromethane

**Online Study Center**

Photo: Crystallization; Videos: Recrystallization; Picking a Solvent

## 2. Diphenyliodonium-2-Carboxylate Monohydrate (4)

K$_2$S$_2$O$_8$
**Potassium persulfate (Potassium peroxydisulfate)**
MW 270.33
*Strong oxidant!*

⚠️

**CAUTION**: Benzene is a mild carcinogen. Carry out this experiment in the hood. Handle concentrated sulfuric acid with care.

Measure 8 mL of concentrated sulfuric acid into a 25-mL Erlenmeyer flask and place the flask in an ice bath to cool. Place 2.0 g of *o*-iodobenzoic acid (*see* Chapter 44) and 2.6 g of potassium persulfate[2] in a 125-mL Erlenmeyer flask, swirl the flask containing sulfuric acid vigorously in the ice bath for 2–3 min, then remove it and wipe it dry. Place the larger flask in the ice bath and pour the chilled acid down the walls to dislodge any particles of solid. Swirl the flask in the ice bath for 4–5 min to produce an even suspension; then remove it and note the time. The reaction mixture foams somewhat and acquires a succession of colors. After it has stood at room temperature for 20 min, swirl the flask vigorously in an ice bath for 3–4 min, add 2 mL of benzene, and swirl and cool until the benzene freezes. Then remove and wipe the flask and note the time at which

---

2. The fine granular material supplied by Fisher Scientific Co. is satisfactory. Persulfate in the form of large prisms should be finely ground prior to use.

the benzene melts. Warm the flask in the palm of your hand and swirl frequently at room temperature for 20 min to promote completion of the reaction in the two-phase mixture.

While the reaction is going to completion, place three 50-mL Erlenmeyer flasks in an ice bath to chill: one containing 19 mL of distilled water, another containing 23 mL of 29% ammonium hydroxide solution, and another containing 40 mL of dichloromethane (bp 40.8°C). At the end of the 20-min reaction period, chill the reaction mixture thoroughly in an ice bath, mount a separatory funnel to deliver into the flask containing the reaction mixture in benzene, and place in the funnel the chilled 19 mL of water. Swirl the reaction flask vigorously while running in the water slowly. The solid that separates is the potassium bisulfate salt of diphenyliodonium-2-carboxylic acid (**3**). Pour the chilled ammonia solution into the funnel and pour the chilled dichloromethane into the reaction flask so that it will be available for efficient extraction of reaction product **4** as it is liberated from **3** on neutralization. While swirling the flask vigorously in the ice bath, run in the chilled ammonia solution during the course of about 10 min. The mixture must be alkaline (pH 9). If not, add more ammonia solution. Pour the mixture into a separatory funnel and rinse the flask with a small amount of fresh dichloromethane. Let the layers separate; then run the lower layer into a tared 125-mL Erlenmeyer flask, through a funnel fitted with a paper containing anhydrous calcium chloride pellets. Extract the aqueous solution with two 10-mL portions of dichloromethane and run the extracts through the drying agent into the tared flask. Evaporate the dried extracts to dryness on a hot water or steam bath (use an aspirator tube) and remove the solvent from the residual cake of solid by connecting the flask, with a rubber stopper, to the aspirator and heating the flask on a hot water or steam bath until the weight is constant; the yield is about 2.4 g.

For recrystallization, dislodge the bulk of the solid with a spatula and transfer it onto weighing paper and then into a 50-mL Erlenmeyer flask. Measure 28 mL of distilled water into a flask and use part of the water to dissolve the residual material in the tared 125-mL flask by heating the mixture to its boiling point over a hot plate or free flame. Pour this solution into the 50-mL flask. Add the remainder of the 28 mL of water to the 50-mL flask and bring the solid into solution at the boiling point. Add a small portion of charcoal for decolorization of the pale-tan solution, swirl, and filter at the boiling point through a funnel, preheated on a steam bath and fitted with moistened filter paper. Diphenyliodonium-2-carboxylate monohydrate (**4**), the benzyne precursor, separates in colorless, rectangular prisms and has a melting point of 219°C–220°C (with decomposition); the yield is about 2.1 g.

**Cleaning Up.** Neutralize the aqueous layer with dilute hydrochloric acid. It should then be diluted with water and flushed down the drain. Allow the dichloromethane to evaporate from the drying agent in the hood and then place the drying agent in the nonhazardous solid waste container along with the decolorizing charcoal. If local regulations do not allow for solvent evaporation in the hood, place the wet drying agent in a special container.

*Online Study Center*

Video: Macroscale Crystallization

# 3. Preparation of 1,2,3,4-Tetraphenylnaphthalene (7)

## Microscale Procedure

Diels–Alder reaction

> **IN THIS EXPERIMENT** the salt from the previous experiment and a cyclic diene are dissolved in a very high boiling solvent. At 200°C the salt decomposes to carbon dioxide, iodobenzene, and benzyne. The latter, in a Diels–Alder reaction, reacts with the cyclic diene. This diene decomposes by losing a molecule of carbon monoxide to give the product, tetraphenylnaphthalene, which is crystallized by adding ethanol to the reaction mixture.

Place 100 mg of the diphenyliodonium-2-carboxylate monohydrate just prepared and 100 mg of tetraphenylcyclopentadienone, prepared according to the procedure in Chapter 51, in a reaction tube. Add 0.6 mL of triethylene glycol dimethyl ether (bp 222°C) so that the solvent rinses the walls of the tube. Support the tube vertically, insert a thermometer, and heat it over a hot sand bath. When the temperature reaches 200°C, remove the tube from the heat and note the time. Maintain the mixture at 200°C–205°C by intermittent heating until the purple color is discharged, the evolution of gas ($CO_2$ and CO) subsides, and a pale-yellow solution is obtained. In case a purple or red color persists after 3 min at 200°C–205°C, add additional small amounts of the benzyne precursor and continue to heat until all the solid is dissolved.

Cool the hot solution to below 100°C and add to it 0.6 mL of ethanol. Bring the solution to boiling, using a wood stick to prevent bumping. If shiny prisms do not separate at once, add water dropwise at the boiling point of the ethanol until crystals begin to separate. Let crystallization proceed at room temperature and then at 0°C. Collect the product using the Pasteur pipette method or on a Hirsch funnel. Wash the crystals with cold methanol. The yield of dry product should be about 80 mg. The pure hydrocarbon (**7**) exists in two crystalline forms (allotropes) and has a double melting point, the first of which is in the range 196°C–199°C. Let the melting point bath cool to about 195°C, remove the sample, and let the sample solidify. Then determine the second melting point, which, for the pure hydrocarbon, is 203°C–204°C.

**Cleaning Up.** The filtrate contains ethanol, iodobenzene, triglyme, and probably small amounts of the reactants and the product. Place the mixture in the halogenated organic solvents container.

Diels–Alder reaction

## Macroscale Procedure

Place 1.0 g of the diphenyliodonium-2-carboxylate monohydrate just prepared and 1.0 g of tetraphenylcyclopentadienone in a 25 × 150-mm test tube. Add 6 mL of triethylene glycol dimethyl ether (bp 222°C) in a way such that the solvent will rinse the walls of the tube. Support the test tube vertically, insert a thermometer,

Use hood; CO is produced.

and heat with a sand bath. When the temperature reaches 200°C, remove from the heat and note the time. Maintain the mixture at 200°C–205°C by intermittent heating until the purple color is discharged, the evolution of gas ($CO_2$ and CO) subsides, and a pale-yellow solution is obtained. In case a purple or red color persists after 3 min at 200°C–205°C, add additional small amounts of the benzyne precursor and continue to heat until all the solid is dissolved. Let the yellow solution cool to 90°C while heating 6 mL of 95% ethanol to its boiling point on a steam or sand bath. Pour the yellow solution into a 25-mL Erlenmeyer flask and use a few drops of the hot ethanol to rinse the test tube. Add the remainder of the ethanol to the yellow solution and heat at the boiling point. If shiny prisms do not separate at once, add a few drops of water at the boiling point of the ethanol until prisms begin to separate. Let crystallization proceed at room temperature and then at 0°C. Collect the product via filtration and wash it with methanol. The yield of colorless prisms is 0.8–0.9 g. The pure hydrocarbon (**7**) exists in two crystalline forms (allotropes) and has a double melting point, the first of which is in the range of 196°C–199°C. Let the bath cool to about 195°C, remove the thermometer, and let the sample solidify. Then determine the second melting point, which, for the pure hydrocarbon, is 203°C–204°C.

**Cleaning Up.** The filtrate contains ethanol, iodobenzene, triglyme, and probably small amounts of the reactants and the product. Place the mixture in the halogenated organic solvents container.

## QUESTIONS

1. To which general class of compounds does potassium persulfate belong?
2. Calculate the volume of carbon monoxide, at standard temperature and pressure, released during the reaction.

# Triptycene via Benzyne

**Online Study Center**

This icon will direct you to techniques, equipment setups, and online resources at http://college.hmco.com/ PIC/williamsonMMOE5e.

**PRELAB EXERCISE:** Outline the preparation of triptycene, writing balanced equations for each reaction, including the reactions that are used to remove excess anthracene.

**1**
**Anthranilic acid**
MW 137.14

$(CH_3)_2CHCH_2CH_2ONO$
**Isoamyl nitrite**
den. 0.87

**2a**

**2b**

$\Delta$
$-N_2, CO_2$

**3**
**Benzyne**

9

10
**Anthracene**
MW 178.22
mp 216°C

**4**
**Triptycene**
MW 254.31, mp 255°C

This interesting cage-ring hydrocarbon results from the 9,10-addition of benzyne to anthracene. In one procedure presented in the literature, benzyne is generated under nitrogen from *o*-fluorobromobenzene and magnesium in the presence of anthracene, but the workup is tedious and the yield only 24%. Diazotization of anthranilic acid to benzenediazonium-2-carboxylate (**2a**) or to the covalent form (**2b**) gives another benzyne precursor, but the isolated substance can be kept only at a low temperature and has a tendency to explode. However, isolation of the precursor is not necessary. On slow addition of anthranilic acid to a solution of anthracene and isoamyl nitrite in an aprotic solvent, the precursor **2a/2b** reacts with anthracene as fast as the precursor is formed. If the anthranilic acid is all present at the start, a side reaction of this substance with benzyne drastically reduces the yield. A low-boiling solvent ($CH_2Cl_2$, bp 41°C) is used in the reaction reported in the literature, in which the desired reaction goes slowly, and a solution of anthranilic acid is added dropwise over a period of 4 h. To bring the reaction time into the limits of a laboratory period, the higher boiling solvent 1,2-dimethoxyethane (bp 83°C, water soluble) is specified in this procedure, and a large excess of anthranilic acid and isoamyl nitrite is used. Treatment of the dark reaction mixture with alkali removes acidic byproducts and most of the color, but the crude product inevitably contains anthracene. However, brief heating with maleic anhydride at a suitable temperature leaves the triptycene untouched and converts the anthracene into its maleic anhydride adduct. Treatment of the reaction mixture with alkali converts the adduct into a water-soluble salt and affords colorless, pure triptycene.

*Aprotic means "no protons"; water (HOH) and alcohols (ROH) are protic.*

*Removal of anthracene*

## EXPERIMENT

## Triptycene

> **IN THIS EXPERIMENT** anthranilic acid is diazotized with isoamyl nitrite to form an unstable diazonium salt that decomposes to benzyne, which reacts with anthracene also present in the solution. This Diels–Alder reaction gives triptycene. Excess anthracene is then reacted in a very similar reaction with maleic anhydride. This material reacts with base to form a water-soluble salt, leaving the triptycene, which crystallizes from the solvent mixture. It is then recrystallized from a different solvent—methylcyclohexane.

*$CH_3OCH_2CH_2OCH_3$*
*Dimethoxyethane (DME)*

*Test dimethoxyethane for peroxides. See Chapter 2 for the peroxides test.*

*Carry out the reaction in a hood.*

*Technique for the slow addition of a solid*

### *Microscale Procedure*

Place 100 mg of anthracene,[1] 0.1 mL of isoamyl nitrite, 1 mL of dimethoxyethane, and a boiling chip in a 10 × 100 mm reaction tube. In a small filter paper, place 130 mg of anthranilic acid and push this about halfway down the reaction tube. Bring the mixture in the tube to a gentle boil; note that the anthracene does not all dissolve. Add to the filter paper 0.5 mL of dimethoxyethane dropwise to leach the acid into the solution below over a period of not less than 20 min. Remove the filter paper, add to the solution 0.1 mL more isoamyl nitrite, add 130 mg more anthranilic acid to the filter paper, replace the paper in the reaction tube, and again leach out the anthranilic acid with 0.5 mL of dimethoxyethane added dropwise

---

1. Practical grade anthracene and anthranilic acid are satisfactory.

Reaction time: 1 h

**Online Study Center**

Video: Microscale Filtration
on the Hirsch Funnel

$$CH_3OCH_2CH_2OCH_2$$
$$|$$
$$CH_3OCH_2CH_2OCH_2$$

**Triethylene glycol
dimethyl ether (triglyme)**
bp 222°C
Miscible with water

**Online Study Center**

Photo: Crystallization;
Video: Recrystallization

$$CH_3OCH_2CH_2OCH_3$$

**1,2-Dimethoxyethane**
bp 83°C

Test dimethoxyethane for
peroxides. *See* Chapter 2
for the peroxides text.

Carry out reaction in hood.

Technique for the slow addition
of a solid

over a 20-min period. Reflux for 10 min more; then add 0.5 mL of ethanol and a solution of 0.15 g of sodium hydroxide in 2 mL of water to produce a suspension of solid in a brown alkaline solvent. Cool the mixture thoroughly in ice and also cool a 4:1 methanol-water mixture for rinsing.

Collect the solid on a Hirsch funnel by vacuum filtration and wash it with the chilled solvent to remove brown mother liquor. Transfer the moist, nearly colorless solid to a tared, clean reaction tube and evacuate while warming until the weight is constant; the anthracene-triptycene mixture (mp about 190°C–230°C) weighs about 100 mg.

Add 50 mg of maleic anhydride and 1 mL of triethylene glycol dimethyl ether (triglyme, bp 222°C) and reflux the mixture for 5 min. Cool to about 100°C, add 0.5 mL of ethanol and a solution of 150 mg of sodium hydroxide in 2 mL of water; then cool in ice, along with 1.25 mL of a 4:1 methanol-water mixture for rinsing. Triptycene separates as nearly white crystals from the slightly brown alkaline solution. The washed and dried product weighs about 75 mg and melts at 255°C. Recrystallize it from methylcyclohexane (23 mL/g). If an insoluble black impurity is seen in the hot solution, transfer the clear solution to a clean tube using a Pasteur pipette, leaving behind the impurity. Slow cooling will produce flat, rectangular, laminated prisms that can be isolated by removing the solvent with a Pasteur pipette.

**Cleaning Up.** Dilute the alkaline filtrate from the reaction with water and flush it down the drain. Methylcyclohexane mother liquor from the crystallization goes in the organic solvents container.

## *Macroscale Procedure*

Place 2 g of anthracene,[2] 2 mL of isoamyl nitrite, and 20 mL of 1,2-dimethoxyethane in a 100-mL round-bottomed flask mounted in a hot sand bath and fitted with a short reflux condenser (Fig. 35.1). Insert a filter paper into a 55-mm short stem funnel, moisten the paper with 1,2-dimethoxyethane, and rest the funnel in the top of the condenser. Weigh 2.6 g of anthranilic acid on a folded filter paper, scrape the acid into the funnel with a spatula, and pack it down. Bring the mixture in the reaction flask to a gentle boil and note that the anthracene does not all dissolve. Measure 20 mL of 1,2-dimethoxyethane into a graduated cylinder, and use a Pasteur pipette to add small portions of the solvent to the anthranilic acid in the funnel—to slowly leach the acid into the reaction flask. If you make sure that the condenser is exactly vertical and the top of the funnel is centered, it should be possible to arrange for each drop to fall free into the flask and not touch the condenser wall. Once dripping from the funnel has started, add fresh batches of solvent to the acid but only 2–3 drops at a time. Plan to complete leaching the first charge of anthranilic acid in a period of not less than 20 min, using about 10 mL

---

2. Practical grade anthracene and anthranilic acid are satisfactory.

■ **FIG. 35.1**
Refluxing a reaction mixture.

Water out

Water in

Reaction time: 1 h

CH₃OCH₂CH₂OCH₂
|
CH₃OCH₂CH₂OCH₂

**Triethylene glycol
dimethyl ether (triglyme)**
bp 222°C
Miscible with water

of solvent. Then add a second 2.6-g portion of anthranilic acid to the funnel, remove from the heat and run in 2 mL of isoamyl nitrite through the condenser by slightly lifting the funnel. Replace the funnel, resume heating, and leach the anthranilic acid as before in about 20 min time. Reflux for an additional 10 min; then add 10 mL of 95% ethanol and a solution of 3 g of sodium hydroxide in 40 mL of water to produce a suspension of solid in a brown alkaline liquor. Cool thoroughly in ice also cool a 4:1 methanol-water mixture for rinsing.

Collect the solid on a small Büchner funnel and wash it with the chilled solvent to remove brown mother liquor. Transfer the moist, nearly colorless solid to a tared 100-mL round-bottomed flask and evacuate on a steam bath until the weight is constant; the anthracene–triptycene mixture (mp about 190°C–230°C) weighs 2.1 g. Add 1 g of maleic anhydride and 20 mL of triethylene glycol dimethyl ether (triglyme, bp 222°C), heat the mixture to the boiling point under reflux (*see* Fig. 35.1 for the apparatus), and reflux for 5 min. Cool to about 100°C, add 10 mL of 95% ethanol and a solution of 3 g of sodium hydroxide in 40 mL of water; then cool in ice, along with 25 mL of a 4:1 methanol-water mixture for rinsing. Triptycene separates as nearly white crystals from the slightly brown alkaline liquor. The washed and dried product weighs 1.5 g and melts at 255°C. It will appear to be colorless, but it contains a trace of black insoluble material.

Dissolve the product in methylcyclohexane (solubility is 23 mL/g), decant from the specks of black material, and allow the solution to cool slowly. The product crystallizes as flat, rectangular, laminated prisms.

**Cleaning Up.** Dilute the alkaline filtrate from the reaction with water and flush it down the drain. Methylcyclohexane mother liquor from the crystallization goes in the organic solvents container.

### For Further Investigation

**Diphenyliodonium-2-carboxylate**

Diphenyliodonium-2-carboxylate is utilized as a benzyne precursor in the synthesis of 1,2,3,4-tetraphenylnaphthalene from tetraphenylcyclopentadienone. For reasons unknown, the reaction of anthracene with benzyne generated in this way proceeds very poorly and gives only a trace of triptycene. What about the converse proposition? Would benzyne generated by diazotization of anthranilic acid in the presence of tetraphenylcyclopentadienone afford 1,2,3,4-tetraphenylnaphthalene in satisfactory yield? If you are interested in exploring this possibility, plan and execute a procedure and see what you can discover.

If benzyne is generated from anthranilic acid in the absence of anthracene, as in this experiment, then dibenzocyclobutadiene (also called biphenylene) is formed.[3]

Devise a procedure for isolating this interesting hydrocarbon.

## Computational Chemistry

Using the AM1 semiempirical molecular orbital program, calculate the heats of formation of isomeric compounds **2a** and **2b** to determine which might be formed in this reaction. Bear in mind that these will be gas phase heats of formation. Repeat the calculation in water, a polar solvent. Does this change the relative stabilities of the two isomers? Explain.

It is instructive to examine the molecular orbitals of benzyne and dibenzocyclobutadiene. Benzyne is made by joining two *ortho* hydrogens into a new bond. Does the central cyclobutadiene ring, which is antiaromatic, have any effect on the appearance of the orbitals when compared with, say, naphthalene? Compare the bond lengths of benzyne to those of benzene.

### QUESTIONS

1. Write equations showing how benzyne might be generated from *o*-fluorobromobenzene and magnesium.
2. Study the mechanism of diazonium salt formation in Chapter 44 and then propose a mechanism for diazotization using isoamyl nitrite.

---

3. Logullo, F. M.; Seitz, A. H.; Friedman, L. *Org. Syn.* **1968**, *48.* Friedman, L.; Logullo, F. M. *J. Org. Chem.* **1969**, *34*, 3089.

**■ FIG. 35.2**
The ¹H NMR spectrum of
anthracene (250 MHz).

PPM (δ)

PPM (δ)

**3.** The ¹H NMR spectrum of anthracene is given in Figure 35.2. Predict the
appearance of the NMR spectrum of triptycene. Which peak in Figure 35.2
will be missing?

# CHAPTER

# 36

# Aldehydes and Ketones

## Online Study Center

This icon will direct you to
techniques, equipment setups,
and online resources at
http://college.hmco.com/
PIC/williamsonMMOE5e.

**PRELAB EXERCISE:** Outline a logical series of experiments designed to identify an unknown aldehyde or ketone with the least
effort. Consider the time required to complete each identification
reaction.

The carbonyl group occupies a central place in organic chemistry. Aldehydes
and ketones—compounds such as formaldehyde, acetaldehyde, acetone, and
2-butanone—are very important industrial chemicals used by themselves and
as starting materials for a host of other substances. For example, more than
10 billion pounds (4.5 billion kilograms) of formaldehyde-containing plastics
are produced in the United States each year.

The carbonyl carbon is $sp^2$ hybridized, the bond angles between adjacent
groups are 120°, and the four atoms R, R′, C, and O lie in one plane:

$$
\begin{array}{c}
\ddot{\text{O}}\colon \\
\parallel \\
\text{C} \\
\diagup \quad \diagdown \\
\text{R} \qquad \text{R}'
\end{array}
$$

The electronegative oxygen polarizes the carbon-oxygen bond, rendering the
carbon electron deficient and hence subject to nucleophilic substitution.

$$
\text{C}=\ddot{\text{O}}\colon \longleftrightarrow \overset{+}{\text{C}}-\ddot{\text{O}}\colon^{-} \quad \text{or} \quad \overset{\delta+}{\text{C}}=\overset{\delta-}{\ddot{\text{O}}}\colon
$$

Geometry of the carbonyl
group

Attack on the $sp^2$ hybridized carbon occurs via the π-electron cloud above and
below the plane of the carbonyl group:

## Reactions of the Carbonyl Group

Many reactions of carbonyl groups are acid catalyzed. The acid attacks the electronegative oxygen, which bears a partial negative charge, to create a carbocation that subsequently reacts with the nucleophile:

The strength of the nucleophile and the structure of the carbonyl compound determine whether the equilibrium lies on the side of the carbonyl compound or the tetrahedral adduct. Water, a weak nucleophile, does not usually add to the carbonyl group to form a stable compound:

In the special case of trichloroacetaldehyde, however, the electron-withdrawing trichloromethyl group allows a stable hydrate to form:

A stable hydrate: chloral hydrate

The compound so formed, chloral hydrate, was discovered by Justus von Liebig in 1832 and was introduced as one of the first sedatives and hypnotics (sleep-inducing substances) in 1869. It is now most commonly encountered in detective fiction as a "Mickey Finn" or "knockout drops."

Reaction with an alcohol; hemiacetals

In an analogous manner, an aldehyde or ketone can react with an alcohol. The product, a hemiacetal or hemiketal, is usually not stable, but in the case of certain cyclic hemiacetals the product can be isolated. Glucose is an example of a stable hemiacetal.

**Hemiacetal**
usually not stable

**Glucose**
a stable cyclic hemiacetal

## Bisulfite Addition

The bisulfite ion is a strong nucleophile but a weak acid. It will attack the unhindered carbonyl group of an aldehyde or methyl ketone to form an addition product:

Because these bisulfite addition compounds are ionic water-soluble compounds and can be formed in up to 90% yield, they serve as a useful means of separating aldehydes and methyl ketones from mixtures of organic compounds. At high sodium bisulfite concentrations these adducts crystallize and can be isolated by filtration. The aldehyde or ketone can be regenerated by adding either a strong acid or base:

## Cyanide Addition

A similar reaction occurs between aldehydes and ketones and hydrogen cyanide, which, like bisulfite, is a weak acid but a strong nucleophile. The reaction is hazardous to carry out because of the toxicity of cyanide, but the cyanohydrins are useful synthetic intermediates:

Cyanohydrin formation and reactions

Amines are good nucleophiles and readily add to the carbonyl group:

The reaction is strongly dependent on the pH. In acid the amine is protonated ($RN^+H_3$) and is no longer a nucleophile. In strong base there are no protons available to catalyze the reaction. But in weak acid solution (pH 4–6) the equilibrium between acid and base (**a**) is such that protons are available to protonate the carbonyl (**b**), and yet there is free amine present to react with the protonated carbonyl (**c**):

(**a**)     $CH_3\ddot{N}H_2 + HCl \rightleftharpoons CH_3\overset{+}{N}H_3 + Cl^-$

(**b**)

(**c**)

## Schiff Bases

The intermediate hydroxyamino form of the adduct is not stable and spontaneously dehydrates under the mildly acidic conditions of the reaction to give an imine, commonly referred to as a *Schiff base*:

**Schiff base**

Imine or Schiff base formation        The biosynthesis of most amino acids proceeds through Schiff base intermediates.

# Oximes, Semicarbazones, and 2,4-Dinitrophenylhydrazones

Three rather special amines form useful stable imines:

$$\text{H}_2\ddot{\text{N}}\text{OH} + \underset{\text{R}'}{\overset{\text{R}}{\diagdown}}\text{C}=\ddot{\text{O}}: \quad \overset{-\text{H}_2\text{O}}{\rightleftharpoons} \quad \underset{\text{R}'}{\overset{\text{R}}{\diagdown}}\text{C}=\text{N}\overset{\ddot{}}{\diagup}\overset{\text{O}-\text{H}}{}$$

**Hydroxylamine**          **Oxime**

$$\text{H}_2\text{NNHCNH}_2 + \underset{\text{R}'}{\overset{\text{R}}{\diagdown}}\text{C}=\text{O} \quad \overset{-\text{H}_2\text{O}}{\rightleftharpoons} \quad \underset{\text{R}'}{\overset{\text{R}}{\diagdown}}\text{C}=\text{N}-\underset{\text{H}}{\text{N}}-\overset{\text{O}}{\overset{\|}{\text{C}}}-\text{NH}_2$$

**Semicarbazide**          **Semicarbazone**

$$\text{H}_2\text{NNH}-\underset{\text{NO}_2}{\bigodot}-\text{NO}_2 + \underset{\text{R}'}{\overset{\text{R}}{\diagdown}}\text{C}=\text{O} \quad \overset{-\text{H}_2\text{O}}{\rightleftharpoons} \quad \underset{\text{R}'}{\overset{\text{R}}{\diagdown}}\text{C}=\text{N}-\underset{\text{H}}{\text{N}}-\underset{\text{NO}_2}{\bigodot}-\text{NO}_2$$

**2,4-Dinitrophenylhydrazine**      **2,4-Dinitrophenylhydrazone**

These imines are solids and are useful for the characterization of aldehydes and ketones. For example, IR (infrared) and NMR (nuclear magnetic resonance) spectroscopies may indicate that a certain unknown is acetaldehyde. It is difficult to determine the boiling point of a few milligrams of a liquid, but if it can be converted to a solid derivative, the melting point *can* be determined with that amount. The 2,4-dinitrophenylhydrazones are usually the derivatives of choice because they are crystalline compounds with well-defined melting or decomposition points, and they increase the molecular weight by 180. Ten milligrams of acetaldehyde will give 51 mg of the 2,4-dinitrophenylhydrazone.

$$\text{CH}_3\overset{\text{O}}{\overset{\|}{\text{C}}}\text{H} + \text{H}_2\text{NNH}-\underset{\text{NO}_2}{\bigodot}-\text{NO}_2 \quad \rightleftharpoons \quad \text{CH}_3\underset{\text{H}}{\text{C}}=\text{N}-\underset{\text{H}}{\text{N}}-\underset{\text{NO}_2}{\bigodot}-\text{NO}_2$$

**Acetaldehyde**    **2,4-Dinitrophenylhydrazine**        **Acetaldehyde 2,4-dinitrophenylhydrazone**
MW 44.05         MW 198.14                    MW 224.19
bp 20.8°C        mp 196°C                     mp 168.5°C

## Tollens' Reagent

Before the advent of NMR and IR spectroscopy and mass spectrometry, the chemist was often called on to identify aldehydes and ketones by purely chemical means. Aldehydes can be distinguished chemically from ketones by their ease of oxidation to carboxylic acids. The oxidizing agent, an ammoniacal solution of silver nitrate, Tollens' reagent, is reduced to metallic silver, which is deposited on the inside of a test tube as a silver mirror.

$$2\ Ag(NH_3)_2OH\ +\ R-\overset{O}{\underset{H}{C}}\ \longrightarrow\ 2\ Ag\ +\ R-\overset{O}{\underset{O^-}{C}}\ \ NH_4^+\ +\ H_2O\ +\ 3\ NH_3$$

## Schiff's Reagent

Another way to distinguish aldehydes from ketones is to use Schiff's reagent. This is a solution of the red dye Basic Fuchsin, which is rendered colorless on treatment with sulfur dioxide. In the presence of an aldehyde, the colorless solution turns magenta.

**Basic Fuchsin, *p*-rosaniline hydrochloride**

**Schiff's reagent, colorless**

**Magenta in color**

## Iodoform Test

A test for methyl ketones

$$CH_3\overset{O}{\overset{\|}{C}}-$$

**Methyl ketone**

Methyl ketones can be distinguished from other ketones by the iodoform test. The methyl ketone is treated with iodine in a basic solution. Introduction of the first iodine atom increases the acidity of the remaining methyl protons, so halogenation stops only when the triiodo compound has been produced. The base then allows the relatively stable triiodomethyl carbanion to leave, and a subsequent proton transfer gives iodoform, a yellow crystalline solid with a melting point of 119°C–123°C. The test is also positive for fragments or compounds easily oxidized to methyl ketones, such as the fragment $CH_3\overset{|}{C}HOH$ or the compound ethanol. Acetaldehyde also gives a positive test because it is both a methyl ketone and an aldehyde.

$$R - \overset{\overset{\displaystyle \ddot{O}:}{\|}}{C} - CH_3 + OH^- \rightleftharpoons R - \overset{\overset{\displaystyle \ddot{O}:}{\|}}{C} - \overset{-}{\ddot{C}}H_2 + H_2O$$

$$R - \overset{\overset{\displaystyle \ddot{O}:}{\|}}{C} - \overset{-}{\ddot{C}}H_2 + I_2 \longrightarrow R - \overset{\overset{\displaystyle O}{\|}}{C} - CH_2I + I^- \xrightarrow{OH^-} \xrightarrow{I_2} \xrightarrow{OH^-} \xrightarrow{I_2} R - \overset{\overset{\displaystyle \ddot{O}:}{\|}}{C} - CI_3$$

$$R - \overset{\overset{\displaystyle \ddot{O}:}{\|}}{C} - CI_3 + OH^- \longrightarrow R - \overset{\overset{\displaystyle :\ddot{O}:^-}{|}}{\underset{:OH}{C}} - CI_3 \longrightarrow R - \overset{\displaystyle :\ddot{O}}{\underset{:O-H}{C}} + :\overset{-}{C}I_3 \longrightarrow R - \overset{\displaystyle :\ddot{O}}{\underset{\ddot{O}:^-}{C}} + CHI_3$$

**Iodoform**
mp 123°C

---

## EXPERIMENTS

## 1. Unknowns

You will be given an unknown that may be any of the aldehydes or ketones listed in Table 36.1. At least one derivative of the unknown is to be submitted to your instructor; but if you first do the bisulfite and iodoform characterizing tests, the results may suggest derivatives whose melting points will be particularly revealing.

You can further characterize the unknown by determining its boiling point, which is best done with a digital thermometer and a reaction tube (*see* Chapter 5). The boiling points of the unknowns are given on this book's website.

Carry out three tests:

Known positive
Known negative
Unknown

In conducting the following tests, you should perform three tests simultaneously: on a compound known to give a positive test, on a compound known to give a negative test, and on the unknown. In this way you will be able to determine whether the reagents are working as they should as well as interpret a positive or a negative test.

## 2. 2,4-Dinitrophenylhydrazones

$$\overset{R}{\underset{R'}{\diagdown}}C = \ddot{O}: + H_2\ddot{N} - \overset{\ddot{N}}{\underset{H}{|}} \overset{NO_2}{\diagdown} - NO_2 \xrightarrow{H^+} \overset{R}{\underset{R'}{\diagdown}}C = \ddot{N} \overset{}{\diagdown} \overset{\ddot{N}}{\underset{H}{|}} \overset{NO_2}{\diagdown} - NO_2$$

An easily prepared derivative
of aldehydes and ketones

To 5 mL of the stock solution[1] of 2,4-dinitrophenylhydrazine in phosphoric acid add about 0.05 g of the compound to be tested. Five milliliters of the 0.1 $M$ solution contains 0.5 mmol (0.0005 mol) of the reagent. If the compound to be tested

---

1. Dissolve 2.0 g of 2,4-dinitrophenylhydrazine in 50 mL of 85% phosphoric acid by heating, cool, add 50 mL of 95% ethanol, cool again, and clarify by suction filtration from a trace of residue.

## TABLE 36.1  •  Melting Points of Derivatives of Some Aldehydes and Ketones

| Compound[a] | Formula | $n_D^{20}$ | MW | Water Solubility | Melting Points (°C) Phenyl-hydrazone | 2,4-dinitro-phenyl-hydrazone | Semi-carbazone |
|---|---|---|---|---|---|---|---|
| Acetone | $CH_3COCH_3$ | 1.3590 | 58.08 | | 42 | 126 | 187 |
| n-Butanal | $CH_3CH_2CH_2CHO$ | 1.3790 | 72.10 | 4 g/100 g | Oil | 123 | 95 (106)[b] |
| 3-Pentanone (diethyl ketone) | $CH_3CH_2COCH_2CH_3$ | 1.3920 | 86.13 | 4.7 g/100 g | Oil | 156 | 138 |
| 2-Furaldehyde (furfural) | $C_4H_3O \cdot CHO$ | 1.5260 | 96.08 | 9 g/100 g | 97 | 212 (230)[b] | 202 |
| Benzaldehyde | $C_6H_5CHO$ | 1.5450 | 106.12 | Insol. | 158 | 237 | 222 |
| Hexane-2,5-dione | $CH_3COCH_2CH_2COCH_3$ | 1.4260 | 114.14 | ∞ | 120[c] | 257[c] | 224[c] |
| 2-Heptanone | $CH_3(CH_2)_4COCH_3$ | 1.4080 | 114.18 | Insol. | Oil | 89 | 123 |
| 3-Heptanone | $CH_3(CH_2)_3COCH_2CH_3$ | 1.4080 | 114.18 | Insol. | Oil | 81 | 101 |
| n-Heptanal | $n\text{-}C_6H_{13}CHO$ | 1.4125 | 114.18 | Insol. | Oil | 108 | 109 |
| Acetophenone | $C_6H_5COCH_3$ | 1.5325 | 120.66 | Insol. | 105 | 238 | 198 |
| 2-Octanone | $CH_3(CH_2)_5COCH_3$ | 1.4150 | 128.21 | Insol. | Oil | 58 | 122 |
| Cinnamaldehyde | $C_6H_5CH{=}CHCHO$ | 1.6220 | 132.15 | Insol. | 168 | 255 | 215 |
| Propiophenone | $C_6H_5COCH_2CH_3$ | 1.5260 | 134.17 | Insol. | About 48 | 191 | 182 |

[a]Visit this book's website for data on additional aldehydes and ketones.

[b]Both melting points have been found, depending on crystalline form of derivative.

[c]Monoderivative or diderivative.

**Online Study Center**

Video: Microscale Crystallization

has a molecular weight of 100, then 0.05 g is 0.5 mmol. Warm the reaction mixture for a few minutes in a water bath and then let crystallization proceed. Collect the product by suction filtration (Fig. 36.1), wash the crystals with a large amount of water to remove all of the phosphoric acid, press a piece of moist litmus paper onto the crystals, and wash them with more water if they are acidic. Press the product between sheets of filter paper until it is as dry as possible and recrystallize from ethanol. Occasionally a high molecular weight derivative will not dissolve in a reasonable quantity (10 mL) of ethanol. In that case cool the hot suspension and isolate the crystals by suction filtration. The boiling ethanol treatment removes impurities so that an accurate melting point can be obtained on the isolated material.

An alternative procedure is applicable when the 2,4-dinitrophenylhydrazone is known to be sparingly soluble in ethanol. Measure 0.5 mmol (0.1 g) of crystalline 2,4-dinitrophenylhydrazine into a 50-mL Erlenmeyer flask, add 15 mL of 95% ethanol, digest on a steam bath until all the solid particles are dissolved, and then add 0.5 mmol of the compound to be tested and continue warming. If there is no immediate change, add, from a Pasteur pipette, 3–4 drops of concentrated hydrochloric acid as a catalyst and note the result. Warm for a few minutes, then

■ **FIG. 36.1**
A Hirsch funnel filtration apparatus.

Filter paper,
12 mm dia.

Polyethylene
filter disk (frit),
10 mm dia.

Hirsch
funnel

To aspirator

25-mL Filter
flask

cool and collect the product. This procedure would be used for an aldehyde like cinnamaldehyde ($C_6H_5CH=CHCHO$).

The alternative procedure strikingly demonstrates the catalytic effect of hydrochloric acid, but it is not applicable to a substance like diethyl ketone, whose 2,4-dinitrophenylhydrazone is much too soluble to crystallize from the large volume of ethanol. The first procedure is obviously the one to use for an unknown.

**Cleaning Up.** The filtrate from the preparation of the 2,4-dinitrophenylhydrazone should have very little 2,4-dinitrophenylhydrazine in it, so after dilution with water and neutralization with sodium carbonate, it can be flushed down the drain. Similarly, the mother liquor from crystallization of the phenylhydrazone should have very little product in it and so should be diluted and flushed down the drain. If solid material is detected, it should be collected by suction filtration, the filtrate flushed down the drain, and the filter paper placed in the solid hazardous waste container because hydrazines are toxic.

## 3. Semicarbazones

**Semicarbazide hydrochloride**    **Pyridine**         **Semicarbazone**         **Pyridine hydrochloride**

Semicarbazide (mp 96°C) is not very stable in the free form and is used as the crystalline hydrochloride (mp 173°C). Because this salt is insoluble in methanol or ethanol and does not react readily with typical carbonyl compounds in alcohol-water mixtures, pyridine, a basic reagent, is added to liberate free semicarbazide.

To 0.5 mL of the stock solution[2] of semicarbazide hydrochloride, which contains 1 mmol of the reagent, add 1 mmol of the compound to be tested and enough methanol (1 mL) to produce a clear solution; then add 10 drops of pyridine (a twofold excess) and warm the solution gently on the steam bath for a few minutes. Cool the solution slowly to room temperature. It may be necessary to scratch the inside of the test tube in order to induce crystallization. Cool the tube in ice, collect the product by suction filtration, and wash it with water followed by a small amount of cold methanol. Recrystallize the product from methanol, ethanol, or ethanol/water. The product can easily be collected on a Wilfilter.

**Cleaning Up.** Combine the filtrate from the reaction and the mother liquor from the crystallization, dilute with water, make very slightly acidic with dilute hydrochloric acid, and flush the mixture down the drain.

*Online Study Center*

Photo: Use of the Wilfilter

## 4. Tollens' Test

Test for aldehydes

$$R-C\overset{\displaystyle O}{\underset{\displaystyle H}{\Big\backslash}} + 2\ Ag(NH_3)_2OH \longrightarrow 2\ Ag + RCOO^-NH_4^+ + H_2O + 3\ NH_3$$

Clean four or five test tubes by adding a few milliliters of 3 $M$ sodium hydroxide solution to each and heating them in a water bath while preparing the Tollens' reagent.

To 2.0 mL of 0.03 $M$ silver nitrate solution, add 1.0 mL of 3 $M$ sodium hydroxide in a test tube. To the gray precipitate of silver oxide, $Ag_2O$, add 0.5 mL of a 2.8% aqueous ammonia solution (10 mL of concentrated ammonium hydroxide diluted to 100 mL). Stopper the tube and shake it. Repeat the process until *almost* all of the precipitate dissolves (3.0 mL of ammonia at most); then dilute the solution with water to 10 mL. Empty the test tubes of sodium hydroxide solution, rinse them, and add 1 mL of Tollens' reagent to each. Add 1 drop (no more) of the substance to be tested by allowing it to run down the inside of the inclined test tube. Set the tubes aside for a few minutes without agitating the contents. If no reaction occurs, warm the mixture briefly on a water bath. As a known aldehyde, try 1 drop of a 0.1 $M$ solution of glucose. A more typical aldehyde to test is benzaldehyde.

At the end of the reaction, promptly destroy any excess Tollens' reagent with nitric acid: It can form an explosive fulminate on standing. Nitric acid can also be used to remove silver mirrors from test tubes.

---

2. Prepare a stock solution by dissolving 1.11 g of semicarbazide hydrochloride in 5 mL of water; 0.5 mL of this solution contains 1 mmol of reagent.

**Cleaning Up.** Place all solutions used in this experiment in a beaker (unused ammonium hydroxide, sodium hydroxide solution used to clean out the tubes, Tollens' reagent from all tubes). Remove any silver mirrors from reaction tubes with a few drops of nitric acid, which is added to the beaker. Make the mixture acidic with nitric acid to destroy unreacted Tollens' reagent and then neutralize the solution with sodium carbonate and add some sodium chloride solution to precipitate silver chloride (about 40 mg). The whole mixture can be flushed down the drain, or the silver chloride can be collected by suction filtration, and the filtrate flushed down the drain. The silver chloride would go in the nonhazardous solid waste container.

## 5. Schiff's Test

Very sensitive test
for aldehydes

Add 3 drops of the unknown to 2 mL of Schiff's reagent.[3] A magenta color will appear within 10 min with aldehydes. As in all of these tests, compare the colors produced by a known aldehyde, a known ketone, and the unknown compound.

**Cleaning Up.** Neutralize the solution with sodium carbonate and flush it down the drain. The amount of *p*-rosaniline in this mixture is negligible (1 mg).

## 6. Iodoform Test

$$R-\overset{\overset{\displaystyle :\ddot{O}:}{\|}}{C}-CH_3 + 3\,I_2 + 4\,OH^- \longrightarrow R-C\overset{\displaystyle \ddot{O}:}{\underset{\displaystyle \ddot{O}:^-}{\diagup}} + CHI_3 + 3\;:\ddot{\underset{..}{I}:}^- + 3\,H_2O$$

**A methyl ketone**                                                                 **Iodoform**
                                                                                    mp 119–123°C

Test for methyl ketones

The reagent contains iodine in potassium iodide solution[4] at a concentration such that 2 mL of solution, on reaction with excess methyl ketone, will yield 174 mg of iodoform. If the substance to be tested is water soluble, dissolve 4 drops of a liquid or an estimated 50 mg of a solid in 2 mL of water in a 20 × 150-mm test tube; add 2 mL of 3 *M* sodium hydroxide, and then slowly add 3 mL of the iodine solution. In a positive test the brown color of the reagent disappears, and yellow iodoform separates. If the substance to be tested is insoluble in water, dissolve it in 2 mL of 1,2-dimethoxyethane, proceed as above, and at the end dilute with 10 mL of water.

Suggested test substances are hexane-2,5-dione (water soluble), *n*-butyraldehyde (water soluble), and acetophenone (water insoluble).

---

3. Schiff's reagent is prepared by dissolving 0.1 g Basic Fuchsin (*p*-rosaniline hydrochloride) in 100 mL of water and then adding 4 mL of a saturated aqueous solution of sodium bisulfite. After 1 h, add 2 mL of concentrated hydrochloric acid.

4. Dissolve 25 g of iodine in a solution of 50 g of potassium iodide in 200 mL of water.

Iodoform can be recognized by its odor and yellow color and, more securely, from its melting point (119°C–123°C). The substance can be isolated by suction filtration of the test suspension or by adding 0.5 mL of dichloromethane, shaking the stoppered test tube to extract the iodoform into the small lower layer, withdrawing the clear part of this layer with a Pasteur pipette, and evaporating it in a small tube on a steam bath. The crude solid is crystallized from a methanol-water mixture (*see* Chapter 4). It can be collected on a Wilfilter.

**Cleaning Up.** Combine all reaction mixtures in a beaker, add a few drops of acetone to destroy any unreacted iodine in potassium iodide reagent, remove the iodoform by suction filtration, and place the iodoform in the halogenated organic waste container. The filtrate can be flushed down the drain after neutralization (if necessary).

## 7. Bisulfite Test

Forms with unhindered carbonyls

$$\underset{H}{\overset{R}{\diagdown}}C\!=\!\ddot{O}\!: \;+\; Na^+SO_3H^- \;\rightleftharpoons\; R\!-\!\underset{H}{\overset{:\ddot{O}H}{\underset{|}{\overset{|}{C}}}}\!-\!SO_3^-Na^+$$

Put 1 mL of the stock solution[5] into a 13 × 100-mm test tube and add 5 drops of the substance to be tested. Shake each tube during the next 10 min and note the results. A positive test will result from aldehydes, unhindered cyclic ketones such as cyclohexanone, and unhindered methyl ketones.

If the bisulfite test is applied to a liquid or solid that is very sparingly soluble in water, formation of the addition product is facilitated by adding a small amount of methanol before the addition to the bisulfite solution.

**Cleaning Up.** Dilute the bisulfite solution or any bisulfite addition products (they will dissociate) with a large volume of water and flush the mixture down the drain. The amount of organic material being discarded is negligible.

## 8. IR and NMR Spectroscopy

IR spectroscopy is extremely useful in analyzing all carbonyl-containing compounds, including aldehydes and ketones (Fig. 36.4 and Fig. 36.11). Refer to the extensive discussion in Chapter 11. In the modern laboratory, spectroscopy has almost completely supplanted the qualitative tests described in this chapter. Figures 36.4 through 36.11 present IR and NMR spectra of typical aldehydes and ketones. Compare these spectra with those of your unknown. Also compare the IR and [1]H NMR spectra of the hydrocarbon fluorene (Fig. 36.2 and Fig. 36.3) with those of the ketone derivative, fluorenone (Fig. 36.4 and Fig. 36.5).

---

5. Prepare a stock solution from 50 g sodium bisulfite dissolved in 200 mL of water with brief swirling.

■ **FIG. 36.2**
The IR spectrum of fluorene in $CS_2$.

■ **FIG. 36.3**
The $^1$H NMR spectrum of fluorene (250 MHz).

**■ FIG. 36.4**
The IR spectrum of fluorenone
(KBr disk).

**■ FIG. 36.5**
The ¹H NMR spectrum
of fluorenone (250 MHz).

A peak at 9.6–10 ppm in the ¹H NMR spectrum is highly characteristic of aldehydes because almost no other peaks appear in this region (Fig. 36.7 and Fig. 36.10). Similarly, a sharp singlet at 2.2 ppm is very characteristic of methyl ketones; beware of contamination of the sample by acetone, which is often used to clean glassware.

■ **FIG. 36.6**
The $^1H$ NMR spectrum of
2-butanone, $CH_3COCH_2CH_3$
(90 MHz).

■ **FIG. 36.7**
The $^1H$ NMR spectrum of
crotonaldehyde, $CH_3CH{=}CHCHO$
(90 MHz).

■ **FIG. 36.8**
The $^{13}$C NMR spectrum of
2-butanone, $CH_3COCH_2CH_3$
(22.6 MHz).

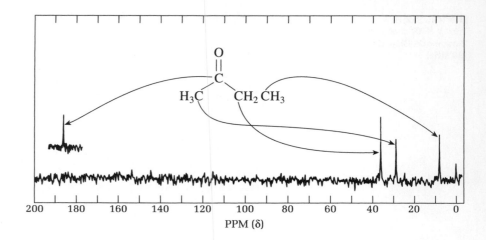

■ **FIG. 36.9**
The $^{13}$C NMR spectrum of
crotonaldehyde (22.6 MHz).

# QUESTIONS

1. What is the purpose of making derivatives of unknowns?
2. Why are 2,4-dinitrophenylhydrazones better derivatives than phenylhydrazones?
3. Using chemical tests, how would you distinguish among 2-pentanone, 3-pentanone, and pentanal?
4. Draw the structure of a compound with the empirical formula $C_5H_8O$ that gives a positive iodoform test and does not decolorize permanganate.
5. Draw the structure of a compound with the empirical formula $C_5H_8O$ that gives a positive Tollens' test and does not react with bromine in dichloromethane.

■ **FIG. 36.10**
The $^1$H NMR spectrum of
benzaldehyde (250 MHz).

■ **FIG. 36.10**
The $^1$H NMR spectrum of
benzaldehyde (250 MHz).

■ **FIG. 36.11**
The IR spectrum of benzaldehyde
(thin film).

6. Draw the structure of a compound with the empirical formula $C_5H_8O$ that reacts with phenylhydrazine, decolorizes bromine in dichloromethane, and does not give a positive iodoform test.

7. Draw the structure of two geometric isomers with the empirical formula $C_5H_8O$ that give a positive iodoform test.

**8.** What vibrations cause the peaks at about 3.6 μm (2940 cm$^{-1}$) in the IR spectrum of fluorene (Fig. 36.2)?

**9.** Locate the carbonyl peak in Figure 36.4.

**10.** Assign the various peaks in the $^1$H NMR spectrum of 2-butanone to specific protons in the molecule (Fig. 36.6).

**11.** Assign the various peaks in the $^1$H NMR spectrum of crotonaldehyde to specific protons in the molecule (Fig. 36.7).

# CHAPTER
# 37

# Dibenzalacetone by the Aldol Condensation

**PRELAB EXERCISE:** Calculate the volumes of benzaldehyde and acetone needed for the macroscale version of this reaction, taking into account the densities of the liquids and the number of moles of each required.

$$2 \text{ } \underset{\substack{\textbf{Benzaldehyde} \\ \text{MW 106.13, bp 178°C} \\ \text{den. 1.04}}}{\boxed{\text{CHO}}} + \underset{\substack{\textbf{Acetone} \\ \textbf{(2-Propanone)} \\ \text{MW 58.08, bp 56°C} \\ \text{den. 0.790}}}{CH_3\overset{O}{\overset{\|}{C}}CH_3} \xrightarrow[\text{MW 40.01}]{\text{NaOH}} \underset{\substack{\textbf{Dibenzalacetone} \\ \textbf{(1,5-Diphenyl-1,4-pentadien-3-one)} \\ \text{MW 234.30}}}{CH=CH\overset{O}{\overset{\|}{C}}CH=CH}$$

The base catalyzed reaction of an aldehyde with a ketone is a mixed aldol condensation, known as the Claisen–Schmidt reaction. Dibenzalacetone is readily prepared by the condensation of acetone with two equivalents of benzaldehyde. The aldehyde carbonyl is more reactive than that of the ketone and, therefore, reacts rapidly with the anion of the ketone to give a β-hydroxyketone, which easily undergoes base-catalyzed dehydration. Depending on the relative quantities of the reactants, the reaction can give either mono- or dibenzalacetone.

Dibenzalacetone is innocuous; its spectral properties (ultraviolet [UV] absorbance) indicate why it is used in sunscreens and sunblock preparations. In the present experiment sufficient ethanol is present as a solvent to readily dissolve the starting material, benzaldehyde, and also the intermediate, benzalacetone. The benzalacetone, once formed, can then easily react with another mole of benzaldehyde to give the product, dibenzalacetone. The mechanism for the formation of benzalacetone is as follows:

$$CH_3\overset{O}{\overset{\|}{C}}CH_3 + OH^- \longrightarrow \left[ CH_3\overset{O}{\overset{\|}{C}}-CH_2{}^- \leftrightarrow CH_3\overset{O^-}{\overset{|}{C}}{=}CH_2 \right] + H_2O$$

$$CH_3\overset{O}{\overset{\|}{C}}CH_2{}^- \overset{H}{\underset{O}{C}} \underset{}{\rightleftharpoons} \ CH_3\overset{O}{\overset{\|}{C}}CH_2\overset{H}{\underset{O^-}{C}} \ \overset{H_2O}{\rightleftharpoons} \ CH_3\overset{O}{\overset{\|}{C}}CH_2\overset{H}{\underset{OH}{C}} + OH^-$$

$$CH_3\overset{O}{\overset{\|}{C}}-\overset{H}{\underset{H}{C}}-\overset{H}{\underset{OH}{C}} \ \overset{^-OH}{\phantom{xx}} \longrightarrow CH_3\overset{O}{\overset{\|}{C}}-C{=}\overset{H}{\underset{H}{C}} + H_2O + OH^-$$

**A β-hydroxyketone**           **Benzalacetone**
                               **(4-Phenyl-3-butene-2-one)**
                                      mp 42°C

---

**EXPERIMENT**

## Synthesis of Dibenzalacetone[1]

### *Microscale Procedure*

> **IN THIS EXPERIMENT** an ethanolic solution of acetone and benzaldehyde is added to aqueous sodium hydroxide. The product, dibenzalacetone, crystallizes after a few minutes. The product is filtered from the mixture, washed, pressed dry, and recrystallized from an ethanol-water mixture. This very important reaction is easily carried out.

If sodium hydroxide gets on the skin, wash until the skin no longer has a soapy feeling.

Into a 10 × 100-mm reaction tube place 2 mL of 3 *M* sodium hydroxide solution. To this solution add 1.6 mL of 95% ethanol and 0.212 g of benzaldehyde.[2] Then add 0.058 g of acetone to the reaction mixture. Alternatively, your instructor may provide a solution that contains 58 mg of acetone in 1.6 mL of ethanol. Cap the tube immediately and shake the mixture vigorously. The benzaldehyde, initially insoluble, goes into solution, resulting in a water-clear, pale-yellow solution. After a minute or so it suddenly becomes cloudy, and a yellow precipitate of the

---

1. Ask your instructor for an alternative procedure for carrying out this experiment, which is found in the *Instructor's Guide*.

2. The benzaldehyde should be pure. Because it is easily air oxidized to benzoic acid on standing in the lab, it should be freshly purified or a newly purchased bottle.

**■ FIG. 37.1**
The Pasteur pipette filtration technique.
Solvent is removed between the pipette
tip and the bottom of tube, which leaves
crystals in the reaction tube.

Pasteur pipette

Reaction tubes

Note square,
flat tip

**■ FIG. 37.2**
A Hirsch funnel filtration apparatus.

Filter paper,
12 mm dia.

Polyethylene
filter disk (frit),
10 mm dia.

Hirsch
funnel

To aspirator

25-mL Filter
flask

**■ FIG. 37.3**
The drying of crystals
under vacuum.

**Online Study Center**

Videos: Microscale Filtration
on the Hirsch Funnel, Formation
of an Oil Instead of Crystals,
Filtration of Crystals Using
the Pasteur Pipette, Microscale
Crystallization

product forms. Continue to shake the tube from time to time for the next 30 min. If the product fails to crystallize, open the tube and scratch the inside of the tube with a glass rod. Remove the liquid from the tube using a Pasteur pipette by squeezing the bulb of the pipette, pressing the tip against the bottom of the tube, and bringing the liquid into the pipette, leaving the crystals in the tube (Fig. 37.1). Add 3 mL of water, cap, and shake the tube vigorously. Remove the wash liquid as before and wash the crystals twice more with 3-mL portions of water.

After the final washing, add 3 mL of water to the tube and collect the crystals on a Hirsch funnel using vacuum filtration. Use the filtrate to complete the transfer of the crystals. Squeeze the product between sheets of filter paper to dry it and then recrystallize the crude dibenzalacetone from a 70:30 ethanol-water mixture. During the recrystallization process (*see* Chapter 4), insert a wooden boiling stick to promote even boiling when heating the solute in the solvent. Remove the tube from the hot sand bath and place it in an insulated container to cool slowly to room temperature. Should the product separate as an oil, obtain a seed crystal, heat the solution to dissolve the oil, and add the seed crystal as the solution cools. If the product continues to oil out, add a bit more ethanol. After cooling the tube for several minutes in ice, collect the product by removing the solvent with a pipette (Fig. 37.1) and washing once with about 0.5 mL of ice-cold 70% ethanol while the tube is in ice. Alternatively, collect the crystals on a Hirsch funnel (Fig. 37.2) or Wilfilter and wash once with ice-cold 70% ethanol. Dry the product under vacuum by attaching the tube to an aspirator (Fig. 37.3). Determine the weight of the dibenzalacetone and its melting point and calculate the percent yield. In a typical experiment, the yield will be 0.10 g (mp 110.5°C–112°C).

**Cleaning Up.** Dilute the filtrate from the reaction mixture with water and neutralize it with dilute hydrochloric acid before flushing down the drain. The ethanol filtrate from the recrystallization should be placed in the organic solvents container.

## *Macroscale Procedure*

If sodium hydroxide gets on the skin, wash until the skin no longer has a soapy feeling.

**Online Study Center**

Video: Macroscale Crystallization

In a 125-mL Erlenmeyer flask containing a magnetic stirring bar, mix 0.05 mol of benzaldehyde with the theoretical quantity of acetone, and add one-half the mixture to a solution of 5 g of sodium hydroxide dissolved in 50 mL of water and 40 mL of ethanol at room temperature (<25°C). After 15 min of stirring, add the remainder of the aldehyde-ketone mixture and rinse the container with a little ethanol to complete the transfer. Stir the mixture for 30 min; then collect the product by suction filtration on a Büchner funnel. Break the suction and carefully pour 100 mL of water on the product. Reapply the vacuum. Repeat this process three times in order to remove all traces of sodium hydroxide. Finally, press the product as dry as possible on the filter using a cork and then press it between sheets of filter paper to remove as much water as possible. Save a small sample for a melting-point determination; then recrystallize the product from ethanol using about 10 mL of ethanol for each 4 g of dibenzalacetone. The yield after recrystallization should be about 4 g.

**Cleaning Up.** Dilute the filtrate from the reaction mixture with water and neutralize it with dilute hydrochloric acid before flushing down the drain. The ethanol filtrate from the recrystallization should be placed in the organic solvents container.

## Molecular Modeling

The name *dibenzalacetone* does not completely characterize the molecule made in this experiment. There are actually three isomeric dibenzalacetones: one melting at 110°C to 111°C, $\lambda_{max}$ = 330 nm, $\varepsilon$ = 34,300; another melting at 60°C, $\lambda_{max}$ = 295 nm, $\varepsilon$ = 20,000; and a third, a liquid with $\lambda_{max}$ = 287 nm, $\varepsilon$ = 11,000.

Both the melting points and the UV spectral data give some hints regarding the structures of these molecules. The first one is very symmetrical and can pack well into a crystal lattice. The long wavelength of the UV light absorption maximum and the high value of the molar absorbance ($\varepsilon$) indicate a long, planar conjugated system (*see* Chapter 14). The other two molecules are increasingly less able to pack nicely into a crystal lattice or to have a planar conjugated system. In the last step of the aldol condensation, the loss of water from the β-hydroxyketone can form molecules in which the alkene hydrogen atoms are either *cis* or *trans* to each other. Write the structures of the three geometric isomers of dibenzalacetone and assign each one to the three molecules described above.

*(handwritten annotations: 200mL Round Bottom Flask, stir bar)*

(a)                              (b)                              (c)

Enter the structures of these three isomers into a molecular modeling program and carry out an energy minimization or semiempirical calculation to find the relative steric energies or heats of formation of each molecule. Note: Once the calculation is complete, the lowest energy conformation of each isomer will be as planar as possible so that there can be maximum overlap of the $p$ orbitals on each $sp^2$ hybridized carbon. To test this idea, calculate the steric energy or heat of formation of benzalacetone (Fig. 37.4a) using the usual energy minimization procedure. The result should be an almost planar molecule. Then deliberately hold the dihedral angle defined by atoms 1, 2, 3, and 4 at 90° (Fig. 37.4b) and again calculate the energy of the molecule. In the latter conformation, the $p$ orbitals of the carbonyl group are orthogonal to the $p$ orbitals of the alkene.

As you may have discovered in calculating the energies of the three geometric isomers of dibenzalacetone, there is still another form of isomerization entering into the conformations of these molecules: single-bond *cis* and *trans* isomers, exemplified by two of the isomers of benzalacetone (Fig. 37.4a and Fig. 37.4c). Both of these conformers are planar in order to achieve maximum overlap of $p$ orbitals, but in 37.4a the carbonyl group is *cis* to the alkene bond, whereas in 37.4c it is *trans*. The barrier to rotation about the single bond is not very high, so these isomers cannot be isolated at room temperature. If your molecular modeling program has a dihedral driver routine, you can calculate the heats of formation of benzalacetone as a function of the dihedral angle defined by atoms 1, 2, 3, and 4 and thus determine the barrier to rotation around this bond in kilocalories per mole.

*Online Study Center*

**General Resources**
Additional Experiments

## QUESTIONS

1. Why is it important to maintain equivalent proportions of reagents in this reaction?
2. Which side products do you expect in this reaction? How are they removed?
3. What evidence do you have that your product consists of a single geometric isomer or a mixture of isomers? Does the melting point give such information?
4. From the $^1$H NMR spectrum of dibenzalacetone (Fig. 37.5), can you deduce which geometric isomer(s) is (are) formed? (*See* Table 12.1 on page 255.)
5. How would you change the procedures in this chapter if you wished to synthesize benzalacetone ($C_6H_5CH$=$CHCOCH_3$) or benzalacetophenone ($C_6H_5CH$=$CHCOC_6H_5$)?

■ **FIG. 37.5**

The $^1H$ NMR spectrum of dibenzalacetone (250 MHz). (An expanded spectrum appears above the normal spectrum.)

■ **FIG. 37.6**

The IR spectrum of dibenzalacetone (KBr disk).

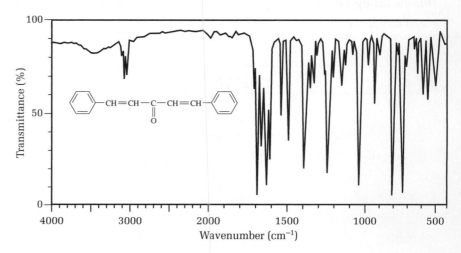

6. Assign the infrared (IR) peak at $1651 \text{ cm}^{-1}$ in Figure 37.6.

7. Write the structures of the three geometric isomers of dibenzalacetone and assign each one to the three molecules described. Disregard single bond *cis* and *trans* isomerism.

8. Draw the structures of all the single bond *cis* and *trans* isomers for each of the three geometric isomers of dibenzalacetone. There are 10 such isomers. Pick out the one you regard as the most stable and calculate its steric energy. Which three or four are represented by the solid with a melting point of 110°C–111°C, the solid with a melting point of 60°C, and the liquid?

9. Carry out a molecular mechanics or semiempirical calculation to determine the relative steric energies or heats of formation of 3 of the 10 possible isomeric dibenzalacetones.

10. (a) Calculate the steric energy or heat of formation for one single bond isomer of *trans*-benzalacetone using the usual energy minimization procedure. The result should be a planar molecule. (b) Then deliberately hold the dihedral angle defined by atoms 1, 2, 3, and 4 at 0°, 90°, and 180° and again calculate the energies of the molecule. What is the approximate barrier to rotation about the single bond?

11. Pick out the isomer you regard as the most stable from Question 7 and calculate its steric energy (if this was not done in Question 8).

# CHAPTER

# 38

# Grignard Synthesis of Triphenylmethanol and Benzoic Acid

**Online Study Center**

This icon will direct you to techniques, equipment setups, and online resources at http://college.hmco.com/PIC/williamsonMMOE5e.

**PRELAB EXERCISE:** Prepare a flow sheet for the preparation of triphenylmethanol. Using your knowledge of the physical properties of the solvents, reactants, and products, show how the products can be purified. Indicate which layer should contain the product in the liquid/liquid extraction steps.

In 1912 Victor Grignard received the Nobel Prize in Chemistry for his work on the reaction that bears his name, a carbon-carbon bond-forming reaction by which almost any alcohol may be formed from appropriate alkyl halides and carbonyl compounds. The Grignard reagent is easily formed by reacting an alkyl halide, in particular a bromide, with magnesium metal in anhydrous diethyl ether. The reaction can be written and thought of as simply

$$R—Br + Mg \longrightarrow R—Mg—Br$$

However, the structure of the material in solution is rather more complex. There is evidence that dialkylmagnesium is present:

$$2\ R—Mg—Br \rightleftharpoons R—Mg—R + MgBr_2$$

Structure of the Grignard reagent

The magnesium atoms, which have the capacity to accept two electron pairs from donor molecules to achieve a four-coordinated state, are solvated by the unshared pairs of electrons on diethyl ether:

$$
\begin{array}{cc}
\overset{\displaystyle Et\diagdown\ddot{\phantom{o}}\diagup Et}{\underset{\displaystyle Et\diagup\ddot{O}\diagdown Et}{\overset{\displaystyle \ddot{O}}{R-\overset{\displaystyle |}{\underset{\displaystyle |}{Mg}}-Br}}} & \qquad and \qquad
\overset{\displaystyle Et\diagdown\ddot{\phantom{o}}\diagup Et}{\underset{\displaystyle Et\diagup\ddot{O}\diagdown Et}{\overset{\displaystyle \ddot{O}}{R-\overset{\displaystyle |}{\underset{\displaystyle |}{Mg}}-R}}}
\end{array}
$$

A strong base and strong
nucleophile

The Grignard reagent is both a strong base and a strong nucleophile. As a base it will react with all protons that are more acidic than those found on alkenes and alkanes. Thus, Grignard reagents react readily with water, alcohols, amines, thiols, and so on to regenerate an alkane:

$$R—Mg—Br + H_2O \longrightarrow R—H + MgBrOH$$

$$R—Mg—Br + R'OH \longrightarrow R—H + MgBrOR'$$

$$R—Mg—Br + R'NH_2 \longrightarrow R—H + MgBrNHR'$$

$$R—Mg—Br + R'C \equiv C—H \longrightarrow R—H + R'C \equiv CMgBr$$
$$\text{(an acetylenic Grignard reagent)}$$

The starting material for preparing the Grignard reagent cannot contain any acidic protons. The reactants and apparatus must be completely and absolutely dry; otherwise the reaction will not start. If proper precautions are taken, however, the reaction proceeds smoothly.

Magnesium metal, in the form of a coarse powder, has a coat of oxide on the outside. A fresh surface can be exposed by crushing the powder under absolutely dry ether in the presence of an organic halide. The reaction will begin at exposed surfaces, as evidenced by a slight turbidity in the solution and evolution of bubbles. Once the exothermic reaction starts, it proceeds easily, the magnesium dissolves, and a solution of the Grignard reagent is formed. The solution is often turbid and gray due to impurities in the magnesium. The reagent is not isolated but reacted immediately with, most often, an appropriate carbonyl compound

$$
R—Mg—Br + R'—\overset{\overset{\displaystyle :\ddot{O}:}{\|}}{C}—R'' \longrightarrow R'—\overset{\displaystyle :\ddot{O}:^-MgBr^+}{\underset{\displaystyle R}{\overset{\displaystyle |}{C}}}—R''
$$

to give, in another exothermic reaction, magnesium alkoxide, a salt insoluble in ether. In a simple acid-base reaction, this alkoxide is reacted with acidified ice water to give the covalent, ether-soluble alcohol and the ionic water-soluble magnesium salt:

$$
R'—\overset{\displaystyle :\ddot{O}:^-MgBr^+}{\underset{\displaystyle R}{\overset{\displaystyle |}{C}}}—R'' \;+\; H^+Cl^- \longrightarrow R'—\overset{\displaystyle :\ddot{O}—H}{\underset{\displaystyle R}{\overset{\displaystyle |}{C}}}—R'' \;+\; Mg^{2+}Br^-Cl^-
$$

A versatile reagent

The great versatility of this reaction lies in the wide range of reactants that undergo it. Thirteen representative reactions are shown in Figure 38.1. In every case except reaction 1, the intermediate alkoxide must be hydrolyzed to give the product. The reaction with oxygen (reaction 2) is usually not a problem because the ether vapor over the reagent protects it from attack by oxygen, but this reaction is one reason why the reagent cannot usually be stored without special precautions. The reaction with solid carbon dioxide (dry ice) occurs readily to produce a carboxylic acid (reaction 3). Reactions 5, 6, and 7 with aldehydes and ketones giving, respectively, primary, tertiary, and secondary alcohols are among the most common. Ring-opening of an epoxide via Grignard addition (reaction 4) yields an alcohol as well. Reactions 8–13 are not nearly so common.

■ **FIG. 38.1**
The versatility of the Grignard reaction is illustrated by the wide range of reactants that undergo it.

In one of the experiments in this chapter, we will carry out another common type of Grignard reaction, the formation of a tertiary alcohol from 2 moles of the reagent and 1 mole of an ester. The ester employed is the methyl benzoate, which can be synthesized in Chapter 40. The initially formed product is unstable and decomposes to a ketone, which, being more reactive than an ester, immediately reacts with more Grignard reagent:

**Bromobenzene**[1]
MW 157.02
bp 156.4°C
den. 1.491

**Magnesium**
At Wt 24.1

**Phenylmagnesium
bromide**

**Methyl benzoate**
MW 136.16
bp 199.6°C
den. 1.09

$-Mg^{2+}Br^-OCH_3^-$

**Triphenylmethanol**
MW 260.34
mp 164.2°C

The primary impurity in these experiments is biphenyl, formed by reacting phenylmagnesium bromide with unreacted bromobenzene. (Figure 38.2 shows the $^{13}C$ NMR spectrum of bromobenzene.) The most effective way to lessen this side reaction is to add the bromobenzene slowly to the reaction mixture so it will

■ **FIG. 38.2**
The $^{13}C$ NMR spectrum of bromobenzene (22.6 MHz).

react with the magnesium and not be present in high concentration to react with previously formed Grignard reagent. The impurity is easily eliminated because it is much more soluble in hydrocarbon solvents than triphenylmethanol.

Biphenyl has a characteristic odor. Triphenylmethanol is odorless.

**Biphenyl**
mp 72°C

Triphenylmethanol can also be prepared from benzophenone.

**Phenylmagnesium bromide**

**Benzophenone**
MW 182.22
mp 48°C

**Triphenylmethanol**
MW 260.34
mp 164.2°C

EXPERIMENTS

# 1. Phenylmagnesium Bromide (Phenyl Grignard Reagent)

Br + Mg → (anhydrous ether) → MgBr

| **Bromobenzene** | **Magnesium** | **Phenylmagnesium bromide** |
| MW 157.02 | At Wt | not isolated, used in situ |
| bp 156°C | 24.31 | |
| den. 1.491 | | |

## Advance Preparation

It is imperative that all equipment and reagents be absolutely dry. The magnesium and the glassware to be used—two reaction tubes, two 1-dram vials (1 dram = 1.78 mL), and a stirring rod—can be dried in a 110°C oven for at least 30 min. Alternatively, if the glassware, syringe, septa, and magnesium appear to be perfectly dry, they can be used without special drying. The plastic and rubber components should be rinsed with acetone if either appears to be dirty or wet with water; then place these components in a desiccator for at least 12 h. Do not place plastic or rubber components in the oven. New, factory-sealed packages of syringes can be used without prior drying. The ether used throughout this reaction must be absolutely dry (absolute ether).

To prepare the Grignard reagent, absolute diethyl ether must be used; elsewhere, ordinary ether (diethyl or *t*-butyl methyl) can be used; for example, ether extractions of aqueous solutions do not need to be carried out with dry ether. It is strongly recommended that *t*-butyl methyl ether be used in all cases except in the preparation of the Grignard reagent itself.

A very convenient container for absolute diethyl ether is a 50-mL septum-capped bottle. This method of dispensing the solvent has three advantages: The ether is kept anhydrous, the exposure to oxygen is minimized, and there is little possibility of its catching fire. Ether is extremely flammable; do not work with this solvent near flames.

To remove ether from a septum-capped bottle, inject a volume of air into the bottle equal to the amount of ether being removed. Pull more ether than needed into the syringe and then push the excess back into the bottle before removing the syringe. In this way there will be no air bubbles in the syringe, and it will not dribble (Fig. 38.3).

⚠ **CAUTION**: Ether is extremely flammable. Extinguish all flames before using ether.

**Online Study Center**

Video: The Grignard Reaction: Removing a Liquid from a Septum-Capped Bottle with Syringe; Photos: Removing a Reagent from a Septum-Capped Bottle, Polypropylene Syringe Containing Ether

## Procedure

**IN THIS EXPERIMENT** magnesium and absolutely dry diethyl ether in a dry, septum-capped tube are reacted with an aromatic halide to give phenylmagnesium bromide. The exothermic reaction is started by crushing the magnesium. If the reaction does not start in 2–3 min, it will be necessary to start again, using equipment that was not used in the first attempt.

Remove a reaction tube from the oven and immediately cap it with a septum. In the operations that follow, keep the tube capped except when it is necessary to open it. After the tube cools to room temperature, add about 2 mmol (about 50 mg) of magnesium powder. Record the weight of magnesium used to the nearest milligram. The magnesium will become the limiting reagent by using a 5% molar excess of bromobenzene (about 2.1 mmol).

Using a dry syringe, add to the magnesium by injection through the septum 0.5 mL of anhydrous diethyl ether. Your laboratory instructor will demonstrate the transfer from the storage container used in your laboratory.

Into an oven-dried vial weigh about 2.1 mmol (about 330 mg) of dry (stored over molecular sieves) bromobenzene. Using a syringe, add to this vial 0.7 mL anhydrous diethyl ether and *immediately*, with the same syringe, remove all the solution from the vial. This can be done virtually quantitatively so you do not need to rinse the vial. Immediately cap the empty vial to keep it dry for later use. Inject

Diethyl ether can be made and kept anhydrous by storing over Linde 5A molecular sieves. Discard diethyl ether within 90 days because of peroxide formation. *t*-Butyl methyl ether does not have this problem.

■ **FIG. 38.3**
A polypropylene syringe (1 mL, with 0.01-mL graduations). The needle is blunt. When there are no air bubbles in the syringe, it will not dribble ether.

■ **FIG. 38.4**
Once the reaction has started, bromobenzene in ether is added slowly from the syringe. The empty needle is for pressure relief, but if condensation is complete (aided by the damp pipe cleaner), it will not be needed. Once the reaction slows down, stir it with a magnetic stirrer and stirring bar.

■ **FIG. 38.5**
To start the Grignard reaction, remove the septum and apply pressure to the stirring rod while rotating the reaction tube on a hard surface that will not scratch the tube, such as a book.

Read volume here

about 0.1 mL of the bromobenzene-ether mixture into the reaction tube and mix the contents by flicking the tube. Pierce the septum with another syringe needle for pressure relief (Fig. 38.4).

It takes much force to crush the magnesium. Place the tube on a hard surface and bear down with the stirring rod while twisting the reaction tube. Do not pound the magnesium.

The reaction will not ordinarily start at this point, so remove the septum, syringe, and empty syringe needle and crush the magnesium with a dry stirring rod. You can do this easily in the confines of the 10-mm-diameter reaction tube while it is positioned on a hard surface. There is little danger of poking the stirring rod through the bottom of the tube (Fig. 38.5). Immediately replace the septum, syringe, and empty syringe needle (for pressure relief). The reaction should start within seconds. The formerly clear solution becomes cloudy and soon begins to boil as the magnesium metal reacts with the bromobenzene to form the Grignard reagent—phenylmagnesium bromide.

**Online Study Center**

Video: The Grignard Reaction: Starting the Reaction

If the reaction does not start within 1 min, begin again with completely different, dry equipment (syringe, syringe needle, reaction tube, etc.). Once the Grignard reaction begins, it will continue. To prevent the ether from boiling away, wrap a pipe cleaner around the top part of the reaction tube. Dampen the pipe cleaner with water or, if the room temperature is very hot, with alcohol.

**Online Study Center**

Video: The Grignard Reaction: Addition of Bromobenzene and Refluxing of Reaction Mixture

To the refluxing mixture add slowly and dropwise over a period of several minutes the remainder of the bromobenzene-ether solution at a rate such that the reaction remains under control at all times. After all the bromobenzene solution is added, spontaneous boiling of the diluted mixture may be slow or become slow. At this point, add a magnetic stirring bar to the reaction tube and stir the reaction mixture with a magnetic stirrer. If the rate of reaction is too fast, slow down the stirrer. The reaction is complete when none or a very small quantity of the metal remains. Check to see that the volume of ether has not decreased. If it has, add more anhydrous diethyl ether. Because the solution of the Grignard reagent deteriorates on standing, Experiment 2 should be started at once. The phenylmagnesium bromide can be converted to triphenylmethanol or to benzoic acid.

## 2. Triphenylmethanol

Phenylmagnesium bromide

Benzophenone
MW 182.22
mp 48°C

Triphenylmethanol
MW 260.34
mp 164.2°C

> **IN THIS EXPERIMENT** the Grignard reagent prepared in Experiment 1 is added to a dry ether solution of benzophenone. Very thorough mixing is required. When the red color disappears, the salt is hydrolyzed by adding hydrochloric acid. More ether is added, the layers separated, and the ether dried and evaporated to give crude product. An impurity, biphenyl, is removed by dissolving it in hexanes, and the product is recrystallized from 2-propanol. The product recrystallizes very slowly, so do not collect the product immediately.

Make *t*-butyl methyl ether anhydrous by storing it over molecular sieves.

*Online Study Center*

Videos: The Grignard Reaction: Dissolution of Benzophenone in Ether, Addition to Grignard Reagent; The Grignard Reaction: Mixing Reaction Mixture by Flicking Reaction Tube

Mixing the reaction mixture is very important.

*Online Study Center*

Videos: The Grignard Reaction: Final Addition of Benzophenone Solution; The Grignard Reaction: Color Change at Completion of Reaction

*Online Study Center*

Videos: The Grignard Reaction: Addition of Hydrochloric Acid; The Grignard Reaction: Solution Treated with Saturated Sodium Chloride Solution

In an oven-dried vial dissolve 2.0 mmol (0.364 g) of benzophenone in 1.0 mL of anhydrous ether by capping the vial and mixing the contents thoroughly. With a dry syringe, remove all the solution from the vial and add it dropwise with *thorough* mixing (magnetic stirring, flicking of the tube, or stirring with a stirring rod near the end of the addition) after each drop to the solution of the Grignard reagent. Add the benzophenone at a rate so as to maintain the ether at a gentle reflux. Rinse the vial with a few drops of anhydrous ether after all the first solution has been added; then add this rinse to the reaction tube.

After all the benzophenone has been added, the mixture should be homogeneous. If not, mix it again, using a stirring rod if necessary. The syringe can be removed but leave the pressure-relief needle in place. Allow the reaction mixture to stand at room temperature. The reaction apparently is complete when the red color disappears.

At the end of the reaction period, cool the tube in ice and add to it dropwise with stirring (use a glass rod or a spatula) 2 mL of 3 *M* hydrochloric acid. A creamy-white precipitate of triphenylmethanol will separate between the layers. Add more ether (it need not be anhydrous) to the reaction tube and shake the contents to dissolve all the triphenylmethanol. The result should be two perfectly clear layers. Remove a drop of the ether layer for thin-layer chromatographic (TLC) analysis. Any bubbling seen at the interface or in the lower layer is leftover magnesium reacting with the hydrochloric acid. Remove the aqueous layer and shake the ether layer with an equal volume of saturated aqueous sodium chloride solution to remove water and any remaining acid. Carefully remove the entire aqueous layer; then dry the ether layer by adding anhydrous calcium chloride pellets to the reaction tube until the drying agent no longer clumps together. Cork the tube and shake it from time to time over 5–10 min to complete the drying.

Using a Pasteur pipette, remove the ether from the drying agent and place it in another tared, dry reaction tube or a centrifuge tube. Use more ether to wash off the drying agent and combine these ether extracts. Evaporate the ether in a hood by blowing nitrogen or air onto the surface of the solution while warming the tube in a beaker of water or in the hand.

After all the solvent has been removed, determine the weight of the crude product. Note the odor of the biphenyl, the product of the side reaction that takes place between bromobenzene and phenylmagnesium bromide during the first reaction.

Trituration (grinding) of the crude product with petroleum ether will remove the biphenyl. Stir the crystals with 0.5 mL of petroleum ether in an ice bath, remove

**■ FIG. 38.6**

An apparatus for drying crystals in a reaction tube under vacuum.

**■ FIG. 38.7**

A semimicroscale, research-type apparatus for the Grignard reaction, with provision for a motor-driven stirrer and an inlet and outlet for dry nitrogen.

Stirrer (connected to motor)

$N_2$ in

Ground glass stopper

$N_2$ out

Reaction pot

*Online Study Center*

Videos: The Grignard Reaction: Crude Product Triturated and Recrystallized—Pure Triphenylcarbinol Isolated; Microscale Filtration on the Hirsch Funnel

*Online Study Center*

Photo: Filtration Using a Pasteur Pipette; Video: Filtration of Crystals Using the Pasteur Pipette

the solvent as thoroughly as possible, add a boiling stick, and recrystallize the residue from boiling 2-propanol (no more than 2 mL). Allow the solution to cool slowly to room temperature and then cool it thoroughly in ice. Triphenylmethanol crystallizes slowly, so allow the mixture to remain in the ice as long as possible. Stir the ice-cold mixture well and collect the product by vacuum filtration on a Hirsch funnel. Save the filtrate. Concentration may give a second crop of crystals.

An alternative method for purifying the triphenylmethanol utilizes a mixed solvent. Dissolve the crystals in the smallest possible quantity of warm ether and add 1.5 mL of hexanes to the solution. Add a boiling stick to the solution and boil off some of the ether until the solution becomes slightly cloudy, indicating that it is saturated. Allow the solution to cool slowly to room temperature. Triphenylmethanol is deposited slowly as large, thick prisms. Cool the solution in ice; after allowing time for complete crystallization to occur, remove the ether with a Pasteur pipette and wash the crystals once with a few drops of a cold 1:4 ether-hexanes mixture. Dry the crystals in the tube under a vacuum (Fig. 38.6).

Determine the weight, melting point, and percent yield of the triphenylmethanol. Analyze the crude and recrystallized product by TLC on silica gel (*see* Chapter 8), developing the plate with a 1:5 mixture of dichloromethane and petroleum ether. An infrared (IR) spectrum can be determined in chloroform solution or by preparing a mull or KBr disk (*see* Chapter 11). Compare the apparatus used in this experiment with the research-type apparatus shown in Figure 38.7.

**Cleaning Up.** Combine the acidic aqueous layer and saturated sodium chloride layers, dilute with water, neutralize with sodium carbonate, and flush down the drain with excess water. Ether is allowed to evaporate from the drying agent in the hood, and the drying agent is then discarded in the nonhazardous solid waste container. If local regulations do not allow for evaporating solvents in a hood, the wet drying agent should be discarded in a special container. The petroleum ether and 2-propanol or ether-hexanes mother liquor are placed in the organic solvents container.

# 3. Benzoic Acid

**IN THIS EXPERIMENT** the Grignard reagent prepared in Experiment 1 is squirted onto a piece of dry ice. The resulting white carboxylate salt is hydrolyzed with hydrochloric acid, releasing benzoic acid, which is extracted into the ether layer. The ether solution could be dried and evaporated to give the product, but a better product is obtained by adding base to make the benzoate salt and then adding acid to this basic solution to cause benzoic acid to crystallize. It can be recrystallized from hot water.

**Phenylmagnesium bromide**

**Carbon dioxide**
MW 44.01
mp − 78.5°C (sublimes)

**Benzoic acid**
MW 122.12
mp 123°C

**CAUTION:** Handle dry ice with a towel or gloves. Contact with the skin can cause frostbite because dry ice sublimes at −78.5°C.

*Online Study Center*
Video: Extraction with Ether

Prepare 2 mmol of phenylmagnesium bromide exactly as described in Experiment 1. Wipe off the surface of a small piece of dry ice (solid carbon dioxide) with a dry towel to remove frost and place it in a dry 30-mL beaker. Remove the pressure-relief needle from the reaction tube; then insert a syringe through the septum, turn the tube upside down, and draw into the syringe as much of the reagent solution as possible. Squirt this solution onto the piece of dry ice; then, using a clean needle, rinse out the reaction tube with 1 mL of anhydrous diethyl ether and squirt this onto the dry ice. Allow excess dry ice to sublime; then hydrolyze the salt by adding 2 mL of 3 $M$ hydrochloric acid.

Transfer the mixture from the beaker to a reaction tube and shake it thoroughly. Two homogeneous layers should result. Add 1–2 mL of acid or ordinary (not anhydrous) ether if necessary. Remove the aqueous layer and shake the ether layer with 1 mL of water, which is removed and discarded. Then extract the benzoic acid by adding to the ether layer 0.7 mL of 3 $M$ sodium hydroxide solution, shaking the mixture thoroughly, and withdrawing the aqueous layer, which is placed in a very small beaker or vial. The extraction is repeated with another 0.5-mL portion of base and finally 0.5 mL of water. Now that the extraction is complete, the ether, which can be discarded, contains primarily biphenyl, the byproduct formed during the preparation of phenylmagnesium bromide.

The combined aqueous extracts are heated briefly to about 50°C to drive off dissolved ether from the aqueous solution and then made acidic by adding concentrated hydrochloric acid (test with indicator paper). Cool the mixture thoroughly in an ice bath. Collect the benzoic acid on a Hirsch funnel and wash it with about 1 mL of ice water while on the funnel. A few crystals of this crude material are saved for a melting-point determination; the remainder of the product is recrystallized from boiling water.

The solubility of benzoic acid in water is 68 g/L at 95°C and 1.7 g/L at 0°C. Dissolve the acid in very hot water. Let the solution cool slowly to room temperature; then cool it in ice for several minutes before collecting the product by vacuum filtration on a Hirsch funnel. Use the ice-cold filtrate in the filter flask to complete the transfer of benzoic acid from the reaction tube. Turn the product out onto a piece of filter paper, squeeze out excess water, and allow it to dry thoroughly. Once dry, weigh it, calculate the percent yield, and determine the melting point along with the melting point of the crude material. The IR spectrum may be determined as a solution in chloroform (1 g of benzoic acid dissolves in 4.5 mL of chloroform) or as a mull or KBr disk (*see* Chapter 11).

**Cleaning Up.** Combine all aqueous layers, dilute with a large quantity of water, and flush the slightly acidic solution down the drain.

## 4. Phenylmagnesium Bromide (Phenyl Grignard Reagent)

| Bromobenzene | Magnesium | Phenylmagnesium bromide |
|---|---|---|
| MW 157.02 | At Wt | not isolated, used in situ |
| bp 156°C | 24.31 | |
| den. 1.491 | | |

Diethyl ether can be kept anhydrous by storing it over Linde 5A molecular sieves. Discard the ether after 90 days because of peroxide formation.

All equipment and reagents must be *absolutely dry*. The Grignard reagent is prepared in a dry, 100-mL round-bottomed flask fitted with a long reflux condenser. A calcium chloride drying tube inserted in a cork that will fit either the flask or the top of the condenser is also made ready (Fig. 38.8a). The flask, condenser, and magnesium (2.00 g = 0.082 mol of magnesium turnings) should be as dry as possible to begin with and then should be further dried in a 110°C oven for at least 35 min. Alternatively, the magnesium is placed in the flask, the calcium chloride tube is attached directly, and the flask is heated gently but thoroughly with a cool luminous flame.[1] Do not overheat the magnesium. It will become deactivated through oxidation or, if strongly overheated, can burn. The flask on cooling pulls dry air through the calcium chloride. Cool to room temperature before proceeding! Extinguish all flames! Ether vapor is denser than air and can travel along bench tops and into sinks. Use care.

**CAUTION:** Ether is extremely flammable. Extinguish all flames before using ether.

---

1. Alternatively, if nitrogen gas is available, the reaction can be run under nitrogen using a bubbler.

■ **FIG. 38.8**

(a) A calcium chloride drying tube fitted with a rubber stopper. Store for future use with a cork in the top and a pipette bulb on the bottom. (b) An apparatus for refluxing the Grignard reaction.

Cotton
or glass wool

8 Mesh
CaCl₂

Cotton

(a)

Calcium chloride tube

Water out

Water in

Round-bottom flask

Thermowell

(b)

**Specially dried ether is required.**

Prepare an ice bath in case control of the reaction becomes necessary, although this is usually not the case. Remove the drying tube and fit it to the top of the condenser. Then pour into the flask through the condenser 15 mL of *absolute* ether (absolutely dry, anhydrous) and 9 mL (13.5 g = 0.086 mol) of bromobenzene. Be sure the graduated cylinders used to measure the ether and bromobenzene are absolutely dry. (More ether is to be added as soon as the reaction starts, but the concentration of bromobenzene is kept high at the outset to promote easy starting.) If there is no immediate sign of reaction, insert a *dry* stirring rod with a flattened end and crush a piece of magnesium firmly against the bottom of the flask under the surface of the liquid, giving a twisting motion to the rod. When this is done properly, the liquid becomes slightly cloudy, and ebullition commences at the surface of the compressed metal. Be careful not to punch a hole in the bottom of the flask. Attach the condenser at once and swirl the flask to provide fresh surfaces for contact. As soon as you are sure that the reaction has started, add an additional 25 mL of absolute ether through the top of the condenser before spontaneous boiling becomes too vigorous (replace the drying tube). Note the volume of ether in the flask. Cool in ice if necessary to slow the reaction but do not overcool the mixture; the reaction can be stopped by too much cooling. Any difficulty in initiating the reaction can be dealt with by trying the following prompts in succession.

**Starting the Grignard reaction**

1. Warm the flask with your hands or in a beaker of warm water. Then see if boiling continues when the flask (condenser attached) is removed from the warmth.
2. Try further mashing of the metal with a dry stirring rod.
3. Add a tiny crystal of iodine as a starter (in this case the ethereal solution of the final reaction product should be washed with sodium bisulfite solution to remove the yellow color).
4. Add a few drops of a solution of phenylmagnesium bromide or methylmagnesium iodide (which can be made in a test tube).
5. Start afresh, taking greater care with the dryness of apparatus, measuring tools, and reagents, and sublime a crystal or two of iodine on the surface of the magnesium to generate Gattermann's activated magnesium before beginning the reaction again.

**Use caution when heating to avoid condensation on the outside of the condenser.**

Once the reaction begins, spontaneous boiling in the diluted mixture may be slow or become slow. If so, mount the flask and condenser in a heating mantle or Thermowell (one clamp supporting the condenser suffices; Fig. 38.8b) and reflux gently until the magnesium has disintegrated and the solution has acquired a cloudy or brownish appearance. The reaction is complete when only a few remnants of metal (or metal contaminants) remain. Check to see that the volume of ether has not decreased. If it has, add more anhydrous ether. Because the solution of Grignard reagent deteriorates on standing, Experiment 5 should be started at once.

## 5. Triphenylmethanol from Methyl Benzoate

$$2 \; \text{C}_6\text{H}_5\text{—MgBr} + \text{C}_6\text{H}_5\text{—C(=O)—OCH}_3 \longrightarrow (\text{C}_6\text{H}_5)_3\text{C—}\ddot{\text{O}}\text{—MgBr} \xrightarrow{\text{H}_3\text{O}^+} (\text{C}_6\text{H}_5)_3\text{C—}\ddot{\text{O}}\text{—H}$$

**Phenylmagnesium
bromide**

**Methyl benzoate**
MW 136.15
bp 198–199°C

**Triphenylmethanol**
MW 260.34
mp 164.2°C

Mix 5 g (0.037 mol) of methyl benzoate and 15 mL of absolute ether in a separatory funnel, cool the flask containing the phenylmagnesium bromide solution briefly in an ice bath, remove the drying tube, and insert the stem of a separatory funnel into the top of the condenser. Run in the methyl benzoate solution *slowly* with only such cooling as is required to control the mildly exothermic reaction, which affords an intermediate salt that separates as a white solid. Replace the calcium chloride tube; swirl the flask until it is at room temperature and the reaction has subsided. Go to Experiment 7.

## 6. Triphenylmethanol from Benzophenone

| Phenylmagnesium bromide | Benzophenone MW 182.22 mp 48°C | | Triphenylmethanol MW 260.34 mp 164.2°C |

Dissolve 6.75 g (0.037 mol) of benzophenone in 25 mL of absolute ether in a separatory funnel and cool the flask containing *half* the phenylmagnesium bromide solution (0.041 mol) briefly in an ice bath. (The other half can be used to make benzoic acid.) Remove the drying tube and insert the stem of the separatory funnel into the top of the condenser. Add the benzophenone solution *slowly* with swirling and only such cooling as is required to control the mildly exothermic reaction, which gives a bright-red solution and then precipitates a white salt. Replace the calcium chloride tube; swirl the flask until it is at room temperature and the reaction has subsided. Go to Experiment 7.

## 7. Completion of Grignard Reaction

This is a suitable stopping point.

The reaction is then completed by either refluxing the mixture for 30 min or stoppering the flask with the calcium chloride tube and letting the mixture stand overnight (subsequent refluxing is then unnecessary).[2]

In this part of the experiment, ordinary (not anhydrous) diethyl ether or *t*-butyl methyl ether may be used.

Pour the reaction mixture into a 250-mL Erlenmeyer flask containing 50 mL of 10% sulfuric acid and about 25 g of ice and use both ordinary ether and 10% sulfuric acid to rinse the flask. Swirl well to promote hydrolysis of the addition compound; basic magnesium salts are converted into water-soluble neutral salts;

---

2. A rule of thumb for organic reactions: A 10°C rise in temperature will double the rate of the reaction.

triphenylmethanol is distributed into the ether layer. An additional amount of ordinary ether may be required. Pour the mixture into a separatory funnel (rinse the flask with ether), shake, and draw off the aqueous layer. Shake the ether solution with 10% sulfuric acid to further remove magnesium salts and wash with saturated sodium chloride solution to remove water that has dissolved in the ether. The amounts of liquid used in these washing operations are not critical. In general, an amount of wash liquid equal to one-third of the ether volume is adequate.

*Saturated aqueous sodium chloride solution removes water from ether.*

To effect final drying of the ether solution, pour the ether layer out of the neck of the separatory funnel into an Erlenmeyer flask, add about 5 g of calcium chloride pellets, swirl the flask intermittently, and after 5 min remove the drying agent by gravity filtration (using a filter paper and funnel) into a tared Erlenmeyer flask. Rinse the drying agent with a small amount of ether. Add 25 mL of 66°C–77°C hexanes and concentrate the ether-hexanes solutions (steam bath or hot plate) in an Erlenmeyer flask under an aspirator tube (*see* Fig. 9.6 on page 206). Evaporate slowly until crystals of triphenylmethanol just begin to separate; then let crystallization proceed, first at room temperature and then at 0°C. The product should be colorless and should melt at no lower than 160°C. Concentration of the mother liquor may yield a second crop of crystals. A typical student yield is 5.0 g. Evaporate the mother liquors to dryness and save the residue for later isolation of the components by chromatography.

Perform a TLC analysis of the first crop of triphenylmethanol and the residue from the evaporation of the mother liquors. Dissolve equal quantities of the two solids (a few crystals) and also biphenyl in equal quantities of dichloromethane (1 or 2 drops). Using a microcapillary, spot equal quantities of material on silica gel TLC plates and develop the plates in an appropriate solvent system. Try a 1:3 dichloromethane–petroleum ether mixture first and adjust the relative quantities of solvent as needed. The spots can be seen by examining the TLC plate under a fluorescent lamp or by treating the TLC plate with iodine vapor. From this analysis decide how pure each of the solids is and whether it would be worthwhile to attempt to isolate more triphenylmethanol from the mother liquors.

Turn in the product in a vial labeled with your name, the name of the compound, its melting point, and the overall percent yield from benzoic acid.

**Cleaning Up.** Combine the acidic aqueous layer and saturated sodium chloride layers, dilute with water, neutralize with sodium carbonate, and flush down the drain with excess water. Ether is allowed to evaporate from the drying agent in the hood, and the drying agent is then discarded in the nonhazardous solid waste container. If local regulations do not allow for evaporating solvents in a hood, the wet drying agent should be discarded in a special container. The ether-hexanes mother liquor is placed in the organic solvents container.

*Dispose of recovered and waste solvents in the appropriate containers.*

 ## 8. Benzoic Acid

Wipe the frost from a piece of dry ice, transfer the ice to a cloth towel, and crush it with a hammer. Without delay (so moisture will not condense on the cold solid), transfer about 10 g of dry ice to a 250-mL beaker. Cautiously pour one-half of the

*Online Study Center*

Video: Macroscale
Crystallization

solution of phenylmagnesium bromide prepared in Experiment 1 onto the dry ice. A vigorous reaction will ensue. Allow the mixture to warm up and stir it until the dry ice has evaporated. To the beaker add 20 mL of 3 *M* hydrochloric acid; then heat the mixture over a steam bath in the hood to boil off the ether. Cool the beaker thoroughly in an ice bath and collect the solid product by vacuum filtration on a Büchner funnel.

Transfer the solid back to the beaker and dissolve it in a minimum quantity of saturated sodium bicarbonate solution (2.8 *M*). Note that a small quantity of a byproduct remains suspended and floating on the surface of the solution. Note the odor of the mixture. Transfer it to a separatory funnel and shake it briefly with about 15 mL of ether. Discard the ether layer, place the clear aqueous layer in the beaker, and heat it briefly to drive off dissolved ether. Carefully add 3 *M* hydrochloric acid to the mixture until the solution tests acidic to pH paper. Cool the mixture in ice and collect the product on a Büchner funnel. Recrystallize it from a minimum quantity of hot water and isolate it in the usual manner. Determine the melting point and the weight of the benzoic acid; calculate its yield based on the weight of magnesium used to prepare the Grignard reagent.

**Cleaning Up.** Combine all aqueous layers, dilute with a large quantity of water, and flush the slightly acidic solution down the drain. The ether-hexanes mother liquor from the recrystallization goes in the organic solvents container. The TLC developer, which contains dichloromethane, is placed in the halogenated organic waste container. Calcium chloride from the drying tube should be dissolved in water and flushed down the drain.

## QUESTIONS

1. Triphenylmethanol can also be prepared by reacting ethyl benzoate with phenylmagnesium bromide and by reacting diethylcarbonate with phenylmagnesium bromide. Write stepwise reaction mechanisms for these two reactions.

$$C_2H_5OCOC_2H_5$$
$$\overset{O}{\overset{\|}{}}$$

2. If the ethyl benzoate used to prepare triphenylmethanol is wet, what byproduct is formed?
3. Exactly what weight of dry ice is needed to react with 2 mmol of phenylmagnesium bromide? Would an excess of dry ice be harmful?
4. In the synthesis of benzoic acid, benzene is often detected as an impurity. How does this come about?
5. In Experiment 3, the benzoic acid could have been extracted from the ether layer using sodium bicarbonate solution. Write equations showing how this might be done and how the benzoic acid would be regenerated. What

practical reason makes this extraction method less desirable than sodium hydroxide extraction?

6. What is the weight of frost (ice) on the dry ice that will react with all of the Grignard reagent used in Experiment 8?

7. How many moles of carbon dioxide are contained in 10 g of dry ice?

8. Just after the dry ice has evaporated from the beaker, what is the white solid remaining?

9. Write an equation for the reaction of the white solid with 3 *M* hydrochloric acid.

10. Write an equation for the reaction of the product with sodium bicarbonate.

11. Would you expect sodium benzoate to have an odor? Why or why not?

12. What odor do you detect after the product has dissolved in sodium bicarbonate solution?

13. What is the purpose of the ether extraction?

14. "Isolate the product in the usual way." What is the meaning of this sentence?

*Online Study Center*

**General Resources**
Web Links

# CHAPTER 39

# The Wittig and Wittig-Horner Reactions

**Online Study Center**

This icon will direct you to techniques, equipment setups, and online resources at http://college.hmco.com/PIC/williamsonMMOE5e.

**PRELAB EXERCISE:** Account for the fact that the Wittig-Horner reaction of cinnamaldehyde gives almost exclusively the *E,E*-butadiene with very little contaminating *E,Z*-product.

$(C_6H_5)_3P$  +  $ClH_2C$—⟨phenyl⟩ ⟶ $(C_6H_5)_3\overset{+}{P}CH_2$—⟨phenyl⟩ $Cl^-$ $\xrightarrow[\text{(50\% NaOH)}]{\text{Base}}$

**Triphenylphosphine**
**(1)**

**Benzyl chloride**
**(2)**

**Benzyltriphenylphosphonium chloride**
**(3)**

$(C_6H_5)_3P{=}CH$—⟨phenyl⟩  $\xleftarrow{\hspace{1cm}}$ **The Wittig reagent, an ylide (4)** $\xrightarrow{\hspace{1cm}}$ $(C_6H_5)_3\overset{+}{P}$—$\overset{-}{CH}$—⟨phenyl⟩ $\longrightarrow$

**9-Anthraldehyde**
**(5)**

$(C_6H_5)_3P$—CH—⟨phenyl⟩
O—C—H ⟨anthracene⟩

$(C_6H_5)_3P$
‖
O

+ ⟨trans product⟩

**4-Anthranyl-2,2,2,3-tetraphenyl-1,2-oxaphosphetane**
**(6)**

**Triphenylphosphine oxide**
**(7)**

*trans*-9-(2-Phenylethenyl)anthracene
**(8)**

The Wittig reaction affords an invaluable method for converting a carbonyl compound to an olefin; for example, the conversion of 9-anthraldehyde (**5**) and benzyl chloride (**2**) by means of the Wittig reagent (**4**) through a four-centered intermediate to the product (**8**). Because the active reagent, an ylide, is unstable, it is generated in the presence of a carbonyl compound by dehydro-halogenation of the phosphonium chloride (**3**) with base. Usually a very strong base, such as phenyllithium in dry ether, is employed, but in the experiments in this chapter, 50% sodium hydroxide is employed. The existence of the four-membered ring intermediate was proved by nuclear magnetic resonance (NMR) in 1984.[1] The phenylethenylanthracene product produces a green fluorescence with Cyalume, a chemiluminescent substance, in the light stick reaction (*see* Chapter 61).

The light stick reaction

Simplified Wittig reaction

When the halogen compound employed in the first step has an activated halogen atom ($RCH{=}CHCH_2X$, $C_6H_5CH_2X$, $XCH_2CO_2H$), a simpler procedure known as the *Horner phosphonate modification* of the Wittig reaction is applicable. When benzyl chloride is heated with triethyl phosphite, chloroethane is eliminated from the initially formed phosphonium chloride with the production of diethyl benzylphosphonate. This phosphonate is stable, but in the presence of a strong base, such as the sodium methoxide used here, it condenses with a carbonyl component in the same way that a Wittig ylide condenses. Thus it reacts with benzaldehyde to give *E*-stilbene and with cinnamaldehyde to give *E,E*-1,4-diphenyl-1,3-butadiene.

**Benzyl chloride**
bp 179°C, MW 126.59
den. 1.10, $n_D^{20}$ 1.5380

**Triethyl phosphite**
bp 156°C, MW 166.16
den. 0.94, $n_D^{20}$ 1.4130

**Benzyltriethylphosphonium chloride**

**Diethyl benzylphosphonate**
bp 156°C/9 torr
MW 228.23, den. 1.045
$n_D^{20}$ 1.4970

**Chloroethane**
bp 12.3°C
MW 64.52, den. 0.891

1. Maryanoff, B. E.; Reitz, A. B.; Mutter, M. S. *J. Am. Chem. Soc.* **1984**, *106*, 1873.

Diethyl benzylphosphonate + $CH_3O^- Na^+$ ⟶

**Sodium methoxide**
MW 52.02

**Ylide**

Ylide +

**Cinnamaldehyde**
MW 132.16, den. 1.11
bp 248°C

⟶

**E,E-1,4-Diphenyl-1,3-butadiene**
mp 153°C, MW 206.27

$+ Na^+ \bar{O}$—P ...

The energy-minimized conformation of *E,E*-1,4-diphenyl-1,3-butadiene. Because of steric hindrance, the molecule is not flat.

≡

## EXPERIMENTS

## 1. Synthesis of *trans*-9-(2-Phenylethenyl)anthracene[2]

### *Microscale Procedure*

> **IN THIS EXPERIMENT** a phosphonium chloride is reacted with base to form the Wittig reagent. An aldehyde in the same solution reacts with the Wittig reagent to give the alkene product. The solvent, dichloromethane, is dried and evaporated to give the product that is, in turn, recrystallized from 1-propanol.

To a reaction tube add 200 mg of benzyltriphenylphosphonium chloride, 115 mg of 9-anthraldehyde, 0.6 mL of dichloromethane, and a magnetic stirring bar. With rapid magnetic stirring, 0.26 mL of 50% sodium hydroxide solution is added

2. For macroscale syntheses of *trans*-9-(2-phenylethenyl)anthracene, see Becker, H. D.; Andersson, K. *J. Org. Chem.* **1983**, *48*, 4552. Merkl, G.; Merz, A. *Synthesis* **1973**, 295. Silversmith, E. F. *J. Chem. Educ.* **1986**, *63*, 645.

**■ FIG. 39.1**
An apparatus for removing
a solvent under a vacuum.

To
aspirator

dropwise from a Pasteur pipette. After stirring vigorously for 30 min, 1.5 mL of dichloromethane and 1.5 mL of water are added; then the tube is capped and shaken. The organic layer is removed and placed in another reaction tube, and the aqueous layer is extracted with 1 mL of dichloromethane. The combined dichloromethane extracts are dried over calcium chloride pellets, the dichloromethane is removed, the drying agent is washed with more solvent, and the solvent is removed under vacuum in the filter flask (Fig. 39.1). To the solid remaining in the filter flask add 3 mL of 1-propanol, heat and transfer the hot solution to an Erlenmeyer flask to crystallize. After cooling spontaneously to room temperature, cool the flask in ice and collect the product on a Hirsch funnel. The triphenylphosphine oxide remains in the propanol solution. The reported melting point of the product is 131°C–132°C.

 ## *Macroscale Procedure*

Into a 10-mL round-bottomed flask place 0.97 g of benzyltriphenylphosphonium chloride, 0.57 g of 9-anthraldehyde, 3 mL of dichloromethane, and a stirring bar. Clamp the flask over a magnetic stirrer and stir the mixture at high speed while adding 1.3 mL of 50% sodium hydroxide solution dropwise from a Pasteur pipette. After the addition is complete, continue the stirring for 30 min; then transfer the contents of the flask to a 50-mL separatory funnel using 10 mL of water and 10 mL of dichloromethane to complete the transfer. Shake the mixture, remove the organic layer, and then extract the aqueous layer once more with 5 mL of dichloromethane. Dry the combined organic layers with anhydrous calcium chloride pellets, transfer the solution to a 50-mL Erlenmeyer flask, evaporate it to dryness on a steam bath, and recrystallize the yellow partially crystalline residue from 15 mL of 2-propanol. The product recrystallizes as thin yellow plates (mp 131°C–132°C). Save the product for use as a fluorescer in the Cyalume chemiluminescence experiment (*see* Chapter 61). Examine a dilute ethanol solution of the product under an ultraviolet (UV) lamp.

# 2. Synthesis of 1,4-Diphenyl-1,3-Butadiene

> **IN THIS EXPERIMENT** benzyl chloride and triethyl phosphite are reacted with each other to give a phosphonate salt. This salt, on reaction with the strong base, sodium methoxide, gives a Wittig reagent-like ylide that reacts with an aldehyde to give the product alkene, diphenylbutadiene. It is isolated by filtration and can be recrystallized, if necessary, from methylcyclohexane.

**CAUTION**: Take care to keep organophosphorus compounds off the skin.

Use freshly prepared or opened sodium methoxide; keep the bottle closed.

**CAUTION**: Handle benzyl chloride in the hood. Severe lachrymator (tear producer) and irritant of the respiratory tract.

**N,N-Dimethylformamide (DMF)**
MW 73.10, bp 153°C
$n_D^{20}$ 1.4310

A highly polar solvent capable of dissolving ionic compounds such as sodium methoxide yet miscible with water.

**CAUTION**: Avoid skin contact with dimethylformamide.

The success of this reaction is strongly dependent on the purity of the starting materials.[3] Benzyl chloride and triethyl phosphite are usually pure enough as received from the supplier. Cinnamaldehyde from a new, previously unopened bottle should be satisfactory, but because it air oxidizes extremely rapidly, it should be distilled, preferably under nitrogen or at reduced pressure, if there is any doubt about its quality. Sodium methoxide, direct from reputable suppliers, has been found on occasion to be completely inactive. No easy method exists for determining its activity; therefore, it is best prepared using the procedure given in *Organic Syntheses*.[4]

To a 10 mm × 100-mm reaction tube add 316 mg (0.0025 mol) of benzyl chloride (α-chlorotoluene), 415 mg (0.44 mL, 0.0025 mol) of triethyl phosphite, and a boiling chip (Fig. 39.2). Because the triethyl phosphite has an offensive odor, obtain this material from the dispenser that has been calibrated to deliver the correct quantity. Place the reaction tube to a depth of about 1 cm in a sand bath that has been preheated to 210°C to 220°C. Reflux the reaction mixture for 1 h. The vapors condense on the cool upper portion of the reaction tube, so a condenser is not necessary. At an internal temperature of about 140°C (which you need not monitor), ethyl chloride is evolved; by the end of the reflux period, the temperature of the reaction mixture will be 200°C to 220°C. At the end of the reaction period, remove the tube from the sand bath, cool it to room temperature, and then add the contents to 160 mg (0.003 mol) of sodium methoxide in a 10-mL Erlenmeyer flask using 1 mL of dry dimethylformamide (DMF) to complete the transfer. Cool the mixture in ice, mix the contents of the flask thoroughly, then add 330 mg (0.0025 mol) of freshly distilled cinnamaldehyde in 1 mL of dry DMF dropwise with thorough mixing in the ice bath. Allow the mixture to come to room temperature. Note the changes that take place over the next few minutes.

After a *minimum* of 10 min, add 2 mL of methanol to the Erlenmeyer flask with thorough stirring. Then almost fill the flask with water and stir the mixture until it is homogeneous in color. The hydrocarbon precipitates from the reaction

---

3. The instructor should try out this experiment before assigning it to a class.

4. For enough sodium methoxide for 50 microscale or 2 macroscale reactions, add 3.5 g of sodium spheres to 50 mL of anhydrous methanol in a 100-mL flask equipped with a condenser. Add the sodium a few pieces at a time through the top of the condenser at a rate so as to keep the reaction under control. It is safest to wait for complete reaction of one portion of sodium before adding the next. After all the sodium has reacted, remove the methanol on a rotary evaporator, first over a steam bath and then over a bath heated to 150°C. The sodium methoxide will be a free-flowing white powder that should keep for several weeks in a desiccator. The yield should be 8.2 g. Source: *Org. Synth.* Col. Vol. IV, **1963**, 651.

**■ FIG. 39.2**
A microscale reflux apparatus.

**■ FIG. 39.3**
A Hirsch funnel for filtration.

Filter paper,
12 mm dia.

Polyethylene
filter disk (frit),
10 mm dia.

Hirsch
funnel

To aspirator

25-mL Filter
flask

Cool part
of tube

Refluxing liquid
(Air condenser)

Sand

Heated
area

Boiling
liquid

Boiling
chip

Electric
flask heater

*:::||||::
**Online Study Center**
*||||||*
Video: Microscale Filtration
on the Hirsch Funnel

mixture after adding the methanol and water. Collect the product by vacuum filtration on a Hirsch funnel (Fig. 39.3) and wash the crystals: first with water to
remove the red color and then with ice-cold methanol to remove the yellow
color. The hydrocarbon is completely insoluble in water and only sparingly soluble in methanol. Weigh the dry product, which should be faint yellow in color,
determine its melting point, and calculate the crude yield. If the melting point is
below 150°C, recrystallize the product from methylcyclohexane (10 mL/g), and
again determine the melting point and yield. Turn in the product in a labeled
vial, giving crude and recrystallized weights, melting points, and yields.

**Cleaning Up.** The filtrate and washings from this reaction are dark, oily, and smell
bad. The mixture contains DMF and could contain traces of all starting materials.
Keep the volume as small as possible and place it in the hazardous waste container
for organophosphorus compounds. If methylcyclohexane was used for recrystallization, place the mother liquor in the organic solvents container.

## 3. Synthesis of 1,4-Diphenyl-1,3-Butadiene

The success of this reaction is strongly dependent on the purity of the starting materials.

**CAUTION:** Keep organophosphorus compounds off the skin. Handle benzyl chloride in the hood. It is a severe lachrymator (tear producer) and respiratory tract irritant.

Use freshly prepared or opened sodium methoxide; keep the bottle closed.

Avoid skin contact with dimethylformamide.

O
‖
C—N
H      CH₃
        CH₃

**N,N-Dimethylformamide (DMF)**
MW 73.10, bp 153°C
$n_D^{20}$ 1.4310

A highly polar solvent capable of dissolving ionic compounds such as sodium methoxide yet miscible with water.

With the aid of pipettes and a pipetter, measure into a $25 \times 150$-mm test tube 5 mL of benzyl chloride ($\alpha$-chlorotoluene) and 7.7 mL of triethyl phosphite. Add a boiling stone, insert a cold finger condenser, and gently reflux the liquid with a flask heater for 1 h. Alternatively, carry out the reaction in a 25-mL round-bottomed flask equipped with a reflux condenser. (Elimination of ethyl chloride starts at about 130°C and, in the time period specified, the temperature of the liquid rises to 190°C–200°C.) Let the phosphonate ester cool to room temperature, pour it into a 125-mL Erlenmeyer flask containing 2.4 g of sodium methoxide, and add 40 mL of dimethylformamide (DMF), using a part of this solvent to rinse the test tube. Swirl the flask vigorously in a water-ice bath to thoroughly chill the contents and continue swirling while running in 5 mL of cinnamaldehyde by pipette. The mixture soon turns deep red; then crystalline hydrocarbon starts to separate. When there is no further change (about 2 min), remove the flask from the cooling bath and let it stand at room temperature for about 10 min. Then add 20 mL of water and 10 mL of methanol, swirl vigorously to dislodge crystals, and finally collect the product on a suction funnel using the red mother liquor to wash the flask. Wash the product with water until the red color of the product is completely replaced by a yellow color. Then wash with methanol to remove the yellow impurity and continue until the wash liquor is colorless. The yield of the crude, faintly yellow hydrocarbon (mp 150°C–151°C) should be about 5.7 g. This material is satisfactory for use in Chapter 50 (1.5 g required). A good solvent for recrystallization of the remaining product is methylcyclohexane (bp 101°C, 10 mL/g; use more if the solution requires filtration). Pure $E,E$-1,4-diphenyl-1,3-butadiene melts at 153°C.

**Cleaning Up.** The filtrate and washings from this reaction are dark, oily, and smell bad. The mixture contains DMF and could contain traces of all starting materials. Keep the volume as small as possible and place it in the hazardous waste container for organophosphorus compounds. If methylcyclohexane was used for recrystallization, place the mother liquor in the organic solvents container.

## QUESTIONS

1. Show how 1,4-diphenyl-1,3-butadiene might be synthesized from benzaldehyde and an appropriate halogenated compound.
2. Explain why the methyl groups of trimethyl phosphite give two peaks in the ${}^1$H NMR spectrum (Fig. 39.4).
3. Write the equation for the reaction between sodium methoxide and moist air.

**■ FIG. 39.4**
The $^1$H NMR spectrum of trimethyl phosphite (90 MHz).

4. The Wittig reaction usually gives a mixture of *cis* and *trans* isomers. Using a molecular mechanics program, calculate the steric energies or heats of formation of both possible products in Experiment 1 or 2. Are these compounds planar? Are the most stable molecules produced in these two experiments? If not, why not?

# Esterification and Hydrolysis

$$\underset{\text{The ester group}}{R-\overset{\overset{\textstyle O}{\|}}{C}-O-R'}$$

**The ester group**

Flavors and fragrances

**PRELAB EXERCISE:** Write the detailed mechanism for the acid-catalyzed hydrolysis of methyl benzoate.

The ester group is an important functional group that can be synthesized in a variety of ways. The low molecular weight esters have very pleasant odors and indeed comprise the major flavor and odor components of a number of fruits. Although a natural flavor may contain nearly 100 different compounds, single esters approximate the natural odors and are often used in the food industry for artificial flavors and fragrances (Table 40.1).

Esters can be prepared by reacting a carboxylic acid with an alcohol in the presence of a catalyst such as concentrated sulfuric acid, hydrogen chloride, *p*-toluenesulfonic acid, or the acid form of an ion exchange resin. For example, methyl acetate can be prepared as follows:

$$\underset{\textbf{Acetic acid}}{CH_3\overset{\overset{\textstyle O}{\|}}{C}-OH} + \underset{\textbf{Methanol}}{CH_3OH} \underset{}{\overset{H^+}{\rightleftharpoons}} \underset{\textbf{Methyl acetate}}{CH_3\overset{\overset{\textstyle O}{\|}}{C}-OCH_3} + H_2O$$

Fischer esterification

The Fischer esterification reaction reaches equilibrium after a few hours of refluxing. The position of the equilibrium can be shifted by adding more of the acid or of the alcohol, depending on cost or availability. The mechanism of the reaction involves initial protonation of the carboxyl group, attack by the nucleophilic hydroxyl, a proton transfer, and loss of water followed by loss of the catalyzing proton to give the ester. Each of these steps is completely reversible, so this process is also, in reverse, the mechanism for the hydrolysis of an ester:

## TABLE 40.1 • Boiling Points and Fragrances of Esters

| Ester | Formula | bp (°C) | Fragrance |
|---|---|---|---|
| 2-Methylpropyl formate | $\overset{O}{\overset{\|}{HC}}-OCH_2\overset{CH_3}{\overset{\|}{CH}}CH_3$ | 98.4 | Raspberry |
| 1-Propyl acetate | $CH_3\overset{O}{\overset{\|}{C}}-OCH_2CH_2CH_3$ | 101.7 | Pear |
| Methyl butyrate | $CH_3CH_2CH_2\overset{O}{\overset{\|}{C}}-OCH_3$ | 102.3 | Apple |
| Ethyl butyrate | $CH_3CH_2CH_2\overset{O}{\overset{\|}{C}}-OCH_2CH_3$ | 121 | Pineapple |
| 2-Methylpropyl propionate | $CH_3CH_2\overset{O}{\overset{\|}{C}}-OCH_2\overset{CH_3}{\overset{\|}{CH}}CH_3$ | 136.8 | Rum |
| 3-Methylbutyl acetate | $CH_3\overset{O}{\overset{\|}{C}}-OCH_2CH_2\overset{CH_3}{\overset{\|}{CH}}CH_3$ | 142 | Banana |
| Benzyl acetate | $CH_3\overset{O}{\overset{\|}{C}}-OCH_2-\phantom{x}$ ⬡ | 213.5 | Peach |
| Octyl acetate | $CH_3\overset{O}{\overset{\|}{C}}-OCH_2(CH_2)_6CH_3$ | 210 | Orange |
| Methyl salicylate | ⬡$\overset{O}{\overset{\|}{C}}-OCH_3$, OH | 222 | Wintergreen |

Other methods are available for synthesizing esters, most of which are more expensive but readily carried out on a small scale. For example, alcohols react with anhydrides and with acid chlorides:

$$CH_3CH_2OH \ + \ CH_3\overset{\displaystyle O}{\overset{\|}{C}}-O-\overset{\displaystyle O}{\overset{\|}{C}}CH_3 \ \longrightarrow \ CH_3\overset{\displaystyle O}{\overset{\|}{C}}-OCH_2CH_3 \ + \ CH_3\overset{\displaystyle O}{\overset{\|}{C}}-OH$$

**Ethanol**    **Acetic anhydride**    **Ethyl acetate**    **Acetic acid**

$$CH_3CH_2CH_2OH \ + \ CH_3\overset{\displaystyle O}{\overset{\|}{C}}-Cl \ \longrightarrow \ CH_3\overset{\displaystyle O}{\overset{\|}{C}}-OCH_2CH_2CH_3 \ + \ HCl$$

**1-Propanol**    **Acetyl chloride**    ***n*-Propyl acetate**

In the latter reaction, an organic base such as pyridine is usually added to react with the hydrogen chloride.

*Other ester syntheses*

Other methods can also be used to synthesize the ester group. Among these are the addition of 2-methylpropene to an acid to form *t*-butyl esters, the addition of ketene to make acetates, and the reaction of a silver salt with an alkyl halide:

$$CH_2{=}\overset{\displaystyle CH_3}{\overset{|}{C}}CH_3 \ + \ CH_3CH_2\overset{\displaystyle O}{\overset{\|}{C}}-OH \ \xrightarrow{H^+} \ CH_3CH_2\overset{\displaystyle O}{\overset{\|}{C}}-O\overset{\displaystyle CH_3}{\underset{\displaystyle CH_3}{\overset{|}{\underset{|}{C}}}}CH_3$$

**2-Methylpropene**    **Propionic acid**    ***t*-Butyl propionate**
**(isobutylene)**

$$CH_2{=}C{=}O \ + \ HOCH_2{-}\langle\bigcirc\rangle \ \longrightarrow \ CH_3\overset{\displaystyle O}{\overset{\|}{C}}-OCH_2{-}\langle\bigcirc\rangle$$

**Ketene**    **Benzyl alcohol**    **Benzyl acetate**

$$CH_3\overset{\displaystyle O}{\overset{\|}{C}}-OAg \ + \ BrCH_2CH_2\overset{\displaystyle CH_3}{\overset{|}{C}}HCH_3 \ \longrightarrow \ CH_3\overset{\displaystyle O}{\overset{\|}{C}}-OCH_2CH_2\overset{\displaystyle CH_3}{\overset{|}{C}}HCH_3$$

**Silver acetate**    **1-Bromo-3-methylbutane**    **3-Methylbutyl acetate**

As noted previously, Fischer esterification is an equilibrium process. Consider the reaction of acetic acid with 1-butanol to give *n*-butyl acetate:

$$CH_3\overset{\displaystyle O}{\overset{\|}{C}}-OH \ + \ HOCH_2CH_2CH_2CH_3 \ \underset{}{\overset{H^+}{\rightleftharpoons}} \ CH_3\overset{\displaystyle O}{\overset{\|}{C}}-OCH_2CH_2CH_2CH_3 \ + \ H_2O$$

The equilibrium constant is as follows:

$$K_{eq} = \frac{[n\text{-BuOAc}][H_2O]}{[n\text{-BuOH}][HOAc]}$$

For primary alcohols reacting with unhindered carboxylic acids, $K_{eq} \approx 4$. If equal quantities of 1-butanol and acetic acid are allowed to react at equilibrium, the theoretical yield of ester is only 67%. To upset the equilibrium we can, by Le Chatelier's principle, increase the concentration of either the alcohol or acid. If either one is doubled, the theoretical yield increases to 85%. When one is tripled, the yield goes to 90%. But note that in the example cited the boiling point of the relatively nonpolar ester is only about 8°C higher than the boiling points of the polar acetic acid and 1-butanol, so a difficult separation problem exists if the product must be isolated by distillation after the starting materials are increased in concentration.

Another way to upset the equilibrium is to remove water. This can be done by adding to the reaction mixture molecular sieves (an artificial zeolite), which preferentially adsorb water. Most other drying agents, such as anhydrous sodium sulfate or calcium chloride pellets, will not remove water at the temperatures used to make esters.

A third way to upset the equilibrium is to preferentially remove the water as an azeotrope (a constant-boiling mixture of water and an organic liquid). The information in Table 40.2 can be found in a chemistry handbook table of ternary (three-component) azeotropes. These data tell us that vapor that distills from a mixture of 1-butanol, *n*-butyl acetate, and water will boil at 90.7°C and that the vapor contains 8% alcohol, 63% ester, and 29% water. The vapor is homogeneous, but when it condenses, it separates into two layers. The upper layer is composed of 11% alcohol, 86% ester, and 3% water, but the lower layer consists of 97% water with only traces of alcohol and ester. If some ingenious way can be devised to remove the lower layer from the condensate and still return the upper layer to the reaction mixture, then the equilibrium can be upset, and nearly 100% of the ester can be produced in the reaction flask.

The apparatus shown in Figure 40.1, modeled after that of Ernest W. Dean and David D. Stark, achieves the desired separation of the two layers. The mixture of equimolar quantities of 1-butanol and acetic acid is placed in the flask along with an acid catalyst. Stirring reduces bumping. The vapor, the temperature of

*Azeotropic distillation*

## TABLE 40.2 • The Ternary Azeotrope of Boiling Point 90.7°C

| Compound | Boiling Point of Pure Compound (°C) | Percentage Composition of Azeotrope | | |
| --- | --- | --- | --- | --- |
| | | *Vapor Phase* | *Upper Layer* | *Lower Layer* |
| 1-Butanol | 117.7 | 8.0 | 11.0 | 2.0 |
| *n*-Butyl acetate | 126.7 | 63.0 | 86.0 | 1.0 |
| Water | 100.0 | 29.0 | 3.0 | 97.0 |

■ **FIG. 40.1**

A microscale Dean-Stark azeotropic esterification apparatus. A cork is used instead of a septum so that layer separation can be observed clearly.

Reflux condenser

Ester + alcohol layer

Water layer

Cork

Reaction mixture

Hot sand bath

Stirring bar and catalyst

■ **FIG. 40.2**

A macroscale azeotropic distillation apparatus, with a Dean-Stark trap where water collects.

which is 90.7°C, condenses and runs down to the sidearm, which is closed with a cork. The layers separate, with the denser water layer remaining in the sidearm and the lighter ester plus alcohol layer running down into the reaction flask. As soon as the theoretical quantity of water has collected, the reaction is over, and the product in the flask should contain an ester of high purity. The macroscale apparatus is illustrated in Figure 40.2.

Esterification using a carboxylic acid and an alcohol requires an acid catalyst. In the first experiment, the acid form of an ion-exchange resin is used. This resin, in the form of small beads, is a cross-linked polystyrene that bears sulfonic acid groups on some of the phenyl groups. It is essentially an immobilized form of *p*-toluenesulfonic acid, an organic-substituted sulfuric acid.

$$\{CH-CH_2-CH-CH_2-CH\}_n$$

$$SO_3^- H^+ \qquad SO_3^- H^+ \qquad SO_3^- H^+$$

An ion-exchange catalyst

This catalyst has the distinct advantage that at the end of the reaction it can be easily removed by filtration. Immobilized catalysts of this type are becoming more and more common in organic synthesis.

If concentrated sulfuric acid were used as the catalyst, it would be necessary to dilute the reaction mixture with ether; wash the ether layer successively with water, sodium carbonate solution, and saturated sodium chloride solution; and then dry the ether layer with anhydrous calcium chloride pellets before evaporating the ether.

## EXPERIMENTS

### 1. *n*-Butyl Acetate by Azeotropic Distillation of Water

$$\underset{\substack{\text{Acetic acid}}}{\overset{O}{\underset{\|}{CH_3C{-}OH}}} + \underset{\substack{\text{1-Butanol}}}{HOCH_2CH_2CH_2CH_3} \overset{H^+}{\rightleftharpoons} \underset{\substack{\textit{n}\text{-Butyl acetate}}}{\overset{O}{\underset{\|}{CH_3C{-}OCH_2CH_2CH_2CH_3}}} + H_2O$$

| Acetic acid | 1-Butanol | *n*-Butyl acetate | |
|---|---|---|---|
| MW 60.05 | MW 74.12 | MW 116.16 | MW 18 |
| bp 117.9°C, den 1.049 | bp 117.7°C, den 0.810 | bp 126.5°C, den 0.882 | |
| $n_D^{20}$ 1.3720 | $n_D^{20}$ 1.3990 | $n_D^{20}$ 1.3940 | |

> **IN THIS EXPERIMENT** acetic acid is esterified with butanol in the presence of an acid catalyst attached to very small polystyrene beads. Because esterification is an equilibrium reaction, the equilibrium is upset by removing the water formed in a unique apparatus patterned after the Dean-Stark water separator. The product is isolated simply by withdrawing it from the reaction flask with a Pasteur pipette, leaving the catalyst behind. It can be further purified by simple distillation.

Ion-exchange resin catalyst

A cork instead of a septum allows the accumulation of water to be observed.

In a 5-mL short-necked, round-bottomed flask, place 0.2 g of Dowex 50X2-100 ion-exchange resin,[1] 0.60 g (0.58 mL; 0.01 mol) of acetic acid, 0.74 g (0.91 mL; 0.01 mol) of 1-butanol, and a stirring bar. Attach the addition port with the sidearm corked and an empty distilling column, as shown in Figure 40.1, and clamp the apparatus at the angle shown. Heat the flask while stirring on a hot sand

---

1. The Dowex resin as received should be washed with water by decantation to remove much of its yellow color. It is then collected by vacuum filtration on a Büchner funnel and returned, in a slightly damp state, to the reagent bottle.

bath and bring the reaction mixture to a boil. Stirring the mixture prevents it from bumping. As an option, you might hold a thermometer just above the boiling liquid and note a temperature of about 91°C. Remove the thermometer and allow the reaction mixture to reflux in a manner such that the vapors condense about one-third of the way up the empty distilling column, which is functioning as an air condenser. Note that the material that condenses is not homogeneous, as droplets of water begin to collect in the upper part of the apparatus. As the sidearm fills with condensate, it is cloudy at first and then two layers separate. When the volume of the lower aqueous layer does not appear to increase, the reaction is over. This will take about 20–30 min. Carefully remove the apparatus from the heat, allow it to cool, and then tip the apparatus very carefully to allow all the upper layer in the sidearm to run back into the reaction flask. Disconnect the apparatus.

Remove the product from the reaction flask with a Pasteur pipette, determine its weight and boiling point, and assess its purity by thin-layer chromatographic (TLC) analysis and infrared (IR) spectroscopy. The product can be analyzed by gas chromatography on a 10-ft (3-m) Carbowax column. At 152°C, the 1-butanol has a retention time of 2.1 min, the *n*-butyl acetate 2.5 min, and the acetic acid 7.5 min. Look for the presence of hydroxyl and carboxyl absorption bands in the IR spectrum. The product can easily be purified by simple distillation (*see* Figure 5.7 on page 97).

**Cleaning Up.** Place the catalyst in the solid hazardous waste container.

## 2. Isobutyl Propionate (2-Methylpropyl propanoate) by Fischer Esterification

$$\underset{\substack{\text{Propanoic acid} \\ \text{MW 74.08} \\ \text{bp 141°C}}}{\text{CH}_3\text{CH}_2\overset{\displaystyle\text{O}}{\overset{\|}{\text{C}}}\text{OH}} + \underset{\substack{\text{2-Methyl-1-propanol} \\ \text{MW 74.12} \\ \text{bp 108°C}}}{\text{HOCH}_2\overset{\displaystyle\text{CH}_3}{\overset{|}{\text{C}}}\text{HCH}_3} \overset{\text{H}^+}{\rightleftharpoons} \underset{\substack{\text{Isobutyl propionate} \\ \text{MW 130.19} \\ \text{bp 136.8°C}}}{\text{CH}_3\text{CH}_2\overset{\displaystyle\text{O}}{\overset{\|}{\text{C}}}\text{OCH}_2\overset{\displaystyle\text{CH}_3}{\overset{|}{\text{C}}}\text{HCH}_3} + \text{H}_2\text{O}$$

**IN THIS EXPERIMENT** a propionate ester is prepared by reacting excess propanoic acid with an alcohol in the presence of an acid catalyst on polystyrene beads (Dowex 50X). The excess acid serves to drive the equilibrium reaction toward formation of the product. The excess propanoic acid is removed by reaction with potassium carbonate, and the water is adsorbed by silica gel when the reaction mixture is chromatographed.

### Procedure

To a reaction tube add 112 mg of 2-methyl-1-propanol (isobutyl alcohol), 148 mg of propanoic acid (propionic acid), 50 mg of Dowex 50X2-100 ion-exchange resin, and a boiling chip. Attach the empty distilling column as an air condenser (Fig. 40.3). Reflux the resulting mixture for 1 h or more, cool it to room temperature, remove the product mixture from the resin with a Pasteur pipette, and chromatograph the product (2-methyl-1-propyl propanoate) on a silica gel column.

■ **FIG. 40.3**
An esterification apparatus.

Wet pipe cleaner —

■ **FIG. 40.4**
A chromatographic column for esters.

Funnel

Glass column

Anhydrous potassium carbonate

Silica gel

Micro Büchner funnel

Polyethylene frit

10-mL Erlenmeyer flask

**CAUTION**: Carry out this part of the experiment in a hood. Dichloromethane is a suspected carcinogen.

## *Chromatography Procedure*

Assemble the column as depicted in Figure 40.4, being sure that it is clamped in a vertical position. Close the valve and fill the column with dichloromethane to the bottom of the funnel. In a small beaker prepare a slurry of 1 g of silica gel in 4 mL of dichloromethane. Stir the slurry gently to remove air bubbles and gently swirl, pour, and scrape the slurry into the funnel, which has a capacity of 10 mL. After some of the silica gel has been added to the column, open the stopcock and allow solvent to drain slowly into an Erlenmeyer flask. Use this dichloromethane to rinse the beaker containing the silica gel. As the silica gel is being added, tap the column with a glass rod or pencil so the adsorbent will pack tightly into the column. Continue to tap the column while cycling the dichloromethane through the column once more; then add 1 g of anhydrous potassium carbonate to the top of the silica gel. The potassium carbonate will remove water from the esterification mixture as well as react with any carboxylic acid present. Allow the solvent to flow until it reaches the top of the potassium carbonate layer.

### Adding the Sample

The solvent is drained just to the surface of the potassium carbonate. Using a Pasteur pipette, add the sample to the column and let it run into the adsorbent, stopping when the solution reaches the top of the potassium carbonate. The flask and ion-exchange resin are rinsed twice with 0.5-mL portions of dichloromethane that are run into the column, with the eluent being collected in a tared reaction tube. The elution is completed with 1 mL more of dichloromethane.

### Analysis of the Product

Evaporate the dichloromethane under a stream of air or nitrogen in the hood and remove the last traces by connecting the reaction tube to a water aspirator. Because the dichloromethane boils at 40°C and the product boils at 137°C, separation of the two is easily accomplished. Determine the weight of the product and its boiling point and calculate the yield. The ester should be a perfectly clear, homogeneous liquid. Obtain an IR spectrum and analyze it for the presence of unreacted alcohol and carboxylic acid. Check the purity of the product by TLC and/or gas chromatography.

**Cleaning Up.** Place the catalyst in the hazardous waste container. Any dichloromethane should be placed in the halogenated organic waste container. The contents of the chromatography column, if free of solvent, can be placed in the nonhazardous solid waste container; otherwise, this material is classified as a hazardous waste and must go in the container so designated.

## 3. Benzyl Acetate from Acetic Anhydride

| Benzyl alcohol | Acetic anhydride | Benzyl acetate | Acetic acid |
|---|---|---|---|
| MW 108.14 | MW 102.09 | MW 150.18 | MW 60.06 |
| bp 205°C | bp 138–140°C | bp 213.5°C | bp 117–118°C |
| $n_D^{20}$ 1.5400 | $n_D^{20}$ 1.3900 | $n_D^{20}$ 1.5020 | $n_D^{20}$ 1.3720 |

To a reaction tube add 108 mg of benzyl alcohol, 102 mg of acetic anhydride, and a boiling chip. Reflux the mixture for at least 1 h, cool the mixture to room temperature, and chromatograph the liquid in exactly the same manner described in Experiment 2. Analyze the product by TLC and by IR spectroscopy as a thin film between salt or silver chloride plates. Note the presence or absence of hydroxyl and carboxyl bands.

This ester cannot be prepared by Fischer esterification using Dowex. Polymerization seems to occur.

**Cleaning Up.** Place the catalyst in the hazardous waste container. Any dichloromethane should be placed in the halogenated organic waste container. The contents of the chromatography column, if free of solvent, can be placed in the nonhazardous solid waste container; otherwise, this material is classified as a hazardous waste and must go in the container so designated.

## Other Esterifications

This experiment lends itself to wide-ranging experimentation. All three methods of esterification previously described can, in principle, be applied to any unhindered primary or secondary alcohol. These methods work well for most of the esters in Table 40.1 and for hundreds of others as well.

## 4. Methyl Benzoate by Fischer Esterification

$$\text{Benzoic acid} + CH_3OH \underset{}{\overset{H_2SO_4}{\rightleftharpoons}} \text{Methyl benzoate} + H_2O$$

| **Benzoic acid** | **Methanol** | **Methyl benzoate** |
|---|---|---|
| mp 122°C | bp 64.6°C | bp 198–199°C |
| MW 122.12 | den. 0.791 | den. 1.094 |
| | MW 32.04 | MW 136.15 |

**IN THIS EXPERIMENT** benzoic acid and excess methanol are refluxed in the presence of a catalytic amount of sulfuric acid. Water and ether are added; the water layer is withdrawn; and the ether layer is washed with water, bicarbonate, and saturated salt solutions. After drying over calcium chloride, the ether is evaporated. Finally, the product, ethyl benzoate, is distilled.

Place 10.0 g of benzoic acid and 25 mL of methanol in a 125-mL round-bottomed flask, cool the mixture in ice, pour 3 mL of concentrated sulfuric acid *slowly and carefully down the walls of the flask*, then swirl to mix the components.

**■ FIG. 40.5**
An apparatus for refluxing
a reaction mixture.

Water out

Water in

Do not use water in the
circulating jacket.

Attach a reflux condenser, add a boiling chip, and reflux the mixture gently for 1 h. (Fig. 40.5 illustrates this apparatus.) Cool the solution, decant it into a separatory funnel containing 50 mL of water, and rinse the flask with 35 mL of ether. Add this ether to the separatory funnel, shake thoroughly, and drain off the water layer, which contains the sulfuric acid and the bulk of the methanol. Wash the ether in the separatory funnel with 25 mL of water followed by 25 mL of 0.5 *M* sodium bicarbonate to remove unreacted benzoic acid. Again shake, with frequent release of pressure by inverting the separatory funnel and opening the stopcock, until no further reaction is apparent; then drain off the bicarbonate layer into a beaker. If this aqueous material is made strongly acidic with hydrochloric acid, unreacted benzoic acid may be observed. Wash the ether layer in the separatory funnel with saturated sodium chloride solution and dry the solution over anhydrous calcium chloride in an Erlenmeyer flask. Add sufficient calcium chloride pellets so that it no longer clumps together on the bottom of the flask. After 10 min decant the dry ether solution into a flask, wash the drying agent with an additional 5 mL of ether, and decant again.

Remove the ether by simple distillation or by evaporation on a steam bath under an aspirator tube. Use the methods illustrated in Fig. 4.22 (on page 79) or Fig. 5.10 (on page 100), or use a rotary evaporator (*see* Fig. 9.8 on page 207). When evaporation ceases, add 2–3 g of anhydrous calcium chloride pellets to the residual oil and heat for about 5 min longer. Then decant the methyl benzoate into a 50-mL round-bottomed flask, attach a stillhead, dry out the ordinary condenser and use it without water circulating in the jacket, and distill. The boiling point of the ester is so high (199°C) that a water-cooled condenser is liable to crack. Use a tared 25-mL Erlenmeyer flask as the receiver to collect material boiling above 190°C. A typical student yield is about 7 g. *See* Chapter 28 for the nitration of methyl benzoate and Chapter 38 for its use in the Grignard synthesis of triphenylmethanol.

IR and nuclear magnetic resonance (NMR) spectra of benzoic acid and methyl benzoate are found at the end of the chapter (Figs. 40.6 to 40.11).

**Cleaning Up.** Pour the sulfuric acid layer into water, combining it with the bicarbonate layer, neutralize it with sodium carbonate, and flush the solution down the drain with excess water. The saturated sodium chloride layer can also be flushed down the drain. If the calcium chloride is free of ether and methyl benzoate, it can be placed in the nonhazardous solid waste container; otherwise it must go into the hazardous waste container. Ether goes into the organic solvents container, along with the pot residues from the final distillation.

# 5. Hydrolysis (Saponification): The Preparation of Soap

In general, the reversal of esterification is called *hydrolysis*. In the case of the hydrolysis of a fatty acid ester, it is called . In this experiment, the saturated fat made from hydrogenated olive oil in Chapter 25 will be saponified to give a soap, which, in this case, will be primarily sodium stearate.

$$
\underset{\textbf{Glycerol tristearate}}{
\begin{array}{c}
\mathrm{CH_2O\overset{\displaystyle O}{\overset{\|}{C}}(CH_2)_{16}CH_3} \\[2mm]
| \\[1mm]
\mathrm{CHO\overset{\displaystyle O}{\overset{\|}{C}}(CH_2)_{16}CH_3} \\[2mm]
| \\[1mm]
\mathrm{CH_2O\overset{\displaystyle O}{\overset{\|}{C}}(CH_2)_{16}CH_3}
\end{array}}
\;\xrightarrow[\mathrm{H_2O}]{\mathrm{NaOH}}\;
\underset{\textbf{Glycerol}}{
\begin{array}{c}
\mathrm{CH_2OH} \\[4mm]
| \\[2mm]
\mathrm{CHOH} \\[4mm]
| \\[2mm]
\mathrm{CH_2OH}
\end{array}}
\;+\;
\underset{\textbf{Sodium stearate (soap)}}{3\,\mathrm{Na^+\,{}^-\!O\overset{\displaystyle O}{\overset{\|}{C}}(CH_2)_{16}CH_3}}
$$

## Microscale Procedure

> **IN THIS EXPERIMENT** a saturated fat is heated with sodium hydroxide in a water-ethanol micture. The resulting soap is precipitated in salt solution and collected on a Hirsch funnel, where it is washed free of base and salt.

Place 0.18 g of the saturated triglyceride prepared in Chapter 25, Experiment 4, in a 5-mL short-necked, round-bottomed flask. Add 1.5 mL of a 50:50 water-ethanol solution that contains 0.18 g of solid sodium hydroxide (weigh this quickly). Add an air condenser and gently reflux the mixture on a sand bath for 30 min, taking care not to boil away the ethanol. At the end of the reaction period, some of the soap will have precipitated. Transfer the mixture to a 10-mL Erlenmeyer flask containing a solution of 0.8 g of sodium chloride in 3 mL of water. Collect the precipitated soap on a Hirsch funnel and wash it free of excess sodium hydroxide and salt using 4 mL of distilled ice water.

Test the soap by adding a very small piece (about 5–15 mg) to a centrifuge or test tube along with 3–4 mL of distilled water. Cap the tube and shake it vigorously. Note the height and stability of the bubbles. Add a crystal of magnesium chloride or calcium chloride to the tube. Shake the tube again and note the results. For comparison, do these same tests with a few grains of a detergent instead of the soap.

## Macroscale Procedure

In a 100-mL round-bottomed flask, place 5 g of hydrogenated olive oil (Chapter 25) or lard or solid shortening (e.g., Crisco). Add 20 mL of ethanol and a hot solution of 5 g of sodium hydroxide in 20 mL of water. This solution, prepared in a beaker, will become very hot as the sodium hydroxide dissolves. Fit the flask with a water-cooled condenser and reflux the mixture on a sand bath for 30 min. At the end of the reaction period, some of the soap may precipitate from the reaction mixture. Transfer the mixture to a 250-mL Erlenmeyer flask containing an ice-cold solution of 25 g of sodium chloride in 90 mL of distilled water. Collect the precipitated soap on a Büchner funnel and wash it free of excess sodium hydroxide and salt using no more than 100 mL of distilled ice water.

Test the soap by adding a small piece to a test tube along with about 5 mL of water. Cap the tube and shake it vigorously. Note the height and stability of the bubbles formed. Add a few crystals of magnesium chloride or calcium chloride to the tube. Shake the tube again and note the results. For comparison, do these same tests with an amount of detergent equal to that of the soap used.

**Online Study Center**

**General Resources**
Additional Experiments

## QUESTIONS

1. In the preparation of methyl benzoate, what is the purpose of (a) washing the organic layer with sodium bicarbonate solution? (b) washing the organic layer with saturated sodium chloride solution? (c) treating the organic layer with anhydrous calcium chloride pellets?
2. Assign the resonances in Figure 40.8 to specific protons in methyl benzoate.
3. Figures 40.9 and 40.10 each have two resonances that are very small. What do the carbons causing these peaks have in common?

■ **FIG. 40.6**
The IR spectrum of benzoic acid (KBr disk).

■ **FIG. 40.7**
The $^1$H NMR spectrum of benzoic acid (250 MHz).

### ■ FIG. 40.8
The $^1$H NMR spectrum of methyl benzoate (250 MHz).

### ■ FIG. 40.9
The $^{13}$C NMR spectrum of benzoic acid (22.6 MHz).

### ■ FIG. 40.10
The $^{13}$C NMR spectrum of methyl benzoate (22.6 MHz).

### ■ FIG. 40.11
The IR spectrum of methyl benzoate.

# CHAPTER

# 41

# Acetylsalicylic Acid (Aspirin)

**PRELAB EXERCISE:** Write detailed mechanisms showing how pyridine and sulfuric acid catalyze the formation of acetylsalicylic acid.

Aspirin is among the most fascinating and versatile drugs known to medicine; it is also among the oldest. The first known use of an aspirin-like preparation can be traced to ancient Greece and Rome. Salicin, an extract obtained from the bark of willow and poplar trees, has been used as a pain reliever (analgesic) for centuries. In the middle of the 19th century, it was found that salicin is a glycoside formed from a molecule of salicylic acid and a sugar molecule. The sodium salt of salicylic acid, sodium salicylate, is easily synthesized on a large scale by heating sodium phenoxide with carbon dioxide at 150°C under slight pressure (the Kolbe synthesis):

**Sodium salicylate**

Unfortunately, however, salicylic acid attacks the mucous membranes of the mouth and esophagus and causes gastric pain that may be worse than the discomfort it was intended to cure. Felix Hoffmann, a chemist for Friedrich Bayer, a German dye company, reasoned that the corrosive nature of salicylic acid could be altered by adding an acetyl group; and in 1893 the Bayer Company obtained a patent on acetylsalicylic acid, despite the fact that it had been synthesized some 40 years previously by Charles Gerhardt. Bayer coined the name *Aspirin* for its new product to reflect its acetyl nature and its natural occurrence in the spirea plant. Over the years Bayer has allowed the term *aspirin* to fall into the public domain so it is no longer capitalized. The manufacturers of Coke (Coca-Cola) and Sanka are working hard to prevent a similar fate befalling their trademarks.

In 1904 Carl Duisberg, the head of Bayer, decided to emulate John D. Rockefeller's Standard Oil Company and formed an *interessen gemeinschaft* (I.G.) of the dye industry (*Farbenindustrie*). This cartel completely dominated the world dye industry before World War I, and it continued to prosper between the wars, even though some of its assets were seized and sold after World War I. Soon after, an American company, Sterling Drug, bought the rights to aspirin for $5.3 million. Sterling was bought by Eastman Kodak in 1988, which was then sold to Smith-Kline Beecham in 1995, which combined in 2000 with Glaxo Wellcome to become GlaxoSmithKline.

Because of their involvement at Auschwitz, the top management of I.G. Farbenindustrie was tried and convicted at the Nuremberg trials after World War II, and the cartel was broken into three large branches—Bayer, Hoechst, and BASF (*Badische Anilin und Sodafabrik*)—each of these companies does more business than DuPont, the largest American chemical company. In 1997 the American rights to the Bayer name and trademark were sold back to Bayer A.G. for $1 billion.

By law, all drugs sold in the United States must meet the purity standards set by the U.S. Food and Drug Administration (FDA), and so all aspirin is essentially the same. Each tablet contains 0.325 g (325 mg) of acetylsalicylic acid held together with a binder. The remarkable difference in price for aspirin is primarily a reflection of the advertising budget of the company that sells it. Bayer has 5% of the painkiller market; lower-priced generic aspirin has 18%.

The gross profit margin on Bayer aspirin is 80%. After advertising costs, the profit margin is 60%.

Aspirin is an analgesic, an antipyretic (fever reducer), and an anti-inflammatory agent. It is the premier drug for reducing fever, a role for which it is uniquely suited. As an anti-inflammatory, it has become the most widely effective treatment for arthritis. Unfortunately, people suffering from arthritis must take so much aspirin (several grams per day) that gastric problems may result. For this reason aspirin is often combined with a buffering agent. Bufferin is an example of such a preparation.

The ability of aspirin to diminish inflammation is apparently due to its inhibition of the synthesis of prostaglandins, a group of $C_{20}$ molecules that exacerbate inflammation. Aspirin alters the oxygenase activity of prostaglandin synthetase by moving the acetyl group to a terminal amine group of the enzyme.

If aspirin were a new invention, the FDA would place many hurdles in the path of its approval. It has been implicated, for example, in Reye's syndrome, a brain disorder that strikes children and young persons under age 18 who take aspirin after contracting flu or chickenpox. It has an effect on platelets, which play a vital role in blood clotting. In newborn babies and their mothers, aspirin can lead to uncontrolled bleeding and circulation problems for the baby—even brain hemorrhages in extreme cases. This same effect can be turned into an advantage, however. Heart specialists urge potential stroke victims to take aspirin regularly to inhibit clotting in their arteries, and it has been shown that taking aspirin will help prevent first heart attacks from occurring.

Over 100 common medications contain aspirin, including, for example, Alka-Seltzer, Anacin, Coricidin, Excedrin, Midol, and Vanquish. Despite its side effects, aspirin is one of the safest, cheapest, and efficacious nonprescription drugs, although acetaminophen (Tylenol, etc.) has 40% and ibuprofen (Advil, etc.) has 26% of the painkiller market in dollar volume ($2.47 billion in 1996). Naproxen (Aleve) has 6% of the market. Aspirin is made commercially employing the same synthesis presented in this chapter.

The mechanism for the acetylation of salicylic acid is as follows:

Acetylsalicylic acid

## EXPERIMENT

# Synthesis of Acetylsalicylic Acid (Aspirin)

| Salicylic acid | Acetic anhydride | Acetylsalicylic acid |
|---|---|---|
| MW 138.12, mp 159°C | MW 102.09, bp 140°C | MW 180.15, mp 128–137°C |

## *Microscale Procedure*

> **IN THIS EXPERIMENT** salicylic acid, acetic anhydride, and an acid catalyst are heated briefly. Water is added to convert excess acetic anhydride to acetic acid and also serves as the crystallization solvent. After the water is removed from the crystals, they are washed, collected, and dried to give pure aspirin.

⚠
Measure acetic anhydride in the hood, as it is very irritating to breathe.

**Online Study Center**

Photo: Filtration Using a Pasteur Pipette; Videos: Filtration of Crystals Using the Pasteur Pipette, Microscale Filtration on the Hirsch Funnel

To a reaction tube, add 138 mg of salicylic acid, a boiling chip, and one small drop of 85% phosphoric acid followed by 0.3 mL of acetic anhydride (the acetic anhydride serves to wash the reactants to the bottom of the tube). Mix the reactants thoroughly; then heat the reaction tube on a steam bath or in a beaker of 90°C water for 5 min. Cautiously add 0.2 mL of water to the reaction mixture to decompose excess acetic anhydride. This will be an exothermic reaction. When the reaction is over, add an additional 0.3 mL of water and allow the tube to cool slowly to room temperature. If crystallization of the product does not occur during the cooling process, add a seed crystal or scratch the inside of the tube with a glass stirring rod. Cool the tube in ice until crystallization is complete (at least 10 min) and then remove the solvent with a Pasteur pipette. If the crystals are too fine for this procedure, collect the product by vacuum filtration on a Hirsch funnel. Complete the transfer of the product to the funnel using a very small quantity of ice water. In either case, turn the product out onto a piece of filter paper and squeeze the crystals between sheets of filter paper to absorb excess water. Allow the product to air-dry thoroughly before determining the weight and calculating the percent yield. Determine the melting point and the infrared (IR) spectrum in chloroform solution. Figure 41.1 shows the IR spectrum of acetylsalicylic acid. Aspirin is hydrolyzed by boiling water, but the reaction is not rapid; if desired, the product may be recrystallized from a small quantity of very hot water.

■ **FIG. 41.1**
The IR spectrum of acetylsalicylic acid in chloroform.

Compare a tablet of commercial aspirin with your sample. Test the solubility of the commercial tablet in water and in toluene and observe if it dissolves completely. Compare its behavior when heated in a melting-point capillary to the behavior of your sample. If an impurity is found, it is probably a substance used as binder for the tablets. Is the binder organic or inorganic?

**Cleaning Up.** Dilute the filtrate with water and flush it down the drain.

## Macroscale Procedure

Place 1 g of salicylic acid in each of four 13 × 100-mm test tubes and add to each tube 2 mL of acetic anhydride. To the first tube add 0.2 g of anhydrous sodium acetate, note the time, stir the mixture with a glass stirring rod, replace the rod with a thermometer, and record the time required for a 4°C rise in temperature. Continue to stir occasionally while starting the next acetylation. Obtain a clean thermometer, put it in the second tube, add 5 drops of pyridine, stir, observe as before, and compare it to the first results. To the third and fourth tubes add 5 drops of boron trifluoride etherate[1] and 5 drops of concentrated sulfuric acid, respectively. What is the order of activity of the four catalysts as judged by the rates of the reactions?

Put all of the tubes in a beaker of hot water for 5 min to dissolve solid material and complete the reactions; then pour all of the solutions into a 125-mL Erlenmeyer flask containing 50 mL of water and rinse the tubes with water. Swirl to aid hydrolysis of the excess acetic anhydride; then cool thoroughly in ice, scratch the side of the flask with a stirring rod to induce crystallization, and collect the crystalline solid; the yield is about 4 g.

Acetylsalicylic acid melts with decomposition at temperatures reported between 128°C to 137°C. It can be crystallized by dissolving it in $t$-butyl methyl ether, adding an equal volume of petroleum ether, and letting the solution stand undisturbed in an ice bath.

Test the solubility of your sample in toluene and in hot water and note the peculiar character of the aqueous solution when it is cooled and when it is then rubbed against the tube with a stirring rod. Note also that the substance dissolves in cold sodium bicarbonate solution and is precipitated by adding concentrated hydrochloric acid.

Compare a tablet of commercial aspirin to your sample. Test the solubility of the commercial tablet in water and in toluene and observe if it dissolves completely. Compare its behavior when heated in a melting-point capillary to the behavior of your sample. If an impurity is found, it is probably a substance used as binder for the tablets. Is the binder organic or inorganic? To interpret your results, consider the mechanism whereby salicylic acid is acetylated.

Note that acetic acid is eliminated during the reaction. What effect would sodium acetate have? How might boron trifluoride etherate or sulfuric acid affect the nucleophilic attack of the phenolic oxygen on acetic anhydride? With what might the pyridine, a base, associate?

---

Measure acetic anhydride in the hood, as it is very irritating to breathe.

Acetylation catalysts:

1. $CH_3COO^- Na^+$

2.

3. $BF_3 \cdot (C_2H_5)_2O$

4. $H_2SO_4$

Pyridine has a bad odor, and boron trifluoride etherate and sulfuric acid are corrosive; work in the hood. Handle with care.

**Online Study Center**

Video: Macroscale Crystallization

$t$-Butyl methyl ether and petroleum ether are very flammable.

Propose mechanisms to interpret your results.

---

1. *To the instructor*: Commercial reagent, if dark, should be redistilled (bp 126°C, water-white).

■ **FIG. 41.2**
The ¹H NMR spectrum of
acetylsalicylic acid (250 MHz).
The COOH proton does not appear
on this spectrum.

The IR and ¹H NMR (nuclear magnetic resonance) spectra of acetylsalicylic acid are presented in Figure 41.1 (on page 572) and Figure 41.2.

**Cleaning Up.** Combine the aqueous filtrates, dilute with water, and flush the solution down the drain.

## QUESTIONS

*Online Study Center*
**General Resources**
Web Links

1. Hydrochloric acid is about as strong a mineral acid as sulfuric acid is. Why would HCl not be a satisfactory catalyst in this reaction?
2. How do you account for the smell of vinegar when an old bottle of aspirin is opened?

# CHAPTER

# 42

# Malonic Ester of a Barbiturate

**Online Study Center**

This icon will direct you to techniques, equipment setups, and online resources at http://college.hmco.com/ PIC/williamsonMMOE5e.

**PRELAB EXERCISE:** Which barbiturate discussed in this chapter cannot be synthesized by the acetoacetic ester reaction?

**B**arbiturates are central nervous system (CNS) depressants used as hypnotic drugs and anesthetics. They are all derivatives of barbituric acid (R = R′ = H), which has no sedative properties. It is called an *acid* because the carbonyl groups render the imide hydrogens acidic:

**Barbituric acid**
R = R′ = H

**Veronal**

Barbituric acid, derived from urea, was first synthesized in 1864 by Adolph von Baeyer. Apparently it was named at a tavern on St. Barbara's day. At the beginning of the 20th century, the great chemist Emil Fischer synthesized the first hypnotic (sleep-inducing) barbiturate, the 5,5-diethyl derivative, at the direction of Joseph von Mering. Von Mering, who made the seminal discovery that removing the pancreas causes diabetes, named the new derivative of barbituric acid Veronal because he regarded Verona as the most restful city on Earth.

Barbiturates are the most widely used sleeping pills and are classified according to their duration of action. Because it is quite easy to put almost any conceivable R group onto diethyl malonate, several thousand derivatives of barbituric acid have been synthesized. Studies of these derivatives have shown that as the alkyl chain (R) gets longer or as double bonds are introduced into the chain, the duration of action and the time of onset of action decreases. The maximum sedation occurs when the alkyl chains contain five or six carbons, as found in amobarbital (R = ethyl, R′ = 3-methylbutyl) and pentobarbital (R = ethyl, R′ = 1-methylbutyl).

**Amobarbital**          **Pentobarbital**          **Phenobarbital**

These two molecules illustrate how subtle changes in molecular structure can affect action. Amobarbital requires 30 min to take effect and sedation lasts for 5–6 h, whereas pentobarbital takes effect in 15 min and sedation lasts only 2–3 h. Phenobarbital (R = ethyl, R′ = phenyl), on the other hand, requires over 1 h to take effect, but sedation lasts for 6–10 h. When the alkyl chains are made much longer, the sedative properties decrease, and the substances become anticonvulsants, which are used to treat epileptic seizures. If the alkyl group is too long or is substituted at one of the two nitrogens, convulsants are produced.

The usual dosage is 10–50 mg per pill. The continuous use of barbiturates leads to physiologic dependence (addiction), and withdrawal symptoms are identical to those experienced by a heroin addict. Replacing the oxygen atom at C-2 with sulfur results in the compound pentothal. Administered intravenously as the sodium salt, pentothal is a fast-acting general anesthetic. Like all general anesthetics, the details of its mode of action are unknown. In low, subanesthetic doses sodium pentothal reduces inhibitions and the will to resist and thus functions as the so-called truth serum, apparently because it takes less mental effort to tell the truth than to prevaricate.

Most barbiturates are made from diethyl malonate. The methylene protons between the two carbonyl groups are acidic ($pK_a$ = 13.3), and resonance will give a highly stabilized enolate anion:

**Sodium pentothal**

The acidic protons can be removed with a strong base, most often sodium ethoxide in dry ethanol. In Experiment 1 carbonate functions as the strong base because, in the absence of water, it is not solvated. In association with the phase-transfer catalyst tricaprylmethylammonium ion, the carbonate is soluble in the organic phase and can react with the diethyl malonate.

**Tricaprylmethylammonium chloride**
Phase transfer catalyst

Soluble in organic phase

The enolate anion of diethyl malonate can be alkylated by an $S_N2$ displacement of bromide to give diethyl *n*-butylmalonate:

Also, because the product still contains one acidic proton, the process can be repeated using the same or a different alkyl halide. The previous product could be hydrolyzed and decarboxylated to give a carboxylic acid:

In Experiment 3 the diethyl *n*-butylmalonate is allowed to react with urea in the presence of a strong base, sodium ethoxide, to give the barbiturate:

*n*-Butylbarbituric
acid

Barbiturates with only one alkyl substituent are rapidly degraded by the body and are therefore physiologically inactive.

## EXPERIMENTS

## 1. Diethyl *n*-Butylmalonate

**Diethyl malonate**
MW 160.17
bp 199.3°C
$n_D^{20}$ 1.4140

**Tricaprylmethylammonium chloride**
MW 404.17

$CH_3CH_2CH_2CH_2Br$
**1-Bromobutane**
MW 137.03
bp 101.6°C
$n_D^{20}$ 1.4390

**Diethyl *n*-butylmalonate**
MW 216.28
bp 235–240°C
$n_D^{20}$ 1.4220

To check reagents, the instructor should run this experiment before assigning it to the class.

**IN THIS EXPERIMENT** potassium carbonate combines with a quaternary ammonium salt to form a strong base that forms an anion from an ester of malonic acid. This anion reacts with bromobutane, also present in the reaction mixture, to give the product, a butyl derivative of the malonic ester. This ester is isolated by adding the reaction mixture to water and extracting with ether. The ether is dried and evaporated to give the product.

⚠ **CAUTION**: Tricapryl-methylchloride (Aliquat 336) is a toxic irritant. Handle with care.

**Online Study Center**

Video: Extraction with Ether

Into a 5-mL long-necked, round-bottomed flask weigh 40 mg of tricapryl-methylammonium chloride, 400 mg of diethyl malonate, 350 mg of 1-bromo-butane, and 415 mg of anhydrous potassium carbonate. Attach the empty distillation column as an air condenser and reflux the mixture for 1.5 h (Fig. 42.1). Allow the mixture to cool somewhat; then cool further in ice. Using 2.5 mL of water, transfer the reaction mixture to a reaction tube, shake the mixture, draw off the organic layer, and extract the aqueous layer with two 1.5-mL portions of ether. Dry the combined organic extracts over anhydrous calcium chloride pellets or sodium sulfate, remove the ether solution from the drying agent, and complete the drying of the ether over Drierite (commercial anhydrous calcium sulfate) or molecular sieves. Place the dry ether solutions in a *dry*, tared 5-mL round-bottomed, long-necked flask and remove the ether by distillation or evaporation in the hood.

■ **FIG. 42.1**
A microscale reflux apparatus.

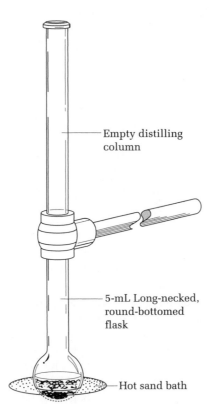

Empty distilling column

5-mL Long-necked, round-bottomed flask

Hot sand bath

Complete the removal of the ether by connecting the flask to the aspirator. Determine the weight and calculate the yield of product. Analyze the product by thin-layer chromatography (TLC) on silica gel using dichloromethane as the eluent. Compare the product to a sample of authentic, pure diethyl *n*-butylmalonate and to the starting materials diethyl malonate and bromobutane. The sample must be pure to be used as a starting material in the next reaction. It must be perfectly clear in appearance and absolutely dry. It can also be analyzed by gas chromatography using a Carbowax column at 160°C.

**Cleaning Up.** The aqueous layer after dilution with water can be flushed down the drain. The drying agents, once free of ether, can be placed in the nonhazardous solid waste container. Ether recovered from the distillation is placed in the organic solvents container, and the dichloromethane goes in the halogenated organic solvents container.

# 2. Diethyl Benzylmalonate

$$C_2H_5OCCH_2COC_2H_5 + K_2CO_3 \xrightarrow{CH_3\overset{+}{N}[(CH_2)_7CH_3]_3Cl^-} C_2H_5O-\overset{O}{\overset{\|}{C}}-\overset{-}{CH}\overset{O}{\overset{\|}{C}}OC_2H_5 + H_2O + CO_2$$

**Diethyl malonate**
MW 160.17
bp 199.3°C
$n_D^{20}$ 1.4140

**Tricaprylmethylammonium chloride**
MW 404.17

CH₂Cl

**Benzyl chloride**
MW 126.59
bp 180°C
$n_D^{20}$ 1.5380

$$C_2H_5O\overset{O}{\overset{\|}{C}}CH\overset{O}{\overset{\|}{C}}OC_2H_5$$
$$|$$
$$CH_2$$
$$|$$
$$C_6H_5$$

**Diethyl benzylmalonate**
MW 250.24
bp 162–163°C/10 mm
$n_D^{20}$ 1.4860

**⚠ CAUTION**: Tricaprylmethylammonium chloride (Aliquat 336) is a toxic irritant. Handle with care.

To check reagents, the instructor should run this experiment before assigning it to class.

Into a 50-mL round-bottomed flask weigh 0.8 g of tricaprylmethylammonium chloride, 8.00 g of diethyl malonate, 13.8 g of benzyl chloride, and 8.3 g of anhydrous potassium carbonate. Attach a water-cooled reflux condenser and reflux the reaction mixture for 1.5 h (Fig. 42.2). Allow the mixture to cool somewhat and then transfer it, using 30 mL of water to rinse out the flask, to a small separatory funnel. Shake the mixture thoroughly and draw off the aqueous layer after the layers have separated. Place the organic layer in a 50-mL Erlenmeyer flask and extract the aqueous layer with two 10-mL portions of ether. Dry the combined extracts over anhydrous calcium chloride pellets (about 10 g), decant the liquid into a 100-mL round-bottomed flask, and evaporate the solvent on a

**■ FIG. 42.2**
A reflux apparatus for macroscale experiments.

Water out

Water in

rotary evaporator. Distill the residue from a 25-mL round-bottomed flask, collecting the material that boils from about 150°C to 165°C at 10 mm pressure. Use the apparatus depicted in Figure 6.7 (on page 119). The product, diethyl benzylmalonate, is reported to boil at 162°C–163°C/10 mm. See the nomograph (Fig. 6.19 on page 128) to determine the boiling point at a different pressure.

**Cleaning Up.** The aqueous layer after dilution with water can be flushed down the drain. Calcium chloride, once free of the ether (evaporate in the hood), can be placed in the nonhazardous solid waste container. Ether recovered from the distillation is placed in the organic solvents container. If local regulations do not allow for the evaporation of solvents in a hood, the wet solid should be disposed in a special container.

$$\underset{\substack{\textbf{Urea}\\ \text{MW 60.06}\\ \text{mp 133–135°C}}}{\text{H}_2\text{NCNH}_2} + \underset{\substack{\textbf{Diethyl \textit{n}-butylmalonate}\\ \text{MW 216.28}\\ \text{bp 235–240°C}}}{\text{CH}_3\text{CH}_2\text{OCCHCOCH}_2\text{CH}_3} \xrightarrow{\text{Na}^+\bar{\text{O}}\text{CH}_2\text{CH}_3} \underset{\substack{\textbf{\textit{n}-Butylbarbituric acid}\\ \text{MW 184.18}\\ \text{mp 209–210°C}}}{}$$

# 3. *n*-Butylbarbituric Acid

## Experimental Considerations

In this experiment, absolute (anhydrous, 100%) ethanol is reacted with sodium metal to prepare a solution of sodium ethoxide in the 5-mL flask. (Commercially prepared sodium ethoxide is often not active.) The empty distilling column functions as an air condenser, and moisture is prevented from diffusing into the apparatus by capping the distilling column with a septum and an empty syringe needle.

The sodium metal may come in the form of small spheres or a large block stored under mineral oil. If in block form, it must be shaved off with a knife and transferred to a beaker containing light mineral oil or hexanes. Use tweezers to handle the sodium, blot it off, and then transfer it to a dry beaker for weighing. Blot off each piece as it is added to the reaction flask down the condenser.

Be extremely careful to have no water near the sodium. If a piece of sodium is accidentally dropped, place it in a beaker containing 1-propanol (see "Cleaning Up" on page 583). Do not dispose of it in a trash can because it can easily start a fire due to oxidation.

## Microscale Procedure

> **IN THIS EXPERIMENT** sodium ethoxide is prepared from sodium and ethanol, which is then reacted with *n*-butylmalonate and urea to produce a barbiturate. The product is isolated by acidification and evaporation until crystallization occurs. The solid is collected on a Hirsch funnel, washed with hexanes, and recrystallized from water.

To a perfectly dry (oven- or flame-dried) 5-mL round-bottomed, long-necked flask add 2.5 mL of absolute ethanol. Add to the flask the empty distilling column capped with a septum and an empty syringe needle. Add to the ethanol 23 mg of sodium metal. Wait for all of the sodium to react; do not proceed until all traces of the metal have disappeared. To the sodium ethoxide solution add 216 mg of dry *n*-butylmalonate (synthesized in Experiment 1 or purchased commercially[1]). Immediately add 60 mg of urea that have been dissolved in 1.1 mL of absolute ethanol. Reflux the resulting mixture on a warm sand bath or a steam bath for 2 h. Which solid separates during this reaction? It may cause the reaction mixture to bump during the first part of the refluxing period, so be sure to clamp the apparatus firmly, as always.

At the completion of the reaction, acidify the solution with 1 mL of 3 *M* hydrochloric acid (check pH with indicator paper) and then reduce the volume in the flask to one-half by distilling the ethanol or by removing the solvent on a rotary evaporator. Cool the solution on ice until no more product crystallizes; then collect it by vacuum filtration on a Hirsch funnel. Empty the filter flask; then wash unreacted diethyl *n*-butylmalonate (detected by its odor) from the crystals with hexanes, press the crystals dry, and then recrystallize the product from boiling

*Online Study Center*

Video: Microscale Filtration on the Hirsch Funnel

---

1. *n*-Butylmalonate is available commercially from Aldrich Chemical Company.

water (20 mL of water per gram of product). The acid crystallizes slowly. Cool the solution for at least 30 min before collecting the product by vacuum filtration. Dry the barbituric acid on filter paper until the next laboratory period; then weigh it, calculate the percent yield, and determine the melting point. Pure *n*-butylbarbituric acid melts at 209°C–210°C.

Wet all paper that has come in contact with sodium metal with alcohol before disposal. This will decompose traces of the sodium metal.

**Cleaning Up.** Add scrap sodium metal to 1-propanol in a beaker. After all traces of the sodium have disappeared, the alcohol can be diluted with water, neutralized with dilute hydrochloric acid, and flushed down the drain. Combine aqueous filtrates and mother liquors from the crystallization, neutralize with sodium carbonate, and flush down the drain. The ligroin wash goes in the organic solvents container.

# 4. Benzylbarbituric Acid

## Experimental Considerations

In this experiment, absolute (anhydrous, 100%) ethanol is reacted with sodium metal to prepare a solution of sodium ethoxide in the 250-mL flask. (Commercially prepared sodium ethoxide is often not active.) Moisture is prevented from diffusing into the apparatus by capping the condenser with a drying tube. The drying tube is prepared by first placing a piece of cotton in the tube, followed by calcium chloride and another firmly placed piece of cotton. At the end of the experiment, dispose of the calcium chloride or cap each end of the drying tube very carefully with corks. The calcium chloride is highly hygroscopic.

The sodium metal may come in the form of small spheres or a large block stored under mineral oil. If in block form, it must be shaved off with a knife and transferred to a beaker containing light mineral oil or hexanes. Use tweezers to handle the sodium, blot it off, and then transfer it to a dry beaker for weighing. Blot off each piece as it is added to the reaction flask down the condenser.

Be extremely careful to have no water near the sodium. If a piece of sodium is accidentally dropped, place it in a beaker containing 1-propanol (see "Cleaning Up" on page 584). Do not dispose of it in a trash can because it can easily start a fire due to oxidation.

## Procedure

To a perfectly dry 250-mL round-bottomed flask add 50 mL of absolute ethanol; then add a water-cooled condenser and a calcium chloride drying tube (Fig. 42.3). Add to the ethanol 0.46 g (0.2 mol) of sodium metal. Warm the flask on a steam bath to dissolve all of the sodium; do not proceed until all traces of the metal have disappeared. To the sodium ethoxide solution add 5.00 g (0.2 mol) of dry *n*-benzylmalonate (synthesized in Experiment 2 or purchased commercially[2]). Immediately add a hot solution of 1.2 g (0.2 mol) of dry urea that has been dissolved in 22 mL of absolute ethanol. Reflux the resulting mixture on a steam bath

2. *n*-Benzylmalonate is available from Aldrich Chemical Company.

■ **FIG. 42.3**
A reflux apparatus with a calcium chloride drying tube.

Water out

Water in

or a heating mantle for 2 h. Which solid separates during this reaction? It may cause the reaction mixture to bump during the first part of the refluxing period, so be sure to clamp the apparatus firmly, as always.

At the completion of the reaction, acidify the solution with 20.0 mL of 3 *M* hydrochloric acid (check pH with indicator paper) and then reduce the volume to one-half by removing the ethanol either by distillation or by evaporation on a rotary evaporator. Cool the solution in an ice bath until no more product crystallizes; then collect it by vacuum filtration on a Büchner funnel. Empty the filter flask; then wash unreacted diethyl benzylmalonate (detected by its odor) from the crystals with hexanes, press the crystals dry, and recrystallize the product from boiling water. The acid crystallizes slowly. Cool the solution for at least 30 min before collecting the product by vacuum filtration. Dry the barbituric acid on filter paper until the next laboratory period; then weigh it, calculate the percent yield, and determine the melting point. Pure benzylbarbituric acid melts at 206°C.

**Cleaning Up.** Add scrap sodium metal to 1-propanol in a beaker. After all traces of sodium have disappeared, the alcohol can be diluted with water, neutralized with dilute hydrochloric acid, and flushed down the drain. Combine aqueous filtrates and mother liquors from the crystallization, neutralize with sodium carbonate, and flush down the drain. The hexanes wash goes in the organic solvents container.

## QUESTIONS

1. The anion of diethyl benzylmalonate is often made by reacting the diester with sodium methoxide. What weight of sodium would be required in Experiment 2 if that method were employed?
2. Outline all of the steps in the synthesis of pentobarbital.
3. What problem might be encountered if 1-bromo-2,2-dimethylpropane, $BrCH_2C(CH_3)_3$, were used as the halide in this synthesis?

# CHAPTER

# 43

# Amines

 **Online Study Center**

This icon will direct you to techniques, equipment setups, and online resources at http://college.hmco.com/PIC/williamsonMMOE5e.

**PRELAB EXERCISE:** Explain why an ordinary amide from a primary amine does not dissolve in alkali, whereas the corresponding sulfonamide of the amine does dissolve.

**A**mines are weak bases and may be regarded as organic substitution products of ammonia. Just as ammonia reacts with acids to form the ammonium ion, so amines react with acid to form the organoammonium ion:

$$\ddot{N}H_3 + H_3\ddot{O}^+ \rightleftharpoons \overset{+}{N}H_4 + H_2\ddot{O}:$$

**Ammonia**    **Ammonium ion**

$$CH_3\ddot{N}H_2 + H_3\ddot{O}^+ \rightleftharpoons CH_3\overset{+}{N}H_3 + H_2\ddot{O}:$$

**Methylamine**    **Methyl-ammonium ion**

When an amine dissolves in water, the following equilibrium is established:

$$R\ddot{N}H_2 + H_2O \rightleftharpoons R\overset{+}{N}H_3 + OH^-$$

and from this the basicity constant can be defined as

$$K_b = \frac{[RNH_3^+][OH^-]}{[RNH_2]}$$

Strong bases have larger values of $K_b$, meaning the amine has a greater tendency to accept a proton from water to increase the concentration of $RNH_3^+$ and $OH^-$. The basicity constant ($K_b$) for ammonia is $1.8 \times 10^{-5}$.

$$\ddot{N}H_3 + H_2O \rightleftharpoons NH_4^+ + OH^-$$

$$K_b = 1.8 \times 10^{-5} = \frac{[NH_4^+][OH^-]}{[NH_3]}$$

Alkyl amines such as methylamine ($CH_3NH_2$) or *n*-propylamine ($CH_3CH_2CH_2NH_2$) are stronger bases than ammonia because the alkyl groups are electron donors and increase the effective electron density on nitrogen. Aromatic amines, on the other hand, are weaker bases than ammonia. Delocalization of the unshared pair of electrons on nitrogen onto the aromatic ring means that the electrons cannot be shared with an acidic proton.

Low molecular weight amines have a powerful fishy odor. Slightly higher molecular weight diamines have names suggestive of their odors: $H_2N(CH_2)_5NH_2$ is cadaverine; and $H_2N(CH_2)_4NH_2$ is putrescine. The lower molecular weight amines with up to about five carbon atoms are soluble in water. Many amines, especially the liquid aromatic amines, undergo light-catalyzed free-radical air oxidation reactions to give a large variety of highly colored decomposition products. The higher molecular weight amines that are insoluble in water will dissolve in acid to form ionic amine salts. This reaction is useful for both characterization and separation purposes. The ionic amine salt on treatment with base will regenerate the amine.

Nitriles and amides are not basic.

Nitriles, $R—C\equiv N$, are not basic and neither are amides or substituted amides. The adjacent electron-withdrawing carbonyl oxygen effectively removes the unshared pair of electrons on nitrogen:

The object of the experiments in this chapter is to identify an unknown amine or amine salt. Procedures for solubility tests are given in Experiment 1. Experiment 2 presents the Hinsberg test for distinguishing between primary, secondary, and tertiary amines; Experiment 3 gives procedures for preparing solid derivatives for melting-point characterizations; and Experiment 4 gives spectral characteristics. You will apply the procedures to known substances along with an unknown substance.

## EXPERIMENTS

# 1. Solubility

**Pyridine**

**p-Toluidine**

**Online Study Center**

Video: Instant Microscale
Distillation

⚠️

Use caution in smelling
unknowns and do so only once.

If necessary, this experiment can be scaled up by a factor of 2 or 3.

The following substances will be tested:

Aniline, $C_6H_5NH_2$ (bp 184°C)

p-Toluidine, $CH_3C_6H_4NH_2$ (mp 43°C)

Pyridine, $C_5H_5N$ (bp 115°C; a tertiary amine)

Methylamine hydrochloride, dec. (salt of $CH_3NH_2$; bp 6.7°C)

Aniline hydrochloride, dec. (salt of $C_6H_5NH_2$)

Aniline sulfate, dec.

Amines are often contaminated with dark oxidation products. Amines can be easily purified by a small-scale distillation (Fig. 43.1).

First, determine if the substance has a fishy, ammonia-like odor; if so, it probably is an amine of low molecular weight. Then test its solubility in water by putting one drop if a liquid or an estimated 10 mg if a solid into a reaction tube or a small glass centrifuge tube, adding 0.1 mL of water, and seeing if the substance dissolves when cooled. If the substance is a solid, rub it well with a stirring rod and break up any lumps before drawing a conclusion.

If the substance is *readily soluble in cold water* and if the odor is suggestive of an amine, test the solution with pH paper and further determine if the odor disappears by adding a few drops of 3 *M* hydrochloric acid. If the properties are more like those of a salt, add a few drops of 3 *M* sodium hydroxide solution. If the solution remains clear, adding a small amount of sodium chloride may cause separation of a liquid or solid amine.

If the substance is not soluble in cold water, see if it will dissolve on heating; be careful not to mistake the melting of a substance for dissolving. If it *dissolves in hot water*, add a few drops of 3 *M* alkali and see if an amine precipitates. (If you are in doubt as to whether a salt has dissolved partially or not at all, pour off the supernatant liquid and make it basic.)

If the substance is *insoluble in hot water*, add 3 *M* hydrochloric acid, heat if necessary, and see if it dissolves. If so, make the solution basic and see if an amine precipitates.

**Cleaning Up.** All solutions should be combined, made basic to free the amines, extracted with an organic solvent such as hexanes. The aqueous layer is flushed down the drain, and the organic layer is placed in the organic solvents container.

**■ FIG. 43.1**
The apparatus for instant microscale distillation. When the refluxing vapor is drawn into the pipette, it will condense to a clear liquid, which is then squirted into the empty reaction tube. Both tubes can be held in one hand.

## 2. Hinsberg Test

If necessary, this experiment can be doubled in scale.

The procedure for distinguishing amines with benzenesulfonyl chloride is to be run in parallel on the following substances:

Aniline, $C_6H_5NH_2$ (bp 184°C)

$N$-Methylaniline, $C_6H_5NHCH_3$ (bp 194°C)

Triethylamine, $(CH_3CH_2)_3N$ (bp 90°C)

Primary and secondary amines react in the presence of alkali with benzenesulfonyl chloride ($C_6H_5SO_2Cl$) to give sulfonamides.

The sulfonamides are distinguishable because the derivative from a primary amine has an acidic hydrogen, which renders the product soluble in alkali (reaction **1**); whereas the sulfonamide from a secondary amine is insoluble (reaction **2**). Tertiary amines lack the necessary acidic hydrogen for the formation of benzenesulfonyl derivatives.

## *Procedure*

To a reaction tube or small centrifuge tube add 50 mg of the amine, 200 mg of benzenesulfonyl chloride, and 1 mL of methanol. Over a hot sand bath or a steam bath, heat the mixture to just below the boiling point, cool, and add 2 mL of 6 *M* sodium hydroxide. Shake the mixture for 5 min; then allow the tube to stand for an additional 10 min with intermittent shaking. If the odor of benzenesulfonyl chloride is detected, warm the mixture to hydrolyze it (*see* Question 7). Cool the mixture and acidify it by adding 6 *M* hydrochloric acid dropwise while stirring. If

Refluxing vapor

Boiling liquid

Boiling chip

Low molecular weight amines have very bad odors; work in the hood.

**CAUTION**: Do not allow benzenesulfonyl chloride to come in contact with the skin.

Distinguishing among primary (1°), secondary (2°), and tertiary (3°) amines

a precipitate is observed at this point, the amine is either primary or secondary. If no precipitate is observed, the amine is tertiary.

If a precipitate is present, remove it by filtration on a Hirsch funnel, wash it with 2 mL of water, and transfer it to a reaction tube. Add 2.5 mL of 1.5 $M$ sodium hydroxide solution and warm the mixture to 50°C. Shake the tube vigorously for 2 min. If the precipitate dissolves, the amine is primary. If it does not dissolve, it is secondary. The sulfonamide of a primary amine can be recovered by acidifying the alkaline solution. Once dry, these sulfonamides can be used to characterize the amine by their melting points.

**Cleaning Up.** The slightly acidic filtrate can be diluted and then flushed down the drain if the unknown is primary or secondary. If the unknown is tertiary, make the solution basic, and extract the amine with hexanes. Place the hexanes layer in the organic solvents container and flush the aqueous layer down the drain.

# 3. Solid Derivatives

Acetyl (CH$_3$C=O) derivatives of primary and secondary amines are usually solids suitable for melting-point characterization and are readily prepared by reaction with acetic anhydride, even in the presence of water. Benzoyl (C$_6$H$_5$C=O) and benzenesulfonyl (C$_6$H$_5$SO$_2$) derivatives are made by reacting the amine with the appropriate acid chloride in the presence of alkali, as in Experiment 2 (the benzenesulfonamides of aniline and of $N$-methylaniline melt at 110°C and 79°C, respectively).

Solid derivatives suitable for the characterization of tertiary amines are the methiodides and picrates:

$$R_3N + CH_3I \longrightarrow R_3\overset{+}{N}CH_3 \; I^-$$

**Methiodide**

2,4,6-Trinitrophenol (Picric acid)   $+ NR_3 \longrightarrow$   Amine picrate

Typical derivatives are to be prepared and, although a determination of melting points is not necessary because the values are given, the products should be saved for possible identification of unknowns.

## Acetylation of Aniline with Acetic Anhydride

$$\text{Aniline} \quad + \quad CH_3-\overset{\displaystyle O}{\overset{\|}{C}}-O-\overset{\displaystyle O}{\overset{\|}{C}}-CH_3 \quad \longrightarrow \quad \text{Acetanilide} \quad + \quad CH_3COOH$$

**Aniline**
MW 93.13
bp 184°C

**Acetic anhydride**
MW 102.09
bp 138–140°C

**Acetanilide**
MW 135.17
mp 114°C

With patience this reaction and the associated recrystallizations can be carried out in a melting point tube on one-tenth the indicated quantities of material using essentially the same techniques; if deemed necessary, this reaction and associated recrystallizations can also be carried out on 10–20 times as much material.

In a dry, small test tube place 46 mg (or 5 drops) of freshly distilled aniline, 51 mg (or 5 drops) of pure acetic anhydride, and a small boiling chip. Reflux the mixture for about 8 min and then add water to the mixture dropwise with heating until all of the product is in solution. This will require very little water. (Note: On a larger scale, exercise caution when adding water to the reaction mixture because the exothermic hydrolysis of acetic anhydride is slow at room temperature but can get out of control at higher temperatures.) Allow the mixture to cool spontaneously to room temperature; then cool the mixture in ice. Remove the solvent with a Pasteur pipette and recrystallize the crude material again from boiling water. Isolate the crystals in the same way: cool the tube in ice water and wash the crystals with a 1–2 drops of ice-cold acetone. Remove the acetone with a Pasteur pipette and dry the product in the test tube. Once dry, determine the weight, calculate the percent yield, and determine the melting point.

**Cleaning Up.** The slightly acidic filtrate and washings can be diluted and then flushed down the drain.

## Acetylation of Aniline Hydrochloride with Acetic Anhydride

If necessary, this experiment can be carried out on 10–20 times as much material.

Dissolve 0.26 g of aniline hydrochloride in 2.5 mL of water; then add 0.21 g of acetic anhydride followed immediately by 0.25 g of sodium acetate. Allow the reaction mixture to stir (using a magnetic stir bar and magnetic stirrer) for 10 min; then cool it in ice. Recrystallize the product from water and wash it with acetone as described in the previous acetylation. *See* Chapter 45 for the mechanism.

**Cleaning Up.** The slightly acidic filtrate can be diluted and then flushed down the drain if the unknown is a primary or secondary amine. If the unknown is a tertiary

*Online Study Center*

Photo: Filtration Using a Pasteur Pipette; Video: Filtration of Crystals Using the Pasteur Pipette amine, make the solution basic and extract the amine with hexanes. Place the hexanes layer in the organic solvents container and flush the aqueous layer down the drain.

## Formation of an Amine Picrate from Picric Acid and Triethylamine

Picric acid
MW 222.11, mp 120–122°C

$+$ $(CH_3CH_2)_3N$

**Triethylamine**
MW 101.19
bp 88.8°C

$\longrightarrow$

**Triethylamine picrate**
MW 323.30, mp 171°C

NOTE: It is preferable to use the moist reagent of picric acid (35% water) and to not allow the reagent to dry out. Your instructor may provide a stock solution containing 3 g of moist picric acid in 25 mL of methanol.

If necessary, this reaction can be doubled in scale.

Dissolve 30 mg of moist (35% water) picric acid in 0.25 mL of methanol; then add 10.1 mg of triethylamine and let the solution stand to deposit crystals of the picrate, an amine salt that has a characteristic melting point. This picrate melts at 171°C.

**Cleaning Up.** The filtrate and the product from this reaction can be disposed of by dilution with a large volume of water and flushing down the drain. Larger quantities of moist picric acid (1 g) can be reduced with tin and hydrochloric acid to the corresponding triaminophenol.[1] Dry picric acid is hazardous and should not be handled.

## 4. The NMR and IR Spectra of Amines

The proton bound to nitrogen can appear between 0.6 ppm and 7.0 ppm on the nuclear magnetic resonance (NMR) spectrum, with the position depending on the solvent, the concentration, and the structure of the amine. The peak is sometimes extremely broad owing to slow exchange and interaction of the proton with the electric quadrupole of the nitrogen. If adding a drop of $D_2O$ (heavy water) to the sample causes the peak to disappear, this is evidence for an amine hydrogen; but alcohols, phenols, and enols will also exhibit this exchange behavior. Figure 43.2 illustrates the $^1H$ NMR spectrum of aniline, in which the amine hydrogens appear as a sharp peak at 3.3 ppm. Infrared (IR) spectroscopy can also be very useful for identification purposes. Primary amines, both aromatic and aliphatic, show a weak doublet between 3300 and 3500 cm$^{-1}$ (N—H stretching) and a strong absorption between 1560 and 1640 cm$^{-1}$ due to N—H bending (Fig. 43.3). Secondary amines show a single peak between 3310 and 3450 cm$^{-1}$. Tertiary amines have no useful IR absorptions. The characteristic ultraviolet (UV) absorption shifts of aromatic amines in the presence and absence of acids are discussed in Chapter 14.

1. Armour, M.-A. *Hazardous Laboratory Chemicals Disposal Guide*; Lewis Publishers: Edmonton, Alberta, Canada, 1996.

**■ FIG. 43.2**
The $^1$H NMR spectrum
of aniline (250 MHz).

**■ FIG. 43.3**
The IR spectrum of aniline.

# 5. Unknowns[2]

**Online Study Center**

**General Resources**
Additional Experiments

Determine first if the unknown is an amine or an amine salt; then determine whether the amine is primary, secondary, or tertiary. Complete identification of your unknown may be required.

## QUESTIONS

1. How could you most easily distinguish between samples of 2-amino-naphthalene and acetanilide?
2. Would you expect the reaction product from benzenesulfonyl chloride and ammonia to be soluble or insoluble in alkali?
3. Is it safe to conclude that a substance is a tertiary amine because it forms a picrate?
4. Why is it usually true that amines that are insoluble in water are odorless?
5. Technical dimethylaniline contains traces of aniline and methylaniline. Suggest a method for eliminating these impurities.
6. How would you prepare aniline from aniline hydrochloride?
7. Write a balanced equation for the reaction of benzenesulfonyl chloride with sodium hydroxide solution in the absence of an amine. What solubility would you expect the product to have in acid and base?
8. Assign the peak at 3347 $cm^{-1}$ in the IR spectrum of aniline (Fig. 43.3).
9. In the proton noise-decoupled $^{13}C$ NMR spectra of benzenesulfonyl chloride (Fig. 43.4) and benzoyl chloride (Fig. 43.5), why is the peak of the carbon bearing the substituent so small?

■ **FIG. 43.4**
The $^{13}C$ NMR spectrum of benzenesulfonyl chloride (22.6 MHz).

The $^{13}$C NMR spectrum of
benzoyl chloride (22.6 MHz).

# The Sandmeyer Reaction: 1-Bromo-4-chlorobenzene, 2-Iodobenzoic Acid, and 4-Chlorotoluene

*Online Study Center*

This icon will direct you to techniques, equipment setups, and online resources at http://college.hmco.com/PIC/williamsonMMOE5e.

**PRELAB EXERCISE:** Using the Sandmeyer reaction as one step, outline the steps necessary to prepare 4-bromotoluene, 4-iodotoluene, and 4-fluorotoluene from benzene.

The Sandmeyer reaction is a versatile means of replacing the amine group of a primary aromatic amine with a variety of different substituents:

$$C_6H_5\overset{+}{N}\equiv N\ BF_4^- \xrightarrow{\Delta} C_6H_5F$$

HBF$_4$

CuCl $\longrightarrow$ C$_6$H$_5$Cl

CuBr $\longrightarrow$ C$_6$H$_5$Br

$$\overset{+}{N}\equiv N \quad Cl^-$$

KI $\longrightarrow$ C$_6$H$_5$I

H$_3$O$^+$, $\Delta$

H$_3$PO$_2$ $\longrightarrow$ C$_6$H$_5$OH

CuCN $\quad$ C$_6$H$_5$H

NaNO$_2$
HCl

$$\longrightarrow NH_2$$

C$_6$H$_5$C$\equiv$N

The benzene diazonium salt is formed by reacting nitrous acid with the amine in acid solution. Nitrous acid is not stable and must be prepared in situ; in strong acid it dissociates to form nitroso ions, $:\overset{..}{N}=O$, which attack the nitrogen of the amine. The intermediate so formed loses a proton, rearranges, and finally loses water to form the resonance-stabilized diazonium ion.

$$NaNO_2 + HCl \rightleftharpoons HO\overset{..}{N}O + Na^+Cl^-$$

**Sodium nitrite**          **Nitrous acid**

$$H_3O^+ + HO\overset{..}{N}O \rightleftharpoons H_2O + H_2\overset{+}{O}\overset{..}{N}O \rightleftharpoons 2\,H_2O + :\overset{+}{N}=O$$

The benzene diazonium ion is reasonably stable in aqueous solution at 0°C; on warming it will form the phenol. A versatile functional group, it will undergo all the reactions depicted on the previous page as well as coupling to aromatic rings activated with substituents such as amino and hydroxyl groups to form the huge class of azo dyes (*see* Chapter 46).

**Benzenediazonium chloride**          *N,N*-**Dimethylaniline**          **A diazo compound**

Diazonium salts are not ordinarily isolated because the dry solid is explosive.

# Three Sandmeyer Reactions

Experiments 1, 2, and 3

$$NaNO_2 + HCl \longrightarrow Na^+Cl^- + HONO$$

**Sodium nitrite**
MW 69.01

**Nitrous acid**

**4-Bromoaniline**
MW 172.03, mp 62–64°C

**4-Bromoaniline hydrochloride**

**4-Bromobenzenediazonium chloride**

**1-Bromo-4-chlorobenzene**
MW 191.46, mp 68°C

In the first microscale procedure (Experiments 1, 2, and 3), after the copper(I) chloride is prepared, 4-bromoaniline is dissolved in the required amount of hydrochloric acid, two more equivalents of acid are added, and the mixture is cooled in ice to produce a paste of the crystalline amine hydrochloride. When this salt is treated with one equivalent of sodium nitrite at 0°C–5°C, nitrous acid is generated and reacts to produce the diazonium salt. The excess hydrochloric acid (beyond the two equivalents required to form the amine hydrochloride and react with sodium nitrite) maintains sufficient acidity to prevent formation of the diazo-amino compound and rearrangement of the diazonium salt. The diazonium salt reacts with copper(I) chloride to give the easily sublimed 1-bromo-4-chlorobenzene.

$$NaNO_2 + HCl \xrightarrow{0°C} HONO + NaCl$$

**Sodium nitrite**                                    **Nitrous**
**MW 69.01**                                          **acid**

Experiments 6, 7, and 8

*p*-Toluidine
MW 107.15, mp 45°C

*p*-Toluidine
hydrochloride

*p*-Methylbenzenediazonium
chloride

**Copper(I) chloride**
MW 98.99

**4-Chlorotoluene**
MW 126.58, bp 162°C

Experiments 4 and 5

**CAUTION**: *p*-Toluidine, like many aromatic amines, is highly toxic. *o*-Toluidine is not only highly toxic but is also a suspected carcinogen.

The second procedure (Experiment 4 microscale, Experiment 5 macroscale) employs potassium iodide, which reacts with a diazonium salt to give an iodobenzoic acid.

The third procedure is a macroscale synthesis of 4-chlorotoluene (Experiments 6, 7, and 8). *p*-Toluidine is dissolved in the required amount of hydrochloric acid, two more equivalents of acid are added, and the mixture is cooled in ice to produce a paste of the crystalline amine hydrochloride. When this salt is treated with one equivalent of sodium nitrite at 0°C–5°C, nitrous acid is liberated and reacts to produce the diazonium salt. The excess hydrochloric acid, beyond the two equivalents required to form the amine hydrochloride and react with sodium nitrite, maintains sufficient acidity to prevent formation of the diazo-amino compound and rearrangement of the diazonium salt. The diazonium salt reacts with copper(I) chloride to give liquid 4-chlorotoluene.

Copper(I) chloride is made by reducing copper(II) sulfate with sodium sulfite (which is produced as required from the cheaper sodium bisulfite). The white solid is left covered with the reducing solution for protection against air oxidation until it is to be used and then dissolved in hydrochloric acid. When the diazonium salt solution is added, a complex forms and rapidly decomposes to give *p*-chlorotoluene and nitrogen. The mixture is very discolored, but steam distillation leaves behind most of the impurities and all salts and yields material substantially pure except for the presence of a trace of yellow pigment, which can be eliminated by distilling the dried oil.

## EXPERIMENTS

# 1. Copper(I) Chloride Solution

$$2\ CuSO_4 \cdot 5\ H_2O + 4\ NaCl + NaHSO_3 + NaOH \rightarrow 2\ CuCl + 3\ Na_2SO_4 + 2\ HCl + 10\ H_2O$$

MW 249.71        MW 58.45   MW 104.97   MW 40.01        MW 99.02

> **IN THIS EXPERIMENT** an aqueous solution of copper sulfate, sodium chloride, sodium bisulfite and sodium hydroxide react to give a precipitate of copper(I) chloride. Immediately before use this white salt is dissolved in hydrochloric acid.

$NaHSO_3$, sodium bisulfite (sodium hydrogen sulfite), not $Na_2S_2O_4$, sodium hydrosulfite

In a reaction tube dissolve 0.30 g of copper(II) sulfate crystals ($CuSO_4 \cdot 5\ H_2O$) in 1.0 mL of water by boiling and then add 140 mg of sodium chloride, which may give a small precipitate of basic copper(II) chloride. Prepare a solution of sodium sulfite from 70 mg of sodium bisulfite and 0.4 mL of 3 *M* sodium hydroxide solution (or use 57 mg of sodium sulfite in 0.4 mL of water). Add this solution dropwise to the hot copper(II) sulfate solution. Rinse the tube that contained the sulfite with a couple drops of water and add it to the copper(I) chloride mixture. Shake the reaction mixture well; then allow it to cool in ice while preparing the diazonium salt (Experiment 2). When you are ready to use the copper(I) chloride (in Experiment 3), remove the water above the solid, wash the white solid once with water, remove the water, and dissolve the solid in 0.45 mL of concentrated hydrochloric acid. This solution is susceptible to air oxidation and should not stand for an appreciable time before use.

**CAUTION**: Do not measure hydrochloric acid in a syringe that has a metal needle.

**Cleaning Up.** Combine the filtrate and washings, neutralize with sodium carbonate, and flush down the drain.

# 2. Diazotization of 4-Bromoaniline

> **IN THIS EXPERIMENT** an ice-cold acidic solution of an aromatic amine is treated with solid sodium nitrite to give a solution of the diazonium salt.

In a 10-mL Erlenmeyer flask place 172 mg of 4-bromoaniline and 0.50 mL of 3 *M* hydrochloric acid. Warm the mixture on a sand bath to dissolve the amine and ensure transformation into hydrochloride. Cool the flask in ice (the hydrochloride will crystallize) and add an ice-cold solution of 70 mg of fresh sodium nitrite in 0.2 mL of water. Use a drop of water to complete the transfer of the nitrite solution. Mix the solid and the sodium sulfite thoroughly. The result should be a homogeneous yellow solid/liquid mixture. Do not allow this mixture to warm above 0°C.

# 3. Sandmeyer Reaction: Synthesis of 1-Bromo-4-Chlorobenzene

**IN THIS EXPERIMENT** the cold copper(I) chloride solution is combined with the cold diazonium salt solution made from 4-bromoaniline in Experiment 2. Nitrogen gas bubbles off, and the diazonium ion is replaced with a chlorine atom to give the product, 1-bromo-4-chlorobenzene.

*Online Study Center*

Video: Microscale Filtration on the Hirsch Funnel; Photo: Sublimation Apparatus

Cool the copper(I) chloride solution (from Experiment 1) in ice and add it to the diazonium chloride solution dropwise with a Pasteur pipette and with thorough mixing. Rinse the tube that contained the copper(I) chloride with a drop of water. Allow the reaction mixture to warm to room temperature and observe the reaction mixture closely. It bubbles as nitrogen gas is evolved. Heat the mixture on a steam or a sand bath to complete the reaction. Cool the mixture thoroughly in ice and collect the dark solid on a Hirsch funnel. Squeeze the product between sheets of filter paper to dry it and then sublime the bromochlorobenzene on a steam or sand bath. After initial sublimation, scrape the residue in the filter flask to sublime more product.

The extraction can be carried out in the microscale separatory funnel (*see* Fig. 7.7 on page 145).

Another way to isolate the product is to extract the product with three 0.5-mL portions of ether and wash the ether extract with 0.2 mL of 3 *M* sodium hydroxide solution, which will remove any 4-bromophenol, and then with 0.5 mL of saturated sodium chloride solution. Dry the ether solution over anhydrous calcium chloride pellets. Add sufficient drying agent so that it no longer clumps together. After 5–10 min, remove the ether with a Pasteur pipette, wash the drying agent with ether, and evaporate the ether solutions in a small, tared filter flask. Use care in this evaporation because the product is volatile. Determine the weight of the crude product; then purify it by sublimation at atmospheric pressure.

*Online Study Center*

Video: Extraction with Ether, Filtration of Crystals Using the Pasteur Pipette; Photo: Filtration Using a Pasteur Pipette

The 15-mL centrifuge tube is filled with ice water while the filter flask is heated on a sand or steam bath (refer to the sublimation apparatus in Fig. 6.21 on page 130). Close the sidearm with a rubber pipette bulb. Roll the flask in the sand and/or use a heat gun to drive product from the sides of the flask to the centrifuge tube. Replace the ice water in the centrifuge tube with room-temperature water before removing it so that moisture will not condense on the tube. Scrape the product onto a piece of weighing paper, determine the weight and the melting point, and analyze it for purity by thin-layer chromatography (TLC) and infrared (IR) spectroscopy in a chloroform solution.

*Online Study Center*

Photo: Sublimation Apparatus

**Cleaning Up.** Combine all aqueous filtrates and washings, neutralize with sodium carbonate, and flush down the drain. Allow the ether to evaporate from the drying agent in the hood before the drying agent is placed in the nonhazardous solid waste container. If local regulations do not allow for the evaporation of solvents in a hood, the wet solid should be disposed in a special container.

# 4. Sandmeyer Reaction: Synthesis of 2-Iodobenzoic Acid

**2-Iodobenzoic acid**
MW 248.02
mp 162–164°C

**2-Aminobenzoic acid**
MW 137.14, mp 144–148°C

**CAUTION:** Do not measure hydrochloric acid in a syringe that has a metal needle.

To a reaction tube containing 137 mg of 2-aminobenzoic acid (anthranilic acid), add 1 mL of water and 0.25 mL of concentrated hydrochloric acid. Heat to dissolve the amino acid and form the hydrochloride; then cool the tube in ice. Cap the tube with a septum; then, with the tube in an ice bath, add a solution of 75 mg of sodium nitrite dissolved in 0.3 mL of water using a syringe equipped with a needle. This addition should be made dropwise with thorough agitation or magnetic stirring of the reaction. Rinse the syringe with 0.1 mL of water, which is also injected into the reaction mixture. After 5 min, a solution of 0.17 g of potassium iodide in 0.25 mL of water is added when a brown complex partially separates.

An empty distilling column is added to the top of the reaction tube to catch any foam. The mixture is allowed to stand without cooling for 5 min and then cautiously warmed to 40°C. At this point a vigorous reaction begins, with nitrogen gas evolution, foaming, and separation of a tan-colored solid. After reacting for 10 min, the mixture is heated in a beaker of boiling water for 10 min and then cooled in ice. A few milligrams of sodium sulfite are added to destroy any iodine present, and the granular tan-colored product is collected and washed with water.

The still-moist product is dissolved in 0.7 mL of ethanol; 0.35 mL of water is added; and the hot, brown solution is treated with enough granular Norit to remove most of the color. The solution is transferred to another reaction tube to remove the decolorizing charcoal, diluted at the boiling point with 0.35–0.40 mL of water, and allowed to stand. 2-Iodobenzoic acid separates in large, slightly yellow needles. The yield should be about 150 mg with a melting point near 164°C.

**Cleaning Up.** The reaction mixture filtrate and mother liquor from the crystallization are combined, neutralized with sodium carbonate, and flushed down the drain. Norit is placed in the nonhazardous solid waste container.

***Online Study Center***

Video: Microscale
Crystallization

# 5. Sandmeyer Reaction: Synthesis of 2-Iodobenzoic Acid

**2-Aminobenzoic acid
(Anthranilic acid)**
MW 137.14, mp 144–148°C

**2-Iodobenzoic acid**
MW 248.02, mp 162–164°C

*Online Study Center*

Video: Macroscale
Crystallization

A 100-mL round-bottomed flask containing 3.4 g of anthranilic acid, 25 mL of water, and 6 mL of concentrated hydrochloric acid is heated until the solid is dissolved. The mixture is then cooled in ice while bubbling in nitrogen to displace the air. When the temperature reaches 0°C–5°C, a solution of 1.8 g of sodium nitrite in 6 mL of water is added slowly. After 5 min, a solution of 4.25 g of potassium iodide in 6 mL of water is added when a brown complex partially separates. This mixture is allowed to stand without cooling for 5 min (under nitrogen) and then warmed to 40°C, at which point a vigorous reaction ensues (gas evolution and separation of a tan-colored solid). After reacting for 10 min, the mixture is heated on a steam or sand bath for 10 min and then cooled in ice. A pinch of sodium bisulfite is added to destroy any iodine present, and the granular tan-colored product is collected and washed with water. The still-moist product is dissolved in 18 mL of 95% ethanol; 9 mL of hot water is added; and the brown solution is treated with decolorizing charcoal, filtered, diluted at the boiling point with 9–10 mL of water, and allowed to stand. 2-Iodobenzoic acid separates in large, slightly yellow needles (mp 164°C); yield is approximately 4.3 g (72%).

**Cleaning Up.** The reaction mixture, filtrate, and mother liquor from the crystallization are combined, neutralized with sodium carbonate, and flushed down the drain with a large excess of water. Decolorizing charcoal is placed in the nonhazardous solid waste container.

# 6. Copper(I) Chloride Solution

$$2\ CuSO_4 \cdot 5\ H_2O + 4\ NaCl + NaHSO_3 + NaOH \rightarrow 2\ CuCl + 3\ Na_2SO_4 + 2\ HCl + 10\ H_2O$$
MW 249.71        MW 58.45   MW 104.97   MW 40.01        MW 99.02

$NaHSO_3$, sodium bisulfite (sodium hydrogen sulfite), not $Na_2S_2O_4$, sodium hydrosulfite

In a 250-mL round-bottomed flask (to be used later for steam distillation) dissolve 15 g of copper(II) sulfate crystals ($CuSO_4 \cdot 5\ H_2O$) in 50 mL of water by boiling; then add 7 g of sodium chloride, which may give a small precipitate of basic copper(II) chloride. Prepare a solution of sodium sulfite from 3.5 g of sodium bisulfite, 2.25 g of sodium hydroxide, and 25 mL of water; add this, not too rapidly, to the hot copper(II) sulfate solution (rinse flask and neck). Shake well, put the flask in a pan of cold water in a slanting position favorable for decantation, and

One may stop here.

let the mixture stand to cool and settle during the diazotization. When you are ready to use the copper(I) chloride for Experiment 8, decant the supernatant liquid, wash the white solid once with water by decantation, and dissolve the solid in 23 mL of concentrated hydrochloric acid. This solution is susceptible to air oxidation and should not stand for an appreciable time before use.

 ## 7. Diazotization of *p*-Toluidine

> **IN THIS EXPERIMENT** a solid amine is dissolved in hydrochloric acid. The ice-cold solution is treated with solid sodium nitrite to diazotize the amine.

**⚠**

**CAUTION**: *p*-Toluidine is toxic. Handle with care.

Put 5.5 g of *p*-toluidine and 7.5 mL of water in a 50-mL Erlenmeyer flask. Measure 12.5 mL of concentrated hydrochloric acid and add 5 mL of it to the flask. Heat over a hot plate and swirl to dissolve the amine to ensure that it is all converted into the hydrochloride. Add the remaining acid, cool thoroughly in an ice bath, and let the flask stand in the bath while preparing a solution of 3.5 g of sodium nitrite in 10 mL of water. To maintain a temperature of 0°C–5°C during diazotization, add a few pieces of ice to the amine hydrochloride suspension and add more later as the first ones melt. Pour in the nitrite solution in portions during 5 min with swirling in the ice bath. The solid should dissolve to a clear solution of the diazonium salt. After 3–4 min, test for excess nitrous acid: Dip a stirring rod in the solution, touch off the drop on the wall of the flask, put the rod in a small test tube, and add a few drops of water. Then insert a strip of starch-iodide paper; an instantaneous deep-blue color due to a starch-iodine complex indicates the desirable presence of a slight excess of nitrous acid. (The sample tested is diluted with water because strong hydrochloric acid alone produces the same color on starch-iodide paper, after a slight induction period.) Leave the solution in an ice bath.

 ## 8. Sandmeyer Reaction: Synthesis of 4-Chlorotoluene

> **IN THIS EXPERIMENT** the ice-cold diazonium salt solution from Experiment 7 is poured into the copper(I) chloride solution prepared in Experiment 6. Nitrogen gas is evolved, and the product, 4-chlorotoluene, separates as oil. It is isolated by steam distillation; the distillate is extracted with ether; and the ether is washed, dried, and evaporated to leave crude product that is purified by simple distillation.

Complete the preparation of copper(I) chloride solution from Experiment 6, cool it in an ice bath, pour in the solution of diazonium chloride through a long-stemmed funnel, and rinse the flask. Swirl occasionally at room temperature for 10 min. You should observe initial separation of a complex of the two components followed by its decomposition with the liberation of nitrogen and the separation of an oil. Arrange for steam distillation (*see* Fig. 6.4 on page 112) or generate steam in situ simply by boiling the flask contents with a Thermowell using the apparatus for simple distillation (*see* Fig. 5.10 on page 100; add more water during the distillation). Do not start the distillation until bubbling in the mixture has practically ceased and an oily layer has separated. Then steam distill and note that *p*-chlorotoluene, although lighter than the solution of inorganic salts in which it was produced, is heavier (den. 1.07) than water. Extract the distillate with a little ether, wash the extract with 3 *M* sodium hydroxide solution to remove any *p*-cresol ($CH_3C_6H_4OH$) present, and then wash with saturated sodium chloride solution. Dry the ether solution over about 2.5 g of anhydrous calcium chloride pellets and filter or decant it into a tared flask, evaporate the ether, and determine the yield and percent yield of product (your crude yield should be about 4.5 g). Pure *p*-chlorotoluene, the IR and nuclear magnetic resonance (NMR) spectra of which are shown in Figures 44.1 through 44.3, is obtained after simple distillation of this crude product. Analyze your crude and distilled product by TLC (using a 3:1 mixture of hexane and dichloromethane as the eluent), IR spectroscopy, and gas chromatography using a Carbowax column. Is it pure?

**Cleaning Up.** Combine the pot residue from the steam distillation with the aqueous washings, neutralize with sodium carbonate, dilute with water, and flush down the drain. Allow the ether to evaporate from the drying agent in the hood; then place the drying agent in the nonhazardous solid waste container.

⚠️ **CAUTION**: Use aspirator tube.

Dispose of copper salts and solutions in the container provided.

■ **FIG. 44.1**
The IR spectrum of *p*-chlorotoluene in carbon disulfide.

■ **FIG. 44.2**
The ¹H NMR spectrum of *p*-chlorotoluene (90 MHz).

■ **FIG. 44.3**
The $^{13}$C NMR spectrum
of *p*-chlorotoluene (22.6 MHz).

## QUESTIONS

1. Nitric acid is generated by the action of sulfuric acid on sodium nitrate. Nitrous acid is prepared by the action of hydrochloric acid on sodium nitrite. Why is nitrous acid prepared in situ rather than obtained from the reagent shelf?

2. Which byproduct would be obtained in high yield if the diazotization of *p*-toluidine were carried out at 30°C instead of 0°C–5°C?

3. How would 4-bromoaniline be prepared from benzene?

# CHAPTER
# 45

# Synthesis and Bioassay of Sulfanilamide and Derivatives

*Online Study Center*

This icon will direct you to techniques, equipment setups, and online resources at http://college.hmco.com/PIC/williamsonMMOE5e.

**PRELAB EXERCISE:** Write balanced equations for each step in the reaction sequence of the synthesis of sulfanilamide. Prepare detailed flow sheets for the three reactions. Look up the solubilities of all reagents and indicate—in separation steps—in which layer you would expect to find the desired product. Calculate the theoretical amount of ammonium hydroxide needed to react with 5 g of *p*-acetaminobenzenesulfonyl chloride to form sulfanilamide.

**P**aul Ehrlich, the father of immunology and chemotherapy, discovered Salvarsan, an arsenical "magic bullet" (a favorite phrase of his) used to treat syphilis. He hypothesized at the beginning of the 20th century that it might be possible to find a dye that would selectively stain (or dye) a bacterial cell and thus destroy it. In 1932 I.G. Farbenindustrie patented Prontosil, a new azo dye that they put through routine testing for chemotherapeutic activity when it was noted that it had a particular affinity for protein fibers like silk.

$$H_2N- \underset{\underset{NH_2}{|}}{\bigcirc} -N=N- \bigcirc -SO_2NH_2$$

**Prontosil**

Prontosil was found to be effective against streptococcal infections in mice, but somewhat surprisingly, it was ineffectual in vitro (outside the living animal). A number of other dyes were tested, but only those having the group

$$=N- \bigcirc -SO_2NH_2$$

607

were effective. French workers hypothesized that the antibacterial activity had nothing to do with the identity of the compounds as dyes but rather with the reduction of the dyes in the body to *p*-aminobenzenesulfonamide, known commonly as sulfanilamide.

$$H_2N-\phantom{}-SO_2NHR$$

**Sulfanilamide, R = H**

On the basis of this hypothesis, sulfanilamide was tested and found to be the active substance.

Because sulfanilamide had been synthesized in 1908, its manufacture was not protected by patents, so the new drug and thousands of its derivatives were rapidly synthesized and tested. When the R group in sulfanilamide is replaced with a heterocyclic ring system—for example, pyridine, thiazole, diazine, merazine, and so on—the sulfa drug so produced is often faster acting or less toxic than sulfanilamide. Although they have been supplanted for the most part by antibiotics of microbial origin, these drugs still find wide application in chemotherapy.

Unlike that of most drugs, the mode of action of the sulfa drugs is now completely understood. Bacteria must synthesize folic acid for growth. Higher animals, like humans, do not synthesize folic acid and hence must acquire it in their food. Sulfanilamide inhibits the formation of folic acid, stopping the growth of bacteria; and because the synthesis of folic acid does not occur in humans, only bacteria are affected.

A closer look at these events reveals that bacteria synthesize folic acid using several enzymes, including one called dihydropteroate synthetase, which catalyzes the attachment of *p*-aminobenzoic acid to a pteridine ring system. When sulfanilamide is present, it competes with the *p*-aminobenzoic acid (note the structural similarity) for the active site on the enzyme.

**Sulfanilamide**

**p-Aminobenzoic acid**

Pteridine part | *p*-Aminobenzoic acid part | Glutamic acid part

**Folic acid**

This activity makes it a competitive inhibitor. Once this site is occupied on the enzyme, folic acid synthesis ceases and bacterial growth stops. Folic acid can also be synthesized in the laboratory.[1]

# PART 1: The Synthesis and Bioassay of Sulfanilamide

**P**art 1 of this chapter represents a multistep synthesis of sulfanilamide, starting with nitrobenzene. Nitrobenzene is reduced with tin and hydrochloric acid to give the anilinium ion, which is converted to aniline with base:

**Anilinium ion**          **Aniline**

The acetylation of aniline in aqueous solution to give acetanilide serves to protect the amine group from reaction with chlorosulfonic acid. Furthermore, if the amine group were protonated, it would be a *meta*-director to substitution. The resulting *meta*-substituted product does not have antibacterial activity. The acetylation takes place readily in aqueous solution. Aniline reacts with acid to give the water-soluble anilinium ion. The anilinium ion reacts with the acetate ion to set up an equilibrium that liberates a small quantity of aniline:

$$C_6H_5-\ddot{N}H_2 + H^+ \longrightarrow C_6H_5-\overset{+}{\ddot{N}}H_3$$

**Aniline**                      **Anilinium ion**

1. Plante, L. T.; Williamson, K. L.; Pastore, E. J. In *Methods in Enzymology*; McCormick, D. B.; Wright, L. D., Eds.; Academic Press: New York, 1990, 66: 533.

This aniline reacts with acetic anhydride to give water-insoluble acetanilide:

**Acetanilide**

This upsets the equilibrium, releasing more aniline, which can then react with acetic anhydride to give more acetanilide.

Acetanilide reacts with chlorosulfonic acid in an electrophilic aromatic substitution:

$$2 \ ClSO_3H \rightleftharpoons O{=}\overset{OH}{\underset{O}{S}}{-}O{-}\overset{O}{\underset{O}{S}}{-}Cl + HCl$$

$$ArH + ClS_2O_6H \rightleftharpoons Ar^+$$

$$Ar^+ \longrightarrow ArSO_3H + ClSO_3H \longrightarrow ArSO_2Cl + H_2SO_4$$

$$CH_3{-}\overset{:\ddot{O}:}{\underset{H}{C}}{-}\ddot{N}{-} \cdots {-}\overset{:\ddot{O}:^-}{\underset{:\ddot{O}:^-}{S}}{-}\ddot{C}l: + 2 \ R\ddot{N}H_2 \longrightarrow CH_3{-}\overset{:\ddot{O}:}{\underset{H}{C}}{-}\ddot{N}{-} \cdots {-}\overset{:\ddot{O}:^- \ H}{\underset{:\ddot{O}:^-}{S}}{-}\overset{}{N}{-}R + R\overset{+}{N}H_2Cl$$

***p*-Acetamidobenzenesulfonamide**

The protecting amide group is removed from the *p*-acetamidobenzenesulfon-amide by acid hydrolysis. The amide group is more easily hydrolyzed than the sulfonamide group:

## The Synthetic Sequence of Sulfanilamide

**Nitrobenzene**
MW 123.11
bp 210.9°C, den. 1.12

**Aniline**
MW 93.12
bp 184.4°C, den. 1.02

**Acetanilide**
MW 135.16
mp 114°C

**p-Acetaminobenzene-
sulfonyl chloride**
MW 233.68

**p-Acetaminobenzene-
sulfonamide**
MW 214.25

**Sulfanilamide
(p-Aminobenzene-
sulfonamide)**
MW 172.20, mp 163–164°C

Nitrobenzene is reduced to aniline by tin and hydrochloric acid. A double salt with tin having the formula $(C_6H_5NH_3)_2SnCl_4$ separates partially during the reaction, and at the end it is decomposed by adding excess alkali, which converts the tin into water-soluble stannite or stannate ($Na_2SnO_2$ or $Na_2SnO_3$) and also liberates aniline.

The aniline is separated from inorganic salts and the insoluble impurities derived from the tin by steam distillation (*see* Chapter 6). The aniline layer from the steam distillate is then dried, distilled, and acetylated with acetic anhydride in aqueous solution to give acetanilide. Treatment of the acetanilide with excess chlorosulfonic acid affects substitution of the chlorosulfonyl group and affords the *para*-substituted product, *p*-acetaminobenzenesulfonyl chloride. The amino group must be protected by acetylation to permit the acid chloride group to be added to the ring. If the amine group were protonated, substitution would occur at the *meta* position. The amide nitrogen of acetanilide is not basic enough to react with chlorosulfonic acid. The next step in the synthesis is ammonolysis of the sulfonyl chloride to give *p*-acetaminobenzenesulfonamide, and the terminal step is removing the protective acetyl group to give sulfanilamide, which can be tested for antibacterial activity.

Use the total product obtained at each step as the starting material for the next step and adjust the amounts of reagents accordingly. Keep an accurate record of your working time. Aim for a high overall yield of pure final product in the shortest possible time. Study the procedures carefully before coming to the laboratory so that your work will be efficient. A combination of consecutive steps that avoids a needless isolation saves time and increases the yield.

## EXPERIMENTS

# 1. Microscale Preparation of Aniline

**Nitrobenzene**          **Aniline**

⚠️ **CAUTION**: Measure nitrobenzene (toxic) in the hood. Do not breathe the vapors. Use a nonmercury thermometer and wear gloves.

Reaction time: 0.5 h

In a 5-mL round-bottomed, long-necked flask place 1.25 g of granulated tin, 600 mg of nitrobenzene, 2.75 mL of concentrated hydrochloric acid, and a magnetic stirring bar. Have an ice bath ready to control the reaction, insert a thermometer, and let the mixture react until the temperature reaches 60°C; then cool it briefly, if necessary, so that the temperature does not rise above 62°C. Swirl the reaction mixture and maintain the temperature between 55°C and 60°C for 15 min. Remove the thermometer, rinse it with water, and fit the flask with an empty fractionating column that will function as an air condenser. Additional cooling can, if

**■ FIG. 45.1**

A microscale apparatus for refluxing a reaction mixture. Use a black Viton connector.

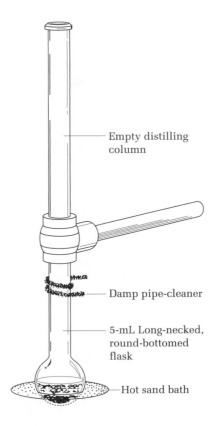

— Empty distilling column

— Damp pipe-cleaner

— 5-mL Long-necked, round-bottomed flask

— Hot sand bath

**CAUTION**: Handle aniline with care. It may be a carcinogen.

**CAUTION**: Handle the hot solution of sodium hydroxide with care; it is very corrosive.

*Online Study Center*

Video: Extraction with Ether

*Online Study Center*

Video: Steam Distillation Apparatus

necessary, be furnished by a damp pipe cleaner (Fig. 45.1). Heat the reaction mixture on a steam bath or a boiling water bath with frequent swirling or magnetic stirring until droplets of nitrobenzene are absent from the condenser and the color due to an intermediate reduction product is gone (about 15 min). During this period, dissolve 2 g of sodium hydroxide in 5 mL of water and cool the solution to room temperature.

At the end of the reduction period, transfer the acid solution to a 15-mL centrifuge tube, cool the mixture in ice, and add the sodium hydroxide solution dropwise with stirring and ice cooling. The alkali neutralizes the aniline hydrochloride, releasing aniline. Gently extract the mixture with three 1.5-mL portions of ether; place the ether extracts in the 5-mL short-necked, round-bottomed flask, which need not be clean or dry. If intractable emulsions form, centrifuge the tube for 1 min to cause the layers to separate. Transfer any emulsion layer to the flask along with the ether layer.

Construct the apparatus shown in Figure 45.2, add a boiling chip, and remove the ether (bp 55°C) by distillation. To the dirty residue of aniline and water add 2 mL of water with a syringe through the addition port. Heat the contents of the flask to boiling. A mixture of water and aniline will steam distill (*see* Chapter 6),

■ **FIG. 45.2**
A small-scale steam distillation
apparatus. Add water via a syringe.

Ice

Hot sand bath

The aniline can be isolated (*see* Experiment 2) or converted directly to acetanilide (*see* Experiment 3).

and the distillate will be cloudy as the two layers, aniline and water, separate. As the distillation proceeds, add water with a syringe through the addition port to keep the volume in the distilling flask constant. Because aniline is fairly soluble in water (3.6 g/100 g of water at 18°C), distillation should be continued somewhat beyond the point at which the distillate has lost its original turbidity (2.5–3 mL more). Make an accurate estimate of the volume of the distillate, for example, by pouring it into a 10-mL graduated cylinder.

The $^{13}$C NMR (nuclear magnetic resonance) spectrum of nitrobenzene (Fig. 45.6) and the infrared (IR; Fig. 45.7) and $^{1}$H NMR spectra of aniline (Fig. 45.8) are found at the end of the chapter.

**Cleaning Up.** Filter the pot residue from the steam distillation through a layer of Celite filter aid on a Büchner funnel. Discard the solid in the nonhazardous solid waste container. The filtrate, after neutralizing with hydrochloric acid, can be flushed down the drain.

# 2. Macroscale Preparation of Aniline

> **IN THIS EXPERIMENT** nitrobenzene is reduced with tin and hydrochloric acid. The reaction mixture is neutralized with strong base, and the aniline is isolated by steam distillation.

**CAUTION:** Measure nitrobenzene and dichloromethane (both are toxic) in the hood. Do not breathe the vapors.

**CAUTION:** Handle aniline with care. It may be a carcinogen.

Reaction time: 0.5 h

**CAUTION:** Handle the hot solution of sodium hydroxide with care; it is very corrosive.

Suitable point of interruption

The reduction of the nitrobenzene is carried out in a 500-mL round-bottomed flask suitable for steam distillation of the reaction product. Put 25 g of granulated tin and 12.0 g of nitrobenzene in the flask, make an ice-water bath ready, add 55 mL of concentrated hydrochloric acid, insert a thermometer, and swirl well to promote reaction in the three-phase system. Let the mixture react until the temperature reaches 60°C; then cool briefly in ice just long enough to prevent a rise above 60°C, so the reaction will not get out of hand. Continue to swirl, cool as required, and maintain the temperature between 55°C and 60°C for 15 min. Remove the thermometer, rinse it with water, fit the flask with a reflux condenser, and heat on a steam bath or heating mantle with frequent swirling until droplets of nitrobenzene are absent from the condenser and the color due to an intermediate reduction product is gone (about 15 min). During this period, dissolve 40 g of sodium hydroxide in 100 mL of water and cool to room temperature.

At the end of the reduction reaction, cool the acid solution in ice (to prevent volatilization of aniline) during gradual addition of the solution of alkali. This alkali neutralizes the aniline hydrochloride, releasing aniline, which will now be volatile in steam. Attach a stillhead with a steam-inlet tube, a condenser, an adapter, and a receiving Erlenmeyer flask (*see* Fig. 6.4 on page 112); heat the flask with a microburner to prevent the flask from filling with water from condensed steam and proceed to steam distill. Because aniline is fairly soluble in water (3.6 g/100 g at 18°C), distillation should be continued somewhat beyond the point at which the distillate has lost its original turbidity (50–60 mL more). Make an accurate estimate of the volume of distillate by filling a second flask with water to the level of liquid in the receiver and measuring the volume of water.

The $^{13}$C NMR spectrum of nitrobenzene (Fig. 45.6) and the IR (Fig. 45.7) and $^1$H NMR (Fig. 45.8) spectra of aniline are found at the end of the chapter.

# 3. Isolation of Aniline–Alternative Choices

Salting out aniline

At this point aniline can be isolated. You could reduce the solubility of aniline by dissolving in the steam distillate 0.2 g of sodium chloride per milliliter, extracting the aniline with three 50-mL portions of dichloromethane, drying the extract, distilling the dichloromethane (bp 41°C), and then distilling the aniline (bp 184°C). Alternatively, the aniline can be converted directly to acetanilide. The procedure calls for pure aniline, but note that the first step is to dissolve the aniline in water and hydrochloric acid. Your steam distillate is a mixture of aniline and water, both of which have been distilled. Are they not both water-white and presumably pure? Hence, an attractive procedure would be to assume that the steam distillate contains the theoretical amount of aniline and to add to it, in turn, appropriate amounts of hydrochloric acid, acetic anhydride, and sodium acetate, calculated from the quantities given in Experiment 2.

**Cleaning Up.** Filter the pot residue from the steam distillation through a layer of Celite filter aid on a Büchner funnel. Discard the tin salts in the nonhazardous solid waste container. The filtrate, after neutralizing it with hydrochloric acid, can be flushed down the drain.

# 4. Acetanilide: Acetylation in Aqueous Solution

Aniline          Acetic anhydride                    Acetanilide

**IN THIS EXPERIMENT** the steam distillate, a mixture of aniline and water, is reacted with acetic anhydride to form crystalline acetanilide that is isolated by filtration. It is carefully dried for the next experiment.

From the volumes of reagents given for this experiment, calculate the numbers of grams, the numbers of moles, and the mole ratios. What is the limiting reagent? Why is sodium acetate found in this mixture?

**CAUTION**: Handle aniline and acetic anhydride in the hood; avoid contact with the skin by wearing gloves.

*Online Study Center*

Video: Microscale Filtration on the Hirsch Funnel

Add to the steam distillate, which consists of a mixture of aniline and water, 0.43 mL of concentrated hydrochloric acid and make up the volume to 13 mL. Add 0.620 mL of acetic anhydride; also prepare a solution of 0.53 g of anhydrous sodium acetate in 3 mL of water. Add the acetic anhydride to the solution of aniline hydrochloride with stirring and then immediately add all of the sodium acetate solution. Stir, cool in ice, and collect the product by suction filtration on a Hirsch funnel. It should be colorless; when dry, the melting point should be close to 114°C. Because the acetanilide *must be completely dry* for use in the next step, it is advisable to put the material in a tared 25-mL Erlenmeyer flask and to heat it on a steam or sand bath under evacuation until the weight is constant (Fig. 45.3). It could then be used to make sulfanilamide.

$^1$H NMR (Fig. 45.9) and IR (Fig. 45.10) spectra of acetanilide are found at the end of the chapter.

**Cleaning Up.** The aqueous filtrate can be flushed down the drain.

**■ FIG. 45.3**
Drying of acetanilide under reduced pressure. Heat the flask on a steam bath or a sand bath.

**CAUTION:** Handle aniline and acetic anhydride in the hood; avoid contact with the skin.

## 5. Acetanilide: Acetylation in Aqueous Solution

Dissolve 5.0 g (0.054 mol) of aniline in 135 mL of water and 4.5 mL (0.054 mol) of concentrated hydrochloric acid; if the solution is colored, filter it by suction through a pad of decolorizing charcoal. Measure out 6.2 mL (0.065 mol) of acetic anhydride; also prepare a solution of 5.3 g (0.065 mol) of anhydrous sodium acetate in 30 mL of water. Add the acetic anhydride to the solution of aniline hydrochloride with stirring and immediately add all of the sodium acetate solution. Stir, cool in ice, and collect the product. It should be colorless, with a melting point close to 114°C. Because the acetanilide *must be completely dry* for use in the next step, it is advisable to put the material in a tared 125-mL Erlenmeyer flask and to heat this on a steam or sand bath under evacuation until the weight is constant (Fig. 45.3).

$^1$H NMR (Fig. 45.9) and IR (Fig. 45.10) spectra of acetanilide are found at the end of the chapter.

**Cleaning Up.** The aqueous filtrate, after neutralizing it with hydrochloric acid, can be flushed down the drain with a large excess of water.

## 6. Sulfanilamide

| Acetanilide | Chloro-sulfonic acid | *p*-Acetaminobenzene-sulfonyl chloride | *p*-Acetaminobenzene-sulfonamide | Sulfanilamide |

**■ FIG. 45.4**
A hydrogen chloride trap.

Polyethylene tubing

Reaction tube

Syringe needle

Damp cotton

25-mL Erlenmeyer flask

**HSO₃Cl**
**Chlorosulfonic acid**

**CAUTION**: Chlorosulfonic acid is a corrosive chemical and reacts violently with water. Withdraw with a pipette and pipette bulb. Neutralize any spills and drips immediately. The wearing of gloves and a face shield is advised.

*Online Study Center*

Photos: Capping a Reaction Tube with a Septum; Placing a Polyethylene Tube through a Septum

**IN THIS EXPERIMENT** perfectly dry acetanilide is treated with chlorosulfonic acid, a highly reactive reagent. The hydrogen chloride evolved is trapped; the reaction mixture is added carefully to water; and the product, *p*-acetaminobenzenesulfonyl chloride, is isolated by filtration. This solid is added to aqueous ammonia to form *p*-acetaminobenzenesulfonamide, which is hydrolyzed with hot hydrochloric acid. The resulting solution is neutralized with sodium bicarbonate to give sulfanilamide.

The chlorosulfonation of acetanilide in the preparation of sulfanilamide is conducted without solvent in the 25-mL Erlenmeyer flask used for drying the precipitated acetanilide from Experiment 3. Alternatively, commercial acetanilide may be employed.

Fit the Erlenmeyer flask with a septum connected to a short length of polyethylene tubing leading into a reaction tube that contains a small piece of damp cotton to trap hydrogen chloride vapors (Fig. 45.4). Add 0.625 mL of chlorosulfonic acid a few drops at a time to 0.25 g of acetanilide from a capped vial in the hood using a Pasteur pipette (not a syringe with metal needle). Connect the flask to the gas trap between additions. The reaction subsides in 5–10 min, with only a few small pieces of acetanilide remaining undissolved.

$$HSO_3Cl + H_2O \rightarrow H_2SO_4 + HCl$$

When this point has been reached, heat the mixture on a steam bath for 10 min to complete the reaction, cool the flask in ice, and deliver the oily product by drops with a Pasteur pipette while stirring it into 3.5 mL of ice water contained in a 10-mL Erlenmeyer flask in the hood. If the material in the flask solidifies (it usually does not), add 3.5 g of ice to the reaction flask. Use extreme caution when adding the oil to ice water and when rinsing out any containers that have held chlorosulfonic acid. Rinse the flask with cold water and stir the precipitated

Do not let the mixture stand
before adding ammonia.

*p*-acetaminobenzenesulfonyl chloride for a few minutes until an even suspension
of granular white solid is obtained. Collect and wash the solid on a Hirsch funnel.

After pressing and draining the filter cake, transfer the solid to the rinsed
reaction flask, add 0.75 mL of concentrated aqueous ammonia solution (ammonium
hydroxide) and 0.75 mL of water, and in the hood heat the mixture over a hot sand
bath to just below the boiling point with occasional swirling. Heat the mixture
in this manner for 5 min. During this treatment, a change can be noted as the
sulfonyl chloride undergoes transformation to a more pasty suspension of the
amide. Cool the suspension well in an ice bath, collect the *p*-acetaminobenzenesul-
fonamide by suction filtration, press the cake on a Hirsch funnel, and allow it to
drain thoroughly. Any excess water will unduly dilute the acid used in the next step.

Transfer the still-moist amide to a well-drained reaction flask, add 0.25 mL of
concentrated hydrochloric acid and 0.5 mL of water, boil the mixture gently until
all the solid has dissolved (5–10 min), and then continue the heating at the boiling
point for an additional 10 min (do not evaporate to dryness). The solution, when
cooled to room temperature, should deposit no solid amide, but if it is deposited,
heating should be continued for a further period. The cooled solution of sulfanilamide hydrochloride is shaken with granulated decolorizing charcoal and filtered
by removing the solution with a Pasteur pipette.

Place the solution in a 30-mL beaker and cautiously add an aqueous solution
of 0.25 g of sodium bicarbonate while stirring to neutralize the hydrochloride.
After the foam has subsided, test the suspension with indicator paper; if it is still
acidic, add more bicarbonate until the neutral point is reached. Cool thoroughly in
ice and collect the granular, white precipitate of sulfanilamide. The crude product
(mp 161°C–163°C) on crystallization from alcohol or water affords pure sulfanilamide (mp 163°C–164°C) with about 90% recovery. Save a small amount
(~10 mg) of sample for the bioassay.

The IR spectrum (Fig. 45.11) of sulfanilamide is found at the end of this chapter.

**Cleaning Up.** Rinse the cotton in the trap with water, add this rinse to the combined aqueous filtrates from all reactions, and neutralize the solution by adding
either 3 *M* hydrochloric acid or sodium carbonate. Flush the neutral solution down
the drain. Any spilled drops of chlorosulfonic acid should be covered with sodium
carbonate; the resulting powder should be collected in a beaker, dissolved in
water, and then flushed down the drain.

$HSO_3Cl$
**Chlorosulfonic acid**

**CAUTION:** Chlorosulfonic acid
is a corrosive chemical and reacts
violently with water. Withdraw
with a pipette and pipette bulb.
Neutralize any spills and drips
immediately. The wearing of
gloves and a face shield is
advised. Work within a hood.

# 7. Sulfanilamide

The chlorosulfonation of acetanilide in the preparation of sulfanilamide
is conducted without solvent in a 125-mL Erlenmeyer flask used for drying the precipitated acetanilide from Experiment 2. Because the reaction is most easily controlled when the acetanilide is in the form of a hard cake, the dried solid is melted
by heating the flask over a hot plate; as the melt cools, the flask is swirled to distribute the material as it solidifies over the lower walls of the flask. Let the flask
cool while making provision for trapping the hydrogen chloride evolved in the
chlorosulfonation. Fit the Erlenmeyer flask with a stopper connected by a section
of rubber tubing to a glass tube fitted with a stopper into the neck of a 250-mL filter

■ **FIG. 45.5**
The chlorosulfonation apparatus
fitted with HCl gas trap.

Solidified acetanilide
in 125-mL flask

250-mL Filter flask

1-cm Clearance

Water

flask half filled with water. The tube should be about 1 cm above the surface of the water and *must not dip into the water* (Fig. 45.5).

$$HSO_3Cl + H_2O \rightarrow H_2SO_4 + HCl$$

Cool the flask containing the acetanilide thoroughly in an ice-water bath, and for 5.0 g of acetanilide measure 12.5 mL of chlorosulfonic acid in a graduated cylinder (supplied with the reagent and kept away from water). Add the reagent in 1–2 mL portions with a capillary dropping tube and connect the flask to the gas trap. The flask is now removed from the ice bath and swirled until a portion of the solid has dissolved and the evolution of hydrogen chloride is proceeding rapidly. Occasional cooling in ice may be required to prevent too brisk a reaction.

The reaction subsides in 5–10 min, with only a few lumps of acetanilide remaining undissolved. When this point has been reached, heat the mixture on a steam or sand bath for 10 min to complete the reaction, cool the flask under the tap, and deliver the oil by drops with a capillary dropper while stirring it into 75 mL of ice water contained in a beaker cooled in an ice bath (in the hood). Use extreme caution when adding the oil to ice water and when rinsing out any containers that have held chlorosulfonic acid. Rinse the flask with cold water and stir the precipitated *p*-acetaminobenzenesulfonyl chloride for a few minutes until an even suspension of granular white solid is obtained. Collect and wash the solid on a Büchner funnel.

After pressing and draining the filter cake, transfer the solid to the rinsed reaction flask, add (for each 5 g of acetanilide) 15 mL of concentrated aqueous ammonia solution and 15 mL of water, and heat the mixture over a flame or sand

Do not let the mixture stand
before adding ammonia.

bath with occasional swirling (in the hood). Maintain the temperature of the mixture just below the boiling point for 5 min. During this treatment a change can be noted as the sulfonyl chloride undergoes transformation to a more pasty suspension of the amide. Cool the suspension well in an ice bath, collect the p-acetaminobenzenesulfonamide by suction filtration, press the cake on the funnel, and allow it to drain thoroughly. Any excess water will unduly dilute the acid used in the next step.

Transfer the still-moist amide to the well-drained reaction flask, add 5 mL of concentrated hydrochloric acid and 10 mL of water (for each 5 g of aniline), and boil the mixture gently until the solid has all dissolved (5–10 min); then continue the heating at the boiling point for an additional 10 min (do not evaporate to dryness). The solution when cooled to room temperature should deposit no solid amide, but, if it is deposited, heating should be continued for a further period. The cooled solution of sulfanilamide hydrochloride is shaken with decolorizing charcoal and filtered by suction.

Place the solution in a beaker and cautiously add an aqueous solution of 5 g of sodium bicarbonate with stirring to neutralize the hydrochloride. After the foam has subsided, test the suspension with litmus; if it is still acidic, add more bicarbonate until the neutral point is reached. Cool thoroughly in ice and collect the granular, white precipitate of sulfanilamide. The crude product (mp 161°C–163°C) on crystallization from alcohol or water affords pure sulfanilamide (mp 163°C–164°C) with about 90% recovery. Determine the melting point and calculate the overall yield from the starting material. Save a small amount (~10 mg) of sample for the bioassay.

The IR spectrum (Fig. 45.11) of sulfanilamide is found at the end of the chapter.

**Cleaning Up.** Add the water from the gas trap to the combined aqueous filtrates from all reactions and neutralize the solution by adding either 10% hydrochloric acid or sodium carbonate. Flush the neutral solution down the drain with a large excess of water. Any spilled drops of chlorosulfonic acid should be covered with sodium carbonate; the resulting powder should be collected in a beaker, dissolved in water, and then flushed down the drain.

## 8. Bioassay of Sulfanilamide

You may bioassay your synthesized sulfanilamide and commercial sulfanilamide with *Bacillus cereus* or *E. coli* to test for its antibiotic behavior. There are many bacterial cultures commercially available, but *Bacillus cereus* is preferred because it is nonpathogenic, requires no special medium for growth, and grows at room temperature in a short time.[2] Cultures of *E. coli* not only have to be handled with care due to their potential pathogenicity, but their optimal growth is at 37°C, which requires an incubator. In addition, *E. coli* is a gram-negative bacteria unlike *Bacillus cereus*, which is gram positive. It would be interesting to test your compound with both bacteria and to compare the two results.

---

2. Other nonpathogens for acceptable use are *Staphylococcus epidermis* and the following strains of *E. coli*: K-12, 1776B, and 1776C.

Agar culture plates are prepared in sterile Petri dishes (equipped with caps) or can be purchased. These plates are stored in the refrigerator. Whenever you handle the sterile agar Petri dishes, try to minimize any exposure to random forms of bacteria or dirt in the laboratory. Always wear gloves when handling the plates and the bacteria because your hands can easily contaminate the agar. Open the cover of the dish a few inches vertically to have just enough room to swab the plate with a known strain of bacteria with a sterilized copper wire. Once you have your sample ready, you will add your sample to the plate.

After preparing your assay, you will observe the growth of bacteria on the plate. If the bacteria do not grow in the area surrounding the sulfanilamide, then it is exhibiting antibiotic behavior.

### *Bioassay Using Agar Medium Bacterial Growth Plates*

Sterilize a copper wire loop by heating it in a flame. Once cool, use the loop to transfer the bacteria from the starter culture plate to a previously prepared sterile agar culture plate.

Using a microspatula cleaned by wiping it with an ethanol-soaked tissue, scoop some sulfanilamide on the tip. Lift the lid on the agar plate just enough to insert the solid. Two students can use one agar plate. Label the bottom of the culture plate with your name and label the synthesized sulfanilamide and commercial sulfanilamide. Be sure to label your plates properly and draw one line down the middle of the plate to separate each student's sample. Place the plate in a quiet area at room temperature or in an incubator oven. The results of the assay *must* be assessed after 24 h. *See* Chapter 6 for an example of an agar plate.

⚠️ **CAUTION**: Wash hands after handling bacterial plates.

**Cleaning Up.** If an autoclave is available, then the agar plates and contaminated materials should be autoclaved for at least 1 h. If an autoclave is not available, soak the agar plates and contaminated materials in bleach for several days.

# PART 2: Synthesis and Bioassay of Sulfanilamide Derivatives[3]

**A**s mentioned at the beginning of this chapter, sulfanilamide and its derivatives can exhibit biological activity. For example, when the R group in sulfanilamide is replaced with a heterocyclic ring system, the sulfa drug so formed is often faster acting or less toxic than sulfanilamide. What if the R group is an alkyl group? What about derivatizing sulfanilamide such that the primary amine is functionalized to an amide? Will these derivatives be more or less active than sulfanilamide? You will be able to answer these questions with the following experiments.

---

3. Note: Some people are allergic to sulfanilamide, so be aware of any individual sensitivity.

## EXPERIMENTS

### 1. Synthesis of 4-Amino-*N*-methylbenzenesulfonamide (*N*-1-methylsulfanilamide)

**Acetanilide**
MW 135.16
mp 114°C

*p*-**Acetaminobenzene-sulfonylchloride**
MW 233.68

*p*-**Acetaminobenzene-*N*-methylsulfonamide**
MW 228.26

*N*-**1-methylsulfanilamide**
MW 186.23
mp 105–107°C

> **IN THIS EXPERIMENT** perfectly dry acetanilide is treated with chlorosulfonic acid, a highly reactive reagent. The hydrogen chloride evolved is trapped, the reaction mixture is added carefully to water, and the product (*p*-acetaminobenzenesulfonyl chloride) is isolated by filtration. This solid is added to a solution of methylamine to form *p*-acetamino-*N*-methylbenzenesulfonamide, which is hydrolyzed with hot hydrochloric acid. The resulting solution is neutralized with sodium bicarbonate to give *N*-1-methylsulfanilamide.

The chlorosulfonation of acetanilide in the preparation of *N*-1-methylsulfanilamide is conducted without solvent in the 25-mL Erlenmeyer flask used for drying the precipitated acetanilide from Experiment 4 in Part 1. Alternatively, commercial acetanilide may be employed.

**CAUTION**: Chlorosulfonic acid is a corrosive chemical that reacts violently with water. Withdraw with a pipette and pipette bulb. Neutralize any spills and drips immediately. The wearing of gloves and a face shield is advised.

*Online Study Center*

Photos: Capping a Reaction Tube with a Septum, Placing a Polyethylene Tube through a Septum

Do not let the mixture stand before addition of methylamine.

Fit the Erlenmeyer flask with a septum connected to a short length of polyethylene tubing leading into a reaction tube that contains a small piece of damp cotton to trap hydrogen chloride vapors (*see* Fig. 45.4 on page 618). Add 0.625 mL of chlorosulfonic acid a few drops at a time to 0.25 g of acetanilide from a capped vial in the hood using a Pasteur pipette (not a syringe with metal needle). Connect the flask to the gas trap between additions. In 5–10 min the reaction subsides, with only a few small pieces of acetanilide remaining undissolved.

When this point has been reached, heat the mixture on a steam or sand bath for 10 min to complete the reaction, cool the flask in ice, and deliver the oily product dropwise with a Pasteur pipette while stirring it into 3.5 mL of ice water contained in a 10-mL Erlenmeyer flask in the hood. If the material in the flask solidifies (it usually does not), add 3.5 g of ice to the reaction flask. Use extreme caution when adding the oil to ice water and when rinsing out any containers that have held chlorosulfonic acid. Rinse the flask with cold water and stir the precipitated *p*-acetaminobenzenesulfonyl chloride for a few minutes until an even suspension of granular white solid is obtained. Collect and wash the solid on a Hirsch funnel.

After pressing and draining the filter cake, transfer the solid to the rinsed reaction flask, add 0.02 mol of methylamine (2.5 mL of a 33% solution of methylamine in absolute ethanol),[4] and heat the mixture over a hot sand bath to just below the boiling point with occasional swirling in the hood. Heat the mixture in this manner for approximately 15 min. Monitor the reaction by thin-layer chromatography (TLC) using ethyl acetate. When complete, remove all excess solvent with a stream of nitrogen gas to leave behind a white solid.

Transfer the still-moist amide to the well-drained reaction flask, add 0.25 mL of concentrated hydrochloric acid and 0.5 mL of water, and boil the mixture gently until all of the solid has dissolved (5–10 min); then continue heating at the boiling point for an additional 10 min (do not evaporate to dryness). The solution, when cooled to room temperature, should deposit no solid amide; if a solid is deposited, heating should be continued for a further period.

Place the solution in a 30-mL beaker and cautiously add an aqueous solution of 0.25 g of sodium bicarbonate while stirring to neutralize the hydrochloride. After the foam has subsided, test the suspension with indicator paper; if it is still acidic, add more bicarbonate until the neutral point is reached. Cool thoroughly in ice and collect the granular, white precipitate of *N*-1-methylsulfanilamide. The crude product can be recrystallized in water to afford pure *N*-1-methylsulfanilamide (mp 105°C–107°C) with about 65% recovery. Save a small amount of sample (~10 mg) for the bioassay.

**Cleaning Up.** Rinse the cotton in the trap with water, add this rinse to the combined aqueous filtrates from all reactions, and neutralize the solution by adding either 3 *M* hydrochloric acid or sodium carbonate. Flush the neutral solution down the drain. Any spilled drops of chlorosulfonic acid should be covered with sodium carbonate; the resulting powder should be collected in a beaker, dissolved in water, and then flushed down the drain.

---

4. The 33% solution of methylamine in absolute ethanol can be purchased from Fluka, catalog no. 65590.

## 2. 4-Amino-*N*-methylbenzenesulfonamide (*N*-1-methylsulfanilamide)

The chlorosulfonation of acetanilide in the preparation of *N*-1-methylsulfanilamide is conducted without solvent in the 125-mL Erlenmeyer flask used for drying the precipitated acetanilide from Experiment 5 in Part 1 (commercial acetanilide can also be used). Because the reaction is most easily controlled when the acetanilide is in the form of a hard cake, the dried solid is melted by heating the flask over a hot plate; as the melt cools, the flask is swirled to distribute the material over the lower walls of the flask as it solidifies. Let the flask cool while making provision for trapping the hydrogen chloride evolved in the chlorosulfonation. Fit the Erlenmeyer flask with a stopper connected by a section of rubber tubing to a glass tube fitted with a stopper into the neck of a 250-mL filter flask half filled with water. The tube should be about 1 cm above the surface of the water and *must not dip into the water* (*see* Fig. 45.5 on page 620).

**CAUTION**: Chlorosulfonic acid is a corrosive chemical that reacts violently with water. Withdraw with pipette and a pipette bulb. Neutralize any spills and drips immediately. The wearing of gloves and a face shield is advised.

Cool the flask containing the acetanilide thoroughly in an ice-water bath. For each 5.0 g of acetanilide, measure 12.5 mL of chlorosulfonic acid in a separate graduated cyclinder (supplied with the reagent and kept away from water). Add the reagent in 1–2 mL portions with a Pasteur pipette and connect the flask to the gas trap. Remove the flask from the ice bath and swirl until a portion of the solid has dissolved and the evolution of hydrogen chloride is proceeding rapidly. Occasional cooling in ice may be required to prevent too brisk a reaction.

The reaction subsides in 5–10 min, with only a few lumps of acetanilide remaining undissolved. When this point has been reached, heat the mixture with a sand bath for 10 min to complete the reaction, cool the flask under the tap, and deliver the oil dropwise with a Pasteur pipette while stirring it into 75 mL of ice water in a beaker cooled in an ice bath (in the hood). Use extreme caution when adding the oil to the ice water and when rinsing out any containers that have held chlorosulfonic acid. Rinse the flask with cold water and stir the precipitated *p*-acetaminobenzenesulfonyl chloride for a few minutes until an even suspension of granular white solid is obtained. Collect and wash the solid on a Büchner funnel. Save a small sample (~6 mg) for bioassay.

Do not let the mixture stand before addition of methylamine.

After pressing and draining the filter cake, transfer the solid to the rinsed reaction flask, add 0.42 mol of methylamine (52 mL of a 33% solution of methylamine in absolute ethanol)[5] for each 5 g of acetanilide and heat the mixture over a flame with occasional swirling (in the hood). Maintain the temperature of the mixture just below the boiling point for approximately 30 min. Monitor the reaction by TLC using ethyl acetate. When complete, remove all excess solvent with a nitrogen gas line to leave behind a white solid. Save a small sample (~6 mg) for bioassay.

Transfer the still-moist amide to the well-drained reaction flask, add 5 mL of concentrated hydrochloric acid and 10 mL of water (for each 5 g of acetanilide), and boil the mixture gently until the solid has all dissolved (5–10 min); then continue the heating at the boiling point for an additional 10 min (do not evaporate to dryness). The solution when cooled to room temperature should deposit no solid amide; if a solid is deposited, heating should be continued for a further period.

---

5. *See* footnote 4.

Place the solution in a beaker and cautiously add an aqueous solution of 5 g of sodium bicarbonate with stirring to neutralize the hydrochloride. After the foam has subsided, test the suspension with indicator paper; if it is still acidic, add more bicarbonate until the neutral point is reached. Cool thoroughly in ice and collect the granular, white precipitate of N-1-methylsulfanilamide. The crude product can be recrystallized in water to afford pure N-1-methylsulfanilamide (mp 105°C–107°C) with about 65% recovery. Determine the melting point and calculate the overall yield from the starting material. Save a small amount (~10 mg) for the bioassay.

**Cleaning Up.** Add the water from the gas trap to the combined aqueous filtrates from all reactions and neutralize the solution by adding either 10% hydrochloric acid or sodium carbonate. Flush the neutral solution down the drain with a large excess of water. Any spilled drops of chlorosulfonic acid should be covered with sodium carbonate; the resulting powder should be collected in a beaker, dissolved in water, and then flushed down the drain.

# 3. Synthesis of Acylated Sulfanilamide Derivatives

**Sulfanilamide**
MW 172.20
mp 163–164°C

**Acid chloride**
R = t-butyl, ethyl,
phenyl, isopropyl

**Acylated
sulfanilamide**

**IN THIS EXPERIMENT** sulfanilamide is treated with an acid chloride to produce an acylated sulfanilamide. The resulting solution is neutralized by sodium bicarbonate.

**CAUTION**: Tetrahydrofuran easily forms peroxides. Discard material exposed to oxygen after 30 days.

The acylation of sulfanilamide to prepare the acylated derivative is conducted in a 25-mL round-bottom flask equipped with a magnetic stirring bar. Transfer 3 mL of tetrahydrofuran to the flask and then add 0.25 g of sulfanilamide (from Experiment 6 or Experiment 7 in Part 1; alternatively, commercial sulfanilamide may be employed) and 0.305 g of sodium bicarbonate. While the mixture is stirring, add 3.63 mol of the desired acid chloride. Attach a condenser to the flask and heat at reflux until the reaction goes to completion. Monitor progress of the reaction by thin-layer chromatography every 15 min using a 20/30/50 mixture of dichloromethane, ethanol, and hexanes to elute.

| TABLE 45.1 • Physical Properties of Acylated Sulfanilamide Derivatives | | | |
|---|---|---|---|
| **Derivative** | **R** | **% Yield[a]** | **mp (°C)** |
| 4-Pivaloylamido benzenesulfonamide | $(CH_3)_3C$ | 60%–67% | 231–233 |
| 4-Propionamido benzenesulfonamide | $CH_3CH_2$ | 66%–76% | 225–227 |
| 4-Benzoylamino benzenesulfonamide | $C_6H_5$ | 60%–86% | 288–290 |
| 4-Isobutanamido benzenesulfonamide | $(CH_3)_2CH$ | 59%–79% | 249–251 |

[a]Percent yield after recrystallization.

When this point has been reached, add 15 mL of cold water to cause complete precipitation of the acylated derivative. Collect the granular white solid and wash with ether on a Hirsch funnel. The crude product can be recrystallized in alcohol to afford the pure acylated sulfanilamide. Save a small amount of pure sample (~10 mg) for the bioassay.

Yield and melting points of four acyl derivatives are displayed in Table 45.1.

## 4. Synthesis of Acylated Sulfanilamide Derivatives

CAUTION: Tetrahydrofuran easily forms peroxides. Discard material exposed to oxygen after 30 days.

The acylation of sulfanilamide is conducted in a 50 mL round-bottom flask equipped with a 1-inch magnetic stirring bar. Add 12 mL of tetrahydrofuran to the flask and then add 1.0 g of sulfanilamide (from Experiment 7 in Part 1; alternatively, commercial sulfanilamide may be employed) and 1.3 g of sodium bicarbonate. While the mixture is stirring, add 14.5 mol of the desired acid chloride. Attach a condenser to the flask and heat at reflux until the reaction goes to completion. Monitor progress of the reaction by thin-layer chromatography every 15 min using a 20/30/50 mixture of dichloromethane, ethanol, and hexanes to elute.

When the reaction is complete, add 30 mL of cold water to cause complete precipitation of the acylated derivative. Collect the granular white solid and wash with ether on a fritted glass funnel. The crude product can be recrystallized from ethanol to afford pure acylated sulfanilamide. *See* Table 45.1 for yields and melting points of four acyl derivatives. Save a small amount of pure sample (~10 mg) for the bioassay.

## 5. Bioassay of Sulfanilamide and Derivatives

**IN THIS EXPERIMENT** sulfanilamide derivatives are tested for antibacterial activity and compared to the activity of sulfanilamide.

The bioassay procedure can be followed as described in Experiment 8 of Part 1 of this chapter. You will need approximately 6 mg of each derivative and of sulfanilamide to test for bioactivity. Be sure to carefully label the agar plates so it is clear what compounds are being tested. The compounds to be tested can be directly placed on the agar/bacteria in a small pile. If a derivative is active toward bacteria, there will be no bacterial growth around the immediate area of where the derivative is placed on the plate. Observe the results from the bioassay and assess whether any of the derivatives are active. Compare the activities of the derivatives with each other and with the sulfanilamide. From these results, what conclusions can you draw about the activity of sulfanilamide derivatives? Which derivatives are antibacterial? Does the primary aromatic amine most likely need to be in this "free" form ($NH_2$) to be antibacterial? Is *N*-methylsulfanilamide more toxic or less toxic toward bacteria than sulfanilamide?

**■ FIG. 45.6**
The $^{13}C$ NMR spectrum of nitrobenzene (22.6 MHz).

**■ FIG. 45.7**
The IR spectrum of aniline.

## QUESTIONS

1. Why is an acetyl group added to aniline (making acetanilide) and then removed to regenerate the amine group in sulfanilamide?
2. What happens when chlorosulfonic acid comes in contact with water?
3. Acetic anhydride, like any anhydride, reacts with water to form a carboxylic acid. How then is it possible to carry out an acetylation in aqueous solution? What is the purpose of the hydrochloric acid and the sodium acetate in this reaction?
4. What happens when *p*-acetaminobenzenesulfonyl chloride is allowed to stand for some time in contact with water?
5. Assign the groups of peaks at 7.0, 7.2, and 7.4 ppm to specific protons in the NMR spectrum of acetanilide (Fig. 45.9).
6. Assign the peaks at 3470 cm$^{-1}$ and 1619 cm$^{-1}$ in the IR spectrum of sulfanilamide (Fig. 45.11).
7. At 98°C the vapor pressure of water is 720 mm Hg and that of aniline is 40 mm Hg. How much aniline steam distills with each gram of water collected?

**■ FIG. 45.8**
The $^1$H NMR spectrum of aniline (250 MHz).

■ **FIG. 45.9**
The $^1$H NMR spectrum of
acetanilide (250 MHz). The amide
proton shows a characteristically
broad peak.

■ **FIG. 45.10**
The IR spectrum of acetanilide
(KBr disk).

■ **FIG. 45.11**
The IR spectrum of sulfanilamide
(KBr disk).

# CHAPTER 46

# Dyes and Dyeing[1]

**Online Study Center**

This icon will direct you to techniques, equipment setups, and online resources at http://college.hmco.com/PIC/williamsonMMOE5e.

**PRELAB EXERCISE:** Operating on the simple hypothesis that the intensity of a dye on a fiber will depend on the number of strongly polar or ionic groups in the fiber molecule, predict the relative intensities produced by Methyl Orange when it is used to dye a variety of different fibers, such as are found in the Multifiber Fabric. Examine the structures of the fiber molecules and try to order them in terms of polarity. Wool is a protein.

**Tyrian purple**
**(6,6′-Dibromoindigo)**

**Perkin's Mauve**
R = H or $CH_3$

$S$ince prehistoric times humans have been dyeing cloth. The "wearing of the purple" has long been synonymous with royalty, attesting to the cost and rarity of Tyrian purple, a dye derived from the sea snail *Murex brandaris*. The organic chemical industry originated in 1856 with William Henry Perkin's discovery of the first synthetic dye, Perkin's Mauve. The correct structure of this dye has only recently been elucidated.[2]

---

1. For a detailed discussion of the chemistry of dyes and dyeing, see Fieser, L. F.; Fieser, M. *Topics in Organic Chemistry*; Reinhold Publishing Corp.: New York, 1963; 357–417.

2. Meth-Cohn, O.; Smith, M. *J. Chem. Soc., Perkin Trans.* 1, **1994**, 5.

A fiber usually absorbs dyes from an aqueous solution. A natural fiber such as cotton has a surface area of about 4.4 hectares per kilogram (5 acres per pound). The dye penetrates the pores in the fiber and is bound to the fiber by electrostatic forces, van der Waals attraction, hydrogen bonding, and covalent bonds (in the case of fiber-reactive dyes). A good dye must be fast to light, heat, and washing; that is, it must not fade, sublime away, or come off during washing.

Among the newest of the dyes are the fiber-reactive compounds, which form a covalent link to the hydroxyl groups of cellulose. This reaction involves an amazing and little-understood nucleophilic displacement of a chloride ion from the triazine part of the molecule by the hydroxyl groups of cellulose, yet the reaction occurs in aqueous solution.

**Chlorantin Light Blue 8G**

■ **FIG. 46.1**
Multifiber Fabric.

Black thread
Acetate rayon
SEF
Arnel
Cotton
Creslan
Dacron 54
Dacron 64
Nylon 6.6
Orlon 75
Silk
Polypropylene
Viscose rayon
Wool

Several dyes are synthesized in this chapter; these and other dyes are used to dye a representative group of natural and synthetic fibers. You will receive several pieces of Multifiber Fabric (Fig. 46.1), which has 13 strips of different fibers woven into it. Below the black thread at the top, the fibers are acetate rayon (cellulose di- or triacetate), SEF (Solutia's modacrylic Self-Extinguishing Fiber), Arnel (cellulose triacetate), cotton, Creslan (polyacrylonitrile), Dacron 54 and 64 (polyester without and with a brightener), nylon 6.6 (polyamide), Orlon 75 (polyacrylonitrile), silk (polyamide), polypropylene, viscose rayon (regenerated cellulose), and wool (polyamide).

Acetate rayon is cellulose (from any source) in which about two of the hydroxyl groups in each unit have been acetylated. This renders the polymer soluble in acetone, from which it can be spun into fiber. The smaller number of hydroxyl groups in acetate rayon compared to cotton makes direct dyeing of rayon more difficult than cotton.

Cotton is pure cellulose. Nylon is a polyamide made by polymerizing adipic acid and hexamethylenediamine. The nylon polymer chain can be prepared with one acid and one amine group at the termini or with both acids or both amines. Except for these terminal groups, there are no polar centers in nylon; consequently, it is difficult to dye. Similarly, Dacron, a polyester made by polymerizing ethylene glycol and terephthalic acid, has few polar centers within the polymer and is also difficult to dye. Even more difficult to dye is Orlon, a polymer of acrylonitrile. Wool and silk are polypeptides cross-linked with disulfide bridges. The acidic and basic amino acids (e.g., glutamic acid and lysine) provide many polar groups in wool and silk to which a dye can bind, making these fabrics easy to dye. In this experiment note the marked differences in shade produced by the same dye on different fibers.

**Cellulose (Cotton, R=H)**
**Acetylated cellulose (Acetate rayon, R=OAc)**

**Wool (R=amino acid residue)**

**Polyethylene glycol terephthalate**
**(Dacron)**

**Nylon**

**Polypropylene**

**Polyacrylonitrile**
**(Orlon)**

# PART 1: Dyes

**Azo group**

The most common dyes are the azo dyes, formed by coupling diazotized amines to phenols. The dye can be made in bulk; as we shall see, the dye molecule can also be developed on and in the fiber by combining the reactants in the presence of the fiber.

One dye, Orange II, is made by coupling diazotized sulfanilic acid with 2-naphthol in alkaline solution; another, Methyl Orange, is prepared by coupling the same diazonium salt with *N,N*-dimethylaniline in a weakly acidic solution. Methyl Orange is used as an indicator because it changes color at pH 3.2–4.4. The change in color is due to the transition from one chromophore (azo group) to another (quinonoid system).

You will prepare one of these two dyes and then exchange samples with a classmate and do the tests with both dyes. Both substances dye wool, silk, and skin, so you must work carefully to avoid getting these on your hands or clothes. The dye will eventually wear off your hands; alternatively, you can clean your hands by soaking them for a few seconds in a warm, slightly acidic ($H_2SO_4$) permanganate solution until heavily stained with manganese dioxide and then removing the stain in a bath of warm, dilute bisulfite solution. (Follow the same procedure if dye is accidentally spilled on your clothing.)

# EXPERIMENTS

## 1. Diazotization of Sulfanilic Acid

### *Microscale Procedure*

**Sulfanilic acid**
MW 173.19

In a $10 \times 100$-mm reaction tube (Fig. 46.2) dissolve, by boiling, 120 mg of sulfanilic acid monohydrate in 1.25 mL of 2.5% sodium carbonate solution (or use 35 mg of anhydrous sodium carbonate and 1.25 mL of water). Cool the solution to room temperature, add 50 mg of sodium nitrite, and stir until it is dissolved. Cool the tube in ice and add to it with thorough stirring a mixture of 0.75 g of ice

**■ FIG. 46.2**
A reaction tube with a boiling stick to promote even boiling.

Wood applicator stick

Cool at this point

Air condenser

Boiling solvent

Temperature controlled by depth in sand

and 0.125 mL of concentrated hydrochloric acid. A powdery white precipitate of the diazonium salt should separate in 1–2 min; the material is then ready for use. The product is not collected but is used in the preparation of Orange II and/or Methyl Orange while in suspension. It is more stable than most diazonium salts and will keep for a few hours.

 ### *Macroscale Procedure*

In a 50-mL Erlenmeyer flask dissolve, if necessary by boiling, 2.4 g of sulfanilic acid monohydrate in 25 mL of 2.5% sodium carbonate solution (or use 0.66 g of anhydrous sodium carbonate and 25 mL of water). Cool the solution under the tap, add 0.95 g of sodium nitrite, and stir until it is dissolved. Pour the solution into a flask containing about 15 g of ice and 2.5 mL of concentrated hydrochloric acid. A powdery white precipitate of the diazonium salt should separate in 1–2 min; the material is then ready for use. The product is not collected but is used in the preparation of Orange II and/or Methyl Orange while in suspension. It is more stable than most diazonium salts and will keep for a few hours.

 ## 2. Orange II (1-*p*-Sulfobenzeneazo-2-Naphthol, Sodium Salt)

**Orange II**

In a 10-mL Erlenmeyer flask dissolve 90 mg of 2-naphthol in 0.5 mL of 3 *M* sodium hydroxide solution and transfer to this solution—with *thorough* stirring—the suspension of diazotized sulfanilic acid prepared in the microscale procedure of Experiment 1. Rinse all of the diazonium salt into the naphthol solution with a few drops of cold water. Coupling occurs very rapidly, and the dye, being a sodium salt, separates easily from the solution because a considerable excess of sodium ion from the carbonate, the nitrite, and the alkali is present. Stir the crystalline paste thoroughly to effect good mixing and, after 5–10 min, heat the mixture in a beaker of boiling water until the solid dissolves. Add 0.25 g of sodium chloride to further decrease the solubility of the product and bring this all into solution by heating and stirring. Allow the flask to cool to near room temperature undisturbed; then cool it in ice. Collect the product on a Hirsch funnel bearing a filter paper (Fig. 46.3). Use saturated sodium chloride solution rather than water to rinse out the flask and to wash the filter cake free of the dark-colored mother liquor. The filtration is somewhat slow, and the transfer to a Hirsch funnel is difficult because it is a paste.

*Online Study Center*

Video: Microscale Filtration on the Hirsch Funnel

Filter paper,
12 mm dia.

Polyethylene
filter disk (frit),
10 mm dia.

Hirsch
funnel

To aspirator

25-mL Filter
flask

The product dries slowly and contains about 20% sodium chloride. Thus the crude yield is not significant, and the material need not be dried before being purified. This particular azo dye is too soluble to be crystallized from water; it can be obtained in a fairly satisfactory form by adding saturated sodium chloride solution to a hot, filtered solution in water and cooling, but the best crystals are obtained from aqueous ethanol.

Transfer the filter cake to a reaction tube, wash the material from the filter paper and funnel with 1 mL of water, and bring all of the solid into solution at the boiling point. It may be necessary to add another 0.25 mL (no more) of water in the course of filtering this hot solution through a Hirsch funnel equipped with a polyethylene frit (Fig. 46.3). Collect the filtrate in a reaction tube, add 2.5–3 mL of ethanol, and allow crystallization to proceed as the tube cools slowly to room temperature. Cool well in ice before collecting the product. Rinse the tube with some of the filtrate and, finally, wash the product with a small quantity of ethanol. The yield of pure crystalline material should be about 150 mg. Orange II separates from aqueous alcohol as the dihydrate, containing two molecules of water of crystallization; allowance for this should be made in calculating the yield. If the water of hydration is eliminated by drying at 120°C, the material will become fiery red.

**Cleaning Up.** The filtrate from the reaction, although highly colored, contains little dye but is highly soluble in water. It can be diluted with a large quantity of water and flushed down the drain. Alternatively, with the volume kept as small as possible, the filtrate can be placed in the aromatic amines hazardous waste container, or it can be reduced with tin(II) chloride (*see* Experiment 5). The crystallization filtrate should go into the organic solvents container. Wash the Hirsch funnel *thoroughly*.

# 3. Orange II (1-*p*-Sulfobenzeneazo-2-Naphthol, Sodium Salt)

**Orange II**

**CAUTION**: Handle 2-naphthol with care, in the hood. Do not breathe the dust or allow skin contact. It is a carcinogen.

In a 250-mL beaker, dissolve 1.8 g of 2-naphthol in 10 mL of cold 3 *M* sodium hydroxide solution; pour into this solution, with stirring, the suspension of diazotized sulfanilic acid prepared in the macroscale procedure of Experiment 1. Rinse the Erlenmeyer flask with a small amount of water and add it to the beaker. Coupling occurs very rapidly, and the dye, being a sodium salt, separates easily from the solution because a considerable excess of sodium ion from the carbonate, the nitrite, and the alkali is present. Stir the crystalline paste thoroughly to effect good mixing and, after 5–10 min, heat the mixture until the solid dissolves. Add 5 g of sodium chloride to further decrease the solubility of the product, bring this all into solution by heating and stirring, set the beaker in a pan of ice and water, and let the solution cool undisturbed. When near room temperature, cool further by stirring and collect the product on a Büchner funnel. Use saturated sodium chloride solution rather than water for rinsing the material from the beaker and for washing the filter cake free of the dark-colored mother liquor. The filtration is somewhat slow.[3]

The product dries slowly; it contains about 20% of sodium chloride. Thus, the crude yield is not significant, and the material need not be dried before being purified. This particular azo dye is too soluble to be crystallized from water; it can be obtained in a fairly satisfactory form by adding saturated sodium chloride solution to a hot, filtered solution in water and cooling, but the best crystals are obtained from aqueous ethanol.

Transfer the filter cake to a beaker, wash the material from the filter paper and funnel with water, and bring the cake into solution at the boiling point. Avoid a large excess of water but use enough to prevent separation of the solid during filtration (use about 25 mL). Filter by suction through a Büchner funnel that has been preheated on a steam bath. Pour the filtrate into an Erlenmeyer flask, rinse the filter flask with a small quantity of water, add it to the flask, estimate the volume, and if greater than 30 mL evaporate by boiling. Cool to 80°C, add 50–60 mL of ethanol, and allow crystallization to proceed. Cool the solution well before collecting the product. Rinse the beaker with the mother liquor and wash finally with a little ethanol. The yield of pure crystalline material is about 3.4 g. Orange II separates from aqueous alcohol with two molecules of water of crystallization; allowance for

---

3. If the filtration must be interrupted, fill the funnel and close the rubber suction tubing (while the aspirator is still running) with a screw pinchclamp placed near the filter flask. Then disconnect the tubing from the trap and set the unit aside. Thus, suction will be maintained, and filtration will continue.

this should be made in calculating the yield. If the water of hydration is eliminated by drying at 120°C, the material will become fiery red.

**Cleaning Up.** The filtrate from the reaction, although highly colored, contains little dye but is highly soluble in water. It can be diluted with a large quantity of water and flushed down the drain. Alternatively, with the volume kept as small as possible, the filtrate can be placed in the aromatic amines hazardous waste container, or it can be reduced with tin(II) chloride (*see* Experiment 6). The crystallization filtrate should go into the organic solvents container.

## 4. Methyl Orange (*p*-Sulfobenzeneazo-4-Dimethylaniline, Sodium Salt)

| **Diazotized sulfanilic acid** 4-Diazobenzenesulfonic acid | **N,N-Dimethylaniline** MW 121.18 bp 194°C | **Methyl Orange** 4-[4-(Dimethylamino)phenylazo] benzenesulfonic acid, sodium salt MW 327.34 $\lambda_{max}$ 507 nm |

Methyl Orange is an acid-base indicator.

**Methyl Orange** (alkali-stable form, pH ≥ 4.4) Yellow

**Methyl Orange** (acid-stable form, pH ≤ 3.2) Red

### Microscale Procedure

In a 10-mL Erlenmeyer flask, mix 75 mg of dimethylaniline and 65 mg of acetic acid. Add the suspension of diazotized sulfanilic acid prepared in the microscale procedure in Experiment 1, with stirring, to the solution of dimethylaniline acetate, rinsing out the last portions with a few drops of water. Stir the mixture thoroughly and, within a few minutes, the red acid-stable form of the dye should separate. A stiff paste should result in 5–10 min, at which time 1 mL of 3 *M* sodium hydroxide

*Online Study Center*

Video: Microscale Filtration
on the Hirsch Funnel

is added to produce the orange sodium salt. Heat the mixture to the boiling point with constant stirring (to avoid bumping), at which time a large part of the dye should dissolve. A better way to do this would be to place the flask in a small beaker of boiling water. Allow the flask to cool slowly to room temperature; then cool it thoroughly in ice before collecting the product by vacuum filtration on a Hirsch funnel bearing a piece of filter paper. Use saturated sodium chloride solution rather than water to rinse the flask and to wash the dark mother liquor from the filter cake.

The crude product does not need to be dried; it can be crystallized from water after performing preliminary solubility tests to determine the proper conditions. The yield should be between 125 mg and 150 mg. Recrystallized material need not necessarily be used for the dyeing process.

**Cleaning Up.** The highly colored filtrates from the reaction and crystallization are very water soluble. After dilution with a large quantity of water, they can be flushed down the drain because the amount of solid is small. Alternatively, the combined filtrates should be placed in the hazardous waste container, or the mixture can be reduced with tin(II) chloride (*see* Experiment 5). Wash the Hirsch funnel *thoroughly*.

 ## *Macroscale Procedure*

In a test tube, thoroughly mix 1.6 mL of dimethylaniline and 1.25 mL of glacial acetic acid. To the suspension of diazotized sulfanilic acid prepared in the macroscale procedure in Experiment 1 contained in a 250-mL beaker add, with stirring, the solution of dimethylaniline acetate. Rinse the test tube with a small quantity of water and add it to the beaker. Stir and mix thoroughly; within a few minutes the red, acid-stable form of the dye should separate. A stiff paste should result in 5–10 min; 18 mL of 3 $M$ sodium hydroxide solution is then added to produce the orange sodium salt. Stir well and heat the mixture to the boiling point, when a large part of the dye should dissolve. Place the beaker in a pan of ice and water and allow the solution to cool undisturbed. When cooled thoroughly, collect the product on a Büchner funnel, using saturated sodium chloride solution rather than water to rinse the flask and to wash the dark mother liquor from the filter cake.

The crude product does not need to be dried; it can be crystallized from water after making preliminary solubility tests to determine the proper conditions. The yield is about 2.5–3 g.

**Cleaning Up.** The highly colored filtrates from the reaction and crystallization are highly water soluble. After dilution with a large quantity of water, they can be flushed down the drain because the amount of solid is small. Alternatively, the combined filtrates should be placed in the hazardous waste container, or the mixture can be reduced with tin(II) chloride (*see* Experiment 5).

 # 5. For Further Investigation

Following the procedure for the synthesis of Orange II, couple diazotized sulfanilic acid with one or more of the following naphthols: 4-amino-5-hydroxy-2,7-naphthalenedisulfonic acid, monosodium salt; 6-amino-4-hydroxy-2-naphthalenesulfonic acid; 7-amino-4-hydroxy-2-naphthalenesulfonic acid; 4-hydroxy-1-naphthalenesulfonic acid, sodium salt; or 6-hydroxy-2-naphthalenesulfonic acid, sodium salt. Despite the apparent complexity of these molecules, they are not expensive; they are used for the commercial synthesis of dyes. The many sulfonic acid groups have been added to make the dyes water soluble; the addition of solid sodium chloride "salts out" the dye, rendering it insoluble.

# 6. Tests

## Solubility and Color

Compare the solubility in water of Orange II to Methyl Orange and account for the difference in terms of structure. Treat the first solution with alkali and note the change in shade due to salt formation; to the other solution alternately add acid and alkali.

## Reduction

A characteristic of an azo compound is the ease with which the molecule is cleaved at the double bond by reducing agents to give two amines. Because amines are colorless, the reaction is easily followed by the color change. The reaction is useful in preparing hydroxyamino and similar compounds in analyzing azo dyes by titration with a reducing agent and in identifying azo compounds from an examination of the cleavage products.

This reaction can, if necessary, be run on five times the indicated scale. Dissolve about 0.1 g of tin(II) chloride in 0.2 mL of concentrated hydrochloric acid, add a small quantity of the azo compound (20 mg), and heat. A colorless solution should result, and no precipitate should form on adding water. The aminophenol or the diamine products are present as the soluble hydrochlorides; the other product of cleavage, sulfanilic acid, is sufficiently soluble to remain in solution.

$$NaO_3S-\text{C}_6H_4-N=N-\text{C}_6H_4-N(CH_3)_2 \xrightarrow[SnCl_2]{HCl} HO_3S-\text{C}_6H_4-NH_2 + Cl^- H_3N^+-\text{C}_6H_4-N(CH_3)_2$$

**Cleaning Up.** Dilute the reaction mixture with water, neutralize with sodium carbonate, and remove the solids by vacuum filtration. The solids go in the aromatic amines hazardous waste container, and the filtrate can be flushed down the drain.

# PART 2: Dyeing

## EXPERIMENTS

*Using good laboratory techniques, your hands will not be dyed; use care.*

The following experiments can, if necessary, be run on up to 10 times the indicated amounts. Wearing disposable gloves (polyethylene or vinyl) might be advisable.

## 1. Direct Dyes

The sulfonate groups on the Methyl Orange and Orange II molecules are polar and thus enable these dyes to combine with polar sites in fibers. Wool and silk have many polar sites on their polypeptide chains and hence bind strongly to a dye of this type. Martius Yellow, picric acid, and eosin are also highly polar dyes and thus dye directly to wool and silk.

**Picric acid**

### Orange II or Methyl Orange

The dye bath is prepared from 50 mg of Orange II or Methyl Orange, 0.5 mL of 3 $M$ sodium sulfate solution, 15 mL of water, and 5 drops of 3 $M$ sulfuric acid in a 30-mL beaker. Place a piece of test fabric, a strip 3/4-in. wide, in the bath for 5 min at a temperature near the boiling point. Remove the fabric from the dye bath, allow it to cool, and then wash it thoroughly with soap under running water before drying it.

Dye untreated test fabric and one or more of the pieces of test fabric that have been treated with a mordant following this same procedure. *See* Experiment 3 in this part for the application of mordants.

**Martius Yellow**

### Picric Acid or Martius Yellow

Dissolve 50 mg of one of these acidic dyes in 15 mL of hot water to which a few drops of dilute sulfuric acid have been added. Heat a piece of test fabric in this bath for 1 min; then remove it with a stirring rod, rinse well, scrub with soap and water, and dry. Describe the results.

### Eosin

**Eosin A**
($\lambda_{max}$ 516, 483 nm)

Dissolve 10 mg of sodium eosin in 20 mL of water and dye a piece of test fabric by heating it with the solution for about 10 min. Eosin is the dye used in red ink. Also dye pieces of mordanted cloth in eosin (*see* Experiment 3 in this part). Wash and rinse the dyed cloth in the usual way.

**Cleaning Up.** In each case dilute the dye bath with a large quantity of water and flush it down the drain.

## 2. Substantive Dyes: Congo Red

Cotton and rayons do not have the anionic and cationic carboxyl and amine groups of wool and silk and hence do not dye well with direct dyes; however, cotton and rayons can be dyed with substances of rather high molecular weight showing colloidal properties. Such dyes probably become fixed to the fiber by hydrogen bonding: Congo Red is a substantive dye.

Dissolve 10 mg of Congo Red in 40 mL of water, add about 0.1 mL each of 3 $M$ solutions of sodium carbonate and sodium sulfate, heat to a temperature just below the boiling point, and introduce a piece of test fabric. At the end of 10 min, remove the fabric and wash in warm water with soap as long as the dye continues to be removed. Place pieces of the dyed material in very dilute hydrochloric acid solution and observe the result. Rinse and wash the material with soap.

**Congo Red**
($\lambda_{max}$ 497 nm)

**Cleaning Up.** The dye bath should be diluted with water and flushed down the drain.

## 3. Mordant Dyes

One of the oldest known methods of producing wash-fast dyes involves using metallic hydroxides, which form a link, or mordant (Latin *mordere*, to bite), between the fabric and the dye. Other substances, such as tannic acid, also function as mordants. The color of the final product depends on both the dye used and the mordant. For instance, the dye Turkey Red (alizarin) is red with an aluminum mordant, violet with an iron mordant, and brownish-red with a chromium mordant. Some commercially important mordant dyes possess a structure based on triphenylmethane, as well as Crystal Violet and Malachite Green.

Chromium functioning as a mordant

**Alizarin**
1,2-Dihydroxyanthraquinone

## Applying Mordants—Tannic Acid, Iron, Tin, Chromium, Copper, and Aluminum

Mordant pieces of test fabric or wool yarn by allowing them to stand in a hot (nearly boiling) solution of 0.1 g of tannic acid in 50 mL of water for 30 min. The tannic acid mordant must now be fixed to the cloth; otherwise it would wash out. For this purpose, transfer the cloth or yarn to a hot bath made from 20 mg of potassium antimonyl tartrate [$K(SbO)C_4H_4O_6$; tartar emetic] in 20 mL of water. After 5 min, wring or blot the cloth and dry it as much as possible over a warm hot plate.

Mordant 1/2-in. strips of test cloth or yarn in the following mordants, which are 0.1 $M$ solutions of the indicated salts: ferrous sulfate, stannous chloride, potassium dichromate, copper sulfate, and potassium aluminum sulfate (alum). (The alum and dichromate solutions should also contain 0.05 $M$ oxalic acid.) Immerse pieces of cloth in the solutions, which are kept near the boiling point, for about 15–20 min or longer. These mordanted pieces of cloth or yarn can then be dyed with alizarin (1,2-dihydroxyanthraquinone, Turkey Red) and either Methyl Orange or Orange II in the usual way.

**Cleaning Up.** Mix the mordant baths. The $Fe^{2+}$ and $Sn^{2+}$ will reduce the $Cr^{6+}$ to $Cr^{3+}$. The mixture can then be diluted with water and flushed down the drain because the quantity of metal ions is extremely small. Alternatively, precipitate the ions as the hydroxides, collect by vacuum filtration, and place the solid in the hazardous waste container.

## Dyeing with a Triphenylmethane Dye: Crystal Violet or Malachite Green

A dye bath is prepared by dissolving 10 mg of either Crystal Violet or Malachite Green in 20 mL of boiling water. Dye the mordanted cloth or yarn in this bath for 5–10 min at a temperature just below the boiling point. Dye another piece of cloth that has not been mordanted and compare the two. In each case allow as much of the dye to drain back into the beaker as possible then, using glass rods, wash the dyed cloth or yarn under running water with soap, blot, and dry.

**Cleaning Up.** The stains on glass produced by triphenylmethane dyes can be removed with a few drops of concentrated hydrochloric acid and washing with water, as HCl forms a di- or trihydrochloride more soluble in water than the original monosalt.

The dye bath and acid washings are diluted with water and flushed down the drain because the quantity of dye is extremely small.

## 4. Developed Dyes

A superior method of applying azo dyes to cotton, patented in England in 1880, is that in which cotton is soaked in an alkaline solution of a phenol and then in an ice-cold solution of a diazonium salt; the azo dye is developed directly on the fiber. The reverse process (ingrain dyeing) of impregnating cotton with an amine,

**CAUTION**: $K(SbO)C_4H_4O_6$ is toxic, and the dust is a suspected carcinogen.

N(CH₃)₂

**Malachite Green**
($\lambda_{max}$ 617 nm)

$+N(CH_3)_2Cl^-$

N(CH₃)₂

$C=$ $=\overset{+}{N}(CH_3)_2Cl^-$

N(CH₃)₂

**Crystal Violet**
($\lambda_{max}$ 591, 540 nm)

which is then diazotized and developed by immersion in a solution of the phenol, was introduced in 1887. The first ingrain dye was Primuline Red, obtained by coupling the sulfur dye Primuline, after application to the cloth and diazotization, with 2-naphthol. Primuline (substantive to cotton) is a complex thiazole prepared by heating *p*-toluidine with sulfur and then introducing a solubilizing sulfonic acid group.

**Primuline**

### *Primuline Red*

**Resorcinol**

**Naphthol AS**

Dye three pieces of cotton cloth in a solution of 20 mg of Primuline and 0.5 mL of sodium carbonate solution in 50 mL of water at a temperature just below the boiling point for 15 min. Wash the cloth twice in about 50 mL of water. Prepare a diazotizing bath by dissolving 20 mg of sodium nitrite in 50 mL of water containing a little ice and, just before using the bath, add 0.5 mL of concentrated hydrochloric acid. Allow the cloth dyed with Primuline to stay in this diazotizing bath for about 5 min. Next, prepare three baths for the coupling reaction. Dissolve 10 mg of 2-naphthol in 0.2 mL of 1.5 *M* sodium hydroxide solution and dilute with 10 mL of water; prepare similar baths from phenol, resorcinol, Naphthol AS, or other phenolic substances.

Transfer the cloth from the diazotizing bath to a beaker containing about 50 mL of water and stir. Put one piece of cloth in each of the developing baths and allow them to remain for 5 min. Primuline coupled to 2-naphthol gives the dye called Primuline Red. Draw the structure of the dye.

### *Para Red, an Ingrain Color*

**4-Nitrobenzene diazonium chloride**

**2-Naphthol**

**Para Red**

A solution is prepared by suspending 300 mg of 2-naphthol in 10 mL of water, stirring well, and adding 3 *M* sodium hydroxide solution, a drop at a time, until

⚠ **CAUTION**: 4-Nitroaniline is toxic. Handle with care.

Para Red is the red dye used for the American flag.

the naphthol just dissolves. Do not add excess alkali. The material to be dyed is first soaked in or painted with this solution and then dried, preferably in an oven.

Prepare a solution of 4-nitrobenzenediazonium chloride as follows: Dissolve 140 mg of 4-nitroaniline in a mixture of 3 mL of water and 0.6 mL of 3 *M* hydrochloric acid by heating. Cool the solution in ice (the hydrochloride of the amine may crystallize), add all at once a solution of 80 mg of sodium nitrite in about 0.5 mL of water, and stir. In about 10 min a clear solution of the diazonium salt will be obtained. Just before developing the dye on the cloth, add a solution of 80 mg of sodium acetate in 0.5 mL of cold water. Stir in the acetate well, add 30 mL of water, and immediately add the cloth. The diazonium chloride solution may also be painted onto the cloth.

Good results can be obtained by substituting Naphthol AS for 2-naphthol; in this case it is necessary to warm the Naphthol AS with alkali and to break the lumps with a flattened stirring rod to bring the naphthol into solution.

**Cleaning Up.** The dye baths should be diluted with water and flushed down the drain because the quantity of dye is extremely small.

##  5. Dyeing with Indigo: The Denim Dye

$$\text{Indigo} \xrightleftharpoons[\text{O}_2\ (\text{air})]{\text{Na}_2\text{S}_2\text{O}_4 + \text{NaOH}} \text{Indigo white}$$

**Indigo**
Insoluble in water
Deep blue

**Indigo white**
The leuco form, soluble in water
Colorless or light yellow

Denim, de Nimes, from Nimes, in southern France

Look around you. No matter where in the world you are carrying out these experiments, you will find people wearing cotton clothes dyed with indigo. Indigo, "the king of dyes," has been used for more than 3500 years, first in India and then in ancient Egypt. It is a unique dye in that it can be abraded from the surface of a fabric, where other dyes penetrate the fiber. This explains why the knees and other parts of blue jeans (dyed exclusively with indigo) will gradually turn white. An advantage of this fact has been taken by manufacturers who sell "stonewashed" denim blue jeans that have been tumbled with pumice to wear off some of the dye (which can then be reused).

This dye originally came from the leaves of the indigo plant, but it is now produced by chemical synthesis. It is one of the *vat dyes*, a term applied to dyes that are reduced to a colorless (leuco) form that is then oxidized on the surface of the fiber. Formerly, the reduction was carried out by fermenting plant leaves; now, chemical reducing agents, most commonly sodium hydrosulfite, are used.

## Experimental Considerations

Indigo is a dark-blue powder that is completely insoluble in water and is not easily wet by water. To disperse the dye in water, a bit of soap is added to the dye bath. The liquid will appear blue, but the dye is not dissolved; it is merely suspended. Sodium hydrosulfite ($Na_2S_2O_4$) is then added to the hot dye bath. Because this reducing agent decomposes on storage, it is difficult to state exactly how much should be used. Add the required amount, stir, and then look through the side of the beaker. If the solution is not transparent and light yellow in color, add more of the reducing agent. From the top the liquid will still appear to be dark blue because of a film of the oxidized dye on the surface of the solution.

## Procedure

Use 100 mg of indigo and a drop of detergent or a pinch of soap powder. Boil the dye with 50 mL of water, 2.5 mL of 3 $M$ sodium hydroxide solution, and about 0.5 g of sodium hydrosulfite until the dye is reduced. At this point a clear solution will be seen through the side of the beaker. Add more sodium hydrosulfite if necessary to produce a clear solution. Introduce a piece of cloth and boil the solution gently for 10 min. Rinse the cloth well in water and then allow it to dry. To increase the intensity of the dye, repeat the process several times with no drying. Describe what happens during the drying process.

Other dyes that can be used in this procedure are Indanthrene Brilliant Violet and Indanthrene Yellow.

**Indanthrene Yellow**

**Cleaning Up.** Add household bleach (5.25% sodium hypochlorite solution) to the dye bath to oxidize it to the starting material. The mixture can be diluted with water and flushed down the drain; alternatively, the small amount of solid can be removed by filtration, with the solid placed in the aromatic amines hazardous waste container.

# 6. Disperse Dyes: Disperse Red

Fibers such as Dacron, acetate rayon, nylon, and polypropylene are difficult to dye with conventional dyes because they contain so few polar groups. These fibers are dyed with substances that are insoluble in water but that, at elevated temperatures

**Celliton Fast Blue B**

**Celliton Fast Pink B**

(pressure vessels), are soluble in the fiber as true solutions. They are applied to the fiber in the form of a dispersion of finely divided dye (hence the name). The Cellitons are typical disperse dyes.

Disperse Red, a brilliant red dye used commercially, is synthesized in this experiment.

## Diazotization of 2-Amino-6-methoxybenzothiazole

**2-Amino-6-methoxybenzothiazole**
MW 180.23
mp 165–167°C

To 135 mg (0.75 mmol) of 2-amino-6-methoxybenzothiazole in 1.5 mL of water in a test tube, add 0.175 mL of concentrated hydrochloric acid; then cool the solution to 0°C–5°C. To this mixture add, dropwise, an *ice-cold* solution of 55 mg of sodium nitrite that has been dissolved in 0.75 mL of water. The reaction mixture changes color, and some of the diazonium salt crystallizes out, but it should not foam. Foaming, caused by the evolution of nitrogen, is an indication the mixture is too warm. Keep the mixture ice-cold until used in the coupling reaction.

## Disperse Red

**N-Phenyldiethanolamine**
MW 181.24
mp 56–58°C

**Disperse Red**

To 135 mg (0.75 mmol) of *N*-phenyldiethanolamine in 0.75 mL of hot water, add just enough 3 *M* hydrochloric acid to bring the amine into solution. This amount is less than 0.5 mL. Cool the resulting solution to 0°C in ice, add to it, dropwise and with *very thorough mixing*, the diazonium chloride solution. Mix the solution well by drawing it into a Pasteur pipette and expelling it into a cold reaction tube. Allow the mixture to come to room temperature over a period of 10 min; then add 225 mg of sodium chloride and heat the mixture to boiling. The sodium chloride decreases the solubility of the product in water. Allowing the hot solution to cool slowly to room temperature should afford easily filterable crystals. Collect the dye on a Hirsch funnel, wash it with a few drops of saturated sodium chloride solution, and press it dry on the funnel.

This reaction often produces a noncrystalline product that looks like purple tar, which is the dye. It can be used to dye the Multifiber Fabric, so do not discard the reaction mixture.

Save the filtrate. Add the material on the filter paper to 50 mL of boiling water and dye a piece of test cloth for 5 min. Do the same with the filtrate. Wash the cloth with soap and water, rinse and dry it, and compare the results.

**Cleaning Up.** The dye baths should be diluted with water and flushed down the drain because the quantity of dye is extremely small.

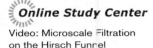

*Online Study Center*

Video: Microscale Filtration on the Hirsch Funnel

## 7. Test Identification Stain

In the dye industry, a proprietary mixture of three dyes called *Test Identification Stain*[4] is used to dye cloth. Prepare a dye bath by dissolving 25 mg of the dye in 25 mL of water, adding 1 drop of acetic acid, and bringing the mixture to a boil. Dye a piece of test cloth in the mixture for 5 min, remove it, immediately wash it under running water, and then scrub it *thoroughly* with soap and water before rinsing and drying. Describe the rather extraordinary result and suggest how this mixture could be used industrially. Several pieces of test cloth can be dyed in this one dye bath. Analyze the dye mixture by thin-layer chromatography, using 40:60 mixture of ethanol and hexane as the eluent.

**Cleaning Up.** Dilute the dye bath with copious amounts of water and flush it down the drain. Dyeing cloth removes much of the dye from the dye bath.

## 8. Fluorescein and Eosin

Fluorescein, as its name implies, is fluorescent; in fact, it is so fluorescent that the sodium salt in water can be detected under ultraviolet (UV) light at concentrations of 0.02 ppm. Because of this property, fluorescein is used to trace the paths of underground rivers such as those found in Mammoth Cave in Kentucky, to trace sources of contamination of drinking water, to find leaks in the huge condensers found in power plants, and so on. It is also used to visualize scratches on the cornea of the eye. The tetrabromo derivative is eosin, a dye used in lipstick, nail polish, red ink, and a biological stain.

---

4. The *Test Identification Stain* is available from Kontes.

The synthesis of fluorescein is very straightforward; however, in this experiment we have deliberately omitted the equations for the synthesis, the mechanism of the reaction, the structures of the molecules in acid and base, and the structures of the tetrabromo derivatives. These are left for you to solve or look up in the library.[5]

## Synthesis of Fluorescein

In a reaction tube heat 100 mg of zinc chloride in a hot sand bath until no more moisture comes from the tube; then add 200 mg of resorcinol and 140 mg of phthalic anhydride. Place a thermometer into the mixture and heat the mixture to 180°C. Stir with a stirring rod and continue heating for 10 min, at which time the mixture should cease bubbling and become stiff. Cool the mixture to below 100°C, add 2 mL of water and 0.2 mL of concentrated hydrochloric acid, then stir and grind the solid (this process is called *trituration*) while heating it on a steam bath or boiling water bath (not on a sand bath because it will bump). Remove the liquid with a Pasteur pipette. After again triturating with another similar portion of water and acid, collect the solid on a Hirsch funnel, wash it well with water, and allow it to dry. A typical yield is 240 mg.

## Fluorescence

Add 50 mg of fluorescein to 0.5 mL of 3 *M* sodium hydroxide and 0.5 mL of water; then dilute the mixture to 100 mL. Examine the solution by transmitted light and then by reflected sunlight or UV light. Make the solution acidic and again examine it under sunlight or UV light.

## Tetrabromofluorescein, Eosin: The Red Ink Dye

In a reaction tube place 100 mg of fluorescein and 0.4 mL of ethanol. To this mixture add 215 mg of bromine in 0.5 mL of carbon tetrachloride, dropwise, mixing thoroughly after each drop. After half the bromine is added, a solution of soluble dibromofluorescein results. When all the bromine has been added, the tetrabromo compound will precipitate. After about 15 min, cool the reaction mixture in ice and remove the solvent with a Pasteur pipette. Wash the product in the reaction tube with a few drops of ice-cold ethanol; then dry it under vacuum. The yield should be about 160 mg.

Prepare the ammonium salt of eosin by exposing the salt to ammonia vapors. Moisten about 80 mg of fluorescein with a few drops of ethanol on a small filter paper. Fold up the paper and push it into a reaction tube above 0.5 mL of ammonium hydroxide. Cap the tube with a septum. After about an hour, a test portion of the derivative should be completely soluble in water. Dye a piece of test cloth with this dye. Eosin is used commercially to dye silk.

---

5. If you would like to read about the original synthesis, see *J. Prakt. Chem.* **1922**, *104*, 123, which will require a reading knowledge of German, and the structure proof in *Comp. Rendu.* **1937**, *205*, 864, which will require a reading knowledge of French. Further searching in the library may disclose other papers in English.

Answer the following questions. The answers for some will come from your experimental observations:

1. Write a balanced equation for the reaction of resorcinol with phthalic anhydride.
2. Calculate the number of moles of starting material, determine the limiting reagent, and calculate the theoretical yield.
3. In view of the balanced equation, why is it necessary to dehydrate zinc chloride, a very hygroscopic substance?
4. To which general category of substances does zinc chloride belong that is relevant to its use in this reaction? What two functions does zinc chloride serve?
5. Write a possible mechanism for the synthesis of fluorescein.
6. Write an equation showing the reaction of fluorescein with sodium hydroxide. *Hint*: A quinoid structure is involved.
7. How do you account for the changes in fluorescence in going from an acidic to a basic medium?
8. Compare the color and fluorescence changes of fluorescein to those of the related compound, phenolphthalein.
9. Write a balanced equation for the bromination of fluorescein to give first the dibromo and then the tetrabromo compound. Explain your choices for the positions of bromination.
10. Write a balanced equation for the reaction of tetrabromofluorescein with ammonia.

## 9. Optical Brighteners: Fluorescent White Dyes

Most modern detergents contain a blue-white fluorescent dye that is adsorbed on the cloth during the washing process. These dyes fluoresce; that is, absorb UV light and reemit light in the visible blue region of the spectrum. This blue color counteracts the pale-yellow color of white goods, which develops because of a buildup of insoluble lipids. The modern-day usage of optical brighteners has replaced a past custom of using bluing (ferriferrocyanide).

**Blankophor B**
An optical brightener

### Dyeing with Detergents

Immerse a piece of test fabric in a hot solution (0.5 g of detergent, 200 mL of water) of a commercial laundry detergent that you suspect may contain an optical brightener (e.g., Tide and Cheer) for 15 min. Rinse the fabric thoroughly, dry, and compare it to an untreated fabric sample under a UV lamp.

**Cleaning Up.** Dilute the solution with water and flush down the drain.

**Online Study Center**

**General Resources**
Additional Experiments

## QUESTIONS

1. Write reactions showing how nylon can be synthesized so that it will react with (a) basic dyes and (b) acidic dyes.
2. Draw the resonance form of dimethylaniline that is most prone to react with diazotized sulfanilic acid.
3. Draw a resonance form of indigo that would be present in base.
4. Draw a resonance form of indigo that has been reduced and is therefore colorless.

# Martius Yellow

**PRELAB EXERCISE:** Prepare a time line for this experiment, clearly indicating which experiments can be carried out simultaneously.

The experiments in this chapter were introduced by Louis Fieser of Harvard University over 70 years ago. Although the initial procedure starts with macroscale quantities of materials, the later steps often become microscale experiments and benefit greatly from using microscale equipment. It has long been the basis for an interesting laboratory competition. Ken Williamson, one of the present authors, was a winner over 50 years ago.

Starting with 5 g of 1-naphthol, a skilled chemist familiar with the procedures can prepare pure samples of the seven compounds in this chapter in 3–4 h. In a first trial of the experiments here, a particularly competent student, one who plans his or her work in advance, can complete the program in two laboratory periods (6 h).

The first compound of the series, Martius Yellow, a mothproofing dye for wool (1 g of Martius Yellow dyes 200 g of wool) discovered in 1868 by Karl Alexander von Martius, is the ammonium salt of 2,4-dinitro-1-naphthol (**1**). This compound is obtained by sulfonating 1-naphthol with sulfuric acid and treating the resulting disulfonic acid with nitric acid in an aqueous medium. The exchange of groups occurs with remarkable ease, and it is not necessary to isolate the disulfonic acid. The advantage of introducing the nitro groups in this indirect way is that 1-naphthol is highly sensitive to oxidation and would be partially destroyed by direct nitration. Martius Yellow is prepared by reacting the acidic phenolic group of compound **1** with ammonia to form the ammonium salt. A small portion of this salt (Martius Yellow) is converted by acidification and crystallization into pure 2,4-dinitro-1-naphthol, a sample of which is saved. The remainder is

**1-Naphthol**
MW 144.16

**1**
MW 234.16

**3**
MW 256.25

**2**
MW 208.65

**4**
MW 258.27

**7**
MW 173.16

**6**
MW 173.16

**5**
MW 215.25

suspended in water and reduced to diaminonaphthol with sodium hydrosulfite according to the following equation:

**Martius Yellow**     $+\ 6\ Na_2S_2O_4\ +\ 8\ H_2O\ \longrightarrow$     **2,4-Diamino-1-naphthol**     $+\ 12\ NaHSO_3$

The diaminonaphthol separates in the free condition, rather than as an ammonium salt. The diamine, unlike the dinitro compound, is a very weak acidic substance that does not readily form salts.

Because 2,4-diamino-1-naphthol is exceedingly sensitive to air oxidation as a free base, it is at once dissolved in dilute hydrochloric acid. The solution of diaminonaphthol dihydrochloride is clarified with decolorizing charcoal and divided into two equal parts. One part on oxidation with iron(III) chloride affords the fiery-red 2-amino-1,4-naphthoquinonimine hydrochloride (**2**). This substance, like many other salts, has no melting point; it is therefore converted for identification to the yellow diacetate (**3**). Compound **2** is remarkable in that it is stable enough to be isolated. On hydrolysis it affords the orange 4-amino-1,2-naphthoquinone (**7**).

The other part of the diaminonaphthol dihydrochloride solution is treated with acetic anhydride and then sodium acetate; the reaction in aqueous solution effects selective acetylation of the amino groups and affords 2,4-diacetylamino-1-naphthol (**4**). Oxidation of compound **4** by $Fe^{3+}$ and oxygen from the air is attended with cleavage of the acetylamino group at the 4-position, and the product is 2-acetylamino-1,4-naphthoquinone (**5**). This yellow substance is hydrolyzed by sulfuric acid to the red 2-amino-1,4-naphthoquinone (**6**), the last member of the series. The reaction periods are brief, and the yields high; however, remember to scale down the quantities of the reagents and the solvents if the quantity of the starting material is less than that called for.[1]

## EXPERIMENTS

If necessary, the quantities in the following seven experiments can be doubled.

# 1. Preparation of 2,4-Dinitro-1-Naphthol (1)

Use care when working with hot concentrated sulfuric and nitric acids.

Place 2.5 g of pure 1-naphthol[2] in a 50-mL Erlenmeyer flask, add 5 mL of concentrated sulfuric acid, and heat the mixture with swirling on a steam bath or a hot plate for 5 min, at which time the solid should have dissolved and an initial red color should be discharged. Cool in an ice bath, add 13 mL of water, and cool the solution rapidly to 15°C. Measure 3 mL of concentrated nitric acid into a test

1. This series of reactions lends itself to a laboratory competition, the rules for which might be as follows: (a) No practice or advance preparation is allowed except the collection of reagents not available at the contestant's bench [ammonium chloride, sodium hydrosulfite, iron(III) chloride solution, and acetic anhydride]. (b) The time scored is the actual working time, including that required for cleaning the apparatus and bench; labels for ziplock bags can be prepared outside of the working period. (c) Time is not charged during an interim period (overnight) when solutions are let stand to crystallize or solids are let to dry, on the condition that during this period no adjustments are made and no cleaning or other work is done. (d) Melting-point and color-test characterizations are omitted. (e) Successful completion of the contest requires preparing authentic and macroscopically crystalline samples of all seven compounds. (f) Judgment of the winners among the successful contestants is based on quality and quantity of samples, technique and neatness, and working time. (Superior performance is 3–4 h.)

2. If the 1-naphthol is dark, it can be purified by distillation at atmospheric pressure in the hood. The colorless distillate is most easily pulverized (also in the hood) before it has completely cooled and hardened.

**1**

⚠

Avoid contact of the yellow
product and its orange
ammonium salt with the skin.

**Martius Yellow**

tube and transfer it with a Pasteur pipette in small portions (0.25 mL each) to the
chilled aqueous solution while maintaining the temperature between 15°C and
20°C by swirling the flask vigorously in the ice bath. When the addition is complete and the exothermic reaction has subsided (1–2 min), warm the mixture
gently to 50°C (1 min), whereupon the nitration product should separate as a stiff
yellow paste. Apply heat (the full heat of the steam bath for an additional 1 min if
using a steam bath), fill the flask with water, break up the lumps and stir to an even
paste, collect the product (**1**) on a small Büchner funnel, wash it well with water,
and then wash it into a 250-mL beaker with water (50 mL). Add 75 mL of hot
water and 2.5 mL of concentrated ammonium hydroxide solution (den. 0.90), heat
to the boiling point, and stir to dissolve the solid. Filter the hot solution by suction
if it is dirty, add 5 g of ammonium chloride to the filtrate to salt out the ammonium
salt (Martius Yellow), cool in an ice bath, collect the orange salt, and wash it with
water containing 1%–2% of ammonium chloride. The salt does not have to be
dried (the dry weight yield is 3.8 g—a percent yield of 88%).

Set aside an estimated 150 mg of the moist ammonium salt. This sample is to
be dissolved in hot water, the solution is acidified with HCl, and the free 2,4-
dinitro-1-naphthol (**1**) is crystallized from methanol or ethanol (use decolorizing
charcoal if necessary); it forms yellow needles (mp 138°C).

The proton nuclear magnetic resonance ($^1$H NMR) and infrared (IR) spectra
of 1-naphthol are shown in Figures 47.1 and 47.2.

■ **FIG. 47.1**
The $^1$H NMR spectrum
of 1-naphthol (250 MHz).

**Cleaning Up.** Combine aqueous filtrates, dilute with water, neutralize with sodium carbonate, and flush the solution down the drain. Recrystallization solvents go in the organic solvents container.

## Preparation of Unstable 2,4-Diamino-1-Naphthol

Na$_2$S$_2$O$_4$
**Sodium hydrosulfite**

**2,4-Diamino-1-naphthol dihydrochloride**

Wash the remaining ammonium salt into a beaker with about 100 mL of water, add 20 g of sodium hydrosulfite, stir until the original orange color has disappeared and a crystalline tan precipitate has formed (5–10 min), and then cool in ice. Prepare two separate solutions: (1) 1 g of sodium hydrosulfite in 50 mL of water for use in washing and (2) a 250-mL beaker containing 3 mL of concentrated hydrochloric acid and 12 mL of water. When collecting the precipitate by suction filtration on a small Büchner funnel, use the hydrosulfite solution for rinsing and washing, be sure to avoid even briefly sucking air through the cake after the reducing agent has been drained away, and immediately wash the solid into the beaker containing the dilute hydrochloric acid and stir to convert all the diamine to the dihydrochloride.

The acid solution, often containing suspended sulfur and filter paper, is clarified using suction filtration through a moist charcoal bed made by shaking 1 g of powdered decolorizing charcoal (carbon) with 13 mL of water in a stoppered flask to produce a slurry and pouring this on the filter paper in a 50-mm Büchner funnel. Pour out the water from the filter flask and then filter the solution of dihydrochloride. Divide the pink or colorless filtrate into approximately two equal parts and immediately add the reagents for converting one part to compound **2** (Experiment 2) and the other part to compound **4** (Experiment 4).

**Cleaning Up.** Neutralize the filtrate with sodium carbonate and dilute with water. In the hood, cautiously add household bleach (aqueous sodium hypochlorite solution) to the mixture until a test with 5% silver nitrate proves that no more hydrosulfite is present (the absence of a black precipitate). Neutralize the solution and filter it through Celite to remove any suspended solids. Dilute the filtrate with water and flush down the drain. The solid residue goes in the nonhazardous solid waste container.

## 2. Preparation of 2-Amino-1,4-naphthoquinonimine Hydrochloride (2)

**2**

To one-half of the diamine dihydrochloride solution (Experiment 1), add 12.5 mL of 1.3 $M$ iron(III) chloride solution,[3] cool in ice, and, if necessary, initiate crystallization by scratching. Rub the liquid film with a glass stirring rod at a single spot slightly above the surface of the liquid. If efforts to induce crystallization are unsuccessful, add more hydrochloric acid until crystallization occurs. Collect the red product and wash with 3 $M$ HCl (aq). The dry weight yield is 1.2–1.35 g.

Divide the moist product into three equal parts; then spread out one part to dry for conversion to compound **3** (Experiment 3). The other two parts can be used while still moist for conversion to compound **7** (Experiment 7) and for recrystallization. Dissolve one part by warming in a little water containing 2–3 drops of hydrochloric acid, shake for a minute or two with decolorizing charcoal, filter by suction, and add concentrated hydrochloric acid to decrease the solubility. Collect the product by suction filtration.

**Cleaning Up.** Neutralize the filtrate with sodium carbonate and collect the iron hydroxide by vacuum filtration through Celite on a Büchner funnel. The solid goes into the nonhazardous solid waste container, whereas the filtrate is diluted with water and flushed down the drain.

## 3. Preparation of 2-Amino-1,4-naphthoquinonimine Diacetate (3)

**3**

A mixture of 0.25 g of the dry quinonimine hydrochloride (**2**; Experiment 2), 0.25 g of sodium acetate (anhydrous), and 1.5 mL of acetic anhydride is stirred in a reaction tube and warmed gently on a sand or steam bath. With thorough stirring, the red salt should soon change into yellow crystals of the diacetate. The solution may appear red, but as soon as particles of red solid have disappeared, the mixture can be poured into about 5 mL of water. Stir until the excess acetic anhydride has either dissolved or become hydrolyzed, collect and wash the product (the dry weight yield is 250 mg) with water, and (drying is unnecessary) crystallize it from ethanol or methanol; yellow needles result (mp 189°C).

---

3. *Preparation of iron(III) chloride solution*: Dissolve 45 g of $FeCl_3 \cdot 6\,H_2O$ (MW 270.32) in 50 mL of water and 50 mL of concentrated hydrochloric acid by warming, cooling, and filtration (produces 124 mL of solution).

**Cleaning Up.** The filtrate should be diluted with water and flushed down the drain. The crystallization solvent goes in the organic solvents container.

## 4. Preparation of 2,4-Diacetylamino-1-Naphthol (4)

To one-half of the diaminonaphthol dihydrochloride solution saved from Experiment 1, add 1.5 mL of acetic anhydride, stir vigorously, and then add a solution of 1.5 g of sodium acetate (anhydrous) and about 50 mg of sodium hydrosulfite in 10–15 mL of water. The diacetate may precipitate as a white powder, or it may separate as an oil that solidifies when chilled in ice and rubbed with a stirring rod. Collect the product and, to hydrolyze any triacetate present, dissolve it in 2.5 mL of 3 M sodium hydroxide and 25 mL of water by stirring at room temperature. If the solution is colored, a few milligrams of sodium hydrosulfite may bleach it. Filter by suction and acidify by gradual addition of well-diluted hydrochloric acid (1 mL of concentrated acid in 19 mL of water). The diacetate tends to remain in a supersaturated solution; hence, either to initiate crystallization or to ensure maximum separation, it is advisable to stir well, rub the walls with a stirring rod, and cool in ice. Collect the product, wash it with water, and divide it into thirds (the dry weight yield is 1–1.3 g).

Two-thirds of the material can be converted without drying into compound **5**, and the other third can be used to prepare a crystalline sample. Dissolve the one-third reserved for crystallization (moist or dry) in enough hot acetic acid to bring about solution, add a solution of a small crystal of tin(II) chloride in a few drops of 10% HCl (aq) solution to inhibit oxidation, and dilute gradually with 5 to 6 volumes of water at the boiling point. Crystallization may be slow, and cooling and scratching may be necessary. The pure diacetate forms colorless prisms (mp 224°C, dec.).

**Cleaning Up.** Combine all filtrates—including the acetic acid used to crystallize the product—dilute with water, neutralize with sodium carbonate, and flush the solution down the drain.

## 5. Preparation of 2-Acetylamino-1,4-Napthoquinone (5)

Dissolve 0.75 g of the moist diacetylaminonaphthol (compound **4**; Experiment 4) in 5 mL of acetic acid (hot), dilute with 10 mL of hot water, and add 5 mL of 0.13 M iron(III) chloride solution. The product separates promptly into flat, yellow needles, which are collected (after cooling) and washed with a little alcohol; the yield is usually 0.6 g. Dry one-half of the product for conversion to compound **6** and crystallize the remaining half from 95% ethanol (mp 204°C).

**Cleaning Up.** Dilute the filtrate with water, neutralize it with sodium carbonate, and flush it down the drain. A negligible quantity of iron is disposed of in this way.

## 6. Preparation of 2-Amino-1,4-Napthoquinone (6)

**6**

To 0.25 g of dry 2-acetylamino-1,4-naphthoquinone (**5**) contained in a reaction tube, add 1 mL of concentrated sulfuric acid and heat the mixture on a steam or sand bath with swirling to promote rapid solution (1–2 min). After an additional 5 min, cool the deep-red solution, dilute it with a large amount of water, and collect the precipitated product. Wash the crude material with water and recrystallize it from alcohol or an alcohol:water mixture[4] while it is still moist. The yield of red needles of the amino quinone (mp 206°C) will be about 200 mg.

**Cleaning Up.** The filtrate is neutralized with sodium carbonate, diluted with water, and flushed down the drain.

## 7. Preparation of 4-Amino-1,2-Naphthoquinone (7)

Compound **2** can be moist.

**7**

Dissolve 0.5 g of the aminonaphthoquinonimine hydrochloride (**2**) reserved from Experiment 2 in 12 mL of water, add 1 mL of concentrated ammonium hydroxide solution (den 0.90), and boil the mixture for 5 min. The free quinonimine initially precipitated is hydrolyzed to a mixture of the aminoquinone (**7**) and the isomeric compound **6**. Cool, collect the precipitate, suspend it in about 25 mL of water, and add 12.5 mL of 3 *M* sodium hydroxide solution. Stir well, remove the small amount of residual 2-amino-1,4-naphthoquinone (**6**) by filtration, and acidify the filtrate with acetic acid. The orange precipitate of compound **7** is collected, washed, and crystallized while still wet from 250–300 mL of hot water (the separation is slow). The yield of orange needles (dec. about 270°C) is about 200 mg.

**Cleaning Up.** The filtrate is neutralized with sodium carbonate, diluted with water, and flushed down the drain.

## QUESTIONS

1. Write a balanced equation for the preparation of 2-amino-1,4-naphthoquinonimine hydrochloride (**2**).
2. What part of the molecule accounts for the peak at 3300 cm$^{-1}$ in the IR spectrum of 1-naphthol (Fig. 47.2)?

---

4. The exact ratio of alcohol:water is determined experimentally. Try different ratios, such as 50:50, 70:30, or 80:20. *See* Chapter 4 on how to choose a recrystallization solvent.

# CHAPTER

# 48

# Diels–Alder Reaction

**Online Study Center**

This icon will direct you to techniques, equipment setups, and online resources at http://college.hmco.com/PIC/williamsonMMOE5e.

**PRELAB EXERCISE:** Describe in detail the laboratory operations, reagents, and solvents you would employ to prepare

Otto Diels and his pupil Kurt Alder received the Nobel Prize in Chemistry in 1950 for their discovery and work on the reaction that bears their names. Its great usefulness lies in its high yield and high stereospecificity. A cycloaddition reaction, the Diels–Alder reaction involves the 1,4-addition of a conjugated diene in the s-*cis* conformation to an alkene in which two new σ (sigma) bonds are formed from two π (pi) bonds.

s-*trans*          s-*cis*

The adduct is a six-membered ring alkene. The diene can have the two conjugated bonds contained within a ring system, as with cyclopentadiene or cyclohexadiene, or the molecule can be an acyclic diene that must be in the *cis* conformation about the single bond before reaction can occur.

The reaction works best when there is a marked difference between the electron densities in the diene and the alkene with which it reacts—the dienophile. Usually the dienophile has electron-attracting groups attached to it, whereas the diene is electron rich, for example, as in the reaction of methyl vinyl ketone with 1,3-butadiene.

**Methyl vinyl ketone     1,3-Butadiene**

Retention of the configurations of the reactants in the products implies that both new σ bonds are formed almost simultaneously. If not, then the intermediate with a single new bond could rotate about that bond before the second σ bond is formed, thus destroying the stereospecificity of the reaction.

**Dimethyl maleate**
**+**
**1,3-Butadiene**

*cis* **Isomer**

The following does not happen:

***trans* Isomer**

A highly stereospecific reaction

This reaction is not polar in that no charged intermediates are formed. Neither is it a radical reaction because no unpaired electrons are involved. It is instead known as a *concerted reaction*, or one in which several bonds in the transition state are simultaneously made and broken. When a cyclic diene and a cyclic dienophile react with each other, more than one stereoisomer may be formed. The isomer that predominates is the one that involves maximum overlap of $\pi$ electrons in the transition state. The transition state for the formation of the *endo* isomer involves a sandwich with the diene directly above the dienophile. To form the *exo* isomer, the diene and dienophile would need to be arranged in a stair-step fashion. Robert B. Woodward and Roald Hoffmann formulated the theoretical rules involving the correlation of orbital symmetry, which govern the Diels–Alder and other electrocyclic reactions.

**Maximum overlap
of $\pi$ electrons**

***endo* Isomer
predominant product**

**π Electron overlap not so large**

***exo* Isomer**

**Woodward's Diels–Alder adduct**

**Cantharidin**

The Diels–Alder reaction has been used extensively in the synthesis of complex natural products because it is possible to exploit the formation of a number of chiral centers in one reaction and also because of the regioselectivity of the reaction. For example, the first step in Woodward's synthesis of cortisone was the formation of a Diels–Alder adduct.

But the reaction is also subject to steric hindrance, especially when the difference between the electron-withdrawing and electron-donating characteristics of the two reactants is not great. When Woodward tried to synthesize cantharidin, the active ingredient in Spanish fly, by the Diels–Alder condensation of furan with dimethylmaleic anhydride, the reaction did not work. The reaction possesses $-\Delta V^*$ (it proceeds with a net decrease in volume). High pressure should overcome this problem, but this reaction will not proceed even at 600,000 lb/in.$^2$ ($4.1 \times 10^{10}$ dynes/cm$^2$). A closely related reaction will proceed at 300,000 lb/in.$^2$ and has been used to synthesize this molecule.[1] Cantharidin is a powerful vesicant (blister former).

Cyclopentadiene is obtained from the light oil from coal tar distillation but exists as the stable dimer, dicyclopentadiene, which is the Diels–Alder adduct from two molecules of the diene. Thus generation of cyclopentadiene by pyrolysis of the dimer represents a reverse Diels–Alder reaction. *See* Figures 48.5 to 48.7 at the end of the chapter for nuclear magnetic resonance ($^1$H and $^{13}$C NMR) and infrared (IR) spectra, respectively, of dicyclopentadiene.

In the Diels–Alder addition of cyclopentadiene and maleic anhydride, the two molecules approach each other in the orientation shown in the top drawing on the previous page because this orientation provides maximal overlap of π bonds of the two reactants and favors formation of an initial π complex and then the final *endo* product. Dicyclopentadiene also has the *endo* configuration.

1. Dauben, W. G.; Kessel, C. R.; Takemura, K. H. *J. Am. Chem. Soc.* **1980**, *102*, 6893.

# EXPERIMENTS

# 1. Cracking of Dicyclopentadiene

Check dicyclopentadiene for peroxides. Commercially available test strips can test for the presence of peroxides. If peroxides are present, the bottle should be discarded. If the bottle is over 12 months old, test for peroxides and/or discard bottle. The presence of peroxides during a distillation can be very dangerous; an explosion may occur.

**Dicyclopentadiene**
den. 0.98
MW 132.20
bp 170°C
$n_D^{20}$ 1.5100

~ 160°C
room temperature

**Cyclopentadiene**
den. 0.80
MW 66.10
bp 41°C

> **IN THIS EXPERIMENT** the dimer of cyclopentadiene is cracked into two molecules of the monomer by dripping it onto very hot mineral oil. The product, which boils at 41°C, is collected in an ice-cooled receiver.

## *Microscale Procedure*

Half fill with mineral oil the 5-mL, short-necked, round-bottomed flask equipped with an addition port bearing a septum on the sidearm and topped with a distillation head and thermometer (Fig. 48.1). Set the controller on the flask heater at half maximum. The temperature can be controlled by piling up or scraping away sand from the flask. You can either start with the thermometer down in the oil and raise it as the temperature approaches 250°C or simply wait until you judge the oil is hot enough.

Place a small, tared collection vial in an ice-filled 30-mL beaker at the end of the distilling head, taking care to keep water out of the vial. Using a syringe, draw 0.6 mL of dicyclopentadiene (Fig. 48.2) from the septum-capped storage container after first injecting 0.6 mL of air into the container to overcome the vacuum. Insert the needle of the filled syringe into a rubber stopper or cork to avoid loss of the contents until they are used.

When the mineral oil is judged to be hot enough (250°C), inject the dicyclopentadiene through the septum on the addition port. Add it dropwise at a rate such that the temperature of the thermometer never exceeds 45°C. The boiling point of cyclopentadiene is 41°C. Add the dimer over a 10-min period. If the dimer is added too slowly, the yield will be lower. Once all the dicyclopentadiene has been added, remove the syringe and weigh the product after closing the vial. Rinse the syringe with acetone in the hood because the dimer has a very

**Online Study Center**

Photo: Capping a Reaction Tube with a Septum; Videos: Steam Distillation Apparatus, The Grignard Reaction: Removing a Liquid from a Septum-Capped Bottle with Syringe

■ **FIG. 48.1**
An apparatus for the microscale
cracking of dicyclopentadiene.
Add dicyclopentadiene dropwise
via a syringe so that the distillate
temperature does not exceed
45°C.

Ice

22-Gauge needle

Sand Bath
no higher
than 200°C = 200°C

Hot sand bath

disagreeable odor. Calculate the percent yield of cyclopentadiene. If the product
is cloudy, add a small quantity of anhydrous calcium chloride pellets to dry it if
the maleic anhydride experiment (Experiment 2) is being done next. It need not be
dry to make ferrocene. Keep this cyclopentadiene on ice and use it the same day
it is prepared for Experiment 2.

The $^1$H NMR, IR, and $^{13}$C NMR spectra of dicyclopentadiene appear in Figures 48.5, 48.6, and 48.7, respectively, at the end of the chapter.

Don't disassemble the apparatus
until you are sure you have
0.3 mL of product.

**Cleaning Up.** Place the mineral oil from the reaction flask and any unused dicyclopentadiene in the organic solvents container. If calcium chloride was used,
free it of cyclopentadiene by evaporation in the hood, and then place it in the nonhazardous solid waste container.

■ **FIG. 48.2**

Dicyclopentadiene has a very bad odor and so is dispensed from a closed container. Remove 0.6 mL from the septum-capped bottle after injecting 0.6 mL of air.

Dicyclopentadiene

Rubber septum

1-mL Syringe

## Macroscale Procedure

Measure 20 mL of dicyclopentadiene into a 100-mL flask and arrange for fractional distillation into an ice-cooled receiver (*see* Fig. 19.2 on page 355). Heat the dimer with an electric flask heater until it refluxes briskly and at a rate such that the monomeric diene begins to distill in about 5 min and soon reaches a steady boiling point between 40°C and 42°C. Apply heat continuously to promote rapid distillation without exceeding the boiling point of 42°C. Distillation for 45 min should provide the 12 mL of cyclopentadiene required for two preparations of the adduct; continued distillation for another half hour produces a total of about 20 mL of monomer.

The $^1$H NMR, IR, and $^{13}$C NMR spectra of dicyclopentadiene appear in Figures 48.5, 48.6, and 48.7, respectively, at the end of the chapter.

**Cleaning Up.** Pour the pot residue of dicyclopentadiene and unused cyclopentadiene into the recovered dicyclopentadiene container. This recovered material can, despite its appearance, be cracked in the future to give cyclopentadiene, but test it for peroxides. If the pot residue is not to be recycled, place it in the organic solvents container.

**Online Study Center**

Video: Macroscale Cracking of Dicyclopentadiene

Reverse Diels–Alder

⚠️

Check dicyclopentadiene for peroxides. Commercially available test strips can test for the presence of peroxides. If peroxides are present, the bottle should be discarded. If the bottle is over 12 months old, test for peroxides and/or discard bottle. The presence of peroxides during a distillation can be very dangerous; an explosion may occur.

# 2. Synthesis of *cis*-Norbornene-5,6-*endo*-Dicarboxylic Anhydride

**Maleic anhydride**
mp 53°C, MW 98.06

*cis*-**Norbornene-5,6-*endo*-dicarboxylic anhydride**
mp 165°C, MW 164.16

> **IN THIS EXPERIMENT** maleic anhydride (the dienophile) is reacted with cyclopentadiene (the diene) in a mixture of solvents that will dissolve the starting material but not the product, which crystallizes. The product is collected by Pasteur pipette filtration.

## Microscale Procedure

*[handwritten margin note: hexanes in after cyclopentadiene]*

Dissolve 0.20 g of powdered maleic anhydride in 1 mL of ethyl acetate in a tared 10 × 100-mm reaction tube; ~~then add 1 mL hexanes (bp 60°C–80°C)~~. This combination of solvents is used because the product is too soluble in pure ethyl acetate and not soluble enough in pure hexanes. To the solution of maleic anhydride add 0.20 mL (0.160 g) of dry cyclopentadiene, mix the reactants, and observe the reaction. Allow the tube to cool to room temperature, during which time crystallization of the product should occur. If crystallization does not occur, scratch the inside of the test tube with a stirring rod at the liquid-air interface. The scratch marks on the inside of the tube often form the nuclei upon which crystallization starts. Should crystallization occur very rapidly at room temperature, the crystals will be very small. If so, save a seed crystal, heat the mixture until the product dissolves, seed it, and allow it to cool slowly to room temperature. You will be rewarded with large platelike crystals. Remove the solvent from the crystals with a Pasteur pipette that is forced to the bottom of the tube, wash the crystals with one portion of cold ligroin, and remove the solvent (*see* Fig. 4.12 on page 73). Scrape the product onto a piece of filter paper, allow the crystals to air-dry, determine their weight, and calculate the yield of the product. Determine the melting point of the product and turn in any material not used in the next experiment. Thin-layer chromatography (TLC) of the product is not necessary because the product is quite pure. The IR spectrum of the anhydride adduct is in Figure 48.8, the $^1$H NMR spectrum is in Figure 48.9, and the $^{13}$C NMR spectrum is in Figure 48.10—all at the end of the chapter.

**Cleaning Up.** Place the crystallization solvent mixture in the organic solvents container. It contains a very small quantity of the product.

*[margin note]* Mixing of the reactants is very important. Pull the reaction mixture into a pipette and then expel it into the reaction tube.

*Online Study Center*

Video: Filtration of Crystals Using the Pasteur Pipette

## *Macroscale Procedure*

Place 6 g of maleic anhydride in a 125-mL Erlenmeyer flask and dissolve the anhydride in 16 mL of ethyl acetate by heating on a hot plate or a steam bath. Add 16 mL of ligroin (bp 60°C–80°C) or hexane, cool the solution thoroughly in an ice-water bath, and leave it in the bath (some anhydride may crystallize).

The freshly distilled cyclopentadiene may be slightly cloudy because of moisture condensation in the cooled receiver and water in the starting material. Add about 1 g of calcium chloride pellets to remove the moisture. Be sure to use the cyclopentadiene immediately because it will dimerize within a few hours of its preparation. Measure 6 mL of dry cyclopentadiene and add it to the ice-cold solution of maleic anhydride. Swirl the solution in an ice bath for a few minutes until the exothermic reaction ceases and the adduct separates as a white solid. Then heat the mixture on a hot plate or a steam bath until the solid is completely dissolved.[2] If you allow the solution to stand undisturbed, you will be rewarded with a beautiful display of crystal formation. The anhydride crystallizes in long spars (mp 164°C–165°C); a typical yield is 8.2 g.[3] The IR spectrum of the anhydride adduct is in Figure 48.8, the [1]H NMR spectrum is in Figure 48.9, and the [13]C NMR spectrum is in Figure 48.10—all at the end of the chapter.

**Cleaning Up.** Place the crystallization solvent mixture in the organic solvents container. It contains a very small quantity of the product. Allow the organic material to evaporate from the drying agent (in the hood); then place it in the nonhazardous solid waste container.

> **CAUTION:** Cyclopentadiene is flammable.

> Rapid addition at 0°C

## 3. *cis*-Norbornene-5,6-*endo*-Dicarboxylic Acid

H₂O +

**endo,cis-Diacid**

---

2. In case moisture has gotten into the system, a little of the corresponding diacid may remain undissolved at this point and should be removed by filtering the hot solution.

3. The student need not work up the mother liquor but may be interested in learning the result. Concentration of the solution to a small volume is not satisfactory because of the presence of dicyclopentadiene, formed by the dimerization of excess monomer; the dimer has high solvent power. Hence, the bulk of the solvent is evaporated on a steam bath, and the flask is connected to a water pump with a rubber stopper and glass tube and heated under vacuum on a steam bath until dicyclopentadiene is removed and the residue solidifies. Crystallization from 1:1 ethyl acetate–ligroin affords 1.3 g of adduct (mp 156°C–158°C); the total yield is 95%.

> **IN THIS EXPERIMENT** the product from Experiment 2 is boiled with water to hydrolyze the anhydride to the corresponding dicarboxylic acid. On cooling the aqueous solution, the product crystallizes.

## Microscale Procedure

To 0.2 g (200 mg) of the anhydride from Experiment 2, add 2.5 mL of water and a boiling stick in a 10 × 100-mm reaction tube. Heat the mixture to boiling by immersing the tube in a hot sand bath (Fig. 48.3). The anhydride may appear to melt and form globules on the bottom of the tube. As the reaction proceeds, the anhydride will react with the water, and the diacid, which is soluble in boiling water, will be formed. Continue to heat for about 2 min after the last globule disappears. Remove the boiling stick from the hot solution and allow the mixture to cool to room temperature.

If crystallization of the diacid does not occur, follow the same procedure used for the anhydride (Experiment 2). On slow cooling with simultaneous crystal growth the solution will deposit long, needlelike crystals. Again, cool the mixture in ice, allow sufficient time for crystal growth to occur, and then collect the product by filtration on a Hirsch funnel. Use the ice-cold filtrate to complete the transfer. Wash the crystals once with a small quantity of ice water and place the product on a piece of filter paper to dry. Do not discard the filtrate until you have weighed the product. More material can be recovered by concentrating the filtrate and allowing it to cool to give a second crop of crystals. This is a general strategy. Weigh the diacid and determine its melting point and percent yield. The melting point depends on the rate of heating as the anhydride reforms and water splits out. The IR and $^1$H NMR spectra of the

■ **FIG. 48.3**
Hydrolysis of the anhydride
to the *endo,cis*-diacid.

Wood applicator
stick

Cool at this point

Air condenser

Boiling solvent

Temperature
controlled by
depth in sand

diacid are presented in Figures 48.11 and 48.12, respectively, at the end of the chapter.[4]

**Cleaning Up.** The aqueous filtrate from the crystallization contains a very small quantity of the diacid. It can be flushed down the drain.

## Macroscale Procedure

For preparation of the *endo,cis*-diacid, place 4.0 g of the anhydride from Experiment 2 and 50 mL of distilled water in a 125-mL Erlenmeyer flask, grasp with a clamp, swirl over a hot plate, and bring the contents to the boiling point, at which time the solid partially dissolves and partly melts. Continue to heat until all the oil is dissolved; then let the solution stand undisturbed. Because the diacid has a strong tendency to remain in supersaturated solution, allow 30 min or more if necessary for the solution to cool to room temperature; then drop in a boiling stone or touch the surface of the liquid once or twice with a stirring rod. Observe the stone and its surroundings carefully, waiting several minutes before applying the more effective method of making one scratch with a stirring rod on the inner wall of the flask at the air-liquid interface. Let crystallization proceed spontaneously to give large needles; then cool the solution in ice and collect the product. The melting point depends on the rate of heating as the anhydride reforms and water splits out. The IR and ${}^1$H NMR spectra of the diacid are found in Figures 48.11 and 48.12, respectively at the end of the chapter.[5]

**Cleaning Up.** The aqueous filtrate from the crystallization contains a very small quantity of the diacid. It can be flushed down the drain.

The temperature of decomposition is variable.

**Online Study Center**

Video: Macroscale Crystallization

---

4. The *endo,cis*-diacid is stable to alkali but can be isomerized to the *trans*-diacid (mp 192°C) by conversion to the dimethyl ester (3 g acid, 10 mL methanol, 0.5 mL concentrated $H_2SO_4$; reflux 1 h). This ester is equilibrated with sodium ethoxide in refluxing ethanol for 3 days and saponified. For an account of a related epimerization and discussion of the mechanism, see Meinwald, J.; Gassman, P. G. *J. Am. Chem. Soc.* **1960**, *82*, 5445. *See also* Williamson, K. L.; Li, Y.-F.; Lacko, R.; Youn, C. H. *J. Am. Chem. Soc.* **1969**, *91*, 6129; Williamson, K. L.; Li, Y.-F. *J. Am. Chem. Soc.* **1970**, *92*, 7654.

5. *See* footnote 4.

# 4. Synthesis of Compound X[6]

> **IN THIS EXPERIMENT** the diacid from Experiment 3 is dissolved in concentrated sulfuric acid and warmed briefly. Water is added very carefully to the mixture, which causes the product to crystallize. A seed crystal is saved, and the product is recrystallized from the aqueous acid mixture. Allow at least 10 min for compound X to crystallize.

## *Microscale Procedure*

General rule: If you do not have the necessary quantity of diacid, scale down the amounts of reactants and solvents to match the quantity of starting material.

Do not put a wood boiling stick into this solution. The sulfuric acid will attack the wood.

**Online Study Center**

Video: Microscale Filtration on the Hirsch Funnel; Photo: Use of the Wilfilter

To a tared 10 × 100-mm reaction tube add 0.15 g of the *endo,cis*-diacid from Experiment 3, followed by 0.25 mL of concentrated sulfuric acid. Warm the mixture on a steam bath or in a beaker of boiling water for about 2 min to allow the anhydride to dissolve/react, cool; then *cautiously* add 0.70 mL of water to the test tube. This should be done dropwise with vigorous mixing of the contents after the addition of each drop. The product will crystallize as a fine powder, often gray in color.

Save a seed crystal and heat the tube on a hot sand bath until the crystals redissolve. Seed the solution and allow it to cool slowly to room temperature. Compound X will crystallize in platelike crystals. The crystallization process for this compound is fairly slow; allow at least 10 min for the solution to come to room temperature and a further 10 min in the ice bath before collecting the crystals on a Hirsch funnel or a Wilfilter (Fig. 48.4). The longer you wait to collect the product, the higher your yield will be. Wash the crystals with one small portion of ice water and scrape the product onto a piece of filter paper. Squeeze the crystals between sheets of filter paper to complete the drying process; then determine the weight, yield, and melting point of the product.

**Cleaning Up.** Dilute the aqueous filtrate with water, neutralize it with sodium carbonate, and flush the resulting solution down the drain. It contains an extremely small quantity of compound X.

## *Macroscale Procedure*

To prepare compound X, place 1 g of the *endo,cis*-diacid and 5 mL of concentrated sulfuric acid in a 50-mL Erlenmeyer flask and heat gently on a hot plate for a minute or two until all the crystals are dissolved. Then cool in an ice bath, add a small piece of ice, swirl to dissolve, and add further ice until the volume is about 20 mL. Heat to the boiling point and let the solution simmer on the hot plate for 5 min. Cool well in ice, scratch the flask to induce recrystallization, and allow for some delay in complete separation. The crystals will often be gray in color. Collect, wash with water, and recrystallize from water. Compound X (about 0.7 g) forms large prisms (mp 203°C). To run an NMR spectrum of compound X, the compound must be dissolved in deuterodimethyl sulfoxide (DMSO-$d_6$).

---

6. The synthesis of compound X was introduced by James A. Deyrup.

■ **FIG. 48.4**
The Wilfilter filtration apparatus.

Crystals

Wilfilter

Filtrate

**Cleaning Up.** Dilute the aqueous filtrate with water, neutralize it with sodium carbonate, and flush the resulting solution down the drain. It contains a small quantity of compound X.

# 5. Structure Determination of Compound X

To determine the formula for compound X, which is an isomer of the diacid, try to answer the following questions: What is meant by "isomer of the starting material"? How many functional groups are in the starting material? What happens when each functional group reacts with a strong acid? What ionic intermediate is formed when the diacid dissolves in concentrated sulfuric acid? Why is the $^1H$ NMR spectrum of X (Fig. 48.13) so much more complex than the spectra of the anhydride and diacid (Fig. 48.9 and Fig. 48.12)? What functional group is missing from compound X (which is seen in Fig. 48.9 and Fig. 48.12)? Write formulas for the possible structures of compound X and devise tests to distinguish among them. What functional group is present in compound X that is not found in the diacid, as determined by an analysis of the IR spectrum of compound X (Fig. 48.14)?

## *Computational Chemistry*

Would you predict that compound X is more or less stable than the isomeric diacid from which it is formed? Test your prediction by carrying out a heat of formation calculation on each of the isomers. To do this, first find the conformation of each isomer with the minimum steric energy using a molecular mechanics program. Then submit each of these to a heat of formation calculation at the AM1 or higher level of calculation.

## QUESTIONS

*Online Study Center*

**General Resources**
Additional Experiments

1. In the cracking of dicyclopentadiene, why is it necessary to distill the product very slowly?
2. Draw the products of the following reactions:

(a)

(b)

(c)

3. What starting material would be necessary to prepare the following compound by the Diels–Alder reaction?

4. If the Diels–Alder reaction between dimethylmaleic anhydride and furan had worked, would cantharidin have been formed?

5. Determine the heats of formation or the steric energies of the *exo-* and *endo-*anhydride adducts using a molecular mechanics program. What do these energies tell you about the mechanism of the reaction?

6. Which molecule, norbornene dicarboxylic acid or compound X, would you predict to be more stable? Why?

**FIG. 48.5**

The $^1$H NMR spectrum of dicyclopentadiene (250 MHz).

PPM ($\delta$)

■ **FIG. 48.6**

The IR spectrum of
dicyclopentadiene (thin film).

■ **FIG. 48.7**

The $^{13}$C NMR spectrum
of dicyclopentadiene.

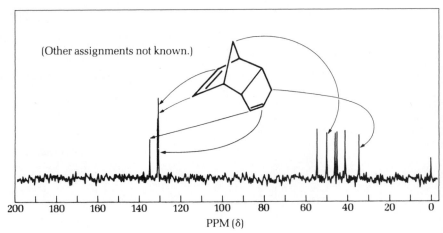

■ **FIG. 48.8**

The IR spectrum of *cis*-
norbornene-5,6-*endo*-
dicarboxylic anhydride.

■ **FIG. 48.9**

The $^1$H NMR spectrum of *cis*-norbornene-5,6-*endo*-dicarboxylic anhydride (250 MHz).

■ **FIG. 48.10**

The $^{13}$C NMR spectrum of *cis*-norbornene-5,6-*endo*-dicarboxylic anhydride (22.6 MHz).

■ **FIG. 48.11**

The IR spectrum of *cis*-norbornene-5,6-*endo*-dicarboxylic acid (KBr disk).

■ **FIG. 48.12**

The $^1$H NMR spectrum of *cis*-norbornene-5,6-*endo*-dicarboxylic acid (250 MHz).

■ **FIG. 48.13**

The $^1H$ NMR spectrum of compound X (250 MHz), dissolved in DMSO-$d_6$.

■ **FIG. 48.14**

The IR spectrum of compound X (KBr disk).

# Ferrocene [Bis(cyclopentadienyl)iron]

**PRELAB EXERCISE:** Propose a detailed outline of the procedure for synthesizing ferrocene, paying particular attention to the time required for each step.

**Dicyclopentadiene**
den. 0.98, MW 132.20

**Cyclopentadiene**
bp 41°C, den. 0.80
MW 66.10

+ KOH ⟶ $\ominus$ K$^+$ + H$_2$O

**Potassium cyclopentadienide**

2 $\ominus$ K$^+$ + FeCl$_2$ · 4 H$_2$O ⟶ Fe + 2 KCl + 4 H$_2$O

**Iron(II) chloride tetrahydrate**
MW 198.81

**Ferrocene**
**Bis(cyclopentadienyl)iron**
MW 186.04, mp 172–174°C

The Grignard reagent is a classic organometallic compound. The magnesium ion, in Group IIA of the periodic table, needs to lose two—and only two—electrons to achieve the inert gas configuration. This metal has a strong tendency to form ionic bonds by electron transfer:

$$RBr + Mg \longrightarrow \overset{\delta^-}{R}-\overset{\delta^+}{Mg}Br$$

In the transition metals, the situation is not so simple. Consider the bonding between iron and carbon monoxide in $Fe(CO)_5$:

The pair of electrons on the carbon atom is shared with iron to form a σ (sigma) bond between the carbon and iron. The π (pi) bond between iron and carbon is formed from a pair of electrons in the $d$ orbital of iron. The π bond is thus formed by the overlap of a $d$ orbital of iron with the $p$-π bond of the carbonyl group. This mutual sharing of electrons results in a relatively nonpolar bond.

Iron has six electrons in the $3d$ orbital, two in the $4s$, and none in the $4p$ orbital. The inert gas configuration requires 18 electrons—ten $3d$, two $4s$, and six $4p$ electrons. Iron pentacarbonyl enters this configuration by accepting two electrons from each of the five carbonyl groups, for a total of 18 electrons. Back-bonding of the $d$-π type distributes the excess electrons among the five carbon monoxide molecules.

Early attempts to form σ-bonded derivatives linking alkyl carbon atoms to iron were unsuccessful, but P. L. Pauson in 1951 succeeded in preparing a very stable substance, ferrocene, $C_{10}H_{10}Fe$, by reacting 2 mol of cyclopentadienyl-magnesium bromide with anhydrous ferrous chloride. Another group of chemists—Geoffrey Wilkinson, Myron Rosenblum, Mark Whiting, and Robert B. Woodward—recognized that the properties of ferrocene (its remarkable stability to water, acids, and air and its ease of sublimation) could be explained only if it had the structure depicted previously and that the bonding of the ferrous iron with its six electrons must involve all 12 of the π electrons on the two cyclopentadiene rings, with a stable 18-electron inert gas structure as the result.

In this chapter ferrocene is prepared by reacting the anion of cyclopentadiene with iron(II) chloride. Abstraction of one of the acidic allylic protons of cyclopentadiene with base gives the aromatic cyclopentadienyl anion. It is considered aromatic because it conforms to the Hückel rule in having $4n + 2\pi$

electrons (where *n* is 1). Two molecules of this anion will react with iron(II) chloride to give ferrocene, the most common member of the class of metal-organic compounds referred to as metallocenes. In this centrosymmetric sandwich-type π complex, all carbon atoms are equidistant from the iron atom, and the two cyclopentadienyl rings rotate more or less freely with respect to each other. The extraordinary stability of ferrocene (stable to 500°C) can be attributed to the sharing of the 12 π electrons of the two cyclopentadienyl rings with the 6 outer shell electrons of iron(II) to give the iron a stable 18-electron inert gas configuration. Ferrocene is soluble in organic solvents, can be dissolved in concentrated sulfuric acid and recovered unchanged, and is resistant to other acids and bases as well (in the absence of oxygen). This behavior is consistent with that of an aromatic compound; ferrocene can undergo electrophilic aromatic substitution reactions with ease.

Cyclopentadiene readily dimerizes at room temperature by a Diels–Alder reaction to give dicyclopentadiene. This dimer can be "cracked" by heating (an example of the reversibility of the Diels–Alder reaction) to give low-boiling cyclopentadiene. In most syntheses of ferrocene, the anion of cyclopentadiene is prepared by reacting the diene with metallic sodium. Subsequently, this anion is allowed to react with anhydrous iron(II) chloride. Here the anion is generated using powdered potassium hydroxide, which functions as both a base and a dehydrating agent.

The anion of cyclopentadiene decomposes rapidly in air and iron(II) chloride; although reasonably stable in the solid state, it is readily oxidized to the iron(III) (ferric) state in solution. Consequently, this reaction must be carried out in the absence of oxygen, which is accomplished by bubbling nitrogen gas through the solutions to displace dissolved oxygen and to flush air from the apparatus. Rather elaborate apparatus is used in research laboratories to carry out the experiment in the absence of oxygen. A very simple apparatus is used in this chapter because no gases are evolved, no heating is necessary, and the reaction is only mildly exothermic.

Ferrocene: soluble in organic solvents, stable to 500°C

## EXPERIMENTS

 # 1. Microscale Synthesis of Ferrocene

**IN THIS EXPERIMENT** one of the reactants and the ionic intermediate will react with atmospheric oxygen; therefore, the reaction is run under a nitrogen atmosphere. Potassium hydroxide is partially dissolved in a solvent, cyclopentadiene is added, and, on shaking and stirring, is converted to the colored anion. To this solution of the cyclopentadienyl anion is added a solution of iron(II) chloride to produce ferrocene. The mixture is poured onto ice-cold acid, and the crystalline product is collected on a Hirsch funnel. Once dry, the product is purified by sublimation.

■ **FIG. 49.1**

The apparatus for flushing air from reaction flask. Check the nitrogen flow rate by allowing it to bubble through an organic solvent before inserting the needle through the septum.

CH₃OCH₂CH₂OCH₃

**1,2-Dimethoxyethane**
(Ethylene glycol dimethyl ether, monoglyme), bp 85°C
Completely miscible with water

 **Online Study Center**

Video: Ferrocene Synthesis

⚠

**CAUTION**: Potassium hydroxide is extremely corrosive and hygroscopic. Immediately wash any spilled powder or solutions from the skin and wipe up all spills. Keep containers tightly closed. Work in the hood.

⚠

**CAUTION**: Dimethyl sulfoxide is rapidly absorbed through the skin. Wash off spills with water. Wear disposable gloves when shaking the apparatus.

To a 5-mL short-necked, round-bottomed flask add a magnetic stirring bar and then quickly add 0.75 g of finely powdered potassium hydroxide,[1] followed by 1.25 mL of dimethoxyethane (DME). The funnel that is a part of the chromatography column makes a convenient addition funnel. Cap the flask with a good septum and pass nitrogen into the flask or, better, through the solution for about 1 min. This is done by connecting a tank of nitrogen via a rubber tube to a 22-gauge needle and adjusting the nitrogen flow to a few milliliters per minute by bubbling it under a liquid such as acetone. With the nitrogen flow adjusted, insert an empty syringe needle through the septum of the flask as an outlet and then insert the nitrogen inlet needle (Fig. 49.1). Remove the needles and shake the flask to dislodge the solid from the bottom and to help dissolve some of it. If possible, stir the mixture magnetically.

To a 10 × 100-mm reaction tube add 0.35 g of finely powdered green iron(II) chloride tetrahydrate and 1.5 mL of dimethyl sulfoxide (DMSO). Cap the tube with a good rubber septum, insert an empty syringe needle through the septum, and pass nitrogen into the tube for about 1 min to displace the oxygen present. Remove the needles and then shake the vial vigorously to dissolve all the iron chloride. Some warming may be needed.

Using an accurate syringe, inject 0.300 mL of freshly prepared cyclopentadiene (*see* Chapter 48 for the preparation of cyclopentadiene) into the flask containing the

---

1. Potassium hydroxide is easily ground to a fine powder in 25-g batches in 1 min, employing an ordinary food blender (e.g., Waring, Osterizer). The finely powdered base is transferred in a hood to a bottle with a tightly fitting cap. Alternatively, grind about 1 g potassium hydroxide in a mortar and transfer it rapidly to the reaction flask.

**Dimethyl sulfoxide, DMSO**
(Methyl sulfoxide), bp 189°C
Completely miscible with
water

Isolation of ferrocene

Purification by sublimation

potassium hydroxide. Do not grasp the body of the syringe because the heat of your hand will cause the cyclopentadiene to volatilize. Stir the mixture vigorously and note the color change as the potassium cyclopentadienide is formed. After waiting about 5 min for the anion to form, pierce the septum with an empty needle for pressure relief and inject the iron(II) chloride solution contained in the reaction tube in six 0.25-mL portions over a 10-min period. Stir the mixture well with a magnetic stirrer. If the stirring bar is immobilized, then between injections remove both needles from the septum and shake the flask vigorously. After all the iron(II) chloride solution has been added, rinse the reaction tube with an additional 0.25 mL of DMSO and add this to the flask. Continue to stir or shake the solution for about 15 min to complete the reaction.

To isolate the ferrocene, pour the dark slurry onto a mixture of 4.5 mL of 6 *M* hydrochloric acid and 5 g of ice in a 30-mL beaker. Stir the contents of the beaker thoroughly to dissolve and neutralize all the potassium hydroxide. Collect the crystalline orange ferrocene on a Hirsch funnel, wash the crystals well with water, press out excess water, squeeze the product between sheets of filter paper to complete the drying, and then purify the ferrocene by sublimation. The filtrate is blue because of dissolved ferrocinium ion. It can be reduced with a mild reducing agent, such as ascorbic acid, to regenerate ferrocene. The amount of ferrocene produced is usually negligible.

To sublime the ferrocene, add the crude dry product to a 25-mL filter flask equipped with a neoprene filter adapter (Pluro stopper) and a 15-mL centrifuge tube that is pushed to within 5 mm of the bottom of the flask (Fig. 49.2). Put a rubber bulb on the side arm of the flask; then add ice to the centrifuge tube and heat the flask on a sand bath to sublime the product. Tilting and rolling the filter flask in the hot sand will help drive ferrocene onto the centrifuge tube. Remove

■ **FIG. 49.2**
The apparatus for the sublimation of ferrocene.

15-mL Centrifuge tube, to be filled with ice

Adapter (Pluro stopper)

Pipette bulb

25-mL Filter flask

Material to be sublimed

Sublimate

Heat source

■ **FIG. 49.3**
Evacuation of a melting point
capillary prior to sealing.

To aspirator

Rubber hose

Glass tube

Rubber septum

Melting point
capillary

Seal here

⚠

**CAUTION**: Potassium
hydroxide is extremely corrosive
and hygroscopic. Immediately
wash any spilled powder or
solutions from the skin and wipe
up all spills. Keep containers
tightly closed. Work in the hood.

**CAUTION**: Dimethoxyethane
can form peroxides. Discard
90 days after opening bottle
because of peroxide formation.

**CAUTION**: Dimethyl sulfoxide
is rapidly absorbed through the
skin. Wash off spills with water.

the flask from the sand bath and use a heat gun to drive the last of the ferrocene
from the sides of the flask to the centrifuge tube. Ferrocene sublimes nicely at atmospheric pressure. Vacuum sublimation is not needed.

When sublimation is complete, cool the flask, remove the ice water from the
centrifuge tube, and replace it with room temperature water (to prevent moisture
from collecting on the tube). Transfer the product to a tared, stoppered vial, determine the weight, and calculate the percent yield. Determine the melting point
in an evacuated capillary because the product sublimes at the melting point
(Fig. 49.3) (*see* Chapter 3 for this technique).

**Cleaning Up.** The filtrate should be slightly acidic. Neutralize it with sodium
carbonate, dilute it with water, and flush it down the drain. Place any unused
cyclopentadiene in the recovered dicyclopentadiene or the organic solvents container. Add 0.4 mL concentrated nitric acid to the sublimation flask to clean it.
After at least 24 hours, dilute the acid, neutralize it with sodium carbonate, and
flush the solution down the drain.

## 2. Macroscale Synthesis of Ferrocene

Following the procedure described in Experiment 1 of Chapter 48, prepare 3 mL
of cyclopentadiene. It need not be dry. While this distillation is taking place,
rapidly weigh 12.5 g of finely powdered potassium hydroxide[2] into a 50-mL
Erlenmeyer flask, add 30 mL of dimethoxyethane (DME), and immediately cool
the mixture in an ice bath. Swirl the mixture in the ice bath for a minute or two;
then bubble nitrogen through the solution for about 2 min. Quickly stopper the
flask and shake the mixture to dislodge the cake of potassium hydroxide from the
bottom of the flask and to dissolve as much of the base as possible (much will remain undissolved).

Grind 3.5 g of iron(II) chloride tetrahydrate to a fine powder, then add 3.5 g of
the green salt to 12.5 mL of dimethyl sulfoxide (DMSO) in a 25-mL Erlenmeyer
flask. Pass nitrogen through the DMSO mixture for about 2 min, stopper the flask,
and shake it vigorously to dissolve all the iron(II) chloride. Gentle warming of the
flask on a steam bath may be necessary to dissolve the last traces of iron(II)
chloride. Transfer the solution rapidly to a 60-mL separatory funnel (serves as an
addition funnel) equipped with a cork or stopper to fit the 50-mL Erlenmeyer flask,
flush air from the funnel with a stream of nitrogen, and stopper it.

Transfer 3 mL of the freshly distilled cyclopentadiene to the slurry of potassium
hydroxide in dimethoxyethane. Shake the flask vigorously and note the color
change as the potassium cyclopentadienide is formed. After waiting about 5 min for
the anion to form, replace the cork or stopper on the Erlenmeyer flask with the separatory funnel quickly (to avoid admission of air to the flask; Fig. 49.4). Figure 49.5
depicts a research-quality apparatus that would be used for this experiment.

2. Potassium hydroxide is easily ground to a fine powder in 75-g batches in 1 min employing an ordinary food
blender (e.g., Waring, Osterizer). The finely powdered base is transferred in a hood to a bottle with a tightly
fitting cap. If a blender is available, crush and then grind 27 g of potassium hydroxide pellets in a large mortar
and quickly weigh 25 g of the resulting powder into a 125-mL Erlenmeyer flask.

**■ FIG. 49.4**

The apparatus for ferrocene synthesis.

FeCl$_2$
in dimethyl sulfoxide,

$$\underset{CH_3SCH_3}{\overset{\overset{\displaystyle O}{\parallel}}{}}$$

Powdered KOH in
dimethoxyethane,
CH$_3$OCH$_2$CH$_2$OCH$_3$

**■ FIG. 49.5**

A research-quality apparatus for the preparation of ferrocene.

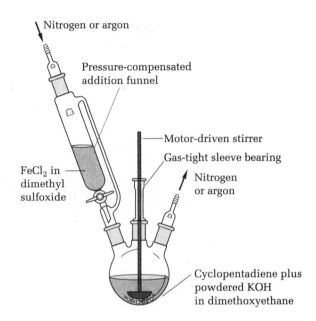

Nitrogen or argon

Pressure-compensated
addition funnel

Motor-driven stirrer

Gas-tight sleeve bearing

Nitrogen
or argon

FeCl$_2$ in
dimethyl
sulfoxide

Cyclopentadiene plus
powdered KOH
in dimethoxyethane

Add the iron(II) chloride solution to the base dropwise over a period of 20 min with vigorous swirling and shaking. Dislodge the potassium hydroxide should it cake on the bottom of the flask. The shaking will allow nitrogen to pass from the Erlenmeyer flask into the separatory funnel as the solution leaves the funnel.[3] Continue to shake and swirl the solution for 10 min after all the iron(II) chloride is added; then pour the dark slurry onto a mixture of 45 mL of 6 *M* hydrochloric acid and 50 g of ice in a 250-mL beaker. Stir the contents of the beaker thoroughly to dissolve and neutralize all the potassium hydroxide. Collect the crystalline orange ferrocene on a Büchner funnel, wash the crystals with water, press out excess water, and allow the product to dry overnight on a watch glass.

Recrystallize the ferrocene from methanol or, better, hexanes. It is also very easily sublimed. In a hood place about 0.5 g of crude ferrocene on a watch glass on a hot plate set to about 150°C. Invert a glass funnel over the watch glass. Ferrocene will sublime in about 1 h, leaving nonvolatile impurities behind. Pure ferrocene melts at 172°C–174°C. Determine the melting point in an evacuated capillary (Fig. 49.3) because the product sublimes at the melting point. Compare the melting points of your sublimed and recrystallized materials.

---

3. If the particular separatory funnel being used does not allow nitrogen to pass from the flask to the funnel, connect the two with a rubber tube leading to a glass tube and stopper at the top of the separatory funnel and to a syringe needle that pierces the rubber stopper in the flask (suggestion of D. L. Fishel).

**Cleaning Up.** The filtrate from the reaction mixture should be slightly acidic. Neutralize it with sodium carbonate, dilute it with water, and flush it down the drain. Place any unused cyclopentadiene in the recovered dicyclopentadiene or the organic solvents container. If the ferrocene has been crystallized from methanol or hexanes, place the mother liquor in the organic solvents container.

## QUESTIONS

1. If ferrocene and all of the reagents in this experiment are stable in air before the reaction begins, why must air be so carefully excluded during the reaction?
2. What special properties do the solvents dimethoxyethane and dimethyl sulfoxide have compared to diethyl ether, for example, that make them particularly suited for this reaction?
3. What is it about ferrocene that allows it to sublime easily whereas many other compounds do not?

# *p*-Terphenyl by the Diels–Alder Reaction

**Online Study Center**

This icon will direct you to techniques, equipment setups, and online resources at http://college.hmco.com/PIC/williamsonMMOE5e.

**PRELAB EXERCISE:** Why does the dicarboxylate (**5**) undergo double decarboxylation to give terphenyl, whereas the diacid formed from the reaction of cyclopentadiene with maleic anhydride (*see* Chapter 48) does not undergo decarboxylation under the same reaction conditions?

**(1)**
***E,E*-1,4-Diphenyl-1,3-butadiene**
MW 206.27

s-*trans* form

s-*cis* form

**(2)**
MW 142.11

**(3)**
MW 348.38, mp 98°C

**(4)**
mp 170°C

**(5)**

**(6)**
***p*-Terphenyl**
MW 230.29, mp 209–210°C

**Diels–Alder reaction**

**E**,*E*-1,4-Diphenyl-1,3-butadiene (**1**) is most stable in the s-*trans* form, but at a suitably elevated temperature the s-*cis* form present in the equilibrium adds to dimethyl acetylenedicarboxylate (**2**) to give dimethyl 1,4-diphenyl-1,4-dihydrophthalate (**3**). This low-melting ester is obtained as an oil that, when warmed briefly with methanolic potassium hydroxide, is isomerized to the high-melting *E*-ester (**4**). The free *E*-acid can be obtained in 86% yield by refluxing the suspension of (**3**) in methanol for 4 h; but in the recommended procedure the isomerized ester is collected, washed to remove the dark mother liquor, and hydrolyzed by brief heating with potassium hydroxide in a high-boiling solvent. The final step, an oxidative decarboxylation, is rapid and nearly quantitative. It probably involves reaction of the oxidant with the dianion (**5**) with removal of two electrons and formation of a diradical, which loses carbon dioxide with the formation of *p*-terphenyl (**6**).

## EXPERIMENTS

# 1. Microscale Synthesis of *p*-Terphenyl

> **IN THIS EXPERIMENT** a previously prepared diene reacts with a dieneophile, an acetylenic diester, in a very high-boiling solvent. The mixture is dissolved in ether, and the water-soluble solvent is removed by extraction with water. The ether solution is dried and evaporated to give the diester intermediate (**3**), which is hydrolyzed in a high-boiling solvent to a dianion. The dianion is oxidized with ferricyanide to the product that is collected on a Hirsch funnel.

⚠️ **CAUTION**: Handle dimethyl acetylenedicarboxylate with care.[1]

$$CH_3OCH_2CH_2OCH_2$$
$$|$$
$$CH_3OCH_2CH_2OCH_2$$

**Triethylene glycol dimethyl ether (triglyme)**
Miscible with water; bp 222°C

Reaction time: 30 min

***Online Study Center***
Video: Extraction with Ether;
Photos: Extraction with Ether,
Drying Crystals Under Vacuum

In a 10 × 100-mm reaction tube, place 100 mg of *E*,*E*-1,4-diphenyl-1,3-butadiene prepared by the Wittig reaction in Chapter 39. Add 74 mg of dimethyl acetylenedicarboxylate (*Caution*: skin irritant), followed by 0.35 mL of triethylene glycol dimethyl ether (triglyme; bp 222°C) and a boiling chip. Reflux the mixture on a hot sand bath for 30 min.

Cool the yellowish solution in cold water, add 2.5 mL of ether (*t*-butyl methyl ether is strongly recommended), and extract the solution three times with 1.5-mL portions of water to remove the high-boiling solvent, shake the ethereal solution once with saturated sodium chloride solution, and dry the solution over calcium chloride pellets. Add the drying agent until it no longer clumps together. Remove the ether solution with a Pasteur pipette and place it in a tared reaction tube. Wash off the calcium chloride pellets with more ether. Evaporate the solvent in the hood under a stream of nitrogen or air, removing the last traces under aspirator vacuum.

While the evaporation is in progress, dissolve 35 mg of potassium hydroxide in 0.35 mL of methanol by heating and stirring. Crystallization of the yellow oil

---

1. This ester is a powerful lachrymator (tear producer) and vesicant (blistering agent) and should be dispensed from a bottle provided with a pipette and a pipetter. Even a trace of ester on the skin should be washed off promptly with methanol, followed by soap and water.

Methanol solutions are strongly fluorescent.

**CAUTION:** Triglyme forms peroxides. Discard 90 days after container is opened.

Rapid hydrolysis of the hindered ester (**4**)

**Online Study Center**

Video: Microscale Filtration on the Hirsch Funnel

containing compound **3** can be initiated by cooling and scratching; this provides assurance that the reaction has proceeded properly. Transfer the methanolic potassium hydroxide to the yellow oil and heat with stirring on a hot sand bath for about 1 min until a stiff paste of crystals of the isomerized ester (**4**) appears. Cool, thin the mixture with methanol, collect the product on a Hirsch funnel, wash it free of the dark mother liquor, and spread it thinly on a filter paper for rapid drying. The yield of pure, white ester (**4**) is about 120 mg. Solutions in methanol are strongly fluorescent.

Place the ester (**4**) in a reaction tube, add 50 mg of potassium hydroxide, and add 0.35 mL of triethylene glycol. Stir the mixture with a stirring rod and heat, raising the temperature to 140°C in the course of about 5 min. By intermittent heating, keep the temperature close to 140°C for an additional 5 min; then cool the mixture under the tap. Pour this cooled mixture into a 4-in. test tube and rinse the reaction tube with about 3.3 mL of water. Heat to boiling; in case there is a small precipitate or the solution is cloudy, add a little pelletized Norit decolorizing charcoal, swirl, and filter the alkaline solution. Then add 225 mg of potassium ferricyanide and heat on a hot sand bath for about 5 min to dissolve the oxidant and to coagulate the white precipitate, which soon separates and is collected by filtration on a Hirsch funnel. The product can be air-dried overnight or dried to constant weight by heating in an evacuated reaction tube on a steam bath. The yield of colorless *p*-terphenyl (mp 209°C–210°C) is 50–60 mg.

**Cleaning Up.** The aqueous layer from the reaction after dilution can be flushed down the drain. Allow the ether to evaporate from the calcium chloride in the hood; then discard the drying agent in the nonhazardous solid waste container. If local regulations do not allow for the evaporation of solvents in a hood, the wet solid should be disposed in a special container. The aqueous reaction mixture after dilution with water is neutralized with dilute hydrochloric acid and flushed down the drain.

## 2. Macroscale Synthesis of *p*-Terphenyl

**CAUTION:** Handle dimethyl acetylenedicarboxylate with care.[2]

**Triethylene glycol dimethyl ether (triglyme)** Miscible with water; bp 222°C

Reaction time: 30 min; reflux on a sand bath or a hot plate

Place 1.5 g of *E,E*-1,4-diphenyl-1,3-butadiene (from the Wittig reaction, Chapter 39) and 1.0 mL (1.1 g) of dimethyl acetylenedicarboxylate (*Caution:* skin irritant) in a 25 × 150-mm test tube and rinse down the walls with 5 mL of triethylene glycol dimethyl ether (triglyme; bp 222°C). Clamp the test tube in a vertical position, introduce a cold finger condenser, and reflux the mixture gently for 30 min. Alternatively, carry out the experiment in a 25-mL round-bottomed flask equipped with a reflux condenser. Cool the yellowish solution under the tap, pour into a separatory funnel, and rinse out the reaction vessel with about 50 mL of ether. Extract twice with water (50–75 mL portions) to remove the high-boiling solvent, shake the ethereal solution with saturated sodium chloride solution, and dry the ether layer over anhydrous calcium chloride pellets. Filter or decant the ether solution into a tared 125-mL Erlenmeyer flask and evaporate the ether on a

---

2. This ester is a powerful lachrymator (tear producer) and vesicant (blistering agent) and should be dispensed from a bottle provided with a pipette and a pipetter. Even a trace of ester on the skin should be washed off promptly with methanol, followed by soap and water.

**CAUTION**: Extinguish all flames before working with ether.

Potassium hydroxide is easily powdered in a blender.

steam bath, eventually with evacuation at an aspirator, until the weight of yellow oil is constant; the yield is about 2.5–2.8 g.

While evaporation is in progress, dissolve 0.5 g of potassium hydroxide (about 5 pellets) in 10 mL of methanol by heating and swirling; this process is greatly hastened by crushing the lumps using a stirring rod with a flattened head. Crystallization of the yellow oil containing compound **3** can be initiated by cooling and scratching; this provides assurance that the reaction has proceeded properly. Pour in the methanolic potassium hydroxide and heat with swirling on a hot plate for about 1 min until a stiff paste of crystals of the isomerized ester (**4**) appears. Cool, thin the mixture with methanol, collect the product, wash it free of the dark mother liquor, and spread it thinly on a filter paper for rapid drying. The yield of pure, white ester (**4**) is 1.7–1.8 g. Solutions in methanol are highly fluorescent.

Rapid hydrolysis of the hindered ester (**4**)

Place the ester (**4**) in a 25 × 150-mm test tube, add 0.7 g of potassium hydroxide (7–8 pellets), and pour in 5 mL of triethylene glycol. Stir the mixture with a stirring rod and heat, raising the temperature to 140°C in the course of about 5 min. By intermittent heating, keep the temperature close to 140°C for an additional 5 min; then cool the mixture under the tap. Pour into a 125-mL Erlenmeyer flask and rinse the tube with about 50 mL of water. Heat to boiling; in case there is a small precipitate or the solution is cloudy, add a little pelletized Norit decolorizing charcoal, swirl, and filter the alkaline solution by gravity. Then add 3.4 g of potassium ferricyanide and heat on a hot plate with swirling for about 5 min to dissolve the oxidant and to coagulate the white precipitate that soon separates. The product can be air-dried overnight or dried to constant weight by heating in an evacuated Erlenmeyer flask on a steam bath. The yield of colorless *p*-terphenyl (mp 209°C–210°C) is 0.7–0.8 g.

**Cleaning Up.** The aqueous layer from the reaction after dilution can be flushed down the drain if local regulations permit. Allow the ether to evaporate from the calcium chloride in the hood; then discard the drying agent in the nonhazardous solid waste container. If local regulations do not allow for the evaporation of solvents in a hood, the wet solid should be disposed in a special container. The aqueous reaction mixture after dilution with water is neutralized with dilute hydrochloric acid and then flushed down the drain.

## Computational Chemistry

Using a molecular mechanics program, calculate the steric energies of all the possible *cis-* and *trans-* isomers as well as all the *s-cis* and *s-trans* forms. There are six possible combinations of these. Which is the most stable?

## QUESTIONS

1. What is the driving force for the isomerization of compound **3** to compound **4**?
2. Why is the *trans-* and not the *cis-* diester (**4**) formed in the isomerization of compound **3** to compound **4**?
3. Why does hydrolysis of compound **4** in methanol require 4 h, whereas hydrolysis in triethylene glycol requires only 10 min?

# Tetraphenylcyclopentadienone

**Online Study Center**

This icon will direct you to
techniques, equipment setups,
and online resources at
http://college.hmco.com/
PIC/williamsonMMOE5e.

**PRELAB EXERCISE:** Write a detailed mechanism for the formation
of tetraphenylcyclopentadienone from benzil and 1,3-diphenyl-
acetone. To which general class of reactions does this condensation
belong?

$$C_6H_5C=O \atop C_6H_5C=O \quad + \quad {C_6H_5 \atop CH_2} - C=O - {CH_2 \atop C_6H_5} \quad \xrightarrow[-2\ H_2O]{C_6H_5CH_2N^+(CH_3)_3OH^-} \quad$$

**Benzil**
MW 210.22
mp 96°C

**1,3-Diphenylacetone**
MW 210.26
mp 35°C

**Tetraphenylcyclopentadienone**
MW 384.45
mp 219°C

**Cyclopentadienone**

**C**yclopentadienone is an elusive compound that has been sought for many
years but with little success. Molecular orbital calculations predict that it
should be highly reactive, and so it is; it exists only as the dimer. The tetraphenyl
derivative of this compound is synthesized in this chapter. This derivative is stable and reacts readily with dienophiles. It is used not only for the synthesis of
highly aromatic, highly arylated compounds but also for examination of the
mechanism of the Diels–Alder reaction itself. Tetraphenylcyclopentadienone has
been carefully studied by means of molecular orbital methods in attempts to understand its unusual reactivity, color, and dipole moment.

The literature procedure for the condensation of benzil with 1,3-diphenyl-
acetone in ethanol with potassium hydroxide as a basic catalyst suffers from the
low boiling point of the alcohol and the limited solubility of both potassium hydroxide and the reaction product in this solvent. Triethylene glycol is a better solvent and permits operation at a higher temperature. In the procedure that follows,

Louis Fieser introduced the idea
of using very high-boiling
solvents to speed up this and
many other reactions.

the glycol is used with benzyltrimethylammonium hydroxide, a strong base readily soluble in organic solvents, which serves as a catalyst.

In Chapter 52 tetraphenylcyclopentadienone will be used to synthesize hexaphenylbenzene and dimethyl tetraphenylphthalate. The mechanism for the formation of tetraphenylcyclopentadienone is as follows:

## EXPERIMENTS

## 1. Tetraphenylcyclopentadienone

**IN THIS EXPERIMENT** benzil and diphenyl acetone are dissolved in a very high-boiling solvent and treated with a strong organic base. The product crystallizes and is collected and washed with methanol. It can, if necessary, be recrystallized from a high-boiling solvent.

**CAUTION:** Triton B is toxic and corrosive.

Short reaction period

*See* Chapter 54 for the preparation of benzil. Into a 10 × 100-mm reaction tube, place 42 mg of pure benzil (free of benzoin), 42 mg of 1,3-diphenylacetone, and 0.4 mL of triethylene glycol, using the solvent to wash the walls of the tube. Clamp the tube over a hot sand bath, insert a thermometer, and heat the solution until the benzil is dissolved. Remove the tube from the heat; then, using a 1-mL syringe, add to the solution 0.20 mL of a 40% solution of benzyltrimethylammonium hydroxide in methanol (Triton B) when the temperature of the solution reaches exactly 100°C. Stir once to mix. Crystallization usually starts in 10–20 s. Let the mixture cool to near room temperature; then cool it in cold water. Add 0.5 mL of methanol, cool the tube in ice, and collect the product.

**■ FIG. 51.1**

The Pasteur pipette filtration technique. Solvent is removed between the pipette tip and the bottom of the tube, leaving crystals in the reaction tube.

Pasteur pipette

Reaction tube

Note square, flat tip

**■ FIG. 51.2**

An apparatus for filtration on a microscale Büchner funnel.

Thermometer adapter

Micro Büchner funnel with fritted disk

To aspirator

Vacuum-tight connector

Reaction tube with filtrate

*Online Study Center*

Photos: Filtration Using a Pasteur Pipette, Micro Büchner Funnel, Use of the Wilfilter; Videos: Filtration of Crystals Using the Pasteur Pipette, Microscale Crystallization

If the crystals are large enough, collection can be done by inserting a Pasteur pipette into the tube and removing the solvent between the tip of the pipette and the bottom of the tube (Fig. 51.1). If the crystals are small, collect them on a Wilfilter, a Hirsch funnel, or a microscale Büchner funnel (Fig. 51.2). In any case, wash the crystals with cold methanol until the washings are purple-pink, not brown. The yield of deep-purple crystals is about 60 mg. If either the crystals are not well formed or the melting point is low, place the material in a reaction tube, add 0.6 mL of triethylene glycol, stir with a stirring rod, and raise the temperature to 220°C to bring the solid into solution. Let it stand for crystallization. (If initially pure material is recrystallized, the recovery is about 90%.)

**Cleaning Up.** Because the filtrate and washings from the reaction contain Triton B, place them in the hazardous waste container. Dilute the crystallization solvent with water and flush it down the drain.

## 2. Tetraphenylcyclopentadienone

*See* Chapter 54 for the preparation of benzil. Measure into a 25 × 150-mm test tube 2.1 g of benzil, 2.1 g of 1,3-diphenylacetone, and 10 mL of triethylene glycol, using the solvent to wash the walls of the test tube. Support the test tube in a hot sand bath, stir the mixture with a stirring rod, heat until the benzil is dissolved, and then remove it from the sand. Measure 1 mL of a commercially available 40% solution of benzyltrimethylammonium hydroxide (Triton B) in methanol into a 10 × 75-mm test tube, adjust the temperature of the solution to exactly 100°C, remove from the heat, add the catalyst, and stir once to mix. Crystallization usually starts in 10–20 s. Let the temperature drop to about 80°C; then cool under the tap, add 10 mL of methanol, stir to a thin crystal slurry, collect the product, and wash it with methanol until the filtrate is purple-pink, not brown. The yield of deep-purple crystals is 3.3–3.7 g. If either the crystals are not well formed or the melting point is low, place 1 g of material and 10 mL of triethylene glycol in a vertically supported test tube, stir with a stirring rod, raise the temperature to 220°C to bring the solid into solution, and let stand for crystallization. (If initially pure material is recrystallized, the recovery is about 90%.)

**Cleaning Up.** Because the filtrate and washings from the reaction contain Triton B, place them in the hazardous waste container. Dilute the crystallization solvent with water and flush it down the drain.

## QUESTION

Draw the structure of the dimer of cyclopentadienone. Why doesn't tetraphenylcyclopentadienone undergo dimerization? This will become much clearer if an energy-minimized structure is generated using a molecular mechanics program.

# Hexaphenylbenzene and Dimethyl Tetraphenylphthalate

**PRELAB EXERCISE:** Explain the driving force behind the loss of carbon monoxide from the intermediates formed in the two reactions in this chapter.

This experiment illustrates two examples of the Diels–Alder reaction, a means of synthesizing molecules that would be extremely difficult to synthesize in any other way. Both reactions employ as the diene the tetraphenylcyclopentadienone prepared in Chapter 51. Although the Diels–Alder reaction is reversible, the intermediate in each of these reactions spontaneously loses carbon monoxide (why?) to form the products.

In the first experiment the dienone is condensed with dimethyl acetylenedicarboxylate using as the solvent 1,2-dichlorobenzene. This solvent is chosen for its solvent properties as well as its high boiling point, which guarantees that the reaction is completed in 1–2 min.

In the second experiment, the dienone is condensed with diphenylacetylene to produce hexaphenylbenzene. Hexaphenylbenzene melts at 465°C without decomposition. Few completely covalent organic molecules have higher melting points. For comparison, lead melts at 327.5°C. As is often the case, a high melting point also means limited solubility. The solvent used to recrystallize hexaphenylbenzene, diphenyl ether, has a very high boiling point (bp 259°C) and has superior solvent power.

# 1. Dimethyl Tetraphenylphthalate

> **IN THIS EXPERIMENT** purple tetraphenylcyclopentadienone undergoes a Diels–Alder reaction with an acetylene diester in a high-boiling solvent. The intermediate adduct spontaneously loses carbon monoxide to give the high-melting product that is purified by crystallization.

## Microscale Procedure

**Tetraphenylcyclopentadienone**
MW 384.45, mp 219°C

**Dimethyl acetylenedicarboxylate**
MW 142.11, bp ~300°C

1,2-Dichlorobenzene
bp 179°C

– CO

**Dimethyl tetraphenylphthalate**
MW 498.58, mp 258°C

⚠️ **CAUTION**: Work in a hood. Dimethyl acetylenedicarboxylate is corrosive and a lachrymator (tear producer). 1,2-Dichlorobenzene is a toxic irritant.

Measure into a reaction tube 50 mg of tetraphenylcyclopentadienone (prepared in Chapter 51), 0.4 mL of 1,2-dichlorobenzene, and 27.5 mg of dimethyl acetylenedicarboxylate. Clamp the tube over a hot sand bath, insert a thermometer, and raise the temperature to the boiling point (180°C–185°C). Boil gently until there is no further color change and let the rim of condensate rise just high enough to wash the walls of the tube. The pure adduct is colorless

and, if the starting ketone is adequately pure, the color changes from purple to
pale tan. A 5-min boiling period should be sufficient. Cool the tube to 100°C,
slowly stir in 0.6 mL of 95% ethanol, and let crystallization proceed. After the
mixture has cooled to near room temperature, cool it in ice. Then collect the
product, either by removing the solvent with a Pasteur pipette or, if the crystals
are too small for this technique, on a Hirsch funnel or a Wilfilter. Wash the
crystals with cold methanol. The yield of dimethyl tetraphenylphthalate will be
about 50 mg.

**Cleaning Up.** Because the filtrate from this reaction contains 1,2-dichlorobenzene,
place it in the halogenated organic solvents container.

⚠️

**CAUTION**: Work in a
hood. Dimethyl acetylenedicarboxylate is corrosive and a
lachrymator (tear producer).
1,2-Dichlorobenzene is a toxic
irritant.

Reaction time: about 5 min

### Macroscale Procedure

Measure into a 25 × 150-mm test tube 2 g of tetraphenylcyclopentadienone,
10 mL of 1,2-dichlorobenzene, and 1 mL (1.1 g) of dimethyl acetylenedicarboxylate. Clamp the test tube in a hot sand bath, insert a thermometer, and raise
the temperature to the boiling point (180°C–185°C). Boil gently until there is
no further color change and let the rim of condensate rise just high enough to
wash the walls of the tube. The pure adduct is colorless and, if the starting ketone is adequately pure, the color changes from purple to pale tan. A 5-min
boiling period should be sufficient. Cool to 100°C, slowly stir in 15 mL of 95%
ethanol, and let crystallization proceed. Cool under the tap, collect the product,
and rinse the tube with methanol. The yield of colorless crystals should be
2.1–2.2 g.

**Cleaning Up.** Because the filtrate from this reaction contains 1,2-dichlorobenzene,
place it in the halogenated organic solvents container.

## 2. Hexaphenylbenzene

**IN THIS EXPERIMENT** tetraphenylcyclopentadienone is heated over a flame with excess diphenylacetylene. Some of the diphenylacetylene is removed with a pipette to
allow the very hot reaction mixture to become even hotter to complete the reaction.
The product, which melts above the melting point of lead, is recrystallized from
diphenyl ether, a high-boiling point solvent, and is collected by filtration.

Diphenylacetylene is a less reactive dienophile than dimethylacetylenedi-
carboxylate; but when heated with tetraphenylcyclopentadienone without solvent
to a high temperature (ca. 380°C–400°C), the reaction will proceed. In the following procedure the dienophile is taken in large excess to serve as solvent. Because
refluxing diphenylacetylene (bp about 300°C) keeps the temperature below the
melting point of the product, removal of the diphenylacetylene causes the reaction
mixture to melt, which ensures completion of the reaction.

**Tetraphenylcyclopentadienone**
MW 384.45, mp 219°C

**Diphenylacetylene**
MW 178.22, mp 61°C

$-CO$

**Hexaphenylbenzene**
MW 534.66
mp 465°C
Energy-minimized conformation

## Microscale Procedure

Tetraphenylcyclopentadienone is a hydrocarbon with a very high melting point. Although the small amount of carbon monoxide produced probably presents no hazard, work in a hood.

**Online Study Center**

Videos: Instant Microscale Distillation, Microscale Filtration on the Hirsch Funnel, Filtration of Crystals Using the Pasteur Pipette; Photo: Filtration Using a Pasteur Pipette

Place 50 mg each of tetraphenylcyclopentadienone (prepared in Chapter 51) and diphenylacetylene (which can be prepared in Chapter 58) in a reaction tube, clamp it upright, and heat the mixture with the flame of a microburner. This is the only experiment in this text that requires a flame to conduct a microscale reaction. Use care to ensure that no flammable solvents are nearby. Soon after the reactants have melted with strong bubbling, the white product becomes visible. Let the diphenylacetylene reflux briefly on the walls of the tube; then remove some of this excess diphenylacetylene by inserting a Pasteur pipette into the vapors above the solid and drawing the hot vapors into the pipette. Repeat this operation until it is possible, through strong heating with the flame, to melt the mixture completely. Then let the melt cool and solidify. Add 0.25 mL of diphenyl ether (bp 259°C), heat the mixture *carefully* to dissolve the solid, then let the product crystallize slowly. Extinguish the flame and, when the tube is cold, add 0.5 mL of toluene to dilute the mixture. Collect the product using the Pasteur pipette method or by vacuum filtration on a Hirsch funnel and wash it with toluene.

This is the only experiment in this text that requires a flame to conduct a microscale reaction. Use care to ensure that no flammable solvents are nearby.

The yield of colorless plates is about 60 mg. Using a Mel-Temp apparatus, *equipped with a 500°C thermometer*, determine the melting point. To avoid oxidation, seal the sample in an evacuated capillary tube. The product should melt at 465°C.

**Cleaning Up.** Place all filtrates in the organic solvents container.

 ## Macroscale Procedure

Tetraphenylcyclopentadienone is a hydrocarbon with a very high melting point. Although the small amount of carbon monoxide produced probably presents no hazard, work in a hood.

Place 0.5 g each of tetraphenylcyclopentadienone (prepared in Chapter 51) and diphenylacetylene (which can be prepared in Chapter 58) in a 25 × 150-mm test tube supported by a clamp and heat the mixture strongly with the free flame of a microburner held in the hand (do not insert a thermometer into the test tube; the temperature will be too high). Soon after the reactants have melted with strong bubbling, white masses of the product become visible. Let the diphenylacetylene reflux briefly on the walls of the tube; then remove some of the diphenylacetylene by letting it condense for a minute or two on a cold finger condenser filled with water but without fresh water running through it (Fig. 52.1) minus the paper thimble. Remove the flame, withdraw the cold finger, and wipe it with a towel. Repeat the operation until you are able, by strong heating, to melt the mixture completely. Then let the melt cool and solidify. Add 10 mL of diphenyl ether (bp 259°C), using it to rinse the walls. Heat *carefully* over a free flame to dissolve the solid; then let the product crystallize. When cold, add 10 mL of toluene to thin the mixture, collect the product, and wash with toluene.

■ **FIG. 52.1**
A Soxhlet-type extractor.

■ **FIG. 52.2**
A research-quality Soxhlet extractor. Solvent vapors from the flask rise through A and up into the condenser. As they condense, the liquid just formed returns to B, where the sample to be extracted is placed. (The bottom of B is sealed at C.) Liquid rises in B to level D, at which time the automatic siphon, E, starts. Extracted material accumulates in the flask as more pure liquid vaporizes to repeat the process.

In case the hexaphenylbenzene is contaminated with insoluble material, crystallization from a filtered solution can be accomplished as follows: Place 10 mL of diphenyl ether in a 25 × 150-mm test tube, pack the sample of hexaphenylbenzene into a 10-mm extraction thimble, and suspend this in the test tube with two nichrome wires, as shown in Figure 52.1. Insert a cold finger condenser supported by an inverted filter adapter and adjust the condenser and the wires so that condensing liquid will drop into the thimble. Let the diphenyl ether reflux until the hexaphenylbenzene in the thimble is dissolved; then let the product crystallize, add toluene, collect the product, and wash with toluene (as described previously). A larger version of this apparatus is shown in Figure 52.2.

The yield of colorless plates is 0.6–0.7 g. The melting point of the product can be determined with a Mel-Temp apparatus and a 500°C thermometer. To avoid oxidation, seal the sample in an evacuated capillary tube. The product should melt at 465°C.

**Cleaning Up.** Place all filtrates in the organic solvents container.

## Computational Chemistry

**Bicyclo[2,2,1]hepta-2,5-dien-7-one**          **Benzene**     **Carbon monoxide**

Calculate, using the AM1 semiempirical method, the heats of formation of bicyclo[2,2,1]hepta-2,5-dien-7-one and of benzene and carbon monoxide in order to confirm theoretically that a reaction might occur.

## QUESTIONS

1. Which two factors probably contribute to the very high melting points of these two hexasubstituted benzenes?
2. What volume of carbon monoxide, measured at standard temperature and pressure (STP), is produced by the decomposition of 38 mg of tetraphenylcyclopentadienone?
3. Is the energy-minimized conformation of hexaphenylbenzene a chiral molecule? Explain why or why not.

# Derivatives of 1,2-Diphenylethane: A Multistep Synthesis[1]

**T**he next eight chapters provide various procedures for the rapid preparation of small samples of twelve related compounds, using benzaldehyde and phenylacetic acid as the initial reactants (see synthetic scheme on the next page). The quantities of reagents specified in the procedures will often provide somewhat more of each intermediate than is required for completing subsequent steps in the sequence of reactions. If the experiments are dovetailed, the entire series of preparations can be completed in a very short working time. For example, one can start the preparation of benzoin (record the time of starting; do not rely on memory) and during the reaction period start the preparation of α-phenylcinnamic acid. The preparation of α-phenylcinnamic acid requires refluxing for 35 min and, while it is proceeding, the benzoin preparation can be stopped when the time is up and the product allowed to crystallize. The α-phenylcinnamic acid mixture can be allowed to stand (and cooled) until one is ready to isolate it. Also, you may want to observe the crystals occasionally during a crystallization process; however, you should utilize most of your time for other operations.

Specific points related to stereochemistry and reaction mechanisms are discussed at the beginning of each individual chapter. Because several of the compounds have characteristic ultraviolet or infrared spectra, pertinent spectroscopic constants are recorded, and brief interpretations of the data are presented. Molecular mechanics calculations give insight into the conformations of several of these compounds.

## QUESTION

Starting with 150 mg of benzaldehyde and assuming an 80% yield on each step, what yield of diphenylacetylene, in milligrams, might you expect in the synthetic sequence on page 702? From the information given and assuming a yield of 80% on the last two reactions, what yield of stilbene dibromide would you expect when starting with 150 mg of benzaldehyde?

---

1. If your work is well organized and proceeds without setbacks, the experiments can be completed in about four laboratory periods. Your instructor may elect to name a certain number of periods in which you are to make as many of the compounds as possible; your instructor may also decide to require only the submission of the end products in each series.

**Benzaldehyde**

Chap. 53 KCN or thiamine

**Benzoin**

Chap. 54 HNO₃

**Benzil**

NaBH₄ Chap. 55

**Hydrobenzoin (Stilbenediol)**

Acetone FeBr₃ Chap. 56

**Stilbenediol acetonide**

Chap. 57 Ac₂O Zn

*E*- and **Z-Diacetates**

Chap. 58 Zn/Hg HCl + EtOH

*E*-**Stilbene**

Chap. 58 Br₂

**meso-Stilbene dibromide**

Chap. 58 KOH

**Diphenylacetylene**

Chap. 59 + CH₂COOH

*E*- and **Z-Phenylcinnamic acid**

Chap. 60 copper chromite catalyst

**Z-Stilbene**

Chap. 60 Br₂

*dl*-**Stilbene dibromide**

702

# The Benzoin Condensation: Catalysis by the Cyanide Ion and Thiamine

*Online Study Center*

This icon will direct you to techniques, equipment setups, and online resources at http://college.hmco.com/ PIC/williamsonMMOE5e.

**PRELAB EXERCISE:** What purposes does sodium hydroxide serve in the thiamine-catalyzed benzoin condensation?

The reaction of 2 mol of benzaldehyde to form a new carbon-carbon bond is known as the *benzoin condensation*. It is catalyzed by two rather different catalysts—the cyanide ion and thiamine (a B-complex vitamin)—which, on close examination, seem to function in exactly the same way.

**Benzaldehyde**
MW 106.12
bp 178°C, den. 1.044

**Benzoin**
MW 212.24
mp 135°C

Consider the reaction catalyzed by the cyanide ion. The cyanide ion attacks the carbonyl oxygen to form a stable cyanohydrin, mandelonitrile, a liquid with a boiling point of 170°C that under the basic conditions of the reaction loses a proton to give a resonance-stabilized carbanion (**A**). The carbanion attacks another molecule of benzaldehyde to give compound **B**, which undergoes a proton transfer and loses cyanide to give benzoin. Evidence for this mechanism lies in the failure of 4-nitrobenzaldehyde to undergo the reaction because the nitro group reduces the nucleophilicity of the anion in compound **A**. On the other hand, a strong electron-donating group in the 4-position of the phenyl ring makes the loss of the proton from the cyanohydrin very difficult; thus 4-dimethylaminobenzalde-hyde also does not undergo the benzoin condensation with itself.

**Benzaldehyde**

**Mandelonitrile**
bp 170°C

**A**

**B**                                 **Benzoin**

The cyanide ion binds irreversibly to hemoglobin, rendering it useless as a carrier of oxygen.

Several biochemical reactions bear a close resemblance to the benzoin condensation but are not, obviously, catalyzed by the highly toxic cyanide ion. Approximately 30 years ago, Ronald Breslow proposed that vitamin $B_1$ (also known as thiamine hydrochloride), in the form of the coenzyme thiamine pyrophosphate, can function in a manner completely analogous to the cyanide ion in promoting reactions like the benzoin condensation. The resonance-stabilized conjugate base of the thiazolium ion, thiamine, and the resonance-stabilized carbanion (**C**), which it forms, are the keys to the reaction. Like the cyanide ion, the thiazolium ion has just the right balance of nucleophilicity, the ability to stabilize the intermediate anion, and good leaving group qualities.

In the reactions that follow, the cyanide ion functions as a fast and efficient catalyst, although in large quantities it is highly toxic. The amount of potassium cyanide used in the first experiment (15 mg) is about eight times lower than the average fatal dose, a difference that underlines the advantage of carrying out organic experiments on a microscale.

The importance of thiamine is evident in that it is a vitamin, an essential substance that must be provided in the diet to prevent beriberi, a nervous system disease. The thiamine-catalyzed reaction is much slower, but the catalyst is completely nontoxic.

**Thiamine hydrochloride**

**Thiamine**

C

B′

+ Thiamine

---

## EXPERIMENTS

## 1. Cyanide-Catalyzed Benzoin Condensation

Reaction time: 30 min

IN THIS EXPERIMENT a tiny amount of potassium cyanide catalyzes the condensation of two molecules of benzaldehyde to a molecule of benzoin in a 30-min reaction. The product crystallizes from the reaction mixture and is isolated by filtration.

⚠️ CAUTION: Potassium cyanide is poisonous. Do not handle if you have open cuts on your hands. Never acidify a cyanide solution (because HCN gas is evolved). Wash your hands after handling cyanide.

### Microscale Procedure

In a reaction tube place 15 mg of potassium cyanide (**poison**!); dissolve it in 0.15 mL of water; add 0.30 mL of 95% ethanol and from a small, accurate syringe add 0.15 mL (157 mg) of pure benzaldehyde[1]; introduce a boiling chip; and reflux

---

1. Commercial benzaldehyde is inhibited against autoxidation with 0.1% hydroquinone. If the material available is yellow or contains benzoic acid crystals, it should be purified. *See also* footnote 2.

the solution gently on a warm sand bath or a steam bath for 30 min (*see* Fig. 39.2 on page 551). Remove the tube and cool it in an ice bath; if no crystals appear within a few minutes, withdraw a drop on a stirring rod and rub it against the inside of the tube to induce crystallization. When crystallization is complete, remove the solvent using a Pasteur pipette and, while keeping the tube on ice, wash the crystals thoroughly with 1 mL of a 1:1 mixture of 95% ethanol and water. Mix the crystals with the wash solvent and then isolate them by filtration on a Hirsch funnel or on a Wilfilter. If pure, this material is usually colorless and has a melting point of 134°C–135°C. The usual yield is 100–120 mg.

**Cleaning Up.** Add the aqueous filtrate to 10 mL of a 1% sodium hydroxide solution. Add 10 mL of household bleach (aqueous sodium hypochlorite solution) to oxidize the cyanide ion. The resulting solution can be tested for cyanide using the Prussian blue test described on the website under Tests for Nitrogen. (Continue this process until no cyanide is present.) When cyanide is no longer present, the solution can be diluted with water and flushed down the drain. Place ethanol used in crystallization in the organic solvents container.

## *Macroscale Procedure*

Place 0.75 g of potassium cyanide (**poison!**) in a 50-mL round-bottomed flask, dissolve it in 7.5 mL of water, add 15 mL of 95% ethanol and 7.5 mL of pure benzaldehyde,[2] introduce a boiling stone, attach a short condenser, and reflux the solution gently on a flask heater for 30 min (Fig. 53.1). Remove the flask and cool it in an ice bath; if no crystals appear within a few minutes, withdraw a drop on a stirring rod and rub it against the neck of the flask to induce crystallization. When crystallization is complete, collect the product and wash it free of the yellow mother liquor with a 1:1 mixture of 95% ethanol and water. Usually this first-crop material is colorless and of satisfactory melting point (134°C–135°C); the usual yield 5–6 g.[3]

**Cleaning Up.** Add the aqueous filtrate to 5 mL of a 1% sodium hydroxide solution. Add 25 mL of household bleach (aqueous sodium hypochlorite solution) to oxidize the cyanide ion. The resulting solution can be tested for cyanide using the Prussian blue test described on the website under Tests for Nitrogen. Continue this process until no cyanide is present. When cyanide is no longer present, the solution can be diluted with water and flushed down the drain. Place ethanol used in crystallization in the organic solvents container.

Photos: Filtration Using a Pasteur Pipette, Use of the Wilfilter; Videos: Filtration of Crystals Using the Pasteur Pipette, Microscale Filtration on the Hirsch Funnel

**CAUTION**: Potassium cyanide is poisonous. Do not handle if you have open cuts on your hands. Never acidify a cyanide solution (because HCN gas is evolved). Wash your hands after handling cyanide.

Video: Macroscale Crystallization

---

2. Commercial benzaldehyde inhibited against autoxidation with 0.1% hydroquinone is usually satisfactory. If the material available is yellow or contains benzoic acid crystals, it should be shaken with equal volumes of a 5% sodium carbonate solution until carbon dioxide is no longer evolved; then dry the upper layer over calcium chloride and distill (bp 178°C–180°C). Avoid exposing the hot liquid to air. The distillation step can be omitted if the benzaldehyde is colorless.

3. Concentration of the mother liquor to a volume of 10 mL gives a second crop (1.4 g, mp 133°C –134.5°C); the best total yield 6.8 g (87%). Recrystallization can be accomplished with either methanol (11 mL/g) or 95% ethanol (7 mL/g) with 90% recovery in the first crop.

■ **FIG. 53.1**
A macroscale reflux apparatus.

Water out

Water in

# 2. Thiamine-Catalyzed Benzoin Condensation

## *Microscale Procedure*

In a reaction tube place 26 mg of thiamine hydrochloride, dissolve it in a drop of water, add 0.30 mL of 95% ethanol, and cool the solution in an ice bath. Add 0.05 mL of 3 *M* sodium hydroxide followed by 0.15 mL (157 mg) of pure benzaldehyde,[4] stir well, add the distilling column as an air condenser, and heat the mixture in a water bath at 60°C for 1–1.5 h. The progress of the reaction should be followed by thin-layer chromatography (TLC). Use benzaldehyde in one lane, benzoin in another, and the reaction mixture in the center. Elute with a 50:50 mixture of petroleum ether and ethyl acetate. Alternatively, the reaction mixture can be stored at room temperature for at least 24 h, although a week would do no harm. (The rate of most organic reactions doubles for each 10°C rise in temperature.)

Cool the reaction mixture in an ice bath. If crystallization does not occur, withdraw a drop of solution on a stirring rod and rub it against the inside surface of the tube to induce crystallization. Remove the solvent using a Pasteur pipette and, while keeping the mixture on ice, wash the crystals with a 1:1 ice-cold mixture of 95% ethanol and water. The product should be colorless and of sufficient purity (mp 134°C–135°C) for use in subsequent reactions; the usual yield is 100–120 mg. If desired, the moist product can be recrystallized from 95% ethanol (7 mL/g) or methanol (11 mL/g) with 90% recovery.

**Cleaning Up.** The aqueous filtrate, after neutralization with dilute hydrochloric acid, is diluted with water and flushed down the drain. Ethanol used in crystallization should be placed in the organic solvents container.

 ## *Macroscale Procedure*

Place 1.3 g of thiamine hydrochloride in a 50-mL Erlenmeyer flask, dissolve it in 4 mL of water, add 15 mL of 95% ethanol, and cool the solution in an ice bath. Add 2.5 mL of 3 *M* sodium hydroxide dropwise with swirling such that the temperature of the solution does not rise above 20°C. To the yellow solution add 7.5 mL of pure benzaldehyde[5] and heat the mixture at 60°C for 1–1.5 h. The progress of the reaction can be followed by thin-layer chromatography (TLC). Alternatively, the reaction mixture may be stored at room temperature for at least 24 h. (The rate of most organic reactions doubles for each 10°C rise in temperature.)

Cool the reaction mixture in an ice bath. If crystallization does not occur, withdraw a drop of solution on a stirring rod and rub it against the inside surface of the flask to induce crystallization. Collect the product by suction filtration and wash it free of the yellow mother liquor with a 1:1 mixture of 95% ethanol and water. The product should be colorless and of sufficient purity (mp 134°C–135°C)

***Online Study Center***
Video: Macroscale Crystallization

---

4. *See* footnote 1.

5. *See* footnote 1.

for use in subsequent reactions; the usual yield is 5–6 g. If desired, the moist product can be recrystallized from 95% ethanol (8 mL/g).

**Cleaning Up.** The aqueous filtrate, after neutralization with dilute hydrochloric acid, is diluted with water and flushed down the drain. Ethanol used in crystallization should be placed in the organic solvents container.

## QUESTIONS

1. Speculate on the structure of the compound formed when 4-dimethyl-aminobenzaldehyde is condensed with 4-chlorobenzaldehyde.
2. Why might the presence of benzoic acid be deleterious to the benzoin condensation?
3. How many π electrons are in the thiazoline ring of thiamine hydrochloride? of thiamine?
4. Locate the CH and OH protons in the proton nuclear magnetic resonance ($^1$H NMR) spectrum of benzoin (Fig. 53.2).

■ **FIG. 53.2**
The $^1$H NMR spectrum of benzoin.

PPM (δ)

# CHAPTER

# 54

# Nitric Acid Oxidation; Preparation of Benzil from Benzoin; and Synthesis of a Heterocycle: Diphenylquinoxaline

*Online Study Center*

This icon will direct you to techniques, equipment setups, and online resources at http://college.hmco.com/ PIC/williamsonMMOE5e.

**PRELAB EXERCISE:** Write a detailed mechanism for the formation of 2,3-dimethylquinoxaline.

**B**enzoin can be oxidized to benzil, an α-diketone, very efficiently by nitric acid or by copper(II) sulfate in pyridine. When oxidized with sodium dichromate in acetic acid, the yield is lower because some of the material is converted into benzaldehyde by cleavage of the bond between two oxidized carbon atoms that is activated by both phenyl groups (**a**). Similarly, when hydrobenzoin is oxidized with dichromate or permanganate, it yields chiefly benzaldehyde and only a trace of benzil (**b**).

Ultraviolet (UV) spectroscopy is used to help characterize aromatic molecules such as benzoin. The absorption band at 247 nm in Figure 54.1 is attributable to the presence of the phenyl ketone group, in which the carbonyl group is conjugated with a benzene ring. Aliphatic α,β-unsaturated ketones, R—CH=CH—C=O, show selective absorption of UV light of comparable wavelength. See the infrared (IR) spectrum of benzoin in Figure 54.2.

**■ FIG. 54.1**
The UV spectrum of benzoin. $\lambda_{max}^{EtOH}$ = 247 nm ($\varepsilon$ = 13,200). Concentration: 12.56 mg/L = 5.92 × $10^{-5}$ mol/L. *See* Chapter 14 for the relationship between the extinction coefficient ($\varepsilon$), absorbance (*A*), and concentration (*C*).

**■ FIG. 54.2**
The IR spectrum of benzoin (KBr disk).

## EXPERIMENTS

# 1. Nitric Acid Oxidation of Benzoin

| Benzoin | Benzil |
|---|---|
| MW 212.24, mp 135°C | MW 210.23, mp 94–95°C |

⚠ **CAUTION:** Handle nitric acid with great care. It is highly corrosive to tissue and a very strong oxidant.

**IN THIS EXPERIMENT** the hydroxyl group of benzoin is oxidized with hot nitric acid to a carbonyl group to give benzil, which is yellow in color. The product is collected on a Hirsch funnel, washed with water, and recrystallized from ethanol.

## Microscale Procedure

Reaction time: 11 min

**Online Study Center**

Photos: Filtration Using a Pasteur Pipette, Use of the Wilfilter; Videos: Filtration of Crystals, Using the Pasteur Pipette, Microscale Filtration on the Hirsch Funnel

Heat a mixture of 100 mg of benzoin and 0.35 mL of concentrated nitric acid on a steam bath or in a small beaker of boiling water for about 11 min. Carry out the reaction in the hood or use an aspirator tube near the top of the tube to remove nitrogen oxides. Be sure all of the benzoin is washed down inside the tube and is oxidized. Add 2 mL of water to the reaction mixture, cool to room temperature, and stir the mixture for 1–2 min to coagulate the precipitated product; remove the solvent with a Pasteur pipette and wash the solid with an additional 2 mL of water. Dissolve the solid in 0.5 mL of hot ethanol and add water dropwise to the hot solution until the solution appears to be cloudy, indicating that it is saturated. Heat to bring the product completely into solution and allow it to cool slowly to room temperature. Cool the tube in ice and isolate the product using a Wilfilter or a Hirsch funnel. Scrape the benzil onto a piece of filter paper, squeeze out excess solvent, and allow the solid to dry. Record the percent yield, the crystalline form, the color, and the melting point of the product.

**Cleaning Up.** The aqueous filtrate should be neutralized with sodium carbonate, diluted with water, and flushed down the drain. Place the ethanol used in crystallization in the organic solvents container.

## Macroscale Procedure

Reaction time: 11 min

Heat a mixture of 4 g of benzoin and 14 mL of concentrated nitric acid on a steam bath for about 11 min. Carry out the reaction in a hood or use an aspirator tube near the top of the flask to remove nitrogen oxides. Add 75 mL of water to the reaction mixture, cool to room temperature, and swirl for 1–2 min to coagulate the precipitated product; collect and wash the yellow solid on a Hirsch funnel, pressing the solid on the filter to squeeze out the water. This crude product (the dry weight yield is 3.7–3.9 g) need not be dried but can be crystallized at once from ethanol. Dissolve the product in 10 mL of hot ethanol, add water dropwise to the cloud point, and set aside to crystallize. Record the yield, the crystalline form, the color, and the melting point of the purified benzil. The IR and $^1$H NMR spectra are found at the end of the chapter (Fig. 54.5 and Fig. 54.6).

### A Test for the Presence of Unoxidized Benzoin

Dissolve about 0.5 mg of crude or purified benzil in 0.5 mL of 95% ethanol or methanol and add 1 drop of 3 *M* sodium hydroxide. If benzoin is present, the solution soon acquires a purplish color due to a complex of benzil with a product of autoxidation of benzoin. If no color develops in 2–3 min, which is an indication that the sample is free of benzoin, add a small amount of benzoin; observe the color that develops; and note that if the test tube is stoppered and shaken vigorously, the color momentarily disappears. When the solution is allowed to stand, the color reappears.

**Cleaning Up.** The aqueous filtrate should be neutralized with sodium carbonate, diluted with water, and flushed down the drain. Ethanol used in crystallization should be placed in the organic solvents container.

## 2. Preparation of Benzil Quinoxaline (Diphenylquinoxaline)

A reaction that characterizes benzil as an α-diketone is a condensation reaction of 1,2-phenylenediamine to the quinoxaline derivative. The aromatic heterocyclic ring formed in the condensation is fused to a benzene ring to give a bicyclic system analogous to naphthalene. The word *heterocyclic* refers to the fact that the ring contains an atom or atoms other than carbon.

**Benzil**
MW 210.23
mp 94–95°C

**1,2-Phenylenediamine**
MW 108.14
mp 103°C

**2,3-Diphenylquinoxaline**
MW 282.33
mp 126°C

⚠️ **CAUTION**: Handle *o*-phenylenediamine with care. It is classified as a suspected carcinogen. Similar compounds (hair dyes) are low-risk carcinogens. Carry out the reaction in the hood.

**IN THIS EXPERIMENT** the two carbonyl groups of benzil are condensed with two amine groups to give an aromatic compound with two nitrogen atoms in the ring. The diamine must be purified by sublimation before being heated with the benzil (no solvent). The product is crystallized from methanol or a mixture of methanol and water. (*See* Chapter 4.)

## *Microscale Procedure*

Commercial 1,2-phenylenediamine (also known as *o*-phenylenediamine) is usually badly discolored (due to air oxidation) and gives a poor result unless it is purified as follows. Place 100 mg of material in a reaction tube, evacuate the tube at full aspirator suction, clamp it, and heat the bottom of the tube with a hot sand bath to distill or sublime colorless 1,2-phenylenediamine from the dark residue into the upper half of the tube (Fig. 54.3). Let the tube cool until the melt has solidified; then scrape out the white solid.

Weigh 105 mg of benzil and 54 mg of your purified 1,2-phenylenediamine into a reaction tube and heat on a steam bath for 10 min, which changes the initially molten mixture to a light tan solid. Dissolve the solid in hot methanol (about 2.5 mL) and let the solution stand undisturbed. If crystallization does not occur within 10 min, reheat the solution and dilute it with a little water to the point of saturation. The crystals should be filtered on a Hirsch funnel as soon as they are formed because brown oxidation products accumulate on standing. The quinoxaline forms colorless needles (mp 125°C–126°C); the yield is about 90 mg.

**Cleaning Up.** The residues from the distillation (sublimation) of 1,2-phenylenediamine (the tube can be rinsed with acetone) and the solvent from the reaction should be placed in the aromatic amines hazardous waste container because the diamine may be a carcinogen.

## *Online Study Center*

Video: Microscale Filtration on the Hirsch Funnel

Reaction time: 11 min

## *Macroscale Procedure*

Commercial *o*-phenylenediamine (also known as 1,2-phenylenediamine) is usually badly discolored (due to air oxidation) and gives a poor result unless it is purified as follows. Place 200 mg of material in a 20 × 150-mm test tube, evacuate the tube at full aspirator suction, clamp it in a nearly horizontal position, and heat the bottom of the tube in a hot sand bath to sublime colorless *o*-phenylene-diamine from the dark residue into the upper half of the tube. Let the tube cool in position until the melt has solidified; then scrape out the white solid.

Weigh 0.20 g of benzil (theoretical = 210 mg) and 0.10 g of your purified *o*-phenylenediamine (theoretical = 108 mg) into a 20 × 150-mm test tube and heat in a steam bath for 10 min, which changes the initially molten mixture to a light tan solid. Dissolve the solid in hot methanol (about 5 mL) and let the solution stand undisturbed. If crystallization does not occur within 10 min, reheat the solution and dilute it with a little water to the point of saturation. The crystals should be filtered as soon as they are formed because brown oxidation products

Reaction time: 11 min

accumulate on standing. The quinoxaline forms colorless needles (mp 125°C–126°C); the yield is 185 mg. The UV spectrum is shown in Figure 54.4.

**Cleaning Up.** Place all the residues from the distillation of 1,2-phenylenediamine and the solvent from the reaction in the aromatic amines hazardous waste container because the diamine may be a carcinogen.

■ **FIG. 54.3**
An apparatus for the vacuum sublimation of 1,2-phenylenediamine.

■ **FIG. 54.4**
The UV spectrum of quinoxaline derivative. $\lambda_{max}^{EtOH} = 244$ nm ($\varepsilon = 37{,}400$); $\lambda_{max}^{EtOH} = 345$ nm ($\varepsilon = 12{,}700$).

■ **FIG. 54.5**
The IR spectrum of benzil (KBr disk).

■ **FIG. 54.6**
The $^1$H NMR spectrum of benzil
(250 MHz).

## QUESTIONS

1. Assign the peak at 1646 cm$^{-1}$ in the IR spectrum of benzil (Fig. 54.5).
2. Assign the peaks at 3410 cm$^{-1}$ and 1664 cm$^{-1}$ in the IR spectrum of benzoin (Fig. 54.2).

# The Borohydride Reduction of a Ketone: Hydrobenzoin from Benzil

**PRELAB EXERCISE:** Compare the reducing capabilities of lithium aluminum hydride to those of sodium borohydride.

Sodium borohydride was discovered in 1943 by H. I. Schlesinger and Herbert C. Brown. Brown devoted his entire scientific career to this reagent, making it and other hydrides the most useful and versatile of reducing reagents. He received a Nobel Prize in Chemistry in 1979 for his work.

Considering the extreme reactivity of most hydrides (such as sodium hydride and lithium aluminum hydride) toward water, sodium borohydride is somewhat surprisingly sold as a stabilized aqueous solution in 14 *M* sodium hydroxide containing 12% sodium borohydride. Unlike lithium aluminum hydride, sodium borohydride is insoluble in ether and soluble in methanol and ethanol.

Sodium borohydride is a mild and selective reducing reagent. In ethanol solution it reduces aldehydes and ketones rapidly at 25°C and esters very slowly, and it is inert toward functional groups that are readily reduced by lithium aluminum hydride, including carboxylic acids, epoxides, lactones, nitro groups, nitriles, azides, amides, and acid chlorides.

The present experiment is a typical sodium borohydride reduction. These same conditions and isolation procedures could be applied to hundreds of other ketones and aldehydes.

## EXPERIMENT

## Sodium Borohydride Reduction of Benzil

The addition of two atoms of hydrogen to benzoin or four atoms of hydrogen to benzil gives a mixture of stereoisomeric diols, of which the predominant isomer is the nonresolvable (1*R*,2*S*)-hydrobenzoin, the *meso* isomer, accompanied by the

**Benzil**
MW 210.22

**(1R,2S)-(meso)-Hydrobenzoin**
mp 137°C, MW 214.25

**(1R,2R) and (1S,2S)-Hydrobenzoin**
mp 120°C

enantiomeric (1R,2R) and (1S,2S) compounds. The reaction proceeds rapidly at room temperature; the intermediate borate ester is hydrolyzed with water to give the product alcohol.

$$4 \ R_2C{=}O + Na^+BH_4^- \longrightarrow (R_2CHO)_4B^-Na^+$$

$$(R_2CHO)_4B^-Na^+ + 2 \ H_2O \longrightarrow 4 \ R_2CHOH + Na^+BO_2^-$$

The procedure that follows specifies the use of benzil rather than benzoin because you can then follow the progress of the reduction by the discharge of the yellow color of the benzil.

Chapter 54 contains the infrared (IR) and nuclear magnetic resonance (NMR) spectra for benzil (Fig. 54.5 and Fig. 54.6, respectively). The ultraviolet (UV) spectrum is shown in Figure 55.2 at the end of this chapter.

## Microscale Procedure

**IN THIS EXPERIMENT** a solid diketone dissolved in ethanol is reduced with solid sodium borohydride to a dialcohol that crystallizes from the reaction mixture and is isolated by filtration. It is easy to follow the progress of the reaction—the diketone is yellow, and the dialcohol is colorless in solution.

*Online Study Center*

Photos: Filtration Using a Pasteur Pipette, Use of the Wilfilter; Videos: Filtration of Crystals Using the Pasteur Pipette. Microscale Crystallization

In a 10 × 100-mm reaction tube, dissolve 50 mg of benzil in 0.5 mL of 95% ethanol; then cool the solution in ice to produce a fine suspension. Add to this suspension 10 mg of sodium borohydride (a large excess). The benzil dissolves, the mixture warms up, and the yellow color disappears in 2–3 min. After a total of 10 min, add 0.5 mL of water, heat the solution to the boiling point, filter the solution in case it is not clear (usually not necessary), and dilute the hot solution with hot water to the point of saturation (cloudiness)—a process that requires about 1 mL of water. (1R,2S)-Hydrobenzoin separates in lustrous thin plates (mp 136°C–137°C) and is best isolated by withdrawing some of the solvent using a Pasteur pipette and then using a Wilfilter (Fig. 55.1). It also can be collected on a microscale Büchner funnel. The yield is about 35 mg.

**Cleaning Up.** Dilute the aqueous filtrate with water, neutralize with acetic acid (to destroy the borohydride), and then flush the mixture down the drain.

### ■ FIG. 55.1

The Wilfilter filtration apparatus.
*See* Chapter 4.

Crystals

Wilfilter

Filtrate

### ■ FIG. 55.2

The UV spectrum of benzil.
$\lambda_{max}^{EtOH} = 260$ nm ($\varepsilon = 19,800$).
One-centimeter cells and 95%
ethanol have been employed
for this spectrum.

## Macroscale Procedure

In a 50-mL Erlenmeyer flask, dissolve 0.5 g of benzil in 5 mL of 95% ethanol and cool the solution under the tap to produce a fine suspension. Then add 0.1 g of sodium borohydride (large excess). The benzil dissolves, the mixture warms up, and the yellow color disappears in 2–3 min. After a total of 10 min, add 5 mL of water, heat to the boiling point, filter in case the solution is not clear, dilute to the point of saturation with more water (10 mL), and set the solution aside to crystallize. *meso*-Hydrobenzoin separates in lustrous thin plates (mp 136°C–137°C); the yield is about 0.35 g. A second crop of material can be obtained by adding solid sodium chloride to the filtrate until no more of the salt will dissolve. Collect the second crop of crystals via filtration and wash the crystals with water.

**Cleaning Up.** Dilute the aqueous filtrate with water, neutralize it with acetic acid (to destroy the borohydride), and flush the mixture down the drain.

## QUESTIONS

1. Using the *R,S* system of nomenclature, draw and name all of the isomers of hydrobenzoin.
2. Calculate the theoretical weight of sodium borohydride needed to reduce 50 mg of benzil.

# The Synthesis of 2,2-Dimethyl-1,5-Dioxolane; The Acetonide Derivative of a Vicinal Diol

**Online Study Center**

This icon will direct you to techniques, equipment setups, and online resources at http://college.hmco.com/ PIC/williamsonMMOE5e.

**PRELAB EXERCISE:** Draw the most stable structures of *meso*- and *racemic*-1,2-stilbenediol using Newman projections. Draw the same two isomers with the hydroxyl groups eclipsed.

In Chapter 55, yellow benzil was reduced with sodium borohydride to the colorless diol. The melting point of the product (137°C) indicates that (1*R*,2*S*)-(*meso*)-hydrobenzoin has been produced. But how do we know that this is the isomer produced? The object of this experiment is to demonstrate, using logic based on nuclear magnetic resonance (NMR) spectral data, that the product is indeed the *meso* isomer and not the *racemic* isomer.

The diol is reacted with acetone to form a crystalline acetonide (a ketal) that will have either structure **1** or structure **2**. An examination of the NMR spectrum of this derivative will reveal whether or not the two methyl groups of the acetonide are equivalent and thus whether the diol is *meso* or *racemic*. Rather than simply explaining each step in this process and the reasoning behind it, you will discover them for yourself by answering the questions at the end of this chapter. By considering each question in turn, you will see how it is possible to determine the stereochemistry of the starting diol.

## EXPERIMENT

**IN THIS EXPERIMENT** a solid diol is dissolved in acetone that is both a solvent and a reactant. The addition of a catalyst, the Lewis acid iron(III) chloride, and heating of the solution causes the formation of a cyclic acetonide. At the end of the reaction, the mixture is poured into water containing a base to react with the acid, and the product is extracted into dichloromethane. This solution is dried and then evaporated to give the solid product, which is then crystallized from hexane.

### Microscale Procedure

Dissolve 100 mg of the diol in 3 mL of acetone and add 30 mg of anhydrous iron(III) chloride. The iron(III) chloride is very hygroscopic and should therefore be weighed from the sealed storage container into a closed vial. Reflux the mixture for 20 min using the apparatus depicted in Figure 56.1.

Transfer the reaction mixture to a 15-mL centrifuge tube containing 4 mL of water and 1 mL of 3 $M$ potassium carbonate solution. Extract the product three times with 1-mL portions of dichloromethane that are transferred to a reaction tube. Wash the combined extracts with 2.5 mL of water and then dry the solution over a few pieces of granular anhydrous calcium chloride. Transfer the dry solution to a clean reaction tube and carefully evaporate it to dryness.

Dissolve the crude product in 0.3–0.5 mL of hexane and cool the solution in ice, whereupon the product will crystallize. Collect the product using the Pasteur pipette method or a Wilfilter (Fig. 56.2). Dry the product under vacuum, if necessary; then record the yield, the melting point, and the infrared (IR) spectrum of the product. Record the $^1$H NMR spectrum of a solution in deuterochloroform (CDCl$_3$) and assign the stereochemistry of the product and thus of the diol. Store the acetonide in a refrigerator before analysis; it slowly decomposes at room temperature.

**Cleaning Up.** Place any excess or recovered dichloromethane in the halogenated waste container. Other organic solutions go in the organic solvents container. The aqueous phases from the extractions can, after dilution with water, be washed down the drain. The drying agent (the calcium chloride pellets), after removal of the solvent in the hood, can be placed in the nonhazardous solid waste container. If local regulations do not allow for the evaporation of solvents in a hood, the wet solid should be disposed in a special container.

*Online Study Center*

Video: Extraction with Dichloromethane; Photo: Extraction with Dichloromethane

*Online Study Center*
Video: Filtration of Crystals Using the Pasteur Pipette; Photos: Filtration Using a Pasteur Pipette, Use of the Wilfilter

■ **FIG. 56.1**
The esterification apparatus.

Wet pipe
cleaner

■ **FIG. 56.2**
The Wilfilter filtration
apparatus. *See* Chapter 4.

Crystals

Wilfilter

Filtrate

## *Macroscale Procedure*

Dissolve 0.30 g of *meso*-hydrobenzoin (*meso*-1,2-stilbenediol), the product of the experiment in Chapter 55, in 9 mL of acetone and add 0.1 g of anhydrous iron(III) chloride. The iron(III) chloride is very hygroscopic and should therefore be weighed from the sealed storage container that can be closed. Reflux the mixture for 20 min using the apparatus depicted in Figure 25.3 (on page 407). Be sure to add a boiling chip to the flask.

Transfer the reaction mixture to a small separatory funnel that contains 30 mL of water and 3 mL of 3 *M* potassium carbonate solution. Extract the product three times with 3-mL portions of dichloromethane that are transferred to a 10-mL Erlenmeyer flask. If you are uncertain which layer is the dichloromethane layer, choose a drop from one layer and add a few drops of water. If two phases can be seen, the layer you chose is the dichloromethane layer. Wash the combined extracts with 8 mL of water and then dry the dichloromethane solution over granular anhydrous calcium chloride pellets. Transfer the dry solution to a clean 25-mL filter flask using more dichloromethane to wash the tube and drying agent. Evaporate the solution to dryness under vacuum (*see* Fig. 25.9 on page 416). Warm the filter flask with your hand to speed the process. The product should crystallize in the filter flask. It may be necessary to scratch or seed the thick liquid that sometimes results to initiate crystallization.

At this point the crude product could be used for NMR analysis. Dissolve about 50 mg of the crystals or oil in deuterochloroform and transfer the solution to a clean, dry NMR tube. Bring the level of the deuterochloroform in the tube to 5.5 cm, cap the tube, and invert it several times to mix the solution.

To purify the crude product, dissolve it in a minimum quantity (0.9–1.5 mL) of dichloromethane, transfer the solution to a reaction tube or a small test tube, evaporate the dichloromethane, seed the resulting oil and then add 0.9–1.5 mL of hexane, stir the mixture, and allow crystallization to proceed. Cool the solution in ice. Collect the acetonide on a small Hirsch funnel. Dry it under vacuum, if necessary (*see* Fig. 4.22 on page 79); then record the yield, the melting point, and the IR spectrum of the product.

Record the $^1$H NMR spectrum of a solution in deuterochloroform and assign the stereochemistry of the product and thus of the diol. Store the acetonide in a refrigerator before analysis; it slowly decomposes at room temperature.

**Cleaning Up.** Place any excess or recovered dichloromethane in the halogenated waste container. Other organic solutions go in the organic solvents container. The aqueous phases from the extractions can, after dilution with water, be washed down the drain. The drying agent (the calcium chloride pellets), after removal of the solvent in the hood, can be placed in the nonhazardous solid waste container. If local regulations do not allow for the evaporation of solvents in a hood, the wet solid should be disposed in a special container.

## QUESTIONS

1. Write the mechanism and discuss the stereochemistry of the sodium borohydride reduction of diphenylethanedione (benzil).
2. Give the mechanism of acetonide formation. What is the role of iron(III) chloride?
3. Carefully draw the structures, using Newman projections, of (1*R*,2*S*)-(*meso*)-hydrobenzoin and (1*R*,2*R*)- and (1*S*,2*S*)-(*racemic*)-hydrobenzoin with the hydroxyl groups eclipsed. Then draw the corresponding acetonides of these two isomers.
4. Is acetonide **1** (on page 719) from the *meso* or the *racemic* diol?
5. Consider the symmetry of the two different acetonides formed. How many methyl peaks would you expect to observe in the NMR spectrum for the acetonide of the *meso* isomer? for the *racemic* isomer?
6. Compare and contrast the IR spectra of the dione, the diol, and the acetonide. Assign the major peaks to functional groups.

# CHAPTER 57

# 1,4-Addition: Reductive Acetylation of Benzil

*Online Study Center*

This icon will direct you to techniques, equipment setups, and online resources at http://college.hmco.com/ PIC/williamsonMMOE5e.

Ac = acetyl group

$$-OAc = \begin{matrix} O \\ \parallel \\ -OCCH_3 \end{matrix}$$

$$\begin{matrix} O & & O \\ \parallel & & \parallel \\ CH_3C-O-CCH_3 \end{matrix}$$

Ac$_2$O is acetic anhydride.

**PRELAB EXERCISE:** Write the complete mechanism for the reaction of acetic anhydride with an alcohol.

In one of the first demonstrations of the phenomenon of 1,4-addition, Johannes Thiele (1899) established that the reduction of benzil with zinc dust in a mixture of acetic anhydride and sulfuric acid involves the 1,4-addition of hydrogen to the α-diketone grouping and acetylation of the resulting enediol before it can undergo ketonization to benzoin. The process of reductive acetylation results in a mixture of the *E*- and *Z*-isomers (**1** and **2**). Thiele and subsequent investigators isolated the more soluble, lower-melting *Z*-stilbenediol diacetate (**2**) only in impure form (mp 110°C). Separation of the two isomers by chromatography is not feasible because they are equally adsorbable on alumina. However, separation is possible by fractional crystallization (described in the following procedures), and both isomers can be isolated in pure condition. In the method prescribed here for the preparation of the isomer mixture, hydrochloric acid is substituted for sulfuric acid because sulfuric acid gives rise to colored impurities and is reduced to sulfur and hydrogen sulfide.[1]

**Benzil**
$$\xrightarrow[\text{Zn, HCl}]{\text{2 H (1,4-addition)}}$$
$$\xrightarrow{\text{Ac}_2\text{O, H}^+}$$

---

1. If acetyl chloride (2 mL) is substituted for the hydrochloric acid–acetic anhydride mixture in the procedure, the Z-isomer is the sole product.

723

*E*-Stilbenediol
diacetate (1)
mp 155°C, $\lambda_{max}^{EtOH}$ 271 nm ($\epsilon = 23,400$)
MW 296.31
**1**

*Z*-Stilbenediol
diacetate (2)
mp 119°C, $\lambda_{max}^{EtOH}$ 265 nm ($\epsilon = 12,800$)
MW 296.31
**2**

Assignment of configuration

The configurations of this pair of geometrical isomers remained unestablished for over 50 years, but the tentative inference that the higher melting isomer has the more symmetrical *E* configuration eventually was found to be correct. Evidence of infrared (IR) spectroscopy is of no avail; the spectra are nearly identical in the interpretable region (4000–1200 cm$^{-1}$) characterizing the acetoxyl groups but differ in the fingerprint region (1200–400 cm$^{-1}$). However, the isomers differ markedly in ultraviolet (UV) absorption (Fig. 57.1); in analogy to *E*- and *Z*-stilbene, the conclusion is justified that the higher melting isomer, because it has an absorption band at a longer wavelength and a higher intensity than its isomer, does indeed have configuration **1**.

■ **FIG. 57.1**

The UV spectra of *Z*- and *E*-stilbene diacetate. *Z*: $\lambda_{max}^{EtOH} = 223$ nm ($\epsilon = 20,500$) and 269 nm ($\epsilon = 10,800$); *E*: $\lambda_{max}^{EtOH} = 272$ nm ($\epsilon = 20,800$). The two phenyl rings cannot be coplanar in the *Z*-diacetate, which prevents overlap of the *p* orbitals of the phenyl groups with those of the central double bond. This steric inhibition of resonance accounts for the diminished intensity of the *Z*-isomer relative to the *E*-isomer. Both spectra were run at the same concentration.

## EXPERIMENTS

# 1. Reductive Acetylation of Benzil

> **IN THIS EXPERIMENT** an ice-cold liquid anhydride reacts with an ice-cold solution of the dienol of a diketone (made by the zinc dust reduction of the diketone). The product, a mixture of solid diacetate isomers, is isolated by filtration. Because the solid mixture is contaminated with zinc dust, it is dissolved in ether to leave the zinc dust behind. The Z- and E-isomers are separated by fractional crystallization. As the ether is evaporated, the E-isomer is the first to crystallize out.

### *Microscale Procedure*

Handle acetic anhydride and hydrochloric acid with care.

Use fresh zinc dust.

*Online Study Center*

Video: Microscale Filtration on the Hirsch Funnel

*Online Study Center*

Photos: Filtration Using a Pasteur Pipette, Use of the Wilfilter; Videos: Filtration of Crystals Using the Pasteur Pipette, Microscale Crystallization

Place one reaction tube containing 0.7 mL of acetic anhydride and another containing 0.1 mL of concentrated hydrochloric acid in an ice bath and, when both are thoroughly chilled, transfer the acid to the anhydride dropwise by means of a Pasteur pipette. Transfer the acidic anhydride solution into a reaction tube containing 100 mg of pure benzil (prepared in Chapter 54) and 100 mg of zinc dust and stir the mixture for 2–3 min in an ice bath. The reaction will give off heat. Remove the tube and allow it to warm up slowly, cooling in ice if the tube becomes too hot to touch.

When there is no further exothermic effect, let the mixture stand for 5 min; then add 2.5 mL of water. Swirl, break up any lumps of product, and allow a few minutes for the hydrolysis of excess acetic anhydride. Next, collect the mixture of product and zinc dust on a Hirsch funnel, wash with water, and press and apply suction to the filter cake until there is no further drip. Digest the solid in a 25-mL Erlenmeyer flask (drying is not necessary) with 7.0 mL of diethyl ether to dissolve the organic material, add about 1 g of anhydrous calcium chloride pellets, and swirl briefly; using a Pasteur pipette, transfer the ether solution into another 10-mL Erlenmeyer flask, wash the drying agent, and concentrate the ether solution (using a steam bath, a boiling stone, and a water aspirator) to a volume of approximately 1.5 mL, transfer to a reaction tube, cork the tube, and let the tube stand undisturbed.

The E-diacetate (**1**) soon begins to separate in prismatic needles; after 20–25 min, crystallization appears to stop. Remove the mother liquor from the reaction tube with a Pasteur pipette or use a Wilfilter (*see* Fig. 4.14 on page 74) or the microscale filtration apparatus shown in Figure 57.2 and evaporate it to dryness in another reaction tube.

Dissolve the white solid remaining in the second reaction tube in 1.0 mL of methanol, let the solution stand undisturbed for about 10 min, and drop in a tiny crystal of the E-diacetate. This should give rise, in 20–30 min, to a second crop of the E-diacetate. Remove the mother liquor from these crystals and concentrate it to 0.7 mL, let it cool to room temperature as before, and again seed with a crystal of the E-diacetate; this usually affords a third crop of this diacetate. The three

Thermometer adapter

Micro Büchner funnel with fritted disk

To aspirator

Vacuum-tight connector

Reaction tube with filtrate

crops might be 30 mg (mp 154°C–156°C), 5 mg (mp 153°C–154°C), and 5 mg (mp 153°C–155°C).

At this point the mother liquor should be rich enough in the more soluble Z-diacetate (**2**) for its isolation. Concentrate the methanol mother liquor and the washings from the third crop of the E-diacetate to a volume of 0.4 mL, stopper the tube, and let the solution stand undisturbed overnight. The Z-diacetate sometimes separates spontaneously in large rectangular prisms of great beauty. If the solution remains supersaturated, adding a seed crystal of the Z-diacetate causes prompt separation of this diacetate in a paste of small crystals [e.g., 20 mg (mp 118°C–119°C) and then a second crop of 7 mg (mp 116°C–117°C)]. This experiment illustrates the great convenience of column chromatography in isomer separations when it can be employed. Separating compounds by fractional crystallization is rarely necessary.

**Cleaning Up.** Spread the zinc out on a watch glass for about 20 min before wetting it and placing it in the nonhazardous solid waste container. Sometimes zinc dust from a reaction like this is pyrophoric (spontaneously flammable in air) because it is so finely divided and has such a large, clean surface area able to react with air. Neutralize the aqueous filtrate with sodium carbonate, remove the zinc salts by vacuum filtration, dilute the filtrate with water, and flush it down the drain. The zinc salts are placed in the nonhazardous solid waste container. Place the filtrate from the fractional crystallization in the organic solvents container. Allow the solvent to evaporate from the calcium chloride pellets in the hood; then place the drying agent in the nonhazardous solid waste container. If local regulations do not allow for the evaporation of solvents in a hood, the wet solid should be disposed in a special container.

## Macroscale Procedure

**Online Study Center**
Video: Macroscale Crystallization

Place one test tube (20 × 150 mm) containing 7 mL of acetic anhydride and another test tube (13 × 100 mm) containing 1 mL of concentrated hydrochloric acid in an ice bath. When both are thoroughly chilled, transfer the acid to the anhydride dropwise in not less than 1 min by means of a Pasteur pipette. Wipe the test tube dry, pour the chilled solution into a 50-mL Erlenmeyer flask containing 1 g of pure benzil and 1 g of zinc dust, and swirl for 2–3 min in an ice bath. Remove the flask and hold it in the palm of your hand; if it begins to warm up, cool further in ice. When there is no further exothermic effect, let the mixture stand for 5 min; then add 25 mL of water. Swirl, break up any lumps of product, and allow a few minutes for the hydrolysis of excess acetic anhydride. Then collect the mixture of product and zinc dust by vacuum filtration, wash with water, and press and apply suction to the cake until there is no further drip. Digest the solid (drying is not necessary) with 70 mL of t-butyl methyl ether to dissolve the organic material, add about 4 g of calcium chloride pellets, and swirl briefly; filter the solution into a 125-mL Erlenmeyer flask, concentrate the filtrate (using a steam bath, a boiling stone, and a water aspirator) to a volume of approximately 15 mL,[2] cork the flask, and let the flask stand undisturbed.

The E-diacetate (**1**) soon begins to separate in prismatic needles; after 20–25 min, crystallization appears to stop. Remove the crystals by filtration on a Hirsch funnel. Wash them once with ether and then evaporate the ether to dryness. Dissolve the residue in 10 mL of methanol and transfer the hot solution to a 25-mL Erlenmeyer flask.

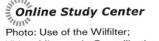
**Online Study Center**
Photo: Use of the Wilfilter;
Video: Microscale Crystallization

Let the solution stand undisturbed for about 10 min, then drop in one tiny crystal of the E-diacetate. This should give rise, in 20–30 min, to a second crop of the E-diacetate. Then concentrate the mother liquor and washings to a volume of 7–8 mL, let cool to room temperature as before, and again seed with a crystal of the E-diacetate; this usually affords a third crop of this diacetate. The three crops might be 300 mg (mp 154°C–156°C), 50 mg (mp 153°C–156°C), and 50 mg (mp 153°C–155°C). The products can also be collected using a Wilfilter or a microscale Büchner funnel.

At this point the mother liquor should be rich enough in the more soluble Z-diacetate (**2**) for its isolation. Concentrate the methanol mother liquor and washings from the third crop of the E-diacetate to a volume of 4–5 mL, stopper the flask, and let the solution stand undisturbed overnight. The Z-diacetate sometimes separates spontaneously in large rectangular prisms of great beauty. If the solution remains supersaturated, adding a seed crystal of the Z-diacetate causes prompt separation of this diacetate in a paste of small crystals [e.g., 215 mg (mp 118°C–119°C) and then a second crop of 70 mg (mp 116°C–117°C)].

**Cleaning Up.** Spread the zinc out on a watch glass for about 20 min before wetting it and placing it in the nonhazardous solid waste container. Sometimes zinc dust from a reaction like this is pyrophoric (spontaneously flammable in air) because it

---

2. Measure 15 mL of a solvent into a second flask of the same size and compare the levels in the two flasks.

is so finely divided and has such a large, clean surface area able to react with air. Neutralize the aqueous filtrate with sodium carbonate, remove the zinc salts by vacuum filtration, dilute the filtrate with water, and flush it down the drain. The zinc salts are placed in the nonhazardous solid waste container. Place the filtrate from the fractional crystallization in the organic solvents container. Allow the solvent to evaporate from the calcium chloride pellets in the hood and then place the drying agent in the nonhazardous solid waste container. If local regulations do not allow for the evaporation of solvents in a hood, the wet solid should be disposed in a special container.

## Computational Chemistry

Construct $Z$- and $E$-stilbene diacetate and carry out an energy minimization on the resulting molecules using a molecular mechanics program. Repeat the calculation using the semiempirical molecular orbital program AM1. From the resulting conformations, can you explain why the UV extinction coefficient $\varepsilon$ for the peak at 269 nm is smaller for the $Z$-isomer than for the $E$-isomer?

# CHAPTER
# 58

# The Synthesis of an Alkyne from an Alkene; Bromination and Dehydrobromination: Stilbene and Diphenylacetylene

**PRELAB EXERCISE:** Calculate the theoretical quantities of thionyl chloride and sodium borohydride needed to convert benzoin to *E*-stilbene.

**E-Stilbene**
mp 125°C, MW 180.24
$\lambda_{max}^{EtOH}$ 301 nm ($\epsilon = 28,500$)
226 nm ($\epsilon = 17,700$)

Heat of hydrogenation, $-20.1$ kcal/mol

**Z-Stilbene**
mp 6°C, MW 180.24
$\lambda_{max}^{EtOH}$ 280 nm ($\epsilon = 13,500$)
223 nm ($\epsilon = 23,500$)

Heat of hydrogenation, $-25.8$ kcal/mol

In this chapter's experiments, benzoin, which was prepared in Chapter 53, is converted to the alkene *trans*-stilbene (*E*-stilbene), which is in turn brominated and dehydrobrominated to form diphenylacetylene, an alkyne.

One method of preparing *E*-stilbene is the reduction of benzoin with zinc amalgam in a mixture of ethanol and hydrochloric acid, presumably through an intermediate:

The procedure that follows is quick and affords very pure hydrocarbon. It involves three steps: (1) replacing the hydroxyl group of benzoin by chlorine to form desyl chloride, (2) reducing the keto group with sodium borohydride to give what appears to be a mixture of the two diastereoisomeric chlorohydrins, and (3) eliminating the elements of hypochlorous acid with zinc and acetic acid. The last step is analogous to the debromination of an olefin dibromide.

**Benzoin**
MW 212.24

**Desyl chloride**
mp 68°C

**Mixture of both diasteromers**

**E-Stilbene**

The minimum energy conformations of *E*- and *Z*-stilbene are shown in Figures 58.1 and 58.2, respectively. These have been calculated using Spartan's

■ **FIG. 58.1**
The minimum energy conformation of *E*-stilbene.

■ **FIG. 58.2**
The minimum energy conformation of *Z*-stilbene.

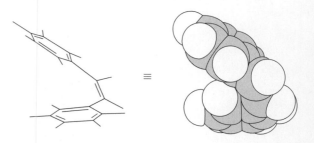

molecular mechanics routine.[1] Note that the *E*-isomer is planar, whereas the *Z*-isomer is markedly distorted from planarity.

---

## EXPERIMENTS

### 1. Stilbene

> **IN THIS EXPERIMENT** the hydroxyl group of solid benzoin is converted to a chloride by heating with liquid thionyl chloride, which acts as both a solvent and a reactant. The hydrogen chloride that is evolved is captured in a gas trap. The thionyl chloride is evaporated under vacuum to leave the oily product—a keto chloride. This material is dissolved in ethanol and reduced to a mixture of isomeric hydroxy chlorides (chlorohydrins) with sodium borohydride. Without isolation this solution of chlorohydrins is reduced again, this time with zinc dust and acetic acid. This process removes the elements of hypochlorus acid (HOCl) to give the product—stilbene.

### *Microscale Procedure*

**Online Study Center**

Photos: Placing a Polyethylene Tube through a Septum, Gas Trap

Place 100 mg of benzoin (crushed to a powder) in a reaction tube, cover it with 0.15 mL of thionyl chloride, and add a boiling chip and a rubber septum bearing a polyethylene tube leading to another reaction tube bearing a plug of damp cotton (Fig. 58.3). This will serve to trap the hydrogen chloride and sulfur dioxide

■ **FIG. 58.3**
A trap for hydrogen chloride and sulfur dioxide. Warm the tubing in a steam bath to bend it permanently.

Polyethylene tubing

Rubber septum

Reaction tubes

Damp cotton

Thionyl chloride and benzoin

---

1. Spartan molecular modeling software is available from Wavefunction, Inc., Irvine, California.

■ **FIG. 58.4**
The Pasteur pipette filtration
technique.

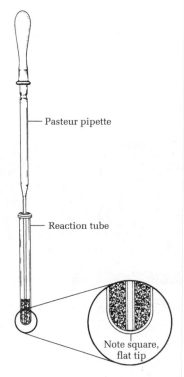

— Pasteur pipette

— Reaction tube

Note square,
flat tip

evolved in this reaction. Warm the reaction mixture gently on a steam bath or on a warm sand bath until the solid has dissolved and then more strongly for 5 min.

*Caution*: If the mixture of benzoin and thionyl chloride is allowed to stand at room temperature for an appreciable time before being heated, an undesired reaction intervenes,[2] and the synthesis of *E*-stilbene is spoiled.

To remove excess thionyl chloride (bp 77°C), connect the reaction tube to an aspirator for a few minutes, add 0.5 mL of petroleum ether (bp 30°C–60°C), boil it off, and evacuate again. Desyl chloride is thus obtained as a viscous, pale-yellow oil (it will solidify on standing). Dissolve it in 1 mL of 95% ethanol, cool under the tap, and add 9 mg of sodium borohydride (an excess is harmful to the success of the reaction). Stir to break up any lumps of the borohydride; after 10 min, add to the solution of chlorohydrins 60 mg of zinc dust and 0.15 mL of acetic acid; then reflux for 1 h and cool under the tap. When white crystals separate, add 2 mL of *t*-butyl methyl ether, wash the ether solution once with an equal volume of water containing 1 drop of concentrated hydrochloric acid (to dissolve basic zinc salts), and then wash with water. Dry the ether over calcium chloride pellets, which is added until it no longer clumps together. Remove the ether from the drying agent with a Pasteur pipette (Fig. 58.4), wash the drying agent with ether, evaporate the ether to dryness

2. $C_6H_5C=O$ $\xrightarrow{SOCl_2}$ $C_6H_5C-O$ $\underset{C_6H_5C-O}{\overset{\|}{}}$ $SO$ $\xrightarrow{NaBH_4}$ $C_6H_5CO$
   |                                                          |
   $C_6H_5CHOH$                         $C_6H_5CH_2$

**Desoxybenzoin**

under the hood, dissolve the residue in a minimum amount of hot 95% ethanol (about 1 mL), and let the product crystallize. *E*-Stilbene separates in diamond-shaped iridescent plates (mp 124°C–125°C); the yield is about 50 mg. *E*-Stilbene can be isolated easily using the Pasteur pipette technique (Fig. 58.4) or a Wilfilter (*see* Fig. 4.14 on page 74).

**Cleaning Up.** Combine the washings from the cotton in the trap and all aqueous layers, neutralize with sodium carbonate, remove zinc salts by vacuum filtration, and flush the filtrate down the drain with excess water. The zinc salts are placed in the nonhazardous solid waste container. Allow the ether to evaporate from the calcium chloride pellets in the hood and then place the drying agent in the nonhazardous solid waste container. If local regulations do not allow for the evaporation of solvents in a hood, the wet solid should be disposed in a special container. Ethanol mother liquor goes in the organic solvents container. Spread out any isolated zinc on a watch glass to dry and air oxidize. Wet it with water and place in the nonhazardous solid waste container.

## *Macroscale Procedure*

Place 2 g of benzoin (crushed to a powder) in a 50-mL round-bottomed flask, cover it with 4 mL of thionyl chloride,[3] warm gently on a steam bath (in the hood) until all the solid has dissolved, and then heat more strongly for 5 min.

   *Caution*: If the mixture of benzoin and thionyl chloride is left standing at room temperature for an appreciable time before being heated, an undesired reaction intervenes,[4] and the synthesis of *E*-stilbene is spoiled.

   To remove excess thionyl chloride (bp 77°C), evacuate on an aspirator for a few minutes, add 5 mL of petroleum ether (bp 30°C–60°C), boil it off (in the hood), and again evacuate. Desyl chloride is thus obtained as a viscous, pale-yellow oil (it will solidify on standing). Dissolve it in 20 mL of 95% ethanol, cool under the tap, and add 180 mg of sodium borohydride (an excess is harmful to the success of the reaction). Stir to break up any lumps of the borohydride; after 10 min, add to the solution of chlorohydrins 1 g of zinc dust and 2 mL of acetic acid and reflux for 1 h. Then cool under the tap. When white crystals separate, add 25 mL of *t*-butyl methyl ether and decant the solution from the bulk of the zinc into a separatory funnel. Wash the solution twice with an equal volume of water containing 0.5–1 mL of concentrated hydrochloric acid (to dissolve basic zinc salts) and then, in turn, with sodium carbonate solution and saturated sodium chloride solution. Dry the ether over anhydrous calcium chloride pellets (2 g), filter

Sometimes zinc dust from a reaction like this is pyrophoric (spontaneously flammable in air) because it is so finely divided and has such a large, clean surface area able to react with air.

Carry out the procedure in the hood.

---

3. The reagent can be dispensed from a burette or measured by pipette; in the latter case the liquid should be drawn into the pipette with a pipette bulb, *not by mouth*.

4.
$$C_6H_5C=O \xrightarrow{SOCl_2} \begin{matrix} C_6H_5C-O \\ \| \\ C_6H_5C-O \end{matrix} SO \xrightarrow{NaBH_4} \begin{matrix} C_6H_5CO \\ | \\ C_6H_5CH_2 \end{matrix}$$

$$\underset{\text{Desoxybenzoin}}{}$$

with the first structure showing $C_6H_5CHOH$ below $C_6H_5C=O$.

■ **FIG. 58.5**

The UV spectra of Z- and E-stilbene. Z: $\lambda_{max}^{EtOH} = 224$ nm ($\varepsilon = 23,300$) and 279 nm ($\varepsilon = 11,100$); E: $\lambda_{max}^{EtOH} = 226$ nm ($\varepsilon = 18,300$) and 295 nm ($\varepsilon = 27,500$). As in the diacetates, steric hindrance and the lack of coplanarity in these hydrocarbons cause the long-wavelength absorption of the Z-isomer to be of diminished intensity relative to the E-isomer.

to remove the drying agent, evaporate the filtrate to dryness under the hood, dissolve the residue in a minimum amount of hot 95% ethanol (15–20 mL), and let the product crystallize. *E*-Stilbene separates in diamond-shaped iridescent plates (mp 124°C–125°C); the yield is about 1 g. The UV spectrum is shown in Figure 58.5.

**Cleaning Up.** Combine the washings from the cotton in the trap and all aqueous layers, neutralize with sodium carbonate, remove zinc salts by vacuum filtration, and flush the filtrate down the drain with excess water. The zinc salts are placed in the nonhazardous solid waste container. Allow the ether to evaporate from the calcium chloride in the hood and then place it in the nonhazardous solid waste container. If local regulations do not allow for the evaporation of solvents in a hood, the wet solid should be disposed in a special container. Ethanol mother liquor goes in the organic solvents container. Any zinc isolated should be spread out on a watch glass to dry and air oxidize. It should then be wetted with water and placed in the nonhazardous solid waste container.

Sometimes zinc dust from a reaction like this is pyrophoric (spontaneously flammable in air) because it is so finely divided and has such a large, clean surface area able to react with air.

## 2. *meso*-Stilbene Dibromide

> **IN THIS EXPERIMENT** an acetic acid solution of stilbene, an alkene, is brominated with a solid bromine donor to give a dibromide that crystallizes and is isolated by filtration.

*E*-Stilbene reacts with bromine predominantly by the usual process of *trans*-addition and affords the optically inactive, nonresolvable *meso*-dibromide; the much lower melting enantiomeric mixture of dibromides is a very minor product of the reaction.

Pyridine + HBr + Br$_2$

↓

Pyridinium hydrobromide
perbromide

In this procedure, the brominating agent will be pyridinium hydrobromide perbromide,[5] a crystalline, nonvolatile, odorless complex of high molecular weight (319.86), which dissociates, in the presence of a bromine acceptor such as an alkene, to liberate 1 mol of bromine. For microscale experiments the perbromide is far more convenient and agreeable to measure and use than free bromine.

| *E*-Stilbene | Pyridinium hydro- | (1*R*,2*S*)-*meso*-Stilbene dibromide |
|---|---|---|
| MW 180.24 | bromide perbromide | mp 236–237°C, MW 340.07 |
| | MW 319.86 | |

## Microscale Procedure

⚠ **CAUTION**: Pyridinium hydrobromide perbromide is corrosive and a lachrymator (tear producer).

Carry out the procedure in the hood.

**Online Study Center**

Photo: Use of the Wilfilter; Video: Microscale Filtration on the Hirsch Funnel

Total time required: 10 min

In a reaction tube, dissolve 50 mg of *E*-stilbene in 1 mL of acetic acid by heating on a steam bath or a hot water bath; then add 100 mg of pyridinium hydrobromide perbromide. Mix by swirling; if necessary, rinse crystals of the reagent down the walls of the flask with a little acetic acid and continue the heating for an additional 1–2 min. The dibromide separates in small plates. Cool this mixture under the tap, collect the product on a Wilfilter (*see* Fig. 4.14 on page 74) or on a Hirsch funnel, and wash it with methanol to remove any color; the yield of colorless crystals (mp 236°C–237°C) is 80 mg. Use this material to prepare the diphenylacetylene after determining the percent yield and the melting point.

**Cleaning Up.** To the filtrate add sodium bisulfite (until a negative test with starch-iodide paper is observed) to destroy any remaining perbromide, neutralize with sodium carbonate, and extract the pyridine released with ether, which goes in the organic solvents container. The aqueous layer can then be diluted with water and flushed down the drain.

---

5. Crystalline pyridinium hydrobromide perbromide suitable for small-scale experiments is available from the Aldrich Chemical Company. Massive crystals commercially available should be recrystallized from acetic acid (4 mL/g). Pyridinium hydrobromide perbromide can also be prepared as follows: Mix 15 mL of pyridine with 30 mL of 48% hydrobromic acid and cool. Add 25 g of bromine gradually with swirling, cool, and collect the product (use acetic acid for rinsing and washing). Without drying the solid, crystallize it from 100 mL of acetic acid. The yield of orange needles should be 33 g (69%).

⚠️ **CAUTION**: Pyridium hydrobromide perbromide is corrosive and a lachrymator (tear producer).

Carry out the procedure in the hood.

Total time required: 10 min

## Macroscale Procedure

In a 50-mL Erlenmeyer flask dissolve 1 g of *E*-stilbene in 20 mL of acetic acid, by heating on a steam bath; then add 2 g of pyridinium hydrobromide perbromide.[6] Mix by swirling; if necessary, rinse crystals of the reagent down the walls of the flask with a little acetic acid; continue the heating for an additional 1–2 min. The dibromide separates almost at once in small plates. Cool the mixture under the tap, collect the product, and wash it with methanol; the yield of colorless crystals (mp 236°C–237°C) is about 1.6 g. Use 0.5 g of this material to prepare the diphenylacetylene and turn in the remainder to your instructor.

**Cleaning Up.** To the filtrate add sodium bisulfite (until a negative test with starch-iodide paper is observed) to destroy any remaining perbromide, neutralize with sodium carbonate, and extract the pyridine released during the reaction with ether, which goes in the organic solvents container. The aqueous layer can then be diluted with water and flushed down the drain.

## 3. Synthesis of Diphenylacetylene

**(1R,2S)-*meso*-Stilbene dibromide**
MW 340.07

**Diphenylacetylene**
mp 61°C, MW 178.22

One method for the preparation of diphenylacetylene involves the oxidation of benzil dihydrazone with mercuric oxide; the intermediate diazo compound loses nitrogen as it is formed to give the hydrocarbon:

**Benzil**          **Hydrazine**          **Benzil hydrazone**                    **Diphenylacetylene**

6. *See* footnote 5.

The method used in the following procedure involves the dehydrohalogenation of *meso*-stilbene dibromide. An earlier procedure in the chemical literature called for refluxing the dibromide with 43% ethanolic potassium hydroxide in an oil bath at 140°C for 24 h. In the following procedure the reaction time is reduced to a few minutes by using high-boiling triethylene glycol as the solvent to permit operation at a higher reaction temperature, a technique introduced by Louis Fieser.

**IN THIS EXPERIMENT** a solution of stilbene dibromide in a very high-boiling solvent reacts with potassium hydroxide to remove 2 mol of HBr to give an alkyne. After the reaction mixture is diluted with water, the product crystallizes, is isolated by filtration, and is recrystallized from ethanol to give pure diphenylacetylene.

**CAUTION:** Potassium hydroxide is corrosive to skin.

Reaction time: 5 min

**Online Study Center**

Photo: Filtration Using a Pasteur Pipette; Video: Filtration of Crystals Using the Pasteur Pipette

## Microscale Procedure

In a reaction tube place 80 mg of *meso*-stilbene dibromide, 40 mg of potassium hydroxide,[7] and 0.5 mL of triethylene glycol. Heat the mixture on a hot sand bath to 160°C, at which point potassium bromide begins to separate. By intermittent heating, keep the mixture at 160°C–170°C for an additional 5 min; then cool to room temperature, remove the thermometer, and add 2 mL of water. The diphenylacetylene that separates as a nearly colorless, granular solid is collected with a Pasteur pipette. The crude product need not be dried but can be crystallized directly from 95% ethanol. Let the solution stand undisturbed to observe the formation of beautiful, very large spars of colorless crystals. After a first crop has been collected, the mother liquor can be concentrated to afford a second crop of pure product; the total yield is 35 mg (mp 60°C–61°C).

**Cleaning Up.** Combine the crystallization mother liquor with the filtrate from the reaction, dilute with water, neutralize with 3 *M* hydrochloric acid, and flush down the drain.

## Macroscale Procedure

**CAUTION:** Potassium hydroxide is corrosive to skin.

Reaction time: 5 min

In a 20 × 150-mm test tube place 0.5 g of *meso*-stilbene dibromide, 3 pellets of potassium hydroxide[8] (250 mg), and 2 mL of triethylene glycol. Insert a thermometer into a 10 × 75-mm test tube containing enough triethylene glycol to cover the bulb and then slip this assembly into the larger tube. Clamp the tube in a vertical position in a hot sand bath and heat the mixture to 160°C, at which point potassium bromide begins to separate. By intermittent heating, keep the mixture at 160°C–170°C for an additional 5 min; then cool to room temperature, remove the thermometer and small tube, and add 10 mL of water. The diphenylacetylene that separates as a nearly colorless, granular solid is collected by suction filtration. The

---

7. Potassium hydroxide pellets consist of 85% KOH and 15% water.

8. *See* footnote 7.

## ■ FIG. 58.6

The UV spectrum of diphenylacetylene. $\lambda_{max}^{EtOH}$ = 279 nm ($\varepsilon$ = 31,400). This spectrum is characterized by considerable fine structure (multiplicity of bands) and a high extinction coefficient.

crude product need not be dried but can be crystallized directly from 95% ethanol. Let the solution stand undisturbed to observe the formation of beautiful, very large spars of colorless crystals. After a first crop has been collected, the mother liquor can be concentrated to afford a second crop of pure product; the total yield is about 0.23 g (mp 60°C–61°C). The UV spectrum is shown in Figure 58.6.

**Cleaning Up.** Combine the crystallization mother liquor with the filtrate from the reaction, dilute with water, neutralize with 3 *M* hydrochloric acid, and flush down the drain.

## QUESTIONS

1. Using a molecular mechanics program, calculate the energy difference or the difference in heats of formation between *Z*- and *E*-stilbene. How do your computed values compare to the difference in heats of hydrogenation of the two isomers (refer to the structures on the first page of this chapter)?
2. Why is *Z*-stilbene not planar? Explain.
3. How do you account for the large difference in the extinction coefficients ($\varepsilon$) for the long-wavelength peaks for *Z*- and *E*-stilbene (*see* Fig. 58.5 on page 734)?

# The Perkin Reaction: Synthesis of α-Phenylcinnamic Acid

**PRELAB EXERCISE:** At the end of the Perkin reaction, the products are present as mixed anhydrides. What are these mixed anhydrides, how are they formed, and how are they converted to the product?

The reaction of benzaldehyde with phenylacetic acid to produce a mixture of the α-carboxylic acid derivatives of Z- and E-stilbene, a form of an aldol condensation known as the *Perkin reaction*, is effected by heating a mixture of the components with acetic anhydride and triethylamine. In the course of the reaction, the phenylacetic acid is probably present both as the anion and as the mixed anhydride resulting from equilibration with acetic anhydride. A 5-h reflux period specified in an early procedure has been shortened by a factor of 10 by restricting the amount of the volatile acetic anhydride, using an excess of the less expensive high-boiling aldehyde component, and using a condenser that permits some evaporation and consequent elevation of the reflux temperature.

E-Stilbene is a byproduct of the condensation, but it has been shown that neither the E- nor Z-acid undergoes decarboxylation under the conditions of the condensation reaction.

At the end of the reaction, the α-phenylcinnamic acids are present in part as the neutral mixed anhydrides, but these can be hydrolyzed by adding excess hydrochloric acid. The organic material is taken up in ether, and the acids are extracted with alkali. Neutralization with acetic acid ($pK_a = 4.76$) then causes precipitation of only the less acidic E-acid (refer to the $pK_a$ values given the structural formulas on the next page); the Z-acid separates when hydrochloric acid is added.

Although Z-stilbene is less stable and has a lower melting point than E-stilbene, the reverse is true of the α-carboxylic acids, and in this preparation the more stable, higher melting E-acid is the predominant product. Evidently, the steric interference between the carboxyl and phenyl groups in the Z-acid is greater than

$$C_6H_5CH_2COOH + CH_3\overset{O}{\overset{||}{C}}O\overset{O}{\overset{||}{C}}CH_3 \longrightarrow C_6H_5CH_2\overset{O}{\overset{||}{C}}O\overset{O}{\overset{||}{C}}CH_3 \xrightarrow{Et_3N} \left[ C_6H_5\overset{-}{C}H\overset{O}{\overset{||}{C}}O\overset{O}{\overset{||}{C}}CH_3 \right] \overset{O}{\overset{||}{C_6H_5C-H}}$$

**Phenylacetic acid**
mp 77°C, bp 265°C
MW 136.14

**Acetic anhydride**
bp 138–140°C
MW 102.09
$n_D^{20}$ 1.3900

**Triethylamine**
bp 89.5°C, den. 0.729
MW 101.19
$n_D^{20}$ 1.4000

**Benzaldehyde**
bp 179°C, den. 1.046
MW 106.12
$n_D^{20}$ 1.5450

$$\left[ \begin{array}{c} C_6H_5CHC-O-C-CH_3 \\ HC-O^- \\ C_6H_5 \end{array} \right] \longrightarrow \left[ \begin{array}{c} :NEt_3 \\ C_6H_5C-C-O^- \\ HC-OAc \\ C_6H_5 \end{array} \right] \xrightarrow{-OAc^-} \left[ \begin{array}{c} C_6H_5CCOO^- \\ HC \\ C_6H_5 \end{array} \right] \xrightarrow{H^+}$$

The $pK_a$ is measured in
60% ethanol.

**E-α-Phenylcinnamic acid[1]**
mp 174°C, $pK_a$ 6.1
MW 224.25

+

**Z-α-Phenylcinnamic acid**
mp 136–137°C, $pK_a$ 4.8
MW 224.25

that between the two phenyl groups in the *E*-acid. Steric hindrance is also evident from the fact that the *Z*-acid is not subject to Fischer esterification (ethanol and an acid catalyst), whereas the *E*-acid is.

## EXPERIMENTS

**IN THIS EXPERIMENT** four reagents react in an aldol-type reaction (the Perkin reaction) between (1) a solution of phenylacetic acid and acetic anhydride to give a mixed anhydride that (2) reacts with an amine to give an anion that then (3) condenses with benzaldehyde and eliminates a molecule of acetic acid to give the product (4), which is a mixture of isomeric phenylacetic acids. The product is isolated by adding hydrochloric acid, which causes the isomeric products to crystallize. The crystals are dissolved in ether and are extracted with aqueous base. The basic solution on acidification with acetic acid deposits the *E*-acid, which is isolated by filtration and recrystallized. The aqueous filtrate on acidification with hydrochloric acid causes the *Z*-acid to crystallize. It is also isolated by filtration.

# 1. Microscale Synthesis of α-Phenylcinnamic Acid

Measure into a reaction tube 250 mg of phenylacetic acid, 0.30 mL of benzaldehyde, 0.2 mL of triethylamine, and 0.2 mL of acetic anhydride. Insert a boiling
stone and reflux the mixture for 35 min (Fig. 59.1). Cool the yellow melt, add
0.4 mL of concentrated hydrochloric acid, and mix; the mixture sets to a stiff
paste. Add 2.5 mL of *t*-butyl methyl ether, warm to dissolve the bulk of the solid,
wash the ethereal solution twice with water, and then extract it five times with
2-mL portions of 0.5 *M* potassium hydroxide solution.[1] Set aside the dark-colored
ethereal solution (discard it after the product is in hand).[2] Acidify the combined,
colorless alkaline extract to pH 6 by adding 0.4 mL of acetic acid, collect the *E*-
acid that precipitates, and save the filtrate and washings. The yield of *E*-acid
(mp 163°C–166°C) is usually about 290 mg. Crystallize 30 mg of the material by

Reflux time: 35 min

**Online Study Center**

Photos: Extraction with Ether,
Use of the Wilfilter; Videos:
Extraction with Ether, Microscale
Filtration on the Hirsch Funnel

■ **FIG. 59.1**
A microscale reflux apparatus.

Cool part
of tube

Refluxing liquid
(Air condenser)

Sand

Heated
area

Boiling
liquid

Boiling
chip

Electric
flask heater

---

1. If stronger alkali is added, the potassium salt may separate.

2. To isolate the stilbene, wash the ethereal solution with saturated sodium bisulfite solution to remove the
benzaldehyde, dry, evaporate, and crystallize the residue from methanol. Large, slightly yellow spars
(mp 122°C–124°C) separate (yield = 9 mg).

dissolving it in 0.6 mL of *t*-butyl methyl ether, adding 0.8 mL of petroleum ether (bp 30°C–60°C), heating briefly to the boiling point, and letting the solution stand. Silken needles form (mp 173°C–174°C). These can be easily collected on a Wilfilter (*see* Fig. 4.14 on page 74).

Adding 0.5 mL of concentrated hydrochloric acid to the aqueous filtrate from the precipitation of the *E*-acid produces a cloudy emulsion, which on standing for about 1/2 h, coagulates to crystals of the *Z*-acid; the yield is 30 mg (mp 136°C–137°C). The *E*-acid can be recrystallized by dissolving 30 mg in 0.5 mL of *t*-butyl methyl ether, filtering (if necessary) from a trace of sodium chloride, adding 1 mL of petroleum ether (bp 30°C–60°C), and evaporating to a volume of 0.5 mL; the acid separates as a hard crust of prisms (mp 138°C–139°C).

**Cleaning Up.** The dark-colored ether solution and the mother liquor from the crystallization are placed in the organic solvents container. After the *Z*-acid has been removed from the reaction mixture, the acidic solution is diluted with water and flushed down the drain.

# 2. Macroscale Synthesis of α-Phenylcinnamic Acid

Reflux time: 35 min

Measure into a 25-mL round-bottomed flask 2.5 g of phenylacetic acid, 3 mL of benzaldehyde, 2 mL of triethylamine, and 2 mL of acetic anhydride. Add a water-cooled condenser and a boiling stone and reflux the mixture for 35 min (Fig. 59.2). Cool the yellow melt, add 4 mL of concentrated hydrochloric acid, and swirl; the mixture sets to a stiff paste. Add *t*-butyl methyl ether, warm to dissolve the bulk of the solid, and transfer to a separatory funnel by using more ether. Wash the ethereal solution twice with water; then extract it with a mixture of 25 mL of water and 5 mL of 3 *M* potassium hydroxide solution.[3] Repeat the extraction two more times and discard the dark-colored ethereal solution (discard it after the product is in hand).[4] Acidify the combined, colorless, alkaline extract to pH 6 by adding 5 mL of acetic acid, collect the *E*-acid that precipitates, and save the filtrate and washings. The yield of *E*-acid (mp 163°C–166°C) is usually about 2.9 g. Crystallize 0.3 g of material by dissolving it in 8 mL of *t*-butyl methyl ether, adding 8 mL of petroleum ether (bp 30°C–60°C) or pentane, heating briefly to the boiling point, and letting the solution stand. Silken needles form (mp 173°C–174°C).

Adding 5 mL of concentrated hydrochloric acid to the aqueous filtrate from the precipitation of the *E*-acid produces a cloudy emulsion, which, on standing for about 1/2 h, coagulates to crystals of *Z*-acid; the yield is about 0.3 g (mp 136°C–137°C). The *E*-acid can be recrystallized by dissolving 30 mg in 0.5 mL of *t*-butyl methyl ether, filtering (if necessary) from a trace of sodium chlo-

---

3. *See* footnote 1.

4. *See* footnote 2.

**■ FIG. 59.2**
A macroscale reflux apparatus.

**■ FIG. 59.3**
The UV spectra of *Z*- and *E*-α-phenylcinnamic acid, run at identical concentrations. *E:* $\lambda_{max}^{EtOH}$ = 222 nm (ε = 18,000), 282 nm (ε = 14,000); *Z:* $\lambda_{max}^{EtOH}$ = 222 nm (ε = 15,500), 292 nm (ε = 22,300).

ride, adding 1 mL of petroleum ether (bp 30°C–60°C), and evaporating to a volume of 0.5 mL; the acid separates as a hard crust of prisms (mp 138°C–139°C).

**Cleaning Up.** The dark-colored ether solution and the mother liquor from the crystallization are placed in the organic solvents container. After the Z-acid has been removed from the reaction mixture, the acidic solution is diluted with water and flushed down the drain.

## Computational Chemistry

Construct *Z*- and *E*-α-phenylcinnamic acids and carry out energy minimizations. Be sure to arrange the carboxyl group to give the lowest steric energy, which you should note. Then carry out AM1 semiempirical molecular orbital heats of formation calculations for the two isomers. Do the calculations confirm that the Z-isomer is more stable than the E-isomer? Would simple inspection of the space-filling model have led to the same conclusion? Can you rationalize the absorbances in the ultraviolet (UV) spectra of the two isomers (Fig. 59.3)?

*Online Study Center*

**General Resources**
Web Links

# CHAPTER 60

# Decarboxylation: Synthesis of *cis*-Stilbene

☀ *Online Study Center*

This icon will direct you to techniques, equipment setups, and online resources at http://college.hmco.com/PIC/williamsonMMOE5e.

**PRELAB EXERCISE:** Calculate the volume (measured at standard temperature and pressure) of the carbon dioxide evolved in this reaction.

$$
\text{E-}\alpha\text{-Phenylcinnamic acid} \xrightarrow[\text{catalyst, quinoline}]{\text{Copper chromite}} \text{Z-Stilbene} \longrightarrow \text{Stilbene dibromide}
$$

**E-α-Phenylcinnamic acid**
MW 224.25

**Z-Stilbene**
mp 4°C, bp 82–84°C/0.4 mm
MW 180.24

**(1*R*,2*R*)- and (1*S*,2*S*)-(±)-Stilbene dibromide**
mp 114°C
MW 340.07

The catalyst for this reaction is copper chromite, 2 CuO · Cr₂O₃, a relatively inexpensive, commercially available catalyst used for both hydrogenation and dehydrogenation as well as decarboxylation.

Decarboxylation of *E*-α-phenylcinnamic acid is effected by refluxing the acid in quinoline in the presence of a trace of copper chromite catalyst; both the basic properties and boiling point (237°C) of quinoline make it a particularly favorable solvent. Z-Stilbene, a liquid at room temperature, can be characterized by the *trans* addition of bromine to give the crystalline (±)-dibromide. A little *meso*-dibromide derived from E-stilbene in the crude hydrocarbon starting material is easily separated by virtue of its sparing solubility.

■ **FIG. 60.1**
The favored conformations
of stilbene dibromide.

(1*R*,2*R*)- or (1*S*,2*S*)-(±)-Dibromide          (1*R*,2*S*)-*meso*-Dibromide

Although free rotation is possible around the single bond connecting the chiral carbon atoms of the stilbene dibromides and hydrobenzoins, evidence from dipole-moment measurements indicates that the molecules tend to exist predominantly in the specific shape or conformation in which the two phenyl groups repel each other and occupy positions as far apart as possible. The optimal conformations of the (±)-dibromide and the *meso*-dibromide are represented in Figure 60.1 by Newman projection formulas, in which the molecules are viewed along the axis of the bond connecting the two chiral carbon atoms. In the *meso*-dibromide the two repelling phenyl groups are on opposite sides of the molecule and so are the two large bromine atoms. Hence, the structure is far more symmetrical than that of the (±)-dibromide. X-ray diffraction measurements of the dibromides in the solid state confirm the conformations indicated in Figure 60.1. The Br-Br distances found are *meso*-dibromide, 4.50 nm and (±)-dibromide, 3.85 nm. The difference in symmetry of the two optically inactive isomers accounts for the marked contrast in properties:

|                 | mp (°C) | Solubility in Ether (18°C) |
|-----------------|---------|----------------------------|
| ±-Dibromide     | 114     | 1 part in 3.7 parts        |
| *meso*-Dibromide | 237     | 1 part in 1025 parts       |

## EXPERIMENTS

**IN THIS EXPERIMENT** phenylcinnamic acid (from Chapter 59) is dissolved in the high-boiling solvent quinoline and treated with the catalyst copper chromite. This mixture is heated under vacuum to remove traces of water and then at high temperature to cause decarboxylation and formation of *cis*-stilbene. The solid catalyst is removed by filtration, the solution is diluted with ether, and the basic quinoline is removed by extraction with aqueous hydrochloric acid. The ether solution is dried and evaporated to give crude stilbene, which is dissolved in acetic acid and brominated (with solid pyridinium hydrobromide perbromide) on mild warming. As the solution cools, a small amount of *meso*-stilbene dibromide crystallizes and is removed by filtration. The acetic acid solution is diluted with water and then extracted with ether. The ether extracts are washed with water and then alkali to remove the acetic acid. The ether is dried and evaporated to give a dark oil that crystallizes. The crystals are purified by recrystallization from methanol to give pure (*R*)- and (*S*)-stilbene dibromide.

$2CuO \cdot Cr_2O_3$
**Copper chromite**

Copper chromite is toxic;
weigh it in the hood.

**Quinoline**, bp 237°C

Reaction time in first step:
10 min

Carry out the procedure in
the hood.

The second step requires
about 30 min.

**Online Study Center**

Video: Extraction with Ether;
Photo: Extraction with Ether,
Use of the Wilfilter

# 1. *cis*-Stilbene and Stilbene Dibromide

Because a trace of moisture causes troublesome spattering, the reactants and catalyst are dried prior to decarboxylation. Place 250 mg of crude, dry $E$-$\alpha$-phenylcinnamic acid and 40 mg of copper chromite catalyst into a reaction tube, and add 0.3 mL of quinoline (bp 237°C; any quinoline that has darkened during storage should be redistilled over a little zinc dust) to wash down the solids. Connect to an aspirator and turn it on full force. Be sure that you have a good vacuum (pressure gauge) and heat the tube strongly on a steam bath with most of the rings removed or heat on a sand bath. Heat and evacuate for 5–10 min to remove all traces of moisture. Then wipe the outside walls of the tube dry, insert a thermometer, clamp the tube over a hot sand bath, raise the temperature to 230°C, and note the time. Next, maintain a temperature close to 230°C for 10 min. Cool the yellow solution containing suspended catalyst to 25°C, add 3 mL of *t*-butyl methyl ether, and filter the solution by gravity (use more ether for rinsing). Transfer the solution to a reaction tube and remove the quinoline by extraction twice with about 1.5 mL portions of water containing 0.35 mL of concentrated hydrochloric acid. Then shake the ethereal solution well with water containing a little sodium hydroxide solution, draw off the alkaline liquor, and acidify it. A substantial precipitate will show that decarboxylation was incomplete, in which case the starting material can be recovered and the reaction repeated. If there is only a trace of precipitate, shake the ethereal solution with saturated sodium chloride solution for preliminary drying, dry the ethereal solution over calcium chloride pellets, remove the ether from the drying agent with a Pasteur pipette, and evaporate the ether. The residual brownish oil (about 150 mg) is crude $Z$-stilbene, which contains a small amount of the $E$-isomer formed by rearrangement during heating.

Dissolve the crude $Z$-stilbene (e.g., 150 mg) in 1 mL of acetic acid and, in subdued light, add double the weight of pyridinium hydrobromide perbromide (e.g., 300 mg). Warm on a steam or sand bath until the reagent is dissolved; then cool under the tap and scratch to effect separation of a small crop of plates of the *meso*-dibromide. Filter the solution by suction if necessary, dilute extensively with water, and extract with ether. Wash the solution twice with water and then with saturated sodium bicarbonate solution (1.3 $M$) until neutral; shake with saturated sodium chloride solution, dry over calcium chloride pellets, and evaporate to a volume of about 1 mL. If a little more of the sparingly soluble *meso*-dibromide separates, remove it and then evaporate the remainder of the solvent. The residual ($\pm$)-dibromide is obtained as a dark oil that readily solidifies when rubbed with a stirring rod. Dissolve it in a small amount of methanol and let the solution stand to crystallize. Isolate the product on a Wilfilter. The ($\pm$)-dibromide separates as colorless prismatic plates (mp 113°C–114°C); the yield is about 60 mg.

**Cleaning Up.** Place the catalyst removed by filtration in the nonhazardous solid waste container. Treat the combined aqueous layers from all parts of the experiment containing quinoline and pyridine with a small quantity of bisulfite to destroy any bromine and neutralize with sodium carbonate. Extract the quinoline and pyridine that are released into ligroin and place the hexanes in the organic solvents container. Dilute the aqueous layer with water and flush it down the drain. After ether is allowed to evaporate from the calcium chloride pellets in the hood, the pellets can be placed in the nonhazardous solid waste container. If local regulations do not allow for the evaporation of solvents in a hood, the wet solid should be disposed in a special container. Methanol from the crystallization goes in the organic solvents container.

## 2. *cis*-Stilbene and Stilbene Dibromide

$2CuO \cdot Cr_2O_3$
**Copper chromite**

Copper chromite is toxic; weigh it in the hood.

**Quinoline**, bp 237°C

Reaction time in first step: 10 min

Carry out the procedure in the hood.

The second step requires about 30 min.

Because a trace of moisture causes troublesome spattering, the reactants and catalyst are dried prior to decarboxylation. The preparations of copper chromite described in the *Journal of the American Chemical Society*[1] are both satisfactory. Place 2.5 g of crude, dry *E*-α-phenylcinnamic acid and 0.2 g of copper chromite catalyst into a 20 × 150-mm test tube, and add 3 mL of quinoline (bp 237°C; any quinoline that has darkened during storage should be redistilled over a little zinc dust) to wash down the solids. Connect with a rubber stopper to an aspirator and turn it on full force. Be sure that you have a good vacuum (check with a pressure gauge) and heat the tube strongly on a steam bath with most of the rings removed or heat on a sand bath. Heat and evacuate for 5–10 min to remove all traces of moisture. Then wipe the outside walls of the test tube dry, insert a thermometer, clamp the tube over a microburner, raise the temperature to 230°C, and note the time. Next, maintain a temperature close to 230°C for 10 min. Cool the yellow solution containing suspended catalyst to 25°C, add 30 mL of *t*-butyl methyl ether, and filter the solution by gravity (use more ether for rinsing). Transfer the solution to a separatory funnel and remove the quinoline by extraction twice with about 15-mL portions of water containing 3–4 mL of concentrated hydrochloric acid. Then shake the ethereal solution well with water containing a little sodium hydroxide solution, draw off the alkaline liquor, and acidify it. A substantial precipitate will show that decarboxylation was incomplete, in which case the starting material can be recovered and the reaction repeated. If there is only a trace of precipitate, shake the ethereal solution with saturated sodium chloride solution for preliminary drying, dry the ethereal solution over anhydrous calcium chloride pellets, remove the drying agent by filtration, and evaporate the ether. The residual brownish oil (1.3–1.8 g) is crude *Z*-stilbene, which contains a small amount of the *E*-isomer formed by rearrangement during heating.

Dissolve the crude *Z*-stilbene (e.g., 1.5 g) in 10 mL of acetic acid and, in subdued light, add double the weight of pyridinium hydrobromide perbromide (e.g., 3.0 g). Warm on a steam or sand bath until the reagent is dissolved; then cool under the tap and scratch to effect separation of a small crop of plates of the *meso*-dibromide (10–20 mg). Filter the solution by suction, dilute extensively with water, and extract with *t*-butyl methyl ether. Wash the solution twice with water and then with 5% sodium bicarbonate solution until neutral; shake with saturated sodium chloride solution, dry over sodium sulfate, and evaporate to a volume of

---

1. *J. Am. Chem. Soc.* **1932**, 54, 1138; *J. Am. Chem. Soc.* **1950**, 72, 2626.

about 10 mL. If a little more of the sparingly soluble *meso*-dibromide separates, remove it by gravity filtration and then evaporate the remainder of the solvent. The residual DL-dibromide is obtained as a dark oil that readily solidifies when rubbed with a stirring rod. Dissolve it in a small amount of methanol and let the solution stand to crystallize. The DL-dibromide separates as colorless prismatic plates (mp 113°C–114°C); the yield is about 0.6 g.

**Cleaning Up.** Place the catalyst removed by filtration in the nonhazardous solid waste container. Treat the combined aqueous layers from all parts of the experiment containing quinoline and pyridine with a small quantity of bisulfite to destroy any bromine and neutralize with sodium carbonate. Extract the quinoline and pyridine that are released into hexanes and place the hexanes in the organic solvents container. Dilute the aqueous layer with water and flush it down the drain. After ether is allowed to evaporate from the calcium chloride pellets in the hood, the pellets can be placed in the nonhazardous solid waste container. If local regulations do not allow for the evaporation of solvents in a hood, the wet solid should be disposed in a special container. Methanol from the crystallization goes in the organic solvents container.

## Computational Chemistry

The Br-Br distances in (1*R*,2*R*)- or (1*S*,2*S*)-stilbene dibromide and (1*R*,2*S*)-stilbene dibromide are 4.50 and 3.85 nm, respectively, as determined by X-ray diffraction of the crystals. Do simple molecular mechanics energy-minimized structures give these same intramolecular distances once the global minimum is found?

## QUESTIONS

1. Draw the mechanism that shows how the bromination of *E*-stilbene produces *meso*-stilbene dibromide.
2. Can nuclear magnetic resonance ($^1$H NMR) spectroscopy be used to distinguish between *meso*- and (±)-stilbene dibromide? Explain why or why not.
3. Why is *Z*-stilbene not planar?
4. Draw a Newman projection of (1*R*,2*S*)-stilbene dibromide (the *meso*-isomer).

# CHAPTER

# 61

# Chemiluminescence: Syntheses of Cyalume and Luminol

**Online Study Center**

This icon will direct you to techniques, equipment setups, and online resources at http://college.hmco.com/ PIC/williamsonMMOE5e.

> **PRELAB EXERCISE:** Write a balanced equation for the reduction of nitrophthalhydrazide to aminophthalhydrazide using sodium hydrosulfite, which is oxidized to bisulfite.

**C**hemiluminescence is a process whereby light is produced via a chemical reaction with the evolution of little or no heat. The periodic flashes of the male firefly in its quest of a mate and the glow of light seen in the wake of a boat, under a rotten log, or among the many organisms found at great depths in the ocean are examples of natural chemiluminescence. The interaction of luciferin from the firefly, the enzyme luciferase, adenosine triphosphate (ATP), and molecular oxygen is the most carefully studied of these reactions. In this chapter, both Cyalume, an invention of the American Cyanamid Company, and luminol are synthesized.

"Light sticks" sold at entertainment events and as toys are solutions of Cyalume. Bending the light stick activates it (by mixing hydrogen peroxide with the Cyalume) to give off a bright glow in a variety of colors. These sticks are also used as emergency lights on life vests, as roadside markers, on fishing lures, and even in golf balls that light up at night! A variety of colors can be produced by allowing synthetic Cyalume to interact with a variety of fluorescers, including one you can prepare by the Wittig reaction. Luminol is not a commercial product. Luminol's chemiluminescence, the mechanism of the reaction, and peroxyoxalate chemistry are presented in several journal articles.[1]

---

1. For the chemiluminescence of luminol, *see* Huntress, E. H.; Stanley, L. N.; Parker, A. S. *J. Chem. Educ.* **1934**, *11*, 142. The mechanism of the reaction has been investigated by White, E. H. et al. *J. Am. Chem. Soc.* **1964**, *86*, 940, 942. Peroxyoxalate chemistry is discussed by Mohan, A. G.; Turro, N. J. *J. Chem. Educ.* **1974**, *51*, 528.

# Cyalume

Cyalumes are the most efficient nonenzyme fluorescing substances known, with quantum efficiencies of up to 26%.

| **2,4,6-Trichlorophenol** | **Oxalyl chloride** | **Triethylamine** |
|---|---|---|
| MW 197.45 | MW 126.93 | MW 101.19 |
| mp 64–66°C | bp 63–64°C | bp 89°C |
| | den. 1.455 | den. 0.726 |
| **1** | **2** | |

**Bis(2,4,6-trichlorophenyl)oxalate**
(A Cyalume)
MW 448.9
mp 162–163°C
**3**

**Triethylamine hydrochloride**

2,4,6-Trichlorophenol (**1**) reacts with the diacid chloride, oxalyl chloride (**2**), in the presence of triethylamine to give the desired oxalic acid ester (**3**) and the amine hydrochloride. Several different oxalic acid esters will work in the luminescent reaction.

**1,2-Dioxetanedione (?)**
**4**

**9,10-Diphenylanthracene**
MW 330.43
mp 245–248°C
**5**

or

Charge transfer complex

$2 CO_2$ + **9,10-Diphenylanthracene**
Singlet excited state

$h\nu$ + **9,10-Diphenylanthracene**
Ground state

*trans*-**9-(2-Phenylethenyl)anthracene**
(from Chapter 39)

In the light-producing reaction, the ester undergoes nucleophilic attack by hydrogen peroxide to produce a peroxyoxalic acid, which, by an intramolecular displacement reaction, generates a small intermediate molecule, perhaps the 1,2-dioxetanedione (**4**). This molecule transmits its energy to a fluorescer, such as 9,10-diphenylanthracene (**5**), through a charge-transfer complex that produces the excited singlet state of the fluorescer when it decomposes to two molecules of carbon dioxide. *trans*-9-(2-Phenylethenyl)anthracene, synthesized by the Wittig reaction in Chapter 39, is also an excellent fluorescer. The observed color depends on the structure of the fluorescer.

The structure of the intermediate molecule (**4**) has not been established with certainty. The intermediate is known to be volatile; carbon dioxide is a product of the reaction, but carbon monoxide is not. Efforts to detect this intermediate by Fourier transform infrared (IR), carbon nuclear magnetic resonance (NMR), and mass spectroscopy have all been unsuccessful.

Cyalume decomposes to a small molecule that transfers its energy to a different fluorescent molecule to produce light. Luminol, on the other hand, is itself fluorescent and gives off light from the excited state. In addition, the luminol reaction is complete in a few minutes, but the peroxyoxalate reaction can last for a day or more.

## Luminol

The oxidation of luminol is attended with a striking emission of blue-green light. An alkaline solution of the compound is allowed to react with a mixture of hydrogen peroxide and potassium ferricyanide. The dianion (**7**) is oxidized to the triplet excited state (two unpaired electrons of like spin) of the amino phthalate ion. This slowly undergoes intersystem crossing to the singlet excited state (two unpaired electrons of opposite spin; **9**), which decays to the ground state ion (**8**) with the emission of one quantum of light (a photon) per molecule. (*See* Chapter 62 for a more detailed discussion of photochemistry.)

Luminol (**6**) is prepared by reducing the nitro derivative (**13**) formed on the thermal dehydration of a mixture of 3-nitrophthalic acid (**11**) and hydrazine (**12**). An earlier procedure for effecting the first step called for adding hydrazine sulfate to an alkaline solution of the acid, evaporating to dryness, and baking the resulting mixture of the hydrazine salt and sodium sulfate at 165°C; this process required 4.5 h for completion. Louis Fieser reduced this working time drastically by adding high-boiling triethylene glycol (bp 290°C) to an aqueous solution of the hydrazine salt, distilling the excess water, and raising the temperature to a point where dehydration to intermediate **13** is complete within a few minutes. Nitrophthalhydrazide (**3**) is insoluble in dilute acid but soluble in alkali by virtue

## CHEMILUMINESCENCE OF LUMINOL

**6**

**Luminol**
**(3-Aminophthalhydrazide)**

**7**

etc.

$H_2O_2$, $K_3Fe(CN)_6$

Ground state

**8**

+ hv ← Singlet excited state **9** ← Intersystem crossing (slow) ← Triplet excited state **10** + $N_2$

## SYNTHESIS OF LUMINOL

**11**
**3-Nitrophthalic acid**
MW 211.13
mp 222°C

**12**
**Hydrazine**

**13**

Sodium hydrosulfite
Na$_2$S$_2$O$_4$
MW 174.10

**6**
**Luminol**
MW 177.16
mp 332°C

**14**

2 OH$^-$

**15**

of enolization; it is conveniently reduced to luminol (**6**) by sodium hydrosulfite (sodium dithionite) in alkaline solution. In dilute, weakly acidic or neutral solution luminol exists largely as the dipolar ion (**14**), which exhibits beautiful blue fluorescence.

An alkaline solution contains the doubly enolized anion (**15**) and displays particularly marked chemiluminescence when oxidized with a combination of hydrogen peroxide and potassium ferricyanide.

---

## EXPERIMENTS

# Cyalume

## 1. Synthesis of Bis(2,4,6-Trichlorophenyl)oxalate

**IN THIS EXPERIMENT** a phenol dissolved in toluene is treated with the diacid chloride oxalyl chloride in the presence of an amine to give a diester. Upon cooling of the hot reaction mixture, the product (along with solid amine hydrochloride) crystallizes. The mixture of solids obtained by filtration is washed with water to remove the amine hydrochloride. The solid diester, a Cyalume, is collected by filtration, dried, and then recrystallized from toluene.

**Online Study Center**

Video: Microscale Filtration on the Hirsch Funnel

Magnetic stirring of this reaction mixture is desirable but not mandatory. The precipitate is viscous and does not stir well.

For a macroscale reaction, triple or quadruple the quantities. Add the thionyl chloride in portions, with cooling if necessary.

Dissolve 0.395 g of 2,4,6-trichlorophenol in 3 mL of toluene in a 5-mL round-bottomed flask equipped with a magnetic stirring bar. Add to this solution 0.28 mL (0.2 g) of triethylamine, cool the container in ice, and add 0.10 mL (0.14 g) of oxalyl chloride. There will be an immediate voluminous precipitate of triethylamine hydrochloride. Add an air condenser to the flask and reflux the mixture gently with stirring for 30 min to complete the reaction. Cool the mixture well in ice and collect the solid on a Hirsch funnel. Complete the transfer of material with a small volume of hexane(s), which is also used to wash the solid on the filter.

Press the material as dry as possible on the filter; then suspend it in about 5–6 mL of water in a 10-mL Erlenmeyer flask. Stir the solid well with the water, using a spatula to break up any lumps; then cap the flask and shake it vigorously. In this process, the amine hydrochloride dissolves in water and leaves the product as a suspension. Empty the filtrate in a filter flask into the flammable waste container; then collect the solid product from the aqueous solution on a Hirsch funnel. Wash the solid well with water and press it as dry as possible on the filter. Spread out the product to dry, determine its weight, and then recrystallize it from the minimum volume of boiling toluene. Allow the product to crystallize slowly; then cool the tube in ice. Collect the product on a Hirsch funnel [use hexane(s) to complete the transfer of material] and wash the crystals with a small volume of hexane(s). Determine the weight and melting point of the dry material and calculate the percent yield.

**Cleaning Up.** Flush the aqueous filtrate down the drain and discard the organic filtrates in the flammable waste container.

## 2. The Chemiluminescent Reaction

Dissolve 50 mg (or less, if that is all you have) of bis(2,4,6-trichloro)oxalate and about 3 mg of 9-(2-phenylethenyl)anthracene (from Chapter 39) or 9,10-diphenylanthracene in 5 mL of diethyl phthalate by warming on a steam or sand bath. In another container, shake 0.2 mL of 30% hydrogen peroxide (or, less satisfactory, 2 mL of 3% peroxide) with 5 mL of diethyl phthalate. Add the peroxide suspension dropwise to the solution of the oxalate ester and the fluorescer in a darkened room. Try cooling and warming the mixture after initial mixing and note the effect of temperature on the rate of a reaction. Also try adding an additional 1–2 mg of the fluorescer (9,10-diphenylanthracene is expensive) and more hydrogen peroxide.

**Cleaning Up.** Dispose of the mixture in the halogenated waste container.

## Computational Chemistry

Using a molecular mechanics program, calculate the energy, after minimization, of the dimer of carbon dioxide, 1,2-dioxetanedione (**4**), the molecule thought to be the intermediate in Cyalume chemiluminescence. Note the bond lengths and angles. This molecule, with only six atoms, is small enough to carry out some of the more advanced calculational methods in a reasonable time frame. Using the energy-minimized structure as input, calculate the optimal geometry of 1,2-dioxetanedione using the AM1 semiempirical molecular orbital method. Again note the bond lengths and angles.

On a high-speed Apple Macintosh computer with the program Spartan,[2] this calculation took less than 8 s. Using this AM1 geometry as input, again calculate the optimal geometry and heat of formation with an ab initio program. At the HF/6-31G*[3] level this calculation took a few minutes. The energy is given by the program in hartrees (1 hartree = 365.5 kcal/mol). Note the changes in energy and in geometry as the calculational method supposedly gets better (and more lengthy). Ab initio calculations are, due to this time constraint, limited to molecules with fewer than 100 atoms. The time needed to carry out the calculation increases at more than a fourth power of the number of atoms in the molecule.

# Luminol

## 1. Synthesis of Luminol

### Microscale Procedure

> **IN THIS EXPERIMENT** nitrophthalic acid is dissolved in an aqueous solution of hydrazine. The high-boiling solvent triethylene glycol is added, and the water is distilled off to allow the reaction to take place at a high temperature. Dilution of the reaction mixture with hot water causes the nitrophthalhydrazide to crystallize. After isolation by filtration, the damp product is reduced with sodium hydrosulfite in basic solution. After heating to complete the reaction, the base is neutralized with acetic acid. On cooling, the product, luminol (3-aminophthalhydrazide), crystallizes and is isolated by filtration.

**CAUTION**: Hydrazine is a carcinogen. Handle with care. Wear gloves and carry out this experiment in the hood.

$HO(CH_2CH_2O)_3H$

**Triethylene glycol**
bp 290°C

First heat a tube containing 3 mL of water on a steam or sand bath. Then heat a mixture of 200 mg of 3-nitrophthalic acid and 0.4 mL of an 8% aqueous solution of hydrazine (dilute 3.12 g of the commercial 64% hydrazine solution to a volume of 25 mL; use *caution*!) in a 10 × 100-mm reaction tube over a hot sand bath until the solid is dissolved. Add 0.6 mL of triethylene glycol and clamp the tube in a vertical position above a hot sand bath. Insert a boiling chip and a thermometer and boil the solution vigorously to distill the excess water. Intermittently remove the thermometer and replace it with an aspirator tube to facilitate this. There will be a period during which the solution will boil at 110°C and then over a 3–4-min period it will rise to 215°C. Lift the tube from the hot sand and, by intermittent gentle heating, maintain a temperature of 215°C–220°C for 2 min. Remove the tube, cool to about 100°C (crystals of the product often appear), add

2. Spartan molecular modeling software is available from Wavefunction, Inc., Irvine, California.

3. The asterisk means that the program has been improved.

*Online Study Center*
Video: Microscale Filtration
on the Hirsch Funnel

The two-step synthesis of a
chemiluminescent substance can
be completed in 25 min.

**CAUTION**: Contact of solid
sodium hydrosulfite with moist
combustible material may cause
a fire.

$$Na_2S_2O_4 \cdot 2\ H_2O$$

**Sodium hydrosulfite dihydrate
(Sodium dithionite)**
MW 210.15

**CAUTION**: Hydrazine is a
carcinogen. Handle with care.
Wear gloves and carry out this
experiment in the hood.

The two-step synthesis of a
chemiluminescent substance can
be completed in 25 min.

**CAUTION**: Contact of solid
sodium hydrosulfite with moist
combustible material may cause
a fire.

the 3 mL of hot water, cool the tube in cold water, and collect the light-yellow
granular nitro compound (**13**) by vacuum filtration on a Hirsch funnel. The reason for adding hot water and then cooling rather than adding cold water is that
the solid is then obtained in more easily filterable form. The dry weight should be
about 140 mg.

The nitro compound need not be dried and can be transferred at once, for reduction, to the uncleaned tube in which it was prepared. Add 1.0 mL of 3 *M*
sodium hydroxide solution and stir with a stirring rod. To the resulting deep
brown-red solution, add 0.6 g of fresh sodium hydrosulfite dihydrate (*not* sodium
hydrogen sulfite or sodium bisulfite). Wash the solid down the walls with a little
water. Heat to the boiling point, stir, and keep the mixture hot for 5 min, during
which time some of the reduction product may separate. Then add 0.4 mL of
acetic acid, cool the tube in a beaker of cold water, and stir; collect the resulting
precipitate of light-yellow luminol (**6**) by vacuum filtration on a Hirsch funnel.
The filtrate on standing overnight usually deposits a further crop of luminol
(20–40 mg).

**Cleaning Up.** Combine the filtrate from the first and second reactions, dilute with
a few milliliters of water, neutralize with sodium carbonate, add 3 mL of household bleach (5.25% sodium hypochlorite solution), and heat the mixture to 50°C
for 1 h. This will oxidize any unreacted hydrazine and hydrosulfite. Dilute the
mixture and flush it down the drain.

### Macroscale Procedure

First heat a flask containing 15 mL of water on a steam bath. Then heat a mixture
of 1 g of 3-nitrophthalic acid and 2 mL of an 8% aqueous solution of hydrazine
(dilute 3.12 g of the commercial 64% hydrazine solution to a volume of 25 mL;
use *caution*!) in a 20 × 150-mm test tube over a Thermowell until the solid is
dissolved, add 3 mL of triethylene glycol, and clamp the tube in a vertical position in a hot sand bath. Insert a thermometer, a boiling chip, and an aspirator tube
connected to an aspirator and boil the solution vigorously to distill the excess
water (110°C–130°C). Let the temperature rise rapidly (3–4 min) until it reaches
215°C. Remove the burner, note the time, and, by intermittent gentle heating,
maintain a temperature of 215°C–220°C for 2 min. Remove the tube, cool to
about 100°C (crystals of the product often appear), add the 15 mL of hot water,
cool under the tap, and collect the light-yellow granular nitro compound (**13**).
The dry weight yield is 0.7 g. The reason for adding hot water and then cooling
rather than adding cold water is that the solid is then obtained in more easily
filterable form.

The nitro compound need not be dried and can be transferred at once, for reduction, to the uncleaned test tube in which it was prepared. Add 5 mL of 3 *M*
sodium hydroxide solution and stir with a stirring rod. To the resulting deep
brown-red solution, add 3 g of fresh sodium hydrosulfite dihydrate (*not* sodium
hydrogen sulfite or sodium bisulfite). Wash the solid down the walls with a little
water. Heat to the boiling point, stir, and keep the mixture hot for 5 min, during
which time some of the reduction product may separate. Then add 2 mL of acetic

acid, cool under the tap, and stir; collect the resulting precipitate of light-yellow luminol (**6**). The filtrate on standing overnight usually deposits a further crop of luminol (0.1–0.2 g).

**Cleaning Up.** Combine the filtrate from the first and second reactions, dilute with a few milliliters of water, neutralize with sodium carbonate, add 40 mL of household bleach (aqueous sodium hypochlorite solution), and heat the mixture to 50°C for 1 h. This will oxidize any unreacted hydrazine and hydrosulfite. Dilute the mixture and flush it down the drain.

## 2. The Light-Producing Reaction

> **IN THIS EXPERIMENT** a basic solution of luminol is mixed with a solution of hydrogen peroxide in the presence of an iron catalyst to produce light.

This reaction can be run on a scale five times larger. Dissolve the first crop of moist luminol (dry weight is about 40–60 mg) in 2 mL of 3 $M$ sodium hydroxide solution and 18 mL of water; this is stock solution A. Prepare a second stock solution (B) by mixing 4 mL of 3% aqueous potassium ferricyanide, 4 mL of 3% hydrogen peroxide, and 32 mL of water. Next, dilute 5 mL of solution A with 35 mL of water and, in a dark place, pour this solution and solution B simultaneously into an Erlenmeyer flask. Swirl the flask and, to increase the brilliance, gradually add further small quantities of sodium hydroxide and ferricyanide crystals.

Ultrasonic sound also can be used to promote this reaction. Prepare stock solutions A and B again but omit the hydrogen peroxide. Place the combined solutions in an ultrasonic cleaning bath (sonicator) or immerse an ultrasonic probe into the reaction mixture. Spots of light are seen where the ultrasonic vibrations produce hydroxyl radicals.

The sanguinary minded can mix solutions A and B, omitting the ferricyanide from solution B. Light can be generated by adding blood dropwise to the reaction mixture.

**Cleaning Up.** Add 2 mL of 3 $M$ hydrochloric acid, dilute the solution with water, and flush the mixture down the drain.

# Photochemistry: The Synthesis of Benzopinacol

**Online Study Center**

This icon will direct you to
techniques, equipment setups,
and online resources at
http://college.hmco.com/
PIC/williamsonMMOE5e.

**PRELAB EXERCISE:** Draw a mechanism for the base-catalyzed cleavage of benzopinacol.

Photochemistry, as the name implies, is the chemistry of reactions initiated by light. A molecule can absorb light energy and then undergo isomerization, fragmentation, rearrangement, dimerization, or hydrogen atom abstraction. The last reaction, hydrogen atom abstraction, is the subject of this chapter. On irradiation with sunlight, benzophenone abstracts a proton from the solvent 2-propanol and becomes reduced to benzopinacol.

$$2\,C_6H_5\overset{\overset{\displaystyle O}{\|}}{C}C_6H_5 \;+\; CH_3\overset{\overset{\displaystyle OH}{|}}{C}HCH_3 \;\xrightarrow{h\nu}\; C_6H_5-\overset{\overset{\displaystyle OH}{|}}{\underset{\underset{\displaystyle C_6H_5}{|}}{C}}-\overset{\overset{\displaystyle OH}{|}}{\underset{\underset{\displaystyle C_6H_5}{|}}{C}}-C_6H_5 \;+\; CH_3\overset{\overset{\displaystyle O}{\|}}{C}CH_3$$

| **Benzophenone** | **2-Propanol** | **Benzopinacol** | **Acetone** |
|---|---|---|---|
| MW 182.21 | MW 60.09 | MW 366.44 | MW 58.08 |
| mp 48°C | bp 82°C | mp 189°C | bp 56°C |
| | $n_D^{20}$ 1.3770 | | $n_D^{20}$ 1.3590 |

The energy of light varies with its frequency according to the following equation:

$$\Delta E = h\nu = h\left(\frac{c}{\lambda}\right)$$

where $h$ is Planck's constant, $\nu$ is the frequency of the light, $\lambda$ is its wavelength, and $c$ is the speed of light. The fact that benzophenone is colorless means it does not absorb visible light, yet irradiation of benzophenone in alcohol results in a chemical change. Pyrex glass is not transparent to ultraviolet (UV) light with a wavelength shorter than 290 nm, so some wavelength between 290 nm and 400 nm (the edge of the visible region) must be responsible. The UV spectrum of benzophenone indicates an absorption band centered at about 355 nm. Light of that wavelength is absorbed by benzophenone.

What, in a general sense, occurs when a molecule absorbs light? A photon is absorbed only if its energy (wavelength) corresponds exactly to the difference between two electronic energy levels in the molecule. In benzophenone the electrons most loosely held and thus most easily excited are the two pairs of nonbonded electrons on the carbonyl oxygen, called the $n$-electrons.

$$\text{C}=\ddot{\text{O}}{:}\quad n\text{-electrons}$$

These electrons have paired spins, $\text{C}=\text{O}\ \uparrow\downarrow$ and only one is excited by the photon of light. The electron goes into the lowest unoccupied excited energy level, $\pi^*$. Each electronic energy level of benzophenone has within it many vibrational and rotational energy levels. An electron can reside in many of these levels in the lower electronic energy level, the ground state $S_0$, and it can be promoted to many vibrational and rotational energy levels in the first excited state ($S_1$). Hence, the UV spectrum does not appear as a single sharp peak but as a band composed of many peaks that arise from transitions between these many energy levels (Fig. 62.1).

The spin of the electron cannot change in going from the ground state $S_0$ to the excited singlet state $S_1$ (conservation of angular momentum). Once in a higher vibrational or rotational state, it can drop to $S_1$ by vibrational relaxation, losing some of its energy as heat. It then undergoes either fluorescence or intersystem crossing. As seen in Figure 62.1, the rate for fluorescence, a light-emitting process

**■ FIG. 62.1**
Electronic energy levels and possible transitions.

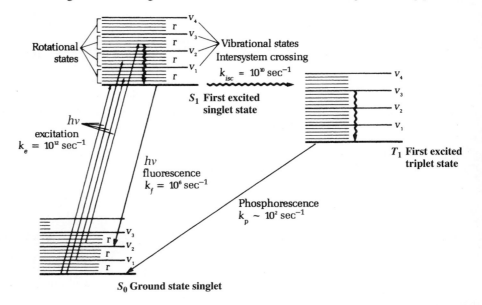

in which the electron returns to the ground state, is $10^4$ times slower than intersystem crossing; so benzophenone does not fluoresce. The electron flips its orientation during intersystem crossing so that it has the same orientation as the electron with which it was paired in the ground state. In this new state, called the *triplet*, it can lose energy as light; but this process, phosphorescence, is a slow one. The lifetime of the triplet state is long enough for chemical reactions to take place. The triplet can also lose energy as heat in a radiationless transition, but the probability of this happening is relatively low. This situation can be represented diagrammatically:

Ground state                     First excited singlet state $S_1$

Intersystem crossing

$(C_6H_5)_2\dot{C}=O\cdot$   $\equiv$   First excited triplet state, $T_1$

This $T_1$ state is a diradical and can abstract a methine proton from the solvent and then a hydroxyl proton to give acetone and diphenyl hydroxy radical, which dimerizes to give the product.

Removal of the methine proton

$(C_6H_5)_2\dot{C}=O\cdot + H-\overset{CH_3}{\underset{CH_3}{C}}-O-H \longrightarrow (C_6H_5)_2\dot{C}-OH + \cdot\overset{CH_3}{\underset{CH_3}{C}}-O-H$

**2-Propanol**

Removal of the hydroxyl proton

$(C_6H_5)_2\dot{C}=O\cdot + H-O-\overset{CH_3}{\underset{CH_3}{C}}\cdot \longrightarrow (C_6H_5)_2\dot{C}-OH + O=C\overset{CH_3}{\underset{CH_3}{}}$

**Acetone**

Dimerization

$2\ (C_6H_5)_2\dot{C}-OH \longrightarrow C_6H_5-\overset{OH}{\underset{C_6H_5}{C}}-\overset{OH}{\underset{C_6H_5}{C}}-C_6H_5$

**Benzopinacol**

## EXPERIMENTS

## 1. Benzopinacol

$$2\ C_6H_5\overset{\displaystyle O}{\overset{\displaystyle \|}{C}}C_6H_5\ +\ \underset{\displaystyle CH_3\overset{\displaystyle OH}{\underset{\displaystyle |}{C}}HCH_3}{}\ \xrightarrow{hv}\ C_6H_5-\underset{\displaystyle C_6H_5}{\overset{\displaystyle OH}{\underset{\displaystyle |}{C}}}-\underset{\displaystyle C_6H_5}{\overset{\displaystyle OH}{\underset{\displaystyle |}{C}}}-C_6H_5\ +\ CH_3\overset{\displaystyle O}{\overset{\displaystyle \|}{C}}CH_3$$

| **Benzophenone** | **2-Propanol** | **Benzopinacol** | **Acetone** |
|---|---|---|---|
| MW 182.21 | MW 60.09 | MW 366.44 | MW 58.08 |
| mp 48°C | bp 82°C | mp 189°C | bp 56°C |

> **IN THIS EXPERIMENT** a tenth of a gram of benzophenone is dissolved in isopropyl alcohol in the presence of a tiny drop of acetic acid. Irradiation by the sun or some other source of UV light results in the photochemical synthesis of benzopinacol, which crystallizes from the reaction mixture.
>
> This experiment should be done when there is good prospect for several hours of bright sun. The benzopinacol is cleaved by alkali to benzhydrol and benzophenone (Experiment 2), and it is rearranged in acid to benzopinacolone (Experiment 3).

### Microscale Procedure

In a 2-mL (1-dram) ampoule dissolve 100 mg of benzophenone in 1 mL of isopropyl alcohol by warming on a sand bath. Add a microdrop of acetic acid, cool the ampoule in dry ice, and seal the neck of the ampoule in a flame (Fig. 62.2). Set the ampoule outside in bright sunlight. On a larger scale, this experiment can take as long as a week due to absorption of UV radiation by the thick glass walls of a flask and by self-absorption of the thick layers of solution. On a microscale, the glass of an ampoule is very thin and the thickness of the entire solution is small, so the reaction proceeds very rapidly. A rough parabolic reflector fashioned from aluminum foil can speed up the reaction even more.

Because benzopinacol is only sparingly soluble in the isopropyl alcohol, its formation is followed by the separation of small, colorless crystals (benzophenone forms large, thick prisms) from around the walls of the ampoule. If the reaction mixture is exposed to direct sunlight, the first crystals separate in less than an hour, and the reaction is complete in a day. In winter the reaction will take longer, and any benzophenone that crystallizes must be brought into solution by warming on a steam bath. When the reaction appears to be over, chill the tube (if necessary) and collect the product. The material should be pure (mp 188°C–189°C). If the yield is low, more material can be obtained by further exposure of the mother liquor to sunlight.

Reaction time: 1 day to 1 week

***Online Study Center***

Photo: Use of the Wilfilter

■ **FIG. 62.2**
Microscale photolysis in a
2-mL (1-dram) ampoule.

Seal here

Score and break
here at end
of reaction

Reaction time: 4 days to
2 weeks

**Cleaning Up.** Dilute the isopropyl alcohol filtrate with water and flush the solution down the drain. Should any unreacted benzophenone precipitate, collect it by vacuum filtration, discard the filtrate down the drain, and place the recovered solid in the nonhazardous solid waste container.

## Macroscale Procedure

In a 50-mL round-bottomed flask dissolve 5 g of benzophenone in 30–35 mL of isopropyl alcohol by warming on a steam bath, fill the flask to the neck with more isopropyl alcohol, and add 1 drop of glacial acetic acid. (If the acid is omitted, enough alkali may be derived from the glass of the flask to destroy the reaction product by the alkaline cleavage described in Experiment 2.) Stopper the flask with a well-rolled, tight-fitting cork or a lightly greased glass stopper, which is then wired in place. Invert the flask in a 100-mL beaker placed where the mixture will be most exposed to direct sunlight for some time. Because benzopinacol is only sparingly soluble in alcohol, its formation can be followed by the separation from around the walls of the flask of small, colorless crystals (benzophenone forms large, thick prisms). If the reaction mixture is exposed to direct sunlight, the first crystals separate in about 5 h, and the reaction is practically complete (with 95% yield) in 4 days. In winter the reaction may take as long as 2 weeks, and any benzophenone that crystallizes must be brought into solution by warming on a steam bath. When the reaction appears to be over, chill the flask (if necessary) and collect the product. The material should be pure (mp 188°C–189°C). If the yield is low, more material can be obtained by further exposure of the mother liquor to sunlight.

**Cleaning Up.** Dilute the isopropyl alcohol filtrate with water and flush the solution down the drain. Should any unreacted benzophenone precipitate, collect it by vacuum filtration, discard the filtrate down the drain, and place the recovered solid in the nonhazardous solid waste container.

# 2. Alkaline Cleavage

**IN THIS EXPERIMENT** you can demonstrate that the alkaline cleavage of benzopinacol produces a low melting solid mixture that can be shown to be equal parts of an alcohol and a ketone. Alternatively, irradiation of benzophenone in an alkaline solution of benzophenone can be shown to produce only the alcohol benzhydrol.

Suspend a small test sample of benzopinacol in alcohol and heat to boiling on a steam bath, being sure that the amount of solvent is not sufficient to dissolve the solid. Add 1 drop of sodium hydroxide solution, heat for 1–2 min, and observe the result. The solution contains equal parts of benzhydrol and benzophenone, which are formed by the following reaction:

$$\underset{\substack{\text{Benzopinacol}\\ \text{mp 189°C}}}{\underset{C_6H_5}{\overset{C_6H_5}{\underset{OH}{\overset{|}{C}}}}\underset{HO}{\overset{C_6H_5}{\overset{|}{C}}}}\quad\xrightarrow{\overset{+}{R\bar{O}Na}}\quad\underset{\substack{\text{Benzhydrol}\\ \text{mp 68°C}}}{\underset{C_6H_5}{\overset{C_6H_5}{\overset{H}{\underset{OH}{\overset{|}{C}}}}}}\;+\;\underset{\substack{\text{Benzophenone}\\ \text{mp 48°C}}}{O=\underset{C_6H_5}{\overset{C_6H_5}{C}}}$$

The low-melting products resulting from the cleavage are far more soluble than the starting material. Analyze the mixture by thin-layer chromatography.

Benzophenone can be converted into benzhydrol in nearly quantitative yield by following the procedure outlined for the preparation of benzopinacol (Experiment 1), modified by adding a *very* small piece of sodium (5 mg) instead of the acetic acid. The reaction is complete when, after exposure to sunlight, the greenish-blue color disappears. To obtain the benzhydrol, the solution is diluted with water, acidified, and evaporated. Benzopinacol is produced as before by photochemical reduction, but it is at once cleaved by the sodium alkoxide. The benzophenone formed by cleavage is converted into more benzopinacol, cleaved, and eventually consumed. Figures 62.3 and 62.4, presented at the end of the chapter, show the infrared (IR) and $^1$H NMR (nuclear magnetic resonance) spectra, respectively, of benzophenone.

## 3. Pinacolone Rearrangement

**Benzopinacol**

(Carbocation)

**Benzopinacolone**
mp 179–180°C

The acid-catalyzed carbonium ion rearrangement is characterized by rapidity and by the high yield.

## Microscale Procedure

In a reaction tube place 50 mg of benzopinacol, 0.25 mL of acetic acid, and one *very* small crystal of iodine (0.0005 g). Heat to boiling for 1–2 min on a sand bath until the crystals are dissolved; then reflux the red solution for 5 min. On cooling, the pinacolone separates as a stiff paste. Thin the paste with alcohol, collect the product by vacuum filtration on a Hirsch funnel, and wash it free of iodine with cold 95% ethanol. The material should be pure; the expected yield is 95%.

**Cleaning Up.** Dilute the filtrate with water, neutralize with sodium carbonate, and flush down the drain.

## Macroscale Procedure

In a 100-mL round-bottomed flask place 5 g of benzopinacol, 25 mL of acetic acid, and two or three very small crystals of iodine (0.05 g). Heat to the boiling point for 1–2 min under a reflux condenser until the crystals are dissolved; then reflux the red solution for 5 min. On cooling, the pinacolone separates as a stiff paste. Thin the paste with alcohol, collect the product, and wash it free from iodine with alcohol. The material should be pure, with a yield of 95%.

**Cleaning Up.** Dilute the filtrate with water, neutralize with sodium carbonate, and flush down the drain.

## QUESTIONS

1. Would the desired reaction occur if ethanol or *t*-butyl alcohol were used instead of isopropyl alcohol in the attempted photochemical dimerization of benzophenone?
2. Assign the peak at 1639 cm$^{-1}$ in the IR spectrum of benzophenone (Fig. 62.3).

■ **FIG. 62.3**
The IR spectrum of benzophenone
(KBr disk).

■ **FIG. 62.3**
The IR spectrum of benzophenone
(KBr disk).

■ **FIG. 62.4**
The ¹H NMR spectrum of
benzophenone (250 MHz).

# Carbohydrates and Sweeteners

**PRELAB EXERCISE:** Draw a flow sheet showing the sequence of tests you would conduct to identify an unknown carbohydrate.

Carbohydrates, the direct product of the photosynthetic combination of carbon dioxide and water, are, by weight, the most common organic compounds on the earth. Because most have the empirical formula $C_nH_{2n}O_n = C_n(H_2O)_n$, it is easy to see why early scientists considered these molecules to be hydrates of carbon. They are polyhydroxyaldehydes and ketones and exist as hemi- and full acetals and ketals. Carbohydrates are classified as mono-, di-, and polysaccharides.

**Saccharin**

**Sodium cyclamate**

**2-Amino-4-nitro-*n*-propoxybenzene**

**Aspartame**
L-Aspartyl-L-phenylalanine methyl ester

**Sucralose**

**Synthetic Sugar Substitutes**

CHO
HO—C—H
HO—C—H
H—C—OH
H—C—OH
CH₂OH
**D-Mannose**

CHO
H—C—OH
HO—C—H
HO—C—H
H—C—OH
CH₂OH
**D-Galactose**

CHO
H—C—OH
HO—C—H
H—C—OH
H—C—OH
CH₂OH
**D-Glucose**

CH₂OH
C=O
HO—C—H
H—C—OH
H—C—OH
CH₂OH
**D-Fructose**

CHO
HO—C—H
H—C—OH
H—C—OH
CH₂OH
**D-Arabinose**

CHO
H—C—OH
HO—C—H
H—C—OH
CH₂OH
**D-Xylose**

**Aldohexoses**         **Ketohexose**         **Aldopentoses**

**Lactose (aldehyde form)**

**Maltose (aldehyde form)**

**Sucrose**

**Disaccharides**

## Sweeteners

The term *sugar* applies to mono-, di-, and oligosaccharides, which are all soluble in water and thereby distinguished from polysaccharides, which are not water soluble. Many natural sugars are sweet, but the data in Table 63.1 show that sweetness varies greatly with stereochemical configuration and is exhibited by compounds of widely differing structural type. Note that D-fructose is 50% sweeter than sucrose, which accounts for the presence of high-fructose corn syrup in many commercial foods and drinks.

The synthesis and sale of the artificial, synthetic sweeteners sucralose (Splenda) and aspartame (NutraSweet) constitute a multibillion-dollar industry. Sucralose is made from sucrose; three hydroxy groups in sucrose are converted to chloro groups with inversion of stereochemistry at one stereocenter.

| TABLE 63.1 • Relative Sweetness of Sugars and Sugar Substitutes | | |
|---|---|---|
| | **Sweetness** | |
| **Compound** | *To Humans* | *To Bees* |
| Monosaccharides | | |
|   D-Fructose | 1.50 | |
|   D-Glucose | 0.55 | |
|   D-Mannose | Sweet, then bitter | —[a] |
|   D-Galactose | 0.55 | — |
|   D-Arabinose | 0.70 | — |
|   D-Xylose | Very sweet | |
| Disaccharides | | |
|   Sucrose (glucose, fructose) | 1.0 | |
|   Maltose (2 glucose) | 0.3 | |
|   Lactose (glucose, galactose) | 0.2 | — |
|   Cellobiose (2 glucose) | Indifferent | — |
|   Gentiobiose (2 glucose) | Bitter | — |
| Polysaccharides | | |
|   Soluble starch suspension | Not sweet | — |
|   Soluble glycogen suspension | Not sweet | — |
| Synthetic sugar substitutes | | |
|   Aspartame (NutraSweet) | 180 | |
|   Saccharin | 550 | |
|   Sucralose (Splenda) | 600 | |
|   2-Amino-4-nitro-1-*n*-propoxybenzene | 4000 | |

[a] —means that is sweeter to humans than to bees.

Sucralose, sold under the name Splenda, is a chlorinated sucrose derivative and is the first synthetic sweetener derived from sugar that offers zero calories and therefore is not metabolized in the body. It is used by diabetics as a sugar substitute. At this time, it appears that there are no side effects or health concerns associated with the consumption of sucralose. Aspartame, a dipeptide, is metabolized in the body and provides 4 cal/g. Aspartame breaks down into methanol, phenylalanine, and aspartic acid. Recent reports have claimed that aspartame can lead to serious health concerns due to its metabolized components—methanol and phenylalanine. For example, methanol can be oxidized to formaldehyde, which can cause severe damage to the neurological and immune systems. In addition, phenylalanine, if not properly metabolized to tyrosine via a required enzyme, can convert to phenylketones, which could lead to mental retardation. Another advantage that sucralose has over aspartame (besides a lack of calories) is that it is heat stable (aspartame is not) and thus can be used for baking and cooking. It is interesting to note that the discovery of all artificial sweeteners has been accidental, which says something about how careless chemists have been.

The exact physiological basis for sweetness, and therefore the relationship between organic structure and sweetness, is not understood.

One gram of sucrose (common table sugar) dissolves in 0.5 mL of water at 25°C and in 0.2 mL at the boiling point, but the substance has marked, atypical crystallizing properties. Despite the high solubility, it can be obtained in beautiful, large crystals (rock candy). More typical sugars are obtainable in crystalline form only with difficulty, particularly in the presence of a trace of impurity, and even then, they give small, ill-formed crystals. Alcohol is often added to a water solution to decrease solubility and thus to induce crystallization. The amounts of 95% ethanol required to dissolve 1-g samples at 25°C are 170 mL for sucrose, 60 mL for glucose, and 15 mL for fructose. Some sugars have never been obtained in crystalline condition and are known only as *viscous syrups*.

## Analysis of Carbohydrates

In this chapter, unknown carbohydrates will be identified using a series of classification tests. The Molisch test (Experiment 1) is a general test for carbohydrates. If the test is positive and the unknown is water soluble (which distinguishes it from polysaccharides such as cellulose), then it can be tested with iodine, which will give a positive test with the partially water-soluble polysaccharide, starch.

If the unknown is a sugar, it can be a mono-, di-, or, less commonly, an oligosaccharide. If it is a monosaccharide, it usually has either five or six carbons, and its carbonyl group can be either an aldehyde or a ketone.

To distinguish aldehydosugars from ketosugars, take advantage of the reducing ability of the aldehyde group. Either the traditional Fehling's or Benedict's test will give a positive reaction with any reducing sugar; in Experiment 2, we will employ the far more sensitive red tetrazolium (RT) test. If the RT test is positive, then Barfoed's test (Experiment 3) allows us to make a distinction between reducing mono- and disaccharides.

To distinguish between pentoses and hexoses, run Bial's test (Experiment 4). Seliwanov's test (Experiment 5) allows us to make a distinction between aldohexoses and ketohexoses.

### Osazones

With phenylhydrazine, many sugars form beautiful crystalline derivatives called *osazones* (Experiment 6). Osazones are far less soluble in water than their parent sugars because the molecular weight is increased by 178 units and the number of hydroxyl groups is reduced by 1. It is easier to isolate an osazone than to isolate the sugar, and syrupy sugars often give crystalline osazones. Osazones of the more highly hydroxylic disaccharides are notably more soluble than those of monosaccharides.

Some disaccharides do not form osazones, but a test for the formation or nonformation of the osazone is ambiguous because the glycosidic linkage may suffer hydrolysis in a boiling solution of phenylhydrazine and acetic acid, with formation of an osazone derived from a component sugar and not from the disaccharide. If a sugar has reducing properties (a positive RT test), it is also capable of osazone formation; hence, an unknown sugar should be tested for reducing properties before preparation of an osazone is attempted.

---

## EXPERIMENTS

 ## 1. Molisch Test for Carbohydrate

> **IN THIS EXPERIMENT** the Molisch reagent is mixed with a dilute solution of a suspected carbohydrate. If the careful introduction of concentrated sulfuric acid produces a purple color at the interface, a carbohydrate is present.

**CAUTION**: Handle concentrated sulfuric acid with care. It is very corrosive. Do not mix the layers in the tube because the heat of mixing can form steam, thus causing the solution to erupt from the tube. Use great care in pouring the reaction mixture into a beaker for disposal.

Label several reaction tubes or 10 × 75-mm culture tubes. To 1 mL of a 1% solution of the carbohydrates in Table 63.1 (or a representative selection chosen by your instructor), add 1 drop of Molisch reagent (a solution of 1.25 g of 1-naphthol in 25 mL of ethanol) and mix thoroughly. Add 1 mL of water in another tube as a control. Then, while tilting each tube, carefully introduce 1 mL of concentrated sulfuric acid down the inside walls. The dense sulfuric acid will form a lower layer. Note the appearance of each tube. Do not mix the layers. A purple color at the interface between the two layers is a positive test for carbohydrates.

**Cleaning Up.** Pour the contents of each tube into a large beaker (*caution!*) and neutralize with sodium bicarbonate. Dilute the neutral solution with water and flush down the drain.

 ## 2. Red Tetrazolium

Red tetrazolium[1], also known as 2,3,5-triphenyl-2H-tetrazolium chloride, is a nearly colorless, water-soluble, light-sensitive substance that oxidizes aldoses and ketoses, as well as other α-ketols, and is thereby reduced. Freshly prepared aqueous solutions should be used in tests. Any unused solution should be acidified and discarded. The reduced form is a water-insoluble, intensely colored pigment—a diformazan.

**Red tetrazolium (RT)**          **RT-Diformazan**

---

1. Red tetrazolium is available from Aldrich Chemical Company.

A sensitive test for reducing sugars

RT affords a highly sensitive test for reducing sugars because it distinguishes between α-ketols and simple aldehydes more sharply than Fehling's and Tollens' tests, two older tests that were used for this purpose.

Put 1 drop of each of the 1% test solutions in clean, marked test tubes and, to each tube, add 1 mL of a 0.5% aqueous solution of red tetrazolium (made by dissolving 0.5 g of red tetrazolium in 100 mL of water) and 1 drop of 3 $M$ sodium hydroxide solution. Put the tubes in a beaker of hot water, note the time, then note the time at which each tube develops color.

To estimate the sensitivity of the test, use the substance that you regard as the most reactive of the five studied. Dilute 1 mL of the 1% solution with water to a volume of 100 mL and run a test with red tetrazolium on 0.2 mL of the diluted solution.

**Cleaning Up.** Dilute the test solutions with water and flush down the drain.

# 3. Barfoed's Test

Barfoed's test is used to distinguish between reducing mono- and disaccharides. In clean, labeled reaction tubes or 10 × 75-mm culture tubes, place 1 mL of 1% solutions of a representative selection of reducing mono- and disaccharide solutions chosen from Table 63.1. To each, add 0.5 mL of Barfoed's reagent (made by dissolving 6.7 g of copper(II) acetate and 0.9 mL of acetic acid in 100 mL of water) and mix. Heat the tubes in a hot water bath for 15 min. Note the time needed for a red precipitate to appear in each tube.

**Cleaning Up.** Dilute the test solutions with water and flush down the drain.

# 4. Bial's Test

Bial's test is used to distinguish between pentoses and hexoses. In clean, labeled reaction tubes or 10 × 75-mm culture tubes, place 1 mL of 1% solutions of a representative selection of pentoses and hexoses chosen from Table 63.1. In one tube, place 1 mL of distilled water to serve as a reference or control solution.

To each tube, add 1 mL of Bial's reagent and a boiling chip. [To make Bial's reagent: dissolve 300 mg of 3,5-dihydroxytoluene in 100 mL of concentrated hydrochloric acid; then add 0.3 mL of a 10% iron(III) chloride solution.] Heat each tube to boiling on a sand bath and note the color that develops. In tubes that have weak or no color, the product of the reaction can be concentrated by diluting the mixture with 2 mL of water, adding 0.5 mL of cyclohexanol, and then extracting the colored product into the alcohol layer by shaking the tube vigorously.

**Cleaning Up.** Dilute the test solutions with water and flush down the drain.

# 5. Seliwanov's Test

Seliwanov's test is used to distinguish between aldohexoses and ketohexoses. To a number of reaction tubes or 10 × 75-mm culture tubes, add 1 mL of the Seliwanov reagent, which is prepared by dissolving 50 mg of 1,3-dihydroxybenzene (resorcinol) in a mixture of 33 mL of concentrated hydrochloric acid and 67 mL of distilled water. To each labeled tube, add 5 drops of a 1% solution of each of the sugars to be tested. After mixing, heat the tubes in a boiling water bath for 2 min. Examine the tubes. The intensity of the color is proportional to the α-ketol concentration.

**Cleaning Up.** Dilute the test solutions with water and flush down the drain.

## 6. Osazones

This experiment can be carried out on a scale two or three times larger than what is specified here.

CAUTION: Conduct this procedure in the hood. Avoid skin contact with phenylhydrazine.

Put 0.33-mL portions of phenylhydrazine reagent [made by neutralizing phenylhydrazine (3 mL) with 9 mL of acetic acid in a 50-mL Erlenmeyer flask, adding 15 mL of water, transfering the mixture to a graduated cylinder, and making up the volume to 30 mL] into each of four cleaned, numbered reaction tubes. Add 1-mL portions of 1% solutions of glucose, fructose, lactose, and maltose and heat the tubes in the beaker of hot water for 20 min. Shake or flick the tubes occasionally to relieve supersaturation. Note the times at which the osazones separate. If, after 20 min, no product has separated, cool and scratch the test tube to induce crystallization.

Collect and save the products for possible use in the later identification of unknowns. Because osazones melt with decomposition, heat the bath at a standard rate (0.5°C/s) when determining the melting point.

**Cleaning Up.** The filtrate can be diluted with water and flushed down the drain. If you must destroy the phenylhydrazine, neutralize the solution and add 2 mL of household bleach (aqueous sodium hypochlorite solution) for each 1 mL of the reagent. Heat the mixture to 45°C–50°C for 2 h to oxidize the amine; then cool the mixture and flush it down the drain.

**Online Study Center**

**General Resources**
Additional Experiments

## QUESTIONS

1. What is the order of relative reactivity in the RT test of the compounds studied?
2. Which test do you regard as the most reliable for distinguishing reducing from nonreducing sugars? Which is most reliable for differentiating an α-ketol from a simple aldehyde?
3. Write a mechanism for the acid-catalyzed hydrolysis of the disaccharide sucrose. Take care to draw the stereochemistry of the products.
4. Will the anomeric forms of glucose give different phenylosazones?

# Biosynthesis of Ethanol

**PRELAB EXERCISE:** List the essential chemical substances, the solvent, and conditions for converting glucose to ethanol.

Fermentation

Humans have been preparing fermented beverages for more than 5000 years. Materials excavated from Egyptian tombs dating to 3000 B.C. demonstrate the operations used in making beer and leavened bread. The history of fermentation, whereby sugar is converted to ethanol by the action of yeast, is also a history of chemistry. The word *gas* was coined by Johannes Baptista van Helmont in 1610 to describe the bubbles produced in fermentation. Antoni van Leeuwenhoek observed and described the cells of yeast with his newly invented microscope in 1680. In 1754 Joseph Black discovered carbon dioxide and showed it to be a product of fermentation, the burning of charcoal, and respiration. In 1789 Antoine Lavoisier showed that sugar gives ethanol and carbon dioxide and made quantitative measurements of the amounts consumed and produced.

Glucose → 2 $C_2H_5OH$ + 2 $CO_2$

Once the mole concept was established (in 1815), Joseph Gay-Lussac could show that 1 mol of glucose gives exactly 2 mol of ethanol and 2 mol of carbon dioxide. But the process of fermentation stumped some great chemists. The little-known Friedrich T. Kützing wrote in 1837, on the basis of microscopic observation, "It is obvious that chemists must now strike yeast off the roll of chemical compounds because it is not a compound but an organized body, an organism." On the other side were chemists such as Jöns Berzelius, who believed that yeast had a catalytic action; and Juston von Liebig, who put forth a "theory of motion of the elements within a compound that caused a disturbance of equilibrium that was communicated to the elements of the substance with which it came in contact thus forming new compounds."

Pasteur's contribution

It remained for Louis Pasteur to show that fermentation was a physiologic action associated with the life processes of yeast. Through his microscope Pasteur could see the yeast cells that grew naturally on the surface of grapes. He showed that grape juice carefully extracted from the center of a grape and exposed to clean air would not ferment. In his classic paper of 1857, Pasteur described fermentation as the action of a living organism; but because the conversion of glucose to ethanol and carbon dioxide is a balanced equation, other chemists disputed his findings. They searched for the substance in yeast that might cause the reaction. The search lasted for 40 years and eventually ended by a clever experiment by Eduard Büchner. He made a cell-free extract of yeast that would still cause the conversion of sugar to alcohol. This cell-free extract contained the catalysts, which we now call *enzymes*, that were necessary for fermentation—a discovery that earned him the 1907 Nobel Prize in Chemistry. In 1905 Sir Arthur Harden discovered that inorganic phosphate added to the enzymes increased the rate of fermentation and was itself consumed. This result led Harden to eventually isolate fructose 1,6-diphosphate. Clearly, the history of biochemistry is intimately associated with the study of alcoholic fermentation.

Ancient peoples discovered many of the essential reactions of alcoholic fermentation completely by accident. That crushed grapes would soon begin to froth and bubble and produce a pleasant beverage is a discovery lost in time. But what of those who lived in colder climates where the grape did not grow? How did they discover that the starch of wheat or barley could be converted to sugar by the enzymes in malt? When grain germinates, enzymes are produced that turn the starch into sugar. The process of malting involves letting the grain start to germinate and then heating and drying the sprouts to stop the process before the enzymes are used up. The color of the malt depends on the temperature of the drying. The darkest is used for stout and porter; the lighter for brown, amber, and pale ale. Because of a discovery made at least 10,000 years ago that the resulting beverage did not spoil as rapidly if hops were added, we now also have beer.

Other sources exist for the amylases that catalyze the conversion of starch to glucose. The Peruvian *campasinos* (peasants) make a drink called "chichi" from masticated wheat, which is dried in small cakes. When water, yeast, and more ground wheat are added, the resulting mixture ferments into a beerlike beverage. The enzyme salivary amylase is the catalyst for this starch-to-glucose conversion.

Bakers make use of fermentation by taking advantage of the gas released to leaven bread. In this chapter baker's yeast is used to convert sucrose, ordinary table sugar, into ethanol and carbon dioxide with the aid of some 14 enzymes as catalysts, in addition to adenosine triphosphate (ATP), phosphate ion, thiamine pyrophosphate, magnesium ion, and reduced nicotinamide adenine dinucleotide (NADH), all present in yeast.

**Sucrose**

The fermentation process—known as the Emden-Meyerhof-Parnas scheme—involves the hydrolysis of sucrose to glucose and fructose, which, as their phosphates, are cleaved to two three-carbon fragments. These fragments, as their phosphates, eventually are converted to pyruvic acid, which is decarboxylated to give acetaldehyde. Acetaldehyde, in turn, is reduced to ethanol in the final step. Each step requires a specific enzyme as a catalyst and often inorganic ions, such as magnesium and, of course, phosphate. Thirty-one kilocalories of heat are released per mole of glucose consumed in this sequence of anaerobic reactions.

    This same sequence of reactions, up to the formation of pyruvic acid, occurs in the human body in times of stress when energy is needed but when not enough oxygen is available for normal aerobic oxidation. The pyruvic acid under these conditions is converted to lactic acid. It is the buildup of lactic acid in the muscles that is partly responsible for the feeling of fatigue.

1. Glucose    2. Glucose-6-phosphate    3. Fructose-6-phosphate    4. Fructose-1,6-*bis*-phosphate    5. Dihydroxyacetone phosphate    6. Glyceraldehyde-3-phosphate

7. 1,3-Diphosphoglyceric acid    8. 3-Phosphoglyceric acid    9. 2-Phosphoglyceric acid    10. Phosphoenolpyruvic acid

14. Lactic acid    11. Pyruvic acid    12. Acetaldehyde    13. Ethanol

The first step in the sequence is the formation of glucose-6-phosphate (**2**) from glucose (**1**). The reaction requires ATP, which is converted to adenosine diphosphate (ADP) by catalysis with the enzyme glucokinase, which requires magnesium ion to function. This conversion is one of the reactions in which energy is released. In the living organism this energy can be used to do work; in fermentation it simply creates heat.

**Adenosine triphosphate (ATP)**

**Adenosine diphosphate (ADP)**

In the next step glucose-6-phosphate (**2**) is converted through the enol of the aldehyde to fructose-6-phosphate (**3**) by the enzyme phosphoglucoisomerase. The fructose monophosphate (**3**) is converted to the *bis*-phosphate (**4**) by the action of ATP under the influence of phosphofructokinase with the release of more energy. This *bis*-phosphate (**4**) undergoes a reverse aldol reaction catalyzed by aldolase to give dihydroxyacetone phosphate (**5**) and glyceraldehyde-3-phosphate (**6**). The latter two are interconverted by means of triosphosphate isomerase. The aldehyde group of glyceraldehyde phosphate (**6**) is oxidized by nicotinamide adenine dinucleotide (NAD) in the presence of another enzyme to a carboxyl group that is phosphorylated with inorganic phosphate. In the next reaction ADP is converted to ATP as 1,3-diphosphoglyceric acid (**7**) loses a phosphate to give 3-phosphoglyceric acid (**8**). A mutase converts the 3-phosphate (**8**) to the

2-phosphate (**9**). An enolase converts 2-phosphoglyceric acid (**9**) to phospho-enolpyruvic acid (**10**), and a kinase converts **10** to pyruvic acid (**11**). A decar-boxylase converts **11** to acetaldehyde (**12**) in the fermentation process. Yeast alcohol dehydrogenase (YAD), a well-studied enzyme, catalyzes the reduction of acetaldehyde to ethanol. The reducing agent is reduced NADH.

**Reduced nicotinamide adenine dinucleotide, NADH**

**Nicotinamide adenine dinucleotide, NAD⁺**

*Enzymes are labile.*
Enzymes are remarkably efficient catalysts, but they are also labile (sensi-tive) to such factors as heat and cold, changes in pH, and various specific inhibitors. In Experiment 1 you will have an opportunity to observe the biosyn-thesis of ethanol and to test the effects of various agents on the enzyme system.

This experiment involves the fermentation of ordinary cane sugar using baker's yeast. The resulting dilute solution of ethanol, after removing the yeast by filtration, can be distilled according to the procedure of Experiment 2 in Chapter 5 (on page 97).

## EXPERIMENTS

## 1. Fermentation of Sucrose

### *Microscale Procedure*

**IN THIS EXPERIMENT** a warm aqueous solution of sucrose and sodium phosphate in the presence of yeast is allowed to ferment to ethanol and carbon dioxide. The enzyme catalysts from the yeast are rendered inactive when the alcohol concentration reaches 10%–12%. Fractional distillation of the reaction mixture will yield 95% pure ethanol, a constant-boiling azeotrope.

Protect the fermenting reaction from exposure to oxygen and contaminating materials. Under aerobic conditions acetobacter bacteria can convert ethanol to acetic acid.

Reaction time: 1 week

To a 5-mL round-bottomed, long-necked flask add 90 mg of dry yeast and 1.25 mL of warm (up to 50°C) water. Shake the mixture thoroughly until it is more or less homogeneous in appearance; then add to it 9 mg of disodium hydrogen phosphate, 1.30 g of sucrose, and an additional 3.75 mL of water, which should be warmed to about 45°C. Shake this mixture to ensure complete mixing and then fit the neck of the flask with a septum that is connected to an 20-cm length of polyethylene tubing (Fig. 64.1). Lead this tubing beneath the surface of about 2 mL of a saturated aqueous solution of calcium hydroxide (limewater) in a reaction tube. The tube in the limewater will act as a seal to prevent air and unwanted enzymes from entering the flask but will allow gas to escape. Place the assembly in a warm spot (the optimal temperature for the reaction is 35°C; an example of a warm spot may be a cupboard near a hot water line) for a week, at which time the evolution of carbon dioxide will have ceased. What is the precipitate in the limewater?

On a small scale, it will be necessary to provide external heat to maintain the fermentation or to group several reactions closely together in an insulated container. On a larger scale, because fermentation is an exothermic reaction, there is enough heat evolved by the reaction to keep the mixture warm and to promote the biosynthesis of ethanol.

On completion of fermentation, add about 0.25 g of Celite filter aid (diatomaceous earth, face powder) to the flask and shake it vigorously. Celite is added to make it possible to filter the solution. Otherwise the yeast cells will clog the filter paper. Filter the mixture on a Hirsch funnel into a 25-mL filter flask that is attached to a water aspirator through a trap by vacuum tubing (*see* Fig. 4.15 on page 74). Because the small filter flask can tip over easily, clamp it to a ring stand. Moisten the filter paper with water, apply gentle suction (water supply to the aspirator turned on full force, valve to trap partly open), and slowly pour the reaction mixture onto the filter. Wash out the flask with 1 mL of water and rinse

■ **FIG. 64.1**

A fermentation apparatus. See Figure 18.2 (on page 347) for the technique of threading a polyethylene tube through a septum.

Polyethylene tubing

Foam

the filter cake with this water. Full vacuum in this filtration will evaporate some of the desired ethanol.

The filtrate, which is a dilute solution of ethanol contaminated with a few bits of cellular material and other organic compounds (acetic acid if you are not careful), is saved in a stoppered container until it is distilled following the procedure outlined in Experiment 2 of Chapter 5.

Refer to the "Cleaning Up" instructions after the following macroscale procedure.

## Macroscale Procedure

Protect the fermenting reaction from exposure to oxygen and contaminating materials. Under aerobic conditions acetobacter bacteria can convert ethanol to acetic acid.

Reaction time: 1 week

Filter aid

Macerate (grind) one-half cake of yeast or half an envelope of dry yeast in 50-mL of water in a beaker, add 0.35 g of disodium hydrogen phosphate, and transfer this slurry to a 500-mL round-bottomed flask. Add a solution of 51.5 g of sucrose in 150 mL of water and shake to ensure complete mixing. Fit the flask with a one-hole rubber stopper containing a bent glass tube that dips below the surface of a saturated aqueous solution of calcium hydroxide (limewater) in a 15-cm test tube (Fig. 64.2). The tube in limewater will act as a seal to prevent air and unwanted enzymes from entering the flask but will allow gas to escape.[1] Place the assembly in a warm spot on your desk (the optimum temperature for the reaction is 35°C; an example of a warm spot may be a cupboard near a hot water line) for 1 week, at which time the evolution of carbon dioxide will have ceased. What is the precipitate in the limewater?

Upon completion of fermentation, add 10 g of Celite filter aid (diatomaceous earth, face powder) to the flask, shake vigorously, and filter. Use a 5.5-cm

1. The tube should be about 0.5 cm below the limewater so that limewater will not be sucked back into the flask should the pressure change.

Büchner funnel placed on a neoprene adapter or Filtervac atop a 500-mL filter flask that is attached to a water aspirator through a trap by vacuum tubing (Fig. 64.3). Celite is added to make it possible to filter the solution. Because the apparatus is top-heavy, clamp the flask to a ring stand. Moisten the filter paper with water, apply gentle suction (water supply to the aspirator turned on full force, clothespin on trap partially closed), and slowly pour the reaction mixture onto the filter. Wash out the flask with a few milliliters of water from your wash bottle and rinse the filter cake with this water.

The filtrate, which is a dilute solution of ethanol contaminated with bits of cellular material and other organic compounds (acetic acid if you are not careful), is saved in a stoppered flask until it is distilled following the procedure outlined in Experiment 2 of Chapter 5.

**Cleaning Up.** Because sucrose, yeast, and ethanol are natural products, all solutions produced in this experiment contain biodegradable material and can be flushed down the drain after dilution with water. The limewater can be disposed of in the same way. The Celite filter aid can be placed in the nonhazardous solid waste container.

# 2. The Effect of Various Reagents and Conditions on Enzymatic Reactions

## *Microscale Procedure*

**CAUTION**: Handle the sodium fluoride solution with great care; it is very poisonous if ingested.

Prepare a fermentation mixture exactly as described in Experiment 1. After about 15 min, when the fermentation should be progressing nicely, divide the mixture into five equal parts in labeled reaction tubes (A, B, C, D, and E). To tube A, add 1.0 mL of water; to tube B, add 1.0 mL of 95% ethanol; to tube C, add 1.0 mL of 0.5 $M$ sodium fluoride solution. Heat tube D for 5 min in a steam bath or boiling water bath and cool tube E for 5 min in ice. Add to each tube 2 drops of mineral oil. The oil will float on the aqueous solution and prevent exposure to oxygen, which is necessary because fermentation is an anaerobic process. Place the tubes in a beaker of water at a temperature of 30°C for 15 min; then connect them one at a time while in the beaker to an inverted 1.0-mL graduated pipette that has been plugged at one end, as shown in Figure 64.4. Record the volume of gas in the pipette every minute for 5 min or count the bubbles evolved per minute. Plot a

**■ FIG. 64.4**
An apparatus for measuring carbon dioxide evolved. See Figure 18.2 (on page 347) for the technique of threading a polyethylene tube through a septum.

Cork

1-mL Graduated pipette

Mineral oil

Test solution

Water

graph of the volume of gas against time for each of the five reactions. What conclusions can you draw from the results of these five reactions?

## Macroscale Procedure

About 15 min after mixing the yeast, sucrose, and phosphate, remove 20 mL of the mixture and place 4 mL in each of five labeled test tubes (A, B, C, D, and E). To tube A, add 1.0 mL of water; to tube B, add 1.0 mL of 95% ethanol; to tube C, add 1.0 mL of 0.5 $M$ sodium fluoride. Heat tube D for 5 min in a steam bath and cool tube E for 5 min in ice. Add 10–15 drops of mineral oil on top of the reaction mixture in each tube (to exclude air because the process is anaerobic). Place the tubes in a beaker of water at a temperature of 30°C for 15 min; then take them one at a time and connect each to the manometer as shown in Figure 64.5. Allow about 30 s for temperature equilibration; then clamp the vent tube and read the manometer. Record the height of the manometer fluid in the open arm of the U-tube every minute for 5 min or until the fluid reaches the top of the manometer. Plot a graph of the height of the manometer fluid against time for each of the five reactions. What conclusions can you draw from the results of these five reactions?

**Cleaning Up.** Combine all of the reaction mixtures, remove the mineral oil from the top, and place it in the organic solvents container. The aqueous solutions, after diluting with about 50 volumes (50 times the volume of the original aqueous solutions) of water, can be flushed down the drain. Solutions containing fluoride ion are highly toxic and should not be placed in the sewer system but neutralized and treated with excess calcium chloride to precipitate calcium fluoride. The latter is separated by filtration and put in the nonhazardous solid waste container.

**CAUTION:** Handle the sodium fluoride solution with great care; it is very poisonous if ingested.

**Online Study Center**

**General Resources**
Additional Experiments

■ **FIG. 64.5**
A manometer gasometer.

Open end manometer

Mineral oil

Water

Test solution

## QUESTIONS

1. Using yeast, can glucose be converted to ethanol? Can fructose be converted to ethanol?
2. Write the equation for the formation of the precipitate formed in a test tube containing calcium hydroxide.
3. In this experiment could 90% ethanol be made by adding more sugar to the fermentation flask?

# Enzymatic Reactions: A Chiral Alcohol from a Ketone and Enzymatic Resolution of DL-Alanine

**Online Study Center**

This icon will direct you to techniques, equipment setups, and online resources at http://college.hmco.com/PIC/williamsonMMOE5e.

**PRELAB EXERCISE:** Study the biochemistry of the conversion of glucose to ethanol (Chapter 64). Which enzyme might be responsible for the reduction of ethyl acetoacetate? Discuss at least three other methods for resolving DL-alanine into its enantiomers. Judging from its name, would you expect acylase to be equally effective on the acetyl derivatives of other amino acids? Explain why the enzyme acylase will hydrolyze only one of a pair of enantiomers and how a mixture of L-alanine and *N*-acetyl-D-alanine can be separated at the end of the reaction.

## PART 1: Enzymatic Reduction of a Ketone to a Chiral Alcohol

$$
\begin{array}{c}
\overset{\displaystyle O}{\underset{\displaystyle R\quad R'}{\|}} \\
\text{Nonselective reduction} \\
\downarrow
\end{array}
$$

50% *R*          50% *S*

ee = 0%

**T**he reduction of an achiral ketone with the usual laboratory reducing agents, such as sodium borohydride or lithium aluminum hydride, will not give a chiral alcohol because the chances for attack on the two sides of the planar carbonyl group are equal. However, if the reducing agent is chiral, a chiral alcohol might be obtained. In recent years organic chemists have devised a number of such chiral reducing agents, but few are as efficient as those found in nature.

In this part of the chapter, we will study the enzymes found in baker's yeast[1] to reduce ethyl acetoacetate to *S*-(+)-ethyl 3-hydroxybutanoate. This compound is a very useful synthetic building block. Many chiral natural product syntheses are based on this hydroxyester.[2]

<br>

$$\underset{\substack{\textbf{Ethyl acetoacetate}\\ \text{MW 130.14}\\ \text{bp 181°C, den. 1.021}\\ n_{\text{D}}^{20}\ 1.4190}}{\text{CH}_3\overset{\text{O}}{\overset{\|}{\text{C}}}\text{CH}_2\overset{\text{O}}{\overset{\|}{\text{C}}}\text{OCH}_2\text{CH}_3} \xrightarrow{\text{yeast}} \underset{\substack{\textbf{\textit{S}-(+)-Ethyl 3-hydroxybutanoate}\\ \text{MW 132.16}\\ \text{bp 180–182°C (71–73°C/12 mm)}\\ n_{\text{D}}^{20}\ 1.4210}}{}$$

Many different enzymes are present in yeast. The primary ones responsible for converting glucose to ethanol are discussed in Chapter 64. While this fermentation reaction is taking place, certain ketones can be reduced to chiral alcohols.

Whenever a chiral product is produced from an achiral starting material, the chemical yield as well as the optical yield is important—in other words, the stereoselectivity of the reaction is critical. The usual method for recording this is to calculate the enantiomeric excess (ee). A sodium borohydride reduction will produce 50% *R* and 50% *S* alcohol with no enantiomer in excess. If 93% *S*-(+) and 7% *R*-(–) isomer are produced, then the enantiomeric excess is 86%. In the yeast reduction of ethyl acetoacetate, several authors have reported enantiomeric excesses ranging from 70% to 97%. This optical yield is distinct from the chemical yield, which depends on how much material is isolated from the reaction mixture.

The use of enzymes to carry out stereospecific chemical reactions is not new, but it is not always possible to predict if an enzymatic reaction (unlike a purely chemical reaction) will occur or how stereospecific it will be if it does take place. Because the yeast reduction of ethyl acetoacetate is easily carried out, it might be an interesting research project to explore the range of possible ketones that yeast will reduce to chiral alcohols. For example, butyrophenone can be reduced to the corresponding chiral alcohol.[3] The following experiments

ee = enantiomeric excess

% ee calculation:

$$\frac{([S] - [R])}{([S] + [R])} \times 100 = \%ee$$

1. Servi, S. *Synthesis* **1990**, 1–25.

2. Amstutz, R.; Hungerbühler, E.; Seebach, D. *Helv. Chim. Acta* **1981**, *64*, 1796; Seebach, D. *Tetrahedron Lett.* **1982**, *159*.

3. Sih, C. J.; Rosazza, J. P. In *Application of Biochemical Systems in Organic Chemistry*, Part 1; Jones, J. B.; Sih, C. J.; Perlman, D. Wiley: New York, 1976; 71–78.

are based on the work of Seebach,[4] Mori,[5] and Ridley.[6] The work of Bucciarelli et al. is also applicable.[7]

## EXPERIMENTS

## 1. Enzymatic Resolution

> **IN THIS EXPERIMENT** ethyl acetoacetate is added to a fermenting sugar solution. After two days the product formed by the chiral reduction of the methyl ketone of the starting material is extracted with ether. The ether is dried and evaporated to give an optically active hydroxyester.

### *Microscale Procedure*

Keep the mixture warm; the reaction is slow at low temperatures.

A 15-mL centrifuge tube can be used for the extraction, and the emulsions can be broken in the centrifuge.

⚠️

**CAUTION**: Extinguish all flames when working with ether.

*Online Study Center*

Videos: Microscale Filtration on the Hirsch Funnel, Extraction with Ether; Photo: Extraction with Ether

In a 25-mL flask dissolve 2.3 g of sucrose and 15 mg of disodium hydrogen phosphate in 8.5 mL of warm (35°C) tap water (*see* Chapter 64). Add 0.5 g of dry yeast and swirl to suspend the yeast throughout the solution. After 15 min, while fermentation is progressing vigorously, add 150 mg of ethyl acetoacetate. Store the flask in a warm place, ideally at 30°C–35°C, for at least 48 h (a longer time will do no harm). At the end of this time, add 0.5 g of Celite filtration aid and remove the yeast cells by filtration on a Hirsch funnel (*see* Chapter 64). Wash the cells with 1.5 mL of water; then saturate the filtrate with sodium chloride to decrease the solubility of the product. Refer to a handbook for the solubility of sodium chloride in water to determine approximately how much to use. Extract the resulting solution five times with 1.5-mL portions of ether in a test tube, taking care to shake hard enough to mix the layers but not so hard as to form an emulsion between the ether and water. Adding a small amount of methanol may help to break up emulsions. Dry the ether layer by adding anhydrous sodium sulfate or calcium chloride pellets until the drying agent no longer clumps together. After approximately 15 min, filter the ether solution into a tared flask and evaporate the filtrate. The remaining residue should weigh about 100 mg. It should, unlike the starting material, give a negative iron(III) chloride test. It can be analyzed by thin-layer chromatography (TLC), using dichloromethane as the solvent, to determine whether unreacted ethyl acetoacetate is present. Infrared (IR) spectroscopy should show the presence of the hydroxyl group and may show unreduced methyl ketone. The nuclear magnetic resonance (NMR) spectrum of the product is easily distinguished from that of the starting material.

4. Seebach, D. *Tetrahedron Lett.* **1982**, *23*, 159–162.

5. Mori, K. *Tetrahedron* **1981**, *37*, 1341.

6. Ridley, D. D.; Stralow, M. *Chem. Commun.* **1975**, *400*.

7. Bucciarelli, M.; Fomi, A.; Moretti, I.; Torre, G. *Synthesis* **1983**, *897*.

**Cleaning Up.** Dilute the aqueous layer with water and flush it down the drain. After the ether evaporates from the drying agent in the hood, place the drying agent in the nonhazardous solid waste container. If local regulations do not allow for the evaporation of solvents in a hood, the wet solid should be disposed in a special container. Dichloromethane (the TLC solvent) goes in the halogenated organic solvents container. Ether distillate goes in the organic solvents container.

## *Macroscale Procedure*

In a 500-mL flask dissolve 80 g of sucrose and 0.5 g of disodium hydrogen phosphate in 300 mL of warm (35°C) tap water (*see* Chapter 64). Add two packets (16 g) of dry yeast and swirl to suspend the yeast throughout the solution. After 15 min, while fermentation is progressing vigorously, add 5 g of ethyl acetoacetate. Store the flask in a warm place, ideally at 30°C–35°C, for at least 48 h (a longer time will do no harm). At the end of this time, add 20 g of Celite filtration aid and remove the yeast cells by filtration on a 10-cm Büchner funnel (*see* Chapter 64). Wash the cells with 50 mL of water; then saturate the filtrate with sodium chloride to decrease the solubility of the product. Refer to a handbook for the solubility of sodium chloride in water to determine approximately how much to use. Extract the resulting solution five times with 50-mL portions of ether, taking care to shake the separatory funnel hard enough to mix the layers but not so hard as to form an emulsion between the ether and water. Adding a small

Keep the mixture warm; the reaction is slow at low temperatures.

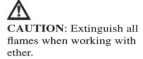

**CAUTION**: Extinguish all flames when working with ether.

■ **FIG. 65.1**
The $^1$H NMR spectrum of ethyl acetoacetate (250 MHz).

**■ FIG. 65.2**
The IR spectrum of ethyl
3-hydroxybutanoate (thin film).

amount of methanol may help to break up emulsions. Dry the ether layer by adding anhydrous sodium sulfate or calcium chloride pellets until the drying agent no longer clumps together. After approximately 15 min, decant the ether solution into a tared distilling flask and remove the ether by distillation or by evaporation. The residue should weigh about 3.5 g. It should, unlike the starting material, give a negative iron(III) chloride test. It can be analyzed by TLC (use dichloromethane as the solvent) to determine whether unreacted ethyl acetoacetate is present. IR spectroscopy should show the presence of the hydroxyl group and may show unreduced methyl ketone. The NMR spectrum of the product is easily distinguished from that of the starting material. NMR and IR spectra of the starting material and *racemic* products are shown in Figures 65.1, 65.2, and 65.3.

**Cleaning Up.** Dilute the aqueous layer with water and flush it down the drain. After the ether evaporates from the drying agent in the hood, place the drying agent in the nonhazardous solid waste container. If local regulations do not allow for the evaporation of solvents in a hood, the wet solid should be disposed in a special container. Dichloromethane (the TLC solvent) goes in the halogenated organic solvents container. Ether distillate goes in the organic solvents container.

## 2. Determination of Optical Purity

The optical purity of the product can be determined by using a polarimeter to measure the optical rotation. The specific rotation, $[\alpha]_D^{25}$, of $S$(+)-ethyl 3-hydroxybutanoate has been reported to vary from +31.3° to +41.7° in chloroform. The specific rotation, $[\alpha]_D^{25}$, of +37.2° (chloroform) corresponds to an enantiomeric excess of 85%.

A better method for determining the optical purity of a small quantity of material is to use an NMR chiral shift reagent.[8] The use of shift reagents is

8. Lipkowitz, K. B.; Mooney, J. L. *J. Chem. Educ.* **1987**, *64*, 985.

The $^1$H NMR spectra of racemic
(±)-ethyl 3-hydroxybutanoate in
0.3 mL of carbon tetrachloride
and 0.2 mL deuterochloroform
(250 MHz). (a) Pure. (b) With 30 mg
of shift reagent. (c) With 50 mg
of shift reagent.

discussed extensively in Chapter 12. A shift reagent will complex with a basic center (the hydroxyl group in ethyl 3-hydroxybutanoate) and cause the NMR peaks to shift, usually downfield. Protons nearest the shift reagent/hydroxyl group shift more than those far away. A chiral shift reagent forms diastereomeric complexes so that peaks from one enantiomer shift downfield more than peaks from the other enantiomer. By comparing the areas of the peaks, the ee can be calculated.

The NMR experiment is easily accomplished by adding 10- to 20-mg increments of tris[3-(heptafluoropropylhydroxymethylene)-(+)-camphorato]-europium(III) shift reagent to a solution of 30 mg of pure dry ethyl 3-hydroxybutanoate in a mixture of 0.3 mL of carbon tetrachloride and 0.2 mL of deuterochloroform.[9]

⚠️ **CAUTION**: Handle carbon tetrachloride in the hood.

The analysis of the NMR spectra is not completely straightforward. For example, in Figure 65.3(a), can you locate the peaks from the proton on the hydroxyl-bearing carbon atom? (Pay attention to the integrals.) Note that the methylene hydrogens on carbon-2 ($\delta = 2.45$ ppm in Fig. 65.3a) are not magnetically (or chemically) equivalent. They are said to be diasteriotopic because diasteriomers could be formed by replacing either one or the other proton with another substituent.

As the optically active shift reagent is added, sets of peaks begin to double as diastereomeric complexes are formed (Fig. 65.3b and Fig. 65.3c). By integrating the pairs of peaks, the ee of one isomer in relation to the other can be determined. In this experiment, only one set of peaks can be seen if the enzymatic reduction gives only one of the two possible enantiomers of the product; thus it is a good idea to compare your product to racemic material, either commercially made or prepared by reducing ethyl acetoacetate by sodium borohydride (follow the procedure in Chapter 26). Also bear in mind that the shift reagent itself will contribute peaks to the spectrum.

The best way to be sure of all the assignments is to run and integrate a series of spectra where the shift reagent is added in 10-mg increments, as is illustrated in Figures 12.6b and 12.6c (on page 263), where 30 mg and 50 mg of reagent have been added.

**Cleaning Up.** Place the contents of the NMR tube in the halogenated solvents waste container.

## 3. Preparation of 3,5-Dinitrobenzoate

3,5-Dinitrobenzoates are common alcohol derivatives; they are easily prepared, the dinitrophenyl group adds considerably to the molecular weight, and they are easily recrystallized.

The 3,5-dinitrobenzoates of a racemic mixture of *R* - and *S*-ethyl 3-hydroxy-butanoate, like most racemates, recrystallize together to give crystals that in this

9. Lipkowitz, K. B.; Mooney, J. L. *J. Chem. Educ.* **1987**, *64*, 985.

**Pyridine**
MW 79.10
bp 115°C
den. 0.978

*S*-(+)-**Ethyl 3-hydroxybutanoate**

**3,5-Dinitrobenzoyl chloride**
MW 230.56
mp 71–74°C

**3,5-Dinitrobenzoate**
mp 154°C

case melt at 146°C. No amount of recrystallization causes one enantiomer to crystallize out while the other remains in solution because they are, after all, mirror images of each other. However, if one enantiomer is in large excess, it is possible to effect a separation by crystallization. In the present case, the *S* enantiomer predominates and it crystallizes out, leaving most of the *R* + *S* racemate in solution. Repeated crystallization increases the melting point and purity to a point at which the optical purity of the crystalline product is almost 100%. At this point the 3,5-dinitrobenzoate has a melting point of 154°C. Treatment of the derivative with excess acidified ethanol regenerates 100% ee *S*-(+)-ethyl 3-hydroxybutanoate.

## Procedure

**IN THIS EXPERIMENT** the hydroxyester from Experiment 1 is reacted with an acid chloride in the presence of base to give 3,5-dinitrobenzoate. The reaction mixture is poured into water, any dinitrobenzoic acid is extracted with bicarbonate, and the crystalline product is recrystallized from ethanol. This material when recrystallized to a constant melting point is enantiomerically pure.

**■ FIG. 65.4**
An apparatus for refluxing reaction mixture.

Syringe needle

Septum

Reaction tube

Refluxing vapors

Hot sand bath

Reactants

To 100 mg of the *S*-(+)-ethyl 3-hydroxybutanoate in a reaction tube (Fig. 65.4), add 175 mg of pure[10] 3,5-dinitrobenzoyl chloride and 1 mL of pyridine. Equip the reaction tube with a septum pierced with a syringe needle. Reflux the mixture for 15 min and then transfer it with a Pasteur pipette to 3.5 mL of water in another reaction tube. Remove the solvent from the crystals and shake the crystals with 2 mL of 0.5 *M* sodium bicarbonate solution to remove dinitrobenzoic acid. Remove the bicarbonate solution and recrystallize the derivative from ethanol. Collect the product on a Wilfilter. Dry a portion and determine the melting point. If it is not near 154°C, repeat the crystallization. Dry the pure derivative and calculate the percent yield.

Treatment of the derivative with excess acidified ethanol will regenerate 100% ee *S*-(+)-ethyl 3-hydroxybutanoate. If several crops of crude product are pooled, distillation at reduced pressure (*see* Chapter 6) can give chemically pure ethyl 3-hydroxybutanoate (bp 71°C–73°C/12 mm). The *R* and *S* enantiomers have the same boiling point, so the optical purity will not be changed by distillation.

**Cleaning Up.** The pyridine solvent goes in the organic solvents container. Dilute the aqueous layer with water and flush it down the drain. Ethanol from the recrystallization goes in the organic solvents container, along with the pot residue if the final product is distilled.

## PART 2: Enzymatic Resolution of DL-Alanine

$$CH_3CHCO^- \quad \longrightarrow \quad CH_3CHCOH \xrightarrow{\text{Acylase}}$$

DL-Alanine (**1**)
mp 295°C
MW 89.10

*N*-Acetyl-DL-alanine (**2**)
mp 137°C
MW 131.13

L-Alanine (**3**)
mp 297°C
$\alpha_D$ +14.4°

*N*-Acetyl-D-alanine (**4**)
mp 125°C
$\alpha_D$ +66.5°

Resolution of DL-alanine (**1**) is accomplished by heating the *N*-acetyl derivative (**2**) in weakly alkaline solution with acylase, a proteinoid preparation from porcine kidney containing an enzyme that promotes rapid hydrolysis of *N*-acyl derivatives of natural L-amino acids but acts immeasurably slowly on the unnatural D-isomers. *N*-Acetyl-DL-alanine (**2**) can thus be converted into a mixture of L-(+)-alanine (**3**) and *N*-acetyl-D-alanine (**4**). The mixture is easily separable into

---

10. Check the melting point of the 3,5-dinitrobenzoyl chloride. If it is below 70°C, it should be recrystallized from dichloromethane.

the components because the free amino acid (**3**) is insoluble in ethanol and the *N*-acetyl derivative (**4**) is readily soluble in this solvent. Note that in contrast to the weakly levorotatory D-(−)-alanine ($\alpha_D = -14.4°$), its acetyl derivative is strongly dextrorotatory ($\alpha_D = +65.5°$).

CH₃CHCOOH ... (structures)

|    |    |    |
| :---: | :---: | :---: |
| (**4**) | (**5**) | (**6**) |

The acetylation of an α-amino acid presents the difficulty that, if the conditions are too drastic, the *N*-acetyl derivative (**4**) is converted in part through the enol (**5**) to the azlactone (**6**).[11] However, under critically controlled conditions of concentration, temperature, and reaction time, *N*-acetyl-DL-alanine can be prepared easily in high yield.

## EXPERIMENTS

## 1. Acetylation of DL-Alanine

| DL-Alanine (**1**) | Acetic anhydride | *N*-Acetyl-DL-alanine (**2**) | |
| :---: | :---: | :---: | :---: |
| MW 89.10 | MW 102.09 | MW 131.13 | |
| mp 295°C | bp 140°C | mp 137°C | |

> **IN THIS EXPERIMENT** a suspension of racemic alanine in acetic acid is reacted with acetic anhydride at 100°C. Water is then added to react with excess anhydride, and the acetic acid and water are removed under vacuum. The last traces of water are removed as an azeotrope with cyclohexane to give the product as a semisolid.

Place 200 mg of DL-alanine and 0.5 mL of acetic acid in a reaction tube, insert a thermometer, and clamp the tube in a vertical position. Measure 0.3 mL of acetic anhydride, which is to be added when the alanine–acetic acid mixture is at

---

11. The azlactone of DL-alanine is known only as a partially purified liquid.

■ **FIG. 65.5**
Drying *N*-acetyl-DL-alanine under reduced pressure.

exactly 100°C. Heat the tube on a hot sand bath with stirring until the temperature of the suspension has risen slightly above 100°C. Stir the suspension and let the temperature gradually fall; when the temperature reaches 100°C, add the 0.3-mL portion of acetic anhydride and note the time. In the course of 1 min the temperature falls (91°C–95°C; cooled by added reagent), rises (100°C–103°C; the acetylation is exothermic), and begins to fall with the solid largely dissolving. Stir to facilitate the reaction of the few remaining particles of solid, let the temperature drop to 80°C, pour the solution into a tared 10-mL Erlenmeyer flask, and rinse the thermometer and test tube with a little acetone. Add 1 mL of water to react with excess anhydride, connect the flask to an aspirator operating at full force (Fig. 65.5), heat gently on a sand bath or put the flask *inside* the rings of a steam bath, and wrap the flask with a towel. Evacuation and heating for about 5–10 min should remove most of the acetic acid and water and leave an oil or thick syrup. Add 2 mL of cyclohexane and evacuate and heat as before for 5–10 min. Traces of water in the syrup are removed as a cyclohexane-water azeotrope. If the product has not yet separated as a white solid or semisolid, determine the weight of the product, add an additional 2 mL of cyclohexane, and repeat the process. When the weight becomes constant, the yield of acetyl DL-alanine should be close to the theoretical amount. The product has a pronounced tendency to remain in supersaturated solution and does not recrystallize readily.

## 2. Enzymatic Resolution of *N*-Acetyl-DL-Alanine

**IN THIS EXPERIMENT** the DL-acetyl alanine from the previous experiment is dissolved in water and neutralized to pH 8 with ammonia. A few milligrams of the enzyme acylase are added; after standing overnight, acetic acid is added, along with Norit pellets. The denatured protein is removed by filtration, and the solution is evaporated to dryness, using cyclohexane to remove the last traces of water. The mixture of the L-amino acid and the D-acetyl derivative is warmed with ethanol to dissolve the D-acetyl alanine, which is discarded; the ethanol-insoluble L-alanine is collected by filtration and then dissolved in warm water. Ethanol is added to the warm aqueous solution until L-alanine begins to recrystallize. After recrystallization ceases, the pure L-alanine is isolated by filtration.

This reaction can be run at 10 times the indicated scale. Add 1.0 mL of distilled water[12] to the Erlenmeyer flask, grasp this with a clamp, swirl the mixture over a hot sand bath to dissolve all of the product, and cool under the tap. Remove a drop of the solution on a stirring rod, add it to 0.5 mL of a 0.3% solution of ninhydrin in water, and heat to boiling. If any unacetylated DL-alanine is present, a purple color will develop and should be noted. Pour the solution into a reaction tube and rinse the flask with a little water. Add 0.15 mL of concentrated ammonia solution,

---

12. Tap water may contain sufficient heavy metal ion to deactivate the enzyme.

**■ FIG. 65.6**
Rotary evaporation.

stir to mix, check the pH with Hydrion paper, and (if necessary) adjust to pH 8 by adding more ammonia dropwise with a pipette. Add 5 mg of commercial acylase, mix with a stirring rod, rinse the rod with distilled water, and make up the volume with distilled water until the tube is about half full. Then stopper the tube, mark it for identification, and let the mixture stand at room temperature overnight or at 37°C[13] for 4 h.

At the end of the incubation period, add 0.3 mL of acetic acid to denature the enzyme; if the solution is not as acidic as pH 5, add more acid. Rinse the cloudy solution into a 10-mL Erlenmeyer flask, add 10 mg of granulated decolorizing carbon, heat and swirl over a hot sand bath for a few minutes to coagulate the protein, and filter the solution. Transfer the solution to a 25-mL round-bottomed flask and evaporate on a rotary evaporator under vacuum (Fig. 65.6); alternatively, add 2 mL of cyclohexane (to prevent frothing) plus a boiling stone and evaporate on a steam bath under vacuum to remove water and acetic acid as completely as possible. Remove the last traces of water and acid by adding 2 mL of cyclohexane and again evaporating to remove water and acetic acid as azeotropes.

The mixture of L-alanine and acetyl-D-alanine separates as a white residue on the walls of the flask. Add 1.5 mL of 95% ethanol, digest on a steam bath, and

Workup time: 30–40 min

---

13. A reasonably constant heating device that will hold 30 or more tubes is made by partially filling a 1-L beaker with water, adjusting the temperature to 37°C, and maintaining this temperature by using a 250-W infrared drying lamp shining horizontally on the beaker from a distance of about 40 cm.

dislodge some of the solid with a spatula. Cool well in ice for a few minutes; then scrape as much of the crude L-alanine as possible onto a suction funnel and wash it with ethanol. Save the ethanol mother liquor.[14] To recover the L-alanine retained by the flask, add 0.2 mL of water and warm on a steam bath until all the solid is dissolved; then transfer the solution to a reaction tube with a Pasteur pipette, rinse the flask with an additional 0.2 mL of water, and transfer in the same way. Add the filtered L-alanine, dissolve by warming, and filter the solution into a reaction tube using a pressure apparatus. Rinse the flask and funnel with 0.1 mL of water and then with 0.5 mL of warm 95% ethanol. Then heat the filtrate on a steam bath and add additional 95% ethanol (in 1–1.5 mL portions) until crystals of L-alanine begin to separate from the hot solution. Let the recrystallization proceed. Collect the crystals and wash with ethanol. The yield of colorless needles of L-alanine ($\alpha_D$ = +13.7° to +14.4°[15] in 1 $N$ hydrochloric acid) varies from 40 mg to 50 mg, depending on the activity of the enzyme.

> Pool several products to make the optical rotation measurement.

**Cleaning Up.** The decolorizing carbon goes in the nonhazardous solid waste container. Dilute the ethanol used in the recrystallization with water and flush it down the drain.

# 3. Acetylation of DL-Alanine

$$\underset{\substack{\text{DL-Alanine (1)} \\ \text{MW 89.10} \\ \text{mp 295°C}}}{\underset{\overset{|}{+\text{NH}_3}}{\overset{\overset{\text{O}}{\|}}{\text{CH}_3\text{CHCO}^-}}} \; + \; \underset{\substack{\text{Acetic anhydride} \\ \text{MW 102.09} \\ \text{bp 140°C}}}{\overset{\overset{\text{O} \quad \text{O}}{\| \quad \|}}{\text{CH}_3\text{COCCH}_3}} \; \xrightarrow{\text{CH}_3\text{COOH}} \; \underset{\substack{\textit{N}\text{-Acetyl-DL-alanine (2)} \\ \text{MW 131.13} \\ \text{mp 137°C}}}{\underset{\overset{|}{\text{NHCOCH}_3}}{\overset{\overset{\text{O}}{\|}}{\text{CH}_3\text{CHCOH}}}} \; + \; \text{CH}_3\text{COOH}$$

Place 2 g of DL-alanine and 5 mL of acetic acid in a 25 × 150-mm test tube, insert a thermometer, and clamp the tube in a vertical position. Measure 3 mL of acetic anhydride, which is to be added when the alanine–acetic acid mixture is at exactly 100°C. Heat the test tube cautiously, with stirring, until the temperature of the suspension has risen slightly above 100°C. Stir the suspension and let the temperature fall gradually; when the temperature reaches 100°C, add the 3-mL portion of acetic anhydride and note the time. In the course of 1 min the temperature falls (91°C–95°C; cooled by added reagent), rises (100°C–103°C; the acetylation is

---

14. In case the yield of L-alanine is low, evaporation of the mother liquor may reveal the reason. If the residue solidifies readily and crystallizes from acetone to give acetyl-DL-alanine (mp 130°C or higher), the acylase preparation is recognized as inadequate in activity or amount. Acetyl-D-alanine is much more soluble and slow to crystallize.

15. Determination of the optical activity can be made in the student laboratory with a Zeiss pocket polarimeter, which requires no monochromatic light source and no light shield. For construction of a very inexpensive polarimeter, see Shaw, W. H. R. *J. Chem. Educ.* **1955**, *32*, 10.

exothermic), and begins to fall with the solid largely dissolving. Stir to facilitate the reaction of a few remaining particles of solid, let the temperature fall to 80°C, pour the solution into a tared 100-mL round-bottomed flask, and rinse the thermometer and test tube with a little acetone. Add 10-mL of water to react with excess anhydride, connect the flask to an aspirator operating at full force (Fig. 65.7), place the flask *inside* the rings of a steam bath, and wrap the flask with a towel. Evacuation and heating for about 5–10 min should remove most of the acetic acid and water and leave an oil or thick syrup. Add 10 mL of cyclohexane and evacuate and heat as before for 5–10 min. Traces of water in the syrup are removed as a cyclohexane-water azeotrope. If the product has not yet separated as a white solid or semisolid, determine the weight of the product, add an additional 10 mL of cyclohexane, and repeat the process. When the weight becomes constant, the yield of acetyl DL-alanine should be close to the theoretical amount. The product has a pronounced tendency to remain in supersaturated solution and does not crystallize readily.

## 4. Enzymatic Resolution of *N*-Acetyl-DL-Alanine

Add 10 mL of distilled water[16] to the reaction flask, grasp this with a clamp, swirl the mixture over a free flame or on a hot sand bath to dissolve all of the product, and cool under the tap. Remove a drop of the solution on a stirring rod, add it to 0.5 mL of a 0.3% solution of ninhydrin in water, and heat to boiling. If any unacetylated DL-alanine is present, a purple color will develop and should be noted. Pour the solution into a 20 × 150-mm test tube and rinse the flask with a little water. Add 1.5 mL of concentrated ammonia solution, stir to mix, check the pH with Hydrion paper, and (if necessary) adjust to pH 8 by adding more ammonia with a Pasteur pipette. Add 10 mg of commercial acylase powder, or 2 mL of fresh acylase solution, mix with a stirring rod, rinse the rod with distilled water, and make up the volume with distilled water until the tube is about half full. Then stopper the tube, mark it for identification (Fig. 65.8), and let the mixture stand at room temperature overnight or at 37°C[17] for 4 h.

Workup time: 30–40 min

At the end of the incubation period, add 3 mL of acetic acid to denature the enzyme; if the solution is not as acidic as pH 5, add more acid. Rinse the cloudy solution into a 125-mL Erlenmeyer flask, add 100 mg of decolorizing carbon (0.5-cm column in a 13 × 100-mm test tube), heat and swirl over a free flame or hot plate for a few moments to coagulate the protein, and filter the solution by suction. Transfer the solution to a 100-mL round-bottomed flask and evaporate on

16. Tap water may contain sufficient heavy metal ion to deactivate the enzyme.

17. A reasonably constant heating device that will hold 15 tubes is made by filling a 1-L beaker with water, adjusting to 37°C, and maintaining this temperature by the heat of a 250-W infrared drying lamp shining horizontally on the beaker from a distance of about 40 cm. The capacity can be tripled by placing other beakers on each side of the first one and a few centimeters closer to the lamp.

**■ FIG. 65.8**
An example of a filter paper
identification marker.

*Online Study Center*

**General Resources**
Additional Experiments

a rotary evaporator under vacuum (*see* Fig. 65.6 on page 796); alternatively, add 20 mL of cyclohexane (to prevent frothing) plus a boiling stone and evaporate on a steam bath under vacuum to remove water and acetic acid as completely as possible. Remove the last traces of water and acid by adding 15 mL of cyclohexane and again evaporating to remove water and acetic acid as azeotropes.

The mixture of L-alanine and acetyl D-alanine separates as a white residue on the walls of the flask. Add 15 mL of 95% ethanol, digest on a steam or sand bath and dislodge some of the solid with a spatula. Cool well in ice for a few minutes; then scrape as much of the crude L-alanine as possible into a suction funnel and wash with ethanol. Save the ethanol mother liquor.[18] To recover the L-alanine retained by the flask, add 2 mL of water and warm on a steam or sand bath until all of the solid is dissolved; then transfer the solution to a 25-mL Erlenmeyer flask with a Pasteur pipette, rinse the flask with an additional 2 mL of water, and transfer in the same way. Add the filtered L-alanine, dissolve by warming, and filter the solution by gravity into a 50-mL Erlenmeyer flask (use a Pasteur pipette to effect the transfer of solution to filter). Rinse the flask and funnel with 1 mL of water and then with 5 mL of warm 95% ethanol. Then heat the filtrate on a steam bath or hot plate and add additional 95% ethanol (in 10–15 mL portions) until crystals of L-alanine begin to separate from the hot solution. Let the recrystallization proceed. Collect the crystals and wash with ethanol. The yield of colorless needles of L-alanine ($\alpha_D = +13.7°$ to $+14.4°$[19] in 1 *N* hydrochloric acid) varies from 0.40 g to 0.56 g, depending on the activity of the enzyme.

**Cleaning Up.** The decolorizing carbon goes in the nonhazardous solid waste container. Dilute the ethanol used in the recrystallization with water and flush it down the drain.

## QUESTIONS

1. Explain the differences in the [1]H NMR spectra shown in Figure 65.3 (on page 790). Are there signals resulting from the shift reagent? If so, identify them.
2. Look up the prices of racemic ethyl 3-hydroxybutanoate and each of the enantiomers in a chemical catalog (e.g., the Aldrich Chemical Company catalog). How do you explain the price differences?
3. Assign the peaks at 3450 cm$^{-1}$ and 1740 cm$^{-1}$ in the IR spectrum of ethyl 3-hydroxybutanoate (Fig. 65.2 on page 789).

---

18. In case the yield of L-alanine is low, evaporation of the mother liquor may reveal the reason. If the residue solidifies readily and crystallizes from acetone to give acetyl-DL-alanine (mp 130°C or higher), the acylase preparation is recognized as inadequate in activity or amount. Acetyl-D-alanine is much more soluble and slow to crystallize.

19. *See* footnote 15.

■ **FIG. 65.9**
The $^{13}C$ NMR spectrum of
L-alanine.

4. The melting point of *N*-acetyl-DL-alanine is 137°C and that of *N*-acetyl-
   D-alanine is 125°C. What would you expect the melting point of *N*-acetyl-
   L-alanine to be? Is the melting point impossible to predict? Explain.
5. Would the $^{13}C$ NMR spectrum of D-alanine differ from that of L-alanine
   (Fig. 65.9)? Explain why or why not.

# CHAPTER

# 66

# The Synthesis of Natural Products: Pseudopellitierene and Camphor

*Online Study Center*

This icon will direct you to techniques, equipment setups, and online resources at http://college.hmco.com/PIC/williamsonMMOE5e.

**PRELAB EXERCISE:** Research Willstätter's synthesis of tropinone and develop a flow sheet for the reaction.

## PART 1: Synthesis of Pseudopellitierene

A principal focus of organic chemistry and certainly of its more specialized branch, biochemistry, is an understanding of the life process. There has long been a quest to understand and then to reproduce in the laboratory the reactions that take place in living organisms without the intervention of catalysts (enzymes) isolated from living organisms.

In the experiment in this part of the chapter, we attempt to reproduce one of the classic syntheses of a complex natural product by a biogenetic-like synthesis: the synthesis of pseudopellitierene. This alkaloid is a naturally occurring substance isolated from the bark of the pomegranate tree. Several isomeric compounds were isolated in 1878; none of the structures were known at the time, hence the names *pellitierene*, *isopellitierene*, and *pseudopellitierene*.

Tropinone is the simplest representative of a very similar class of alkaloids that includes atropine (used by ophthalmologists to dilate the pupil of the eye), scopolamine (used to combat seasickness), and cocaine (a component of Coca-Cola at the turn of the 20th century).

Tropinone was synthesized by Willstätter in 1903 in a classic 15-step synthesis. It also was the object of the first biogenetic-like synthesis of a naturally occurring molecule. Robert Robinson (later Sir Robert), while at the University of Liverpool in 1917, synthesized tropinone by the Mannich-type condensation of succindialdehyde, methylamine, and acetone dicarboxylic acid. For his many contributions to synthetic organic chemistry, Robinson was awarded the Nobel Prize in Chemistry in 1974. Willstätter received the Nobel Prize in Chemistry in 1920. We will reproduce Robinson's synthesis in this experiment.

**Tropinone**  **Cocaine**  **Atropine**  **Scopolamine**

## EXPERIMENT

## The Biomimetic Synthesis of Pseudopellitierene

In this experiment, we will synthesize pseudopellitierene using essentially the same procedure employed by Robinson: simply stirring a mixture of the three re-actants at room temperature. Pseudopellitierene, with a higher melting point, is somewhat easier to isolate than tropinone is.

| **Glutaraldehyde** MW 100.12 | **Methylamine hydrochloride** MW 67.52 mp 232–234 °C | **1,3-Acetone-dicarboxylic acid** MW 146.10 mp 133 °C (dec) | **Pseudopellitierene** |

The mechanism for the glutaraldehyde precursor hydrolysis and for the biogenetic-like synthesis of pseudopellitierene is given here. As you can see, this is a complex series of reactions:

**3,4-Dihydro-2-ethoxy-2H-pyran**
MW 128.17

**Glutaraldehyde**

(continued on next page)

(continued from previous page)

## Procedure

**IN THIS EXPERIMENT** an acidic solution of a pyran is converted over a 1-h period to a dialdehyde. After methylamine and acetone dicarboxylic acid are added in a buffered basic solution, a complex series of reactions takes place. A day later, hydrochloric acid is added, and decarboxylation occurs; after neutralization, the product is extracted eight times with dichloromethane. The solvent is dried and evaporated, and the product is sublimed to give pseudopelliteriene.

To a 250-mL round-bottomed flask equipped with a magnetic stirring bar, add 6.6 mL of deoxygenated water, 0.88 mL of concentrated hydrochloric acid, and 2.56 g of 3,4-dihydro-2-ethoxy-2H-pyran. The mixture is magnetically stirred rapidly for 20 min and then allowed to stand for 1 h. To the yellow solution, add 14.0 mL of water, 2.0 g of methylamine hydrochloride dissolved in 20 mL of water, 3.32 g of 1,3-acetonedicarboxylic acid (freshly recrystallized from ethyl acetate) dissolved in 32 mL of water and 1.42 g of disodium hydrogen phosphate, and 0.44 g of potassium hydroxide dissolved in 8 mL of water by heating. Measure the pH. The solution is stirred slowly under a nitrogen atmosphere for at least 24 h (again measure the pH); then 1.32 mL of concentrated hydrochloric acid is added, and the mixture is heated on a steam bath for 1 h. After cooling to room temperature, 4.4 g of potassium hydroxide dissolved in 4 mL of water is added. The mixture is extracted eight times with 12-mL portions of dichloromethane in the hood, and the extract is dried with anhydrous sodium sulfate (not calcium chloride). The combined dichloromethane extract is carefully evaporated, and the residue is sublimed under aspirator (or oil pump) vacuum to give pseudopellitierene. Resublimation may be necessary to give purer material. The melting point reported for pseudopellitierene is 54°C.

**Cleaning Up.** Neutralize the aqueous layer with dilute hydrochloric acid; then flush the resulting solution down the sink with lots of water.

# PART 2: The Conversion of Camphene to Camphor

In the following experiments, 2,2-dimethyl-3-methylenebicyclo[2.2.1]heptane, otherwise known as *camphene* (**1**), will be converted to 1,7,7-trimethylbicyclo-[2.2.1]heptan-2-one, which is camphor (**8**), through the intermediate isomeric

**1**
**Camphene**
MW 136.24
mp 48–52°C, bp 159–160°C

**2**

**5**
**Bornyl acetate**
MW 196.29
bp 223–224°C

**4**
**Isobornyl acetate**
MW 196.29
bp 220–224°C

**3**

KOH/ethanol

**6**
**Borneol**
MW 154.25

**7**
**Isoborneol**
MW 154.25

Jones' reagent

**8**
**Camphor**
MW 152.24
mp 175°C, bp 204°C

alcohols borneol (**6**) and isoborneol (**7**) and *endo-* and *exo-*1,7,7-trimethylbicyclo-[2.2.1]heptan-2-ol.

Camphene (**1**), as its odor will indicate, can be isolated from turpentine and a number of other naturally occurring oils. Upon reaction with acetic acid and a catalytic amount of sulfuric acid, it forms a carbocation (**2**) that undergoes rearrangement (Wagner–Meerwein rearrangement) to the secondary carbocation (**3**), with subsequent solvolysis (reaction with the solvent) to form predominantly the *exo*-acetate (**4**) with some of the *endo*-acetate (**5**). Even though **2** is a tertiary carbocation and more stable than **3**, the acetate (**4**) is formed because the acetate ion can more easily attack the unhindered cation (**3**) than the hindered cation (**2**). In Experiment 3, computational chemistry is used to predict and/or confirm that the *exo*-acetate (**4**) will be formed in preference to the *endo*-isomer (**5**).

Isobornyl acetate (**4**), with the acetate group in the *exo* position, is accompanied by a small amount of the isomeric *endo* compound, bornyl acetate (**5**). These isomers are liquids and not easily purified. They can be hydrolyzed to the corresponding alcohols, borneol (**6**) and isoborneol (**7**), which are solids.

This mixture of isomers can be oxidized to camphor (**8**). All of these compounds have a roughly spherical shape. Intermolecular interactions are relatively small; therefore, the compounds sublime easily. When you look up the physical properties of isoborneol, note the small difference between its melting point and its boiling point.

In carrying out this series of reactions, bear in mind that one of the goals is to prepare enough of the final product to purify and to characterize by thin-layer chromatography (TLC), melting point, and infrared (IR) and nuclear magnetic resonance (NMR) spectroscopy. For characterization you will need about 20 mg. With skill, you should end up with a lot more than that.

## EXPERIMENTS

> **IN THIS EXPERIMENT** camphene is dissolved in acetic acid, which is both the solvent and a reactant. A catalytic amount of sulfuric acid is added, and the mixture is heated briefly and then diluted with ether. Several extractions with water remove the acetic acid and sulfuric acid. The ether is dried and evaporated under vacuum to give primarily isobornyl acetate, which also contains some bornyl acetate.

## 1. Laboratory Conversion of Camphene to Camphor

### *Microscale Procedure: Camphene (1) to Isobornyl Acetate (4)*

In a 5-mL round-bottomed flask, place a stirring bar, 3 mL of acetic acid, and 1.36 g of camphene (0.01 mol); to the resulting solution, add a solution of 0.25 g of concentrated sulfuric acid in 0.3 mL of water. Because acetic acid is a solvent and sulfuric acid is a catalyst, the amounts of these reagents need not be measured with great accuracy. Heat the mixture in a boiling water bath for 15 min with stirring. Alternatively, the mixture can simply be heated over a steam bath with frequent swirling of the contents. Do not reflux the mixture directly on a sand bath because the sand bath would be too hot and cause undesired side reactions. The reaction mixture will turn dark in color.

After the heating period, cool the reaction mixture and place it in a 15-mL centrifuge tube, complete the transfer with about 2 mL of *t*-butyl methyl ether, and extract the mixture with two 4-mL portions of water. This extraction will remove most of the acetic acid. The separations are achieved by drawing off the aqueous layer using a 9-in. Pasteur pipette. Wash the ether layer with about 2 mL of 10% sodium carbonate or bicarbonate solution and dry the solution over calcium chloride pellets for about 5 min. Transfer the solution to the tared 25-mL filter flask and remove the ether under reduced pressure (Fig. 66.1). This will take just a minute or so because of the ether's low boiling point. Save about 50 mg of the product for analysis. It will be a brown liquid mixture of isobornyl acetate with some bornyl acetate. In this product isolation, the exact amounts of ether, water, and carbonate solutions are not important, so there is no need to measure them precisely.

**Cleaning Up.** Neutralize the aqueous washes with bicarbonate and then flush the resulting solution down the drain with lots of water; alternatively, combine these washes with the filtrate from the next experiment, neutralize, and flush down the drain. Place the calcium chloride in the nonhazardous waste container.

## Microscale Procedure: Isobornyl Acetate (4) [and Bornyl Acetate (5)] to Isoborneol (7) [and Borneol (6)]

> **IN THIS EXPERIMENT** the acetates made in the previous experiment are hydrolyzed by heating in a solution of potassium hydroxide dissolved in a mixture of water and ethanol. The reaction mixture is poured onto ice, and the crystallized alcohols are collected by filtration.

In a 5-mL round-bottomed flask place a boiling chip and a solution of 0.45 g of potassium hydroxide pellets (the pieces are 85% potassium hydroxide and 15% water) dissolved in 0.7 mL water to which is added 2.3 mL of ethanol. To this basic solution add 1.0 g of isobornyl/bornyl acetate, apply an air condenser, and reflux the mixture on a sand bath for 1 h. With careful adjustment of the amount of heat, it is easy to reflux this mixture using an air condenser (*see* Fig. 25.1 on page 406). Just make sure that the top of the condenser is not hot, which would indicate that ethanol may be escaping. Wrap a wet pipe cleaner around the air condenser if more cooling is needed (*see* Fig. 40.3 on page 561). At the end of the

*Online Study Center*

Video: Extraction with Ether;
Photo: Extraction with Ether

■ **FIG. 66.1**
Evaporation of a solvent under reduced pressure. Heat and swirling motion is supplied by one hand; the vacuum is controlled by thumb of other.

To aspirator

**Online Study Center**
Video: Microscale Filtration on the Hirsch Funnel;
Photo: Sublimation Apparatus

hour, pour the solution onto about 5 g of ice in a 10-mL Erlenmeyer flask. Swirl the mixture until the product is completely solidified; then collect it on a Hirsch funnel. Wash it well with water; then carefully press the product onto the funnel using a spatula. It will have the consistency of light-brown sugar. Try to squeeze out as much of the water as possible. Save about 60 mg of this sample for analysis. It sublimes easily and will give pure white crystals. Determine the melting point of the sublimed product in a sealed melting-point capillary (*see* Fig. 3.6 on page 49 for this technique). Explain the melting point in terms of the expected composition of the product.

**Cleaning Up.** Neutralize the filtrate with dilute hydrochloric acid or combine it with the washings from the previous experiment, neutralize, and flush down the drain with excess water.

## Microscale Procedure: Borneols [(5) and (6)] to Camphor (7)

> **IN THIS EXPERIMENT** the mixture of alcohols from the previous experiment is dissolved in acetone and oxidized with Cr(VI) reagent. On dilution with water, the product (camphor) separates as a solid and is isolated by filtration. It is then purified by sublimation.

⚠ **CAUTION**: Jones' reagent is very corrosive because it is a powerful oxidizing agent prepared in concentrated sulfuric acid. Wash up spills immediately. If spilled on the skin, flush with water until the skin is neutral. $Cr^{6+}$ as a dust is a carcinogen if inhaled. It is unlikely that you would inhale this reagent; however, exercise care when preparing the reagent from solid chromium trioxide.

In a 15-mL centrifuge tube dissolve 0.6 g of the borneols in 1.2 mL of acetone, add a magnetic stirring bar, and then add to the solution 1 mL of Jones' reagent[1] dropwise with thorough stirring and cooling. After all of the reagent is added, the mixture should be homogeneous in appearance. If not, stir it some more. After about 30 min, add water dropwise to the reaction mixture, diluting it to about 12 mL. Collect the solid by filtration on a Hirsch funnel, squeeze it as dry as possible, and then sublime the camphor at atmospheric pressure (Fig. 66.2). Determine the melting point, analyze the product by TLC and IR and NMR spectroscopy, and then submit the product to your instructor in a labeled vial (not in a plastic bag).

**Online Study Center**
Photo: Sublimation Apparatus

**Cleaning Up.** Complete the reduction of any remaining chromic ion in the filtrate by adding solid sodium thiosulfate until the solution becomes cloudy and blue. Neutralize with sodium carbonate and filter the flocculent precipitate of $Cr(OH)_3$ on a Hirsch funnel. Dilute the filtrate with water and flush down the drain. Place the precipitate on the filter paper in the heavy metals hazardous waste container.

---

1. Jones' reagent is prepared by dissolving 13.5 g of chromium trioxide in 11.5 mL of concentrated sulfuric acid and then adding, with great care, enough water to bring the volume up to 50 mL.

■ **FIG. 66.2**
A sublimation apparatus.

15-mL Centrifuge tube, to be filled with ice

Adapter (Pluro stopper)

Pipette bulb

25-mL Filter flask

Material to be sublimed

Sublimate

Heat source

## 2. Laboratory Conversion of Camphene to Camphor

### *Macroscale Procedure: Camphene (1) to Isobornyl Acetate (4)*

In a 25-mL round-bottomed flask place a stirring bar, 12 mL of acetic acid, and 5.44 g of camphene (0.01 mol); to the resulting solution add a solution of 1.0 g of concentrated sulfuric acid in 1.2 mL of water. Because acetic acid is a solvent and sulfuric acid is a catalyst, the amounts of these reagents need not be measured with great precision. Heat the mixture in a boiling water bath for 15 min with stirring. Alternatively, the mixture can simply be heated over a steam bath with frequent swirling of the contents. Do not reflux the mixture directly on a sand bath because the sand bath would be too hot and cause undesired side reactions. The reaction mixture will turn dark in color.

After the heating period, cool the reaction mixture and place it in a small separatory funnel, complete the transfer with about 8 mL of *t*-butyl methyl ether, and extract the mixture with two 15-mL portions of water. This extraction will remove most of the acetic acid. Wash the ether layer with about 8 mL of 10% sodium carbonate or bicarbonate solution and dry the solution over calcium chloride pellets for about 5 min. Transfer the solution to a tared 100-mL round-bottomed flask and remove the ether with a rotary evaporator. This will take just a minute or so because of ether's low boiling point. Save about 200 mg of the product for analysis. It will be a brown liquid mixture of isobornyl acetate with some bornyl acetate. In this product isolation, the exact amounts of ether, water, and carbonate solutions are not important, so there is no need to measure them precisely.

**Cleaning Up.** Neutralize the aqueous washes with bicarbonate and then flush the resulting solution down the sink with lots of water; alternatively, combine these washes with the filtrate from the next experiment, neutralize, and flush down the sink. Place the calcium chloride in the nonhazardous waste container.

## Macroscale Procedure: Isobornyl Acetate (4) [and Bornyl Acetate (5)] to Isoborneol (7) [and Borneol (6)]

In a 25-mL round-bottomed flask place a boiling chip and a solution of 1.8 g of potassium hydroxide pellets (the pieces are 85% potassium hydroxide and 15% water) dissolved in 2.8 mL water to which is added 9.2 mL of ethanol. To this basic solution add 4.0 g of isobornyl–bornyl acetate, apply a condenser, and reflux the mixture on a sand bath for 1 h. At the end of the hour, pour the solution onto about 20 g of ice in a 50-mL Erlenmeyer flask. Swirl the mixture until the product is completely solidified; then collect it on a Hirsch funnel. Wash it well with water, then carefully press the product onto the funnel using a spatula. It will have the consistency of light-brown sugar. Try to squeeze out as much water as possible. Save about 250 mg of this sample for analysis. It sublimes easily and will give pure white crystals. Determine the melting point of the sublimed product in a sealed melting-point capillary. Explain the melting point in terms of the expected composition of the product.

**Cleaning Up.** Neutralize the filtrate with dilute hydrochloric acid, or combine it with the washings from the previous experiment, neutralize, and flush down the drain with excess water.

## Macroscale Procedure: Borneols [(5) and (6)] to Camphor (8)

⚠️ **CAUTION**: Jones' reagent is very corrosive because it is a powerful oxidizing agent prepared in concentrated sulfuric acid. Wash up spills immediately. If spilled on the skin, flush with water until the skin is neutral. $Cr^{6+}$ as a dust is a carcinogen if inhaled. It is unlikely that you would inhale this reagent; however, exercise care when preparing the reagent from solid chromium trioxide.

In a 25-mL Erlenmeyer flask, dissolve 2.4 g of the borneols in 4.8 mL of acetone, add a magnetic stirring bar, and then add to the solution 4.0 mL of Jones' reagent[2] dropwise with thorough stirring and cooling. After all of the reagent is added, the mixture should be homogeneous in appearance. If not, stir it some more. After about 30 min, add water dropwise to the reaction mixture, diluting it to about 50 mL. Collect the solid by filtration on a Hirsch funnel, squeeze it as dry as possible, and then sublime the camphor at atmospheric pressure (*see* Fig. 66.2 on page 809). Determine the melting point, analyze the product by TLC and IR and NMR spectroscopy (Fig. 66.3), and then submit the product to your instructor in a labeled vial (not in a plastic bag).

**Cleaning Up.** Complete the reduction of any remaining chromic ion in the filtrate by adding solid sodium thiosulfate until the solution becomes cloudy and blue. Neutralize with sodium carbonate and then filter the flocculent precipitate of

---

2. Jones' reagent is prepared by dissolving 13.5 g of chromium trioxide in 11.5 mL of concentrated sulfuric acid and then adding, with great care, enough water to bring the volume up to 50 mL.

$Cr(OH)_3$ on a Hirsch funnel. Dilute the filtrate with water and flush down the drain. Place the precipitate on the filter paper in the heavy metals hazardous waste container.

## 3. Computational Chemistry: The Theoretical Conversion of Camphene to Camphor

Molecular mechanics calculations have been dealt with previously (*see* Chapter 15), but these calculations represent just a part of a much larger topic, computational chemistry—an important new part of organic chemistry. Just a few years ago this area of chemistry was relegated to the specialist, the "theoretical chemist," who laboriously encoded the Cartesian coordinates of all the atoms in a molecule and then submitted the calculation to a large mainframe computer for solution. The data returned in enormous piles of digital printout, which was interpretable only by the expert. In the past few years, however, workstations and even graphically oriented desktop computers can handle these computations, while the input and output can be manipulated and displayed on a computer monitor as a drawing of the molecule. For example, the distribution of charge density can be color encoded and superimposed on the space-filling model of the molecule. Thus computational chemistry has passed from the hands of the specialist to the practicing organic chemist.

In the example presented here, we try to predict by computation which isomer—isobornyl acetate (**4**) or bornyl acetate (**5**)—is favored when camphene is treated with acetic acid and a catalytic amount of sulfuric acid.

As seen in the reaction scheme, camphene (**1**), on protonation, will give a carbocation (**3**), which is attacked by the nucleophile, acetic acid, to give either **4** or **5**, the isomeric acetates. The carbocation, an $sp^2$-hybridized carbon, is planar and has an empty $p$ orbital. We know that the acetic acid must attack this from "above" or "below," along the axis of the $p$ orbital. The reactants must come within bonding distance for the product to form, but preventing this is steric hindrance from other parts of the molecule. It is not obvious from scrutiny of the drawing whether there is more steric hindrance from the "top" or the "bottom" of the ion in **3**. Even when molecular models are held in the hand, it looks as if the bridge methyl group is going to get in the way of the acetic acid, so we would predict that attack by acetic acid would take place from the bottom side of the cation.

To approach this problem from a calculational standpoint, we first assemble the framework of the cation on a computer and then minimize its energy. We then proceed to carry out a molecular orbital calculation. In this case we use a semi-empirical method because it is fast and will provide the type of information we need. Among the well-known methods are modified neglect of diatomic overlap (MNDO), Austin Method 1 (AM1),[3] and MNDO parameterization method 3 (MNDO/PM3 or simply PM3).

Using a computer program such as Spartan,[4] the lowest unoccupied molecular orbital (LUMO) of the cation (**3**) can be calculated and superimposed on the space-filling model of the cation. Through color coding, you can see on the computer monitor that there is more of the empty molecular orbital exposed on the top (*exo*) side of the cation than on the bottom (*endo*) side. If you have the facilities, try to reproduce this calculation. This same procedure is used to predict the direction of attack of the borohydride anion on 2-methylcyclohexanone (*see* Chapter 26).

*Online Study Center*

**General Resources**
Additional Experiments

## QUESTIONS

1. The molal freezing point depression constant for camphor is 39.7°C. What does this mean, and what effect might it have on your observed melting point for camphor?

2. Determine, using a molecular mechanics program, whether the most stable conformation of pseudopellitierene is shown at the bottom of page 806. Pay particular attention to the locations of the carbonyl group and the methyl group.

---

3. Its originator, Michael J. S. Dewar, was a professor at the University of Texas at Austin.

4. Spartan molecular modeling software is available from Wavefunction, Inc., Irvine, California.

## CHAPTER

# 67

# Polymers: Synthesis and Recycling

*Online Study Center*

This icon will direct you to techniques, equipment setups, and online resources at http://college.hmco.com/PIC/williamsonMMOE5e.

**PRELAB EXERCISE:** In the preparation of nylon by interfacial polymerization, sebacoyl chloride is synthesized from decanedioic acid and thionyl chloride. What volume of hydrogen chloride is produced in this reaction? What volume of sulfur dioxide is produced?

## PART 1: Synthesis of Polymers

We use more synthetic polymers than steel, aluminum, and copper combined.

**P**olymers are ubiquitous. Natural polymers such as proteins (polyamino acids), DNA (polynucleotides), and cellulose (polyglucose) are the basic building blocks of plant and animal life. Synthetic organic polymers, or plastics, are now among our most common structural materials. In the United States we make and use more synthetic polymers than we do steel, aluminum, and copper combined—at present worth $310 billion in annual shipments in an industry employing 1.4 million people.

*Polymers*, from the Greek meaning "many parts," are high molecular weight molecules composed of repeating units of smaller molecules. Most polymers consist of long chains held together by hydrogen bonds, van der Waals forces, and the tangling of the long chains. When heated, the covalent bonds of some polymers, which are thermoplastic, do not break; the chains slide over one another to adopt new shapes. These shapes can be films, sheets, extrusions, or molded parts in a myriad of forms.

Nitrocellulose

The first synthetic plastic was nitrocellulose, made in 1862 by nitrating cellulose, a natural polymer. Nitrocellulose, when mixed with a plasticizer such as camphor to make it more workable, was originally used as a replacement for ivory in billiard balls and piano keys and to make celluloid collars. This material, from which the first movie film was made, is notoriously flammable.

Cellulose acetate

Cellulose acetate, made by treating cellulose with acetic acid and acetic anhydride, was originally used as a waterproof varnish to coat the fabric of airplanes during World War I. It later became important as a photographic film base and as acetate rayon.

The first completely synthetic organic polymer was Bakelite, named for its discoverer Leo Baekeland, a Belgian chemistry professor who invented Velox, the first successful photographic paper. He immigrated to America at age 35 and sold his invention to George Eastman for $1 million in 1899. Baekeland then turned his attention to finding a replacement for shellac, which comes from the Asian lac beetle. At the time, shellac was coming into great demand in the fledgling electrical industry as an insulator. The polymer Baekeland produced (Bakelite) is still used for electrical plugs, switches, and the black handles and knobs on pots and pans. It has superior electrical insulating properties and very high heat resistance. It is made by the base-catalyzed reaction of excess formaldehyde with phenol. In a low molecular weight form it is used to glue together the layers of plywood or mixed with a filler such as sawdust. When it is heated to a high temperature, cross-linking occurs as the polymer "cures."

Most polymers are amorphous, linear macromolecules that are thermoplastic and soften at high temperature. In Bakelite the polymer cross-links to form a three-dimensional network, and the polymer becomes a dark, insoluble, infusible substance. Such polymers are said to be *thermosetting*. Natural rubber is thermoplastic. It becomes a thermosetting polymer when heated with sulfur, as Charles Goodyear discovered. With 2% sulfur the rubber becomes cross-linked but is still elastic; at 30% sulfur the rubber can be made into bowling balls. Some other important thermosetting polymers are urea-formaldehyde resins and melamine-formaldehyde resins. The latter are among the hardest of polymers and take on a high-gloss finish. Melamine is used extensively to manufacture plastic dinnerware.

Even though vinyl chloride was discovered in 1835, poly(vinyl chloride) (PVC) was not produced until 1912. It is now one of our most common polymers; production in 1997 was over 20 million metric tons. The monomer is made by the pyrolysis of 1,2-dichloroethane, which is formed by chlorination of ethylene. Free radical polymerization follows Markovnikov's rule to give the head-to-tail polymer with high specificity:

$$n\text{CH}_2{=}\text{CHCl} \longrightarrow -(\text{CH}_2\text{CHCl})_n-$$

**Vinyl chloride**          **Polyvinyl chloride**

Pure PVC is an extremely hard polymer. PVC, along with some other polymers, can be modified by the addition of plasticizers; the greater the ratio of plasticizer to polymer, the greater the flexibility of the polymer. PVC pipe is rigid and contains little plasticizer, whereas shower curtains contain a large percentage of plasticizers. The most common plasticizer is di(2-ethylhexyl) phthalate, which can be added in concentrations up to 50%.

**Melamine,
a thermosetting polymer**

**Di(2-ethylhexyl) phthalate (Dioctyl phthalate)**

PVC is used for raincoats, house siding, and artificial leather (e.g., handbags, briefcases, and inexpensive shoes). It is found in garden hose, floor covering, swimming pool liners, and automobile upholstery. When vinyl upholstery is exposed to high temperatures, as in the interior of an automobile, the plasticizer distills out. The result is an opaque, difficult-to-remove film on the insides of the windows, and upholstery that is hard and brittle.

Monomers can be polymerized in the gas phase, in bulk, as suspensions and as emulsions. The most common method of making PVC is by emulsifying the vinyl chloride monomer in water with surfactants (soaps), water-soluble catalysts, and heat. The monomer is polymerized to solid particles, which are suspended in the aqueous phase. This product can be centrifuged and dried or used as such. Chemists can control the average molecular weight, which can become very high in emulsion polymerization. A high molecular weight means a more rigid and stronger polymer but also one that is more difficult to work with because of greatly increased viscosity. An emulsion of polyvinyl acetate is sold as latex paint. When the vehicle of emulsion (water) evaporates, the polymer is left as a hard film. A thicker emulsion of polyvinyl acetate is an excellent adhesive, the familiar white glue. When vinyl acetate and vinyl chloride are polymerized together, a copolymer results that has properties all its own. This copolymer is particularly good for detailed moldings and is used to make phonograph records.

*Polymerization methods*

Poly(vinylidene chloride) is primarily extruded as a film that has low permeability to water vapor and air and is therefore used as the familiar clinging plastic food wrap, Saran Wrap.

$$—(CH_2CCl_2)_n—$$

**Poly(vinylidene chloride)**

Poly(tetrafluoroethylene), also known as Teflon, is another halogenated polymer and has a number of unique properties. It has a very high melting point ($327°C$), it does not dissolve in any solvent, and nothing sticks to it. It is also an excellent electrical insulator. A product of the DuPont company, Teflon is one of the densest polymers and also one of the most expensive. The surface of the polymer must be etched with metallic sodium to form free radicals to which glue can adhere. The polymer has an extremely low coefficient of friction, which makes it useful for bearings. Its chemical inertness makes it an ideal liner for chemical reagent bottle caps, and its no-stick property is ideal for the coating on the inside of frying pans. At $380°C$ Teflon is still so viscous that it cannot be injection molded. Instead it is molded by pressing the powdered polymer at high temperature and pressure, a process called *sintering*.

$$—(CF_2CF_2)_n—$$

**Poly(tetrafluoroethylene)
Teflon**

The polymer produced in highest volume is polyethylene. Invented by the British, who call it *polythene*, and put into production in 1939, for a long time it could only be produced by the oxygen-catalyzed polymerization of ethylene at pressures near $2.81 \times 10^7$ kg/m$^2$ (40,000 lb/in.$^2$). Such pressures are expensive and dangerous to maintain on an industrial scale. The polyethylene produced has a low density and is used primarily to make film for bags of all types—from sandwich bags to trash can liners. The opaque appearance of polyethylene is due to crystallites, regions of order in the polymer that resemble crystals in an otherwise highly branched material. In the 1950s Karl Ziegler and Giulio Natta developed catalysts composed of titanium chloride ($TiCl_4$), alkyl aluminum, and transition

*High-density polyethylene*

$$—(CH_2CH_2)_n—$$

**Polyethylene**

metal halides with which ethylene can be polymerized at pressures of just $3.16 \times 10^5$ kg/m$^2$ (450 lb/in.$^2$). The resulting product has a higher density and a higher softening temperature (20°C) than the low-density material. The catalysts that Ziegler and Natta developed, and for which they received the 1963 Nobel Prize in Chemistry, cause linear polymerization and thus a crystalline product. High-density polyethylene is as rigid as polystyrene and yet has high-impact resistance. It is used to mold very large articles such as luggage, the cases for domestic appliances, trash cans, and soft drink crates.

**Polystyrene**

Polystyrene is a brilliantly clear, high-refractive-index polymer familiar in the form of disposable drinking glasses. It is brittle and produces sharp, jagged edges when fractured. It softens in boiling water and burns readily with a very smoky flame. But it can be foamed readily and makes a very good insulator—witness the disposable white hot-drink cup. It is used extensively for insulation when properly protected from ignition. The addition of a small quantity (7%) of poly(butadiene) to the styrene makes a polymer that is no longer transparent but that has high impact resistance. Blends of acrylonitrile, poly(butadiene), and styrene (ABS) have excellent molding properties and are used to make automobile bodies. One formulation can be chrome plated for automobile grills and bumpers.

$$\text{CH}_2\!=\!\text{CHCN} \quad \text{C}_6\text{H}_5\text{CH}\!=\!\text{CH}_2 \quad \text{CH}_2\!=\!\text{CH}\!-\!\text{CH}\!=\!\text{CH}_2$$

**Acrylonitrile**          **Styrene**          **1,3-Butadiene**

**Rubber**

Joseph Priestley, the discoverer of oxygen, named *rubber* for its ability to remove lead pencil marks. Rubber is an *elastomer*, defined as a substance that can be stretched to at least twice its length and return to its original size. The Germans, cut off from a supply of natural rubber, began manufacturing synthetic rubber during World War I. Called "buna" for *bu*tadiene and sodium (*Na*), the polymerization catalyst, it was not the ideal substitute. Cars with buna tires had to be jacked up when not in use because their tires would develop flat spots. The addition of about 25% styrene greatly improved the qualities of the product; styrene-butadiene synthetic rubber now dominates the market, a principal outlet being automobile tires. Adding 30% acrylonitrile to butadiene produces nitrile rubber, which is used to make conveyor belts, storage tank liners, rubber hoses, and gaskets.

**Synthetic rubber; a styrene-butadiene copolymer**

The chemist classifies polymers in several ways. There are thermosetting plastics such as Bakelite and melamine and the much larger category of thermoplastic materials, which can be molded, blown, and formed after polymerization. There are the arbitrary distinctions made among plastics, elastomers, and fibers. And there are the two broad categories formed by the polymerization reaction itself: (1) *addition polymers* (e.g., vinyl polymerizations), in which a double bond of a monomer is transformed into a single bond between monomers; and (2) *condensation polymers* (e.g., Bakelite), in which a small molecule, such as water or alcohol, is split out as the polymerization reaction occurs.

**Addition polymers**

**Condensation polymers**

Nylon

One of the most important condensation polymers is nylon, a name so ingrained into our language that it has lost its trademark status. It was developed by Wallace Carothers, the director of organic chemicals research at DuPont, and was the outgrowth of his fundamental research into polymer chemistry. Introduced in 1938, nylon was the first totally synthetic fiber. The most common form of nylon is the polyamide formed by the condensation of hexamethylene diamine and adipic acid:

$$n\,H_2N(CH_2)_6NH_2 \;+\; n\,HOOC(CH_2)_4COOH$$

| Hexamethylene diamine | Adipic acid |
|---|---|

$$n\,H_3\overset{+}{N}(CH_2)_6\overset{+}{N}H_3 \;+\; n\,\overset{-}{O}\overset{\displaystyle O}{\overset{\|}{C}}(CH_2)_4\overset{\displaystyle O}{\overset{\|}{C}}\overset{-}{O}$$

Pressure, 280°C

$$\boldsymbol{+}NH(CH_2)_6NH\overset{\displaystyle O}{\overset{\|}{C}}(CH_2)_4\overset{\displaystyle O}{\overset{\|}{C}}\boldsymbol{+}_n OH \;+\; (2n-1)H_2O$$

**Nylon 6.6**

The reactants are mixed together to form a salt that melts at 180°C. This is converted into the polyamide by heating to 280°C under pressure, which eliminates water. Nylon 6.6 is used to make textiles, whereas nylon 6.10, from the 10-carbon diacid, is used for bristles and high-impact sports equipment. Nylon can also be made by interfacial and ring-opening polymerization, both of which are used in the following experiments.

The condensation polymer made by reacting ethylene glycol with 1,4-benzene dicarboxylic acid (terephthalic acid) produces a polymer that is converted in large part into the fibers such as Dacron. The polymerization is run as an ester interchange reaction using the methyl ester of terephthalic acid:

$$H_3CO\overset{\displaystyle O}{\overset{\|}{C}}\!-\!\!\left\langle\!\bigcirc\!\right\rangle\!-\!\overset{\displaystyle O}{\overset{\|}{C}}OCH_3 \;+\; HOCH_2CH_2OH$$

$$H_3CO(\overset{\displaystyle O}{\overset{\|}{C}}\!-\!\!\left\langle\!\bigcirc\!\right\rangle\!-\!\overset{\displaystyle O}{\overset{\|}{C}}OCH_2CH_2O)_nH \;+\; 2n\,CH_3OH$$

**Polyethylene glycol terephthalate (Dacron)**

The structure of a polymer is not simple. For example, the polymerization of styrene produces a chiral carbon at each benzyl position. We can ask whether the phenyl rings are all on the same side of the long carbon chain, whether they alternate positions, or whether they adopt some random configuration. In the case of copolymers we can ask whether the two components alternate (ABABABABAB . . .), whether they adopt a random configuration (AABBBABAAB . . .), or whether they polymerize as short chains of one and then the other (AAAABBBBBAAA BBBB . . .). These questions are important because the physical properties of the resulting polymer depend on the configuration. The polymer chemist is concerned with finding the answers and discovering catalysts and reaction conditions that can control these parameters.

In the experiments that follow, you will find that the preparation of nylon by interfacial polymerization is a spectacular and reliable experiment easily carried out in one afternoon. The synthesis of Bakelite works well; it requires overnight heating in an oven to complete the polymerization. Nylon by ring-opening polymerization requires skill and care because of the high temperatures involved. The polymerization of styrene also requires care but is somewhat easier to carry out.

In the recycling experiments in Part 2 of this chapter, you will depolymerize a PET bottle to short-chain polyols and then use that material to make fiberglass and polyurethane foam.

## EXPERIMENTS

# 1. Nylon by Interfacial Polymerization[1]

$$2\ SOCl_2\ +\ HOOC(CH_2)_8COOH\ \longrightarrow\ ClC(CH_2)_8CCl\ +\ 2\ HCl\ +2\ SO_2$$

| **Thionyl chloride** | **Sebacic acid** | **Sebacoyl chloride** |
|---|---|---|
| MW 118.97 | MW 202.25 | MW 239.14 |
| bp 79°C | mp 137°C | bp 168°C/12 mm |

$$n\ ClC(CH_2)_8CCl + n\ H_2N(CH_2)_6NH_2\ \longrightarrow\ +C(CH_2)_8CNH(CH_2)_6NH+_n\ +\ 2n\ HCl$$

**Hexane-1,6-diamine**          **Nylon 6.10**

MW 116.21
mp 45–46°C

---

1. Morgan, P. W.; Kwolek, S. L. *J. Chem. Educ.* **1959**, *36*, 182.

**IN THIS EXPERIMENT** a 10-carbon diacid is converted to the corresponding 10-carbon diacid chloride with thionyl chloride. The hydrogen chloride gas given off during the reaction is trapped. The diacid chloride product is dissolved in dichloromethane and transferred to a beaker. An aqueous basic solution of a 6-carbon diamine is floated on top of the dense organic layer. Where the solutions come in contact, nylon 6.10 is formed and can be lifted from the beaker and wrapped on some type of cylinder.

In this experiment a diamine dissolved in water is carefully floated on top of a solution of a diacid chloride dissolved in an organic solvent. Where the two solutions come in contact (the interface), an $S_N2$ reaction occurs to form a film of a polyamide. The reaction stops there unless the polyamide is removed. In the case of nylon 6.10, the product of this reaction, the film is so strong that it can be picked up with a wire hook and continuously removed in the form of a rope.

This reaction works because the diamine is soluble in both water and dichloromethane, the organic solvent used. As the diamine diffuses into the organic layer, a reaction occurs immediately to give the insoluble polymer. The hydrochloric acid produced reacts with the sodium hydroxide in the aqueous layer. The chloride does not hydrolyze before reacting with the amine because it is not very soluble in water. The acid chloride is conveniently prepared using thionyl chloride.

## Microscale Procedure

In a reaction tube fitted with a gas trap (Fig. 67.1), place 0.25 g of sebacic acid (1,8-octane dicarboxylic acid, otherwise known as decanedioic acid), 0.25 mL of thionyl chloride and 12 mg (a small drop) of *N,N*-dimethylformamide. Heat the tube to 60°C–70°C in a water bath in the hood. This is best accomplished by putting very hot water in a beaker and placing the beaker on a steam bath or hot plate to maintain it at 60°C–70°C. As the reaction proceeds, the product forms a liquid layer on the bottom of the tube. Flick the tube to use this liquid to wash down unreacted acid (if necessary) as the reaction proceeds. When all of the acid has reacted and gas evolution has ceased (about 10–15 min), the product should be a clear liquid. Add some dichloromethane to the reaction tube and transfer the solution to a 50-mL beaker using a total of 12 mL of dichloromethane. To prevent the polymer from sticking to the beaker, coat the inside with a thin layer of silicone vacuum grease or silicone oil. Carefully pour onto the top of this dichloromethane solution 0.25 g of hexane-1,6-diamine (hexamethylenediamine) that has been dissolved in 6 mL of water containing 1 mL of 3 *M* sodium hydroxide. Immediately pick up the polymer film at the center with a copper wire hook and lead it over the outer surface of a bottle, beaker, or graduated cylinder as it is removed. Wind up as much of the polymer as possible, wash it thoroughly in water, and press it as dry as possible. After the polymer has dried, determine its weight and calculate the percent yield. Try dissolving the polymer in two or three solvents. Attach a piece of the polymer to your laboratory report.

**■ FIG. 67.1**

An apparatus for acid chloride synthesis. HCl and SO₂ are trapped in the damp cotton.

**Cleaning Up.** Add the cotton from the trap to the used reaction mixture and stir the mixture vigorously to cause nylon to precipitate. Decant the water and dichloromethane and then squeeze the solid as dry as possible; place the solid in the nonhazardous solid waste container. Place the dichloromethane in the halogenated organic solvents container. After neutralization, dilute the aqueous layer with water and flush down the drain.

 ### *Macroscale Procedure*[2]

In a 15-mL tapered 14/20 flask, fitted with a condenser and gas trap like the one shown in Fig. 18.3 (on page 348), place 2 g of sebacic acid (1,8-octane dicarboxylic acid, otherwise known as decanedioic acid), 2 mL of thionyl chloride, and 0.1 mL of *N,N*-dimethylformamide. Heat the tube to 60°C–70°C in a water bath in the hood. This is best accomplished by putting very hot water in a beaker and placing the beaker on a steam bath or hot plate to maintain it at 60°C–70°C. As the reaction proceeds, the product forms a liquid layer on the bottom of the tube. Use this liquid to wash down unreacted acid as the reaction proceeds. When all of the

2. East, G. C.; Hassell, S. *J. Chem. Educ.* **1983**, *60*, 69.

acid has reacted and gas evolution has ceased (about 10–15 min), transfer the product, which should be a clear liquid at this point, to a 250-mL beaker using 50 mL of dichloromethane. Carefully pour onto the top of this dichloromethane solution 2 g of hexane-1,6-diamine (hexamethylenediamine) that has been dissolved in 50 mL of water containing 1 g of sodium hydroxide. Pick up the polymer film at the center with a copper wire and lead it over the outer surface of a bottle, beaker, or graduated cylinder as it is removed. Remove as much of the polymer as possible, wash it thoroughly in water, and press it as dry as possible. After the polymer has dried, determine its weight and calculate the percent yield. Try dissolving the polymer in two or three solvents. Attach a piece of the polymer to your laboratory report.

**Cleaning Up.** Add the cotton from the trap to the used reaction mixture and stir the mixture vigorously to cause nylon to precipitate. Decant the water and dichloromethane and then squeeze the solid as dry as possible; place the solid in the nonhazardous solid waste container. Place the dichloromethane in the halogenated organic solvents container. After neutralization, dilute the aqueous layer with water and flush down the drain.

# 2. The Condensation Polymerization of Phenol and Formaldehyde: Bakelite

The condensation of phenol with formaldehyde is a base-catalyzed process in which one resonance form of the phenoxide ion attacks formaldehyde, yielding a mixture of methoyl phenols. The resulting dimethoyl phenol is then cross-linked by heat, presumably by dehydration with the intermediate formation of benzyl-carbocations. The resulting polymer is Bakelite. Because the cost of phenol is relatively high and the polymer is somewhat brittle, it is common practice to add an extender such as sawdust to the material before cross-linking. The mixture is placed in molds and heated to form the polymer. The resulting polymer, like other thermosetting polymers, is not soluble in any solvent and does not soften when heated. It is the plastic used to make the black handles on kitchen pots and pans that withstand the heat of an oven.

 *Microscale Procedure*

**IN THIS EXPERIMENT** phenol and formaldehyde are heated in the presence of ammonium hydroxide to produce a low molecular weight polymer. This material is poured into a disposable tube, separated from the aqueous layer, acidified with acetic acid, and heated overnight to complete the polymerization, producing Bakelite.

**Phenol**
MW 94.11
mp 40–42°C, bp 182°C

*p*-**Methoyl phenol**    **Dimethoyl phenol**     **Trimethoyl phenol**

**Bakelite**

**CAUTION**: Formaldehyde solution is a suspected carcinogen. Handle the solution with care. Make measurements and carry out the experiment in the hood. The oven should be well ventilated or in a hood. Phenol is very corrosive to the skin.

In a 5-mL short-necked, round-bottomed flask place 0.75 g of phenol and 2.5 mL of 37% by weight aqueous formaldehyde solution. The formaldehyde solution contains 10%–15% methanol, which has been added as a stabilizer to prevent the formaldehyde from polymerizing. Add 0.37 mL of concentrated ammonium hydroxide to the solution, attach an empty distilling column as an air condenser (Fig. 67.2), and reflux it for 5 min beyond the point at which the solution turns cloudy; the total reflux time is about 10 min. In the hood pour the warm solution

■ **FIG. 67.2**
An apparatus for Bakelite synthesis.

Empty distilling column

5-mL Short-necked, round-bottomed flask

Hot sand bath

■ **FIG. 67.3**
To free the polymer, turn the thumbscrew to break the tube. Wear safety glasses. You may wish to wrap the tube in a piece of cloth.

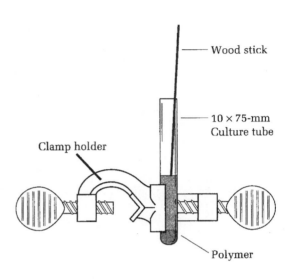

Wood stick

$10 \times 75$-mm Culture tube

Clamp holder

Polymer

into a small, disposable tube, for example, a $10 \times 75$-mm soft glass test tube or vial. *Immediately* clean the flask with a small amount of acetone. Heat the tube and contents inside a steam bath or on a sand bath for 10 min to complete the reaction. Allow the mixture to stand at room temperature for 10–20 min. As the mixture cools, droplets of product will sink to the bottom, giving a clear upper layer that is then poured off. If the upper layer is poured off prematurely, the yield of product will be much lower. Warm the yellow-orange lower layer on a steam bath and add acetic acid dropwise (0.5 mL maximum) with thorough mixing until the layer is clear. It should remain clear even when the polymer is cooled to room temperature. Heat the tube on a water bath at 60°C–65°C for 30 min; after placing a wood stick in the polymer to use as a handle, leave the tube, labeled with your name, in an 85°C oven overnight or until the next laboratory period. To free the polymer, the tube may need to be broken (Fig. 67.3). A filler such as sawdust or a small piece of cloth can be used to make a stronger polymer. Attach a piece of the polymer to your laboratory report.

**Cleaning Up.** Place the acetone wash in the organic solvents container. Should it be necessary to destroy formaldehyde, add 25 mL of household bleach for each milliliter of the solution. After 20 min, flush the solution down the drain. Discard the used glass tube in the nonhazardous solid waste container.

**CAUTION**: Formaldehyde solution is a suspected carcinogen. Handle the solution with care. Make measurements and carry out the experiment in the hood. The oven should be well ventilated or in a hood. Phenol is very corrosive to the skin.

### Macroscale Procedure

In a 25-mL round-bottomed flask place 3.0 g of phenol and 10 mL of 37% by weight aqueous formaldehyde solution. The formaldehyde solution contains 10%–15% methanol, which has been added as a stabilizer to prevent the formaldehyde from polymerizing. Add 1.5 mL of concentrated ammonium hydroxide to the solution and reflux it for 5 min beyond the point at which the solution turns cloudy; the total reflux time is about 10 min. In the hood pour the warm solution into a disposable test tube or culture tube and draw off the upper layer. Immediately clean the flask with a small amount of acetone. Warm the viscous milky lower layer on a steam or sand bath and add acetic acid dropwise with thorough mixing until the layer is clear. The layer should remain clear even when the polymer is cooled to room temperature. Heat the tube on a water bath at 60°C–65°C for 30 min. Then, after placing a wood stick in the polymer to use as a handle, leave the tube, labeled with your name, in an 85°C oven overnight or until the next laboratory period. To free the polymer, the tube may need to be broken. Attach a piece of the polymer to your laboratory report.

**Cleaning Up.** Place the acetone wash in the organic solvents container. Should it be necessary to destroy formaldehyde, add 25 mL of household bleach for each milliliter of the solution. After 20 min, flush the solution down the drain. Discard the used glass tube in the nonhazardous solid waste container.

## 3. Nylon by Ring-Opening Polymerization[3]

**IN THIS EXPERIMENT** a cyclic 6-carbon amide is heated to a very high temperature with sodium hydride and poly(ethylene glycol) in the presence of a polymerization coinitiator, *N*-acetylcaprolactam. A very smooth, tough nylon is produced.

Interfacial polymerization in the manner described in Experiment 1 is not a commercial process; the ring-opening of caprolactam, however, is a commercial process. The nylon produced, nylon 6.6, is used extensively in automobile tire cords and for gears and bearings in small mechanical devices.

The catalyst used in this reaction is sodium hydride; therefore, this is referred to as an *anionic polymerization*. The sodium hydride removes the acidic lactam proton to form an anion (**1**) that attacks the coinitiator, acetylcaprolactam (**2**), which has an electron-attracting acetyl group attached to the nitrogen. The ring of the acetylcaprolactam is attacked by the anion, and the acetylcaprolactam ring opens, forming a substituted caprolactam (**3**) that still has an electron-attracting group attached to nitrogen. A proton transfer reaction occurs, generating a new caprolactam anion (**4**), and **4** then attacks **3**, and so on.

Ordinarily anionic polymerizations must be run in the absence of oxygen, but the addition of poly(ethylene glycol) serves to complex with the sodium ion just as 18-crown-6 (crown ether) does and enhances the catalytic activity of the sodium hydride.

##  *Microscale Procedure*

Set the heater to its maximum setting on your sand bath. The sand must be quite hot before beginning this experiment. Heating can also be done over a small Bunsen burner flame.

Into a disposable 10 × 75-mm test tube place 1 g of caprolactam, 60 mg of poly(ethylene glycol), and 1 small drop of *N*-acetylcaprolactam. Heat the mixture. As soon as it has melted, remove it from the heat and add 20 mg of gray (not white) sodium hydride (50% dispersion in mineral oil). Mix the catalyst with the reactants by stirring with a Pasteur pipette and heat the mixture *rapidly* to boiling (200°C–230°C). This should take place over a 2-min period. Polymerization takes place rapidly, as indicated by an increase in viscosity. If polymerization has not occurred within 3 min, remove the tube from the heat, cool it somewhat, and add another 15 mg of sodium hydride. When the solution is so viscous that it will barely flow, insert a wood stick and with help from a partner draw fibers from the melt. After it cools, the nylon 6.6 can usually be removed from the tube as one cylindrical piece. Try dissolving a piece of the nylon or the fibers in various solvents. Test the physical properties of the fibers by stretching them to the breaking point. Describe your observations and attach a piece of fiber to your laboratory report. Do not forget to turn off the heater.

Active sodium hydride is gray. If it is white, it has probably decomposed on contact with a slight amount of moisture.

**Cleaning Up.** Place the used tube, if coated with polymer, in the nonhazardous solid waste container. Should it be necessary to destroy sodium hydride, add it to excess 1-butanol (38 mL/g of hydride). After the reaction has ceased, cautiously add water; then dilute with more water and flush the solution down the drain.

## *Macroscale Procedure*

Set the heater to its maximum setting on your sand bath. The sand must be quite hot before beginning this experiment. Heating can also be done over a small Bunsen burner flame.

Into a disposable 4-in. test tube place 4 g of caprolactam, 0.25 g of poly-(ethylene glycol), and 2 drops of *N*-acetylcaprolactam. Heat the mixture. As soon as it has melted, remove it from the heat and add 50 mg of gray (not white) sodium hydride (50% dispersion in mineral oil). Mix the catalyst with the reactants by stirring with a Pasteur pipette and heat the mixture *rapidly* to boiling (200°C–230°C). This should take place over a 2-min period. Polymerization takes place rapidly, as indicated by an increase in viscosity. If polymerization has not occurred within 3 min, remove the tube from the heat, cool it somewhat, and add another 50 mg of sodium hydride. When the solution is so viscous that it will barely flow, insert a wood stick and with help from a partner draw fibers (mono-filament line) several feet long from the melt. After it cools, the nylon-6.6 can usually be removed from the tube as one cylindrical piece. Try dissolving a piece of the nylon or the fibers in various solvents. Test the physical properties of the fibers by stretching them to the breaking point. Describe your observations and attach a piece of fiber to your laboratory report. Do not forget to turn off the heater.

**Cleaning Up.** Place the used tube, if coated with polymer, in the nonhazardous solid waste container. Should it be necessary to destroy sodium hydride, add it to excess 1-butanol (38 mL/g of hydride). After the reaction has ceased, cautiously add water; then dilute with more water and flush the solution down the drain.

# 4. Polystyrene by Free-Radical Polymerization

Polystyrene, the familiar crystal-clear brittle plastic used to make disposable drinking glasses and, when foamed, lightweight white cups for hot drinks, is usually made by free-radical polymerization. An initiator is not used commercially because polymerization begins spontaneously at elevated temperatures. At lower temperatures a variety of initiators could be used [e.g., 2,2′-azobis-(2-methyl-propionitrile), which was used in the free-radical chlorination of 1-chlorobutane]. In this experiment we use benzoyl peroxide as the initiator. On mild heating it splits into two benzoyloxy radicals

$$\underset{\substack{\textbf{Benzoyl peroxide}\\ \textbf{MW 242.23}\\ \textbf{dec 106°C}}}{\text{Ph–C(O)–O–O–C(O)–Ph}} \xrightarrow{\Delta} 2\ \underset{\textbf{Benzoyloxy radical}}{\text{Ph–C(O)–O·}} \equiv R\cdot$$

that react with styrene through initiation, propagation, and termination steps to form polystyrene:

*Initiation*:

$$R \cdot + CH_2 = \underset{\underset{C_6H_5}{|}}{CH} \longrightarrow R CH_2 \underset{\underset{C_6H_5}{|}}{CH} \cdot$$

*Propagation*:

$$RCH_2 \underset{\underset{C_6H_5}{|}}{CH} \cdot + CH_2 = \underset{\underset{C_6H_5}{|}}{CH} \longrightarrow RCH_2 \underset{\underset{C_6H_5}{|}}{CH} - CH_2 \underset{\underset{C_6H_5}{|}}{CH} \cdot + CH_2 = \underset{\underset{C_6H_5}{|}}{CH} \longrightarrow RCH_2 \underset{\underset{C_6H_5}{|}}{CH} - CH_2 \underset{\underset{C_6H_5}{|}}{CH} - CH_2 \underset{\underset{C_6H_5}{|}}{CH} \cdot \text{, etc.}$$

*Termination*:

$$2 R \cdot \longrightarrow R - R, \quad RCH_2 \underset{\underset{C_6H_5}{|}}{CH} \cdot + R \cdot \longrightarrow RCH_2 \underset{\underset{C_6H_5}{|}}{CH} - R, \quad 2 RCH_2 \underset{\underset{C_6H_5}{|}}{CH} \cdot \longrightarrow RCH_2 \underset{\underset{C_6H_5}{|}}{CH} - \underset{\underset{C_6H_5}{|}}{CH} CH_2 R$$

**Styrene**
MW 104.15
bp 145–146°C

**Polystyrene**
MW 300,000–25,000,000

**4-*t*-Butylcatechol**

In the termination sequence the first two reactions are rarely observed; termination occurs almost entirely by the third reaction. The final polymer has about 3000 monomer units in a single chain. The lifetime of the radical is about 1 s.

To prevent styrene from polymerizing in the bottle in which it is sold, the manufacturer adds 10–15 ppm of 4-*tert*-butylcatechol, a radical inhibitor (a particularly good chain terminator). The inhibitor must be removed by passing the styrene through a column of alumina before the styrene can be polymerized.

## Synthesis of Polystyrene

Styrene is flammable, an irritant, and has a bad odor. Work with it in the hood.

**IN THIS EXPERIMENT** the polymerization inhibitor is removed from commercial styrene by filtering it through alumina into a disposable culture tube. A small amount of peroxide catalyst is added, and the mixture is heated to initiate polymerization. A clear, very brittle plastic is produced.

In a Pasteur pipette loosely place a very small piece of cotton; then fill the pipette half full with alumina. Add to the top of the pipette 1.5 mL of styrene and

**CAUTION:** Benzoyl peroxide is flammable and may explode on heating or on impact. There is no need for more than a gram or two in the laboratory at any one time. It should be stored in and dispensed from waxed paper containers, not in metal or glass containers with screw-cap lids.

***Online Study Center***

Photo: Filtration Using a Pasteur Pipette

collect 1 mL in a disposable 10 × 75-mm test tube. Add to the tube 50 mg of benzoyl peroxide and a wood boiling stick; then heat the tube over a hot sand bath. When the temperature reaches about 135°C, polymerization begins; because it is an exothermic process, the temperature rises. Keep the reaction under control by cautious heating. The temperature rises, perhaps to 180°C, well above the boiling point of styrene (145°C); the viscosity also increases. Pull the wood stick from the melt from time to time to form fibers; when a cool fiber is found to be brittle, remove the tube from the heat. Alternatively, heat the tube on a steam bath for 90 min with the wood stick in place. Allow the tube to cool; then cool it in ice. Often the polymer will shrink enough so that it can be pulled from the tube; otherwise break the tube. If the polymer is sticky, the polymerization can be completed in an oven overnight at a temperature of about 85°C.

**Cleaning Up.** Shake the alumina out of the pipette and place it in the styrene/alumina hazardous waste container. Clean up spills of benzoyl peroxide immediately. It can be destroyed by reacting each gram with 1.4 g of sodium iodide in 28 mL of acetic acid. After 30 min neutralize the brown solution with sodium carbonate, dilute with water, and flush down the drain.

# PART 2: Recycling Plastics

2.2 lb = 1 kg

**G**arbage. Trash. Hardly the concern of the organic chemistry lab. Or is it? In the United States 320 billion pounds of trash is generated each year. Eighty-five percent of this trash is dumped into landfills, which are filling up fast. Plastics make up 11% of this waste. In any given year we produce about 55 billion pounds of plastic, of which 22 billion pounds are discarded. More than half of this plastic has been used in packaging. By volume, this waste plastic makes up 30% of municipal solid waste. Some 28% of milk bottles and soft drink containers are recycled. Some plastic can be recycled simply by melting it and remolding it; other polymers must be treated chemically.

Plastics must be carefully sorted before recycling because some are incompatible with others. The Society of the Plastics Industry has adopted a coding system in which manufacturers place an alphanumeric identification code at or near the bottom of each container to identify the type of plastic resin being used. Poly(ethylene terephthalate) (PET) is the leading recycled plastic.

In the experiments in this part of the chapter, a plastic soft-drink bottle made of PET is converted to fiberglass-reinforced polyester laminates and polyurethane foam.

Polyethylene terephthalate  PETE

PET is the plastic from which clear plastic bottles for soft drinks are made by blow-molding. PET is made of repeating units of ethylene glycol and terephthalic acid (benzene 1,4-dicarboxylic acid). Each repeating unit has a molecular weight of 192; with the number of repeating units more than 100, the polymer has a molecular weight of about 20,000. When this polymer is formed into sheets, it is called *Mylar*, and when it is spun into fibers, it is called *Dacron*. In the United Kingdom it is called *Terylene*. Millions of kilograms of PET are produced each year, and most of this ends up in waste dumps wherever it is not recycled. As we

High-density polyethylene  HDPE

Polyvinyl chloride  V

Low-density polyethylene  LDPE

Polypropylene

Polystyrene

Other

become increasingly aware of the environmental toll such waste entails, we are turning to recycling. This experiment demonstrates the commercial processes by which PET bottles are chemically recycled.

The process is fairly straightforward. The polymer is broken down (depolymerized) into small fragments that can then be used as the starting material for making a new polymer. In the case of PET, the new polymer could be fiberglass or polyurethane foam. It cannot be blow-molded again to make new beverage bottles because of possible contamination.

The chemical structure of PET is as follows:

$$\left[ CH_2CH_2O\overset{O}{\overset{\|}{C}} - \bigcirc - \overset{O}{\overset{\|}{C}}O \right]_n$$

PET is an ester. Low-molecular-weight esters are prepared and hydrolyzed according to the following equilibrium reaction, which can be either acid- or base-catalyzed (*see* Chapter 40):

$$R\overset{O}{\overset{\|}{C}}OH + R'OH \underset{}{\overset{H^+}{\rightleftharpoons}} R\overset{O}{\overset{\|}{C}}OR' + H_2O$$

If you put small pieces of PET in a flask with either aqueous acid or base and reflux the mixture, there will be no apparent reaction. It is remarkably resistant to hydrolysis. Try it for yourself. Esters undergo a related reaction called *transesterification* that can be used to depolymerize PET:

$$R\overset{O}{\overset{\|}{C}}OR' + R''OH \underset{}{\overset{cat}{\rightleftharpoons}} R\overset{O}{\overset{\|}{C}}OR'' + R'OH$$

When applied to PET, the reaction produces a mixture containing three esters when the alcohol is diethylene glycol:

$$\left[ CH_2CH_2O\overset{O}{\overset{\|}{C}} - \bigcirc - \overset{O}{\overset{\|}{C}}O \right]_n + HOCH_2CH_2OCH_2CH_2OH \rightleftharpoons$$

$$HOCH_2CH_2OCH_2CH_2O\overset{O}{\overset{\|}{C}} - \bigcirc - \overset{O}{\overset{\|}{C}}OCH_2CH_2OCH_2CH_2OH +$$

$$HOCH_2CH_2OCH_2CH_2O\overset{O}{\overset{\|}{C}} - \bigcirc - \overset{O}{\overset{\|}{C}}OCH_2CH_2OH + HOCH_2CH_2O\overset{O}{\overset{\|}{C}} - \bigcirc - \overset{O}{\overset{\|}{C}}OCH_2CH_2OH$$

At the end of the reaction, ethylene glycol and unreacted diethylene glycol are also present. This reaction is catalyzed by a mixture of manganese acetate and manganese carbonate.

## Polyurethane Foam from Plastic Beverage Bottles

The familiar foam-rubber cushions that you sit on are most likely made of polyurethane. It is also widely used in the building trades as a rigid insulating material in the form of lightweight boards.

A urethane (also called a carbamate) is a functional group that looks like an amide on one side and an ester on the other. It is made by reacting an isocyanate with an alcohol:

$$R-N{=}C{=}O + R'-OH \;\rightleftharpoons\; R-NH-\overset{\displaystyle O}{\overset{\|}{C}}-O-R'$$

**An isocyanate**            **A urethane**

Polyurethanes, like many other polymers, are made from difunctional starting materials:

$$O{=}C{=}N-R-N{=}C{=}O + H-O-R'-O-H \;\rightleftharpoons\; O{=}C{=}N\!\left[\!R-NH-\overset{\displaystyle O}{\overset{\|}{C}}-O-R'\!\right]_{n}\!\!O-H$$

**A polyurethane**

Isocyanates react with water to form an amine and carbon dioxide:

$$R-N{=}C{=}O + H_2O \longrightarrow R-\overset{\displaystyle H}{\overset{|}{N}}-\overset{\displaystyle O}{\overset{\|}{C}}-OH \longrightarrow R-\overset{\displaystyle H}{\overset{|}{N}}-H + CO_2$$

For this reason, a small amount of water can be added during the polymerization step. The carbon dioxide is one means of forming the bubbles in the foam.

As noted earlier, commercial polyurethane can be rigid or flexible, depending on the nature of the R and R' group. In this experiment, the foamed polymer will be rigid. The most common diisocyanate is tolylene diisocyanate.

**Tolylene 2,4-diisocyanate**

# Fiberglass-Reinforced Polyester from Plastic Beverage Bottles

PET resin in the form of shredded particles of the clear plastic from beverage bottles is refluxed in the presence of a catalyst to effect an ester interchange reaction with propylene glycol. This glycolysis produces a mixture of three esters, each of which is terminated with a hydroxyl group.

$$\left[CH_2CH_2OC\text{—}\!\!\bigcirc\!\!\text{—}CO\right]_n + CH_2CHCH_3 \overset{cat}{\rightleftharpoons} CH_3CHCH_2OC\text{—}\!\!\bigcirc\!\!\text{—}COCH_2CHCH_3 +$$

$$CH_3CHCH_2OC\text{—}\!\!\bigcirc\!\!\text{—}COCH_2CH_2OH + HOCH_2CH_2OC\text{—}\!\!\bigcirc\!\!\text{—}COCH_2CH_2OH + HOCH_2CH_2OH$$

Alcohols react with anhydrides to give esters. In a very simple example, acetic anhydride reacts with ethyl alcohol to give the ester ethyl acetate and acetic acid:

$$CH_3COCCH_3 + CH_3CH_2OH \longrightarrow CH_3COCH_2CH_3 + CH_3COH$$

If the anhydride is maleic anhydride, then the resulting ester terminates in a carboxyl group, which, in an equilibrium reaction, can react with another molecule of alcohol to give an unsaturated ester and water:

$$\bigcirc + HOCH_2CH_3 \longrightarrow HOCCH\!=\!CHCOCH_2CH_3$$

$$CH_3CH_2OH + HOCCH\!=\!CHCOCH_2CH_3 \rightleftharpoons CH_3CH_2OCCH\!=\!CHCOCH_2CH_3 + H_2O$$

This reaction is employed with the mixture of diols produced in the glycolysis of PET. When this mixture reacts with maleic anhydride and water is allowed to distill from the reaction mixture, the result is an unsaturated polymeric ester:

$$\text{CH}_3\overset{\text{OH}}{\underset{}{\text{CHCH}_2\text{O}}}\overset{\text{O}}{\underset{}{\text{C}}}\!\!-\!\!\bigcirc\!\!-\!\!\overset{\text{O}}{\underset{}{\text{C}}}\overset{\text{OH}}{\underset{}{\text{OCH}_2\text{CHCH}_3}} + \text{(maleic anhydride)} \longrightarrow$$

$$\left[\!\!\begin{array}{c} \overset{}{\underset{\text{CH}_3\text{CHCH}_2\text{O}\overset{}{\text{C}}}{-\text{O}}}\!\!-\!\!\bigcirc\!\!-\!\!\overset{\text{O}}{\underset{}{\text{C}}}\overset{\text{O}}{\underset{}{\text{OCH}_2\text{CHCH}_3}}\overset{\text{O}\quad\text{O}}{\underset{}{\text{OCCH}=\text{CHC}}} \end{array}\!\!\right]_n + \text{H}_2\text{O}$$

This unsaturated polyester is of fairly low molecular weight and thus is a liquid at room temperature. It can be cross-linked into a rigid, high molecular weight ester with styrene in a radical-catalyzed reaction:

$$\text{RO}\overset{\text{O}}{\underset{}{\text{C}}}\text{CH}=\text{CH}\overset{\text{O}}{\underset{}{\text{C}}}\text{OR} + \bigcirc\!\!-\!\!\text{CH}=\text{CH}_2 \longrightarrow$$

This polyester, when reinforced with fiberglass, can be molded and polymerized in place to form such things as automobile bodies and boats up to 100 ft in length.

## EXPERIMENTS

# 1. Depolymerization of PET with Diethylene Glycol

**IN THIS EXPERIMENT** shredded PET is heated with diethylene glycol, a 4-carbon dialcohol, and a catalyst. Depolymerization results in the formation of three esters in a clear solution. This material is used to prepare polyurethane foam.

## Microscale Procedure

Into a 13 × 100-mm test tube, weigh 1.74 g of PET from a clear soft-drink bottle that has been cut into 6-mm ($\frac{1}{4}$-in.) squares or shredded material from a commercial shredder.[4] Add 3 mg of manganese acetate [Mn(OAc)$_2$] and 3 mg of manganese carbonate (MnCO$_3$) followed by 1.26 g of diethylene glycol. Add a boiling chip and reflux the mixture. Note the time at which all the chips of PET disappear. If the mixture is cooled at this point, it will have a cloudy white appearance because the material still has a high molecular weight and comes out of solution. Continue to reflux for a total of 45 min; at the end of this time, the mixture should remain clear after cooling. At this point, the glycolysis (transesterification) is finished, and the tube contains the mixture of three esters shown on page 829. This material will be used for making fiberglass and polyurethane foam in the next two experiments.

Look up information on PET. How much is produced each year? Are Mylar and Dacron simply different forms of this polymer? See if you can discover anything about the use of manganese acetate and manganese carbonate to catalyze a reaction of this type.

## Macroscale Procedure

Into a 20 × 150-mm test tube, weigh 7.5 g of PET from a clear soft-drink bottle that has been cut into 6-mm ($\frac{1}{4}$-in.) squares or shredded material from a commercial shredder.[5] Add 18 mg of manganese acetate [Mn(OAc)$_2$] or antimony acetate [Sb(OAc)$_3$] and 18 mg of manganese carbonate (MnCO$_3$) followed by 5.0 g of diethylene glycol. Add a boiling chip and reflux the mixture. Poke the shredded plastic down into the liquid as the reaction proceeds. Note the time at which all the chips of PET disappear. If the mixture is cooled at this point, it will have a cloudy white appearance because the material still has a high molecular weight and comes out of solution. Continue to reflux for a total of 45 min; at the end of this time, the mixture should remain clear after cooling. At this point, the glycolysis (transesterification) is finished, and the tube contains the mixture of three esters shown on page 829. This material will be used for making fiberglass and polyurethane foam in the next two experiments.

Look up information on PET. How much is produced each year? Are Mylar and Dacron simply different forms of this polymer? See if you can discover anything about the use of manganese acetate or antimony acetate and manganese carbonate to catalyze a reaction of this type.

---

4. In places where a deposit on bottles is required, there is often a machine that will shred the bottles and return the deposit. Obtain this shredded material from the store and treat it first with toluene to remove the label glue and then, after drying, separate it by flotation in water. The paper and black polyethylene base of the bottle will float, the aluminum cap will sink, and the PET (density about 1) will be suspended in the water.

5. *See* footnote 4.

## 2. Prepolymer for Polyurethane Foam

**Castor oil**
Mostly the triglyceride
of ricinoleic acid

Into a 100-mL round-bottomed flask, place 21.5 g of castor oil and 7.5 g of trans-esterified PET from the preceding experiment. Add a magnetic stirring bar and thermometer and, with vigorous stirring in the hood, add 27.5 g of tolylene 2,4-diisocyanate all at once from an Erlenmeyer flask. Over about the next 15 min, the temperature of the mixture rises to 80°C. Some foaming takes place. Once the temperature starts to drop, place the flask in a Thermowell and heat the mixture at 120°C for 1 h. Remove it from the heat and cool it in a water bath to room temperature. This prepolymer should contain approximately 15% free isocyanate. It will keep overnight or longer, if refrigerated.

## 3. Polyurethane Foam

To prepare foamed polyurethane, in a small paper cup mix 0.175 g of buffered catalyst (prepared from 7 g of diethylaminoethanol, 3.32 mL of concentrated hydrochloric acid, and 6.6 mL of water), 1 drop of Dow Corning silicone oil, and 0.55 mL of water. To this, add 5 g of the prepolymer mix (from the preceding experiment) and stir the mixture thoroughly with a metal spatula. Carry out this operation in the hood. Leave the cup with your name on it in the hood until the next lab period.

## 4. Conversion of PET into Fiberglass-Reinforced Polyester

### Microscale Procedure

To 3 g of PET cut into 6-mm ($\frac{1}{4}$-in.) squares or shredded material from a commercial shredder,[6] add 1.4 g of propylene glycol, 5 mg of manganese acetate tetrahydrate, and 2 mg of manganese carbonate. Insert a thermometer into the

---

6. *See* footnote 4.

mixture as far as possible and heat the mixture on a sand bath until the propylene glycol boils (187°C). A rapid increase in the temperature is taken as an indication that the glycolysis reaction is over. When all the particles have disappeared and the temperature reaches 200°C, cool the mixture to about 150°C and add 1.56 g of maleic anhydride. Heat the mixture to boiling and allow about 0.3 mL of water to escape (you probably will not see it). Again monitor the temperature. When it reaches 215°C, the reaction is probably over. Cool the mixture to 140°C and, with stirring, add to it 4.6 g of styrene in the hood.

The mixture of styrene and unsaturated polyester resin can now be polymerized. To do this, add 30 mg of cobalt acetate, stir well with a stirring rod, and then add 100 mg of methyl ethyl ketone peroxide (the radical catalyst) with thorough stirring. The mixture can be poured onto aluminum foil to polymerize; alternatively, the mixture can be poured onto 2.5 g of fiberglass on aluminum foil, the air squeezed out, and then allowed to polymerize in the hood. This fiberglass-reinforced polyester should be about 3-mm ($\frac{1}{8}$-in.) thick. The original procedure says that these castings were allowed to cure at room temperature for 16 h, followed by a post-cure at 70°C for 1 h and a 100°C cure for 1 h. You could simply place the unpolymerized material in a well-ventilated oven at 80°C overnight.

## *Macroscale Procedure*

Into a 20 × 150-mm test tube, place 6 g of PET cut into 6-mm ($\frac{1}{4}$-in.) squares or material from a commercial shredder.[7] Add 2.8 g of propylene glycol, 10 mg of manganese acetate tetrahydrate, and 4 mg of manganese carbonate. Insert a thermometer into the mixture as far as possible and heat the mixture on a sand bath until the propylene glycol boils (187°C). A rapid increase in the temperature is taken as an indication that the glycolysis reaction is over. When all the particles have disappeared and the temperature reaches 200°C, cool the mixture to about 150°C and add 3.12 g of maleic anhydride. Heat the mixture to boiling and allow about 0.6 mL of water to escape (you probably will not see it). Again monitor the temperature. When it reaches 215°C, the reaction is probably over. Cool the mixture to 140°C and, with stirring, add to it 9.2 g of styrene in the hood.

The mixture of styrene and unsaturated polyester resin can now be polymerized. To do this, add 60 mg of cobalt acetate, stir well with a stirring rod, and then add 0.20 g of methyl ethyl ketone peroxide (the radical catalyst) with thorough stirring. This mixture can be poured onto aluminum foil to polymerize; alternatively, the mixture can be poured onto 5 g of fiberglass on aluminum foil, the air squeezed out, and then allowed to polymerize in the hood. This fiberglass-reinforced polyester should be about 3-mm ($\frac{1}{8}$-in.) thick. The original procedure says that these castings were allowed to cure at room temperature for 16 h, followed by a post-cure at 70°C for 1 h, and a 100°C cure for 1 h. You could simply place the unpolymerized material in a well-ventilated oven at 80°C overnight.

CAUTION: Cobalt acetate is a suspected carcinogen and a mutagen. Handle with care.

CAUTION: Cobalt acetate is a suspected carcinogen and a mutagen. Handle with care.

---

7. *See* footnote 4.

## QUESTIONS

1. What might the products be from an explosion of smokeless gunpowder? How many moles of carbon dioxide and water would come from 1 mol of trinitroglucose? Does the molecule contain enough oxygen for the production of these two substances?
2. Write a balanced equation for the reaction of sebacoyl chloride with water.
3. In the final step in the synthesis of Bakelite, the partially polymerized material is heated at 85°C for several hours. What other product is produced in this reaction?
4. Give the detailed mechanism of the first step of the ring-opening polymerization of caprolactam.

**Online Study Center**

**General Resources**
Web Links

# Searching the Chemical Literature

**Online Study Center**

This icon will direct you to
techniques, equipment setups,
and online resources at
http://college.hmco.com/
PIC/williamsonMMOE5e.

**I**n planning a synthesis or investigating the properties of compounds, the
organic chemist is faced with the problem of locating information on the chemical and physical properties of substances that have been previously prepared as
well as the best methods and reagents for carrying out a synthesis. With tens of
millions of compounds known and new ones being reported at the rate of more
than 700,000 per year, this task might seem formidable. However, the field of
chemistry has developed one of the best information retrieval systems of all the
sciences. In a university chemistry library or over the Internet (at an academic institution that has subscribed to the proper databases), it is possible to locate information on almost any known compound in a few minutes. Even a modest library
will have information on several million different substances in printed form.

The ultimate source of information is the primary literature: articles written
by individual chemists and published in journals. The secondary literature
consists of compilations of information taken from these primary sources. For
example, to find the melting point of benzoic acid, one would naturally turn to a
secondary reference containing a table of physical constants of organic compounds. If, on the other hand, one wished to have information about the phase
changes of benzoic acid under high pressure, one would need to consult the primary literature. If a piece of information, such as a melting point, is crucial to an
investigation, the primary literature should be consulted because, for instance,
transcription errors can occur in the preparation of secondary references.

In this chapter we will first consider the traditional chemistry library and then
the resources available on the Internet, although a distinction between the real and
virtual library is becoming less clear each day.

## Handbooks

For rapid access to information such as melting point, boiling point, density, solubility, optical rotation, $\lambda_{max}$, and crystal form, one turns first to the *Handbook of
Chemistry and Physics*, now in its 86th edition.[1] This book contains information

1. Lide, D. R., Ed. *CRC Handbook of Chemistry and Physics,* 86th ed.; CRC Press: Boca Raton, FL, 2005.

on some 15,000 organic compounds, including the Beilstein reference to each compound. The compounds presented are well known and completely characterized. Most are commercially available. *The Merck Index*[2] contains information on more than 10,000 compounds, especially those of pharmaceutical interest. In addition to the usual physical properties and information and literature references to synthesis and isolation, medicinal properties, such as toxicity data, are found. Approximately one-third of *The Merck Index* is devoted to such items as a long cross index of names (which is very useful for looking up drugs), a table of organic name reactions, an excellent section on first aid for poisons, a list of chemical poisons, and a list of the locations of many poison control centers. *The Merck Index* is also available online for a fee.

In nearly every organic chemistry laboratory throughout the world can be found a copy of the *Aldrich Handbook*, a comprehensive catalog of chemicals and equipment related to organic chemistry.[3] Not only does this catalog list the prices of more than 48,000 chemicals (primarily organic) as well as their molecular weights, melting points, boiling points, and optical rotations, it also gives references to infrared (IR) and nuclear magnetic resonance (NMR) spectral data and to *Beilstein's Handbüch der organischen Chemie, The Merck Index*, and *Fiesers' Reagents for Organic Synthesis*. In addition, the catalog provides the reference to the *Registry of Toxic Effects of Chemical Substances* (RTECS #) for each chemical[4] and, if appropriate, the reference to *Sax's Dangerous Properties of Industrial Materials*.[5] Special hazards are noted (e.g., "severe poison," "lachrymator," and "corrosive"). The Aldrich catalog also provides pertinent information (uses, physiological effects, etc.) with literature references for many compounds. Material Safety Data Sheets (MSDS) are also available.

A complete list of all chemicals sold commercially and the addresses of the companies making them is found in *Chem Sources*,[6] which is published yearly.

After consulting these three single-volume references, one would turn to more comprehensive multivolume sources such as the *Dictionary of Organic Compounds*.[7] This dictionary, still known as "Heilbron," the name of its former editor, now comprises nine volumes of specific information, with primary literature references on the syntheses, reactions, and derivatives of more than 50,000 compounds. *Rodd's Chemistry of Carbon Compounds*,[8] another valuable

---

2. O'Neill, M. J.; Smith, A.; Heckelman, P. E.; Budavari, S., Eds. *The Merck Index,* 13th ed.; Merck and Co., Inc.: Rahway, NJ, 2001.

3. The *Aldrich Handbook* is free from the Aldrich Chemical Company. Obtain your copy on the Internet at http://www.sigmaaldrich.com/Brands/Aldrich.html.

4. RTECS is a compendium of toxicity data extracted from the open scientific literature. Formerly maintained by the National Institute for Occupational Safety and Health, it is now managed by MDL Information Systems, Inc., a wholly owned subsidiary of Elsevier Science, Inc.

5. Lewis, R. J. *Sax's Dangerous Properties of Industrial Materials*; Wiley-Interscience: New York, 2004.

6. *Chem Sources* is recompiled annually and published by Chemical Sources International, Inc.

7. Buckingham, J. *Dictionary of Organic Compounds*, 6th ed.; Chapman & Hall/CRC Press: Boca Raton, FL, 1996.

8. *Rodd's Chemistry of Carbon Compounds,* 2nd ed.; Elsevier Science: Amsterdam, 2002.

multivolume work with primary literature references, is organized by functional group rather than in dictionary form.

The ultimate reference for chemistry as a whole, including articles on individual substances, environmental aspects, industrial processes, pharmaceuticals, polymers, food additives, and so on, is the 27-volume *Kirk-Othmer Encyclopedia of Chemical Technology.*[9] It is also available, for a fee, at Kirk-Othmer Online.[10]

*Beilstein's Handbüch der organischen Chemie* is certainly not a "handbook" in the traditional sense—it can occupy an entire alcove of a chemistry library! This reference covers every well-characterized organic compound that has been reported in the literature up to 1979—syntheses, properties, and reactions. Included with the main reference set are three supplements covering the periods 1910–1919, 1920–1929, and 1930–1949; later supplements are available as well. Although written in German, this reference can provide much information even to those who possess no knowledge of the language. Physical constants and primary literature references are easy to pick out. A German-English dictionary for Beilstein is available on the Internet as well as a number of guides and leading references to its use.

The *Handbüch* is organized around a complex classification scheme that is explained on the Internet and in *The Beilstein Guide,*[11] but the casual user can gain access through the *Handbook of Chemistry and Physics* and the Aldrich catalog. In the *Handbüch* the reference for 2-iodobenzoic acid is listed as **B9**$^2$, 239, which means the reference will be found on p. 239 of Vol. (Band) 9 of the second supplement (Zweites Ergänzungwerk, EII). In the Aldrich catalog the reference is given as Beil **9**, 363, which indicates information can be found in the main series on p. 363, Vol. 9. The system number assigned to each compound can be traced through the supplements. The easiest access to the *Handbüch* itself is through the formula index (General Formel register) of the second supplement. A rudimentary knowledge of German will enable one to pick out, for example, iodo (Jod) benzoic acid (Saure).

A modern, up-to-date English version of the *Handbüch* is available as CrossFire Beilstein for a fee on the Internet.[12] Some 100 properties of 9 million compounds and 9.5 million reactions are available for searching.

## Chemical Abstracts

Secondary references are incomplete; some suffer from transcription errors, and the best—*Beilstein's Handbüch der organischen Chemie*—is not completely up-to-date in its survey of the organic literature, so one must often turn to the primary chemical literature. However, there are more than 14,000 periodicals

9. *Kirk-Othmer Encyclopedia of Chemical Technology,* 5th ed.; John Wiley and Sons: New York, 2004.

10. The online version is at http://www.MRW.interscience.wiley.com/kirk/.

11. Weissback, O.; Hoffmann, H. M. R.; Weissback, O. *The Beilstein Guide: A Manual for the Use of Beilstein's Handbuch Der Organischen Chemie*; Springer: New York, 1976.

12. CrossFire Bielstein can be fouind online at http://www.mdl.com/products/knowledge/crossfire_beilstein/.

where chemical information might appear. Chemists are fortunate to have the *Chemical Abstracts*, an index that covers this huge volume of literature very promptly and publishes biweekly abstracts of each article. The information in each abstract is then compiled into author, chemical substance, general subject, formula, and patent indexes.

The Chemical Abstracts Service (CAS) and STN International, a database vendor, are owned by the American Chemical Society. Since publication of *Chemical Abstracts* began in 1907, chemical nomenclature has changed. Now, no trivial (common) names are used, so acetone appears as 2-propanone and *o*-cresol appears as benzene, 1-methyl, 2-hydroxy. It takes some experience to adapt to these changing names; even now, *Chemical Abstracts* does not exactly follow the rules of the International Union of Pure and Applied Chemistry (IUPAC). The formula index is useful because it provides not only reference to specific abstracts but also a correct name that can be found in the chemical substances index.

## Current Contents

As long as a year can elapse between the time a paper appears and the time an abstract is published. *Current Contents* and *Chemical Titles* help fill the gap by providing much of the information found in an abstract: a list of the titles, authors, and keywords that appear in titles of papers recently published. (In some cases, these may include papers about to be published.)

### Science Citation Index

The *Science Citation Index* is unique in that it allows for a type of search not possible with any other index—a search forward in time. For example, if one wanted to learn what recent applications have been made of the coupling reaction first reported by Stansbury and Proops in 1962, one would look up their paper [*J. Org. Chem.*, **27**, 320 (1962)] in a current volume, say 1998, of *Science Citation Index* and find there a list of those articles published in 1998 in which an author cited the 1962 work of Stansbury and Proops in a footnote. This database is available, by subscription, on the Internet, where it is known as the ISI Web of Knowledge.[13]

# Planning a Synthesis

In planning a synthesis, the organic chemist is faced with at least three overlapping considerations: the chemical reactions, the reagents, and the experimental procedure to be employed. For students, a good place to start might be an advanced textbook such as *Advanced Organic Chemistry*[14] or *March's Advanced Organic Chemistry*,[15] both containing literature references. Often a good lead is

---

13. The database is available online at http://isiwebofknowledge.com.

14. Carey, F. J.; Sundberg, R. J. *Advanced Organic Chemistry: Structure and Mechanisms* (part A) and *Reaction and Synthesis* (part B), 4th ed.; Plenum: New York, 2004.

15. Smith, M. B.; March, J. *March's Advanced Organic Chemistry: Reactions, Mechanisms, and Structure*, 5th ed.; Wiley-Interscience: New York, 2001.

through named reactions, for which *Name Reactions and Reagents in Organic Chemistry*[16] is a useful reference.

It is instructive for the beginning synthetic chemist to read about some elegant and classical syntheses that have been carried out in the past. For this purpose, see *Art in Organic Synthesis*[17] for some of the most elegant syntheses carried out in recent years. Similarly, *Creativity in Organic Synthesis*[18] and *Selected Organic Syntheses*[19] are compendia of elegant syntheses. At a more advanced level, see *The Logic of Chemical Synthesis.*[20] For natural product synthesis, see *The Total Synthesis of Natural Products,*[21] an excellent series in nine volumes, with the most recent published in 1992. *Modern Synthetic Reactions*[22] is a more comprehensive source of information with several thousand references to the original literature. It has been brought further up-to-date at about the same level in *Some Modern Methods of Organic Synthesis.*[23]

An extremely useful reference is the *Compendium of Organic Synthetic Methods,*[24] presently in eight volumes, the the most recent published in 1995. The material is organized by reaction type. In addition, *Organic Reactions,*[25] published annually, should be consulted. Over 100 preparative reactions, with examples of experimental details, are covered in great detail in 59 volumes. *Annual Reports in Organic Synthesis*[26] is an excellent review of new reactions in a given year, organized by reaction type. *Stereoselective Synthesis*[27] treats one of the newest and most important areas of organic chemistry.

The late Mary Fieser was responsible for publishing the first 17 volumes of *Feisers' Reagents for Organic Synthesis.*[28] The last three volumes have been edited by Tse-Lok Ho.[29] Begun in 1967, this series critically surveys the reagents employed to carry out organic synthesis. Included are references to the original

16. Mundy, B. P.; Ellerd, M. G.; Favaloro, F. G. *Name Reactions and Reagents in Organic Chemistry*, 2nd ed.; Wiley-Interscience: New York, 2005.

17. Anand, N.; Bindra, J. S.; Randanathan, S. *Art in Organic Synthesis*, 2nd ed.; Wiley-Interscience: New York, 1988.

18. Bindra, J. S. *Creativity in Organic Synthesis*; Academic Press: San Diego, 1975.

19. Fleming, I. *Selected Organic Syntheses: A Guide for Organic Chemists*; University Microfilms International: Ann Arbor, MI, 1991.

20. Corey, E. J.; Cheng, Xue-Min. *The Logic of Chemical Synthesis,* new ed.; Wiley-Interscience: New York, 1995.

21. ApSimon, J. *The Total Synthesis of Natural Products*; Wiley-Interscience: New York.

22. House, H. O. *Modern Synthetic Reactions,* 2nd ed.; Benjamin-Cummings: San Francisco,1972.

23. Carruthers, W. *Some Modern Methods of Organic Synthesis,* 3rd ed.; Cambridge University Press: Oxford, UK, 1987.

24. Smith, M. B. *Compendium of Organic Synthetic Methods*; John Wiley and Sons: New York, 1995; e-book, 2004.

25. Paquette, L. A. *Organic Reactions*; Wiley: New York.

26. *Annual Reports in Organic Synthesis*; Academic Press: San Diego; Volume *100* was published in 2004.

27. Atkinson, R. S. *Stereoselective Synthesis,* new ed.; John Wiley and Sons: New York, 1995.

28. Fieser, M. *Fiesers' Reagents for Organic Synthesis*; Wiley-Interscience: New York.

29. Ho, Tse-Lok. *Fiesers' Reagents for Organic Synthesis*; Wiley-Interscience: New York.

literature and suppliers of reagents, and to *Organic Syntheses* (see below), as well as the critical reviews that have been written about various reagents. The index of reagents according to type of reaction is very useful when planning a synthesis. On a much larger scale is the comprehensive *Encyclopedia of Reagents for Organic Synthesis*,[30] which gives a critical assessment of most of the reagents one would use in carrying out a synthesis.

To carry out a reaction on one functional group without affecting another functional group, it is often necessary to put on a protective group. This rather specialized subject is admirably covered in *Protective Groups in Organic Synthesis*.[31]

Before carrying out the synthesis itself, one should consult *Organic Syntheses*,[32] an annual series published since 1921 that is grouped in nine collective volumes with reaction and reagent indexes. *Organic Syntheses* gives detailed laboratory procedures for carrying out more than 1000 different reactions. Each synthesis is submitted for review, and then the procedure is sent to an independent laboratory for checking and verification. The reactions, unlike many that are reported in the primary literature, are carried out a number of times and can therefore be relied on to work. This valuable series is now available at no cost on the Internet.

Laboratory techniques are covered in *Techniques of Organic Chemistry*,[33] an uneven, multiauthored compendium that is outdated yet unique in many respects. An excellent advanced text on the detailed mechanisms of many representative reactions is *Mechanism and Theory in Organic Chemistry*.[34]

*Methoden der Organischen Chemie* (*Methods of Organic Chemistry*,[35] often called Houben-Weyl) is a multivolume work, published in English since 1990. Detailed experimental procedures are presented for the syntheses of thousands of compounds, making this a valuable resource. *Organic Reactions*,[36] as the name implies, has long chapters or even an entire volume devoted to one reaction. Each article of this 59-volume series is written by an authority.

## The Modern Chemistry Library

Although there will probably always be a chemistry library, more and more of the resources of the traditional library are becoming available on the Internet. All 2.5 million pages (a stack of paper 81 stories high) of all the journals published by

30. Paquette, L. A. *Encyclopedia of Reagents for Organic Synthesis,* 8 vol.; John Wiley and Sons: New York, 1995.

31. Greene, T. W.; Wuts, P. G. M. *Protective Groups in Organic Synthesis,* 3rd ed.; Wiley-Interscience: New York, 1999.

32. *Organic Syntheses* is a publication of the American Chemical Society. All past and present procedures are available online at http://www.orgsyn.org.

33. Weissberger, A., Ed. *Techniques of Organic Chemistry,* 14 vol.; Wiley-Interscience: New York; 1955.

34. Lowry, T. H.; Richardson, K. S. *Mechanism and Theory in Organic Chemistry,* 3rd ed.; Benjamin Cummings: San Francisco, 1987.

35. Houben, J.; Weyl, T. *Methoden der Organischen Chemie* (*Methods of Organic Chemistry*); Georg Thieme: Stuttgart, Germany.

36. *See* footnote 25.

the American Chemical Society (ACS) since it began in 1879 are now online and searchable. No longer will it be necessary to go to the library stacks and pull off, for example, volume 26 of the *Journal of Organic Chemistry*, turn to page 4563, and write down the details of the syntheses of steroidal α-acetoxyketones. If your institution subscribes to the ACS database, you can simply go to a computer terminal and get the same information in a fraction of the time.

All of the ACS journals are published both on paper and electronically. A subscription to the database might seem expensive until you consider how many cubic feet those volumes would occupy in a library and what it costs to build and maintain that space. There are a number of commercial vendors of databases, of which the ACS is but one. STN International, owned by the American Chemical Society, is connected to more than 200 of these databases. The *Chemical Abstracts* database is available only on STN; you can carry out structural searches on the 40 million substances in this index. CrossFire Beilstein is the Web version[37] of this comprehensive database of organic compounds.

Many journals are now available by electronic subscription. Most of the ACS journals and even secondary references are now available on the Internet. Very little of this information is free to the outsider, but the economics and usefulness of computer literature searching and retrieval make it extremely attractive for the institution. The chemical information available to the casual Internet surfer is of varying quality, but certain governmentally supported sites, such as the Protein Data Bank,[38] are invaluable.

Using a computer, it is now possible to conduct substructure searches of a database that are impossible via the printed page. For example, one can ask for a list of all three-membered rings bearing a single hydroxyl group, even if the ring is buried in a large, complex molecule.

It is not always easy to navigate this wealth of information for an online search, especially a structure search. The command language has a steep learning curve; searches are often the province of the trained professional librarian. SciFinder Scholar, published by the Chemical Abstracts Service of the American Chemical Society and purchased by libraries, eases this searching problem for the novice.[39]

Although the journals may disappear from the shelves, books, monographs, and treatises, including many of the secondary references, will remain in the chemistry libraries of the future.

*Online Study Center*

**General Resources**
Web Links

---

37. *See* footnote 12.

38. The Protein Data Bank is available online at www.rcsb.org/pdb.

39. The SciFinder Scholar can be found online at http://www.cas.org/SCIFINDER/SCHOLAR/index.html.

# Index